ASTRONOMY AND ASTROPHYSICS LIBRARY

For further volumes:
http://www.springer.com/series/848

Gennady S. Bisnovatyi-Kogan

Stellar Physics

2: Stellar Evolution and Stability

Second Edition

Springer

Dr. Gennady S. Bisnovatyi-Kogan
Space Research Institute
Russian Academy of Sciences
Profsoyuznaya Str. 84/32
117997 Moskva
Russia
gkogan@iki.rssi.ru

Translators of the 1st edition:
Dr. A. Y. Blinov and Dr. M. Romanova

ISSN 0941-7834
ISBN 978-3-642-26648-5 ISBN 978-3-642-14734-0 (eBook)
DOI 10.1007/978-3-642-14734-0
Springer Heidelberg Dordrecht London New York

Credit: NASA, The Hubble Heritage Team (STScI/AURA)

Cover design: eStudio Calamar S.L.

Printed on acid-free paper

Springer is part of Springer Science+Business Media (www.springer.com)

To my teacher, Yakov Borisovich Zeldovich

Preface to the Second English Edition

During about 10 years since the publication of the first edition, considerable progress was made in different branches of stellar astrophysics. The physical background did not change to visible extend, so only the second volume of the book, dealing with stellar structure and stability needed serious updates. The main updates concern the following topics.

1. The theory of line-driven stellar winds, where general relativistic effects, in particular, line red-shift in the gravitational field were included into consideration, leading to important enhancement of the wind acceleration.
2. The influence of Urca shells on the convective presupernova model has been under discussion for more than 30 years, with principally different answers. A physically relevant approach to this problem is presented in more details in the updated book.
3. Core-collapse supernovae explosion mechanisms. Several new mechanisms have been suggested during last 10 years. Major progress in this field is connected with the development of a magnetorotational mechanism of the explosion. Two and three dimensional calculations are performed in many groups all over the world. The review of all these mechanisms is included, with the discussion of the development of the magnetorotational instability during a magnetorotational explosion.
4. Strange quark stars. This topic was developing in last years in connection with observational search of strange quark stars. The establishment of existence or non-existence of quark stars would be very important for the study of the properties of superdense matter, and physics of strong interactions.
5. Theory of accretion disks around black holes. The progress here is connected with construction of self-consistent models of advective accretion disks at large luminosity, which do not exist in the frame of a standard model. The new solutions represent disks with variable vertical optical depth, large far from the black hole, and low inside. The high temperature region in the central part of the disk may explain the observed hard tails in the luminous X-ray sources. Magnetic field generation by battery effects in accretion disks is analyzed, and a low efficiency of this mechanism is shown. Another addition is connected with the

consideration of a large scale magnetic field amplification in Keplerian accretion disks, which is very important for modeling of a jet formation.
6. A discussion of jet formation and magnetic jet collimation is added, because jets are closely connected with the accretion disks.
7. The most powerful explosions in the universe have now been related to gamma ray bursts, for which the cosmological origin has been established. The review of the observational properties and theoretical models for these mysterious objects is added in the second edition. All models have a hypothetical character, because the nature of the central machine, as well as radiation mechanism, remain unclear. Therefore, the main attention is given to the analysis of the observational data. Observationally similar objects, called soft gamma repeators (SGR), should have quite different nature, and are related to neutron stars inside our galaxy. The observational data and modeling of SGR is also added.
8. New results, obtained for stabilization of stellar collapse in non-spherical case, have been included in this edition. It was shown that in Newtonian gravity an unlimited collapse to a point may happen only in a spherically symmetric case. Any deviations from the symmetry lead to formation of a dynamically stabilized configuration, which does not collapse.

Many new references have been added. The total list now contains more than a thousand items. Misprints found in the first edition have been corrected here.

Moscow, May 2010 *G.S. Bisnovatyi-Kogan*

Preface to the First English Edition

The development of the theory of stellar evolution has been relatively rapid since 1989, which is when the Russian edition of this book was published. Progress in the field concerned mainly a better understanding of the physical background of stellar processes, in particular the improvements made in calculating new opacity tables. The latter led to a better description of some observational phenomena, such as Cepheid oscillation models, but otherwise led mainly to quantitative corrections of previously known results. The field that may be strongly influenced by the increase of opacity according to the new tables is the mass loss from massive evolved stars. This latter phenomenon has not yet been investigated, however. Many new results have been obtained in helioseismology, the theory of supernova explosions, accretion theory, and 2-D calculations of different phenomena, such as star formation, explosions of rotating magnetized stars and numerical simulations of stellar convection.

This book has been updated and now includes over 150 new references. New material has also been added to otherwise well established sections. This includes the CAK theory of mass loss from hot luminous stars, the description of the Eggleton method of stellar evolution, and a more detailed consideration of the accretion disk structure around black holes.

The revisions and additions of new material substantially increased the number of pages, making it desirable to produce two essentially self-contained volumes. The first volume, "Fundamental concepts and stellar equilibrium," contains the material related to the first six chapters of the Russian edition. The second volume, "Stellar evolution and stability," includes the material of the other chapters of the Russian edition. While both volumes retain the structure of the Russian edition, each of these two volumes now has a self-contained character and could be interesting for different kinds of readers. The first one contains a detailed description of physical processes in stars and the mathematical methods of evolutionary calculations. Thus this volume will be of interest for physicists and specialists in numerical mathematics regardless of the level of their actual involvement in the work on stellar evolution. The second volume contains both the qualitative and the quantitative descriptions of stellar evolution, explosions, stability and oscillations. This will be of interest for the wider astronomical community, observers and theoreticians alike, working or interested in astrophysical phenomena related to stellar formation and evolution.

Finally, those who directly work in the theory of stellar evolution, or want to study this field in depth, will find that both volumes provide them with a comprehensive introduction to and survey of the present state of the art in this field.

I am very grateful to my colleagues from the Cambridge Institute of Astronomy, D.O. Gough, P. Eggleton, and C. Tout, and from Queen Mary and Westfield College, I.W. Roxburgh, A.G. Polnarev, and S.V. Vorontsov, for their hospitality during my stay in these places, many discussions and help with the work on different parts of the book, as well as improvements to the English. I am also grateful to M.M. Romanova, A. Blinov, and S.V. Repin, who translated the Russian edition of the book into English and prepared the TeX file of this translation.

Moscow, August 2000 *G.S. Bisnovatyi-Kogan*

Preface to the Russian Edition

The desire of astrophysicists to gain insights into the mysteries of the birth and death of stars has required the application of almost all branches of modern physics. The results of atomic physics are necessary for studies of stellar birth out of the interstellar medium; and knowledge of the structure of white dwarfs and neutron stars requires the use of liquid- and solid-state theory, and the theory of phase transitions, superconductivity, and superfluidity. Between these extremes, in the area where stars mostly exist, the laws of nuclear physics and weak interactions, and the theories of matter and radiative transfer, are at work.

The equilibrium of a star is described by the equations of hydrodynamics supplemented by general relativity and electromagnetic field theory. The problem of turbulence and convection, not yet completely resolved in terrestrial applications, is even more important and difficult in problems of stellar evolution. This book deals with many of these problems, trying to develop the theory of stellar evolution from a physical standpoint. In this regard, I have followed D.S. Frank-Kamenetski's excellent *Physical Processes in Stellar Interiors* [375]; however, spectacular achievements in the field during the last 25 years have considerably reduced the overlap with this book. An essential part of the items treated here has been considered more qualitatively in Ya.B. Zel'dovich's lectures published in *Physical Grounds of Stellar Structure and Evolution* [1077].

The astrophysical results given in Part II have much of the descriptive character typical of astronomy. The reason for this is that the major results are obtained here through numerical experiments which, just as in a real astronomical situation, can only be described rather than reproduced in a book. I have also tried to clarify, whenever possible, the physical sense of the results.

The material in this book is to some extent presented according to my personal preference, as particular attention is often paid to items connected to my scientific interests. Nonetheless, I have tried to preserve a general understanding of the problems discussed and to give results which are basic, as I see them, to the theory of stellar evolution.

I have tried to select from a large variety of papers those reviews and monographs either representing an important advance in solving some astrophysical problem, or dealing with interesting physical problems which are not necessarily important (or are not regarded as such) for the development of the theory of stellar evolution.

I have considerably softened the selection rules for my own results. Some problems in the book remain unanswered; I have included them in the hope that some reader will succeed in finding their solution.

The book is concerned only with the evolutionary paths of single stars. The theory of stellar evolution for close binaries, in which there is a considerable increase in the evolutionary paths, is treated in the recently published monograph by A.G. Masevich and A.V. Tutukov *Stellar Evolution: Theory and Observations* [673]. The relationship between theory and observations is also considered in this book.

I gratefully acknowledge the assistance of and useful discussions with S.I. Blinnikov, S.A. Lamzin and A.F. Illarionov, and express my particular thanks to E.V. Bugaev and D.G. Yakovlev, who have read several chapters of the book and made many helpful remarks.

Moscow, September 1989 *G.S. Bisnovatyi-Kogan*

Contents of Volume 1

Contents of Volume 2

Chapter 7
Star Formation

7.1 Observations of the Regions of Star Formation

7.1.1 Introduction

The matter in the Universe (its barionic component) is concentrated mainly in stars. Inside galaxies, stars contain more than 90% of the matter, and in galactic clusters, due to the existence of intercluster gas, stars contain more than 70% of the matter. The presence of heavy elements (heavier than carbon) in the intercluster gas, with an abundance of the order of one third of solar gas, indicates that almost all barionic matter in the Universe went through a stellar stage. According to modern views, the enrichment of intercluster gas by heavy elements happens due to outflow of matter from galaxies, where the production of heavy elements takes place due to stellar evolution. It follows from the cosmological models of a hot Universe that only hydrogen and helium, with very small additions of lithium, beryllium and boron, were produced in the Big Bang. All heavier elements, starting from carbon, are produced as a result of stellar evolution (see Sect. 4.4, Vol. 1).

The gas inside galaxies is divided into three phases: hot with $T \sim 10^6 - 10^7$ K, warm with $T \sim 10^4$ K, and cold with $T \leq 100$ K. This division takes place because the thermally stable conditions of the galactic gas, which is approximately in the state of thermal equilibrium under actions of cooling and heating, correspond to these particular temperature intervals. In the static equilibrium galactic gas has an almost uniform pressure, and low temperature regions have larger densities. According to the Jeans criterion of gravitational stability of a uniform gas cloud with density ρ_0, and temperature T_0, masses exceeding Jeans mass M_J are unstable with respect to contraction, leading to star formation. We have [1082]

$$M_J = \frac{4\pi^{3/2}}{3}\left(\frac{\gamma\mathscr{R}T}{G}\right)^{3/2}\rho_0^{-1/2} \approx 10^4 M_\odot T_{100}^{3/2} n_{10000},$$

where $T_{100} = T_0/100\,\mathrm{K}$, $n_{10000} = n_0/10^4\ \mathrm{g/cm}^{\ 3}$, $n_0 \approx \rho_0/m_\mathrm{p}$. For $n_0 = 10^6$, $T = 4\,\mathrm{K}$ we obtain $M_J = 8M_\odot$. The limit of stable masses M_J in the cold phase is

G.S. Bisnovatyi-Kogan et al., *Stellar Physics: 2: Stellar Evolution and Stability*, Astronomy and Astrophysics Library, DOI 10.1007/978-3-642-14734-0_1,

considerably less than the mass of the cold cloud. In this situation the development of instability, accompanied by energy losses due to radiation, leads to a collapse of parts of a gas cloud and the formation of new stars.

7.1.2 Observational Data

Observational surveys have been carried out in infrared ($\lambda = 1 - 100\,\mu$m) and submillimeter ($\lambda = 0.1 - 1\,\mu$m) ranges which have made it possible to study regions of star formation. According to theoretical concepts [1049], the star formation results from instability (thermal or gravitational) development in a dense interstellar cloud leading to hydrodynamical contraction of the cloud or its part, and to formation of a relatively dense, optically thick core. A core surrounded by a dense gas-dust envelope accreting or (and) outflowing outward is likely to have a large size $R \geq 10^3 R_\odot$, a low photosphere temperature $T \leq 500$ K, and be available for surveys only at long wavelengths ($\lambda \geq 2\,\mu$m).

Ground-based observations in the 2.0–20 μm range of regions of the expected star formation in Orion, Monoceros and other constellations have yielded significant results. These regions include young star clusters (associations), HII regions, dark clouds,[1] central condensations of molecular clouds, maser and infrared sources. Observations have provided data on 30 objects that exhibit properties similar to those of a protostar after opaque core formation [1055]. The luminosity of these protostars is $\sim 10^3 L_\odot$ on average, and sometimes reaches $2 \times 10^5 L_\odot$ or $25 L_\odot$. Most of the flux is radiated at wavelengths $\lambda > 2\,\mu$m, and the optical flux observed in a few cases has been well below the infrared one. The energy distribution of one source (NGC 2264) is shown in Fig. 7.1 from [1055]. All these sources are associated with molecular clouds, most of them include H_2O-masers. Some sources are located within compact HII-zones, which, in turn, are embedded in molecular clouds. Source sizes range from 10^3 to $10^5 R_\odot$. Dust and various molecules (CO and others) are important contributors to the observed emission from these sources.

The available observational data suggest that the above 30 objects have characteristics resulting from effects connected with current or recent accretion. These data, however, are not sufficient for making a definite conclusion that the objects are in an accretion stage (see Sect. 7.2) and have a gravitational, rather than thermonuclear, inner energy source.

The observed outflow velocities achieving 10–100 km s^{-1} represent an important and not fully understood property of these objects. These are far higher than the characteristic free-fall velocities of $\sim$1 km s^{-1} in clouds, but well below the observed outflow velocities of about 1,000 km s^{-1} in hot stars [711]. The mass flow from these objects exceeds by far the observed flow from hot stars. It is possible that such an outflow is typical for massive stars in the stage of transition from protostar

[1] In dark gas-dust clouds the absorption in the optical range is $\leq 10^m$.

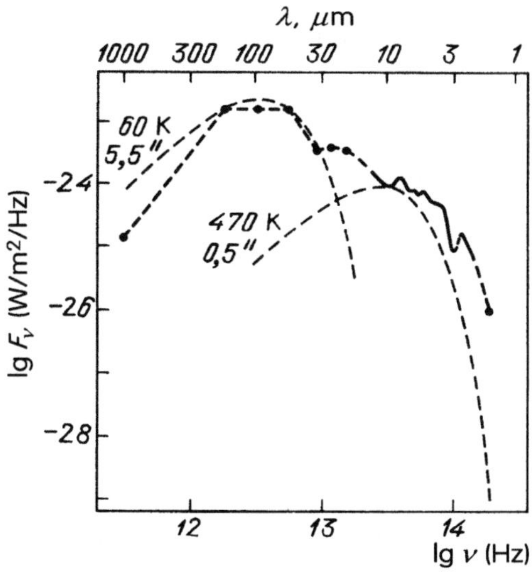

Fig. 7.1 The 1.6/1000 μm energy distribution of the infrared source NGC 2264. *Solid line* represents spectrophotometry, *dashed* segments are interpolations. The *thin dashed lines* are theoretical blackbody emission from circular sources with the indicated temperature and angular diameter

to the main sequence. It may be accounted for by a large optical thickness with $\tau_c \gg 1$ near the critical point. The photosphere radius exceeds here the critical radius ($r_{ph} > r_c$), contrary to the hot stars, where the acceleration of matter is associated with the radiation pressure in lines while the mean optical thickness of the flow is small ($\tau_c \ll 1$). The absence of any adequate theoretical explanation of this phenomenon has given rise to a hypothesis on an essential role in matter outflow and protostar formation [330] of some processes inside stars that are either unknown to date or not addressed by current theoretical models. This hypothesis resembles ideas suggested by Ambarzumian [22]. One of the possible reasons for the absence of observed accretion in the above 30 sources is that they may be fairly massive, $\geq 3 - 5M_\odot$, so that the accretion stage with a gravitational energy source predominance is too short to be observable. However, less heavy stars of mass $\sim 1M_\odot$ with a sufficiently long accretion stage may be cooler and emit mostly at wavelengths $\lambda > 20$ μm. Observational data providing evidence for this assumption have been obtained in ground-based and airplane surveys in the far-infrared ($\lambda > 60$ μm) and submillimeter ranges [559], and in the infrared IRAS satellite surveys that allow us to make measurements in the range of 10–100 μm.

Observations of the Bok globule B 335 in the range 0.06 − 1 mm have resulted in the discovery at its centre of a very cold compact source with a total luminosity of $\geq 5.3L_\odot(D/400 \text{ pc})^2$, which exceeds 70% of the total luminosity of the globule [559]. The H_2 density in the central peak is $\geq 10^6$ cm$^{-3}(D/400 \text{ pc})^2$, the mass of the central core is $6.5M_\odot(D/400 \text{ pc})^2$, and a star with a mass of less than $2M_\odot$ in a stage of gravitational contraction may represent the embedded energy source. Curiously, the central source has not been discovered either at radio or at near-infrared wavelengths $\lambda \leq 10$ μm. The spectrum of this source is given in Fig. 7.2

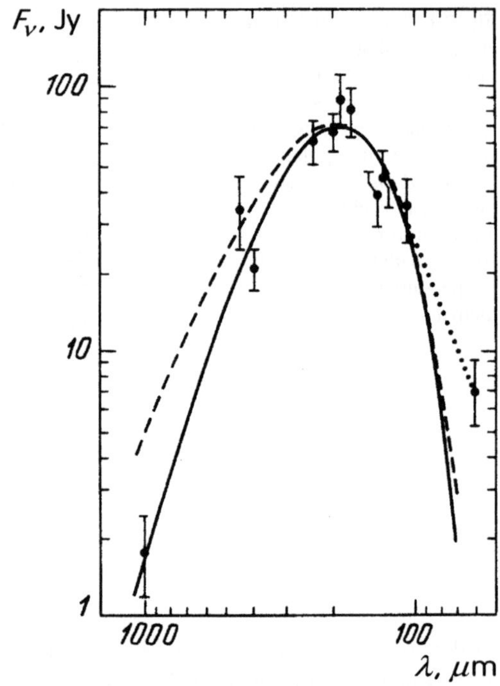

Fig. 7.2 Spectrum of the source in B335. On the experimental dots are superimposed the lines: $\nu^2 B_\nu(15\text{K})$ *solid line*, $\nu B_\nu(18\text{K})$ *dashed line*, $\nu^{-2.5}$ for $\lambda < 110\,\mu\text{m}$ *dotted line*, $1\,\text{Jy} = 10^{-26}$ $\text{Wt} \times \text{m}^{-2} \times \text{Hz}^{-1}$, $B_\nu(T)$ is the Planck energy distribution (see Sect. 2.1, Vol. 1)

from [559]. The absence of emission in these ranges provides evidence for the existence of a very thick dust envelope. This source is the first example from a possibly wide class of protostars of a low luminosity seen only in the far-infrared and submillimeter ranges.

Radio observations of five rotational transition line profiles of H_2CO and CS in B335 provide direct, kinematic evidence of collapse [1083]. The best fit model gave an age of 1.5×10^5 yr, corresponding to an infall radius of about 0.04 pc and a total mass 0.4 $M_\odot$ for the central star and disc. Outside the infall radius there is a static envelope with a r^{-2} density distribution, an average temperature of 13 K and turbulent velocity of 0.13 km/s.

Forming stars of a mass of about the solar mass have been found in the IRAS satellite surveys in Barnard 5 (B5) dark cloud [86], the Chameleon 1 dark cloud [76] and also in the dust cloud L 1551 [356]. These papers report the discovery of more than 10 sources of this type in all. The luminosity of most of them does not exceed $10L_\odot$ and is, in a few cases, $\sim 1L_\odot$. The observed dust temperature is low, $T_d = 20-60$ K, in some sources, and the luminous region length approaches 10^{17} cm. The observed sources are likely to be at different stages of star formation, including the onset of gravitational collapse (IRS2 from [86]). The spectra of the sources IRS1 and IRS2 are shown in Fig. 7.3, see also [87].

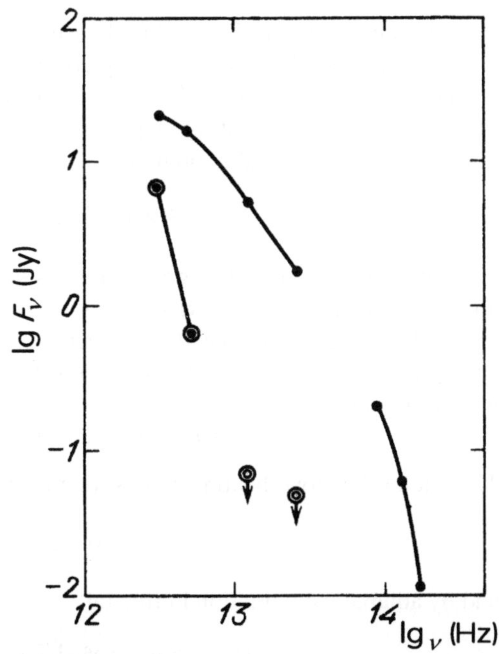

Fig. 7.3 Energy distribution for two infrared sources in B5 dark cloud, the IRAS data cover the frequency range $\nu < 3 \times 10^{13}$ Hz, *filled circle* refers to the source IRS1, *encircled dot* to the source IRS2

7.2 Spherically Symmetric Collapse of Interstellar Clouds

Consider physical processes in a cloud that contracts and transforms into a protostar. At the initial stage, the cloud is transparent for radiation.

7.2.1 Heat Balance of an Optically Thin Cloud

Collisional excitations of the fine structure of CII ion and OI atom, the energy transfer to the dust in its collisions with atoms and molecules are the principal mechanisms of gas cooling in a cloud of molecular hydrogen. CII ions in a molecular cloud result from the action of cosmic rays and background ultraviolet radiation. Defining τ_{CR} as the thickness for penetration of cosmic rays and hard radiation inside the cloud, we have for the rates of molecular cloud cooling via various mechanisms [622]

$$\Lambda_{CII} = 9.0 \times 10^{19} e^{-\tau_{CR}} \rho e^{-92/T} \ \ (\text{erg g}^{-1}\ \text{s}^{-1}), \tag{7.1}$$

$$\Lambda_{OI} = 2.5 \times 10^{20} \rho T^{0.33} e^{-228/T} \ \ (\text{erg g}^{-1}\ \text{s}^{-1}), \tag{7.2}$$

$$\Lambda_{d} = 1.1 \times 10^{14} \rho T^{1/2} (T - T_d) \ \ (\text{erg g}^{-1}\ \text{s}^{-1}). \tag{7.3}$$

Here T is measured in K, $1/10$th of carbon is assumed to be in the gas phase, the opacity corresponding to cosmic ray absorption and included in the $\tau_{\rm CR}$ definition is taken to be $\kappa_{\rm CR} = 300\,{\rm cm}^2\,{\rm g}^{-1}$, T_d is the dust temperature determined by equilibrium between dust heating by gas at the rate Λ_d from (7.3), and thermal cooling of dust grains by the radiation flux

$$j = 2.3 \times 10^{-4} k_P T_d^4 \ ({\rm erg\,m^{-2}\,s^{-1}}), \tag{7.4}$$

from their surface. The dust grain radius is taken to be $r_d = 2 \times 10^{-5}$ cm, and the number of dust grains per gram of gas $N_d = 2 \times 10^{11}$. The mean Planck opacity of the dust matter, calculated (see Sect. 2.2, Vol. 1) as an average of the spectral absorption coefficient over the frequency, with a Planckian spectral distribution as weight function, is

$$k_P = 3 \times 10^{-5}\, T_d^3 \ ({\rm cm^2\,g^{-1}}). \tag{7.5}$$

The rate of the cloud heating by cosmic rays is

$$\Gamma_{\rm CR} = 2.5 \times 10^{-3} e^{-\tau_{\rm CR}} \ ({\rm erg\,g^{-1}\,s^{-1}}) \tag{7.6}$$

and by adiabatic contraction in free fall

$$\Gamma_f = 3.8 \times 10^4 \rho^{1/2} T \ ({\rm erg\,g^{-1}\,s^{-1}}). \tag{7.7}$$

Solving the balance equation

$$\Gamma_{\rm CR} + \Gamma_f = \Lambda_{\rm CII} + \Lambda_{\rm OI} + \Lambda_d, \quad \Lambda_d = j \tag{7.8}$$

gives the gas and dust temperature in the cloud. For a cloud of uniform density, with $\tau_{\rm CR}$ referring to the cloud center, we obtain equilibrium values for the gas temperature of clouds of various masses and densities given in Table 7.1. We see from Table 7.1 that in a wide range of densities the cloud temperature varies weakly, remaining between 5 and 11 K at $\rho = 10^{-19}$–10^{-13} g cmof dense interstellar clouds are analyzed in [576] in more detail.

Table 7.1 Equilibrium temperatures of dense gas clouds

	$M = M_\odot$		$M = 10^3\,M_\odot$	
$\lg \rho$, g cm^{-3}	τ_{CR}	T, K	τ_{CR}	T, K
−22	4.2(−2)	50.6	4.2(−1)	48.1
−21	2.0(−1)	25.4	2.0(0)	24.6
−20	9.1(−1)	15.6	9.1(0)	16.3
−19	4.2(0)	8.4	4.2(+1)	6.6
−18	2.0(+2)	5.1	2.0(+2)	5.1
−17	9.1(+1)	5.4	9.1(+2)	5.4
−16	4.2(+2)	6.3	4.2(+3)	6.3
−15	2.0(+3)	7.5	2.0(+4)	7.5
−14	9.1(+3)	9.1	9.1(+4)	9.1
−13	4.2(+4)	11.0	4.2(+5)	11.0
−12	2.0(+5)	13.3	2.0(+6)	13.3

7.2.2 Equations for Cloud Collapse

In a spherically symmetric approximation, with radiation and heat transfer processes taken into account, the gas motion is described by a set of time-dependent hydrodynamic equations in a Lagrangian system with the mass coordinate m, radiative (L^{rad}) and convective (L^{conv}) heat transfer, and volume energy losses Λ (see Sect. 6.1, Vol. 1). The values r, ρ, P, T and v represent radius, density, pressure, temperature, and radial velocity of the matter. We have the following equations

$$\frac{\partial^2 r}{\partial t^2} + 4\pi r^2 \frac{\partial}{\partial m}(P + \Pi) + \frac{Gm}{r^2} = 0 \text{ (the equation of motion)}, \tag{7.9}$$

$$\frac{\partial r}{\partial t} = v, \tag{7.10}$$

$$\frac{\partial r}{\partial m} = \frac{1}{4\pi\rho r^2} \quad \text{(the continuity equation)}, \tag{7.11}$$

$$\frac{\partial E}{\partial t} - \frac{P + \Pi}{\rho^2}\frac{\partial \rho}{\partial t} + \frac{\partial L_r}{\partial m} = -\Lambda \quad \text{(the energy equation)}, \tag{7.12}$$

$$L_r^{\rm rad} = -\frac{64\pi^2 acr^4 T^3}{3\kappa}\frac{\partial T}{\partial m}\left(1 + \frac{4}{3}\frac{1}{\kappa\rho T}\frac{\partial T}{\partial r}\right)^{-1}, \tag{7.13}$$

$$L_r^{\rm conv} = f\pi r^2 c_p \rho \left(\frac{Gm}{Tr^2}\right)^{1/2} l^2 (\Delta\nabla T)^{3/2}, \tag{7.14}$$

$$\Lambda = \Lambda_{\rm CII} + \Lambda_{\rm OI} + \Lambda_d - \Gamma_{\rm CR}. \tag{7.15}$$

Here, P is the gas pressure P_g for an optically thin region, $\tau \ll 1$, and a sum of a gas and radiation pressure $P_g + P_r$ for an optically thick region $\tau \gg 1$ (see Sect. 1.1, Vol. 1)

$$P_g = \rho \mathcal{R} T, \quad P_r = \frac{aT^4}{3}. \tag{7.16}$$

Here, a is the constant of the radiation energy density, c is the light velocity, c_p is a heat capacity at constant pressure, $\mathcal{R}$ is the gas constant, τ is an optical depth of a corresponding mass layer, $\Delta\nabla T$ is an excess of the temperature gradient over an adiabatic one in a convectively unstable region, f is a parameter determined below. At large heat flows the dynamical effect of radiation on matter may be important in optically thin regions as well. To allow for this effect, the last term in the left-hand side of (7.9) should be rewritten as

$$\frac{Gm}{r^2}\left(1 - \frac{L}{L_{\rm cr}}\right), \quad L_{\rm cr} = \frac{4\pi cGM}{\kappa}, \tag{7.17}$$

where L_{cr} is the critical Eddington luminosity. The form (7.17) is used in [910–912]. In [31, 576–578, 622, 1037] the dynamical effect of the heat flow is ignored, and the form (7.9) is used. The relation for the radiative energy flux L_r^{rad} (7.13) is such that at $\kappa\rho r \gg 1$ it turns into the well-known relation (2.2.32) or (6.1.4a) relation from Vol. 1

$$L_r^{rad} = -\frac{4acT^3}{3\kappa\rho}\frac{dT}{dr}4\pi r^2 \tag{7.18}$$

for the radiative heat transfer and, for the case of a small optical thickness, $\kappa\rho r \ll 1$, will tend to the free radiation flux

$$L_r^{rad} = 4\pi acr^2 T_0^4, \tag{7.19}$$

with a temperature $T = T_0$ at $\tau = 0$ (see Sect. 2.2, Vol. 1). In [622] and [1037] the relation (7.13) with $\Lambda = 0$ has been used for describing the heat transfer process at all optical thicknesses. The condition $\Lambda = 0$ has been also used in [31] with, however, the radiation heat transfer approximation (7.18) for L_r^{rad}, corresponding to $\kappa\rho r \gg 1$. In [577, 578] the same relation with $\Lambda = 0$ has been applied to optically thick layers, while for optically thin layers $\Lambda \neq 0$ has been adopted and the term $\partial L_r/\partial m$ was omitted. Calculations in [31, 1037] take account of convection. The coefficient f is taken there to be

$$f = \left(1 + \left|\frac{\Delta\nabla T}{\nabla T}\right| f_1\right)^{-1}, \quad f_1 = 10^3 \tag{7.20}$$

in order to avoid an overestimate of L_r^{conv} for not very small values of $|\Delta\nabla T/\nabla T|$. The opacity κ_d due to the scattering off dust has been studied at $T \leq 2{,}000$ K in [419, 560]:

$$\kappa_d = Q\frac{\pi r_d^2 x_d}{m_d} = Q\frac{\pi r_d^2 \alpha_d}{\mu_N m_u}, \tag{7.21}$$

where x_d is the weight fraction of the dust, α_d is the number of dust grains per gas particle, μ_N is the mean molecular weight of the gas

$$\mu_N = \left(\sum_i \frac{x_i}{A_i}\right)^{-1}, \tag{7.22}$$

where x_i is a weight concentration, and A_i is a mass number of the i-th element. The sizes of dust grains are [577]

$$\begin{aligned} r_d &= 2\times 10^{-5}\ \text{cm for ice}, \\ r_d &= 6\times 10^{-6}\ \text{cm for SiO}_2, \\ r_d &= 6\times 10^{-6}\ \text{cm for graphite}. \end{aligned} \tag{7.23}$$

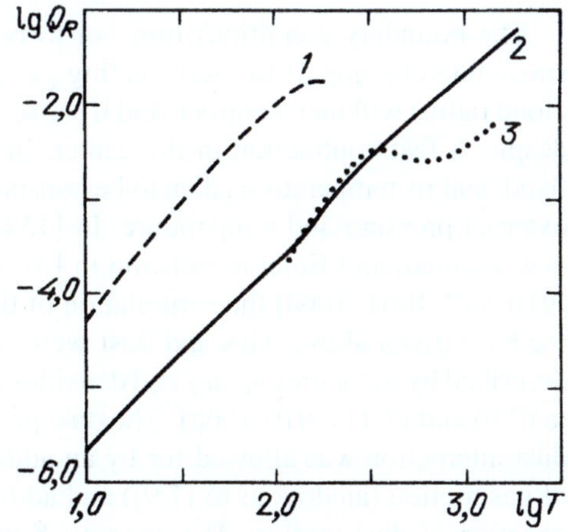

Fig. 7.4 The Rosseland mean dust absorption coefficient for diverse matter: 1,ice; 2, graphite; 3, SiO_2

The Rosseland mean absorption coefficient $Q = Q_R(T)$ is plotted in Fig. 7.4 from [560]. In [577, 578] these values have been used in calculations with $\alpha_d = 10^{-12}$, $x_H = 0.7$, $x_{He} = 0.28$ for $T < 1{,}500$ K. For $T > 1{,}700$ K, data from [316] (see also Table 2.1, Vol. 1) have been used, the range of $1{,}500 < T < 1{,}700$ was covered with the aid of interpolation. The ice component was assumed to sublimate at $T \geq 200$ K, the mineral components (C, SiO_2) at $T \geq 1{,}500$ K. In [1037] the value $\kappa_d = 0.01\,\text{cm}^2\,\text{g}^{-1}$ was specified at low temperatures, tables from [316] were used for $T > 4{,}000$ K, tables from [55] were used at $2{,}000 < T < 4{,}000$ K, and calculation in the range between κ (2,000 K) and κ at a low temperature were made with the aid of a smooth interpolation. In [31] tables from [316] were used for $T > 5{,}000$ K, tables similar to [55] for $1{,}300 < T < 5{,}000$, while for $T < 1{,}300$ K data were taken from [410]. In calculations from [622] it was adopted that $\kappa = \kappa_d = 0.15\,\text{cm}^2\,\text{g}^{-1}$.

A term with artificial viscosity Π has been used in [31, 578, 1037] in a standard form (see, for example, [861]) to produce the possibility for a through numerical calculation of shocks

$$\Pi = \mu\rho\left(\frac{\partial(r^2 v)}{\partial m}\right)^2, \quad \text{where } \mu \text{ is a coefficient of artificial viscosity.} \tag{7.24}$$

A discussion on initial and boundary conditions needed for calculations is given in [622]. As an initial condition, we can take either an equilibrium isothermal sphere of radius

$$R_{\text{sph}} = 0.41\frac{GM}{\mathcal{R}T}, \quad \mathcal{R} \text{ is the gas constant}, \tag{7.25}$$

or a homogeneous cloud with a mass not less than the Jeans mass, unstable to collapse. As noted in [622], an initial density distribution has almost no effect on collapse process.

The boundary conditions turn out to be more important. An isolated cloud is surrounded by gas. If the surrounding gas pressure P_s is taken to be constant, the cloud radius will then contract, and if we set $P_s = 0$, the outer boundary will expand despite a fast contraction in the center. In [31, 622] the outer boundary has been fixed, and its temperature taken to be constant, while in [577, 1037] constants are the external pressure and temperature. In [134, 577] a Lagrangian calculation scheme was adopted, and Eulerian scheme in [31, 622, 910–912, 931–933, 1061–1065]. In [910–912, 1061–1064] the formulation of the problem differs somewhat from what we have given above. Gas and dust were treated separately. Each component was described by velocities u_g, u_d and densities ρ_g, ρ_d inserted into equations analogous to (7.9) and (7.11) [910, 1064]. The dust pressure was assumed to be zero. The gas–dust interaction was allowed for by an additional term $\beta(u_g - u_d)$ in the equation of gas motion (analogous to (7.9)) and additional term $[-\beta(u_g - u_d)\rho_g/\rho_d]$ in the equation of dust motion. The quantity β was estimated in [1064]. The dynamical effect of radiation on dust was also taken into account in the form (7.17) with $\kappa = \kappa_d$ from (7.21). In order to determine temperatures at the stage of opaque core and falling dusted envelope (cocoon), the following procedure was applied instead of solving the energy equations for gas and dust. The luminosity at a given radius $r > r_s$ was taken to be

$$L = L_s + \dot{M}\left(\frac{GM_s}{2r_s} - \frac{GM_s}{r_1} + \frac{u_1^2}{2} - \epsilon\right), \tag{7.26}$$

where L_s, M_s, r_s are the luminosity, mass and radius of the central core, r_1, u_1 are the radius and velocity at the shock, ϵ is the H_2 dissociation energy, $\dot{M}$ is the mass flow onto the central core. In considering massive objects the central core is, for most of the time, a main-sequence star. On finding L, the radiation temperature T_r is obtained from the radiative heat transfer equation (7.18). The dust temperature T_d is set equal to the gas temperature T_g. In the optically thick region $T_r = T_d$, while far from the core

$$T_d^5 = \frac{T_r L}{4\pi acr^2}. \tag{7.27}$$

This procedure has been applied only to investigations of the collapse of massive clouds with $M \geq 3M_\odot$ with use of Eulerian coordinates.

The results of all authors are qualitatively consistent with each other, nevertheless, quantitative discrepancies between them are fairly large.

7.2.3 Calculational Results

A contraction of an infinite homogeneous medium does not alter its homogeneity. The boundary conditions in all the above versions give rise to a rarefaction wave travelling towards the centre with the sound velocity. As the matter velocity does not exceeding the free-fall velocity, the rarefaction wave has time to reach the centre, while the density decreases outward. During further collapse the density contrast increases, the outer layers contract slowly, and a dense core forms inside. The isothermal collapse of a cloud with a self-similar solution

$$\rho = 0.705\frac{\mathscr{R}T}{G}r^{-2}, \quad u = 3.28\sqrt{\mathscr{R}T}, \tag{7.28}$$

is studied best of all. The numerical calculation of this problem [622] shows that a self-similar regime sets in for the major part of the cloud mass even at a fixed outer boundary. Calculations have been made with $M = 1M_\odot$, $T_i = 10\,\mathrm{K}$, $R_i = 1.63 \times 10^{17}\,\mathrm{cm}$, $\rho_i = 1.1 \times 10^{-19}\,\mathrm{g\,cm^{-3}}$. In order for the collapse to proceed, the initial radius must not exceed

$$R_{i,m} = 0.46\frac{GM}{\mathscr{R}T}. \tag{7.29}$$

In all numerical calculations [31, 576–578, 622, 1037, 1061, 1063, 1064] the initial stage of collapse is isothermal and tends to a self-similar regime. Because of a strong non-homogeneity, with density increasing by six orders of magnitude a static optically thick core with hydrogen in the molecular phase forms at the centre. The initial core mass is $10^{-3} - 10^{-2}$ of the cloud mass. The remaining matter of the cloud keeps contracting hydrodynamically and, on passing through the shock front, joins the static core. In a short time of several to several hundred years, when the temperature in the core reaches $\sim$1,900 K, the molecular hydrogen starts to dissociate. The central part of the core undergoes the stage of hydrodynamical collapse again, and its central density increases by an additional four or five orders of magnitude. Some years later the dense secondary core grows to such an extent that all traces of the primary core vanish, and its central temperature becomes about $(1\text{–}2) \times 10^4$ K. This core is a veritable stellar seed and, after the surrounding envelope vanishes upon accreting onto the core or flying away, it transforms into a star.

The fate of the surrounding envelope depends on the relation between two time scales: τ_{ac}, the time of accretion, and τ_{KH}, the Kelvin–Helmholtz time, determined by the core contraction due to energy losses. For low-mass stars with $M <\sim 3M_\odot$, $\tau_{\mathrm{ac}} < \tau_{\mathrm{KH}}$, and the protostar accretes all the surrounding envelope and appears in the optical range before nuclear reactions ignite in the centre. The quasistatic contraction of the protostar (T Tauri stage, see Chap. 8) places it on the main sequence, where hydrogen ignites.

For protostars with $M >\sim 3M_\odot$, $\tau_{\mathrm{ac}} > \tau_{\mathrm{KH}}$, and the core transforms into a main-sequence star as early as at the stage of envelope accretion. A part of its lifetime on the main sequence the star is completely closed by the envelope and seen only as an infrared object. The luminosity of a central star with $M \leq\sim 9M_\odot$ is insufficient to eject the surrounding envelope, and the mass M_* of the star arisen in the optical range is of about the cloud mass M. If $M >\sim 9M_\odot$, a part of the envelope is swept away by the stellar radiation so that

$$M_* \approx 3M^{1/2} \text{ at } M >\sim 9M_\odot. \tag{7.30}$$

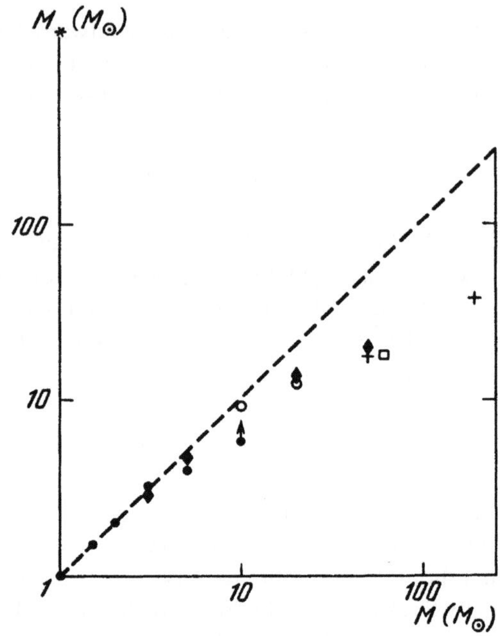

Fig. 7.5 The mass M_* of an optical star as a function of the initial cloud mass M after a spherical collapse. Data are taken from [622] (*filled circle*), [1037] (*filled diamond*), [31] (*open square*), [1064] (*plus*), [1061] (*open circle*)

This $M_*(M)$ dependence based on computations of various authors is shown in Fig. 7.5 from [1063]. Stars with $M >\sim 3M_\odot$ appear in the optical range immediately on the main sequence. Parameters of protostars with $M < 3M_\odot$ appearing in the optical range differ strongly in computations of different authors (see Table 7.2). It is not quite understood now whether this is a consequence of differing initial conditions or of a difference in applied numerical schemes.

In the accretion process the envelope contains several layers with different physical properties resulting from the dust presence [1063]. Around a static and optically opaque central core of radius $\sim 10^{12}$ cm there is an accretion shock front, and above, up to $r_R \approx 10^{14}$ cm, there is an optically transparent zone of matter fall, where dust is entirely absent. The silicate dust sublimation due to the interaction with the core radiation takes place near r_R so that at $r > r_R$ the falling matter is essentially opaque in the optical and ultraviolet ranges. There occurs in this range a conversion of core radiation into an infrared one. On radius $r_p \approx 10^{15}$ cm the optical thickness for the outgoing infrared radiation is of the order of unity, therefore r_p is called, in [1063], the artificial photosphere radius. Up to $r_v \approx 10^{16}$ the infrared radiation field is strong enough to sublimate the ice dust which can exist at the surface of cold mineral dust grains with $T \leq 150$ K. At $r > r_v$ the falling matter has unperturbed initial composition incorporating ice dust which interacts with radiation more effectively than the silicate dust (see (7.21) and (7.24)) and affects the spectrum emitted by the infrared star (see Figs. 7.1–7.3). The structure of accreting gas–dust envelopes (cocoons) has been also studied in [910–912, 1062, 1065].

As may be seen from Table 7.2, the properties of a new born optical star of not large mass are not quite clear. According to [577, 1037], these stars have large

Table 7.2 Initial states of clouds and properties of equilibrium low-mass protostars of hydrostatical formation

$\frac{M}{M_\odot}$	R_0, cm	ρ_0, g cm^{-3}	T_0, K	t, yr	$\frac{R_*}{R_\odot}$	ρ_c, g cm^{-3}	T_c, K	T_{ef} K	$\frac{L}{L_\odot}$	Ref.
0.1	1.71(+16)	1.0(−17)	3	2.94(+4)	10.2	5.0(−2)	1.5(+5)	1.6(+3)	0.6	[1037]
0.2	3.42(+16)	2.5(−18)	3	5.56(+4)	19.0	7.0(−3)	2.0(+5)	1.7(+3)	2.6	[1037]
0.5	8.56(+16)	4.0(−19)	3	1.34(+5)	43.7	3.4(−3)	2.5(+5)	2.1(+3)	32	[1037]
1.0	1.71(+17)	1.0(−19)	3	2.67(+5)	90.0	3.0(−4)	3.5(+5)	2.2(+3)	165	[1037]
0.25	4.1(+16)	1.8(−18)	10		1.6			3.7(+3)	0.5	[622]
1.0	1.63(+17)	1.1(−19)	10	1.2(+6)	2			4.4(+3)	1.5	[622]
1.5	2.0(+17)	4.9(−19)	10		2.5			4.7(+3)	3.0	[622]
2.0	3.26(+17)	2.53(−20)	10	1.4(+6)	3.7			6.8(+3)	30	[622]
1.0	1.63(+16)	1.1(−16)	10	3.5(+4)	6			4.0(+3)	9	[622]
1.0	1.46(+17)	1.5(−19)	10	2.46(+5)	139	0.3	1(+5)	560	2.4	[577]
1.0		1.1(−19)	10		4.69				6.2	[931–933]

radii and developed convective zones and so continue their way towards the main sequence along the Hayashi convective track (see Chap. 8). The model with $1M_\odot$ from [577] shown in Table 7.2 is radiative, nevertheless, a rapid convection development in it leads to an increase in the luminosity L and temperature $T_{\rm ef}$, causing them to approach the values from the model [1037]. On the contrary, the convection in models from [622] is weaker, and they appear with small radii and luminosities. The model with $1M_\odot$ from [622] has a radiative core, and half the mass belongs to the convective zone. The convection is significantly weaker in the model with $1.5M_\odot$, so the star appears in the very end of the Hayashi track, whereas for $M \geq 2M_\odot$ the Hayashi track is absent since the star appears immediately in the radiative state.

Detailed calculations of formation of a protostar with $M = 1M_\odot$ have been carried out in [931–933]. The applied method was analogous to that in [622] which included radiative transfer in moving matter to the Eddington approximation (see Sect. 2.5, Vol. 1), and deuterium burning as well. The opacity tables from [19, 316, 419] were used. The initial and boundary conditions were similar to those in [622]. The optical star appears with $R = 4.69R_\odot$, $L = 6.20L_\odot$. These calculations underlie the birthline for low-mass stars with $0.2M_\odot \leq M \leq 1M_\odot$ obtained in [930], where stars appear in the optical range and continue to evolve towards the main sequence along the Hayashi track (Fig. 7.6a). The value of $\dot{M} = 10^{-5}M_\odot\,{\rm yr}^{-1}$ obtained in [931–933] for $1M_\odot$ was used for all protostars in an accretion stage. As shown in [930], this stellar birthline is in a good agreement with observations (Fig. 7.6b).

It was suggested in [788], that stars end their accretion phase more quickly, through the action of a powerful wind. The estimated mass of the star, passing the contraction pre-main-sequence phase as an optical object have been increased from $3M_\odot$ up to $8M_\odot$. This conclusion was based on the existence of relatively massive Herbig Ae and Be stars, which are supposed to have larger radii than main sequence (MS) stars, indicating their pre-main-sequence nature, see also [854]. The line of birth of stars with masses 0.6–6 $M_\odot$ and their evolutionary tracks to the MS are given in Fig. 7.6b from [788].

7.3 Collapse of Rotating Clouds

Rotation of collapsing interstellar clouds requires two- and three-dimensional calculations. Both Eulerian [197, 198, 218, 226–230, 732, 975, 986] and Lagrangian [35, 42, 553, 754] difference schemes are used in studies of the hydrodynamic axisymmetric contraction. Despite the resemblance of physical assumptions and initial conditions, disagreement in the results of diverse authors is not only quantitative, but also qualitative. The discrepancies are especially large for the case of a rapid rotation. In Eulerian schemes the collapse of rapidly rotating clouds results in the formation of a toroidal figure with a lack of density at the centre. Using Lagrangian schemes always results in obtaining disk-like figures with a maximum central density. As noted in [553, 754], Eulerian schemes yield a large scheme displacement

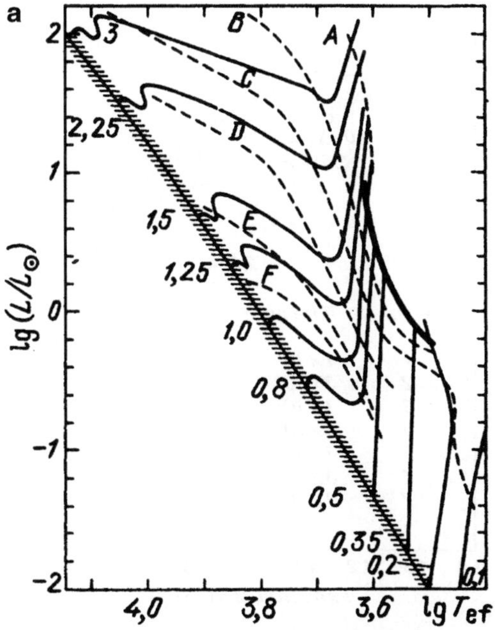

Fig. 7.6a The theoretical birthline *heavy solid curve* for stars with masses between 0.2 $M_\odot$ and 1 $M_\odot$ in the Hertzsprung–Russell diagram. Shown also are isochrones for quasistatic contraction to the given state (Kelvin–Helmholtz age) *dashed curves*. Evolutionary tracks of motion towards the main sequence for stars with various masses taken from [485, 930] are shown by solid curves. The main-sequence line is hatched. Isochrones: 3×10^4 yr (A), 3×10^5 yr (B), 10^6 yr (C), 3×10^6 yr (D), 10^7 yr (E), 2×10^7 yr (F). Chemical composition: $X_H = 0.708$, $X_{^3\mathrm{He}} = 0.0$, $X_{^4\mathrm{He}} = 0.272$, $X_{^{12}\mathrm{C}} = 0.00361$, $X_{^{14}\mathrm{N}} = 0.00120$, $X_{^{16}\mathrm{O}} = 0.00108$. The mixing length is determined by the density scale height: $l = \rho/2|\nabla\rho|$ (see Sect. 3.1, Vol. 1)

viscosity leading to an unphysical angular momentum inflow to the centre which results in torus formation. The results of Lagrangian calculations seem more reliable. We present here the results obtained in [35, 42].

7.3.1 Set of Equations and Difference Scheme Properties

The set of equations used for the two-dimensional collapse investigations has the form

$$\frac{d\mathbf{r}}{dt} = \mathbf{u}, \tag{7.31}$$

$$\frac{d\rho}{dt} + \rho\nabla \cdot \mathbf{u} = 0, \tag{7.32}$$

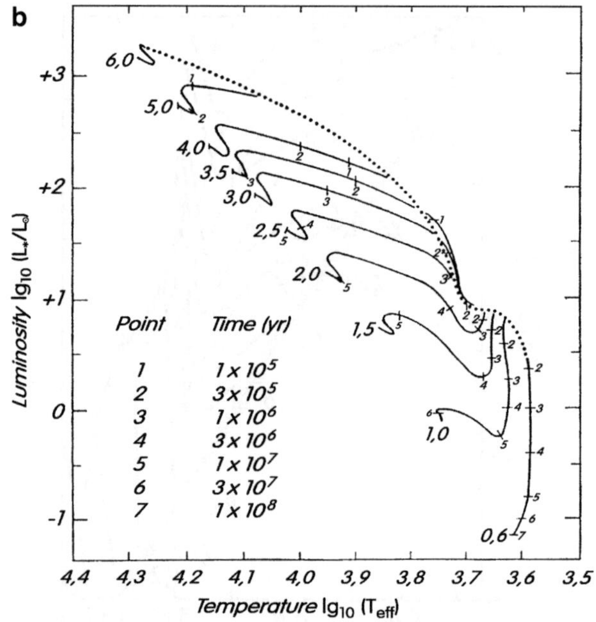

Fig. 7.6b Evolutionary tracks in the H-R diagram for masses between 0.6 and $6M_{\odot}$. Each track is labeled by the corresponding mass in solar units. Tick marks indicate evolutionary times, as given in the table. Each track starts at the birthline (*dotted curve*) and ends at the zero age MS

$$\rho \frac{d\mathbf{u}}{dt} + \nabla P + \rho \nabla \Phi = 0, \tag{7.33}$$

$$\rho \frac{dE}{dt} + P \nabla \cdot \mathbf{u} = 0. \tag{7.34}$$

Here (d/dt) is the substantial (Lagrangian) time derivative related to the Eulerian derivatives by

$$\frac{d}{dt} = \frac{\partial}{\partial t} + \mathbf{u} \cdot \frac{\partial}{\partial \mathbf{r}}, \tag{7.35}$$

and Φ is a solution of a Poisson's equation

$$\nabla^2 \Phi = 4\pi G \rho. \tag{7.36}$$

In all the reported two-dimensional calculations, except [197,198], the polytropic equation of state

$$P = K\rho^{\gamma}, \quad E = \frac{1}{1-\gamma} \frac{P}{\rho} \tag{7.37}$$

has been considered with a K related to the entropy and changing only in the presence of shocks, in accordance with (7.34).[2] In [197, 198], similarly to [622], radiative heat transfer has been taken into account. In [35, 42], a solution of (7.31–7.37) has been searched for in cylindrical coordinates (r, φ, z) with zero density and pressure specified on the outer boundary, and a perfect gas equation of state with

$$\gamma = 5/3, \quad P = \rho \mathscr{R} T. \tag{7.38}$$

The problem is solved in dimensionless variables with use of the scale factors

$$\begin{aligned} \rho_0 &= 1.492 \times 10^{-17}\ \text{g cm}^{-3}, \quad r_0 = z_0 = 3.81 \times 10^{16}\ \text{cm}, \\ t_0 &= 5 \times 10^{11}\ \text{s}, \quad P_0 = \rho_0 r_0^2/t_0^2, \\ u_{0\varphi} &= u_{0r} = u_{0z} = u_0 = r_0/t_0, \quad \omega_0 = u_0/r_0 = 1/t_0, \\ \Phi_0 &= 4\pi G \rho_0 r_0^2, \quad T = u_0^2/\mathscr{R}, \quad \epsilon_0 = u_0^2 = P_0/\rho_0. \end{aligned} \tag{7.39}$$

Here, $\mathscr{R}$ is the gas constant. The angular velocity of rotation is

$$\omega = u_\varphi / r. \tag{7.40}$$

Using (7.39), we can rewrite (7.31–7.37) on substituting

$$F \Rightarrow F \cdot F_0 \tag{7.41}$$

in dimensionless variables. This leads to transformations only in the equation of motion (7.33), Poisson's equation (7.36), and thermodynamic relations, which become

$$\rho \frac{d\mathbf{u}}{dt} + \nabla P + q\rho \nabla \Phi = 0, \tag{7.42}$$

$$P = \rho T, \quad E = \frac{T}{\gamma - 1}, \tag{7.43}$$

$$\nabla^2 \Phi = \rho. \tag{7.44}$$

The dimensionless parameter q is

$$q = 4\pi G \rho_0 t_0^2 = 3.127. \tag{7.45}$$

A sphere of radius R characterized by dimensionless quantities

$$\begin{aligned} (1)\ \rho^{(0)} = 1; \quad P^{(0)} = 1/14; \quad R^{(0)} = 1; \quad \omega^{(0)} = 0.502\,; \\ u_\varphi^{(0)} = r\omega^{(0)}, \quad u_r^{(0)} = u_z^{(0)} = 0\,; \end{aligned} \tag{7.46}$$

[2] For a polytrope with $\gamma = 1$, $P = K\rho$ (isotherm) we have $E = K \ln(\rho/\rho_0)$ for $\rho > \rho_0$, $\rho_0 \to 0$.

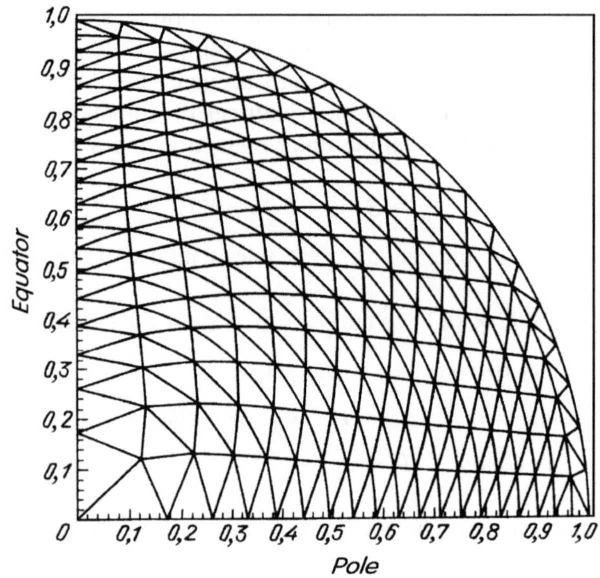

Fig. 7.7 Computation region broken up into triangular cells at initial time. The number of knots is 227, number of cells 396

$$(2)\ \rho^{(0)} = 1; \quad P^{(0)} = 1/560; \quad R^{(0)} = 1; \quad \omega^{(0)} = 1.004;$$
$$u_\varphi^{(0)} = r\omega^{(0)}, \quad u_r^{(0)} = u_z^{(0)} = 0 \tag{7.47}$$

has been taken in [35, 42] for the initial conditions. Similar initial conditions have been adopted in [229]. A difference scheme on an irregular triangular grid of arbitrary structure in a Lagrangian coordinate system was applied in calculations [35,42] (Fig. 7.7). A completely conservative difference scheme of two-dimensional hydrodynamics underlies the calculation techniques from [38]. The inclusion of gravitation causes the numerical scheme as a whole to be no longer completely conservative, but the angular momentum conservation does hold both locally and globally. Theoretical studies of implicit difference schemes on triangular grids have been carried out in [32, 33]. Stability of such a difference scheme is proved to the linear approximation in [39] on an example of a model problem.

7.3.2 Calculational Results

We present calculational results in a dimensionless form. The boundary pressure at the initial moment has been taken to be equal to $P^{(0)}$, then to decrease linearly in t to zero, and from $t = 1$ up to the calculation termination $P_s = 0$. A smooth density decrease is adopted in order to eliminate the rarefaction wave otherwise arising at initial time, which distorts boundary cells and reduces the calculation accuracy. At the beginning, the cloud compression along the z axis is higher than in the equatorial plane. In the first variant (7.46) from [35] the maximum compression is achieved at $t = 1.3$, when the central density $\rho_c = 45$. A shock arises at this moment, and its subsequent propagation causes the outer regions to expand. The spheroid axis ratio is, at this moment, in good agreement with calculations in [229]. At $t = 3$ the cloud occupies a sphere of radius $R = 1.5$, and the expansion turns into a new contraction. The second contraction proceeds more smoothly than the first one because of increasing internal energy of the cloud.

By the time $t = 6.6$ the cloud almost achieves an equilibrium state with weak fluctuations around it. At the last time point of calculations the kinetic energy of rotation is $\sim$1% of the total energy, the central density $\rho_c = 10$, the outer size $R \approx 1.3$. Changes in the cloud shape and velocity field during contraction are shown in Fig. 7.8a–d, and the time evolution of the parameters $\alpha = E_{\text{inner}}/|E_{\text{grav}}|$ and $\beta = E_{\text{rot}}/|E_{\text{grav}}|$ in Fig. 7.9. The quantity $\alpha + \beta$ tends to an equilibrium value of $1/2$ according to the virial theorem [617].

In the second variant (7.47) the initial pressure has been reduced by a factor of 40, and the rotation velocity doubled as compared to the first version. Here, the initial values of α and β are

$$\alpha_0 = 0.00425, \qquad \beta_0 = 0.324. \tag{7.48}$$

This variant was calculated in [42], where grid restructuring was utilized. Average numbers of knots of about 6,000 and cells of about 12,000 were used. Maximum contraction occurs at $t = 1.2856$ with $\rho_c^{max} \approx 195$. Density contours and their distributions along the rotational axis and the equatorial plane are given in Fig. 7.10. At this moment the cloud has an oblate spheroid shape with an axis ratio $\sim 1 : 5$. The inner part of the cloud, containing $\sim$90% of its mass has a much flatter shape with an axis ratio $\sim 1 : 100$. At the time of maximum contraction the shock wave reflects from the equatorial plane, propagates outwards, reaching the boundary of the cloud at $t = 1.3149$, with the density in central cells at $\rho_c \approx 93$. At this intermediate stage, the maximum density is situated outside the centre of the cloud, resembling the final density distribution of torus shape from Eulerian scheme calculations in [229]. The Mach number of the shock at its beginning is about 30 near the centre, and increases during the outward shock propagation. The shock disappears when it reaches the outer boundary in the equatorial plane at $t = 1.4046$. The cloud proceeds to expand until $t = 1.4992$, when the central density $\rho_c \approx 10$ is a maximal one, and the central part of the cloud restores its disc-like shape. The second contraction begins at that time, with density slowly growing near the centre. The light and very stretched envelope continues to expand, part of the matter ($\sim$5%) which has kinetic energy exceeding the potential one, flies away to infinity. The calculations were finished at $t = 1.77425$ with $\rho_c \approx 32$, and the distribution of parameters

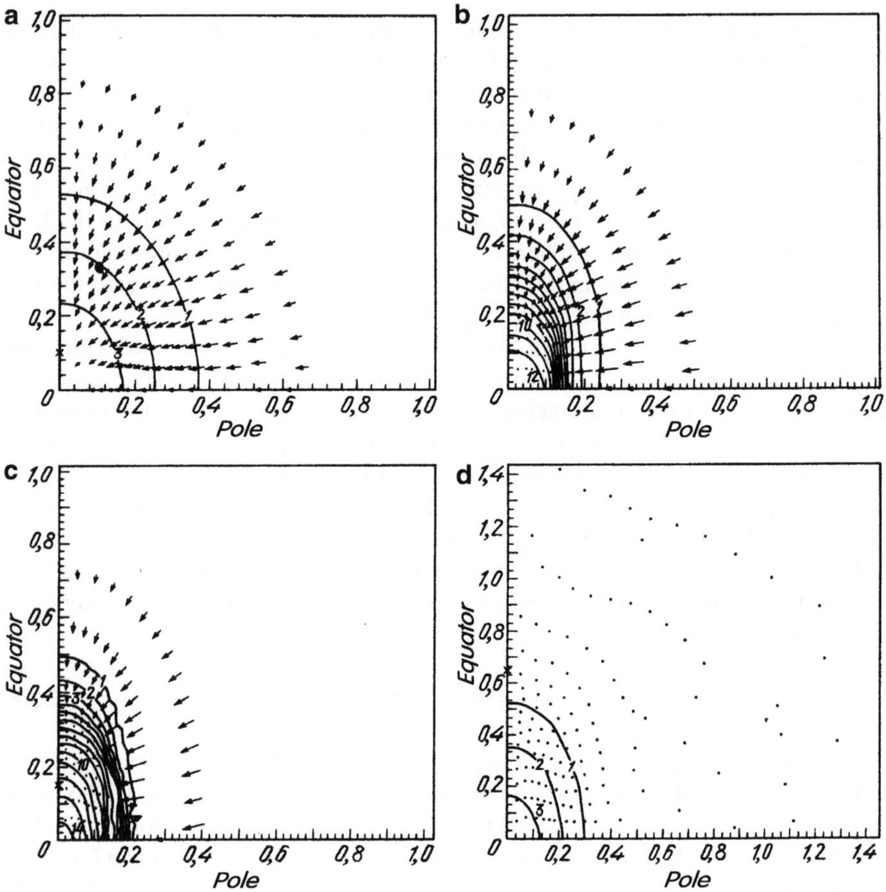

Fig. 7.8 Changes in the cloud shape during collapse in the variant (7.46): (**a**) $t = 1.0$; (**b**) $t = 1.2$; (**c**) $t = 1.3$; (**d**) $t = 6.6$. The numbers mark lines of equal density with $\rho_k = 3k$, the arrows determine a velocity field, their lengths are proportional to the velocities

Fig. 7.9 Time dependence of α, β and $\alpha + \beta$ in the version (7.46). At the last point at $t = 6.6$ one has $\alpha = 0.4048$, $\beta = 0.09451$

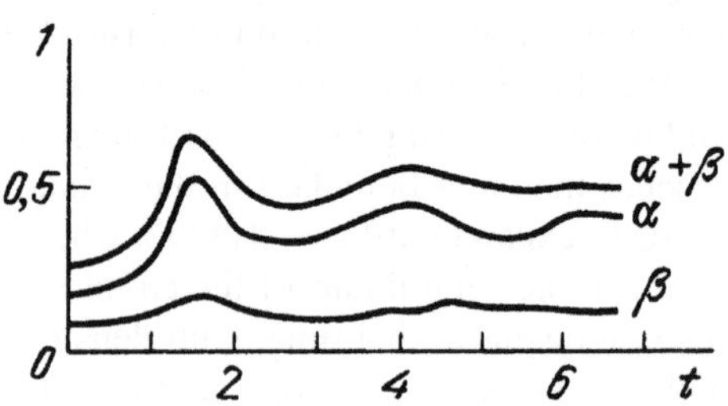

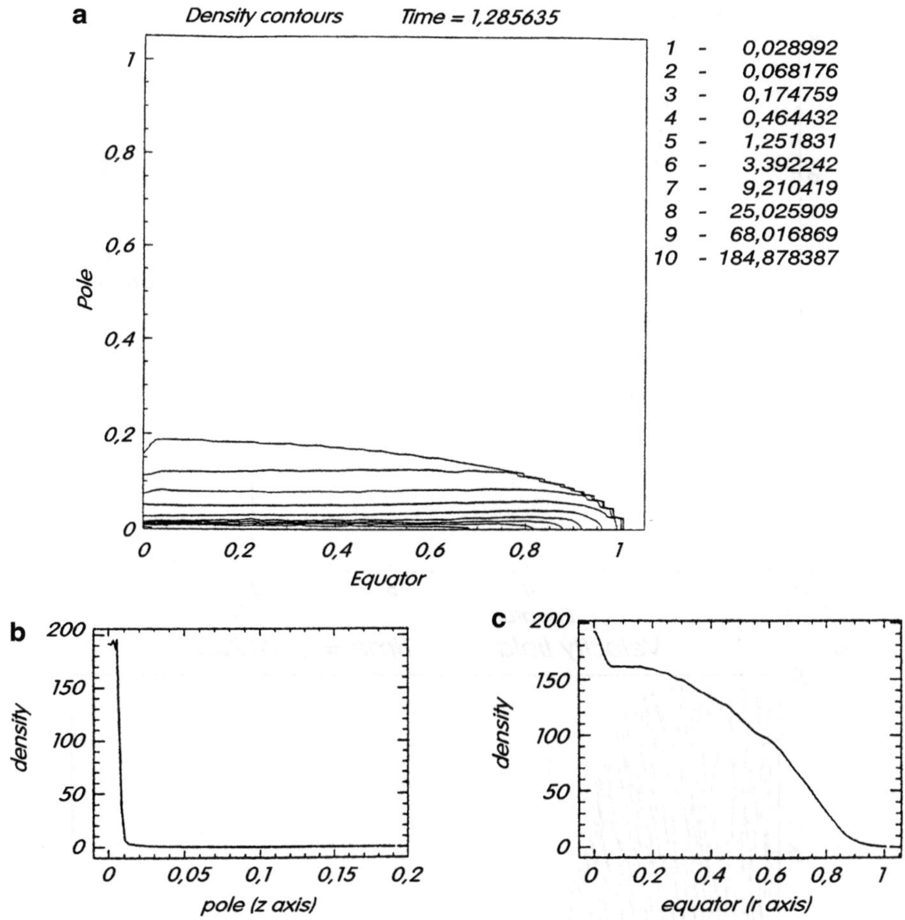

Fig. 7.10 Density contours (**a**), density distributions along z-axis (**b**), and along r-axis (**c**) for t =1.276856

presented in Fig. 7.11. The time dependence of the parameters α and β are given in Fig. 7.11 g. The second contraction starts at $\alpha_1 = 0.08$, $\beta_1 = 0.34$, so the equilibrium figure will definitely have a disc-shaped form. This result differs from [229], where a torus formation was obtained for higher initial pressures with $\alpha_0 = 0.0085$, and the same rotational energy $\beta = 0.324$ (see (7.48)). The star formation during collapse of rotating clouds is strongly influenced by heat processes considered in Sect. 7.2. Nonetheless, they have not been effectively included in rotational model calculations.

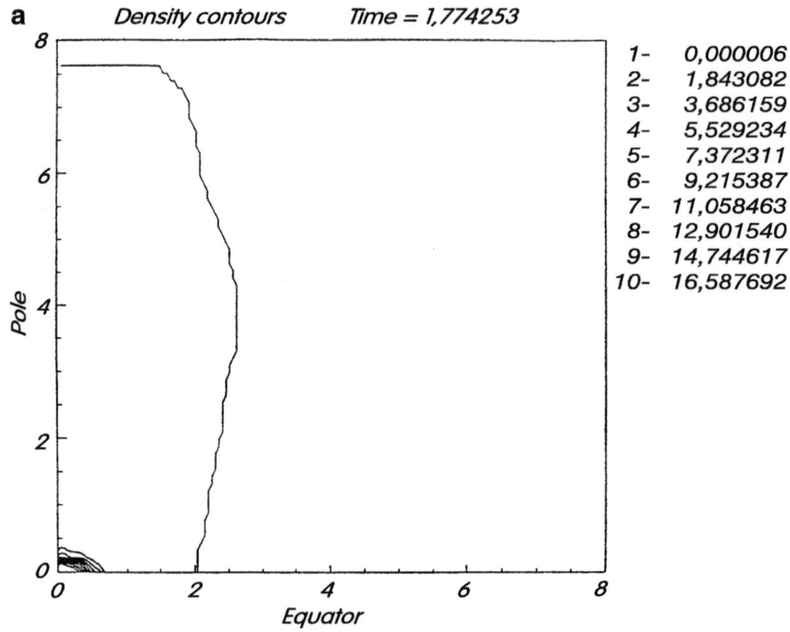

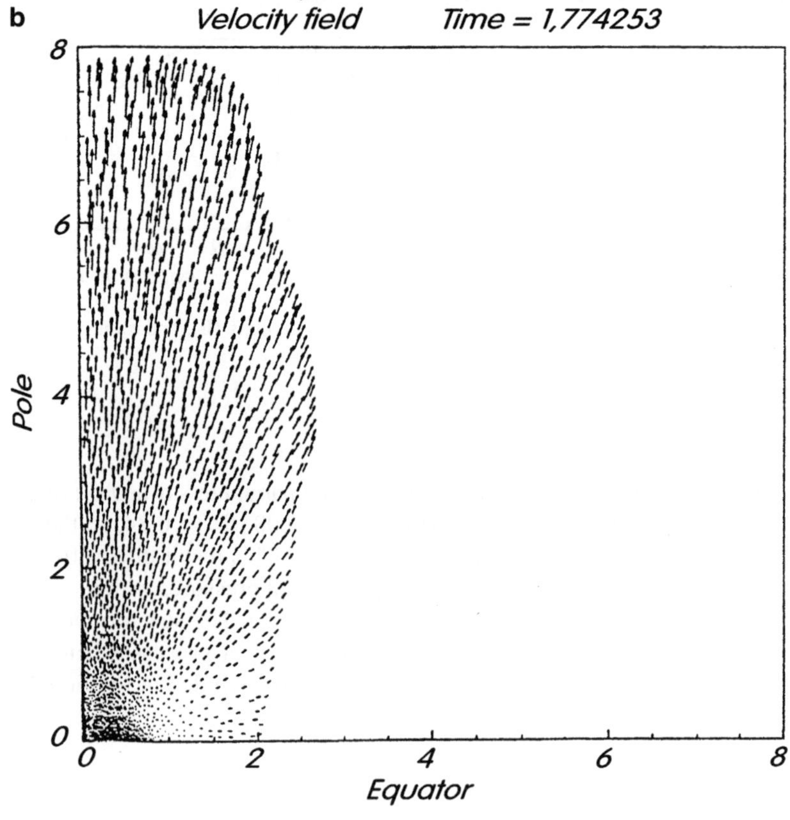

Fig. 7.11 a–b. Density contours (**a**), velocity field (**b**). For (**c**)–(**g**) see next page

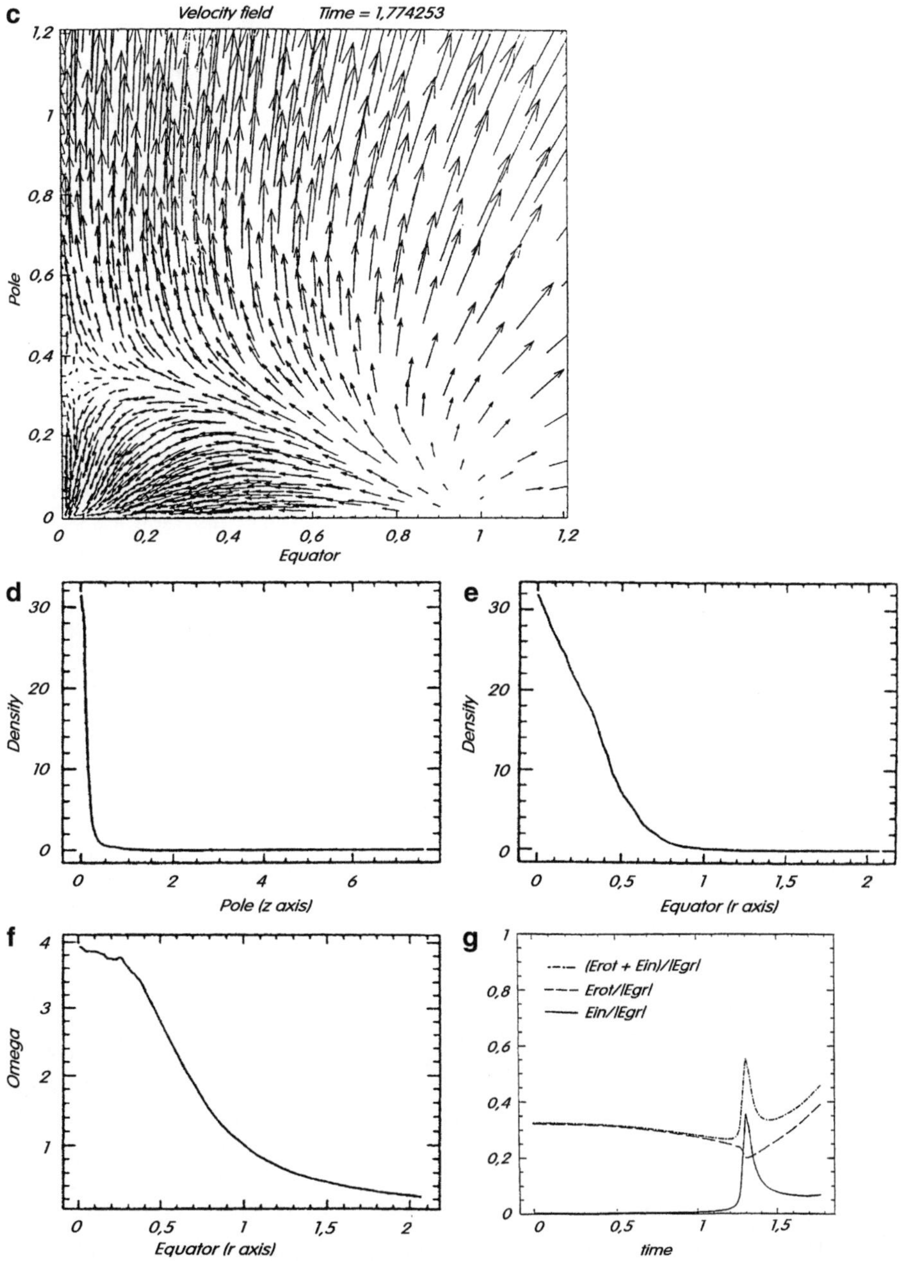

Fig. 7.11 c–g. Central part of the velocity field (**c**), density distributions along z-axis (**d**) and along r-axis (**e**), and distribution of angular velocity along r axis (**f**) for $t = 1.774253$, Time variation of parameters $\alpha = E_{inner}/|E_{grav}|$, $\beta = E_{rot}/|E_{grav}|$, and $\alpha + \beta$ (**g**)

Chapter 8
Pre-Main Sequence Evolution

As can be seen from the calculations of cloud collapse described in Sect. 7.2, stars with mass $M > 3M_\odot$ appear in the optical range immediately near to the main sequence. Objects of lower mass exist for part of the time as optical stars with radiation due to gravitational contraction energy. Consider evolution of a star from its appearance in the optical region until it reaches the main sequence, and its central temperature becomes sufficiently high to initiate a nuclear reaction converting hydrogen into helium (see review [105]).

8.1 Hayashi Phase

The pre-main sequence evolution of stars takes place at not very high temperatures, when a non-full ionization of matter and large opacity cause such stars to be almost convective. This fact was first established by Hayashi [461, 462], who took into account a convection in constructing evolutionary tracks for contracting stars in the HR diagram (see Fig. 7.6 from [930] plotted with the aid of calculations from [485]). Evolutionary calculations in the papers listed below have been made by the Henyey method. This method is based on the division of the stellar mass into J intervals, and writing the differential equations describing static equilibrium, mass conservation, and heat transfer in a difference form. The solution of linearized difference equations is obtained by the "back substitution method", which is very convenient for numerical computations (see Sect. 6.1, Vol. 1).

8.1.1 Nuclear Reactions

Though reaction rates are insufficient for establishing a thermal equilibrium of the star, some contribution into the heat balance of the star in the phase of gravitational contraction to the main sequence can nevertheless be done by reactions with light elements. Calculations in [462] include the burning of ^{2}D, ^{7}Li, ^{9}Be, small amounts of which were formed at the beginning of the Universe expansion. The rate of energy

G.S. Bisnovatyi-Kogan et al., *Stellar Physics: 2: Stellar Evolution and Stability*, Astronomy and Astrophysics Library, DOI 10.1007/978-3-642-14734-0_2,

release in the ^{2}D(p, γ)^{3}He reaction is given in Sect. 4.2, Vol. 1. For the two other reactions, we have from [258, 393]

$$
\begin{aligned}
&{}^7\mathrm{Li}(\mathrm{p},\alpha)^4\mathrm{He}, \qquad Q_6 = 17.346,\\
N_A\langle{}^7\mathrm{Li\,p}\rangle_\alpha &= 1.096\times 10^9\ T_9^{-2/3}\exp\left(-\frac{8.472}{T_9^{1/3}}\right)\\
&\quad - 4.830\times 10^8\ T_{9A}^{5/6}\ T_9^{-3/2}\exp\left(-\frac{8.472}{T_{9A}^{1/3}}\right)\\
&\quad + 1.06\times 10^{10}\ T_9^{-3/2}\exp\left(-\frac{30.442}{T_9}\right),\\
T_{9A} &= T_9/(1+0.759\,T_9), \qquad (8.1)\\
\epsilon_{^7\mathrm{Li\,p}\alpha} &= 2.391\times 10^{18} x_{^7\mathrm{Li}}\, x_H\, \rho N_A\langle{}^7\mathrm{Li\,p}\rangle_\alpha;
\end{aligned}
$$

and

$$
\begin{aligned}
&{}^9\mathrm{Be}(\mathrm{p},\alpha)^6\mathrm{Li}, \qquad Q_6 = 2.126,\\
N_A\langle{}^9\mathrm{Be\,p}\rangle_\alpha &= 2.11\times 10^{11}\ T_9^{-2/3}\exp\left[-\frac{10.359}{T_9^{1/3}} - \left(\frac{T_9}{0.520}\right)^2\right]\\
&\quad\times\Big(1 + 0.040\ T_9^{1/3} + 1.09\ T_9^{2/3} + 0.307\,T_9\\
&\quad\quad + 3.21\ T_9^{4/3} + 2.30\ T_9^{5/3}\Big)\\
&\quad + \frac{4.51\times 10^8}{T_9}\exp\left(-\frac{3.046}{T_9}\right) + \frac{6.70\times 10^8}{T_9^{3/4}}\exp\left(-\frac{5.160}{T_9}\right), \qquad (8.2)\\
\epsilon_{^9\mathrm{Be\,p}\alpha} &= 2.279\times 10^{17} x_{^9\mathrm{Be}}\, x_H\, \rho N_A\langle{}^9\mathrm{Be\,p}\rangle_\alpha.
\end{aligned}
$$

Detailed calculations for the pre-main sequence evolution of stars with masses of 0.5, 1.0, 1.25, 1.5, 2.25, 3.0 $M_\odot$ have been performed in [485]. They include reactions of the pp-cycle and the ^{12}C, ^{14}N, and ^{16}O burning in the CNO cycle. The corresponding reactions and the energy release per reaction Q_6 (in MeV) are

$$
\begin{aligned}
&\mathrm{p}\,(\mathrm{p},\mathrm{e}^{+\,\circ})^2\mathrm{D}, \quad \text{with } Q_6 = 1.192;\\
&{}^3\mathrm{He}\,({}^3\mathrm{He},2\mathrm{p})^4\mathrm{He}, \quad \text{with } Q_6 = 12.860;\\
&{}^{12}\mathrm{C}\,(\mathrm{p},\gamma)^{13}\mathrm{N}, \quad \text{with } Q_6 = 1.944;\\
&{}^{14}\mathrm{N}\,(\mathrm{p},\gamma)^{15}\mathrm{O}, \quad \text{with } Q_6 = 7.297;\\
&\text{and } {}^{16}\mathrm{O}\,(\mathrm{p},\gamma)^{17}\mathrm{F}, \quad \text{with } Q_6 = 0.600.
\end{aligned}
$$

The reaction rates and energy release rates of the above reactions are given in Sect. 4.2, Vol. 1. The notations used in (8.1)–(8.3) are: $T_9 = T/10^9$ K, $\langle 01\rangle \equiv \langle\sigma v\rangle_{01}$ ($cm^3\ s^{-1}$) is the reaction rate, σ is a cross-section of the reaction, v is a relative velocity, $\langle\rangle$ denotes averaging of the corresponding values of reacting nuclei over the energy distributions (Maxwellian), $N_A = 6.02252 \times 10^{23}\ g^{-1}$ is the Avogadro number. These notations are common in the literature, following [392], see also Sect. 4.1, Vol. 1.

From [258, 459] one has the reaction rate

$$^4\mathrm{He}(^3\mathrm{He},\gamma)^7\mathrm{Be}, \qquad Q_6 = 1.588,$$

$$N_A\langle^4\mathrm{He}\,^3\mathrm{He}\rangle_\gamma = 5.61\times 10^6\, T_9^{-3/2}\, T_{9A}^{5/6} \exp\left(-\frac{12.826}{T_{9A}^{1/3}}\right), \tag{8.3}$$

$$T_{9A} = T_9/(1+0.0495\,T_9),$$

$$\epsilon_{^4\mathrm{He}\,^3\mathrm{He}\gamma} = 1.277\times 10^{17} x_{^4\mathrm{He}}\, x_{^3\mathrm{He}}\, \rho N_A\langle^4\mathrm{He}\,^3\mathrm{He}\rangle_\gamma.$$

For other reactions of pp- and CNO-cycles, the heat release is taken into account in [485], but they are assumed to proceed instantaneously. For example, the reaction $p\,(p, e^+\nu)^2D$ is assumed to produce ^3He yielded by the fast $^2D(p,\gamma)^3$ He reaction with $Q_6 = 5.494$, rather than ^2D, while the $^{12}C(p,\gamma)^{13}N$ reaction eventually produces ^{14}N, because the reactions $^{13}N \rightarrow {}^{13}C + e^+ + \nu$ with $Q_6 = 1.51$, and $^{13}C(p,\gamma)^{14}N$ with $Q_6 = 7.551$ proceed more rapidly, (see Sect. 4.2, Vol. 1). The $^6Li\,(p,\alpha)^3He$, $^{10}B\,(p,\alpha)^7Be$ and $^{11}B\,(p,\alpha)^8Be$, reactions, which are of little importance in the energy release, are also considered in [444]. All the reaction rates used in the calculations have been multiplied by a screening factor increasing the reaction rates due to a reduction in the electric repulsion between nuclei (see Sect. 4.5, Vol. 1). The times of approach to the main sequence t_{ms} for stars of diverse masses are [485]:

t_{ms}, yr	2.514(6)	5.855(6)	1.821(7)	2.954(7)	5.016(7)	1.550(8)
$M/M_\odot$	3.0	2.25	1.5	1.25	1.0	0.5

We now clarify the reasons for the appearance of extrema on evolutionary tracks [485] (Fig. 7.6). The first minimum to the right is caused by the growth of the radiative core. Further contraction of stars with $M \geq 0.8\,M_\odot$ is accompanied by a rapid accumulation of ^3He produced in the H and ^2D burning in the reactions $p(p,e^+\nu)^2D$, $^2D(p,\gamma)^3He$, and by a diminution of ^{12}C in the reaction $^{12}C(p,\gamma)^{13}N$ (see Sect. 4.2, Vol. 1) that enhance the pressure gradient slowing the star contraction and lead to convection development in the centre. As a result, the star passes over the peak of its luminosity; a star with $M \leq M_\odot$ reaches the main sequence after the disappearance of the convective core.

For $M = 1.5\,M_\odot$, a decrease in the concentration of ^{12}C reduces the fractional contribution of the nuclear energy into the luminosity. The gravitational energy begins to play a more important role when the star has not reached the main sequence, which causes the luminosity to reach a second minimum and to increase subsequently. After the concentration of ^{12}C has reached an equilibrium value, the role of the nuclear energy becomes more important for the second time, the contraction slows down, the luminosity reaches a maximum once again, and the star reaches the main sequence with a non-zero mass of its convective core. The appearance of new extrema near the main sequence for $M \geq 1.25\, M_{\odot}$ is due to an increased role of the CNO-cycle of hydrogen burning in stars of large mass compared to the pp-cycle.

Calculations of pre-main sequence evolution for masses $M = (0.5–2.5)\, M_{\odot}$ for two different metallicities ($Z = 0.02$ and 0.04) and different convection factors α_p from (8.6) have shown [390] a high sensitivity of results to the input parameters. It was concluded that it is quite difficult to determine masses of T Tauri stars with a precision of better than $0.3 M_{\odot}$. The same conclusion holds, to a lesser extent, for age determination, especially if the metallicity of the observed star is not accurately measured.

8.1.2 *Non-Ideality of Matter*

In studies of the evolution of low-mass stars, $M \leq 0.2 M_{\odot}$, Coulomb corrections to the equation of state and ionization by pressure have been taken into account ([479], see Sect. 1.4, Vol. 1). Simplified techniques have been used in [380, 443–446] in order to include the ionization by pressure. The level shift had been taken into account by using a factor $\phi(r_0) = \exp(r_p/r_0)^3$ in the Saha equation [617], which is written in this case as

$$\frac{y_{i,j-1}}{y_{ij}} = n_e \frac{g_{i,j-1}}{2g_{ij}} \left(\frac{2\pi\hbar^2}{m_e kT} \right)^{3/2} e^{I_{ij}/kT} e^{(r_p/r_0)^3} \equiv n_e K(T). \tag{8.4}$$

Here $I_{ij} = \epsilon_{i,j-1} - \epsilon_{ij}$ is the ionizaton energy (potential) of the j-th electron, and $I_{i0} = 0$; y_{ij} is a fraction of a j-folded ionization of the i-th element; g_{ij} is a statistical weight of the i-th element in the ionization state j. Moreover,

$$r_0 = \left(\frac{3}{4\pi n_i} \right)^{1/3}$$

is the mean interion separation, and

$$r_p = \gamma \frac{\hbar^2}{m_e e^2}$$

is the quantity close to the Bohr radius and given in Table 8.1 from [443].

The degeneracy of electrons has been included in $K(T)$ in the right-hand side of (8.4) by using, as in [479], the relevant expression for the chemical potential μ_{te} of

Table 8.1 Radii r_p

Element	r_p, A	Element	r_p, A	Element	r_p, A
H	0.795	N	0.53	Ni	1.07
H^-	2.12	O	0.45	Ca	2.03
H_2	1.18	C	0.66	Al	1.21
H_2^+	1.30	S	0.82	Na	1.55
He	0.475	Si	1.06	K	2.06
He^+	0.400	Fe	1.22		
Ne	0.32	Mg	1.32		

electrons in the basic relation determining the ionization equilibrium

$$\mu_{t,i,j-1} = \mu_{t,ij} + \mu_{te}, \tag{8.5}$$

where $\mu_{t,ij}$ is the chemical potential of the i-th element in the ionization state j (see also Sect. 1.2, Vol. 1).

8.1.3 *Evolution of Low-Mass Stars, Minimum Mass of a Star on the Main Sequence, Role of Various Factors*

Evolutionary calculations of low-mass stars in approach to the main sequence have been made by the Henyey method. Not only diverse methods of allowing for the non-ideality have been used in these calculations but a variety of chemical compositions and coefficients α determining the mixing length l have also been considered. Here, $\alpha \equiv \alpha_p$ is connecting the pressure scale height H_p with l, so that

$$l = \alpha_p H_p = \alpha_p \frac{P}{|dP/dr|} \tag{8.6}$$

(see Sect. 3.1, Vol. 1). Evolutionary tracks as functions of α have been studied in [444]. The effect of specified boundary conditions, deuterium burning and an applied method of allowing for the non-ideality on the evolutionary track of a star contracting to the main sequence have been investigated in [445, 446].

The evolutionary tracks of stars calculated in [380] from the Hayashi boundary are shown in Fig. 8.1. The chemical composition corresponds to $x_H = 0.7$, $x_{He} = 0.27$, $x_Z = 0.03$. The atmosphere and envelope are calculated similarly to [769] by solving equations for the static atmosphere with an approximate description of optically thin layers with $\tau < 1$. This description gives a smooth transition to the equation for the radiative heat conductivity, valid for deep layers inside the photosphere (see Sects. 2.2.3 and 6.1, Vol. 1). The envelope mass equals 3% of the stellar mass. The convection is included according to the mixing length theory with $l = P/\nabla P$ ($\alpha = 1$), the non-ideality similarly to [443], see (8.4); the hydrogen burning reactions in the proton cycle with the inclusion of screening are also considered in this paper.

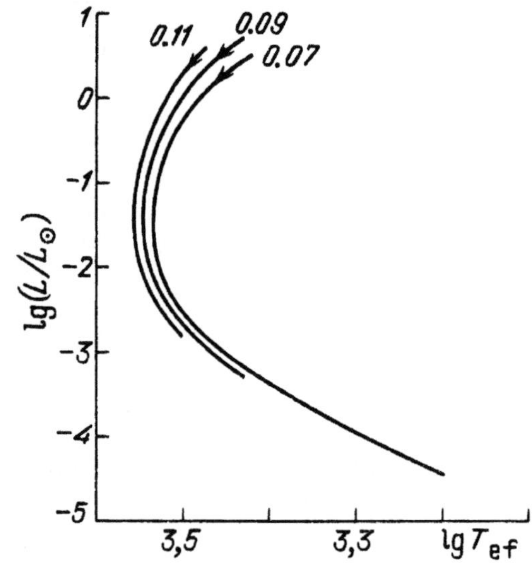

Fig. 8.1 Tracks for spherically symmetric stars with masses of 0.07, 0.09 and 0.11 $M_{\odot}$ in the HR diagram

Table 8.2 Parameters of stars with maximum central temperatures T_c

$\frac{M}{M_{\odot}}$	t, yr	$\lg T_c$	$\lg \rho_c$	$(\beta-\alpha)_c$	$\frac{L_{grav}}{L}$	$\lg T_{ef}$	$\lg \frac{L}{L_{\odot}}$	$\lg \frac{R}{R_{\odot}}$
0.07	2.2(8)	6.448	2.5	4.2	0.96	3.41	−3.27	−0.94
0.08	3.3(8)	6.524	2.62	4.2	0.78	3.42	−3.28	−0.96

Table 8.3 Parameters of stars on the main sequence

$\frac{M}{M_{\odot}}$	t, yr	$\lg T_c$	$\lg \rho_c$	$(\beta-\alpha)_c$	$\lg T_{ef}$	$\lg \frac{L}{L_{\odot}}$	$\lg \frac{R}{R_{\odot}}$
0.09	1.6(9)	6.596	2.74	4.41	3.42	−3.30	−0.27
0.10	8.9(8)	6.645	2.67	3.35	3.46	−3.07	−0.93
0.11	5.7(8)	6.684	2.59	2.57	3.49	−2.88	−0.89

According to the calculations, stars with masses of $0.09M_{\odot}$, $0.1M_{\odot}$ and $0.11M_{\odot}$ reach the main sequence, whereas stars with mass $0.08M_{\odot}$ and less reach a state with a maximum central temperature T_c during contraction and subsequently cool, transforming into degenerate hydrogen dwarfs.

Table 8.2, from [380], gives parameters of stars with maximum T_c, while Table 8.3 presents parameters of stars on the main sequence, when $L_{\text{grav}}/L = 0.01$.

The central values of

$$\alpha = \frac{M_e C^2}{\text{kT}} \qquad \beta = \frac{M_{te}}{\text{kT}} \tag{8.7}$$

are used to determine a level of electron degeneracy. Large positive values of $(\beta-\alpha)$ correspond to highly degenerate, and large negative ones determine non-degenerate electrons (see Sect. 1.2, Vol. 1).

Fig. 8.2 lg T_c as a function of lg ρ_c for a star with initial mass 0.07 $M_\odot$ and the mass increase rate $10^{-12} M_\odot$/yr near the main sequence. The *dashed curves* represent spherically symmetric stars with constant masses of 0.07, 0.08, 0.09, 0.10 and 0.11 $M_\odot$

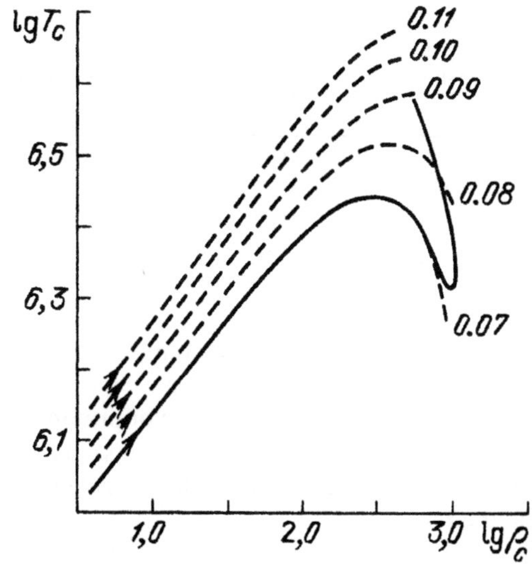

The minimum mass of a star on the main sequence is in the interval $0.08 M_\odot \leq M \leq 0.09 M_\odot$. The low-mass stars reaching the main sequence (Table 8.2) conserve the full convectivity state. The line in the HR diagram along which a fully convective star of a given mass evolves in the absence of degeneracy is obtained in [461, 462] and is called the Hayashi boundary or Hayashi track. It is shown in [461,462] that the presence of a radiative core shifts the star to the left of this boundary (see Fig. 7.6). In the presence of accretion, a low-mass star with $M < M_{\min}$ may turn back to the main sequence upon going over the T_c minimum. A typical evolutionary track is shown in Fig. 8.2 for the initial mass $0.07 M_\odot$ and accretion rate $10^{-12} M_\odot$/yr.

Models for low-mass stars with a small-scale magnetic field included by adding the magnetic pressure term

$$P_m = C\rho^{4/3} \tag{8.8}$$

to the equation of state have been calculated in [379]. The magnetic pressure reduces the temperature needed for equilibrium of a star of the same mass and its density and increases the minimum mass of the star reaching the main sequence: $M_{\min}^{(H)} > M_{\min}$. Calculations with $C = 2.4 \times 10^{13}$ in (8.8) have yielded the value $M_{\min}^{(H)} = 0.12 M_\odot$ (see Table 8.4). Note that a star with $M = 0.12 M_\odot$ reaches the main sequence upon going over the central temperature maximum. The rotation effect on $M_{\min}$ is considered in Sect. 8.2.

8.1.4 Evolutionary Role of the Mass Loss

Observational evidence for the mass loss from T Tauri stars identified with young contracting stars has been obtained by Kuhi [595], see also [297]. There are

Table 8.4 Characteristics of stars at the time when T_c is maximum (two lines first) or upon reaching the main sequence (last two lines), $\beta - \alpha$ is defined in (8.6), $\beta_m = P_m/(P + P_m)$, P and P_m are the matter and magnetic pressure, respectively

$\frac{M}{M_\odot}$	t, yr	$\frac{E_M}{E_{grav}}$	$(\beta_m)_c$	$\lg T_{ef}$	$\lg \frac{L}{L_\odot}$	$\lg \frac{R}{R_\odot}$	$\lg T_c$	$\lg \rho_c$	$(\beta - \alpha)_c$	$\lg \frac{L_{grav}}{L}$
0.11	2.9(8)	0.28	0.22	3.41	−3.00	−0.80	6.454	2.52	4.2	0.96
0.12	3.8(8)	0.27	0.20	3.42	−3.01	−0.82	6.520	2.60	4.2	0.85
0.12	7.6(9)	0.27	0.20	3.30	−3.72	−0.94	6.457	2.94	8.43	0.01
0.13	2.9(9)	0.25	0.19	3.40	−3.20	−0.88	6.575	2.81	5.13	0.01

observational data (see, for example, [251]) providing evidence for the matter outflow from some young stars to have the form of bipolar flows (see reviews in [366]). This may be due to the presence of protoplanetary disks [686, 972].

An empirical inclusion of mass loss in evolutionary calculations of stars contracting to the main sequence has been made in [367]. The mass flux has been specified in the form

$$\frac{dM}{dt} = -\frac{\alpha R^3}{M}. \tag{8.9}$$

We measure the stellar mass M and radius R in solar units, time t in years, α in $M_\odot$/yr. For the solar wind $\alpha = 3 \times 10^{-14}\ M_\odot$/yr. For T Tauri stars the mass flux may be greater by several orders of magnitude. The evolutionary tracks of stars with initial masses of 2.93 and 2.31 $M_\odot$, various α, and with chemical composition $x_{\mathrm{H}} = 0.739$, $x_{\mathrm{He}} = 0.24$, $x_Z = 0.021$ are shown in Fig. 8.3, from [367]. Contrary to [485], where $l = \rho/2|\nabla\rho|$, the mean free path of the convective element in the mixing length theory is twice the pressure scale height: $l = 2P/|\nabla P|$. Deuterium burning is also examined, with initial concentration of deuterium taken to be equal to the earth concentration: $x_{2\mathrm{D}}/x_{\mathrm{H}} = 1.4 \times 10^{-4}$ [575]. As shown in [367], the lines of equal age (see Fig. 7.6) in the HR diagram are not very sensitive to the mass loss rate determined by α.

8.2 Evolution of Rapidly Rotating Stars on Gravitational Contraction Stages

The rotation velocities of T Tauri stars are difficult to determine by reason of enhanced broadening of emission lines. An estimate of rotation velocity for such stars has been made from narrow absorption lines [468] and interpreting observations of the FeI fluorescence line with $\lambda\lambda$ 4063, 4132, [1039]. This estimate turned out to be $\langle v \sin i \rangle = 20 - 65\,\mathrm{km\,s^{-1}}$.

The polytropic equation of state $P = K\rho^\gamma$ with $\gamma = 4/3$ for stars with mass 3–12 $M_\odot$ in [216] and $\gamma = 5/3$ for $M \leq 1\ M_\odot$ in [712] has been used in theoretical studies of the evolution of rapidly rotating stars at the stage of gravitational contraction. A fixed distribution of angular momentum has been studied in [216],

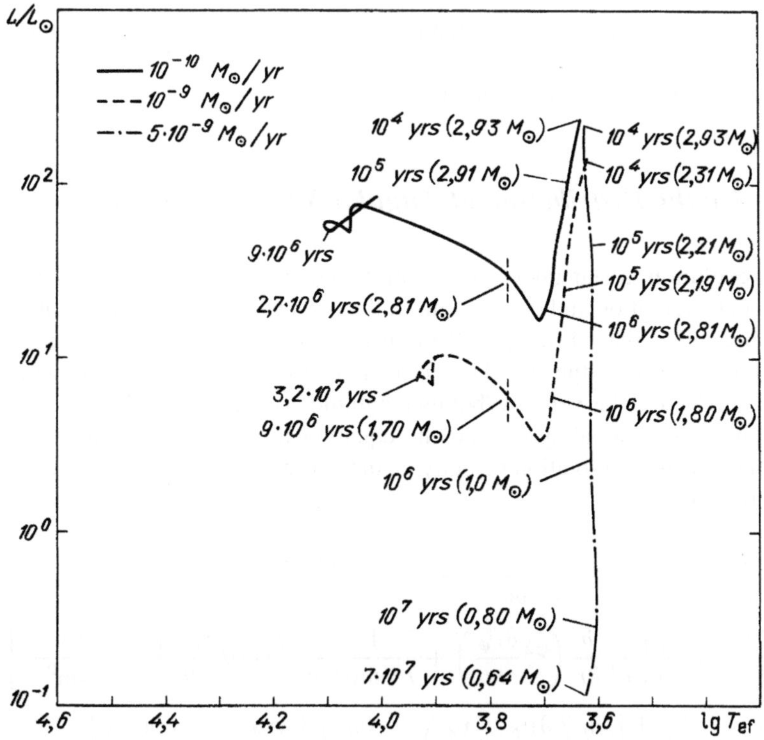

Fig. 8.3 Evolutionary track in the HR diagram for contracting stars with mass loss including the deuterium burning. The initial masses are 2.93 $M_{\odot}$ and 2.31 $M_{\odot}$. The *dashed vertical line* indicates the point where the star stops having outer convection zone and continues to evolve with a constant mass. The masses and ages of the models indicated by pointer are also given in the figure. The pointers on the left of the figure (without mass indication) mark the zero-age main sequence

while in [712] the rotation has been taken to be solid-body and critical[1] throughout the evolution. In this section we follow [167, 168], where an exact equation of state for normal chemical composition is used with a full description of ionization states and radiation pressure (see Sect. 1.1, Vol. 1). A distribution of the effective temperature over the stellar surface is obtained by fitting a thin radiative envelope to the convective core on the poles and at the equator separately. Evolutionary stages of fully convective stars are considered, with entropy and angular velocity being constant over the star. The total angular momentum is taken to be constant during the evolution. A constant entropy under conditions of well-developed convection is obvious, because convection in deep stellar layers is carrying heat very effectively.

[1] For a critically rotating star, the centrifugal force at the equator counterbalances the surface gravity.

Thus, a very small excess of the temperature gradient over the isentropic state is enough to carry almost all heat flux over the star (see Sect. 3.1, Vol. 1). The constancy of the angular velocity remains to be proved.

8.2.1 *On the Distribution of Angular Velocity of Rotation*

During evolution, a star undergoes contraction, expansion and mixing due to the meridional circulation and convection. In the presence of differential rotation a turbulent instability may develop. All these phenomena result in a redistribution of angular momentum in the star. The viscosity due to microscopic phenomena tends to equalize the angular velocity, but its magnitude is usually small.

For the case of axial symmetry, the equation in spherical coordinates $(r, 0, \varphi)$ describing changes in rotation velocity v_ϕ in the presence of isotropic viscosity have the form [591]

$$\begin{aligned}\frac{\partial v_\phi}{\partial t} + v_r\frac{\partial v_\phi}{\partial r} + \frac{v_\theta}{r}\frac{\partial v_\phi}{\partial \theta} + \frac{v_\phi(v_r + \cot\theta\, v_\theta)}{r} \\ = \frac{\eta}{\rho}\left[\frac{1}{r^2}\frac{\partial}{\partial r}\left(r^2\frac{\partial v_\phi}{\partial r}\right) + \frac{1}{r^2\sin\theta}\frac{\partial}{\partial\theta}\left(\sin\theta\frac{\partial v_\phi}{\partial\theta}\right) - \frac{v_\phi}{r^2\sin^2\theta}\right] \\ + \frac{1}{\rho}\left[\frac{\partial\eta}{r}\left(\frac{\partial v_\phi}{\partial r} - \frac{v_\phi}{r}\right) + \frac{\partial\eta}{r\partial\theta}\left(\frac{1}{r}\frac{\partial v_\phi}{\partial\theta} - \frac{v_\phi\cot\theta}{r}\right)\right].\end{aligned} \tag{8.10}$$

Here, $\eta(\rho, \mathrm{T})$ is the viscosity coefficient. For convenience, we shall write hereinafter (8.10) in terms of specific angular momentum j and angular velocity Ω,

$$j = r v_\phi \sin\theta,\ \Omega = \frac{v_\phi}{r\sin\theta}. \tag{8.11}$$

Using (8.11), we rewrite (8.10) as

$$\begin{aligned}\frac{\partial j}{\partial t} + v_r\frac{\partial j}{\partial r} + \frac{v_\theta}{r}\frac{\partial j}{\partial\theta} = \frac{\sin\theta}{\rho r^2}\frac{\partial}{\partial r}\left(r^4\eta\sin\theta\frac{\partial\Omega}{\partial r}\right) \\ + \frac{1}{\rho\sin\theta}\frac{\partial}{\partial\theta}\left(\sin^3\theta\,\eta\frac{\partial\Omega}{\partial\theta}\right).\end{aligned} \tag{8.12}$$

The viscosity of matter is usually negligible, and at $\eta = 0$ the angular momentum conservation follows from (8.12) for all fluid rings.

The physical picture is much more complicated in the convectively unstable region. The laminar convection in a rotating fluid has been examined both theoretically and experimentally in many papers (see [429, 1035]). Various forms of angular velocity distribution may correspond to the steady state in which a convectively unstable rotating medium tends to settle down. As can be seen from (8.12), the condition j = const. reduces the left side to zero, while Ω = const. does the same for the right side. These conditions cannot be simultaneous at υ_r, $\upsilon_\theta \neq 0$ so that the steady rotation law is intermediate between the conditions j = const. and Ω = const. The greater the viscosity coefficient, the more solid-body the rotation. The dimensionless parameter characterizing the rotation law,

$$\alpha = \eta/\rho r \upsilon_r \tag{8.13}$$

is the inverse Reynolds number corresponding to the circular velocity $(v_r,\ v_\theta)$. The Reynolds number

$$Re = \frac{v L_d}{\nu} = \frac{\rho v L_d}{\eta} \tag{8.14}$$

is a dimensionless parameter, determining the character of the hydrodynamic motion, which usually becomes turbulent at large Re (see [591], and Sect. 3.1.3, Vol. 1); η and ν are coefficients of the dynamic and the kinematic viscosity, respectively, related by $\eta = \rho\nu$. L_d is the characteristic scale of the flow. For a steady state, we have Ω = const. at $\alpha \gg 1$, and we obtain j = const. at $\alpha \ll 1$. According to the numerical calculations in [1035], the rotation is almost uniform at $\alpha = 5$, while at $\alpha = 0.04$ the angular momentum per unit mass is constant over almost the total volume, except for a central region, where $r \to 0$, $\alpha \to \infty$ and Ω = const.

Convection is always turbulent in stars due to a low viscosity and large characteristic scales (see Sect. 3.1.3, Vol. 1). Also, this scale is usually much smaller than the characteristic core size. Application of numerical modeling is almost impossible here because even for the case of a laminar convection, the computations can incorporate only a small number of convective cells. The microscopic viscosity in stars is so small that a state with j = const. is likely to set in over one convective cell. The averaged Ω distribution over the convective core rather than the instantaneous distribution of parameters inside the cell should nevertheless be taken into consideration for a small-scale turbulent convection. The large-scale distribution results from interactions between convective cells which take the form of convective, or turbulent, viscosity.

With the turbulent viscosity coefficient

$$\eta_{\mathrm{T}} = \rho \upsilon_{\mathrm{T}} l, \tag{8.15}$$

where υ_{T} is the mean turbulent velocity, l is the turbulence length scale and α from (8.13) becomes

$$\alpha = \left(\frac{\upsilon_{\mathrm{T}}}{\upsilon_r}\right)\frac{l}{r}. \tag{8.16}$$

The r/l ratio is nearly equal to the number of convective cells over the length of the convective zone in the star. In the convective core of a star with mass $30M_\odot$ on the main sequence $l/r \approx 0.1$, $\upsilon_{\mathrm{T}} \approx 2 \times 10^5\,\mathrm{cm\,s^{-1}}$ [144, 724]. Estimates for the circulation velocity obtained from similarity relations and the theory of solar rotation [335] give $\upsilon_r < 10^3\,\mathrm{cm\,s^{-1}}$ for a rotation velocity below the rotation limit. Hence, $\alpha \gg 1$, and in calculations Ω = const. is adopted for the convective core.

Note that the solid-body criterion $\alpha \gg 1$ holds only for $l \ll r$. The notion of turbulent viscosity otherwise has no meaning, and several convective cells arise, all of them having a steady rotation law close to $j = \text{const.}$ Nonetheless, just as in a convectively stable zone, a steady state may never be reached here so that evolutionary calculations taking into account the secular change of the angular velocity (see Sect. 6.3.2, Vol. 1) are needed for the determination of the angular velocity distribution at each evolutionary step.

The above considerations hold only under simple assumptions on the axial symmetry, isotropy of turbulent viscosity, absence of magnetic field. All these assumptions may no longer be valid: instabilities break the axial symmetry [334], the presence of a magnetic field gives rise to additional forces acting in the ϕ direction and complicates (8.10) and (8.12). In several convection models an anisotropic viscous-stress tensor arises, differing in principle from a normal viscosity tensor and not becoming zero for the solid-body rotation [564]. All these complicated effects have been studied with the aid of only simplified models [335, 1010]. For $\alpha \gg 1$, the condition $\Omega = \text{const.}$ seems to be the most reasonable in evolutionary calculations.

Observations of the Sun show a slight (15–20%) equatorial acceleration. This property seems to be inherent in all rotating convective regions of stars and planets and arises from interactions of convective vortices with the overall star rotation. The buoyant force of a vortex with angular velocity ω having a positive component in the direction of the overall angular velocity Ω exceeds the same force of a vortex with a negative component. This gives rise to a constant angular momentum flow outward. In a stationary state a differential rotation sets in, when the angular velocity increases outward (along the cylindrical radius) and the angular momentum flux outward is balanced by the flux inward due to the turbulent viscosity effect (see Problem 1). Other types of interaction between convection, rotation and circulation leading to solar equatorial acceleration have been considered in [335, 851, 1011].

8.2.2 Method for Evolutionary Calculations

The star is divided into an isentropic core rotating as a solid body, and a thin envelope consisting of a radiative outermost region and an underlying non-adiabatic convective zone. The equation of state $P(\rho, T)$ and isentropes $T = T_S(\rho)$ have been derived for the composition $x_{\rm H} = 0.7$, $x_{\rm He} = 0.28$, $x_Z = 0.02$ by use of the approximate formulae for thermodynamic functions from [769]. An error in these formulae approximating the tables [1025] has been corrected according to [1026].

The equations describing the equilibrium of a rigidly rotating ($\Omega = \text{const.}$) self-gravitating stellar core with a barotropic equation of state $P = P(\rho)$ have been solved using a variant of a self-consistent field method, described in [204]. This method is based on an iterative procedure for finding the equilibrium solution by subsequent use of the equilibrium equation in the integral form, and the integral representation of the gravitational potential (see Sect. 6.2.3, Vol. 1). The equation of state is determined by accepting a constant specific entropy S along the core

$P = P_S(\rho)$, and the total angular momentum of the core J was suggested to remain constant during evolution, as well as its mass M. The mass and angular momentum of the envelope had been taken as negligibly small. In order to find the luminosity of the star and its location in the HR diagram, an envelope, taken to the plane approximation with a thickness h well below the radius R, has been fitted to the core. The envelope is taken to be in mechanic and heat equilibrium in the effective gravitational field

$$g_{\rm ef} = |\nabla(\chi - \phi)|, \quad \text{where} \quad \chi = \int_0^r \Omega^2 r' dr' \tag{8.17}$$

is the centrifugal potential.

For a rigid rotation we have $\chi = (\Omega^2 r^2/2)$. The envelope is described by equations

$$\frac{dP}{dx} = -\rho g_{\rm ef}, \quad F = F_{\rm rad} + F_{\rm conv}, \quad F_{\rm rad} = -\frac{4\alpha c T^3}{3\kappa\rho}\frac{dT}{dx}. \tag{8.18}$$

Here, the convective heat flux is calculated according to the mixing length model (see Sect. 3.1.3, Vol. 1),

$$F_{conv} = c_p \rho \left[-\left(\frac{\partial\rho}{\partial T}\right)_P \frac{g_{ef}}{\rho} \right]^{1/2} \frac{l^2}{4} (\Delta\nabla T)^{3/2}, \tag{8.19}$$

where the derivative at constant pressure $(\partial\rho/\partial T)_P$ takes account of the variable ionization states. The variable molecular weight $\mu(\rho, T)$ is found from the condition of ionization equilibrium for an ideal non-degenerate gas (8.4) with $r_p = 0$, x is the coordinate in the envelope in the direction of $\nabla(\chi - \phi)$. The calculations for the envelope have been performed with $l = P/|dP/dx| = H_P$. The core model has been constructed for the outer boundary condition $P = \rho = 0$. The resulting error $\sim h/R$ is small for a thin envelope. The opacity κ for (8.18) has been taken from the tables [316], while for $10^{-3} < \tau < 2/3$ the second and third relations in (8.18) have been replaced by the approximate relation from [769], obtained on the basis of a solution of the radiative transfer equation in the Eddington approximation

$$\begin{aligned} \frac{dT}{dr} &= -\frac{3\kappa\rho F}{4acT^3} - \frac{f}{2} T_0 R_0^{1/2} r^{-3/2}, \\ f &= 1 - \frac{3}{2}\tau \quad \text{for} \quad \tau < 2/3 \\ &= 0 \quad \text{for} \quad \tau \geq 2/3. \end{aligned} \tag{8.20}$$

Here, $T_0^4 = (F/ac)$, and the calculations in [769] have shown that the distribution in a stellar atmosphere depends only weakly on the choice of R_0 and T_0, corresponding to zero optical depth $\tau = 0$ (see also Sect. 2.2.3, Vol. 1). The envelope model is determined uniquely at known $T_{\rm ef}$ and $g_{\rm ef}$. All envelopes are radiative at $\tau = 2/3$, $T = T_{\rm ef}$, but soon with increasing τ the radiative gradient begins to exceed the adiabatic gradient, and convection breaks out. The density is small in the outer layers of the convective envelope so that the heat transfer by convection has a small efficiency, hence, there is a strong non-adiabaticity, and the entropy increases from the surface inwards. On penetrating downwards from the surface, the convective energy transfer grows increasingly effective, and the temperature gradient approaches the adiabatic gradient. The adiabatic regime of the solution is assumed to set in when the condition

$$|\nabla - \gamma_2| \leq 10^{-3}, \quad \nabla = d \ln T / d \ln P \tag{8.21}$$

is satisfied, where

$$\begin{aligned} \gamma_2 &= \left(\frac{\partial \ln T}{\partial \ln P}\right)_S \\ &= \left[\left(\frac{\partial \ln P}{\partial \ln T}\right)_\rho - \left(\frac{\partial \ln P}{\partial \ln \rho}\right)_T \left(\frac{\partial S}{\partial \ln T}\right)_\rho \Big/ \left(\frac{\partial S}{\partial \ln \rho}\right)_T\right]^{-1}. \end{aligned} \tag{8.22}$$

The quantity $T_{\rm ef} = (4\,F/ac)^{1/4}$ has been selected in such a way as to make the envelope entropy S_0 at the point $\nabla \approx \gamma_2$ equal to the core entropy S. First, the dependence $S_0(g_{\rm ef}, T_{\rm ef})$ has been tabulated, then the values of $T_{\rm ef}(g_{\rm ef}, S_0)$ have been found by interpolation of these tables. Note that a similar method for fitting envelopes has been applied to constructing models for non-rotating convective low-mass stars [479]. The calculations reveal a weak dependence $T_{\rm ef}(g_{\rm ef})$ (Fig. 8.4) and a fair accuracy of the relation $T_{\rm ef} \sim g_{\rm ef}^{0.08}$ obtained in [653].

Using the distribution $T_{\rm ef}(\theta)$ over the stellar surface $R = R(\theta)$ allows as to find the total luminosity

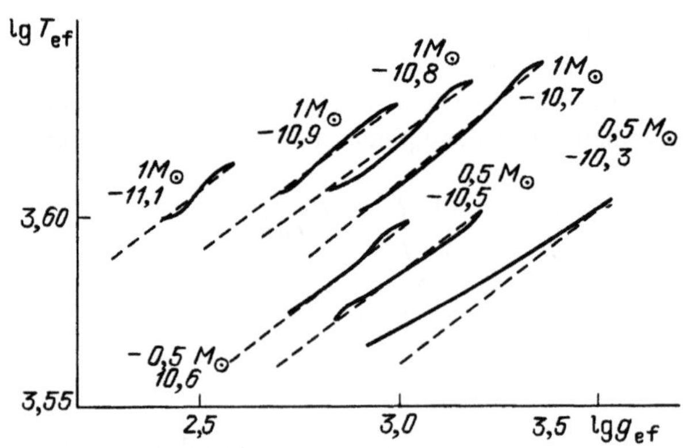

Fig. 8.4 Effective temperature T_{ef} as a function of gravity g_{ef} along stellar surface for models with $0.5\,M_\odot$, $J = 4 \times 10^{50}\,{\rm g\,cm^2\,s^{-1}}$ and with $1\,M_\odot$, $J = 14.2 \times 10^{50}\,{\rm g\,cm^2\,s^{-1}}$. *Dashed lines* represent the slope for the dependence obtained in [653]. Each curve is labeled by a number indicating the value of the corresponding parameter $\lg \rho_0$ (see Tables 8.5–8.7)

$$L = \frac{1}{2}\pi \alpha c \int_0^{\pi} T_{\mathrm{ef}}^4 R \left[R^2 + \left(\frac{dR}{d\theta}\right)^2 \right]^{1/2} \sin\theta \, d\theta. \tag{8.23}$$

With the luminosity and model energy $\mathscr{E}$, we are in a position to determine the age difference between the two convective models

$$\Delta t = 2\frac{|\mathscr{E}_1 - \mathscr{E}_2|}{L_1 + L_2}, \tag{8.24}$$

which may be determined from the mean luminosity between two times "1" and "2". The total energy $\mathscr{E}$ is negative, incorporates thermal, rotational and gravitational energies and has been obtained from the core equilibrium computations (the envelope energy being ignored),

$$\begin{aligned} \mathscr{E} &= -\frac{1}{2}\int \phi dm + \frac{1}{2}\Omega^2 \int r^2 \sin^2\theta dm + \int E(\rho, S) dm \\ dm &= 2\pi\rho r^2 \sin\theta \, dr d\theta. \end{aligned} \tag{8.25}$$

Relations for the specific internal energy $E(\rho, T)$ for the mixture of an ideal gas with radiation, taking into account incomplete ionization, are given in Sect. 1.1, Vol. 1.

8.2.3 Calculation Results

Calculation of non-rotating star models, comparison with the results [367,462,485] together with a test by the Henyey method under the same physical assumptions, reveals that the accuracy of the above method is within 5%, providing the envelope thickness does not exceed $0.3R$, and the radiative core mass is not above 25% of the stellar mass. For stars with masses of 10, 2, 1, $0.5M_{\odot}$, the envelopes are thin if $L < 4 \times 10^4$, 2000, 100, $50L_{\odot}$, respectively. The main calculational results are shown in Figs. 8.5 and 8.6, and Tables 8.5–8.7 taken from [167]. For comparison, note that the angular momentum of the Sun in solid-body rotation is 1.6×10^{48} g cm^2 s^{-1} [15]. The entropy has been characterized by the matter density $\rho_0 (\lg \rho_0)$ at $T = T_0$, $\lg T_0 = 3.3$.

The results of evolutionary calculations for stars with $M = 0.5, 1, 2M_{\odot}$ are given in Tables 8.5–8.7, respectively. Tabulated are time dependences of the polar R_p and equatorial R_e radii, central temperature T_c and central density ρ_c, luminosity L, effective temperature at the pole T_p and at the equator T_e. Given also are equatorial rotation velocities v_{eq} and relative envelope thicknesses h/R_e. A rotating star turns out to have a lower temperature and a higher luminosity than a non-rotating star with the same entropy. For the last model in Table 8.5 with $M = 0.5M_{\odot}$ and the maximum oblateness $R_e/R_p = 1.6$ (compare Fig. 8.7 for $1M_{\odot}$) the increase in luminosity is 25%. The effective acceleration is, for this model, ~24 times lower

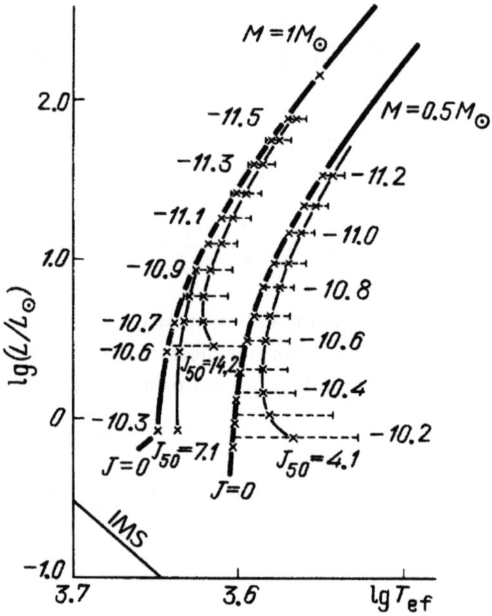

Fig. 8.5 Evolutionary tracks for models of contracting stars with $M = 0.5\,M_\odot$ and $1\,M_\odot$ at diverse values of the angular momentum J_{50} (in 10^{50} g cm^2 s^{-1}) plotted for mean effective temperatures of stars. *The heavy line* represents the results of the calculations obtained using the Henyey method for non-rotating stars, the crosses correspond to models calculated by the described method. The numbers indicate the values of $\lg \rho_0$ parameterizing the core entropy. *The dashed horizontal lines* mark the effective temperature dispersion over the rotating star surface. *The line labeled* IMS is the initial main sequence for a non-rotating star with $X_H = 0.70$ and $X_Z = 0.02$

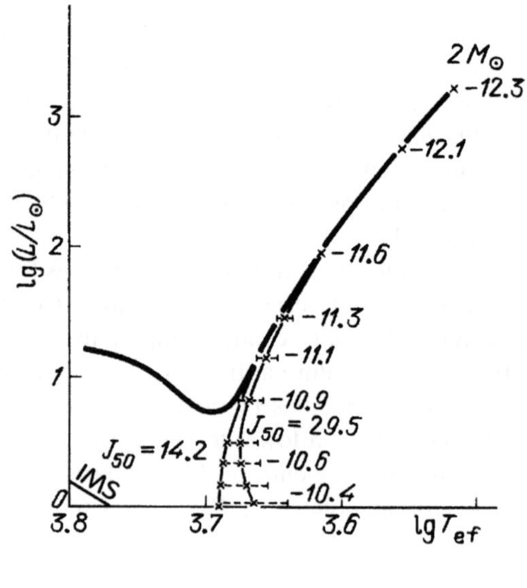

Fig. 8.6 Tracks for models with $2\,M_\odot$. The notation is the same as in Fig. 8.5. The radiative core formation causes the star to pass onto the horizontal branch of the track

Table 8.5 Results of evolutionary calculations for a star of 0.5 $M_\odot$

$\lg \rho_0$	$\frac{R_e}{R_\odot}$	$\frac{R_p}{R_\odot}$	T_c, K	ρ_c, $\mathrm{g\,cm^{-3}}$	$\frac{L}{L_\odot}$	T_e, K	T_p, K	t, yr	v_{eq}, $\mathrm{km\,s^{-1}}$	$\frac{h}{R_e}$	J $\mathrm{g\,cm^2\,s^{-1}}$
−11.3	20.4		2.47(5)	1.16(−3)	53.8	3475		0	0	0.35	
−11.0	9.03		4.87(5)	8.6(−3)	14.4	3750		8.06(3)	0	0.12	
−10.7	4.44		9.35(5)	5.96(−2)	4.14	3920		6.0(4)	0	0.045	0
−10.4	2.36		1.72(6)	0.36	1.28	4020		3.70(5)	0	0.020	
−10.2	1.62		2.50(6)	1.07	0.609	4020		1.10(6)	0	0.012	
−11.3	21.8	20.3	2.4(5)	1.12(−3)	55.5	3400	3470	0	24.5	0.39	
−11.0	9.86	8.94	4.66(5)	8.05(−3)	14.9	3640	3750	7.50(3)	41.6	0.16	
−10.8	5.09	4.35	8.57(5)	0.0523	4.39	3740	3930	5.48(4)	72.9	0.068	4.10(50)
−10.4	2.97	2.28	1.49(6)	0.292	1.45	3720	4010	3.04(5)	127	0.045	
−10.2	2.47	1.54	2.08(6)	0.832	0.773	3365	4040	8.13(5)	193	0.25[a]	

[a] h is large only near equator owing to a decrease in g_{ef}.

at the equator than at the pole. The deviation of the evolutionary tracks to the left (Figs. 8.5 and 8.6, see also Figs. 7.6a, 7.6b and 8.3) arises from the growth of the radiative core and cannot be described by this method. A radiative core treatment should include the non-adiabatically, non-stationary circulation, redistribution of angular momentum and use of the full form of the evolutionary equations, where changes in the distribution of circulation velocities and angular momentum distribution are calculated at each evolutionary step in the same way as changes of the chemical composition (see Sect. 6.3.2, Vol. 1).

The time for a rotating star to contract to the state with a given entropy somewhat decreases because of increasing the luminosity of models with the same S. A rotating star, however, has a lower central temperature and density than a non-rotating star with the same S, so it reaches the main sequence having less entropy and luminosity than the non-rotating star. Approximate calculations of uniformly rotating stars on the main sequence show that the difference in luminosity increases with decreasing mass [370, 794]. If we extrapolate the results [794], we shall have for $M = 0.5 M_\odot$ that a critically rotating star on the main sequence turns out to have a luminosity of 1.6 times lower than the non-rotating star. In addition, a rotating star radiates more heat in its pre-main-sequence phase, so the critical rotation nearly doubles its time of approach to the main sequence.

8.3 Models for the Matter Outflow from Young Stars

The mechanisms of quasistationary matter outflow observed in various stars [289, 595] may be divided into four categories:

(1) Outflow under the effect of the light pressure at small and large optical thicknesses of the outflowing region of the envelope

Table 8.6 Evolution of a star of $1 M_\odot$

$\lg \rho_0$	$\frac{R_e}{R_\odot}$	$\frac{R_p}{R_\odot}$	T_c, K	ρ_c, g cm^{-3}	$\frac{L}{L_\odot}$	T_e, K	T_p, K	t, yr	v_{eq}, km s^{-1}	$\frac{h}{R_e}$	J g cm^{-2} s^{-1}
−11.5	21.7		4.15(5)	1.34(−3)	83.6	3760		0	0	0.21	
−11.2	10.5		7.88(5)	9.23(−3)	26.2	4040		1.5(4)	0	0.10	
−10.9	5.23		1.54(6)	0.0669	8.02	4270		1.1(5)	0	0.034	0
−10.6	2.73		2.91(6)	0.444	2.52	4420		6.5(5)	0	0.015	
−10.3	1.51		5.24(6)	2.52	0.80	4450		3.6(6)	0		
−11.5	22.0	21.7	4.12(5)	1.33(−3)	84.3	3740	3760	0	15.8		
−11.2	10.75	10.5	7.80(5)	9.03(−3)	26.4	4010	4040	1.5(4)	27.8		
−10.9	5.44	5.20	1.50(6)	0.0642	8.13	4180	4270	1.1(5)	51.3		7.1(50)
−10.6	2.92	2.69	2.76(6)	0.412	2.55	4285	4410	6.4(5)	93.4		
−10.3	1.70	1.48	4.81(6)	2.22	0.83	4260	4440	3.3(6)	162		
−11.5	23.05	21.6	4.03(5)	1.29(−3)	85.7	3680	3760	0	31.7	0.30	
−11.2	11.5	10.4	7.46(5)	8.50(−3)	27.1	3920	4040	1.4(4)	55.1	0.12	
−10.9	6.12	5.12	1.39(6)	0.0575	8.57	4040	4260	9.7(4)	100	0.054	14.2(50)
−10.7	4.25	3.26	2.04(6)	0.192	4.04	3990	4365	3.1(5)	150	0.044	
−10.6	3.66	2.62	2.44(6)	0.341	2.79	3860	4395	5.4(5)	184	0.044	

Table 8.7 Evolution of a star of $1 M_\odot$

$\lg \rho_0$	$\frac{R_e}{R_\odot}$	$\frac{R_p}{R_\odot}$	T_c, K	ρ_c, g cm^{-3}	$\frac{L}{L_\odot}$	T_e, K	T_p, K	t, yr	v_{eq}, km s^{-1}	J g cm^{-2} s^{-1}
−12.3	127		1.79(5)	2.86(−5)	1680	3280		0		
−12.1	61.4		3.09(5)	1.43(−4)	567	3600		5.14(2)		
−11.6	18.0		9.15(5)	3.62(−3)	85.9	4147		1.26(4)		0
−11.3	9.22		1.73(6)	2.43(−2)	29.4	4436		7.28(4)		
−11.0	4.69		3.55(6)	0.175	9.6	4704		4.31(5)		
−10.7	2.43		6.43(6)	1.22	3.05	4903		2.53(6)		
−10.4	1.33		1.18(7)	7.40	0.94	4953		1.40(7)		
−12.3	128	127	1.78(5)	2.86(−5)	1680	3276	3280	0	4.34	
−12.1	61.6	61.4	3.11(5)	1.44(−4)	563	3590	3601	4.70(2)	6.34	
−11.6	18.2	18.0	9.12(5)	3.59(−3)	86.1	4139	4147	1.28(4)	16.0	1.419(51)
−11.3	9.32	9.20	1.72(6)	2.40(−2)	29.5	4422	4436	7.24(4)	29.3	
−11.0	4.78	4.67	3.30(6)	0.171	9.67	4674	4703	4.23(5)	55.3	
−10.7	2.52	2.41	6.25(6)	1.17	3.08	4844	4901	2.45(6)	104	
−10.4	1.41	1.31	1.13(7)	6.86	0.98	4867	4949	1.16(7)	186	
−12.3	129	127	1.78(5)	2.84(−5)	1680	3260	3280	0	9.08	
−12.1	62.3	61.4	3.09(5)	1.43(−4)	566	3577	3600	4.66(2)	13.2	
−11.6	18.5	18.0	8.99(5)	3.52(−3)	86.8	4111	4146	1.22(4)	33.2	2.953(51)
−11.3	9.65	9.14	1.67(6)	2.31(−2)	27.5	4373	4434	7.18(4)	60.2	
−11.0	5.10	4.61	3.15(6)	0.159	9.90	4574	4699	4.09(5)	112	
−10.7	2.84	2.36	5.75(6)	1.03	3.22	4632	4891	2.22(6)	207	
−10.4	1.78	1.27	9.80(6)	5.55	1.06	4345	4932	1.10(7)	371	

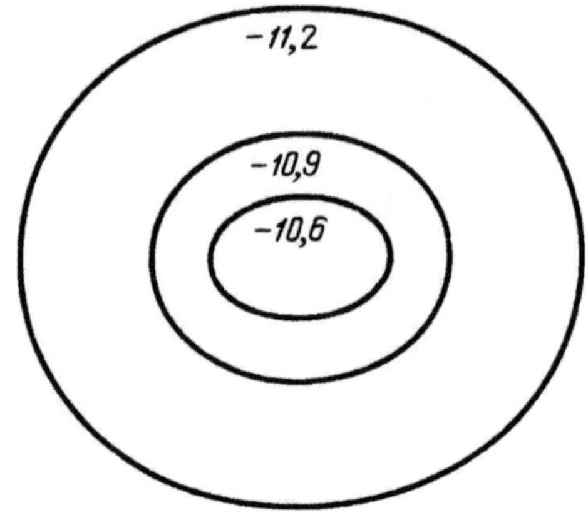

Fig. 8.7 The surface shape for a contracting star with $1\,M_\odot$ and $J_{50} = 14.2$ at various times of its evolution. Indicated are $\lg \rho_0$ values (see Table 8.6)

(2) Non-thermal heating and expansion of corona
(3) Rotation mechanism of outflow
(4) Outflow from stellar envelopes owing to the energy release in recombination of atoms and molecules

The role of the rotational mechanism is far from clear because the observable angular velocities are below their critical values. This mechanism seems to be the most important in non-stationary stages of evolution (see Chap. 10). The first of these four mechanisms is due to a large luminosity of the star and is important for very bright blue massive stars and for giants or supergiants at late evolutionary stages (see Chap. 9). Stars of low mass and luminosity, including young contracting stars, have convective envelopes that give rise to non-thermal heating of corona and to its solar wind-like outflow. Recombination effects reduce the adiabatic index down to low values $\gamma_1 < 4/3$ because of transformations between the kinetic energy of particles, and an energy of recombination and ionization, and may thus become important for the mass loss during the earliest stages of star formation. The adiabatic index γ_1 is defined as follows

$$\gamma_1 = \left(\frac{\partial \ln P}{\partial \ln \rho}\right)_S$$

$$= \left(\frac{\partial \ln P}{\partial \ln \rho}\right)_T - \left(\frac{\partial \ln P}{\partial \ln T}\right)_\rho \left(\frac{\partial S}{\partial \ln \rho}\right)_T \Big/ \left(\frac{\partial S}{\partial \ln T}\right)_\rho, \qquad (8.26)$$

see Sect. 1.1, Vol. 1. It should be noted that the mechanism of mass loss under the effect the of the light pressure is very important for forming massive stars with $M > 9\,M_\odot$ and leads to the dependence (7.30). Consider models of outflowing stars losing mass via the second and fourth of the four mentioned mechanisms.

8.3.1 Outflowing Bipolytropic Models [170]

Consider a star characterized by an equation of state of the form

$$P = K_2 \rho^{1+1/n_2} \text{ for } \rho < \rho_a, n_2 \gg 1; \tag{8.27}$$

$$P = K_1 \rho^{1+1/n_1} \text{ for } \rho > \rho_a, \tag{8.28}$$

where

$$K_1 \rho_a^{1/n_1} = K_2 \rho_a^{1/n_2}.$$

A similar equation of state is determined by the gas properties including recombination. The specific thermal energy of this gas is

$$E = K_2 n_2 \rho^{1/n_2} \text{ for } \rho < \rho_a, \tag{8.29}$$

$$E = K_1(n_2 - n_1)\rho_a^{1/n_1} + K_1 n_1 \rho^{1/n_1} \text{ for } \rho > \rho_a. \tag{8.30}$$

Consider a star with mean density $\overline{\rho}$ such that $\rho_a \ll \overline{\rho}$. For a purely polytropic star of mass M and radius R, characterized by the equation of state (8.28) for all densities, the total energy ϵ is determined by [617] (see Sect. 10.1)

$$\epsilon = -\frac{3 - n_1}{5 - n_1}\frac{GM^2}{R}. \tag{8.31}$$

Here, the zero energy corresponds to the stellar matter energy at zero density with no gravitational interaction. If the mass of the envelope with $\rho < \rho_a$ is negligible, then the total energy of the star characterized by the bipolytropic equation of the state (8.27) and (8.28) will be written as

$$\epsilon = (n_2 - n_1)K_1 \rho_a^{1/n_1} M - \frac{3 - n_1}{5 - n_1}\frac{GM^2}{R}. \tag{8.32}$$

It can be seen from (8.32) that at a sufficiently high n_2 the total energy becomes positive, and a run-away to infinity becomes energetically allowable. If $n_1 < 3$, then most of the mass will be in stable equilibrium even at a positive total energy (see Chap. 12), while for the matter at $\rho < \rho_a$ the outflow is still possible. When the ratio $\rho_a/\overline{\rho}$ is sufficiently small, the mass loss rate $\dot{M}$ is so low that the major part of the stellar mass is in static equilibrium, and the outflow proceeds in a quasistationary regime with velocity distribution

$$u = -\frac{\dot{M}}{4\pi \rho r^2}. \tag{8.33}$$

In a stellar envelope at $\rho < \rho_a$ with a stationary radial outflow the Bernoulli integral holds [591]

$$H = \frac{u^2}{2} + E + \frac{P}{\rho} - \frac{GM}{r} = \frac{u^2}{2} + (n_2 + 1) K_2 \rho^{1/n_2} - \frac{GM}{r}. \tag{8.34}$$

In order for the condition $\rho \to 0$ as $r \to \infty$ to be satisfied, the solution (8.33), (8.34) has to go through a critical sonic point where the relations

$$u_{\rm cr}^2 = u_{\rm s,cr}^2 = \frac{n_2 + 1}{n_2} K_2 \rho_{\rm cr}^{1/n_2},$$

$$u_{\rm cr}^2 = \frac{GM}{2r_{\rm cr}}, \tag{8.35}$$

are valid. Here, the sound velocity $u_s = (\partial P / \partial \rho)_{K_2}^{1/2}$. To construct a model of a static star with envelope outflow, the solution for the core should be fitted to the solution (8.33) and (8.34) for the envelope in such a way that the density ρ and pressure P are continuous, and the velocity u is very low.

For the case of a barotropic equation of state $P(\rho)$, the equilibrium equation for a static core has an integral yielded by the Bernoulli integral (8.34) with $u \to 0$. The integral H is to be evaluated on the boundary of the static core $r = R$, where we set approximately $\rho = \rho_a$, $u \ll u_s$, and

$$H = (n_2 + 1) K_2 \rho_a^{1/n_2} - \frac{GM}{R}. \tag{8.36}$$

Inserting (8.35) into (8.34), we express H in terms of the critical point radius $r_{\rm cr}$:

$$H = \frac{2n_2 - 3}{4} \frac{GM}{r_{\rm cr}}. \tag{8.37}$$

Combining (8.36) and (8.37) gives the radius $r_{cr}(M, R)$ which, in turn, allows as to find, using (8.35) and (8.33), the mass flux

$$\dot{M} = -4\pi \rho_{\rm cr} r_{\rm cr}^2 u_{\rm cr} = -4\pi r_{\rm cr}^{1.5-n_2} \left[\frac{n_2}{2(1+n_2)} \right]^{n_2} \frac{(GM)^{n_2+0.5}}{\sqrt{2} K_2^{n_2}}$$

$$\approx -\frac{4\pi}{e} r_{\rm cr}^{1.5-n_2} \left(\frac{GM}{2K_2} \right)^{n_2+0.5} \sqrt{K_2}, \; n_2 \gg 1, \tag{8.38}$$

$$r_{\rm cr} = \frac{2n_2 - 3}{4} GM \left[(n_2 + 1) K_2 \rho_a^{1/n_2} - \frac{GM}{R} \right]^{-1}. \tag{8.39}$$

From the theory of polytropic stars we have [269] (see Sect. 10.1)

$$R = \left[\frac{(n_1 + 1) K_1}{4\pi G} \right]^{(n_1/(3-n_1))} \left(\frac{M}{4\pi M_{n_1}} \right)^{(1-n_1)/(3-n_1)} \xi_{n_1}, \tag{8.40}$$

where ξ_n, M_n are dimensionless quantities depending only on n. With (8.39) and (8.40), the function $\dot{M}(M, R)$ becomes

$$\dot{M} = -\pi\left(\frac{e}{K_2\rho_a^{1/n_2}}\right)^{3/2} \times (GM)^2\rho_a\left\{1 - \frac{G(4\pi M_{n_1})^{(1-n_1)/(3-n_1)}}{n_2 K_1 \rho_a^{1/n_1}\xi_{n_1}} \times \left[\frac{4\pi G}{(n_1+1)K_1}\right]^{n_1/(3-n_1)} M^{2/(3-n_1)}\right\}^{n_2-1.5}. \quad (8.41)$$

The outflow from the stellar envelope requires that the condition $H > 0$ must be satisfied, which is equivalent to the condition $(\partial\epsilon/\partial M)_K \geq 0$ at $n_2 \gg 1$, if (8.40) and (8.42) are taken into account. At $H = 0$ the energy ϵ is positive and $\partial\epsilon/\partial M = 0$:

$$\epsilon = \epsilon_0 = \frac{2GM^2}{(5-n_1)R} \quad \text{at } H = 0;\ M = M_2. \quad (8.42)$$

At fixed $n_1 < 3$ and $n_2 \gg 1$ bipolytropic models fall into three types depending on the stellar mass M

$$(1)\ M > M_1,\ \epsilon < 0,\ \partial\epsilon/\partial M < 0,\ H < 0 \quad (8.43)$$

are stable static models with negative total energy;

$$(2)\ M_2 < M < M_1,\ 0 < \epsilon < \epsilon_0,\ \frac{\partial\epsilon}{\partial M} < 0,\ H < 0 \quad (8.44)$$

are static models with positive total energy. The run-away to infinity is energetically allowable for the matter, but models are stable against small perturbations (metastable as a whole), and the run-away implies penetrating through a potential barrier;

$$(3)\ M < M_2,\ 0 < \epsilon_0 < \epsilon,\ \frac{\partial\epsilon}{\partial M} > 0,\ H > 0 \quad (8.45)$$

are quasisteadily outflowing models (there are no strictly static models here, but most of the mass is in almost static equilibrium since $\dot{M}$ is low at low ρ_a).

The value of M_1 may be found from (8.32) with $\epsilon = 0$, and M_2 from (8.36) with $H = 0$, using (8.40). The condition that the critical radius r_{cr} be outside the star

$$r_{cr}/R > 1 \quad (8.46)$$

is necessary for applying the approximate evaluation of $\dot{M}$ in (8.41). It follows from (8.39) and (8.40) that (8.46) is valid at

$$y = \left(\frac{M}{M_2}\right)^{\kappa} = \frac{GM/R}{(n_2+1)K_2\rho_a^{1/n_2}} > \frac{2}{n_2+0.5}, \quad \kappa = \frac{2}{3-n_1}. \quad (8.47)$$

With y from (8.47) and (8.36), we may rewrite (8.41) in the form

$$\dot{M} = -aM^2(1-y)^{n_2-1.5} = -aM_2^2 y^{2/\kappa}(1-y)^{n_2-1.5},$$
$$a = \pi\left(\frac{e}{K_2\rho_a^{1/n_2}}\right)^{3/2} G^2\rho_a,$$
$$aM_2^2 = \pi e^{3/2}(n_2+1)^{3/2}\rho_a R_2^2\sqrt{\frac{GM_2}{R_2}}. \tag{8.48}$$

Obviously, the maximum in $\dot{M}$ from (8.48) occurs at

$$y = \frac{2/\kappa}{n_2 - 1.5 + 2/\kappa}, \qquad \left(\text{with} \quad \frac{dy}{dM} = \kappa\frac{y}{M}\right). \tag{8.49}$$

At $\kappa > 1$ ($n_1 > 1$) the condition (8.47) stops being valid before $\dot{M}$ reaches a maximum. On the boundary of the validity condition (8.47) the mean stellar density is of order ρ_a and so, as the adiabatic index γ_1 from (8.27) and (8.26), ($\gamma_1 = 1 + (1/n_1)$) is close to unity, the star turns out to be unstable and runs away on the dynamical time scale. This run-away takes place somewhat earlier than the equality in the condition (8.47) is reached. For outflowing stars, just near the boundary $\partial\epsilon/\partial M = 0$, $y = 1$, the value of $\dot{M}$ is small but increasing as the outflow proceeds and M decreases. As the stellar mass reduces by a factor of $(n_2/2)^{1/\kappa}$, the outflow turns into run-away of the whole star. Note that during the mass loss the outflow and run-away velocities remain always of the order of the sound velocity u_s from (8.35) corresponding to the density ρ_a, with the exception of stars with masses $M \approx M_2$, for which outflowing velocities are small due to a large critical radius in (8.39).

For a star with mass $M < M_2$ the dissipation time is, according to (8.48),

$$\tau \approx \frac{M}{\dot{M}} \approx \tau_{h2}\frac{\overline{\rho}_2}{\rho_a}\left(\frac{4}{3e^{3/2}n_2^{3/2}y^{1/\kappa}(1-y)^{n_2-1.5}}\right) =$$
$$= \tau_h\frac{\overline{\rho}}{\rho_a}\left(\frac{4}{3e^{3/2}n_2^{3/2}y^{3/2}(1-y)^{n_2-1.5}}\right), \tag{8.50}$$

where $\tau_{h2} = R_2/\sqrt{GM_2/R_2}$ is the time of hydrodynamical run-away of the core, $\overline{\rho}_2 = 3M_2/4\pi R_2^3$ is the mean density of a core with mass M_2 and radius R_2; τ_h and $\overline{\rho}$ are the same for $M < M_2$. The minimum in the dimensionless quantity in the latter parentheses (8.50) occurs at $y = 3/2n_2$ and equals $8/9\sqrt{2/3} \approx 0.73$ for $n_2 \gg 1$. The dissipation time of a bipolytopic star with $n_2 \gg 1$ and $\rho_a \ll \overline{\rho}$ is thus always well above the hydrodynamical time of the core, and the applied approximation is therefore correct.

A more complicated 3-polytropic model was used in [303] to mimic the disruption of the neutron star with mass slightly less than the minimal one (see Fig. 11.1).

Note that the behaviour of a 3-polytropic star during this disruption, as well as of the one with a more realistic equation of state, calculated in [303], is well reproduced by the picture of disruption of bi-polytropic stars, described above [170].

8.3.2 *Outflowing Models for Isentropic Hydrogen Stars [179]*

In the ionization and dissociation region the equation of state $P(\rho)$ has $\gamma_1 < 4/3$ at a constant entropy and even $\gamma_1 < 1$ if the radiation is taken into account. This makes possible the existence of steadily outflowing stable configurations similar to the bipolytropic models (8.45) described above. Fully convective contracting stars in the Hayashi stage are characterized by a constant entropy (Sect. 8.1).

Thermodynamic functions for hydrogen along isentropes are shown in Figs. 8.8–8.11 according to calculations [179]. The entropy values are taken to be $S = 20, 24, 30, 36$ in units

$$\mathscr{R} = k/m_p = 8.317 \times 10^7 \text{ erg g}^{-1}\text{ K}^{-1}. \tag{8.51}$$

A gas has been considered in thermodynamic equilibrium with radiation (see Sect. 1.1, Vol. 1), with ionized, atomic, molecular hydrogen in the ground state taken into account, together with rotational and vibrational excitations of the molecular fundamental term. According to the methods described in [426,617], a correction is

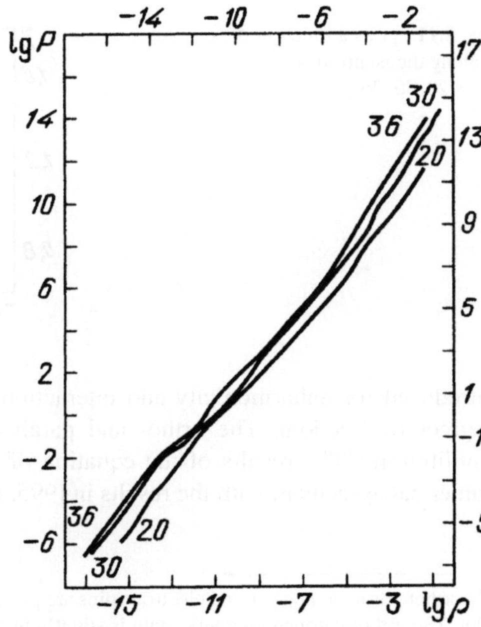

Fig. 8.8 P as a function of ρ along the isentropes $S = 20, 30, 36$

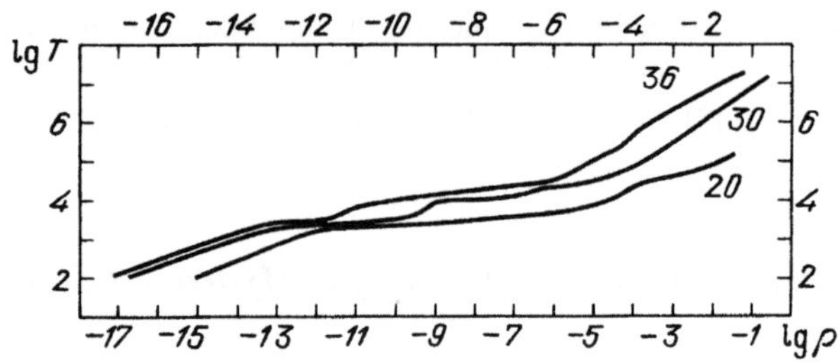

Fig. 8.9 T as a function of ρ along the isentropes $S = 20, 30, 36$

Fig. 8.10 β_g as a function of ρ along the isentropes $S = 20, 30, 36$

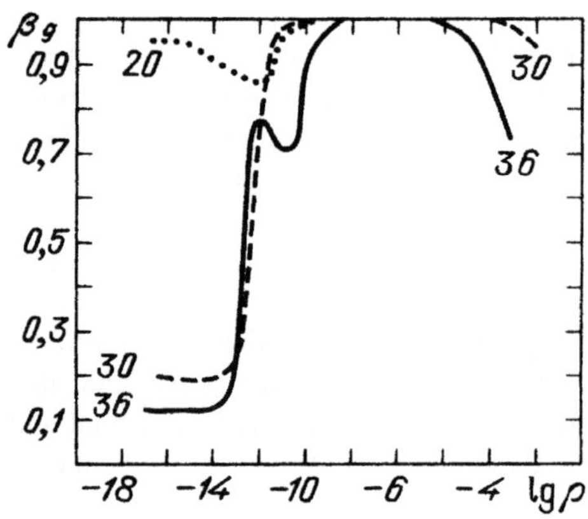

Fig. 8.11 γ_1 as a function of ρ along the isentropes $S = 20, 30, 36$

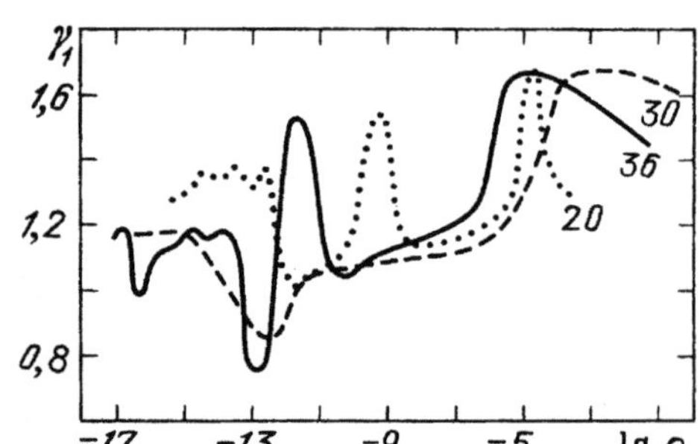

introduced for anharmonicity and interactions between rotational and vibrational degrees of freedom. The ortho- and parahydrogen contents are assumed to be equilibrium.[2] The results of the equation of state calculations [179] are in good numerical agreement with the results in [995, 1025, 1026].

[2] In orthohydrogen, nuclear and electron spins are parallel, while in parahydrogen they are antiparallel. The orthohydrogen energetic state is slightly higher than the parahydrogen state; transitions between these states give rise to an emission line of 21 cm in the radio range.

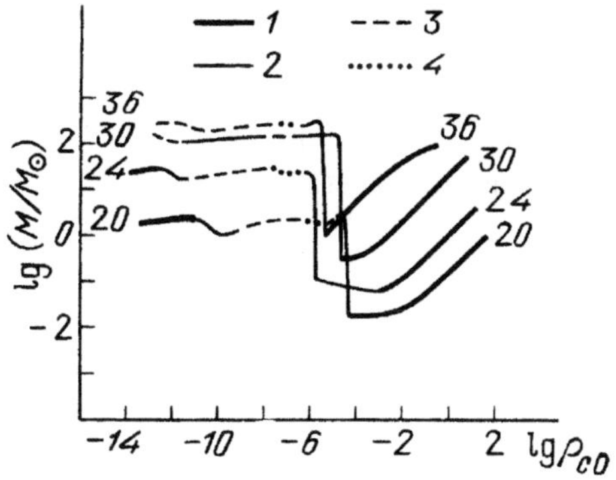

Fig. 8.12 M as a function of ρ_{c0} for static models with $S = 20, 30, 36$. 1, stable branches; 2, intervals with one unstable mode; 3, two; 4, three unstable modes

Taking the equilibrium radiation into account, we obtain $\gamma_1 < 1$ for the dissociation region (Fig. 8.11). For a perfect gas with no radiation always $\gamma_1 > 1$ since the temperature increases along isentropes with increasing density.

Static models for hydrogen stars at $S =$ const. and their stability have been examined in [995] with the aid of numerical solution of the equilibrium equations (1a) in Problem 1, Sect. 9.2, and (9.97). Similar calculations have also been made in [179], from where Fig. 8.12 is taken, which represents the mass as a function of central density $M(\rho_{c0})$. In addition to the stability and instability regions obtained in [995], the number of unstable modes is found in [179] by applying the condition of extremal intersection [1075], see also Chap. 12. The function $M(\rho_{c0})$ has a particularity consisting of the presence of a steep fall appearing at $S > 14$ in the region where the major part of the star is in a partial ionization state. No static model has been constructed in [179,995] for the fall interval. After the fall, the number of unstable modes either decreases by unity, and the model becomes stable (for $S = 20,\ 30,\ 36$ in Fig. 8.12), or does not change, and the instability remains (for $S = 24$).

Besides static models, models with stationary outflow analogous to (8.45) have been constructed in [179] to the left of the fall, with a curve $M(\rho_{c0})$ for outflowing models continuously fitted to the static curve $M(\rho_{c0})$ at the low endpoint of the fall (Fig. 8.13). The techniques used for constructing models with outflow are the same as for analogous bipolytropic models and require that the Bernoulli integral H from (8.34) be positive if the zero-point energy is the energy of the molecular hydrogen state. With the known H, writing the relation (8.34) at the critical pont (8.35) where

$$u_{\rm cr}^2 = (\partial P/\partial \rho)_S = \gamma_{\rm cr}\frac{P_{\rm cr}}{\rho_{\rm cr}} = \frac{GM}{2r_{\rm cr}}, \tag{8.52}$$

we can find numerically the critical point parameters and the mass-loss rate $\dot{M}$ from (8.33).

For finding the integral H, the following procedure has been used in [179]. The structure of static cores is calculated for a given entropy by integrating equations of static equilibrium (1a) in Problem 1, Sect. 9.2, and mass conservation (9.97) from

Fig. 8.13 M as a function of ρ_{c0} for outflowing models. The *dashed line* confines the region of stationary outflow, the isentropes $S = 14, 50$ are taken from Fig. 1 in [995]

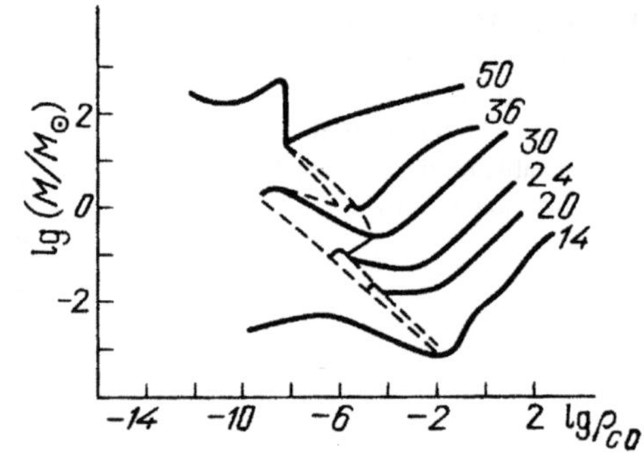

the centre outwards at a given central density ρ_{c0}. The quantity H from (8.34) is evaluated at each step. The possibility to construct a model with static core and stationary envelope outflow analogous to (8.45) requires the following conditions:

1. There must be a shell extended along the radius and having $H \approx$ const. > 0 and a mass much lower relative to the static core mass. The reason is that in a model with outflow there is an intermediate region with $u \ll u_s$ where hydrostatic equations are valid with fair accuracy. The structure of the outflowing star must not depend on the fitting point of two solutions. The constancy of H and a low value of the shell mass will determine the outflowing model unambiguously at given S and ρ_{c0}.
2. The critical radius $r_{\rm cr}$ from (8.52) must exceed the fitting point radius r_b (see (8.46)).
3. The mass m_l of the layer $r_b < r < r_{\rm cr}$ must be much smaller than the core mass M. This allows use of the Bernoulli integral (8.34) without self-gravitation. The same condition determines a quasistationary character of outflow with the dissipation time of the star $\tau \gg \tau_h$ (see (8.50)).
4. The velocity at the fitting point must be low, $u_b^2/H \ll 1$, which is analogous to the condition 1.
5. Besides condition 3, neglecting the self-gravitation of matter in the flow requires that the inequality

$$\Delta H = G \int_{r_b}^{r_{\rm cr}} \frac{dm}{r} \ll H \qquad (8.53)$$

be satisfied. The outflowing models presented in Fig. 8.13 satisfy all the above conditions. The dependence $H(r)$ for static models with various central densities r_{c0} at $S = 30$ shown in Fig. 8.14 illustrates the outflowing model appearance. The curve $H(r)$ has a plateau for models with

$$\rho_{c,\,\rm min} = 8.0 \times 10^{-10} < \rho_{c0} < 3.4 \times 10^{-5}\ {\rm g\ cm^{-3}} = \rho_{c,\,\rm max} \qquad (8.54)$$

and the value $\rho_{c0} = \rho_{c,\,\rm max}$ corresponds to the fall of the static curve $M(\rho_{c0})$. Outflowing models do not exist to the right of this fall, while to the left an outflowing solution exists together with much more massive static unstable solutions.

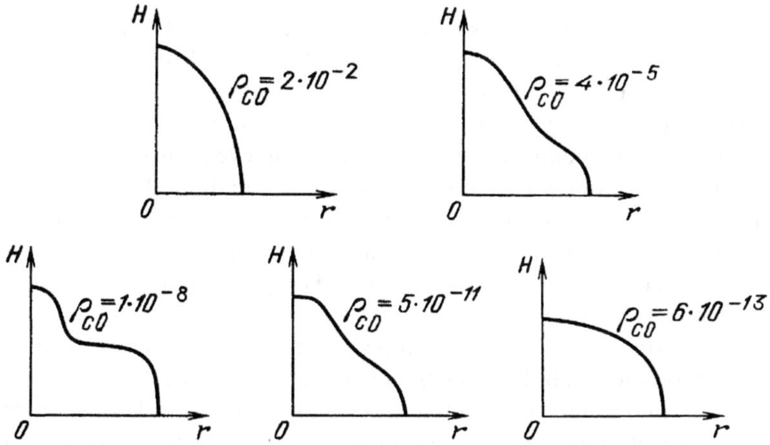

Fig. 8.14 Successive shapes of the dependence of $H = E + P/\rho - Gm/r$ on the current radius of static model along the isentrope $S = 30$ for various ρ_{c0} (qualitative picture). Shown are only portions of $H(r)$ for which $H > 0$

Table 8.8 Models of stationary outflowing isentropic stars

$\frac{S}{\mathfrak{R}}$	ρ_{c0}, g cm^{-3}	$\frac{M}{M_\odot}$	$\frac{r_b}{r_\odot}$	$\frac{H_b}{\mathfrak{R}}$	ρ_b, g cm^{-3}	T_b, K	μ_b
30	8.7(−6)	0.358	629	1.2(4)	7.8(−12)	2100	1.13
30	2.7(−9)	1.594	3178	9300	7.2(−13)	1820	1.27
30	8.05(−10)	1.460	4227	1.3(4)	1.0(−12)	1840	1.74
24	1.8(−6)	0.103	218	2503	9.5(−12)	1897	1.47
24	1.0(−6)	0.107	274	4000	6.6(−11)	1860	1.49
24	5.1(−7)	0.094	365	8800	2.6(−11)	1990	1.42

The subscript "*b*" denotes quantities at the fitting-point, μ is the molecular weight, $\beta_g = P_g/P$ (see (Sect. 1.1, Vol. 1)), ϵ/M is the specific total energy with respect to molecular hydrogen.

Table 8.8

β_{gb}	$\frac{r_{cr}}{R_\odot}$	$\frac{\dot{M}}{M_\odot/yr}$	$\frac{10^3 m_l}{M}$	$\frac{10^3 \Delta H}{H}$	$\frac{10^3 u_b^2}{H}$	E/M 10^{12} erg g^{-1}
0.963	1400	4.5(−5)	3	0.8	0.3	4.16
0.759	4720	3.8(−4)	6	1	5	1.50
0.770	6700	1.1(−3)	30	4	7	1.61
0.969	420	7.5(−6)	0.4	0.2	0.8	1.50
0.958	571	9.3(−6)	0.8	0.4	1	1.18
0.987	882	2.5(−4)	20	4	8	1.39

Parameters of some outflowing models for which conditions 1–5 hold are given in Table 8.8. Minimizing the quantity

$$\eta = \frac{m_l}{M} + \frac{\Delta H}{H_0} + \frac{u_b^2}{H_b} \tag{8.55}$$

yields the fitting point for models from Table 8.8 and Fig. 8.13, though the results are not very sensitive to the fitting-point shift along the plateau portion of $H(r)$. The dissociation energy is 2.15×10^{12} erg g^{-1} for hydrogen, the ionization energy is 13×10^{12} erg g^{-1}; all the models except the first in Table 8.8 have a negative energy with respect to atomic hydrogen. Comparison reveals that the outflow region along the ρ_{c0} axis in Fig. 8.13 is proportional to the height of the fall on the curve $M(\rho_{c0})$ along the M axis (Fig. 8.12). The fall height is maximized at $S = 30$, when the region of outflowing models is maximum. The disappearance of the fall on the curves $M(\rho_{c0})$ from [995] determines the upper and lower boundaries for outflowing models in Fig. 8.13.

The cores of stars with stationary outflow seem to be dynamically stable, though rigorous methods for stability studies are not yet available. An outflowing star may transform into a static one with a lower mass and higher density, contrary to bipolytropic models where, for the case of a star with stationary outflow, ρ always decreases in time ($\overline{\rho} \sim M/R^3 \sim M^{2n_1/(3-n_1)}$, $n_1 < 3$, see (8.40)), and the star eventually dissipates.

The realization of such outflowing models in nature requires that the protostar matter be additionally heated at sufficiently early phases of protostar formation by cosmic rays or radiation emitted by hot stars forming in the neighbourhood.

8.3.3 Models for Outflowing Coronae of Young Stars

The observed shapes of emission lines H and K of ionized calcium and Balmer lines of hydrogen from T Tauri stars correspond to outflow velocities above 100 km s^{-1}. These values exceed by far the thermal velocities at temperatures of possible formation of these lines [594]. To account for this fact, together with other observable properties of young stars, a model of outflowing corona where cold and dense condensations form owing to the thermal instability development has been examined in [181]. After being formed, the condensations undergo a slowdown under the effect of stellar gravity and produce the observed line emission.

It is assumed that the corona formation is due to the conversion into heat of the mechanic energy flow from the convective envelope. The magnetic field plays an important role in this conversion. In analogy with the most simple models for the solar corona [798], we assume the corona of a young star to be isothermal inside and adiabatic outside. The quantity $\int dP/\rho$ in the Bernoulli integral (8.34) has the form [181, 617]

$$\int \frac{dP}{\rho} = E + P/\rho - TS \quad \text{at } T = T_0 = \text{const.}$$
$$= E + P/\rho \quad \text{at } S = S_1 = \text{const.} \tag{8.56}$$

Accordingly,

$$H = H_i = (u^2/2) + E + (P/\rho) - T_0 S - (GM/r) \quad \text{at } T = T_0 \tag{8.57}$$
$$= H_a = (u^2/2) + E + (P/\rho) - (GM/r) \quad \text{at } S = S_1. \tag{8.58}$$

The continuity condition for u, ρ and T at the point $r = r_1$, where the transition from isothermal flow to an adiabatic one occurs, yields the relation

$$H_a = H_i + T_0 S_1. \tag{8.59}$$

The critical point is assumed to be located in the isothermal region where, instead of (8.52), we have

$$u_{\rm cr}^2 = \left(\frac{\partial P}{\partial \rho}\right)_T = \frac{GM}{2r_{\rm cr}}. \tag{8.60}$$

With the temperature of the isothermal corona T_0 and the integral H_i from (8.58), using (8.60), we may find the critical point parameters $u_{\rm cr}$, $\rho_{\rm cr}$, $r_{\rm cr}$ and the mass flow $\dot{M}$ from (8.33). In order to determine the matter velocity at infinity, the entropy S_1 should be specified at the fitting point $r = r_1$:

$$u_\infty^2 = 2(H_i + T_0 S_1). \tag{8.61}$$

The integral H_i is determined by the specified corona luminosity. To calculate the luminosity when H_i is known, we have to specify a mechanism for corona acceleration and find the temperature, density and velocity distribution in it.

The rapid conversion of the mechanical energy flow $Q(\mathrm{erg\ s^{-1}})$ into heat is assumed to take place at radius r_b nearly equal to the stellar photosphere radius r_{ph}. The plasma parameters are almost discontinuous at this point, and the temperature is assumed to increase from the photosphere value T_{ph} to the corona value T_0. The conservation laws hold at the heat-discontinuity point [151]:

$$\rho_{b-} u_{b-} = \rho_{b+} u_{b+}, \tag{8.62}$$
$$P_{b-} + \rho_{b-} u_{b-}^2 = P_{b+} + \rho_{b+} u_{b+}^2, \tag{8.63}$$
$$\frac{u_{b-}^2}{2} + E_{b-} + \frac{P_{b-}}{\rho_{b-}} + \frac{Q}{4\pi r_b^2 \rho_{b-} u_{b-}} = \frac{u_{b+}^2}{2} + E_{b+} + \frac{P_{b+}}{\rho_{b+}}. \tag{8.64}$$

The quantities to the left of the discontinuity point (closer to the star) are labeled here by the subscript ($_b-$), to the right by ($_b+$). From (8.58) and (8.33) we have two other equations:

$$\rho_{b+} u_{b+} = \frac{\dot{M}(H_i, T_0)}{4\pi r_b^2}, \tag{8.65}$$

$$\frac{u_{b+}^2}{2} + E_{b+} + \frac{P_{b+}}{\rho_{b+}} = H_i + T_0 S_b + \frac{GM}{r_b}. \tag{8.66}$$

With the thermodynamic functions $P(\rho,T), E(\rho,T), S(\rho,T)$ for the ideal gas in equilibrium with radiation, taking into account the incomplete ionization states (see Sect. 1.1, Vol. 1) and given values

$$T_{b-} = T_{ph}, \quad T_{b+} = T_0, \quad r_b = r_{ph}, \quad H_i \tag{8.67}$$

the relations (8.62–8.66) represent five equations for five unknowns

$$\rho_{b-}, \rho_{b+}, u_{b+}, u_{b-}, Q. \tag{8.68}$$

The solutions to this system for hydrogen plasma transparent for radiation and characterized by the above thermodynamic functions (see [179]) have been obtained in [181, 182] for several sets of the parameters (8.67) and are given in Table 8.9.

A part (a half or less) of the corona X-ray emission L_{cor} strikes the stellar surface and heats it, forming a hot chromosphere layer with $T_{ch} = (5\text{–}10) \times 10^4$ K in addition to the "cold" chromosphere layer with $T_{ch0} \approx 10^4$ K owing its origin to heat conducting effects. The hot chromosphere parameters: temperature T_{ch}, density ρ_{ch}, luminosity L_{ch} and thickness h_{ch} are calculated either from the energy balance of a chromosphere layer with uniform density under the assumption of minimum chromosphere density providing the given cooling rate ([162], first two lines in Table 8.9), or from the condition of equilibrium between the chromosphere and corona pressures ([163], the final three lines in Table 8.9).

The hot chromosphere of T Tauri stars examined in [181] has been discovered by ultraviolet observations from the IUE satellite, and its luminosity has been found to be $\sim 0.3 L_\odot$ in the 1,150–3,100 $_\text{A}$ range [408] for RU Lupi. Chromosphere and corona parameters and the expected X-ray emission of this star have been evaluated in [182] (last three lines in Table 8.9). Observations in the soft X-ray range

Table 8.9 Model parameters of outflowing **T Tauri** stars

$\frac{r_b}{R_\odot}$	T_{ph}, 10^3 K	T_0, 10^6 K	ρ_{b+}, 10^{-13} g cm^{-3}	u_{b+}, km s^{-1}	$\frac{r_{cr}}{R_\odot}$	u_{cr}, km s^{-1}	ρ_{cr}, 10^{-14} g cm^{-3}	$\dot{M}$, 10^{-8} $M_\odot$/yr
3.5	4.9	1.0	7	67	5.7	129	10	6
2.0	4.9	1.6	6	72	3.6	163	8	2
2.4	4.43	2	1					0.9
2.4	4.43	2.4	1.2					2
2.4	4.43	2.4	1.7					3

L_{ph} is the photosphere luminosity, M_{cor} is the corona mass at $r_b < r < r_c$

Table 8.9

$\frac{L_{ph}}{L_\odot}$	$\frac{L_{cor}}{L_\odot}$	$\frac{M_{cor}}{M_\odot}$	T_{ch}, 10^4 K	ρ_{ch}, 10^{-11} g cm^{-3}	$\frac{L_{ch}}{L_\odot}$	h_{ch}, km	t_1, s	t_2, s
6.36	108	3(−11)	8.9	2	54	1.6	41	9(4)
2.06	14.6	5(−12)	4.6	2	7.3	1.85	108	4(4)
2	0.6		6.0	0.33	0.3	10		
2	2		5.8	0.5	0.7	10		
2	4		6.0	0.67	1.0	6		

from the Einstein [381, 407] and ROSAT [739] satellites have revealed a moderate luminosity $L_X \leq 10^{31}$ erg s^{-1} of T Tauri stars which is well below the values of $L_{\rm cor}$ given in Table 8.9 (see reviews [692,789]). Observations from the Astron satellite provide evidence for a strong RU Lupi variability in the X-ray range of 2–7 KeV [420]. The weak observed values of the X-ray luminosity of young stars might be due to the strong X-ray absorption in the envelope around them, and their real X-ray luminosity $L_{\rm cor}$ might exceed by far the observed values so that $L_{\rm cor} \geq L_{\rm ch}$, in accordance with [182].

The heat discontinuity (8.62–8.64) for models from Table 8.9 occurs at densities well below the photosphere density $\rho_{b-} \ll \rho_{\rm ph}$ and optical thicknesses $\tau = 10^{-2}$–10^{-3} [181]. In the temperature range of $T = (1\text{–}3) \times 10^6$ K the cooling function $\Lambda(T)$ erg cm^3 s^{-1} for transparent plasma with normal composition decreases with temperature [313], giving rise to instability with respect to the formation of condensations with lower temperature and higher density.[3] The time for a condensation to form is

$$t_1 \approx \frac{E}{n^2 \Lambda(T)}. \tag{8.69}$$

The minimum size of condensation is determined by the heat conductivity and equals

$$l_{\rm min} \approx \frac{1}{n}\sqrt{1.8 \times 10^{-6} T^{7/2}/\Lambda(T)} \approx 4 \times 10^7 \text{ cm}$$
$$\text{for } n = 10^{11} \text{ cm}^{-3},\ T = 10^6 \text{ K}. \tag{8.70}$$

Thermal instability in the solar corona does not develop because of a low density and large value of $l_{\rm min}$. The time of thermal instability development t_1 is given in Table 8.9 along with the time t_2 for outflowing gas to travel through the characteristic size of a corona [181]. The condition $t_2 \gg t_1$ means that the thermal instability has the time to develop, and cold dense condensations emerge.

The existence of dense chromospheres and coronae in T Tauri stars requires a theoretical explanation for formation in stars with large convective envelopes of a powerful mechanical energy flow comparable with photosphere flow. This phenomenon may have an explanation from the thermodynamical standpoint under the assumption of a high temperature T_m in the region where the mechanical energy flow forms. The efficiency of conversion of the heat flow into mechanical energy flow in a star treated as a heat engine [105] does not exceed

$$\eta < \frac{T_m - T_{ph}}{T_m}, \tag{8.71}$$

and at $T_m \gg T_{ph}$ may be close to unity.

[3] The quantity $\Lambda(T) n_e (n_e + n_H)$ erg cm^{-3} s^{-1} represents the energy amount emitted by 1 cm^3 of plasma in 1 s; $\Lambda(T)$ is evaluated in [313].

If the coronal emission of T Tauri stars is equal to or greater than the photosphere luminosity, it may resolve, in part, a contradiction appearing in determining the age of young stellar clusters [105, 167]. The age determined from the turning point of massive stars off the main sequence (see Sects. 9.2 and 9.3) turns out to be less than the age determined as the time of contraction of low-mass stars. These ages are $(4\text{–}8)\times 10^8$ and 2×10^9 yr for Hyades; 6×10^6 and 2×10^7 yr for NGC 2264, respectively [588]. When the total bolometric luminosity is several times greater than the photospherical one, this must be taken into account in evolutionary calculations, and the second quantity will then decrease in such a way as to weaken or even remove this contradiction.

8.3.4 On the Phenomenon of Fuor

Several stars are known at present which rapidly increase their luminosity by about a factor $\sim$100 and keep this high level for many years. FU Ori was the first star of this kind and gave its name to the phenomenon of fuor [23]. The latter may be attributed to the birth of a young star with sufficiently high mass $M > 3M\odot$ which completes the accretion stage, evaporates its dust envelope after hydrogen ignition, and appears immediately near the main sequence ([622, 1063], see Sect. 7.2). This simple explanation encounters difficulties when applying to the fuor of V1057 Cyg where a T Tauri star had been observed before the outburst [469]. The masses of T Tauri stars are well below the estimates of stellar masses for FU Ori and V1075 Cyg [800] after outburst. To avoid these difficulties, one may assume that V1075 Cyg is a binary containing a T Tauri star and a young star with significantly higher mass which produces the phenomenon of fuor [105, 180]. At the stage of dust envelope the massive star was emitting mostly in the infrared range, while in the optical range its emission was weaker than that of the neighbouring T Tauri star. At present the T Tauri star is essentially, by a factor $\sim$100, weaker than its companion, but we may still expect its discovery which is facilitated by strong emission lines, ultraviolet excesses and other peculiarities of T Tauri stars [800]. With the aid of speckle-interferometry observations the T Tauri star has been found to be a binary itself, its companion being an infrared star with insufficiently studied properties [337, 455].

Other fuor models are based on phenomena of non-stationary disc accretion, strongly increasing the star luminosity. In this case, we may expect that the star will return in reasonable time to its previous state.

Problem. Find the equatorial acceleration of a convective star due to buoyancy of eddies.

Solution 8.1. [107, 332]. The eddies, whose projection $\omega_\Omega = \omega \sin\theta$ on the direction of angular velocity $\Omega \parallel OZ$ is positive, are under the action of a larger centrifugal force F_{cf} than those with negative projection

$$F_{cf} \approx \frac{(\Omega r + \Omega r_0)^2}{r} \sin\theta, \tag{8.72}$$

where r is a spherical radius, r_0 is the radius of the eddy.

Larger F_{cf} require smaller pressure P inside the eddy as well as density in order to be in equilibrium with the surrounding medium. Smaller density means larger buoyancy force for rising and smaller sinking force for descending convective eddies. All this leads to a net flux of angular momentum in the direction of entropy decrease, i.e., to the surface of a star with a convective envelope. The part of the pressure deficit inside the eddy, linear to ω, is connected with the part of the centrifugal potential $\Delta\phi_{cf} = \Omega\omega r_0^2 \sin^2\theta$. The density difference in eddies rotating in opposite directions is obtained from the Bernoulli integral taking into account the enthalpy, the gravitational and centrifugal potentials (see Sect. 6.2.3, Vol. 1), and is equal to

$$\frac{\Delta\rho}{\rho} = \frac{2\Delta\phi_{cf}}{v_s^2} = 2\sin^2\theta\,\frac{\Omega\omega r_0^2}{v_s^2}. \tag{8.73}$$

The density is smaller for eddies with positive projection of rotational velocity on the direction of stellar rotation velocity Ω, $v_s^2 = 5P/3\rho$ is the sound velocity squared. The difference of convective velocities v, corresponding to the eddies with opposite rotation follows from (8.73), and the relations for a convective velocity, taking into account the buoyancy acceleration (see Sect. 3.1, Vol. 1)

$$\frac{1}{2}\rho v^2 = \frac{1}{2}\Delta\nabla\rho\, dr^2 g + \frac{1}{2} g\Delta\rho dr, \tag{8.74}$$

with $\overline{dr} = l/2$, and is given by

$$2v\,\Delta v = l\frac{Gm}{r^2}\frac{\Omega\omega r_0^2}{v_s^2}\sin^2\theta. \tag{8.75}$$

The velocity difference (8.75) corresponds to the projection on the equatorial plane of the flux of angular momentum to stellar surface $\dot{J}_{out}$ through the area unit

$$\dot{J}_{\text{out}} \approx \Delta v\rho\omega r_0^2 \sin\theta. \tag{8.76}$$

Differential rotation produced by angular momentum flux (8.76) is smoothed by convective viscosity leading to the angular momentum flux J_{in} inside

$$\dot{J}_{\text{in}} \approx \eta_{\text{conv}} r^2 \sin^2\theta\, |\nabla\Omega|. \tag{8.77}$$

In the stationary state $\dot{J}_{out} = \dot{J}_{in}$. Supposing that the radiative core inside the star (Sun) rotates rigidly with $\Omega = \Omega_0$ and $|\nabla\Omega| \approx \Delta\Omega/h$, where h is the thickness of the convective zone, we obtain the following estimation for the equatorial acceleration of the star δ

$$\delta = \frac{\Omega_{\rm eq} - \Omega_{\rm pole}}{\Omega} \approx \left(\frac{Gm}{rv_s^2}\right)\left(\frac{r_0}{r}\right)^2 \frac{h}{r}\left(\frac{\omega r_0}{v}\right)^2 . \tag{8.78}$$

This estimation is sensitive to the choice of v_s and gives, for the Sun [107], $\delta = 2$–20% for $v_s^2 = 10^{14}$–$10^{13}\,\mathrm{cm^2/s^2}$.

Chapter 9
Nuclear Evolution of Stars

After seminal studies [878], the calculations of stellar evolution have been performed by various groups of researchers in a wide mass range with use of increasingly more powerful computers. There is presently (2001) a general understanding of the nuclear evolution of a star from the main sequence (MS) to a white dwarf, neutron star, or black hole formation. However, although much effort has gone into solving these problems, we now have but a crude evolutionary scheme, and many details are not sufficiently reliable. The results of calculations made by diverse authors, though qualitatively similar, differ in detail. A major reason is the uncertainty in most of the physical grounds of the stellar evolution theory, such as convection, mixing, rates of nuclear reactions at low energies, and others. This might also be due to computational difficulties arising from the accumulation of numerical errors which cause different numerical schemes for solving equations of static equilibrium, thermal evolution, and heat transfer (see Chap. 6, Vol. 1) to give different results sometimes. With regard to this last point, the situation is less dramatic than in the theory of two-dimensional collapse where the use of different numerical schemes, e.g., Lagrangian or Eulerian, leads to qualitatively different results (see Sect. 7.3).

Detailed evolutionary calculations have been made by Japanese researchers [462], Schwarzschild and Harm [456–458, 883–887], Iben [485–497, 500–502, 504–508], Paczynski [769–776, 778–781], and Maeder with co-authors [276, 663–665, 867]. Important results have been also obtained by Kippenhahn et al., see review [468], Stothers et al. [937–939], Soviet authors [134, 672, 724, 1004–1006, 1017, 1018], and others; see also [569, 870].

Observational data for stellar parameters come mainly from binary stars, where absolute values of masses and radii can be determined [27].

G.S. Bisnovatyi-Kogan et al., *Stellar Physics: 2: Stellar Evolution and Stability*, Astronomy and Astrophysics Library, DOI 10.1007/978-3-642-14734-0_3,

9.1 Sources of Uncertainty in Evolutionary Calculations

9.1.1 Convection

Since there is no strict quantitative theory of convection (see Chap. 3, Vol. 1) to date, a phenomenological description based on the mixing-length theory is usually used in evolutionary calculations.[1] Several authors specify differently the uncertain parameter in this theory, the mixing length l. Iben [485, 492, 507, 943] defined the mixing length by the scale of nonuniformity in the density distribution

$$l = \alpha|\rho/\nabla\rho| = \alpha H_\rho,\ \alpha = 1/2. \tag{9.1}$$

In his later studies [85, 493, 495, 496, 608], he used the pressure scale height

$$l = \alpha|P/\nabla P| = \alpha H_{\rm p},\ 0.4 < \alpha < 1.2. \tag{9.2}$$

In [937], the relation (9.1) was used with various α. For some cases (low densities, high luminosity and opacity), using (9.1) overestimates the effectiveness of convective transport. In [770, 773, 778], the relation (9.2) was used with $\alpha = 1$. In [134], the convection was described in the same way. If (9.2) is applied, an inverse density gradient may arise in the outer envelopes of stars with $T_{\rm ef} \leq 6{,}000$ K from a limited capacity of the heat transfer by convection. The above methods for describing convection have been used only in convective envelopes of stars. In convective cores, where densities are sufficiently high, the approximation of adiabatic convection has been used. As pointed out in [134], this approximation may be no longer valid, for example, near the end of the hydrogen burning and formation of a radiative helium core. The approximation of adiabatic convection is usually used for burning shells (see below), where it is too crude because the energy flow may vary by orders of magnitude within the limits of the pressure scale height. The effect of the non-adiabaticity in this region on evolutionary tracks is not yet studied.

9.1.2 Semiconvection

A star's departure from the MS, where it has been chemically homogeneous, results for stars with $M > 10 M_\odot$ in formation above the convective core of a zone characterized by a very small excess of the temperature gradient $\gamma_{\rm rad} = {\rm d}\ln T/{\rm d}\ln P$ over the adiabatic gradient $\gamma_2 = (\partial \ln T/\partial \ln P)_{\rm S}$ (8.22). This zone is called the intermediate convective or semiconvective zone. The reason for its formation with increasing stellar mass is a more important role of the radiation pressure which

[1] The description of convection in [462] is even more simplified.

lowers ∇_a in comparison with what we see in the case of a gas without radiation. The small difference between γ_2 and γ_{rad} and the proximity of this zone to the convective core results in significant uncertainties in computations. Agreement is not yet achieved on the criteria for convective stability in a chemically inhomogeneous medium. Both criteria are used and considerably affect the mixing between the core and semiconvective zone. For a better description of this zone, more complicated models based on the consideration of physical processes inside the semiconvective zone are probably needed [926] (see Sect. 3.1.3, Vol. 1). Calculations made in [942] using both Schwarzschild and Ledoux criteria, and comparison with observations including test for SN 1987 A, gave support for the better validity of Ledoux criterion of convection, which takes into account a stabilizing influence of the gradient of chemical composition, so that instability sets in when (see Sect. 3.1.2, Vol. 1)

$$\frac{dS}{dr} - \left(\frac{\partial S}{\partial \mu}\right)_{\rho,P} \frac{d\mu}{dr} < 0. \tag{9.3}$$

For the case of the Ledoux criterion (9.3), a radiative shell that impedes the inflow of additional matter into the convective core arises between this core and the semiconvective zone [724]. The Schwarzschild criterion (9.2), which does not take into account the stabilizing influence of the chemical gradients, and where the instability sets in at (see Sect. 3.1.1, Vol. 1)

$$\frac{dS}{dr} < 0, \tag{9.4}$$

implies a contact between the semiconvective zone and convective core, but the full mixing approximation cannot be applied, and a concentration gradient is supposed to be established with $\gamma_2 = \gamma_{rad}$ [672]. Calculations based on diverse criteria exhibit uncertainties within $\sim$10% at the hydrogen-burning phase and essentially larger differences on later evolutionary stages. This is true, in particular, with regard to the position in the HR diagram of massive stars with $10 \leq M \leq 40 M_\odot$ in the core helium-burning phase: for the case of the criterion (9.3), the major part of helium is consumed when the star is a red supergiant with $T_{ef} \leq 5000$ K, while at (9.4) the star spends most of the time of helium burning in the region of blue supergiants with $T \geq 10^4$ K. Stars with mass $M \geq 40 M_\odot$ always exhaust their helium while in the red supergiant region, and the choice of criterion turns out to be of little importance here [672].[2] The subsequent evolution of massive stars, from the carbon burning onwards, continues in the red supergiant region and is independent of the choice of convection criterion [672].

Chemical mixing in convective regions, including the semiconvective zone, may be consistently treated by diffusive description (P. Eggleton, Programme STAR. Private communication, 1994), [347], see also Chap. 6, Vol. 1.

[2] See, however, [938], Fig. 9.18 and Table 9.8.

9.1.3 Convective Non-Locality and Overshooting

The convective elements are produced in an unstable region, being accelerated by the buoyancy force. They penetrate into the convectively stable region beyond (overshooting), thereby extending the mixing zone.

The overshooting length, however, depends strongly on how it has been determined. Calculations have yielded $d \leq 0.15\, H_p$ in [663, 901] (see also Chap. 3, Vol. 1), whereas in [296], where a statistical model of turbulent diffusion is used, d/H_p approaches 0.7. Diverse cases with $0 \leq d/H_p \leq 0.7$ have been considered in calculations [939] treating the evolution of massive stars with $15 \leq M/M_\odot \leq 120$ before the exhaustion of hydrogen in the centre. The major effects of overshooting and additional mixing should be expected for the thin shell-burning phase. The mixing-length theory is not good enough for a description of convective penetration and overshooting, and a nonlinear approach is needed [1067], see also Chap. 3, Vol. 1.

9.1.4 Opacity and Nuclear Reactions

The development of atomic and nuclear physics has allowed both theoretical and experimental determination of the opacity of matter, and nuclear reaction rates (see Chaps. 2 and 4, Vol. 1). Contrary to convection, treating the opacity implies dealing with fairly rigorous theories. The computations resulting from a large number of lines are nevertheless so voluminous that corrections to opacity tables are regularly published [19, 254, 312, 315, 316]. Revision of these tables [20, 512–514, 837] led to a considerable increase of opacity in the regions of heavy element line formation.

Theoretical studies are of little help for accurate determination of nuclear reaction rates, and the absence of strong interaction theory makes it necessary to use experimental data. These are often difficult to obtain in the low-energy range which is of interest for astrophysicists. Obtaining new data leads to changes in rates of many reactions (see reviews [392, 393, 459]). Large uncertainties arise also in calculations of nuclear reaction screening at high densities because of the necessity to solve a many-body problem, for which there exist only approximate and not very accurate approaches (see Chap. 4., Vol. 1). Such errors in combination with the very strong dependence of the reaction rates on parameters may be the cause of the contradiction between the theoretical and observational neutrino flux from the Sun, which as claimed in [319, 320, 340] could be removed by improving the input nuclear data.

The choice of formulae and tables for opacity and also for nuclear reaction rates has often to do with the individual approach of the researcher making the evolutionary calculations.

9.1.5 Methods for Calculating Envelope

If the envelope is assumed to be in thermal equilibrium, it is usually calculated separately from the core, and a fitting procedure is used to obtain a fully self-consistent stellar model (see Chap. 6, Vol. 1). The envelope mass and fitting techniques may be adopted in various ways. The static envelope is taken to be 3% of the stellar mass in [134, 570], yet in [570] the envelope models are calculated beforehand and their parameters are obtained with the aid of interpolation, while in [134] the envelope models are calculated for each stellar model separately. The envelope mass is 5% of the stellar mass in [1017, 1018], and 10% and even more in [770, 771, 773, 778]. Calculations of ionization equilibrium differ in detail as well, which results in differences in the equation of state even for the same chemical compositions. All these details affect the evolutionary tracks.

9.1.6 Other Factors

Differences in initial chemical compositions and masses of stars in diverse models create difficulties in comparing different authors' results. In various papers, masses are taken to be

$$
\begin{aligned}
M/M_\odot = {} & 0.8, 1.5, 3, 7, 10, 15 && \text{in } [770],\\
& 1.25, 1.5, 2.25, 3, 5, 9, 15 && \text{in } [492],\\
& 16, 32, 64 && \text{in } [1018],\\
& 30 && \text{in } [724],\\
& 9, 30 && \text{in } [134],\\
& 15, 30 && \text{in } [937],\\
& 3, 5, 7 && \text{in } [85],\\
& 15, 25 && \text{in } [608],\\
& 0.8, 0.9, , 1, 1.25, 1.5, 1.7, 2, 2.5, 3, 4, 5, 7, &&\\
& 9, 12, 15, 20, 25, 40, 60, 85, 120 && \text{in } [276],\\
& 1, 2, 4, 8, 16, 32, 64 && \text{in } [809],
\end{aligned}
\tag{9.5}
$$

and initial chemical compositions

$$
\begin{aligned}
(x_{\rm H}, x_{\rm He}, x_{\rm Z}) = {} & (0.708, 0.272, 0.02) && \text{in } [486]\text{–}[492],\\
& (0.602, 0.354, 0.044) && \text{in } [570],\\
& (0.7, 0.27, 0.03) && \text{in } [770],\\
& (0.75, 0.22, 0.03) && \text{in } [134, 724],
\end{aligned}
$$

$$\begin{array}{ll} (0.7, 0.28, 0.02) & \text{in [608]}, \\ (0.68, 0.3, 0.02) & \text{in [867]}, \\ (0.756, 0.243, 0.001) & \text{in [867]}, \\ 0.62 \le x_{\rm H} \le 0.739, 0.021 \le x_{\rm Z} \le 0.044 & \text{in [937]}, \\ 0.71 \le x_{\rm H} \le 0.78, 0.001 \le x_{\rm Z} \le 0.02 & \text{in [85]}. \end{array} \tag{9.6}$$

The stellar rotation and magnetic field may significantly influence the nuclear evolution, giving rise to meridional circulation and consequent additional mixing. Calculations incorporating phases from the MS to the final evolutionary stages do not deal with these factors, in general.

The discrepancies caused by using different calculation schemes (see Chap. 6, Vol. 1) or versions of the same scheme may be small for chemically homogeneous models on the MS. Accumulated numerical errors may become significant, and the results become sensitive to the applied numerical scheme in late evolutionary stages, when $\geq$1,000 evolutionary steps are required. This problem has not been studied so far. The chaotic appearance of loops in the HR diagram for evolutionary tracks of massive stars impose an element of stochasticity in the stellar evolution problem.

The mass loss is an important factor that influences the evolution of stars. In the absence of well-developed theory for stellar outflow, the mass losses have been included phenomenologically (with the exception of an attempt to obtain a self-consistent solution in [134]) in evolutionary calculations for massive [276, 664, 665, 867, 938, 1021] and intermediate-mass stars [457, 501, 874, 876], giving rise to numerous uncertainties.

9.2 Evolution of Stars in Quiescent Burning Phases

The results of evolutionary calculations are usually represented in the form of evolutionary tracks in the Hertzsprung–Russell (HR) diagram where the logarithm of the effective temperature $\lg T_{\rm ef}$ is plotted on the abscissa, the luminosity logarithm $\lg L$ on the ordinate. By definition, the effective temperature is connected with L and the stellar radius R, corresponding to a total stellar mass M, as

$$L = \pi a c R^2 T_{\rm ef}^4. \tag{9.7}$$

Stars with mass $M > 0.8\,M_\odot$ for which the nuclear evolution length does not exceed the cosmological time of $\sim 2 \times 10^{10}$ years are not degenerated on the MS of chemically homogeneous stars with normal composition, corresponding to the solar surface (Table 1.1, Vol. 1), or depleted in heavy elements composition. The evolution of a star is accompanied by an increase in its central density and approach to a state of degeneracy. For the case of low-mass stars with $M \leq 2.25\,M_\odot$, the helium core, which forms after hydrogen in the centre, has been exhausted and the star has left the MS, turns out to be degenerate. For stars of intermediate mass $2.25 < M/M_\odot \leq 8$, the helium core is not degenerated, but a carbon core forming after helium has been exhausted becomes degenerated; for $M = (8\text{–}10)\,M_{\odot}$, the onset of degeneracy is in the phase of oxygen–neon–magnesium core. In massive stars with $M > 13M_{\odot}$, the degeneracy arises about only at final evolutionary phases with a large neutrino luminosity $L_{\nu} \gg L_{\rm opt}(\equiv L)$ [494, 1032, 1033].

After the degenerate core formation, thermal instabilities develop in stars, leading to strong and rapid changes in energy release rates and to essential, though not very pronounced, changes in $T_{\rm ef}$ and L. These instabilities include helium outbursts in the degenerate core of low-mass stars [458, 883, 887], outbursts in nondegenerate helium shell source in the presence of degenerate carbon core in low- and middle-mass stars [456, 884, 885, 1034]. A particular role belongs to outbursts in carbon-degenerate cores that may result in supernova explosions. A variety of simplifying conditions is often used in calculations of these evolutionary stages [772, 838, 953, 1004] because the conventional schemes which gave good results for early stages of stellar evolution (see Chap. 6, Vol. 1) prove inefficient or very time-consuming.

An instability, probably of a thermal origin, manifested in a nonregular pattern of loops in the HR diagram may occur in nondegenerate stars as well ([271, 522, 651]; see below). As the development times are essentially larger and overfalls in energy release smaller in that case, this instability does not create any serious difficulties in calculations.

9.2.1 Iben's Calculations

The first large series of evolutionary calculations for stars with diverse masses was published by Iben in 1964–1967 [486–492]. Calculations have been made by the Henyey method with fitted envelope (see Chap. 6, Vol. 1), for initial chemical composition, $x_{\rm H} = 0.708$, $x_{\rm He} = 0.272$, and $x_{\rm Z} = 0.02$. The convection in the envelope has been calculated with a mixing length equal to half of the density scale height $l = H_{\rho}/2$. The Schwarzschild condition (3.1.3) that does not include the chemical composition gradient has been adopted as the convection criterion. The main results of these calculations are given in Figs. 9.1–9.5 and Tables 9.1 and 9.2. Models with $M > 3M_{\odot}$ have been calculated for stages preceding helium exhaustion in the centre, and with $M < 3M_{\odot}$ for stages before to core helium ignition. Calculations for the evolution of a $5M_{\odot}$ star are extended further along the shell helium-burning phase. For stars with $M \leq 9M_{\odot}$, portions between dots in Figs. 9.1–9.4 correspond to the following evolutionary phases:

1–3 Core hydrogen burning (MS)

3–4 Overall gravitational contraction; this phase is absent for the star with $M = 1M_{\odot}$

4–5 Formation of hydrogen-burning shell

5–6 Hydrogen burning in a thick shell

6–9 The diminution of the hydrogen-burning shell thickness

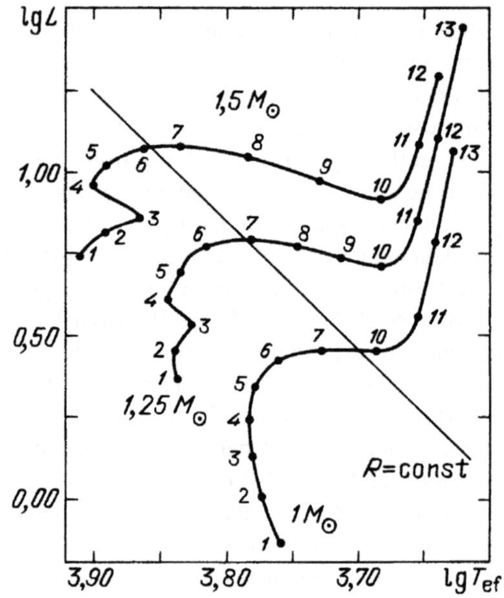

Fig. 9.1 Evolutionary tracks for stars of mass 1, 1.25, 1.5 $M_\odot$ and chemical composition from (9.6). Luminosity L is in $L_\odot$ and effective temperature is in K. Each *number* represents corresponding evolutionary lifetime from Table 9.1. The time t_1 corresponds to the stage of gravitational contraction to the main sequence (MS). The *straight line* is that of constant radius R, from [560]

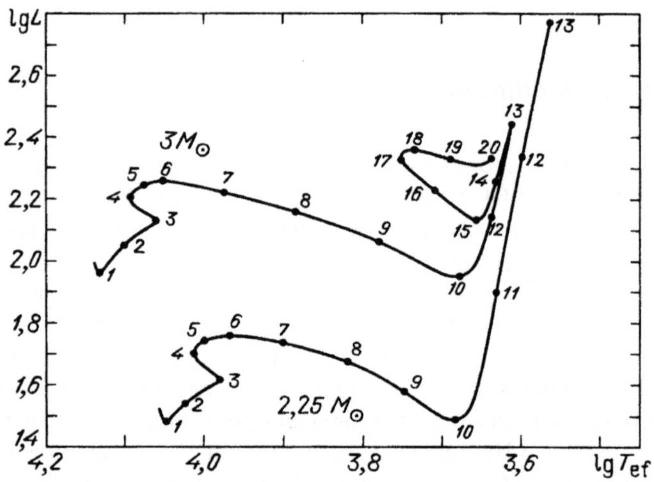

Fig. 9.2 The same as in Fig. 9.1 for $M = 2.25, 3\, M_\odot$, from [486,491]

9–10 Quick propagation of convection from the surface shells inward

10–13 Red-giant phase

13 Onset of the $3^4\text{He} \rightarrow {}^{12}\text{C}$ reaction in the core

13–15 The first phase of core helium burning

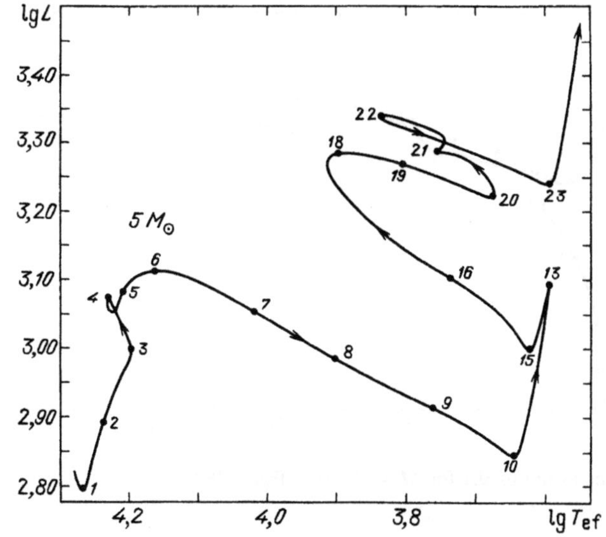

Fig. 9.3 The same as in Fig. 9.1 for $M = 5\,M_\odot$, from [487]

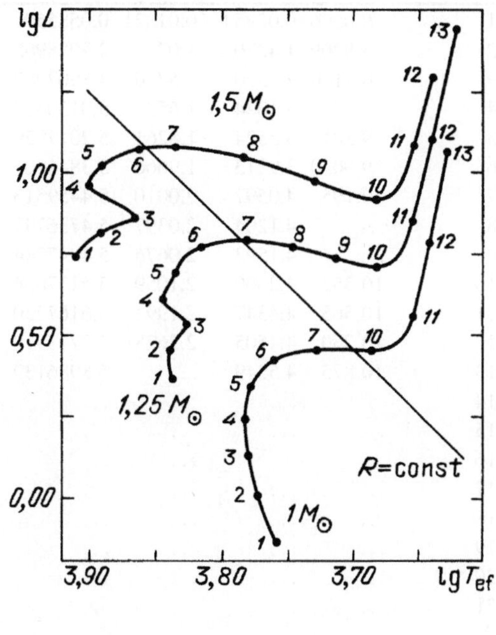

Fig. 9.4 The same as in Fig. 9.4 for $M = 9\,M_\odot$, from[488]

15–16 Disappearance of the deep convective envelope, rapid contraction
16–18 Main core helium-burning phase
20–21 Overall contraction with helium exhaustion in the centre

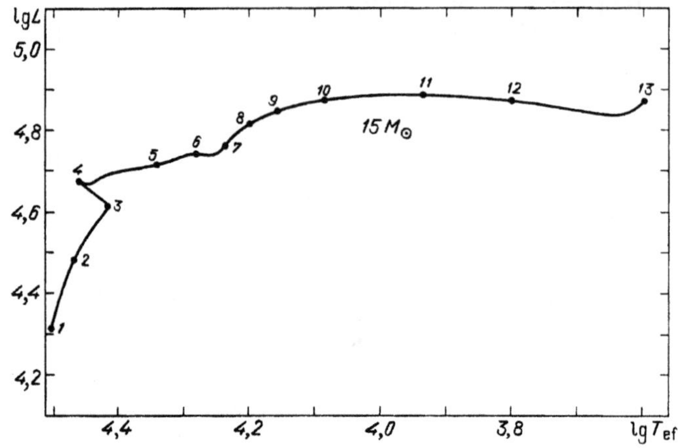

Fig. 9.5 The same as in Fig. 9.1 for $M = 15\,M_\odot$, from [489]

Table 9.1 Evolutionary lifetimes

	$1\,M_\odot$	$1.25\,M_\odot$	$1.5\,M_\odot$	$2.25\,M_\odot$	$3\,M_\odot$	$5\,M_\odot$	$9\,M_\odot$	$15\,M_\odot$
Point	10^9 yr			10^8 yr		10^7 yr		
1	0.0506	0.02954	0.01821	0.058550	0.024586	0.15800	0.0232171	0.0138224
2	3.8209	1.4220	1.0277	2.7988965	1.38921	4.01899	1.435125	0.663253
3	6.7100	2.8320	1.5710	4.8502987	2.23669	6.60443	2.129274	1.024045
4	8.1719	3.1044	1.652	5.0150323	2.34089	6.82168	2.189700	1.046745
5	9.2012	3.5524	1.8261	5.2017959	2.40119	6.83608	2.193710	1.048983
6	9.9030	3.9213	1.9666	5.3846801	2.44420	6.95886	2.198813	1.050644
7	10.195	4.0597	2.0010	5.4459513	2.47004	7.00750	2.206125	1.054302
8	...	4.1204	2.0397	5.4736797	2.47865	7.02016	2.209479	1.113604
9	...	4.1593	2.0676	5.4947244	2.48429	7.02709	2.212819	1.154207
10	10.352	4.2060	2.1059	5.5157054	2.48925	7.03418	2.213585	1.192760
11	10.565	4.4347	2.1991	5.6167250	...	...	...	1.208043
12	10.750	4.4505	2.2628	5.7773918	2.50728	...	2.215236	1.210199
13	10.875	4.5349	...	5.8986139	2.53163	7.08275	2.220137	1.211368
14	...	...	...	...	2.55850	...	2.243431	...
15	...	...	...	...	2.78295	7.57595	2.267412	...
16	...	...	...	...	2.94233	7.77057	2.273715	...
17	...	...	...	...	3.06968	...	2.277173	...
18	...	...	...	...	3.19043	8.68987	2.314993	...
19	...	...	...	...	3.23566	...	2.567444	...
20	...	...	...	...	3.26323	8.78291	2.623007	...
21	...	...	...	...	...	8.79060	2.625870	...
References	[490]	[490]	[490]	[491]	[486]	[487]	[488]	[489]

21–22 Helium burning in a thick shell

23 Neutrino emission out of the core, helium burning in a thin shell

Table 9.2 Principal parameters of stars on evolutionary tracks in Fig. 9.2–9.5 and Table 9.1

$\frac{M}{M_\odot}$	ρ_c, g cm^{-3}	T_c, 10^6 K	$\frac{L}{L_\odot}$	$\frac{R_S}{R_\odot}$	ρ_c, g cm^{-3}	T_c, 10^6 K	$\frac{L}{L_\odot}$	$\frac{R_S}{R_\odot}$	$\frac{M_1}{M_\odot}$	$\frac{M_2}{M_\odot}$	$\frac{M_{c,ms}}{M_\odot}$	$\frac{M_{c,max}}{M_\odot}$	t_{ms}, yr	Ref.
	Main sequence				Last model in Table 9.1									
1	90.0	13.9	0.73	0.87	9.12(4)	27.4	11.4	6.18	0.2	...	...	...	1.02(10)	[490]
1.25	93.5	16.6	2.3	1.08	1.3(5)	31.0	29.0	10.3	0.23	...	0.012	0.058	4.03(9)	[490]
1.5	87.7	18.8	5.4	1.18	1.14(5)	29.9	19.6	7.72	0.22	...	0.056	0.12	1.98(9)	[490]
2.25	58.8	22.2	30	1.45	3.2(5)	92.5	590	59.2	0.38	...	0.26	0.33	5.32(8)	[491]
3	40.4	24.1	95	1.75	4.14(4)	158	210	26.4	0.59	0.14	0.46	0.53	2.21(8)	[486]
5	17.5	27.3	620	2.40	2.16(4)	184	1.94(3)	44	1.08	0.3	1.0	1.1	6.44(7)	[487]
9	10.5	31.0	4.5(3)	3.40	1.33(4)	262	1.6(4)	37	1.9	0.74	2.4	2.7	2.11(7)	[488]
15	6.17	34.4	2(4)	4.40	3.9(3)	300	7.8(4)	530	4.7	2.7	5.9	5.9	1.01(7)	[489]

ρ_c and T_c are the central density and temperature, L and R_S the luminosity and radius of the star, M_1 is the mass of the core where hydrogen is exhausted, M_2 the mass of the C–O core, $M_{c,ms}$ the mass of the convective core on the main sequence (model 1 in Table 9.1), $M_{c,max}$ the maximum mass of the convective core, t_{ms} the lifetime of the star on the main sequence

The tracks between points 21 and 23 in Fig. 9.3 are topologically equivalent to those obtained from similar calculations [565, 566].

Intervals 1–4 in Fig. 9.5 for $M = 12M_\odot$ correspond to the same evolutionary phases as for lower masses, other phases are:

4–7	Shell hydrogen burning
7–10	Helium burning in the core and hydrogen burning in the shell
11–13	Indent rapid expansion of the envelope

The relevant evolutionary lifetimes are listed in Table 9.1. Point 1 is determined by the time of gravitational contraction to the MS.

With stellar mass increasing from 1 to 1.25 $M_\odot$, the radiative core with burning hydrogen transforms into a convective core. This is caused by the transition from the proton–proton reaction of hydrogen burning, to its burning in the carbon cycle (see Chap. 4, Vol. 1), the latter having a steeper temperature dependence. In the presence of a radiative core, the overall contraction phase is absent for $M = 1M_\odot$, and a smooth transition from core H to shell H burning occurs.

As can be seen from Table 9.2, for $M = 15M_\odot$ the convective core mass decreases monotonically with time and for lower masses passes through a maximum. The ^{4}He burning results in the formation of ^{12}C and ^{16}O in the core with almost all ^{4}He converted into ^{16}O for $M \geq 5M_\odot$. This result, however, is not reliable by reason of an uncertainty in the $^{12}\mathrm{C}(\alpha, \gamma)^{16}\mathrm{O}$ reaction rate used in [486–491].

For stars with $M \leq 2.25\,M_\odot$, the helium core turns out to be degenerate, and the subsequent evolution results in a helium flash in the core. In more massive stars the helium burning begins smoothly, without flash. As pointed out in [492], an essential contribution in the luminosity is made near point 11 by the nitrogen-burning reaction $^{14}\mathrm{N}(\alpha, \gamma)\ ^{18}\mathrm{F}(\beta^+\nu)\ ^{18}\mathrm{O}$.

A total of 320 mass shells have been used in calculations, each track covered by 500–660 models.

The loops on evolutionary tracks for stars with $M = 3$–$9\,M_\odot$ arise from the above-mentioned thermal-like instability. This also explains an ambiguity in calculating stellar models with given masses and chemical composition distributions found in [776, 838, 839]. The ambiguity arises beyond the point where the determinants obtained from solving evolutionary equations by the Schwarzschild method or, upon linearizing the system of difference equations appearing during solving differential equations by the Henyey method (see Chap. 6, Vol. 1), become zero. In the case of instability, small changes in initial data drastically distort evolutionary tracks in the region of loops. Figure 9.6 from [85] shows a strong dependence of the loop shapes on chemical composition. Also, essential differences of evolutionary lifetimes may be seen from Table 9.3. As shown in [771], the loop shapes depend not only on physical assumptions but also on the type of computer used for running the same program, that is, on such particulars as the number of significant digits, rounding schemes and other details. All these facts provide evidence for the stochastic nature of the loop shapes. For massive stars with $M \geq 15M_\odot$, the appearance of loops is associated in [1086] with the hydrogen shell outgoing beyond the boundary of the

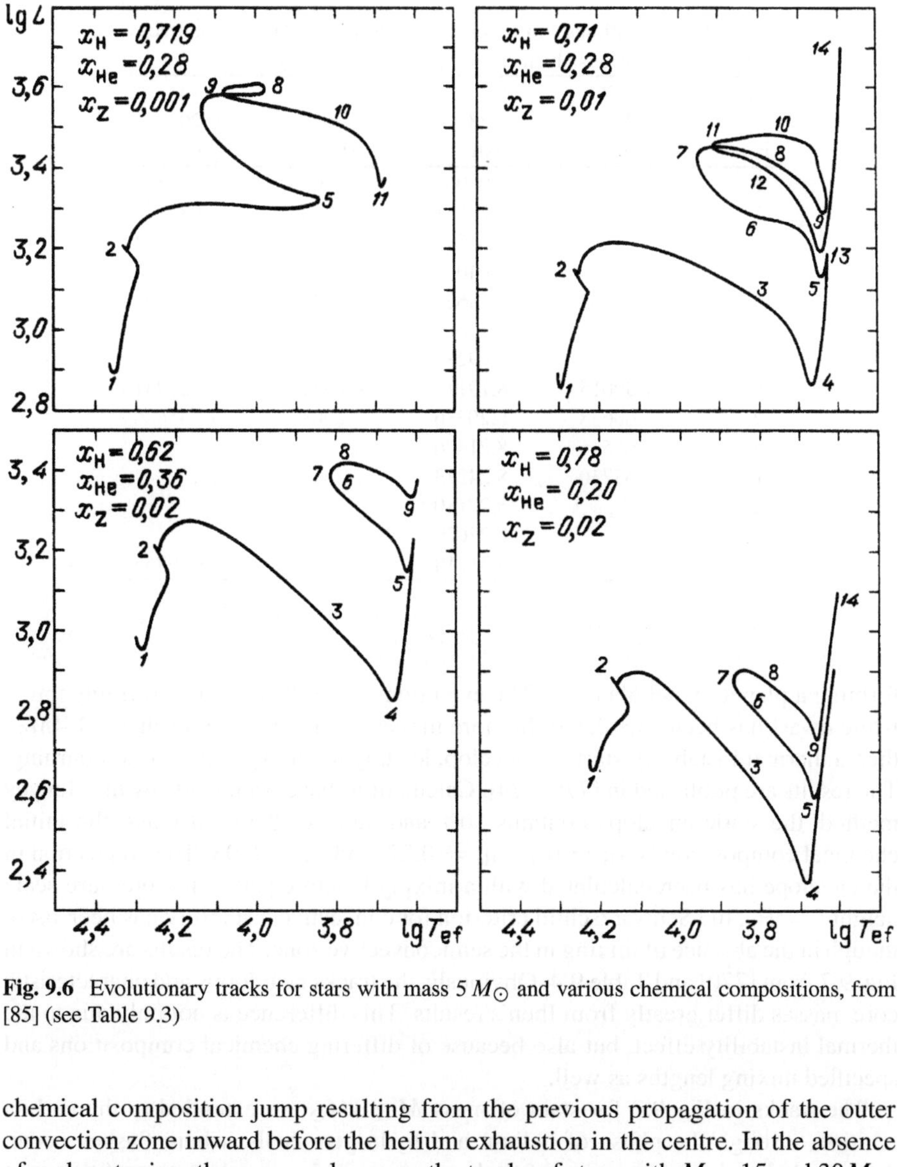

Fig. 9.6 Evolutionary tracks for stars with mass 5 $M_\odot$ and various chemical compositions, from [85] (see Table 9.3)

chemical composition jump resulting from the previous propagation of the outer convection zone inward before the helium exhaustion in the centre. In the absence of such outgoing, there are no loops on the tracks of stars with $M = 15$ and $30 M_\odot$.

9.2.2 Paczynski's Calculations

In papers published in 1970–1971, Paczynski calculated the evolution of stars for the mass range from 0.8 to $15 M_\odot$ up to the formation of a degenerate carbon core. The stars with $M <\sim 8 M_\odot$ are then assumed to lose a significant part of their mass, thus

Table 9.3 Evolutionary times of stars with $M = 5M_{\odot}$ on the tracks in Fig. 9.6 for various chemical compositions (in 10^7 yr)

x_{H}	0.719	0.71	0.62	0.78
x_{He}	0.28	0.28	0.36	0.20
x_{Z}	0.001	0.01	0.02	0.02
1	0.02174	0.06008	0.07035	0.21501
2	6.26784	6.44761	4.55190	9.83129
3	...	6.63593	4.72257	10.22774
4	...	6.64071	4.72685	10.24225
5	6.44162	7.14636	5.28979	11.48509
6	...	7.35031	5.82833	11.88947
7	...	7.99281	5.98230	12.06863
8	7.59015	8.19118	6.09349	13.27731
9	7.60638	8.20386	6.20085	13.45606
10	7.65546	8.21476	...	...
11	7.65730	8.24283	...	...
12	...	8.27180	...	...
13	...	8.27629	...	...
14	...	8.31318	...	13.69765

forming a planetary nebula (PN). The evolution of the PN core transforming into a white dwarf has been calculated. In more massive stars the core reaches $\sim 1.4M_{\odot}$, then a thermal instability starts to develop, leading to an explosive carbon burning. The results are published in [770–774]. Calculations have been made by the Henyey method, the static envelope contains 10% and more of the total mass, the initial chemical composition is $x_{\mathrm{H}} = 0.7$, $x_{\mathrm{He}} = 0.27$, and $x_{\mathrm{Z}} = 0.03$. The convection in the envelope has been calculated with a mixing length equal to the pressure scale height $l = H_{\mathrm{p}}$, the Schwarzschild criterion for convective instability has been used, though in the absence of mixing in the semiconvective zone. The results are shown in Fig. 9.7 from [770] and Table 9.4. Obviously, the tracks near loops and characteristic core masses differ greatly from Iben's results. This difference is not only due to the thermal instability effect, but also because of differing chemical compositions and specified mixing lengths as well.

The tracks in Fig. 9.7 for stars of mass $M \geq 3M_{\odot}$ are extended to the carbon ignition, through the phase of helium and hydrogen shells, where flashes occur. A method allowing suppression of flashes and approximate calculation of the averaged evolution has been used in [770, 772].[3] Helium flashes occur if a degenerate carbon core is present and will be discussed in detail in the next section. As may be

[3] A modified version of Paczynski's program allowing stable running in rapid evolutionary phases, based on a special choice of the calculation of the derivatives for different functions in difference equations, has been used in [779], see also Chap. 6, Vol. 1; input into luminosity due to traversing of the boundary of chemical composition jump by the convective zone is correctly calculated in [781].

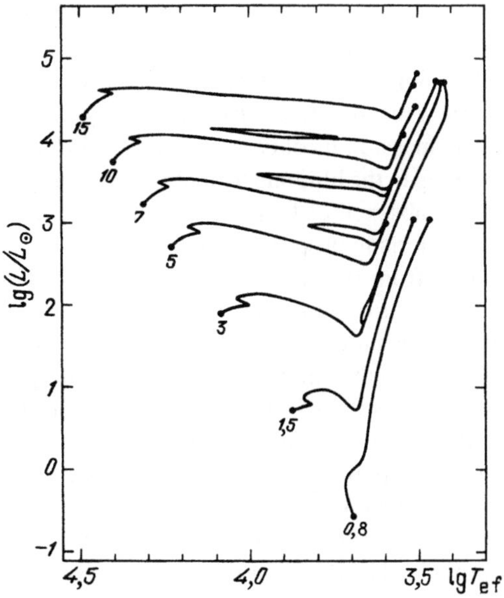

Fig. 9.7 Evolutionary tracks for stars with initial composition $x_H = 0.7$, $x_{He} = 0.27$, $x_Z = 0.03$, from the MS to helium flash for $M = 0.8\,M_\odot$ and $1.5\,M_\odot$ or to carbon ignition for $M = 3$–$15\,M_\odot$. The *numbers* indicate stellar masses in solar units, the *dots* represent the MS and models at times of helium or carbon ignition in the core, from [770]

Table 9.4 Core masses as a function of stellar masses at various evolutionary phases [770]

$\frac{M}{M_\odot}$	$M_{core}/M_\odot$ (inside hydrogen burning shell)		
	Helium ignition	Helium exhaustion in the core	Carbon ignition
0.8	0.39	...	...
1.5	0.40	...	...
3.0	0.35	0.51	1.39
5.0	0.56	0.95	1.39
7.0	0.83	1.45(1.02)	1.39
10.0	1.35	2.32	2.32
15.0	2.54	3.89	3.91

seen from Table 9.4, after helium exhaustion in a star with $M = 7M_\odot$, the convective envelope penetrates the carbon core region, thereby reducing its mass from 1.45 to $1.02\,M_\odot$.

An extended review of the evolution of single and binary stars is given in [503].

9.2.3 *Evolution of Massive Stars*

The core of a massive star with $M \geq 13M_\odot$ remains non-degenerate up to the final evolutionary stages, when considerable neutrino losses cause the central region of the star to contract.

Evolutionary tracks of stars with $M = 15$ and $30M_{\odot}$ are calculated in [937] for diverse chemical compositions (see (9.6)), the Ledoux criterion for convection (9.3) and a variety of mixing lengths $l = \alpha H_{\rho}$, $0.4 \leq \alpha \leq 10$. The opacity and nuclear reaction rates also vary. Many tracks have the same topology for stars of the same mass. In most cases, a star of $15M_{\odot}$ spends 17–62% of the lifetime of helium burning in the blue supergiant region owing to the loop formation. In a star of $30M_{\odot}$, in the absence of loops, all helium exhausts in the red supergiant phase. This is a characteristic property of tracks calculated with use of the Ledoux criterion. Some tracks from [937] are shown in Fig. 9.8. For stars with $15M_{\odot}$, no loop occurs at certain sets of input parameters, and almost all helium exhausts in the red supergiant phase. This might be due to the fact that the outer convective zone does not penetrate inside, and the chemical composition jump does not form [1086].

For the tracks in Fig. 9.8, the lifetimes of hydrogen burning are $\tau_{\rm H} = 1.2 \times 10^7$ and 6.1×10^6 yr, helium burning $\tau_{\rm He}/\tau_{\rm H} = 0.10$ and 0.08, the lifetime of ^{4}He burning in the region of blue supergiants with $\lg T_{\rm e} \geq 4.1$ is $\tau_{\rm bl}/\tau_{\rm He} = 0.37$ and 0.03 for $M = 15$ and $30M_{\odot}$, respectively.

Evolutionary calculations for stars with $M = 15, 30, 60M_{\odot}$ have been performed in [1086] for initial chemical composition $x_{\rm H} = 0.7$, $x_{\rm He} = 0.27$, $x_{\rm Z} = 0.03$, mixing length $l = H_{\rm p}$ and static envelope mass of 5–15% of the stellar mass. The Schwarzschild criterion (9.4) for convection has been used, although the mixing in the semiconvective zone has not been allowed. This treatment is argued in [1086] to be equivalent to using the Ledoux criterion (9.3). The results of calculations are given in Fig. 9.9 and Table 9.5. The presence of loops in the HR diagram causes stars with $M = 15$ and $30M_{\odot}$ to spend most of the lifetime of core helium burning in the blue supergiant region. Calculations have been performed until the exhaustion of helium in the core. The fraction of oxygen relative to carbon

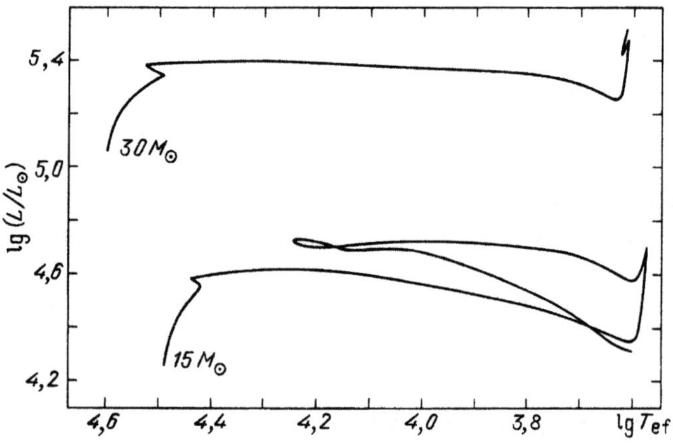

Fig. 9.8 Evolutionary tracks for stars of 15 and 30 $M_{\odot}$ from the MS to the end of core helium burning for initial chemical composition $x_{\rm H} = 0.739$, $x_{\rm He} = 0.24$, $x_{\rm Z} = 0.021$, the Ledoux criterion for convection, and mixing length $l = 0.4\,{\rm H}_{\rho}$, from [937]

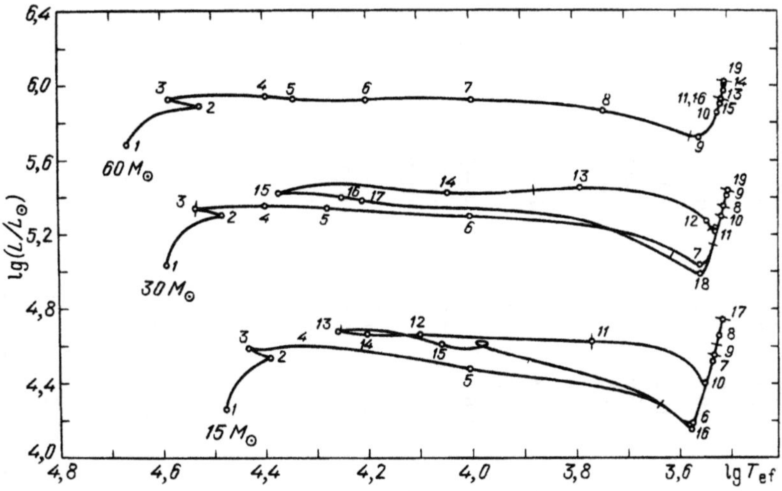

Fig. 9.9 Evolutionary tracks for stars of 15, 30, and 60 $M_\odot$ from the MS to the carbon ignition in the centre for initial chemical composition $x_H = 0.7$, $x_{He} = 0.27$, and $x_Z = 0.03$, the Ledoux criterion and $l = 0.4\,H_P$, from [1086]

increases with the increase of the stellar mass in the course of helium burning. For $M = 15, 30, and 60 M_\odot$, after the exhaustion of helium in the core, the weight fraction of carbon is $x_{^{12}C} = 0.4, 0.3$, and 0.2, respectively. A more accurate values for the $^{12}C(\alpha, \gamma)^{16}O$ reaction rate was used here in comparison with [486–490], where $x_{^{12}C}$ was considerably lower (further refinements for this reaction rate are given in [393], see also Chap. 4, Vol. 1).

As the envelope expands and its matter is progressively penetrated by the convective zone moving inward, the latter passes over a chemical composition jump due to the previous hydrogen burning in the convective core. Such a penetration leads to an increase in luminosity accounted for by the gravitational energy release in the jump [781, 1086]

$$\Delta L = \frac{dM_{CE}}{dt}\left(\Delta E - \frac{P}{\rho^2}\Delta\rho\right) = \frac{dM_{CE}}{dt}\frac{kT}{m_p}\left(\frac{4}{\beta} - 1.5\right)\Delta\left(\frac{1}{\mu}\right), \tag{9.8}$$

where M_{CE} is the core mass inside the jump, the molecular weight μ is the number of nucleons per gas particle and $\beta \equiv \beta_g$ is the ratio of the gas pressure to the total pressure. The penetration of convection from the envelope into the burning zone has been examined in [1006].

The evolutionary calculations of massive stars have been extended to a stage where the star loses its hydrodynamical stability, that is, to the presupernova model, primarily, in [1033] (see also [1032]). Stars with masses $M = 15$ and $25 M_\odot$, initial chemical composition $x_H = 0.7$, $x_{He} = 0.28$, $x_Z = 0.02$ and $l = H_p$ have been

Table 9.5 Evolutionary parameters of stars in Fig. 9.9 taken from [1086]

Point	T_c, K	L_c g cm^{-3}	$\frac{L_{opt}}{L_\odot}$	$\frac{L_\nu}{L_\odot}$	$\frac{L_H}{L_\odot}$	$\frac{L_{He}}{L_\odot}$	$\frac{L_C}{L_\odot}$	$\frac{L_g}{L_\odot}$	T_{ef}, K	$\frac{R_{ph}}{R_\odot}$	$\frac{M_{conv}}{M_\odot}$	$x_{H,c}$ [$x_{^{12}C,c}$]	t_{ev} yr
	3.2(7)	5.3(0)	1.8(4)	...	1.8(4)	0	0	0	2.99(4)	5.1	5.8	0.7	0
1	3.6(7)	3.0(0)	1.1(5)	...	1.1(5)	0	0	0	3.92(4)	7.2	15.8	0.7	0
	3.9(7)	2.0(0)	4.7(5)	...	4.7(5)	0	0	0	4.71(4)	10.3	40.1	0.7	0
	4.1(7)	9.3(0)	3.4(4)	...	3.1(4)	0	0	3.8(3)	2.46(4)	10.2	2.5	0.034	9.6(6)
2	4.6(7)	5.1(0)	2.0(5)	...	1.8(5)	0	0	2.4(4)	3.06(4)	15.9	8.2	0.030	4.8(6)
	4.9(7)	3.3(0)	7.7(5)	...	6.7(5)	0	0	1.0(5)	3.43(4)	25.0	21.8	0.024	3.2(6)
	5.4(7)	2.5(1)	3.8(4)	...	3.5(4)	0	0	3.5(3)	2.70(4)	8.9	1.1	1.2(−5)	2.8(5)
3	6.7(7)	1.9(1)	2.2(5)	...	1.7(5)	0	0	4.3(4)	3.43(4)	13.3	4.3	4(−6)	1.4(5)
	7.9(7)	1.6(1)	8.3(5)	...	6.2(5)	0	0	2.2(5)	3.88(4)	20.3	15.2	...	7.8(4)
	1.6(8)	2.2(3)	1.5(4)	...	3.1(4)	7.5(3)	0	−2.4(4)	3.73(3)	296	0.7	[4(−3)]	6.6(4)
6	1.8(8)	7.0(2)	1.9(5)	...	1.6(5)	9.2(4)	0	−5.9(4)	1.01(4)	145	3.7	[0.011]	2.2(4)
	2.0(8)	4.0(2)	8.1(5)	...	4.2(5)	4.0(5)	0	−1.4(4)	1.58(4)	120	9.4	[0.005]	1.2(4)
	1.8(8)	1.5(3)	2.5(4)	...	1.1(4)	1.4(4)	0	1.0(2)	3.52(3)	428	1.7	[0.37]	5.8(5)
10	1.9(8)	6.2(2)	1.9(5)	...	8.2(4)	1.1(5)	0	1.5(2)	3.24(3)	1404	5.6	[0.13]	6.6(4)
	2.0(8)	3.7(2)	7.0(5)	...	9.5(5)	6.6(5)	0	−5.1(4)	3.30(3)	2579	14.0	[0.021]	4.5(3)
	2.2(8)	2.3(3)	4.7(4)	...	2.7(4)	2.0(4)	0	5.5(2)	1.59(4)	28.7	1.8	[0.56]	5.5(5)
14	2.0(8)	6.8(2)	2.6(5)	...	4.6(4)	1.4(5)	0	7.0(4)	1.10(4)	140	6.5	[0.41]	1.8(5)
	2.1(8)	3.5(2)	9.9(5)	...	2.6(5)	7.2(5)	0	4.9(3)	3.19(3)	3267	16.2	[0.033]	2.3(3)
	6.4(8)	5.0(5)	5.3(4)	7.2(4)	0	5.5(4)	9.0(3)	6.1(4)	3.22(3)	742	3.0(−4)	[0.41]	1.2(5)
17	3.6(8)	4.0(3)	2.3(5)	3.5(3)	8.6(4)	2.4(3)	0	1.5(5)	1.62(4)	61.7	2.7	[0.30]	2.2(5)
	3.0(8)	9.1(2)	8.6(5)	2.1(3)	3.4(4)	8.0(5)	0	3.0(4)	3.21(3)	3004	20.6	[0.247]	3.2(5)
	...	...	...	...	...	...	...	...	...	...	...	...	...
19	8.4(8)	1.2(5)	2.6(5)	5.9(6)	0	7.5(5)	4.5(6)	8.8(5)	3.14(3)	1742	0.41	[0.30]	1.2(4)
	4.2(8)	2.8(3)	1.0(6)	2.5(4)	1.6(5)	1.1(4)	0	8.7(5)	3.18(3)	3324	16.8	[0.194]	1.3(4)

t_{ev} is the lifetime of the evolution between the given and previous points in Fig. 9.9, the upper lines give values for $15M_\odot$, middle lines for $30M_\odot$, lower lines for $60M_\odot$, the dashes indicate data absence in [1086]

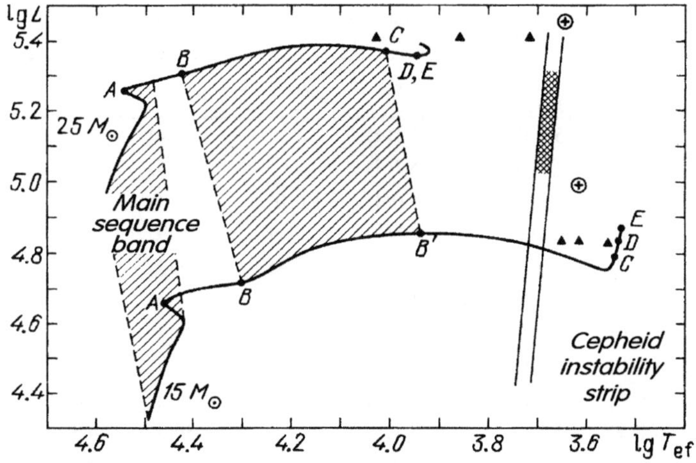

Fig. 9.10 Evolutionary tracks for 15, 25 $M_\odot$ models from [608], BB′ and BC are the paths during the core helium-burning phase in the 15 $M_\odot$ and 25 $M_\odot$ case, respectively, CD is the path during the double H–He shell burning, core carbon burning occurs between points D end E. Evolutionary tracks from [1033] coincide for the major part with [608]. The *triangles* indicate the results of evolutionary calculations from [1033], where a difference may be seen between these and the results from [608]. The calculations in [1033] have been carried through to the onset of instability, that is, the presupernova model (marked by the *circled cross*)

taken under consideration. The initial data coincide[4] with [608], where evolutionary calculations have been carried out through the stage of carbon exhaustion in the centre and formation of a carbon shell along with helium and hydrogen shells. Tracks from [608] are given in Fig. 9.10, those of model stars from [1033] which have a marked difference from [608] being shown by triangles. For the rest, these tracks are indistinguishable. If convection arises according to the Schwarzschild criterion, the helium burning occurs in the blue supergiant region, and no loops arise. After the carbon-burning phase, the neutrino losses cause the core evolution to accelerate to such a degree that the outer radius and optical luminosity of the star remain almost unaltered up to the onset of collapse (see Table 9.6). The distribution of concentration of elements and parameters of star before collapse are given in Figs. 9.11 and 9.12 for $15M_\odot$, while in Figs. 9.13 and 9.14 the same quantities are shown for $M = 25M_\odot$. One may see in the centre the effect of the endothermic reaction of neutron and alpha-particle detachment from iron peak elements. The tracks at late evolutionary stages have been slightly changed in later calculations [1040, 1053]. More than 100 isotopes of a variety of elements have been taken into account in these calculations.

[4] It is asserted in [1033] that the calculations use the Ledoux criterion for convection. However, there is a good coincidence between evolutionary tracks from [1033] and [608], where the Schwarzschild criterion has been used (see Fig. 9.10). In both calculations helium burning is in the blue supergiant region.

Table 9.6 Major phases of evolution of stars with $M = 15$ and $25\, M_\odot$ with initial composition $x_{\rm H} = 0.7, x_{\rm He} = 0.28, x_Z = 0.02$ (from [1033])

Burning phase	T_c, K	ρ_c, g cm^{-3}	$\frac{L_{\rm opt}}{L_\odot}$	$\frac{L_\nu}{L_\odot}$	$T_{\rm ef}$, K	$\frac{R_{\rm ph}}{R_\odot}$	$t_{\rm ev}$ yr
Hydrogen	3.4(7)	5.9(0)	2.1(4)	...	3.26(4)	4.6	1.2(7)
	3.7(7)	3.8(0)	8.1(4)	...	3.98(4)	6.0	7.3(6)
Helium	1.6(8)	1.3(3)	6.0(4)	1.0(0)	1.59(4)	31.6	1.3(6)
	1.8(8)	6.2(2)	2.5(5)	1.9(1)	1.58(4)	67.5	6.7(5)
Carbon	6.2(8)	1.7(5)	8.6(4)	8.9(4)	4.26(3)	532	6.3(3)
	7.2(8)	6.4(5)	3.1(5)	2.6(6)	4.36(3)	963	165
Neon	1.3(9)	1.6(6)	9.7(4)	1.8(8)	4.28(3)	560	7.0
	1.4(9)	3.7(6)	3.1(5)	2.0(9)	4.36(3)	963	1.2
Oxygen	1.9(9)	9.7(6)	9.7(4)	2.1(9)	4.28(3)	560	1.7
	1.8(9)	1.3(7)	3.1(5)	6.0(9)	4.36(3)	963	0.51
Silicon	3.1(9)	2.3(8)	9.7(4)	8.9(10)	4.28(3)	560	1.6(−2)
	3.4(9)	1.1(8)	3.1(5)	9.9(11)	4.36(3)	963	3.8(−3)
Collapse	8.3(9)	6.0(9)	9.7(4)	1.8(15)	4.28(3)	560	9.5(−9)
	8.3(9)	3.5(9)	3.1(5)	2.1(15)	4.36(3)	963	1.1(−8)

All the parameters, with the exception of $t_{\rm ev}$, define the conditions immediately after ignition of each nuclear fuel, $t_{\rm ev}$ is the time until the ignition of the next fuel. The upper lines give values for $15 M_\odot$, lower lines for $25 M_\odot$, the neutrino luminosity during hydrogen burning has not been evaluated

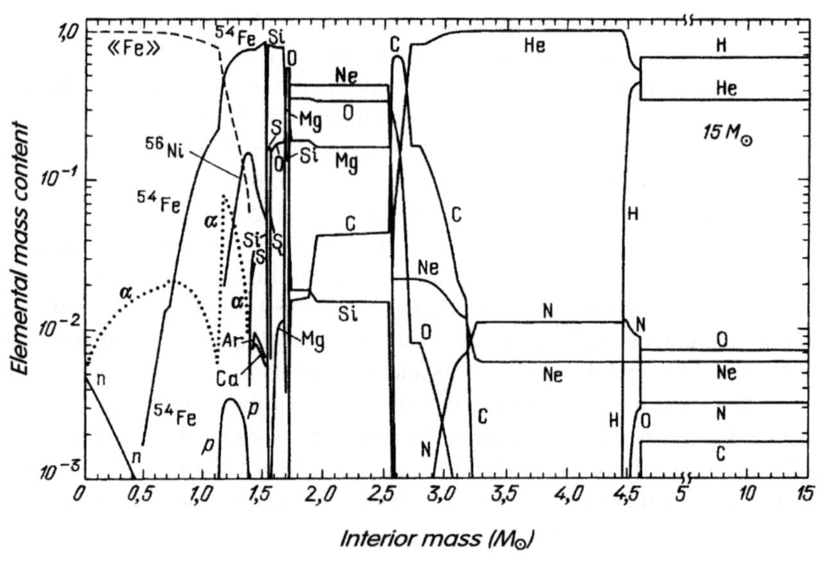

Fig. 9.11 Weight concentrations of various elements in the $15\, M_\odot$ model star before core collapse. The nuclear quasiequilibrium (NQE) has been used inside $M = 1.56\, M_\odot$. Weak interaction reactions have been treated kinetically under condition of a full neutrino escape (see Sect. 1.3, Vol. 1). In this region, the curve for ^{56}Ni represents all iron peak elements with $A = 2Z$, for ^{54}Fe all elements with $A = 2(Z + 1)$, and the curve labeled <<Fe>> represents all other NQE elements with $Z \geq 22$, from [1033]

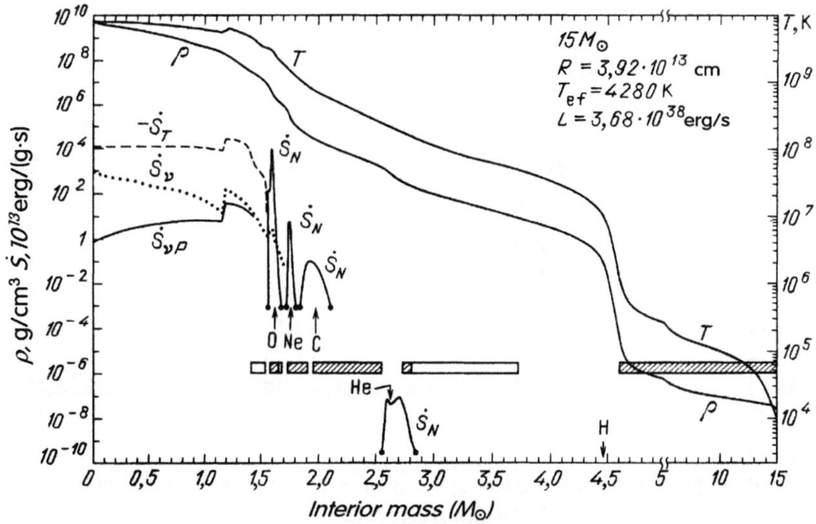

Fig. 9.12 The distribution of parameters of 15 $M_\odot$ model before core collapse. Scales for density ρ and temperature T are selected in such a way as to conserve the distance between the curves at $\rho \sim T^3$. The quantity $-\dot{S}_T$ gives the rate of the total energy loss due to neutrino emission and heavy element photo-disintegration, $\dot{S}_\nu$ is the rate of total neutrino loss, $\dot{S}_{\nu P}$ is the rate of neutrino losses due to emission of plasma neutrinos, from [82] (see Sect. 5.2, Vol. 1). The energy release rates in nuclear burning are denoted by $\dot{S}_N$ with indication of the main nuclear fuel. All the quantities $\dot{S}_i$ have the same scale $\dot{S}$. Zones of strong convection are represented by the *hatched strip*, semiconvective zones by the *blank strip*. The quantities R, T_{ef}, and L are the photosphere radius, effective temperature and optical luminosity, respectively, from [1033]

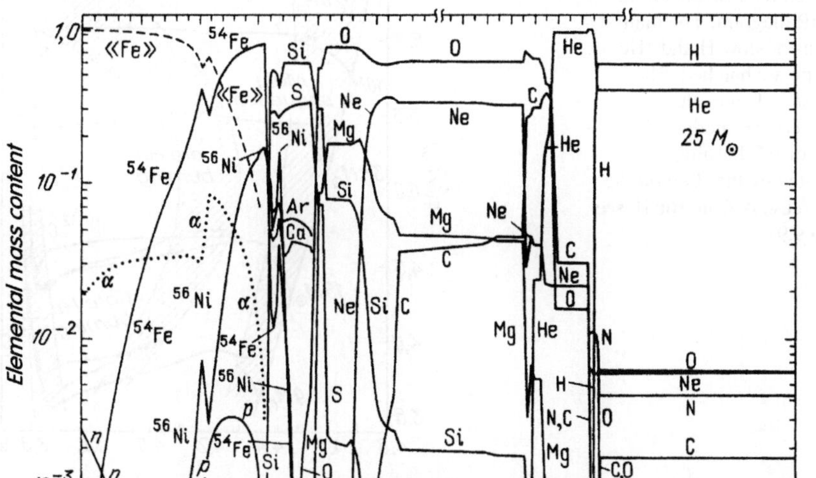

Fig. 9.13 Composition of supernova with $M = 25\,M_\odot$. NQE has been adopted inside $M = 1.61\,M_\odot$, the other notation is the same as in Fig. 9.11 (from [1033])

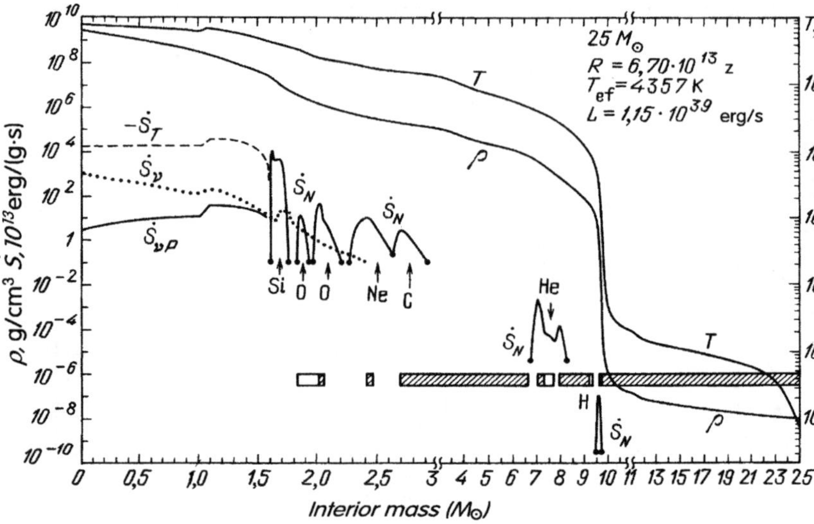

Fig. 9.14 Parameters of supernova with $M = 25\,M_\odot$. The same notation as in Fig. 9.12 (from [1033])

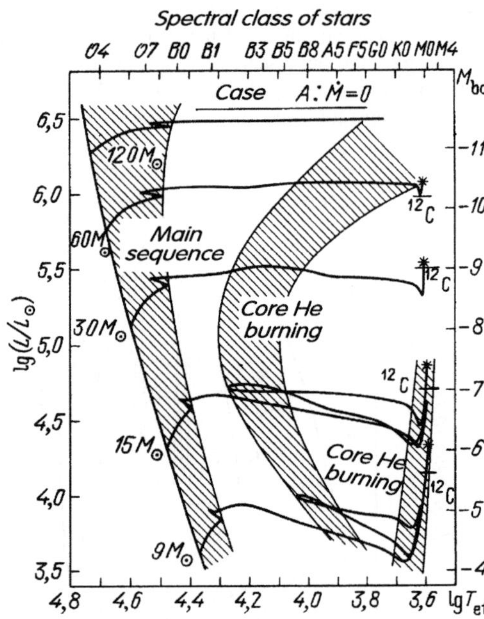

Fig. 9.15 Evolutionary tracks for stars with constant mass and initial chemical composition $x_{\rm H} = 0.7$, $x_{\rm Z} = 0.03$, Schwarzschild criterion for convection, in the HR diagram. The main regions of slow H and ^{4}He burning are hatched. The *asterisk* indicates the presupernova model. The onset of ^{12}C burning is indicated by the *dash on the track* (case A from [665], see Table 9.9)

The evolution of stars with masses of 9, 15, 30, 60, and 120 $M_\odot$ has been investigated in [664, 665] through to the end of the core carbon-burning phase and for various regimes of mass loss. The evolutionary tracks for stars of a constant mass are shown in Fig. 9.15; the results of calculations are given in Table 9.9. The initial composition $x_\mathrm{H} = 0.7$, $x_\mathrm{He} = 0.27$, $x_\mathrm{Z} = 0.03$ and Schwarzschild criterion for convection have been adopted. Contrary to [608, 1048], the track of a $15M_\odot$ model exhibits the presence of loops. The presupernova locations on tracks are indicated by asterisks (Fig. 9.15) and agree with the results of [1033] (Fig. 9.10).

9.2.4 Evolution of Massive Stars with Mass Loss

Both direct observations [711] and the existence of single helium Wolf–Rayet stars that have lost their envelope [829] provide evidence for mass outflow during evolution. The theory does not allow, in the general case, to find the dependence of the mass flux $\dot{M}$ on stellar parameters L, R, x_i; therefore empirical dependences obtained from observations [610] are mostly used in evolutionary calculations. In the stages leading to the Wolf–Rayet star formation, $\dot{M}$ is so large that the flow may accelerate in optically thick layers. A method for performing self-consistent evolutionary calculations with theoretical derivation of $\dot{M}(L, R, x_i)$ has been developed in [134, 185].

To describe a steady-state outflow of an optically thick atmosphere, we use the hydrodynamic equations for spherically symmetric stationary flow of a perfect gas in equilibrium with radiation [149, 185]

$$\dot{M}\left(E + \frac{P}{\rho} - \frac{GM}{r} + \frac{u^2}{2}\right) + 4\pi\lambda r^2 \frac{\mathrm{d}T}{\mathrm{d}r} = -L, \tag{9.9}$$

$$u\frac{\mathrm{d}u}{\mathrm{d}r} = -\frac{1}{\rho}\frac{\mathrm{d}P}{\mathrm{d}r} - \frac{GM}{r^2}, \tag{9.10}$$

$$4\pi\rho u r^2 = -\dot{M}, \tag{9.11}$$

$$P = \rho\mathscr{R}T + \frac{aT^4}{3}, \qquad E = \frac{3}{2}\mathscr{R}T + \frac{aT^4}{\rho} + \epsilon_i,$$

$$\lambda = \frac{4acT^3}{3\kappa\rho}, \quad \mathscr{R} = \frac{k}{\mu m_\mathrm{u}}, \quad \mu = \mu(\rho, T), \tag{9.12}$$

$$\kappa = \kappa(\rho, T), \quad \epsilon_i = \epsilon_i(\rho, T).$$

Here, λ is the radiative heat transfer coefficient, κ (cm^2 g $^{-1}$) is the opacity of the matter, determining the absorption of photons (see Chap. 2, Vol. 1), μ is the molecular weight, ϵ_i is the specific internal energy, consisting of excitation and ionization energies, and the dissociation energy of atoms and molecules. The first term to the left in (9.9) represents the kinetic energy flux L_k, the second one is the radiative energy flux L_th. To derive (9.9), the stationary energy equation in the differential form [614] has been used

$$\rho u\left(\frac{\mathrm{d}E}{\mathrm{d}r}-\frac{P}{\rho^2}\frac{\mathrm{d}\rho}{\mathrm{d}r}\right)=\frac{1}{r^2}\frac{\mathrm{d}}{\mathrm{d}r}\left(\lambda r^2\frac{\mathrm{d}T}{\mathrm{d}r}\right)$$

and (9.10) in the form

$$\dot{M}\left[\frac{\mathrm{d}}{\mathrm{d}r}\left(\frac{u^2}{2}-\frac{GM}{r}\right)+\frac{1}{\rho}\frac{\mathrm{d}P}{\mathrm{d}r}\right]=0.$$

Multiplying the former by $4\pi r^2$, using (9.11), combining with the second equation and integrating gives [149] the energy equation in the integral form (9.9).

During the radial outflow, the matter passes through an isothermal critical point where the relations are obtained by substituting (9.9) and (9.11) into (9.10) and writing an explicit expression for the derivative $\mathrm{d}\rho/\mathrm{d}r$. We obtain

$$u^2=\left(\frac{\partial P}{\partial \rho}\right)_T=u_{\mathrm{T}}^2,$$

$$\frac{GM}{r}=2u^2+\frac{1}{4\pi\rho r\lambda}\left(\frac{\partial P}{\partial T}\right)_\rho\left[L+\dot{M}\left(E+\frac{P}{\rho}-\frac{GM}{r}+\frac{u^2}{2}\right)\right]. \tag{9.13}$$

A theory of outflowing atmosphere with an arbitrary relation between L_{th} and L_{k} is developed in [185]. The evolutionary calculations in [134] are based on the approximation $L_{\mathrm{th}} \gg L_{\mathrm{k}}$, when the first term in (9.9) is neglected. Taking the values ρ_{cr}, T_{cr}, and r_{cr} at the critical point (9.13) for a scale, we introduce dimensionless variables. We then obtain from (9.9–9.12) two equations for the dimensionless variables T and ρ at $L_{\mathrm{th}} \gg L_{\mathrm{k}}$:

$$\frac{\mathrm{d}T}{\mathrm{d}x}=A_4\frac{\rho}{T^3}\frac{\kappa}{\kappa_{\mathrm{cr}}}, \tag{9.14}$$

$$\frac{\mathrm{d}\rho}{\mathrm{d}x}=\left[A_3-2\gamma_{\rho,\mathrm{cr}}\frac{x^3}{\rho^2}-\left(A_1\frac{T^3}{\rho}+\gamma_{\mathrm{T}}\frac{\mu_{\mathrm{cr}}}{\mu}\right)\frac{\kappa}{\kappa_{\mathrm{cr}}}\frac{\rho}{T^3}A_4\right]$$
$$\times\left(\frac{T}{\rho}\frac{\mu_{\mathrm{cr}}}{\mu}\frac{\gamma_\rho}{\gamma_{\rho,\mathrm{cr}}}-\frac{x^4}{\rho^3}\right)^{-1}\frac{1}{\gamma_{\rho,\mathrm{cr}}}. \tag{9.15}$$

Here,

$$x=\frac{r_{\mathrm{cr}}}{r},\quad \gamma_\rho=1-\left(\frac{\partial\ln\mu}{\partial\ln\rho}\right)_T,\quad \gamma_{\mathrm{T}}=1-\left(\frac{\partial\ln\mu}{\partial\ln T}\right)_\rho \tag{9.16}$$

and the dimensionless parameters

$$A_1=\frac{4}{3}\frac{aT_{\mathrm{cr}}^3}{3\rho_{\mathrm{cr}}(k/\mu_{\mathrm{cr}}m_{\mathrm{u}})},\qquad A_3=\frac{GM}{r_{\mathrm{cr}}T_{\mathrm{cr}}(k/\mu_{\mathrm{cr}}m_{\mathrm{u}})},$$
$$A_4=\frac{L}{4\pi}\frac{3\kappa_{\mathrm{cr}}\rho_{\mathrm{cr}}}{4acT_{\mathrm{cr}}^4r_{\mathrm{cr}}}=\frac{L}{L_{\mathrm{c,cr}}}\frac{A_3}{A_1},\qquad L_{\mathrm{C}}=\frac{4\pi cGM}{\kappa}, \tag{9.17}$$

are used. The subscript 'cr' denotes values at the critical point characterized by the condition (9.13) that yields the relation between the parameters (9.17)

$$\frac{L}{L_{c,cr}}\frac{A_3}{A_1} = \frac{A_3 - 2\gamma_{\rho,cr}}{A_1 + \gamma_{T,cr}}. \tag{9.18}$$

It follows from (9.18) that $L/L_{c,cr} < 1$. At a given stellar mass M, an atmosphere with steady-state outflow, similarly to a static one, is determined by two parameters for which it is convenient to take L and ρ_{cr}. The equation for A_4 in (9.17) combined with (9.18) gives

$$r_{cr} = \frac{L}{4\pi}\frac{3\kappa_{cr}\rho_{cr}}{4acT_{cr}^4}\frac{A_1 + \gamma_{T,cr}}{A_3 - 2\gamma_{\rho,cr}}. \tag{9.19}$$

Using (9.19) and the expression for A_1 and A_3 from (9.17), we obtain

$$A_3 = 2\gamma_{\rho,cr}\left[1 - (A_1 + \gamma_{T,cr})\frac{\mathscr{R}_{cr}T_{cr}}{GM}\frac{L}{4\pi}\frac{3\kappa_{cr}\rho_{cr}}{4acT_{cr}^4}\right]^{-1} \tag{9.20}$$

$$= f(\rho_{cr}T_{cr}).$$

Specifying T_{cr} at given ρ_{cr} and L, we obtain all the parameters (9.17). Integrating the system (9.14–9.15) outward for $x < 1$, we choose T_{cr} that will satisfy physical conditions far from the star. We then find all the parameters, including $\dot{M}$, of the model with steady-state outflow with the aid of the iteration scheme for evolutionary calculations of the Henyey type, or any other (see Chap. 6, Vol. 1).

To leave the critical point when integrating (9.14) and (9.15), we have to use the asymptotic relations [134, 185]

$$\rho = 1 - \alpha(1-x), \quad T = 1 - \beta(1-x), \quad \beta = \frac{A_3 - 2\gamma_{\rho,cr}}{A_1 + \gamma_{T,cr}},$$

$$\alpha = \frac{Z_2}{2Z_1} + \sqrt{\left(\frac{Z_2}{2Z_1}\right)^2 - \frac{Z_3}{Z_1}}, \qquad Z_1 = 1 + \gamma_{\rho,cr} + \left(\frac{\partial \ln \gamma_\rho}{\partial \ln \rho}\right)_{cr},$$

$$Z_2 = 8\gamma_{\rho,cr} - \beta\left\{\gamma_{T,cr}\left[\gamma_{\rho,cr} + \left(\frac{\partial \ln \gamma_T}{\partial \ln \rho}\right)_{cr}\right]\right.$$

$$\left. + \gamma_{\rho,cr}\left[\gamma_{T,cr} + \left(\frac{\partial \ln \gamma_\rho}{\partial \ln T}\right)_{cr}\right] + (A_1 + \gamma_{T,cr})\left(\frac{\partial \ln \kappa}{\partial \ln \rho}\right)_{cr}\right\}, \tag{9.21}$$

$$Z_3 = \beta^2\left\{(A_1 + \gamma_{T,cr})\left(\frac{\partial \ln \kappa}{\partial \ln T}\right)_{cr}\right.$$

$$\left. - \gamma_{T,cr}\left[4 - \gamma_{T,cr} - \left(\frac{\partial \ln \gamma_T}{\partial \ln T}\right)_{cr}\right]\right\} + 6\gamma_{\rho,cr}.$$

The most complicated problem of all is that of the outer boundary conditions determining $T_{cr}(\rho_{cr}, L)$. Instead of the equations of radiation hydrodynamics (9.9–9.12), one should use for outer shells at small optical thicknesses the hydrodynamic equations describing the gas component simultaneously with the radiation transfer equation. All types of opacity should be taken into account in the flow region. The major contribution to opacity belongs, at low temperatures, to the molecules and dust.

The description must give a smooth transition from an optically thick case with thermodynamic equilibrium between matter and radiation, to an optically thin one where the gas may be considered to be in local thermodynamic equilibrium with a temperature T_g, and radiation at small τ is represented by free radial flow of quanta whose energies approximately retain their values at $\tau \sim 1$. The simple Eddington approximation with a constant ratio of radiation pressure P_{rad} to the energy density of radiation S, as $P_{rad} = S/3$ at any optical depth, cannot be applied to the extended outflow. The description similar to (8.20) may be used (see also Chap. 2, Vol. 1). The best variant of such a description with a proper account of energy and momentum exchange between gas and radiation, valid at all τ, is not yet established, and the problem is now under consideration.

The variant of approximate description of the flow with a smooth transition from optically thin to optically thick layers has been suggested in [124]. The expressions for radiation pressure, radiation energy density, and radiative heat flux have been chosen in the form

$$P_{rad} = \frac{aT^4}{3}(1 - e^{-\tau}) + \frac{L_{th}(r)}{4\pi r^2 c}, \quad \rho E_{rad} = aT^4(1 - e^{-\tau}) + \frac{L_{th}(r)}{4\pi r^2 c},$$

$$L_{th} = -\frac{4\pi c r^2}{\kappa\rho}\left(\frac{dP_{rad}}{dr} - \frac{\rho E_{rad} - 3P_{rad}}{r}\right), \tag{9.22}$$

where

$$\tau = \int_r^{\infty} \kappa\rho dr. \tag{9.23}$$

These expressions have been used in (9.9)–(9.12) instead of relative expressions there, which are valid only for conditions of local thermodynamic equilibrium

$$P_{rad}^{eq} = \frac{aT^4}{3}, \quad \rho E_{rad}^{eq} = aT^4, \quad L_{th}^{eq} = -\frac{16\pi acT^3r^2}{3\kappa\rho}\frac{dT}{dr}. \tag{9.24}$$

The expression for L_{th} in (9.22) follows exactly from the radiative transfer equation. The relations (9.22) give equilibrium equations in the optically thick limit $\tau \to \infty$, and in the optically thin limit $\tau \to 0$ describe a radial free photon flux with the anisotropic pressure tensor with one non-zero component

$$P_{rr} \equiv P_{rad} = E_{rad} = \frac{L_{th}}{4\pi c r^2}. \tag{9.25}$$

The Euler equation (9.10) is used in the form

$$u\frac{\mathrm{d}u}{\mathrm{d}r} = -\frac{1}{\rho}\frac{\mathrm{d}P_{\mathrm{g}}}{\mathrm{d}r} - \frac{GM}{r^2}\left(1 - \frac{L_{\mathrm{th}}}{L_{\mathrm{c}}}\right), \tag{9.26}$$

where

$$L_{\mathrm{c}} = \frac{4\pi c GM}{\kappa} \tag{9.27}$$

is the critical Eddington luminosity, at which the radiation force balances the gravity, $P_{\mathrm{g}} = \rho\mathscr{R}T$. Equations (9.26) and (9.10) are identical in the case of a local thermodynamic equilibrium. While anisotropic terms in P_{rad} and E_{rad} are important only at $\tau < 1$, where $L_{\mathrm{th}} \approx L_{\mathrm{th}}^{\infty} = \mathrm{const.}$, the solution in [124] was looked for at $L_{\mathrm{th}} = L_{\mathrm{th}}^{\infty}$ in the expressions for P_{rad} and E_{rad}. Equations (9.9)–(9.12) with corrections from (9.22)–(9.27) have been solved in [124], and parameters describing a solution which satisfies the conditions $\rho = T = 0$ at infinity were found numerically for the simplified case of constant κ and μ. This description gives a promising possibility to calculate stellar evolution with self-consistent treatment of the mass loss.

To satisfy the outer boundary conditions for a gas temperature and density $T_{\mathrm{g}} = \rho = 0$ at $r = \infty$, the set of equations must be integrated outside, and the value of the temperature at the critical point T_{cr}, which at $\tau_{\mathrm{cr}} \gg 1$ is the same for the gas and the radiation, must be adjusted at given L, M and ρ_{cr}. This procedure replaces the integration of (8.20) together with the corresponding equation for P,

$$\frac{\mathrm{d}P}{\mathrm{d}\tau} = \frac{Gm}{\kappa r^2}\left(1 + f\,\frac{2aT^3 r^{1/2} T_0 R^{1/2}}{3Gm\rho}\right) \tag{9.28}$$

with specified $\rho(\tau = 0)$ as suggested in [769] for the description of extended static atmospheres (see also Sect. 6.1, Vol. 1). The solution of a full set of equations for gas and radiation will, in principle, give the value of the optical depth $\tau = \int_r^{\infty} \kappa\rho \mathrm{d}r$ on the level of the photosphere where $L = 4\pi\sigma R^2 T_{\mathrm{ef}}^4$. To avoid this complicated procedure, two approaches have been used up to now.

If the optical thickness is sufficiently high in the critical point region, then the transparent shells are of little importance for determining $T_{\mathrm{cr}}(\rho_{\mathrm{cr}}, L)$. Under these circumstances, the conditions $T = \rho = 0$ at $r = \infty$ and (9.9–9.12) have been used in [134, 185] throughout all the regions.

To find the $T_{\mathrm{cr}}(\rho_{\mathrm{cr}}, L)$ dependence, the approximate solution obtained in [172] for $\kappa = \mathrm{const}$ and $\mu = \mathrm{const}$ in the supersonic region with $L > L_{\mathrm{c}}$ is used in [134]. Equations (9.9–9.12) written for the variables u and T have at $L_{\mathrm{th}} \gg L_{\mathrm{c}}$ and $u \gg u_{\mathrm{T}}$ the solution [172] satisfying, at $r = \infty$, the conditions $T = \rho = 0$:

$$\frac{u^2}{2} + \left(\frac{L}{L_{\mathrm{c}}} - 1\right)\frac{GM}{r} = \left(\frac{L}{L_{\mathrm{c}}} - 1\right)\frac{GM}{r_{\mathrm{cr}}}, \quad L_{\mathrm{c}} = \mathrm{const.},$$

$$T^4 = -\frac{3\kappa\dot{M}L}{16\pi^2 ac}\int_r^{\infty}\frac{\mathrm{d}r}{r^4}\left[2\left(\frac{L}{L_{\mathrm{c}}} - 1\right)GM\left(\frac{1}{r_{\mathrm{cr}}} - \frac{1}{r}\right)\right]^{-1/2}, \tag{9.29}$$

$$\dot{M} = -4\pi\rho_{\rm cr} r_{\rm cr}^2 u_{\rm cr}, \quad u_{\rm cr}^2 = \frac{k}{\mu_{\rm cr} m_{\rm u}} T_{\rm cr} = u_{T,\rm cr}^2.$$

Combining the three equations (9.29) in such a way as to eliminate $\dot{M}$, we obtain the searching relationship $T_{\rm cr}(\rho_{\rm cr}, r_{\rm cr}, L)$. It is convenient to write it in the form of the relation between the dimensionless variables (9.17)

$$A_1 = \frac{64}{15\sqrt{2}} \sqrt{A_3} \frac{L}{L_{\rm c}} \left(\frac{L}{L_{\rm c}} - 1\right)^{-1/2}. \qquad (9.30)$$

For the condition $T = \rho = x = 0$ to be valid at low A_1, L and $\kappa = \text{const}$, we must take [149] $A_3 = 3$. The relation $T_{\rm cr}(\rho_{\rm cr}, r_{\rm cr}, L)$ is equivalent to a relation similar to (9.30) between the dimensionless parameters (9.17). The following relations have been used in [134] instead of (9.30) for the variable κ and μ:

$$A_3 = (A+3)\frac{\mu_0}{\mu_{\rm cr}} \frac{\max \kappa(T)}{3\kappa_{\rm L}}\bigg|_{\rho_{\rm cr}} \quad \text{for } A > 7 \text{ or } A > A_1, \qquad (9.31)$$

$$A_3 = (A_1+3)\frac{\mu_0}{\mu_{\rm cr}} \frac{\max \kappa(T)}{3\kappa_{\rm L}}\bigg|_{\rho_{\rm cr}} \quad \text{for } A < 7 \text{ and } A < A_1. \qquad (9.32)$$

Here,

$$A = A_1^2 \left(\frac{L}{L_{\rm m}} - 1\right)\left(\frac{L_{\rm m}}{L}\right)^2 \frac{450}{4096}$$
$$(= A_3 \text{ from (9.30) at } L_{\rm m} = L_{\rm c}),$$
$$\kappa_{\rm L} = \min\left[\frac{4\pi c GM}{L}, \frac{\max \kappa(T)}{3}\bigg|_{\rho_{\rm cr}}\right], \qquad (9.33)$$
$$L_{\rm m} = \frac{4\pi c GM}{\max \kappa(T)|_{\rho_{\rm cr}}} \quad \text{at } \max \kappa(T)|_{\rho_{\rm cr}} < 3\frac{4\pi c GM}{L},$$
$$L_{\rm m} = \frac{L}{3} \quad \text{at } \max \kappa(T)|_{\rho_{\rm cr}} > 3\frac{4\pi c GM}{L},$$

where $\mu_0 = \max(\mu)$ corresponds to the neutral gas.

The procedure for determining the critical point is the following. On finding $\kappa_0 = 4\pi c GM/L$ at given L and $\rho_{\rm cr}$, we obtain $T_{\rm cr0}$ from tables $\kappa(\rho, T)$ and obtain A_1 and A_3 from (9.17) and (9.2.17). Thermodynamic relations yield μ, γ_ρ and $\gamma_{\rm T}$. Then, $L_{\rm c,cr}$ and κ_1 are determined from (9.18) to the next approximation, and the procedure is iterated until the critical point parameters are obtained with some predetermined accuracy. The next step is to find $r_{\rm cr}$ from (9.19) or from the equation for A_3 in (9.17), and the values of $u_{\rm cr} = (\gamma_{\rho,\rm cr} kT/\mu_{\rm cr} m_{\rm u})^{1/2}$ and $\dot{M} = -4\pi\rho_{\rm cr} u_{\rm cr} r_{\rm cr}^2$. Upon leaving the critical point using the expansion (9.21), the integration inward of the star of (9.14) and (9.15) in [134] includes 3% of the stellar mass. On the boundary of the core which is calculated by the Henyey method, the velocity is so small ($u \ll u_{\rm T}$)

that the parameters are fitted in the same way as in the static case (see Chap. 6, Vol. 1). In the region of incomplete H and He ionization, the relations for opacities from [460] have been used in calculations [134].

The static evolution of stars with $M = 9$ and $30M_{\odot}$ has been calculated in [134] to the onset of convective penetration inside the core ($9M_{\odot}$), or to the point where the static models are no longer possible to construct by the adopted method ($30M_{\odot}$). For $30M_{\odot}$, this point is assumed to represent the onset of outflow with corresponding evolutionary phase calculated in accordance with the above method. The adopted initial composition is $x_{\rm H} = 0.75$, $x_{\rm He} = 0.22$ and $x_{\rm Z} = 0.03$. The convection is calculated with the mixing length $l = H_{\rm p}$ and the Ledoux criterion for convection; therefore, mixing between the convective core and semiconvective zone is absent owing to the presence of a radiative shell between them.

The results of calculations are given in Table 9.7 and Figs. 9.16 and 9.17. It has not been possible to construct a static model under the adopted physical conditions for the portion of the track with $30M_{\odot}$ beyond the point E; this point has therefore been taken for the onset of outflow. The dashed line indicates an approximate track for the outflow phase. Twenty-three outflowing models in all have been constructed for a 15-year evolutionary period. The stellar mass has decreased over this period down to $23M_{\odot}$, thus yielding a huge mass flux $\dot{M} \approx 0.5M_{\odot}/\text{year}$. Such a flux is no doubt a result of an inadequate choice of the form of the relationship (9.2.17). A real $\dot{M}$ may be 2–4 orders of magnitude less. The ranges of parameter variations of the critical point over the evolution time are:

$$\begin{aligned} \dot{M} &= (0.5 - 0.4)M_{\odot}/\text{yr}, \quad u_{\rm cr} = 18\,\text{km s}^{-1}, \\ T_{\rm cr} &= (2.5 - 2.4) \times 10^4\,\text{K}, \quad \rho_{\rm cr} = 8 \times 10^{-10}\,\text{g cm}^{-3}, \\ R_{\rm cr} &= (610 - 550)R_{\odot}. \end{aligned} \tag{9.34}$$

Table 9.7 Evolutionary parameters of stars shown in Fig. 9.16

Point	Model	T_c, K	ρ_c, g·cm^{-3}	$\frac{L}{L_{\odot}}$	$\frac{L_{\rm H}}{L_{\odot}}$	$\frac{L_{\rm He}}{L_{\odot}}$
A	0	3.1(7)	1.1(1)	3.5(3)	3.5(3)	0
	0	3.8(7)	3.7(0)	1.1(5)	1.1(5)	0
B	20	4.0(7)	2.0(1)	6.6(3)	6.6(3)	0
	133	5.1(7)	7.5(0)	2.2(5)	2.2(5)	0
C	28	4.5(7)	5.3(1)	7.5(3)	6.7(3)	0
	244	8.2(7)	3.6(1)	2.5(5)	1.9(5)	0
D	64	7.8(7)	1.9(3)	8.1(3)	8.5(3)	0
	342	1.8(8)	5.9(2)	2.4(5)	1.6(5)	1.3(5)
E	132	1.4(8)	5.1(3)	5.0(3)	6.7(3)	1.1(3)
	452	1.8(8)	5.8(2)	2.3(5)	1.6(5)	1.3(5)

Here $t_{\rm ev}$ is the time of the evolution between the given and precedent points on tracks in Fig. 9.16, the values for $M = 9M_{\odot}$ are given in the upper lines, for $M = 30\,M_{\odot}$ in the lower lines, the number in parentheses in the column $R_{\rm ph}/R_{\odot}$ is the critical radius of the first outflowing model

Table 9.7 (end)

Point	$\frac{L_g}{L_\odot}$	T_{ef}, K	$\frac{R_{ph}}{R_\odot}$	$\frac{M_{conv}}{M}$	$x_H[x_{^{12}C}]$	t_{ev}, yr
A	0	2.43(4)	3.4	0.29	0.75	0
	0	4.07(4)	6.7	0.54	0.75	0
B	2.3	2.08(4)	6.3	0.11	0.036	2.6(7)
	4.5(2)	3.10(4)	16.4	0.28	0.02	5.9(6)
C	7.5(2)	2.24(4)	5.7	0.037	2.8(−4)	5.6(5)
	5.8(4)	3.44(4)	14.0	0.12	0	8.0(4)
D	−3.8(2)	1.57(4)	12.2	0	0	9.3(4)
	−5.1(4)	1.54(4)	68.1	0.15	[6.8(−3)]	1.5(4)
E	−2.8(3)	5.16(3)	89.0	0.035	[4.4(−3)]	8.5(4)
	−5.5(4)	5.16(3)	607 (611)	0.18	[1.3(−2)]	3.0(3)

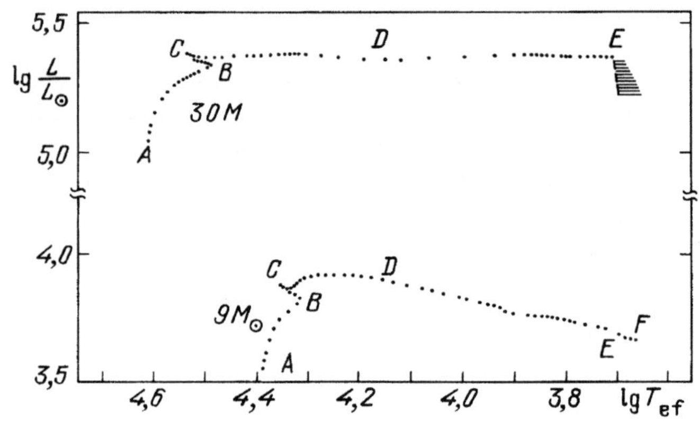

Fig. 9.16 Evolutionary tracks for stars with masses of 9 and 30 $M_\odot$ from the MS to the initial phase of ^{4}He burning in a core with initial composition $x_H = 0.75$, $x_{He} = 0.22$, $x_Z = 0.03$, Ledoux criterion and $l = H_P$, from [134]. The point E on the track with $M = 30\,M_\odot$ is assumed to represent the onset of strong outflow, the phase with outflow is indicated by a *dashed line*

The small change in critical parameters during the outflow shows that it will last until the star loses its hydrogen envelope, and the residual helium core becomes a Wolf–Rayet star. Such a mechanism for the Wolf–Rayet star formation has been suggested in [134]. A check test is proposed there as well: a search for an extended gas envelope with a mass comparable to the mass of a helium star.

Such envelopes have been found around many stars of this type that exhibit no evident indication of binaries [649]. As also noted in [134], at initial phases of outflow the optical depth of the outflowing envelope may reach very large values, and all the star emission will then be re-emitted in the infrared range, thereby giving a bright infrared object. Objects of this type are likely to have been already observed from the IRAS satellite [478]. The distribution of parameters in the envelope of the last static model and the first outflowing model of the same mass is shown in Fig. 9.17.

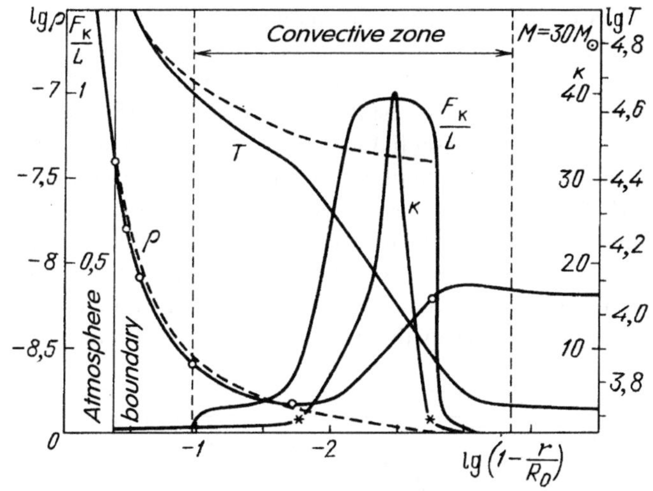

Fig. 9.17 The structure of static (*solid line*) and outflowing (*dashed line*) envelopes at the point E for 30 $M_{\odot}$. The outflowing envelope is calculated with the aid of the relation (9.2.17) instead of the outer boundary condition. Shown are the distributions for density ρ, temperature T, opacity κ (the luminosity to critical luminosity ratio $k = L/L_c$), the flux fraction transferred by convection F_k/L, R_0 is the radius of the static star. The *asterisks* label points with $k = 1$ ($\kappa = \kappa_0$). Shown are only parts of outflowing models adjacent to the static core. A close coincidence of the curves $\rho(r)$ and $T(r)$ referring to static and outflowing models may be seen near the core, $\kappa_0 = 4\pi c GM/L = 1.7$

The outflowing model exhibits the absence of a density inversion, a more rapid ρ decrease and a slower T reduction compared to the static envelope. The solution for static and outflowing envelopes is in good consistency near the core, which proves that the atmosphere processes have only a slight effect on the static core structure. The results of evolutionary calculations with these boundary conditions [134] produced values for the mass-loss rate, which were too large what may be connected with the crudeness of this approximation.

Another approximation used by several authors [557, 1087, 1088] was suggested in [385]. Instead of integrating the equations in the region with $\tau \ll 1$, the position of the photosphere was specified by a relation [557] $\tau_{ph} \equiv (\kappa\rho r)_{ph} = 8/3$, and it was demanded that at this level the effective temperature T_{ef} is reached. This is a rather rough condition, because it ignores the distributions $\rho(r)$ and $\kappa(r)$ in the outflowing region and deals only with local values in the definition of the photosphere. It is equivalent to applying this condition to the static atmosphere instead of solving the equations (8.20) and (9.28) at $0 < \tau < 2/3$ for finding the conditions at the photosphere in the extended envelopes. The self-consistent models of Wolf–Rayet stars have been constructed in such a way in [557], but with a somewhat artificial enhancement of the opacity. At $T > 2 \times 10^5$ K, the enhanced opacity coincides with the value from new Livermore tables, and at lower temperatures the correction coefficient is entirely arbitrary and may be as large as 7 compared to the old Los–Alamos tables.

The evolution of massive stars with outflow at an empirically specified flux $\dot{M}(L,R)$ has been studied by several authors. Evolutionary calculations of stars with initial mass 15, 30, and $60 M_\odot$ have been performed in [938] through the helium exhaustion in the centre. The results are given in Fig. 9.18 and Table 9.8. Tracks with a constant mass differ essentially from tracks of the same mass obtained in other studies with use of the same Ledoux criterion for convection (Fig. 9.8 [937] and Fig. 9.9 [1086]) in that they have no loops at all. For $M = 15$ and $30 M_\odot$, almost all helium burns in the red supergiant region (see Table 9.8). The difference from [937] is in the mixing length only: $l = H_p$ instead of $l = 0.4 H_\rho$ in [937]. The difference from [1086] is in chemical composition and some details of the criterion for convection. Comparing these results suggests that some random elements

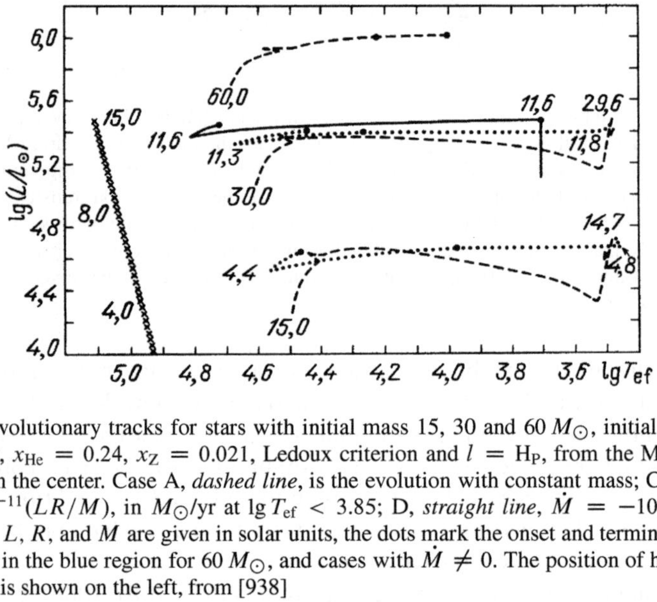

Fig. 9.18 Evolutionary tracks for stars with initial mass 15, 30 and 60 $M_\odot$, initial composition $x_H = 0.739$, $x_{He} = 0.24$, $x_Z = 0.021$, Ledoux criterion and $l = H_P$, from the MS to the ^{4}He exhaustion in the center. Case A, *dashed line*, is the evolution with constant mass; C, *dotted line*, $\dot{M} = -10^{-11}(LR/M)$, in $M_\odot$/yr at $\lg T_{ef} < 3.85$; D, *straight line*, $\dot{M} = -10^{-2}\, M_\odot$/yr at $\lg T_{ef} < 3.7$; L, R, and M are given in solar units, the dots mark the onset and termination of slow ^{4}He burning in the blue region for 60 $M_\odot$, and cases with $\dot{M} \neq 0$. The position of homogeneous helium stars is shown on the left, from [938]

Table 9.8 Evolutionary parameters of stars in Fig. 9.18

Initial mass, $M_\odot$	Case	$\lg T_{ef}$, K	τ_H, yr	τ_{He}/τ_H	τ_{bl}/τ_{He}	τ_y/τ_{He}	Finite mass, $M_\odot$
15	A	...	1.2(7)	0.092	0.040	0.002	15
15	C	4.56	1.2(7)	0.092	0.760	0.009	4.4
30	A	...	5.8(6)	0.082	0.034	0.005	30
30	C	4.67	5.8(6)	0.082	0.831	0.015	11.3
30	D	4.81	5.8(6)	0.082	1.000	...	11.6
60	A,C,D	4.22	3.7(6)	0.086	1.000	...	60

T_{ef} is the maximum effective temperature during core ^{4}He burning, τ_H the time of core H burning, τ_{He} the time of core ^{4}He burning, τ_{bl} the time of ^{4}He burning in the blue phase, τ_y the time of ^{4}He burning in the yellow phase underlie the mechanism of loop formation. Including the mass losses into calculations (under the same physical assumptions) causes the track already in the core helium-burning phase to turn back into the blue giant region, where the star spends most of its lifetime in this phase (see Table 9.8). Both variants of mass-loss rate and the rapid law $\dot{M} = -10^{-2} M_\odot/\text{yr}$ at $\lg T_{\text{ef}} < 3.7$ (case D), analogously to [134], give qualitatively similar tracks (L, R, M are expressed in solar units).

$$\dot{M} = -\frac{LR}{M} \times 10^{-11} \frac{M_\odot}{\text{yr}} \quad \text{at} \quad \lg T_{\text{ef}} < 3.85 \text{ (case C)}, \tag{9.35}$$

A variety of mass-loss laws have been studied in [664,665]. The adopted physical conditions have been discussed above. The results are shown in Figs. 9.19 and 9.20a and Table 9.9. The law used here differs from [938]: $\dot{M} = -(NL/c^2) M_\odot$ / yr, $N = 65$–150. For the case C with a largest $\dot{M}$, the tracks coincide qualitatively with

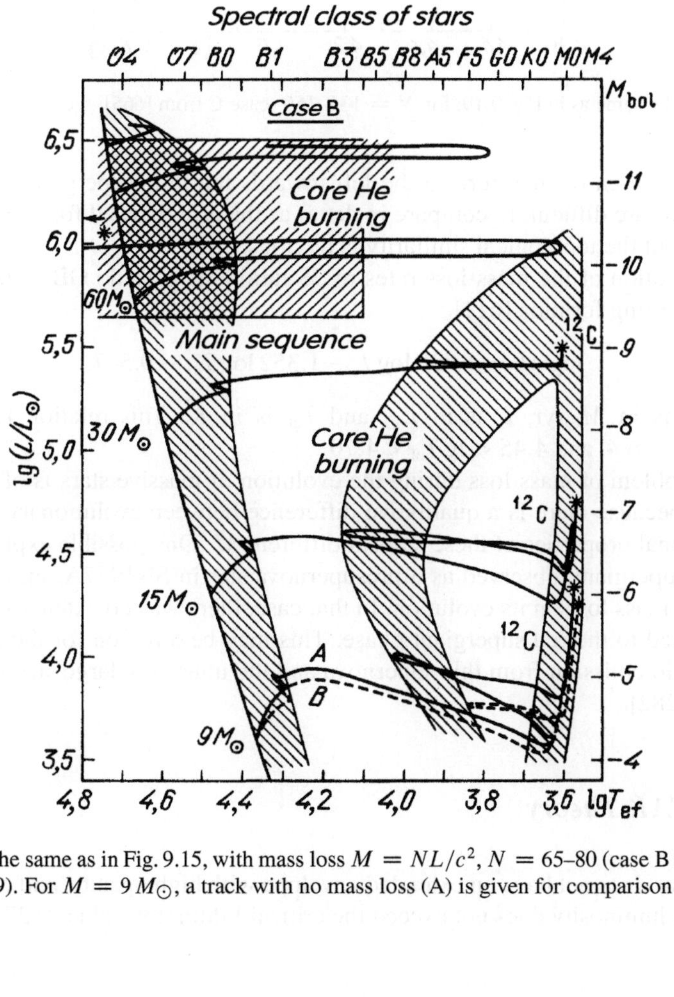

Fig. 9.19 The same as in Fig. 9.15, with mass loss $M = NL/c^2$, $N = 65$–80 (case B from [665], see Table 9.9). For $M = 9\,M_\odot$, a track with no mass loss (A) is given for comparison

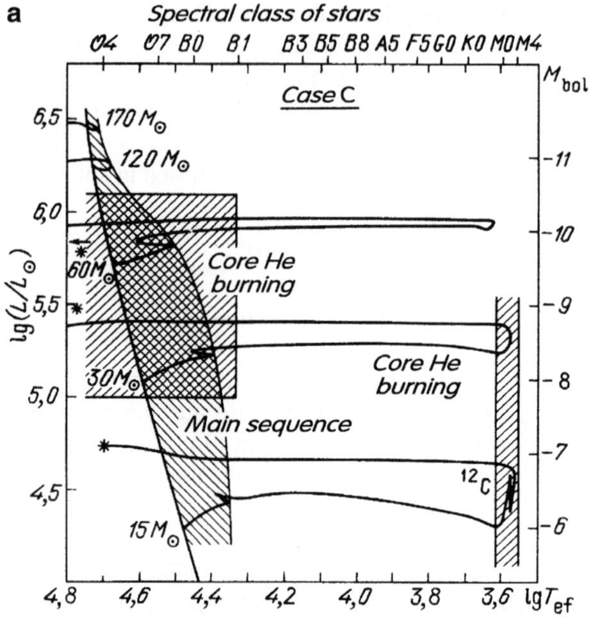

Fig. 9.20 The same as in Fig. 9.19, for $N = 135$–150 (case C from [665], see Table 9.9)

[938]: in both cases they turn to the left towards the blue supergiant region. The calculations are difficult to compare in detail as the criteria used for convection are different, but the topological similarity is obvious.

Investigation of the mass-loss rates of 28 luminous galactic OB stars led to an empirical fitting formula [611]

$$\log \dot{M} = 1.738 \log L - 1.352 \log T_{\rm ef} - 9.547. \tag{9.36}$$

Here, $\dot{M}$ is in $M_\odot$/yr, L is in $L_\odot$ and $T_{\rm ef}$ is in K. This relation is valid for $5.0 < \lg L < 6.4$, and $4.45 < \lg T_{\rm ef} < 4.70$.

The problem of mass loss during the evolution of massive stars is of major importance because there is a qualitative difference between evolutionary tracks and observational properties of these stars at different $\dot{M}$. One possible explanation for the blue supergiant, observed as a presupernova star in SN1987A, may be the influence of mass loss on its evolution. In that case, there was no extensive mass-loss stage related to the red supergiant phase. This may be a reason for the absence of strong radio emission from this supernova and the unusually large size of its radio remnant [282].

9.2.5 CAK Theory

When a star does not have an extended envelope with high opacity, and everywhere at $\tau > 1$ a luminosity does not exceed the critical Eddington value (9.27), the mass

Table 9.9 Principal nuclear burning phases for evolution of the stars shown in Figs. 9.15, 9.19, 9.20 (from [664,665])

Burning phase	Case	$\frac{M}{M_\odot}$	$\dot{M}$, $M_\odot$/yr	T_c, K	ρ_c, g·cm^{-3}	$\frac{L}{L_\odot}$	T_{ef}, K	t_{ev}, yr
Hydrogen	A	9	0	2.9(7)	8.9(0)	3.9(3)	2.32(4)	2.3(7)
	B	9	1.3(−8)	2.9(7)	8.9(0)	3.9(3)	2.32(4)	2.4(7)
Helium	A	9	0	1.4(8)	4.3(3)	8.8(3)	4.04(3)	3.7(6)
	B	8.55	6.8(−7)	1.4(8)	4.6(3)	8.3(3)	3.97(3)	3.8(6)
^{12}C and ^{16}O	A	9	0	6.3(8)	3.5(5)	2.1(4)	3.77(3)	3.6(5)
	B	6.2	1.9(−6)	6.1(8)	3.2(5)	1.8(4)	3.65(3)	6.4(5)
Hydrogen	A	15	0	3.2(7)	5.2(0)	1.9(4)	3.00(4)	1.1(7)
	B	15	7.1(−8)	3.2(7)	5.2(0)	1.9(4)	3.00(4)	1.2(7)
	C	15	1.3(−7)	3.2(7)	5.2(0)	1.9(4)	3.00(4)	1.2(7)
Helium	A	15	0	1.6(8)	1.5(3)	4.3(4)	3.82(3)	1.2(6)
	B	13.7	2.2(−7)	1.5(8)	1.9(3)	2.8(4)	7.66(3)	1.2(6)
	C	12.2	1.0(−5)	1.6(8)	1.8(3)	4.1(4)	3.76(3)	1.3(6)
^{12}C and ^{16}O	A	15	0	7.3(8)	2.1(5)	6.1(4)	3.78(3)	7.6(4)
	B	11.1	8.5(−6)	7.2(8)	2.2(5)	5.5(4)	3.72(3)	3.8(4)
	C	3.5	4.2(−7)	7.1(8)	1.6(5)	5.2(4)	5.30(4)	9.6(4)
Hydrogen	A	30	0	3.6(7)	2.9(0)	1.2(5)	3.93(4)	5.7(6)
	B	30	5.1(−7)	3.6(7)	2.9(0)	1.2(5)	3.93(4)	6.8(6)
	C	30	2.9(−6)	3.6(7)	2.9(0)	1.2(5)	3.94(4)	6.8(6)
Helium	A	30	0	1.9(8)	5.6(2)	3.0(5)	1.76(4)	4.6(5)
	B	24.8	3.1(−6)	1.8(8)	6.1(2)	2.5(5)	9.55(3)	5.0(5)
	C	21.25	6.6(−5)	1.8(8)	6.7(2)	1.8(5)	3.94(3)	5.1(5)
^{12}C and ^{16}O	A	30	0	9.1(8)	8.2(4)	3.5(5)	3.83(3)	7.9(3)
	B	12.5	7.2(−5)	9.3(8)	1.2(5)	3.2(5)	4.06(3)	4.6(3)
	C	10.15	2.9(−6)	9.7(8)	2.3(5)	3.0(5)	5.98(4)	8.3(3)
Hydrogen	A	60	0	3.9(7)	1.9(0)	5.3(5)	4.72(4)	3.7(6)
	B	60	2.7(−6)	3.9(7)	1.9(0)	5.3(5)	4.72(4)	4.2(6)
	C	60	5.4(−6)	3.9(7)	1.9(0)	5.3(5)	4.72(4)	4.2(6)
Helium	A	60	0	2.0(8)	3.2(2)	1.1(6)	1.02(4)	2.8(5)
	B	45.2	2.6(−4)	1.9(8)	2.7(2)	8.8(5)	4.23(3)	3.1(5)
	C	34.8	4.6(−4)	1.9(8)	3.4(2)	8.1(5)	4.43(3)	3.2(5)
^{12}C and ^{16}O	A	60	0	9.4(8)	6.9(4)	1.1(6)	3.95(3)	1.3(2)
	B	24.8	4.0(−5)	9.5(8)	6.8(4)	1.1(6)	2.16(5)	1.2(2)
	C	14.6	1.0(−4)	2.1(8)	5.6(2)	3.4(5)	1.24(5)	...

The parameters determine conditions near the onset of hydrogen and helium burning and at the end of carbon burning, t_{ev} is the lifetime before ignition of the next element. The last line of the Table gives the last model of the given case, where the carbon has not yet ignited

outflow may originate in the optically thin region due to large radiation pressure in the spectral lines. The outflow of such a type is probably observed in the luminous O and B stars. The self-consistent theory of radiation-driven winds in such stars was developed in [256] and is called the CAK theory.

The important starting point in this theory is the fact that in both static and rapidly expanding atmospheres, the force due to an optically thick line is proportional to the line strength, but inversely proportional, approximately, to the line optical depth, and therefore independent of the line strength overall. For the prevalent ions of an abundant element such as C, N and O, there are large numbers of lines that are optically thick over most of the atmosphere of a hot star, and they all contribute equally to the force, or, to be more accurate, in proportion to the continuum flux at their respective frequencies.

The force due to a single line per unit mass of material is

$$f_{\mathrm{rad,L}} = \frac{\kappa_{\mathrm{L}} F_{\mathrm{c}} \Delta\nu_{\mathrm{D}}}{c} \min(1, 1/\tau_{\mathrm{L}}), \tag{9.37}$$

where κ_{L} is the monochromatic line absorption coefficient, per unit mass, divided by the line profile factor, and the function is assumed to be normalized on a scale of frequency expressed in thermal Doppler units. The line optical depth τ_{L} differs in static and expanding cases and is given by

$$\tau_{\mathrm{L}} = \int_r^{\infty} \kappa_{\mathrm{L}} \rho \mathrm{d}r \tag{9.38}$$

in the static and

$$\tau_{\mathrm{L}} = \kappa_{\mathrm{L}} \rho v_{\mathrm{th}} \left| \frac{\mathrm{d}v}{\mathrm{d}r} \right|^{-1} \tag{9.39}$$

in the expanding atmosphere (Sobolev approximation [920]). In the latter case, τ_{L} counts only the absorbers in a section of the column across which the velocity changes by v_{th}, the random velocity of atoms. The factor F_{c} is the continuum flux per interval of frequency at the frequency of the line. We introduce a variable t which, in the static case, is an optical depth connected with the scattering by free electrons (Thomson scattering) by

$$t = \int_r^{+\infty} \sigma_{\mathrm{T}} \rho \mathrm{d}r \text{ (static)} = \sigma_{\mathrm{T}} \rho v_{\mathrm{th}} \left| \frac{\mathrm{d}v}{\mathrm{d}r} \right|^{-1} \text{ (expanding)}. \tag{9.40}$$

Here, the Thomson opacity is given by

$$\sigma_{\mathrm{T}} = \frac{8\pi}{3} \left(\frac{\mathrm{e}^2}{m_{\mathrm{e}} c^2} \right)^2 \frac{1}{\mu_{\mathrm{i}} m_{\mathrm{u}}} (\mathrm{cm}^2\ \mathrm{g}^{-1}), \tag{9.41}$$

where m_{u} is the atomic mass unit and μ_i is the number of nucleons per free electron. The total radiation force due to an ensemble of lines is then simply the sum of (9.37) over all the lines and may be expressed as (F is the total flux)

$$f_{\mathrm{rad}} = \frac{\sigma_{\mathrm{T}} F}{c} M(t), \tag{9.42}$$

where the force multiplier is given by

$$M(t) = \sum_{\text{lines}} \frac{F_c \Delta\nu_D}{F} \min\left(\frac{1}{\beta}, \frac{1}{t}\right). \tag{9.43}$$

The quantity $\beta = t/\tau_L$ is related to the atomic populations and the oscillator strength, and it is calculated in [256].

For t as large as unity, $M(t)$ is proportional to the fraction of the continuum blocked by lines with thermal Doppler widths. This fraction is of the order of 1%, so the line force is a small perturbation to the continuum flux force. In the limit of small optical depth $\tau \ll 1$, the value of $M(t)$ may become very large. For the typical value of $F/F_c \simeq 2.5kT_{ef}/h$ (h is the Planck constant), the maximum value of the force multiplier is

$$M_{max} = M(0) \approx 0.7 \times 10^8 \left(\frac{10^4 K}{T_{ef}}\right) X, \tag{9.44}$$

where X is the fraction of all atoms in the gas which are able to absorb effectively in the spectral region where most of the flux is emerging. In O stars, X may be about 10% of the abundance of C, N and O, so X is about 10^{-4} and $M_{max} \sim 10^3$. The radiation force for electron scattering is about 0.05–0.5 of the gravitational, so at small optical depths the force exerted in the lines will exceed gravity by about two orders of magnitude in an O star, making impossible the existence of the static atmosphere of such a star. The dependence $M(t)$ is calculated in [256] for different T_{ef} and may be expressed as

$$\begin{gathered} M(t) = kt^{-\alpha}, \quad \text{which reads as} \\ 0.0076\,t^{-0.742};\ 0.0026\,t^{-0.737};\ 0.0021\,t^{-0.811} \\ \text{for } T_{ef} = 3\times 10^4;\ 4\times 10^4;\ 5\times 10^4, \quad \text{respectively.} \end{gathered} \tag{9.45}$$

This is valid for $X = 10^{-4}$, and for other abundances of absorbing ions the value $M(t)$ may be calculated from

$$M(X', v'_{th}, t) = \frac{X'}{X} M\left(X, v_{th}, \frac{X' v_{th}}{X v'_{th}} t\right). \tag{9.46}$$

About 10% of the total force (9.42) is given by resonance lines.

The equation of motion describing the optically thin winds from O stars is written, for a given distribution $T(r)$, and contains only gas characteristics: u, $P_g = \rho\mathscr{R}T$, and the radiation force, which substitute the radiation pressure gradient in (9.10). We have instead

$$v\frac{dv}{dr} = -\frac{1}{\rho}\frac{dP_g}{dr} - \frac{GM}{r^2} + \frac{\sigma_T L}{4\pi r^2 c}[1 + M(t)]. \tag{9.47}$$

This is solved together with the continuity equation (9.11) at given M (mass) and L. Combining (9.42)–(9.47) and (9.11), we obtain an equation

$$\left(v - \frac{\mathcal{R}T}{v}\right)\frac{\mathrm{d}v}{\mathrm{d}r} = -\frac{GM}{r^2}(1-\Gamma) + \frac{2\mathcal{R}T}{r} - \mathcal{R}\frac{\mathrm{d}T}{\mathrm{d}r} + \frac{\Gamma GMk}{r^2}\left[\frac{4\pi}{\sigma_{\mathrm{T}} v_{\mathrm{th}}(-\dot{M})}\right]^{\alpha}\left(r^2 v\frac{\mathrm{d}v}{\mathrm{d}r}\right)^{\alpha}, \tag{9.48}$$

where $\Gamma = L/L_{\mathrm{cT}}$, and $L_{\mathrm{cT}} = 4\pi c GM/\sigma_{\mathrm{T}}$ is the Eddington luminosity in the presence of Thomson scattering only (see (9.41)). Equation (9.48) is strongly non-linear and has a singular point. Its solution has been investigated in [256] where it has been used for constructing a model envelope for the O5f star (see also [654]).

Location of the singular point of (9.48) is determined by solution of three equations

$$F(x, w, w') = 0\,, \tag{9.49}$$

$$\frac{\partial F(x, w, w')}{\partial w'} = 0\,, \tag{9.50}$$

$$\frac{\partial F}{\partial x} + w'\frac{\partial F}{\partial w} = 0. \tag{9.51}$$

Here, (9.49) represents (9.48) written in the coordinates

$$w = \frac{1}{2}v^2, \quad x = -\frac{1}{r}, \quad w' \equiv \frac{\mathrm{d}w}{\mathrm{d}x}, \tag{9.52}$$

where it reads

$$F(x, w, w') \equiv \left(1 - \frac{\mathcal{R}T}{2w}\right)w' - h(x) - C(w')^{\alpha}, \tag{9.53}$$

with

$$h(x) = -GM(1-\Gamma) - 2\frac{\mathcal{R}T}{x} - \frac{\mathcal{R}}{x^2}\frac{\mathrm{d}T}{\mathrm{d}r},$$

$$C = \Gamma GMk\left[\frac{4\pi}{\sigma_{\mathrm{T}} v_{\mathrm{th}}(-\dot{M})}\right]^{\alpha}. \tag{9.54}$$

For isothermal flow with no line force $T =$ const., $C = 0$, we obtain the singular point from (9.1.32) to (9.1.37)

$$w = \frac{\mathcal{R}T}{2}, \quad \mathcal{R}T = -\frac{GM(1-\Gamma)x}{2}, \quad w' = -\frac{2w}{x}. \tag{9.55}$$

The first two relations in (9.55) coincide with (9.13) at $\lambda = \infty$ and with a reduced mass $M(1-\Gamma)$. The third relation in (9.55) determines the first term in an expansion of the equation of motion at the critical point.

The problem of stability of solutions in CAK theory has been investigated by different methods [768, 805, 852]. The flow is unstable to formation of clumps and shocks, producing a hot gas responsible for the observed soft X-ray emission from blue luminous stars [768] with $L_x \simeq 10^{-7} L_{\rm bol}$.

9.2.6 *Line-Driven Winds in the Presence of Strong Gravitational Fields*

An important generalization of the CAK model was done by Dorodnitsyn [326], which includes the gravitational redshifts of lines. This generalized model could describe correctly the formation of fast outflows in the vicinity of strong gravitational field sources. The gradient of the gravitational potential $\mathrm{d}\phi/\mathrm{d}r$ plays the same role as $\mathrm{d}v/\mathrm{d}r$ plays in acceleration of line-driven winds from hot stars. It is shown that due to the gravitational redshift of frequency, the gradient of the gravitational potential allows a line to shift out from behind of its own shadow. The line is shifted to the extent where the continuum flux is not reduced by absorption, hence $\mathrm{d}\phi/\mathrm{d}r$ exposes the line to powerful radiation.

Thus, in [326] the resultant wind is termed "Gravitationally Exposed Flow" (GEF). In this chapter, the following toy model is adopted. Consider a wind accelerated by the radiation pressure in lines at the vicinity of BH. The wind is assumed to be spherically symmetric, with a strongly simplified ionization problem. These crude assumptions are accepted in [326] just to compare a solution for GEF with a pure CAK solution.

If a wind is accelerated in a strong gravitational field, we may expect that a photon emitted by the disc may become resonant with the opacity of some line not only because of the Doppler effect but also because of the gravitational redshifting. The redshifting of the photon's frequency is the only effect of general relativity (GR) taken into account in this approximate treatment. A photon with the frequency $\nu_{\rm D}$, emitted at the point with the gravitational potential ϕ_1, will be registered at the point where $\phi = \phi_2$ ($|\phi_1| > |\phi_2|$, $\phi < 0$) reddened according to the well-known approximate formulae [615]:

$$\Delta\nu \equiv \nu_2 - \nu_{\rm D} = \frac{\phi_1 - \phi_2}{c^2}\nu_{\rm D}. \tag{9.56}$$

If there is a velocity difference between these two points, the photon will be additionally red-shifted due to the Doppler effect. The resultant frequency, seen by the absorber at his rest frame, reads:

$$\nu \simeq \nu_{\rm D} + \nu_{\rm D}\left(-\frac{v}{c} + \frac{\Delta\phi}{c^2}\right). \tag{9.57}$$

To derive a relation for the radiation force, we need to calculate the corresponding Sobolev optical depth, but now taking into account the gravitational redshifting. Assuming that absorption occurs in a line with the delta-function profile, for the optical depth in a radial direction we obtain, instead of (9.39),

$$\tau_{\mathrm{L}} = \int_r^{\infty} \kappa_{\mathrm{L}} \rho \, \mathrm{d}r \simeq \frac{\rho v_{\mathrm{th}} \kappa_{\mathrm{L}}}{\mathrm{d}v/\mathrm{d}r + \frac{1}{c}\mathrm{d}\phi/\mathrm{d}r}. \tag{9.58}$$

A characteristic length: $\delta r \sim v_{\mathrm{th}}/(\mathrm{d}v/\mathrm{d}r + \frac{1}{c}\mathrm{d}\phi/\mathrm{d}r)$ gives a thickness of a shell where the absorption due to a single line take place. Introducing an optical depth parameter, which is analogous to (9.40), we have:

$$t \equiv \frac{\sigma_{\mathrm{e}} \rho v_{\mathrm{th}}}{\mathrm{d}v/\mathrm{d}r + \frac{1}{c}\mathrm{d}\phi/\mathrm{d}r}. \tag{9.59}$$

It is assumed here that τ must be calculated taking into account redshifting. No assumptions about the particular physical mechanism that produces redshifting are made. By substituting (9.59) into (9.42) and (9.45), the radiation force is obtained as

$$f_L = \frac{L\sigma_{\mathrm{e}} k}{4\pi r^2 c}\left(\frac{4\pi}{|\dot{M}|\sigma v_{\mathrm{th}}}\right)^{\alpha}\left[v r^2\left(\frac{\mathrm{d}v}{\mathrm{d}r} + \frac{1}{c}\frac{\mathrm{d}\phi}{\mathrm{d}r}\right)\right]^{\alpha}, \tag{9.60}$$

where the radiation force was transformed using the continuity equation

$$|\dot{M}| = 4\pi\rho v r^2. \tag{9.61}$$

Taking into account the gravitational line redshift, we obtain from (9.47) the following equation of motion:

$$v\frac{\mathrm{d}v}{\mathrm{d}r} = -\frac{1}{\rho}\frac{\mathrm{d}P_{\mathrm{g}}}{\mathrm{d}r} - \frac{\mathrm{d}\phi}{\mathrm{d}r} + \frac{L\sigma}{4\pi r^2 c} + \frac{L\sigma_{\mathrm{e}} k}{4\pi r^2 c}\left(\frac{4\pi}{|\dot{M}|\sigma v_{\mathrm{th}}}\right)^{\alpha}\left[v r^2\left(\frac{\mathrm{d}v}{\mathrm{d}r} + \frac{1}{c}\frac{\mathrm{d}\phi}{\mathrm{d}r}\right)\right]^{\alpha}. \tag{9.62}$$

As the equation of motion of CAK, equation (9.62) is nonlinear with respect to the velocity gradient. Instead of $(\mathrm{d}v/\mathrm{d}r)^{\alpha}$ term, equation (9.62) includes a combination: $(\mathrm{d}v/\mathrm{d}r + \frac{1}{c}\mathrm{d}\phi/\mathrm{d}r)^{\alpha}$. This property changes the structure of the transonic solution, going through the critical point, which differs from the sonic one. Both the Newtonian potential and the potential of Paczynski and Wiita (PW) [785]

$$\phi = -\frac{GM}{r - r_{\mathrm{g}}}. \tag{9.63}$$

have been considered in [326]. For the Schwarzschild black hole, the PW potential correctly reproduces the positions of both the last stable orbit and the marginally

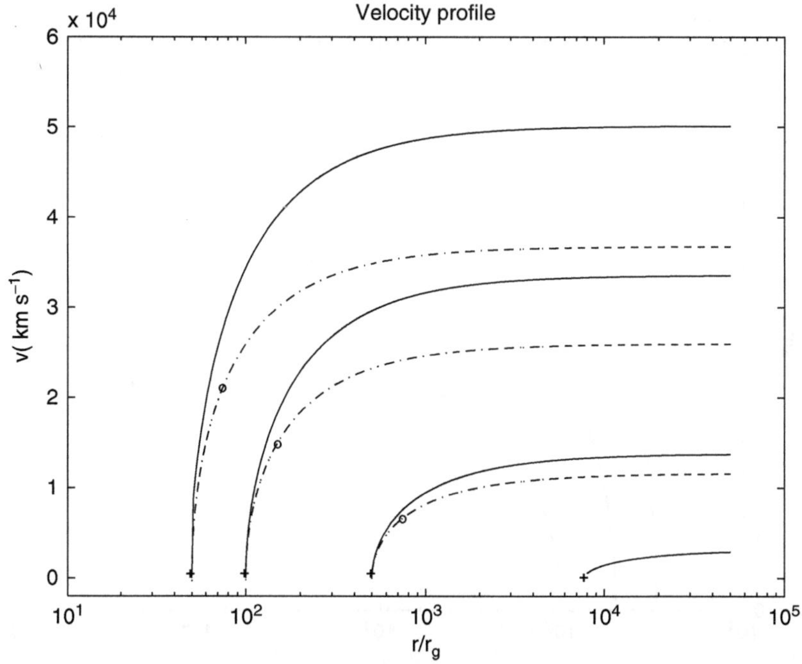

Fig. 9.21 Solution of the wind equation (9.62) for the Newtonian potential. *Solid line* – GEF solution, *dashed line* – CAK solution. Crosses indicate GEF critical points, *circles* – CAK critical points. Curves for $r/r_g = 8{,}000$ are graphically indistinguishable, from [326]

bound orbit. The numerical solutions for the isothermal wind, in the GEF model in comparison with CAK solutions, obtained in [326] are given in Figs. 9.21 and 9.22.

Exact general relativistic calculations in GR have been done in [327] for the GEF model in the Schwarzschild metric, representing a spherically symmetric black hole. A considerable gain was found in the wind velocity in comparison with both the CAK solution and the semiclassical solution of [326]. The position of the GEF critical point is found to be closer to the black hole than the CAK critical point. As follows from [327], even when the solution originates sufficiently far from the black hole (in fact, it is determined by the position of the critical point), there exists a large gain in the flow velocity in GEF in comparison with the CAK case.

The shapes of spectral lines from a plasma that is rapidly moving in the vicinity of a neutron star or a black hole have been studied in [328], taking into account gravitational redshifting. In the well-studied case of winds from normal hot stars, a rapidly moving wind interacts with the continuum radiation of a star and produces a P Cygni profile. From the theory of stellar winds, it is known that in the case of an arbitrary spherically symmetrical distribution of plasma that is moving with a gradually increasing velocity, an absorption line that is blueshifted with respect to the emission line is observed. This is widely interpreted as a fingerprint of moving plasma in a variety of astrophysical situations.

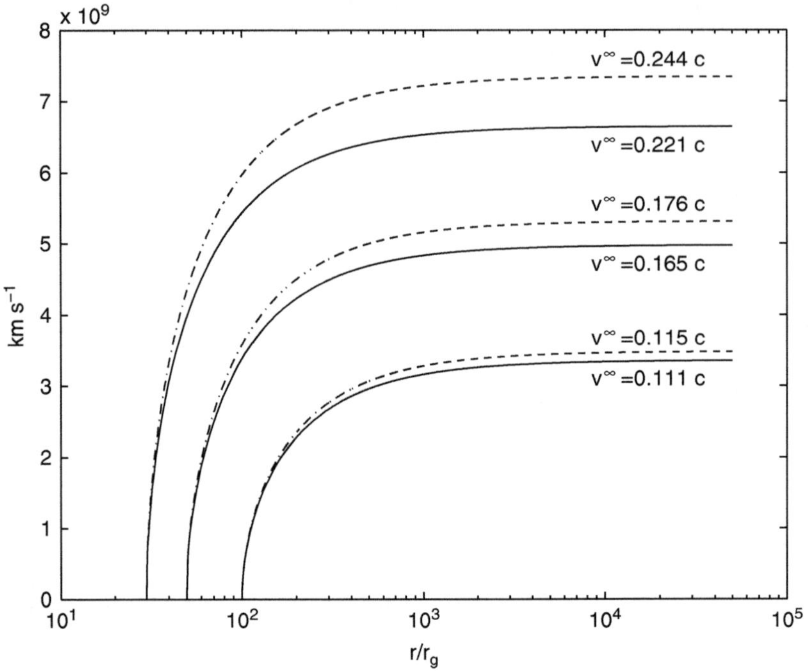

Fig. 9.22 *Solid line* indicates GEF solution for the Newtonian potential, *dashed line* GEF solution for Paczynski – Wiita potential, from [326]

Strong gravitational redshifting helps the radiation to escape efficiently and to interact with matter only locally. Gravitationally redshifted narrow absorption features in many AGN spectra as well as gravitationally redshifted absorption lines in the X-ray burst spectra of neutron stars have been observed, see observations of the gravitationally redshifted absorption lines in the X-ray burst spectra of EXO0748.676 [308]. If the interpretation is correct, these features are produced close to the compact object, where extreme conditions are coupled with high amplitude fluctuations of the radiation field and short dynamical time-scale. As a result, these lines may be highly variable or/and transient.

The calculations are made for spherically symmetric outflow exposed by continuum radiation of a spherical core. The regions of the flow under consideration were located at radii larger $10r_g$. In the calculations [328], the idea of equal frequency surfaces (EFS) plays a major role. The strong gravitational field changes the shape and locus of such surfaces. Their topology is complex; the gravitational field strongly distorts the P Cygni profile. Some branches of the EFS in the redshifted part of the spectra are found to be in front of the core, meaning there is the possibility for the absorption component to be observed as redshifted with respect to the emission. From numerical calculations, which are second-order accurate in v/c, it is established that a superposition of Doppler and gravitational shifting of

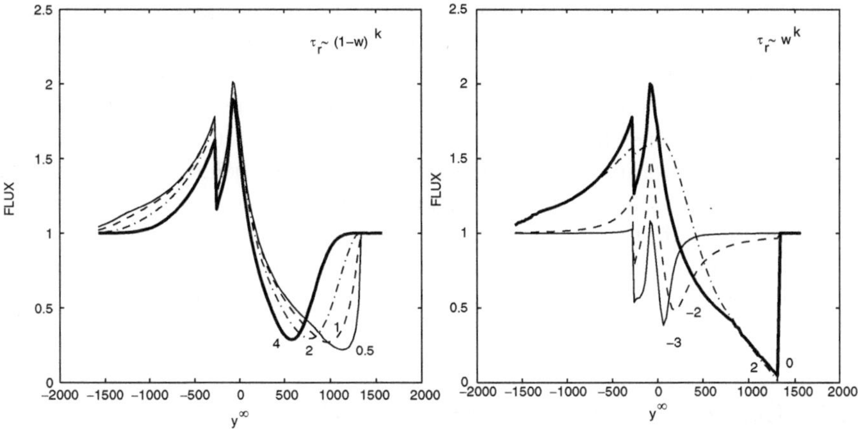

Fig. 9.23 Line profiles for different absorption laws: $\tau_r \sim (1-w)^k$ (9.64) (*left*) and $\tau_r \sim w^k$ (9.65) (*right*). Profiles for $R^* = 10r_g$, $V^\infty = 0.3\,c$ and $T_0 = 4$ and the linear velocity law (9.66). Curves are labeled by the parameter k from absorption laws. Vertical axes: flux. Horizontal axes: frequency shift, y^∞, from [328]

frequency can distort the P Cygni profile in such a way that blue- and redshifted absorption features are observed simultaneously. Often the redshifted absorption line is superimposed on the emission wing. Profiles with more than one emission and two absorption features are possible, (Fig. 9.23).

The radial optical depth $\tau_r = t(\mu = 1)$ is parameterized, as a function of the velocity. The following dependencies are considered:

$$\tau_r(w) = T_0(k+1)\frac{(1-w)^k}{(1-w_c)}, \tag{9.64}$$

and

$$\tau_r(w) = T_0(k+1)\frac{w^k}{(1-w_c^{k+1})}, \tag{9.65}$$

where $w \equiv v/V^\infty$ is the non-dimensional velocity, $w_c = v(R^*)/V^\infty = 0.01$ and k is a free parameter. The parameter T_0 is related to the total optical depth at the line centre. The following velocity laws were adopted in [328]: the linear velocity law

$$u(x) = U^\infty(x-1)/(x_t-1), \tag{9.66}$$

where $U^\infty = V^\infty/v_{th}$, V^∞ is a terminal velocity, $x_t = R_t/R^*$, and R_t is the wind terminating radius, $R_t = 100R^*$. Another was the "stellar" type velocity law:

$$w = w_c + (1-w_c)\left(1 - \frac{1}{x^{\alpha_1}}\right)^{\alpha_2}, \tag{9.67}$$

where α_1 and α_2 determine the slope and shape of the velocity profile.

The separation between red- and blueshifted absorption features is a function only of the dynamics and relative importance of gravity in the line-forming region. The particular shape and intensity of the emission component that separates the absorption lines depend sensitively on the assumed parameters. Gravitationally redshifted absorption lines form in places close to the compact object. The emission component is necessarily formed in plasma occupying a much larger volume. Thus, these features are formed in places that are possibly strongly separated in space. It is suggested in [328] that this property of line profiles to have complicated narrow absorption and emission components in the presence of strong gravity may help researchers to study spectroscopically the innermost parts of an outflow.

9.2.7 Calculations with New Opacity Tables

New calculations of evolutionary tracks for a wide range of masses have been performed using new estimations of the input physics: opacities [512–514], nuclear reactions [258] and new neutrino losses from plasma processes [231]. Of these, the most important are changes in opacities. The new ones (see Vol. 1) typically amount to a factor of 3 at 300,000 K for Solar metallicity with respect to the Los Alamos [312] values. Extensive and detailed calculations have been performed in [276, 867].

The masses in the range (0.8–120) $M_\odot$ (21 values) have been investigated. The metallicity composition varied from $Z = 0.001$ to $Z = 0.02$. The change of opacity leads from $X_{He} = 0.279$ and $X_Z = 0.0195$ to revision of the initial helium content of the Sun to $X_{He} = 0.299$ and $X_Z = 0.0188$. The mixing length theory of convection with pressure gradient height H_p and $\alpha_p = 1.6$ was used; the overshoot length d_{over} was taken to be equal to 0.2 H_p. These choices have been made on the basis of a best fit on T_{ef} of the Sun and red giants, and also on the basis of investigations into different parameters of 65 stellar clusters. The calculations have been performed for constant mass evolution, and for the case with a mass loss. The standard mass-loss rates according to the semi-empirical formulae

$$\log(-\dot{M}\,[M_\odot/\mathrm{yr}]) = -7.93 + 1.64\,\log(L/L_\odot) + 0.16\log(M/M_\odot) - 1.61\log T_{ef}[\mathrm{K}] \tag{9.68}$$

have been taken throughout the HR diagram until the formation of the Wolf–Rayet (WR) stars. The scaling of mass-loss rate with metallicity, $\dot{M} \sim Z^{0.5}$, was adopted. For WNL stars (WR stars with surface hydrogen abundance $X_{He} \leq 0.4$ and a blue location in the HR diagram, $\lg T_{ef} \geq 4$), the mass-loss rate $\dot{M} = -4\times 10^{-5} M_\odot/\mathrm{yr}$ was adopted, and for two types of WR stars, WNE (no hydrogen and still no traces of products of He burning) and WC (no hydrogen, and show these products) the dependence was assumed to be

$$\dot{M} = -(0.6\text{–}1.0)\,10^{-7}(M/M_\odot)^{2.5}\ M_\odot/\mathrm{yr}\,. \tag{9.69}$$

To strengthen the effect of mass loss on the evolution, calculations with mass-loss rate increased by a factor of 2 in post-MS stages have been performed for M in the range (20–120) $M_{\odot}$. Calculational results for $Z = 0.02$ are presented in Figs. 9.24–9.26.

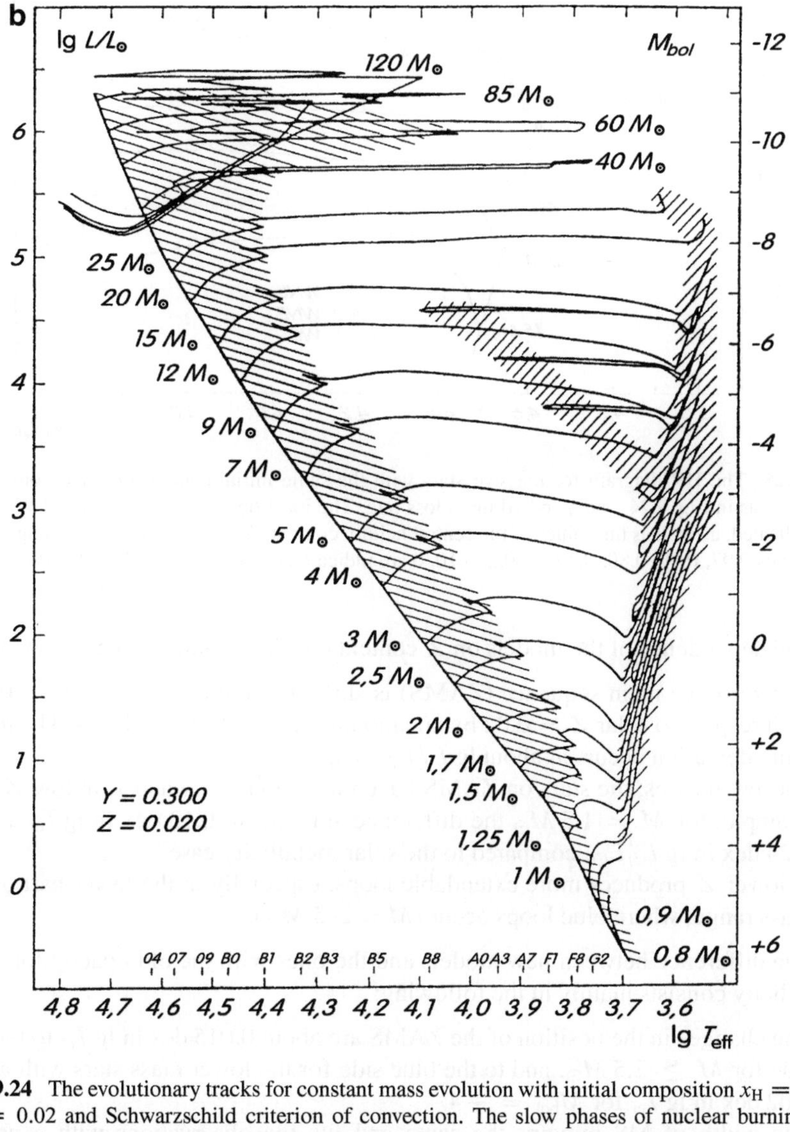

Fig. 9.24 The evolutionary tracks for constant mass evolution with initial composition $x_{\rm H} = 0.68$, $x_{\rm Z} = 0.02$ and Schwarzschild criterion of convection. The slow phases of nuclear burning are indicated by hatched areas. At 1.25 $M_{\odot}$, the model without overshooting is shown [867]

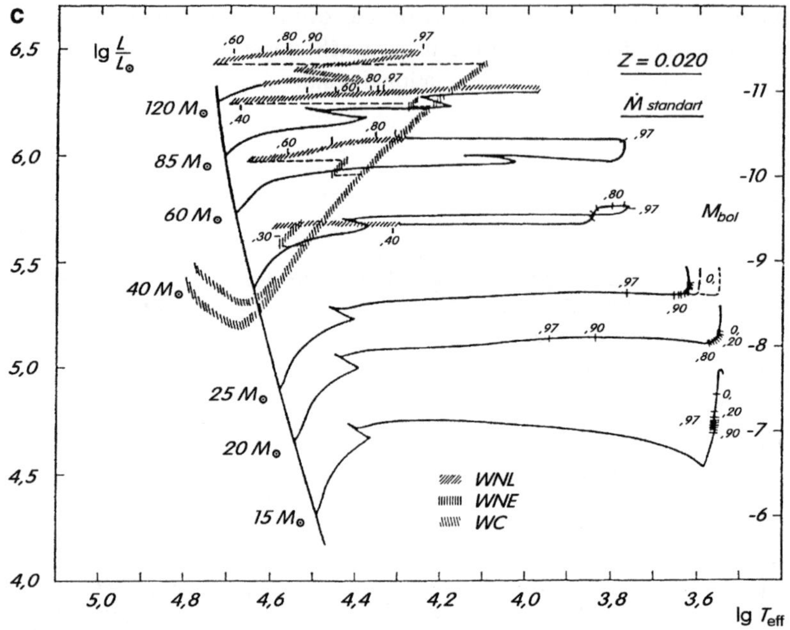

Fig. 9.25 The HR diagram for massive stars with the same initial composition and convection criterion, as in Fig. 9.24, and standard mass loss rate. The locations of WNL, WNE and WC stars are indicated, as well as the values of the central helium content. X_{He} during the He-burning phase; the values 0.97, 0.90, 0.80, 0.70, 0.60,... 0.10, 0 are indicated by small bar on the track [867]

The dependence of the models on Z content consists mainly in the following:

1. The zero-age main sequence (ZAMS) is shifted to the blue at low $Z = 0.001$ with respect to solar $Z = 0.02$ by an amount (0.04–0.1) dex in lg T_{ef}. The maximum deviation occurs at about lg $L/L_\odot = 2$.
2. For low masses, the stars on ZAMS have much higher T_{ef} and L at low Z. For example, for $M = 1.5\,M_\odot$ the difference amounts to 0.13 dex in lg T_{ef} and to 0.24 dex in lg $L/L_\odot$, compared to the solar metallicity case.
3. A lower Z produces more extendable loops, especially in the lower part of the mass range, where blue loops occur ($M = 2$–$5\,M_\odot$).

The difference between new models and the ones with the old opacity for solar metallicity consists mainly in the following:

1. The changes in the position of the ZAMS are about 0.015 dex in lg T_{ef} to the red side for $M \geq 2.5\,M_\odot$, and to the blue side for the lower mass stars with about 0.02 dex in lg T_{ef} for $M_{bol} = +4$.
2. The width of MS remains the same and fits the observations with a proper choice of overshoot parameters: 0.2 for the new and 0.25 for the old Los Alamos opacities.
3. With increasing opacities in the new tables, the blueward extension of loops on the helium burning phase is a little shorter for initial masses between 2 and 12

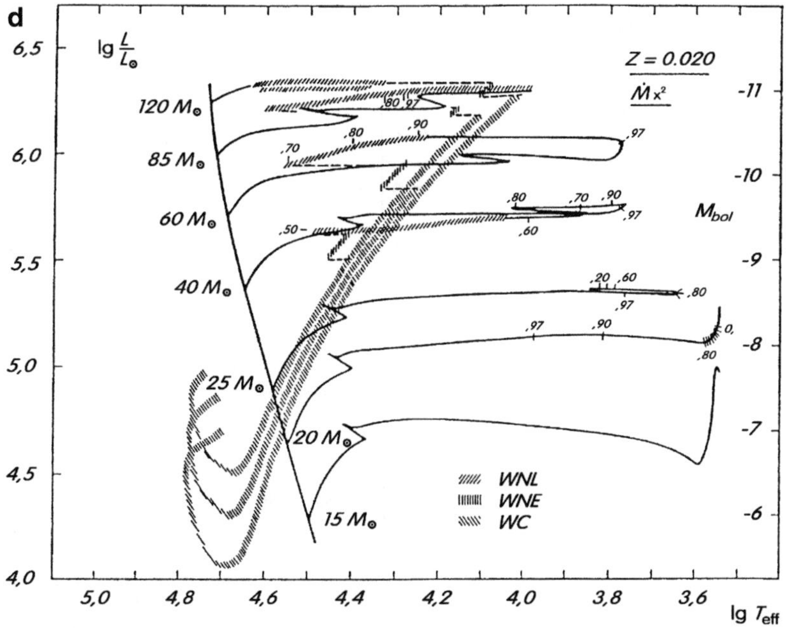

Fig. 9.26 The HR diagram for massive stars with same input parameters as in Fig. 9.24, and mass loss rates increased by a factor of 2 in post-MS phases. Same remarks as in the Fig. 9.25 are used [867]

$M_\odot$. The largest difference occurs for $M = 5\,M_\odot$, for which the extension has been reduced by a factor of two.

4. The position of the RGB and asymptotic giant branch (AGB) (see Sect. 9.3) presents no particular discrepancies. The position of the Core Helium Flashes (CHFs) are bluer in new models by about 0.04–0.07 dex in lg $T_{\rm ef}$. The $T_{\rm ef}$ of red supergiants (RSG) is greater by about 0.04 dex in lg $T_{\rm ef}$ for 15 $M_\odot$ and the difference increases up to 25 $M_\odot$.

Hence, the differences between the models with the old and new opacities are rather modest, and there is great similarity between the internal structures of the two sets of models. Larger changes occur with the ratios between ^{12}C and ^{16}O at the end of the helium-burning phase. These ratios change from the interval $^{12}{\rm C}/^{16}{\rm O} = 0.015$–$0.14$ in old to $^{12}{\rm C}/^{16}{\rm O} = 0.13$–$0.46$ in the new models. We may summarize that the differences between evolutionary tracks for different $Z = (0.02, 0.004$ and $0.001)$ are substantially larger than the differences between models with the old and new opacities and the same Z.

Evolutionary calculations with the new opacities, investigating the dependence of evolutionary tracks on $X_{\rm Z}$ and on iron abundance, have been made in [941]. Masses in the range 1.5–60 $M_\odot$ have been calculated. A mixing length theory with $\alpha_{\rm p} = 1.4$ was used. Convective core overshooting and semiconvection above the hydrogen-burning layers have been ignored, being unimportant in these stars. The evolutionary

Fig. 9.27 Evolutionary tracks running from ZAMS to the end of helium burning for initial compositions $x_H = 0.7$, $x_Z = 0.02$ (*solid lines*), $x_Z = 0.03$ (*dash-dotted lines*) and high iron composition [941]

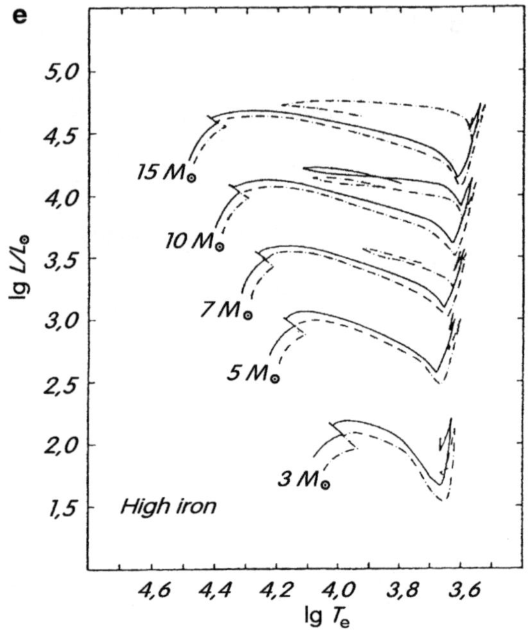

tracks run from ZAMS to the end of the core helium burning in the case of stellar masses 3, 5, 7, 10 and 15 $M_\odot$. Otherwise, only the MS phase was covered. The results of calculations for high iron abundance 0.1 X_Z (compared to the normal 0.007 X_Z) are presented in Fig. 9.27.

New opacity tables have been used in [351] in an attempt to construct a model and an evolutionary status of P Cygni star, which is one of the brightest in the Galaxy with $\lg L/L_\odot = 5.86 \pm 0.1$ and $T_{ef} = 19300 \pm 700$ K. Archival data have shown that $\lg T_{ef}$ changes by -0.027 ± 0.004 per century at constant luminosity. Evolutionary calculations made in [351] lead to the conclusion that P Cygni has already terminated core hydrogen burning. Its luminosity is provided by hydrogen burning in the shell, and the star is on its way to ignite core helium burning.

The relaxation code [353] has been used, which takes into account the inertia terms. Mixing length theory with $\alpha_p = 1.6$ for a convection in the envelope and the Schwarzschild criterion were used, while overshooting has not been included. The mass-loss rate according to (9.68) has been used. The mass-loss rates following from the theory of radiation driven winds, which are lower by a factor of two, were also used to study the sensitivity of the results to the details of mass-loss prescription. The evolution of stars with six initial masses between 30 and 70 $M_\odot$ was calculated. Models with $T_{ef} = 20{,}000$ K were compared to observations of luminosity and evolutionary speed. The observations (within the error bars) may be fitted by stars with initial masses between 55 and 65 $M_\odot$, which have final masses between 47 and 53 $M_\odot$. Calculations using the Ledoux criterion and with the semiconvective zone taken according to [1033], which is the most efficient of all descriptions of semiconvection, have shown much worse agreement with the observations. This favours the use of the Schwarzschild criterion for convection in massive stars; see, however, the opinion expressed in [942]. Mass loss strongly affects the results. The observations seem to suggest that P Cygni has lost $\sim 10\,M_\odot$ during its hydrogen-burning period in accordance with (9.68).

Similar calculations for 60 $M_\odot$ using another prescription for the mass-loss rate have been carried out in [621]. Instead of (9.68), the relation (9.36) was used during the O star phase. For luminous blue giants (LBG) which are more pulsationally unstable with new opacity tables, the enhanced mass-loss rate

$$\begin{aligned} \dot{M} &= 3.358 \times 10^{-4} f(4.63 - \log T_{\mathrm{ef}}/\mathrm{K})\; M_\odot/\mathrm{yr} \\ &\qquad \text{for } 4.63 > \log T_{\mathrm{ef}}/\mathrm{K} > 4.535 \\ \dot{M} &= 1.636 \times 10^{-4} f(\log T_{\mathrm{ef}}/\mathrm{K} - 4.34)\; M_\odot/\mathrm{yr} \\ &\qquad \text{for } 4.535 > \log T_{\mathrm{ef}}/\mathrm{K} > 4.34 \end{aligned} \tag{9.70}$$

was used, essentially with $f = 1$, which is proportional to the pulsational instability increment (see Sect. 14.3). As a result, a lower mass for P Cygni star was obtained, and some changes in the evolutionary scenario of very massive stars were suggested.

Analysis of red–blue loop formation and its dependence on the semiconvective mixing and overshooting was done with new opacities in [350], by evolutionary calculations of stars in mass range 1.5–20 $M_\odot$. The same code as in [351, 353], with various criteria for convective instability, mixing prescription in semiconvective zone and core overshooting length was used. Using the Schwarzschild criterion, loops become more extended with increasing stellar mass and disappear abruptly at $M = 13\,M_\odot$. An intermediate convection zone is formed here, and the convective envelope is prevented from downward penetration into a hydrogen-burning shell. When using the Ledoux criterion, the blue loops exist till $M = 19\,M_\odot$ and are absent at 20 $M_\odot$. Their shape is very sensitive to the details of the semiconvective mixing, probably due to instability action. It is found that semiconvective mixing promotes the formation of the loops, but core overshooting strongly inhibits their extension.

Extended evolutionary calculations of helium stars have been performed in [1052] for the mass range 4–20 $M_\odot$ with account of mass loss. As a result, the final masses converge to a narrow range of small values 2.26–3.55 $M_\odot$ for all stars under consideration. The formation of helium stars is related in [1052] to belonging to binary systems, and products of their evolution were considered as progenitors of Type Ib and Ic supernovae.

Evolutionary calculations with new opacities using an updated variant of the Eggleton method have been performed in [809]. The results of these calculations are presented in Fig. 9.28.

Problem 1. Given the opacity

$$\begin{aligned} \kappa &= \kappa_0 \quad \text{for } T > T_c, \\ \kappa &= \kappa_1 \left(\frac{T}{T_c}\right)^n, n > 0, \quad \text{for } T < T_c, \quad \kappa_1 \gg \kappa_0, \end{aligned} \tag{9.71}$$

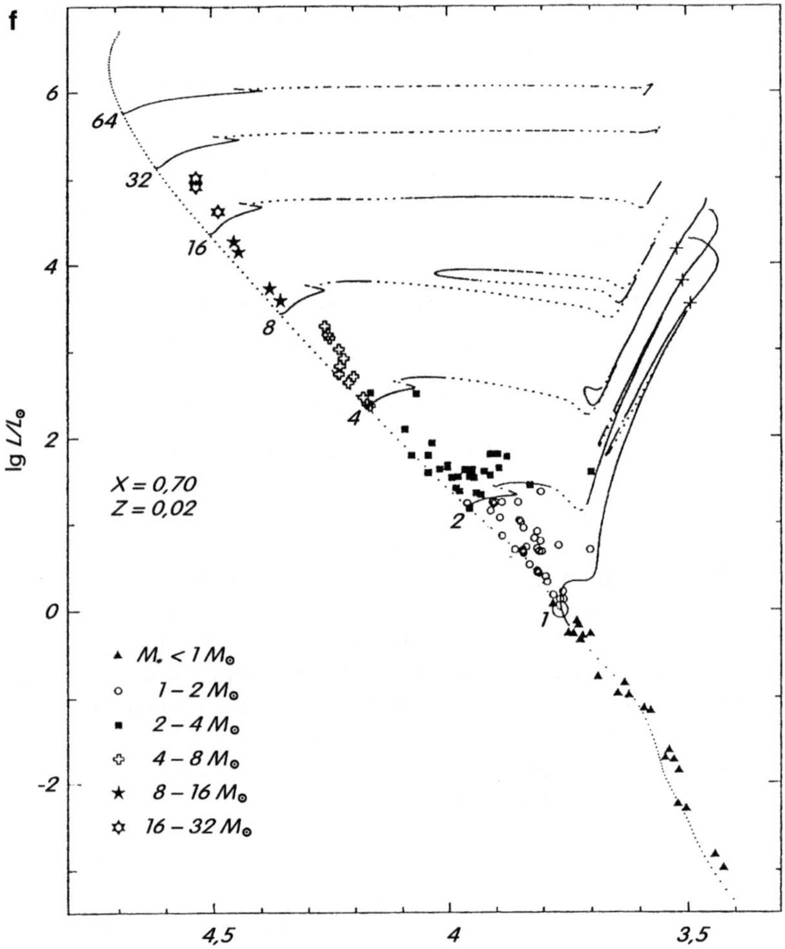

Fig. 9.28 HR diagram of ZAMS and evolutionary tracks, calculated in [809]. The masses of the models are indicated at the starts of the tracks, in solar units. The *solid* portions of the tracks indicate where evolution is on a relatively slow nuclear time-scale, the *dotted* parts show evolution on a thermal time-scale and the *dashed* parts show an intermediate time-scale. The different symbols indicate the positions of binary components with well-determined masses, radii and luminosities. The position of the Sun is indicated by a solar symbol $\odot$. The crosses (+) show the position of the giant branches of 1-, 2- and 4- $M_\odot$ models, where the binding energy of the envelope becomes positive

find the boundary between static and outflowing models in the absence of convection.

Solution. [899]. Taking the equilibrium equation

$$\frac{\mathrm{d}P}{\mathrm{d}r} = -\rho \frac{Gm}{r^2} \tag{9.72}$$

at $m = M$ =const., and the radiative heat conductivity equation

$$L_r = L_r^{rad} = -\frac{4acT^3}{3\kappa\rho} 4\pi r^2 \frac{dT}{dr} \tag{9.73}$$

at L_r =const., divide (9.72) by (9.73) and, using $P(\rho, T)$ from (9.12) and the boundary condition $P = 0, \rho = 0$ at $T = 0$, obtain the solution in the form

$$P = \frac{L_{c0}}{L} \frac{a(T^4 - T_c^4)}{3} + \frac{aT_c^4}{3} \frac{L_{c1}}{L} \frac{4}{4-n} \quad \text{for } T > T_c, \tag{9.74}$$

$$P = \frac{L_{c1}}{L} \frac{aT^4}{3} \frac{4}{4-n} \left(\frac{T_c}{T}\right)^n \quad \text{for } T < T_c,$$

$$\rho(T) = \left(P - \frac{aT^4}{3}\right) \Big/ \mathscr{R}T, \tag{9.75}$$

$$L_{c0} = \frac{4\pi cGM}{\kappa_0}, \quad L_{c1} = \frac{4\pi cGM}{\kappa_1}. \tag{9.76}$$

For T and P from (9.74) to become simultaneously zero, we must have

$$n < 4. \tag{9.77}$$

For T, ρ and P from (9.75) to become simultaneously zero, we must have

$$n < 3, \quad \frac{L_{c1}}{L} \frac{4}{4-n} > 1. \tag{9.78}$$

On substituting (9.74), (9.75) into (9.73), the solution reads

$$\frac{1}{r} \frac{GM}{4\mathscr{R}} + C_1 = \int \left[1 - \frac{L}{L_{c0}} + \left(\frac{T_c}{T}\right)^4 \left(\frac{4}{4-n} \frac{\kappa_0}{\kappa_1} - 1\right)\right]^{-1} dT,$$
$$\text{for } T > T_c, \tag{9.79}$$

$$\frac{1}{r} \frac{GM}{4\mathscr{R}} + C_2 = \int \left[\frac{4}{4-n} - \alpha \left(\frac{T}{T_c}\right)^n\right]^{-1} dT.$$

for $T < T_c$, where $\alpha = L/L_{c1}$.

The constants C_1 and C_2 are to be found from the condition $T = T_c$ at $r = R_c$. For $n = 1$, we have

$$T = \frac{T_c}{\alpha} \left\{\frac{4}{3} - \left(\frac{4}{3} - \alpha\right) \exp\left[\frac{\alpha}{T_c} \frac{GM}{4\mathscr{R}} \left(\frac{1}{R_c} - \frac{1}{r}\right)\right]\right\}, \tag{9.80}$$

$$T < T_c.$$

Reducing T to zero at a finite radius R requires that

$$\frac{GM}{4\mathscr{R}R_c} > -\frac{T_c}{\alpha}\ln\left(1-\frac{3\alpha}{4}\right),\ \alpha < \frac{4}{3}. \tag{9.81}$$

The conditions (9.78) with $n = 1$ and (9.81) are required for the static envelope of the star to exist. When the inequality (9.81) no longer holds, the radiative envelope must be outflowing.

Problem 2. Show that when conditions (9.78) and (9.81) are no longer valid, there is a region with simultaneous existence of a static solution for convective, and an outflowing solution for radiative envelopes.

Solution. As the convection (say, according to the mixing-length theory) effectively reduces α, the static solution still exists at formal breaking of condition (9.81) or analogous to it at $n \neq 1$. Simultaneously, when these conditions are broken, we have for the case of a radiative envelope a solution outflowing to infinity and obtained numerically in [899] from (9.9 to 9.12) with the use of (9.71) for $n = 1, \alpha = 5$, $L/L_{c0} = 1/2$, and $n = 1, \alpha = 2, L/L_{c0} = 1/2$.

9.3 Evolution with Degeneracy, Thermal Flashes

In low-mass stars with $M \leq 2.25M_\odot$, there forms a degenerate helium core where the onset of helium burning is accompanied by a thermal flash. This flash results from a thermal instability due to the fact that an increase in temperature in degenerate matter leads to almost no increase in pressure. The heat capacity of the star is positive in this process, and contrary to ordinary stars with a negative heat capacity, no stabilizing feedback is present here.

The thermal flash makes the helium core non-degenerate, and a quiet evolutionary phase with further core helium burning returns. This phase proceeds qualitatively in the same way as in the case of intermediate-mass stars with $2.25 \leq M/M_\odot \leq \sim 8$, where the degeneracy occurs primarily after formation of a carbon core and two shell sources: helium and hydrogen.

The evolution of low- and intermediate-mass stars (LI-stars) in the shell-burning phase proves surprisingly similar. In this phase, the observed properties of a star – its position in the HR diagram – only slightly depend on the total stellar mass and are mainly determined by the carbon core mass. All stars on average follow the same track, called convergent. The stars reaching this track are also referred to as the AGB stars. The motion along the convergent track is accompanied by thermal flashes in helium shells, the number of flashes increasing with stellar mass. Their nature differs from that of CHFs since the matter in the helium shell is non-degenerate. A positive heat capacity is a common property for both these kinds of flash, but in the latter case it is due to the form of the thin-shell energy source. The evolution along the convergent track is accompanied by a mass loss leading eventually to the white dwarf formation out of the degenerate carbon core. Immediately before this formation, the star rapidly ejects the residual envelope which is subsequently observed as a nebula illuminated by the radiation of a hot central star PN. The carbon core of the most massive LI-stars may approach the mass $M = 1.39 M_\odot$ by the time a thermal instability, which is usually thought to result in observable Type I supernova explosion, develops in its interiors. Consider the above evolutionary phases of LI-stars in more detail.

9.3.1 Core Helium Flash

The thermal instability development in a degenerate core was predicted by Mestel [689] and first calculated by Schwarzschild and others [456, 458, 883, 887]. A population II star (depleted in metals) with $M = 1 M_\odot$, initial chemical composition $x_H = 0.9$, $x_{He} = 0.099$, $x_Z = 0.001$ and Schwarzschild criterion for convection has been examined in [456]. The core convection has been taken to be adiabatic at any time, including the flash peak. The maximum central temperature during the flash is 2×10^8 K, and the maximum helium burning rate $\sim 10^{11} L_\odot$. The luminosity of the star experiences little changes during the flash: from $2720 L_\odot$ before the onset to $2740 L_\odot$ in the peak, and in 3.5×10^5 yr falls off to $170 L_\odot$. The respective values of the effective temperature are 4500, 4060 and 4610 K, and of the radius 105, 106 and 20 $R_\odot$. The last numbers correspond to a model with helium burning in a non-degenerate core. During the flash peak, x_Z approaches 0.0077 in the centre; hence, 0.7% of the helium in the core centre burns out. The energy release rate is such that the core remains in a state close to the static equilibrium. The convection that develops in the core cannot reach the hydrogen-burning shell having two pressure scale heights of separation from it.

From that time onward, the helium flash has been repeatedly calculated (see review [866]), and never in these calculations could the core convection reach the hydrogen-burning shell. The non-stationary character of convection during the flash has been a major and nonetheless non-calculated factor. A stabilizing role of the chemical composition gradient $\nabla\mu$ is also unclear under these conditions. The second type of mixing which may occur during the flash is due to the penetration of the outer convective envelope inside the core. The results of various authors are contradictory on this point, being strongly dependent on the adopted form of neutrino losses due to plasma neutrinos which differ from each other (see [82, 866]). The major result common for all calculations is that the flash is followed by the appearance of a star having a non-degenerate core with helium burning, and the amount of helium burnt out over the flash does not exceed 1%. The duration of the flash is 5×10^4 yr, and the location of the star in the HR diagram does not undergo significant changes during this time. Interpolation formulae for the mass of the helium core M_{CHe} before the flash (in $M_\odot$) have been obtained in [494] using numerical calculations:

$$M_{C\mathrm{He}} = 0.475 + 0.23(0.3 - x_{\mathrm{He}}) \\ - 0.01(\lg x_{\mathrm{Z}} + 3) + 0.035(0.8 - M). \qquad (9.82)$$

These calculations also give the lifetime t_{RG} of the red giant phase of stellar evolution from the time of luminosity L_{RG} to CHF:

$$\lg(t_{\mathrm{RG}}/10^7\,\mathrm{yr}) = 2.351 - 0.84\lg(L_{\mathrm{RG}}/L_\odot) \\ - 0.04(\lg x_{\mathrm{Z}} + 3) + 1.36\lg(1 - x_{\mathrm{He}}) - 0.27\lg M, \qquad (9.83)$$

where M is the stellar mass in $M_\odot$, and x_{He} and x_{Z} are the initial concentrations. Write out also the relation determining the lifetime of the star t_{t} before the turning point of the MS (see Fig. 9.7) as a function of luminosity L_{t} at the turning point and initial concentrations x_{He} and x_{Z}:

$$\lg\left(\frac{t_{\mathrm{t}}}{10^{10}\,\mathrm{yr}}\right) = 0.42 - 1.1\lg L_{\mathrm{t}} + 0.59(0.3 - x_{\mathrm{He}}) \\ -0.14(\lg x_{\mathrm{Z}} + 3), \qquad -4 \le \lg x_{\mathrm{Z}} \le -3. \qquad (9.84)$$

This relation is useful to estimate the age of clusters from their HR diagram.

9.3.2 *Horizontal Branch*

Low-mass stars with a non-degenerate helium core and a hydrogen envelope, resulting from a helium flash, are located near the line in the HR diagram which is called the horizontal giant branch (HB). As the mass of the helium core $M_{C\mathrm{He}}$ during the flash depends slightly on stellar mass, the models of stars on the horizontal branch represent helium cores of almost equal masses which are surrounded by hydrogen envelopes of different masses. Evolution of models with $M_{C\mathrm{He}} = 0.475 M_\odot$ and various hydrogen envelopes is calculated in [943]. The results are given in Fig. 9.29. The initial models are located on the zero-age horizontal branch (ZAHB). These models burn helium in the core and hydrogen in the shell. As helium burns out, the models evolve off the ZAHB and are situated on the HB. After the helium has been exhausted, the phase of rapid core contraction starts (dashed portions in Fig. 9.29) and lasts until helium shell ignition. Subsequently, the stars take positions in the HR diagram that correspond to what is called the upper horizontal branch (UHB). The motion along the UHB is accompanied by an increase in the hydrogen shell importance, an decrease in the separation of one shell from another, a increase in the model luminosity, and results in arriving on the AGB. Almost all stars from the HB reach the AGB, with the exception of stars with a small hydrogen envelope, located on the left edge adjacent to the helium MS. These stars transform into white dwarfs without visiting the AGB. Stars with $M \ge M_\odot$ do not leave, after the helium flash,

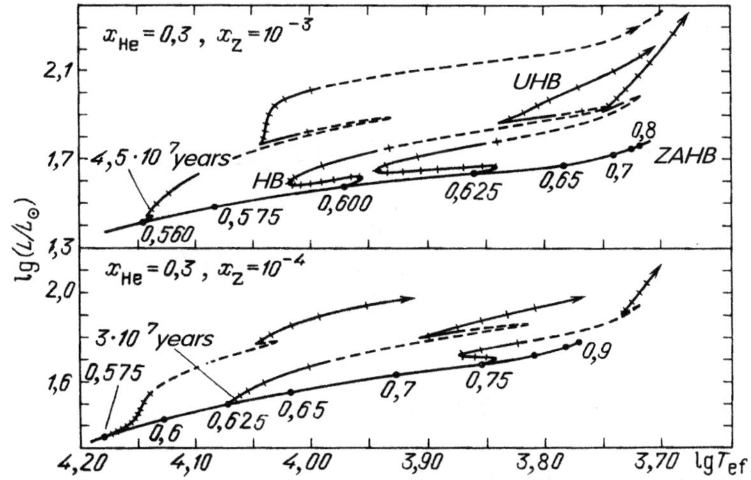

Fig. 9.29 Evolutionary tracks for the core helium burning and double shell source phases of models with two envelopes chemical compositions. ZAHB is the zero-age horizontal branch, total model masses are indicated below it, the helium core has a constant mass of 0.475 $M_{\odot}$. The *solid curves* represent the HB, horizontal branch departing from the ZAHB, and UHB, the upper horizontal branch. Rapid phases of evolution are represented by dashed portions. The time interval between neighbouring tick marks is 5×10^6 yr. After UHB, the stars move to a position on the AGB, asymptotic giant branch, from [943]

the red giant branch (RGB) from which they get onto AGB almost continuously, as is the case of intermediate-mass stars.

The location of stars at various evolutionary stages mentioned above is indicated in a colour $(B - V)$ versus luminosity $(m_{\rm ph})$[5] diagram for the globular cluster $M13$ (Fig. 9.30 from [943]), which is similar to the HR diagram. The time of evolution of a star on the HB is approximately given by the expression [494]

$$\lg(t_{\rm HB}/10^7\,{\rm yr}) = 0.74 - 2.2(M_{C\rm He} - 0.5), \qquad (9.85)$$

where the dependence $M_{C\rm He}(M, x_{\rm He}, x_{\rm Z})$ is determined by (9.82).

[5] The stellar magnitude M is defined as a logarithm of the stellar luminosity. The bolometric (total) absolute magnitude $M_{\rm bol} = 4.74 - 2.5\lg(L/L_{\odot})$. The visual magnitude (interstellar absorption is not taken into account) is $m = M - 5 + 5\lg d_{\rm pc}$, $d_{\rm pc}$ being the distance to the star in parsecs. For crude estimates of the spectrum, stellar magnitudes in separate spectral ranges are used; the latter are determined by the following optical filters: m_U around $\lambda = 3650_{\rm A}$, m_B around $\lambda = 4400_{\rm A}$, m_V around $\lambda = 5500_{\rm A}$ with $\Delta\lambda \approx 800_{\rm A}$. Photo-visual magnitude is $m_{\rm ph} \approx m_V$. The colour index $B - V$ $(\equiv m_B - m_V)$ does not depend on the distance to the star and corresponds to its temperature, $B - V \approx (7300\ {\rm K}/T_{\rm ef}) - 0.60$. More accurate definitions of stellar magnitudes taking into account the transparency curves for filters, energy distributions in the stellar spectrum and interstellar absorption are given in [15].

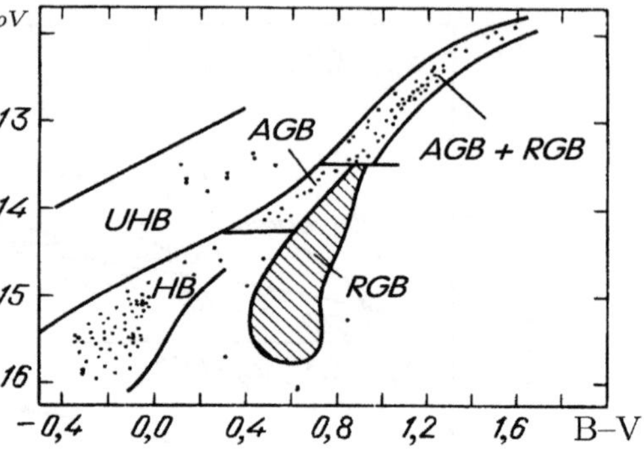

Fig. 9.30 Colour (B–V) – luminosity (m_{pV}) diagram (sketch) for the globular cluster M13. Marked are the following groups of stars: RGB, red giant branch; HB, horizontal branch; UHB, upper horizontal branch; AGB, asymptotic giant branch (from [943])

9.3.3 Asymptotic Giant Branch

For stars with a degenerate carbon core and two very close helium- and hydrogen-burning shells, their location in the HR diagram is nearly independent of the total stellar mass and is determined mainly by the carbon core mass. The AGB represents a convergent track on which all LI-stars arrive. The existence of a convergent track has been established by Paczynski [770, 772] and Uus [1004, 1005]. The convergence of all stars towards this track may be seen, for example, in Fig. 9.7 from [770]. The presence of two thin, closely located shells causes serious computational difficulties in calculations of the AGB evolution by conventional methods like Schwarzschild or Henyey (see Chap. 6, Vol. 1). The situation is still more complicated because the helium-burning shell is unstable on the major part of the AGB and gives rise to thermal flashes (ThF). During ThF, as in the case of CHF, the star remains in static equilibrium. The calculation of one flash requires several thousand models, while the number of flashes for intermediate-mass stars is of the order of 1,000; so direct evolutionary calculations become very time-consuming.

For calculations of AGB evolution, several methods have been suggested which allow suppression of ThF and an averaged evolutionary track to be obtained. The method used by Paczynski and Uus in the discovery of the convergent track consists in adopting a quasistationary approximation for the material flow through the two thin shells. The approximation was first applied in [343] to compute a thin hydrogen-burning shell. It is based on a small magnitude of Eulerian $(\partial/\partial t)_r$ relative to Lagrangian $(\partial/\partial t)_m$ time derivatives. The reason for this is that all parameters change drastically when passing through the burning shell, but time changes for all parameters are small on both sides of the shell. We have [772]

$$\left(\frac{\partial}{\partial t}\right)_m = \left(\frac{\partial}{\partial t}\right)_r + \left(\frac{\partial r}{\partial t}\right)_m \left(\frac{\partial}{\partial r}\right)_t$$
$$= \left(\frac{\partial}{\partial t}\right)_r - \left(\frac{\partial m}{\partial t}\right)_r \left(\frac{\partial}{\partial m}\right)_t . \qquad (9.86)$$

On the core boundary[6] $(\partial m/\partial t)_r \approx \mathrm{d}M_c/\mathrm{d}t = \dot{M}_c$; so, using the small value of $(\partial/\partial t)_r$, we obtain from (9.86)

$$\left(\frac{\partial}{\partial t}\right)_m = -\dot{M}_c\left(\frac{\partial}{\partial m}\right)_t. \tag{9.87}$$

The relations

$$\left(\frac{\partial}{\partial r}\right)_t = \left(\frac{\partial}{\partial m}\right)_t\left(\frac{\partial m}{\partial r}\right)_t \quad \text{and}$$
$$\frac{dm}{dt} = \left(\frac{\partial m}{\partial t}\right)_r + \left(\frac{\partial m}{\partial r}\right)_t\left(\frac{\partial r}{\partial t}\right)_m = 0 \tag{9.88}$$

have been used to derive (9.86). The relation (9.87) allows us to reduce the stellar evolution equations with partial derivatives in equations for chemical composition and gravitational energy, given in Chap. 6, Vol. 1, to ordinary differential equations. The chemical evolution equations are represented by [1004]

$$\frac{dx_{\mathrm{H}}}{dm} = 4m_{\mathrm{p}}\frac{\epsilon_{\mathrm{CNO}}}{Q_{\mathrm{CNO}}}\frac{1}{\dot{M}_c}, \quad \frac{dx_{\mathrm{He}}}{dm} = -\frac{dx_{\mathrm{H}}}{dm}, \quad \frac{dx_{\mathrm{C}}}{dm} = 0 \tag{9.89}$$

in the hydrogen-burning zone and by

$$x_{\mathrm{H}} = 0, \quad \frac{\mathrm{d}}{\mathrm{d}m}(x_{\mathrm{He}} + x_{^{12}\mathrm{C}} + x_{^{16}\mathrm{O}} + x_{^{20}\mathrm{Ne}}) = 0,$$
$$\frac{\mathrm{d}x_{\mathrm{He}}}{\mathrm{d}m} = \left(\frac{3m_\alpha\epsilon_{3\alpha}}{Q_{3\alpha}} + \frac{m_\alpha\epsilon_{^{12}\mathrm{C}\alpha}}{Q_{^{12}\mathrm{C}\alpha}} + \frac{m_\alpha\epsilon_{^{16}\mathrm{O}\alpha}}{Q_{^{16}\mathrm{O}\alpha}}\right)\frac{1}{\dot{M}_c}, \tag{9.90}$$
$$\frac{\mathrm{d}x_{^{12}\mathrm{C}}}{\mathrm{d}m} = \left(-\frac{3m_\alpha\epsilon_{3\alpha}}{Q_{3\alpha}} + \frac{m_{^{12}\mathrm{C}}\epsilon_{^{12}\mathrm{C}\alpha}}{Q_{^{12}\mathrm{C}\alpha}}\right)\frac{1}{\dot{M}_c},$$
$$\frac{\mathrm{d}x_{^{16}\mathrm{O}}}{\mathrm{d}m} = \left(\frac{m_{^{16}\mathrm{O}}\epsilon_{^{16}\mathrm{O}\alpha}}{Q_{^{16}\mathrm{O}\alpha}} - \frac{(m_\alpha + m_{^{12}\mathrm{C}})\epsilon_{^{12}\mathrm{C}\alpha}}{Q_{^{12}\mathrm{C}\alpha}}\right)\frac{1}{\dot{M}_c}$$

in the helium-burning zone. The pp-cycle reaction is omitted in (9.89), the $^{16}\mathrm{O}(\alpha,\gamma)^{20}\mathrm{Ne}$ reaction is taken into account in (9.90), which is not important at earlier evolutionary stages. The term with gravitational energy is written as

$$\epsilon_{\mathrm{gr}} = -\frac{\partial E}{\partial t} + \frac{P}{\rho^2}\frac{\partial\rho}{\partial t} = \left[\left(\frac{\partial E}{\partial\rho} - \frac{P}{\rho^2}\right)\frac{\mathrm{d}\rho}{\mathrm{d}m} + \frac{\partial E}{\partial T}\frac{\mathrm{d}T}{\mathrm{d}m}\right]\dot{M}_c. \tag{9.91}$$

[6] The core mass M_c is defined as if the core boundary is in the middle of the hydrogen-burning shell [770].

Models with double-burning shells have been calculated in [1005] with no inclusion of gravitational energy, at $\epsilon_{gr} = 0$. Write the first equation (9.89) and the third (9.90) in the form

$$\frac{dx_H}{dm} = \frac{\epsilon_H}{E_H} \frac{1}{\dot{M}_c}, \tag{9.92}$$

$$\frac{dx_{He}}{dm} = \frac{\epsilon_{He}}{E_{He}} \frac{1}{\dot{M}_c}. \tag{9.93}$$

Here, $\epsilon_H = \epsilon_{CNO}$, $E_H = Q_{CNO}/4m_p$, $\epsilon_{He} = \epsilon_{3\alpha}$, and $E_{He} = Q_{3\alpha}/3m_\alpha$; the latter quantity determines the energy released by conversion of 1 g of helium into carbon. From the energy equation

$$\frac{dL_r}{dr} = 4\pi\rho r^2 \left(\epsilon_n - \epsilon_\nu - \frac{\partial E}{\partial t} + \frac{P}{\rho^2}\frac{\partial \rho}{\partial t}\right), \tag{9.94}$$

neglecting neutrino emission and production of gravitational energy, that is $\epsilon_\nu = \epsilon_{gr} = 0$, and using (9.3.11), we have upon integrating

$$\begin{aligned} x_H &= x_{H0} + \frac{L(m) - L}{E_H \dot{M}_c}, \qquad L_{He} = L - x_{H0} E_H \dot{M}_c\,, \\ x_{He} &= x_{H0} + x_{He0} + \frac{L(m) - L_{He}}{E_{He}\dot{M}_c}, \\ L_i &= L_{He} - (x_{H0} + x_{He0})\, E_{He}\dot{M}_c. \end{aligned} \tag{9.95}$$

Here, L_{He} is the luminosity on the boundary between the shell sources, L is the outer luminosity and L_i is the luminosity of the carbon core. We obtain from (9.95) the relationship between L, L_i and $\dot{M}_c$

$$\dot{M}_c = \frac{L - L_i}{x_{H0}E_H + (x_{H0} + x_{He0})\, E_{He}}. \tag{9.96}$$

Using (9.96) and (9.95) gives the x_H and x_{He} distribution in shells as a function of luminosity $L(m)$ with total luminosity L as a parameter. The procedure for constructing an equilibrium model with two quasistationary shells with $\epsilon_{gr} = 0$ is as follows [1005]. First, we choose a fitting point inside the carbon core such that $m = m_f$. For a given L and effective temperature T_{ef}, the equations of stellar structure (9.94), (9.72), and (9.73) in Problem 1 after Sect. 9.2, with continuity equation

$$\frac{dm}{dr} = 4\pi\rho r^2, \tag{9.97}$$

the relation for the total heat flux (including the convective one from (8.19)),

$$L_r = L_r^{rad} + 4\pi r^2 F_{conv}, \tag{9.98}$$

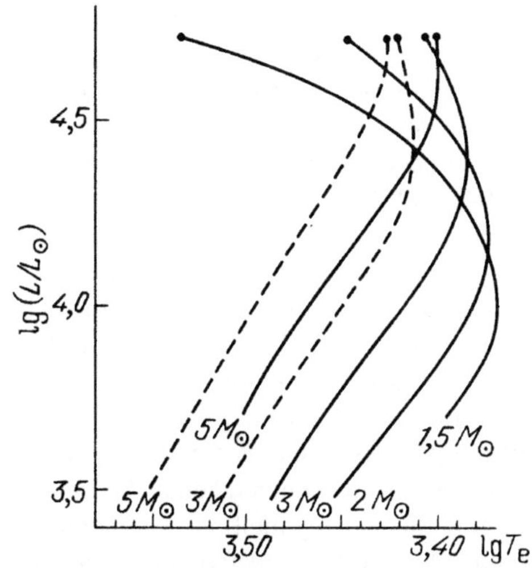

Fig. 9.31 The AGB tracks for intermediate-mass stars, from [1005]. For comparison, the *dashed lines* represent tracks from [770]

and combined with (9.95) and (9.96) are integrated inward to $m = m_f$, where $\rho = \rho_f$, $T = T_f$ and $r = r_f$. Taking the carbon core isothermal with zero luminosity $L_i = 0$, we perform the integration from the centre with given ρ_c and $T_c = T_f$. We then obtain the mismatch with respect to ρ and r at the fitting point m_f. Constructing a model with given L and M reduces to a search of T_{ef} and ρ_c satisfying the fitting conditions for ρ and r. The method for finding T_{ef} and ρ_c is similar to the Schwarzschild method, but only two parameters are to be found here instead of four in the Schwarzschild method (see Chap. 6, Vol. 1).

The initial composition adopted in [1005] is $x_{H0} = 0.7$, $x_{He0} = 0.26$, $x_{Z0} = 0.04$, the mixing length $l = H_p$. Evolutionary tracks are shown in Fig. 9.31. The obtained models have a strongly rarefied, almost totally convective envelope with a large region of inverse gradient of density. All evolutionary tracks turn to the left at a high luminosity, and $T_{ef,min} = 2{,}300 - 2{,}400$ K. A definite linear dependence of the stellar luminosity on the carbon–oxygen core mass M_{CO}, obtained in [1005] in the form

$$L/L_\odot = 59100\,(M_{CO}/M_\odot - 0.51), \tag{9.99}$$

is a remarkable peculiarity of these models. Substituting the numerical values of constants into (9.96) gives

$$\begin{aligned}\dot{M}_c &= 0.47 \times 10^{20}(M_{CO}/M_\odot - 0.51)\,\text{g cm}^{-1}\\ &= 0.71 \times 10^{-6}(M_{CO}/M_\odot - 0.51)M_\odot/\,\text{yr}.\end{aligned} \tag{9.100}$$

The core mass and stellar luminosity increase exponentially with time, the duration of the core growth phase before ^{12}C ignition being several million years. The mass of the matter between shells decreases on the AGB with time from $0.015M_\odot$

at $M_{CO} = 0.6 M_\odot$ to $5 \times 10^{-6} M_\odot$ at $M_{CO} = 1.39 M_\odot$. A calculation scheme for evolution at the AGB phase with suppression of thermal flashes, based on neglecting the gravitational energy, has been considered in [953].

The AGB evolution has been calculated in [770, 772] using the Henyey method. For an envelope with $T_m \leq 10^6$–10^7 K, the system of equations of stellar structure (9.94), (9.97), (9.98), (9.72), and (9.73) in Problem 1 after Sect. 9.2 has been integrated inward to $T = T_m$ for $L =$ const. and $\epsilon_{gr} = 0$ similar to [1005]. At $T = T_m$, this system has been supplemented with (9.89) and (9.90). Contrary to [1005], ϵ_{gr} has been taken here from (9.91). In such a more complete form, the system is integrated further to the fitting point in the core at $m = m_f$. The quantity $\dot{M}_c$ here is not a one-to-one function of L, as is the case in (9.96), and has to be obtained from fitting to the core which is not treated as isothermal. The core temperature at the fitting point is found as a result of this procedure as well. The results of envelope integrations have been taken as boundary conditions in determining the core structure by the Henyey method. Another simplification based on a weak dependence of parameters on the effective temperature in the range $T > T_m$ has been introduced in [772] because of a large luminosity of AGB stars. The results of calculations of tracks for initial composition $x_{H0} = 0.7$, $x_{He0} = 0.27$, $x_{Z0} = 0.03$ at $l = H_p$ are given in Fig. 9.7 [770], compare with Fig. 9.31 [1005]. The difference in tracks is likely to be due to the difference in initial compositions, because the computational accuracy is nearly the same in both cases. The dependence of the luminosity on the core mass is determined by the relation [770]

$$L/L_\odot \approx 59250\,(M_c/M_\odot - 0.52). \tag{9.101}$$

It does not depend on the total mass M and is very close to (9.99) from [1005]. For intermediate-mass stars with $M \geq 2.25 M_\odot$, the gravitational energy release becomes important, and the luminosity between flashes is given by the relation [505] (M in $M_\odot$)

$$L \approx (L_H + L_g) = 6.34 \times 10^4 (M_c - 0.44)(M/7)^{0.19}, \tag{9.102}$$

where a weak dependence on the total mass M (in $M_\odot$) is taken into account. The averaged change of the radius of AGB star is given by the approximate formula [501]

$$R = 312 (L/10^4 L_\odot)^{0.68} \left(\frac{1.175}{M}\right)^{0.31S} \left(\frac{x_Z}{0.001}\right)^{0.088} (l/H_p)^{-0.52}, \tag{9.103}$$

where

$$S = 0 \text{ for } M \leq 1.175$$
$$1 \text{ for } M > 1.175. \tag{9.104}$$

9.3.4 Thermal Flashes in Helium-Burning Shell

Consecutive evolutionary computations for a star with $M = 1M_{\odot}$ without simplifications introduced in the preceding subsection have resulted [884] in the discovery of the thermal instability of a helium-burning shell with no degeneracy. The presence of this instability is a common property for stars on the major part of the AGB. A detailed theoretical explanation for this phenomenon is given in [884] (see also Sect. 13.2). The possibility for thermal instabilities to develop in thin degenerate burning shells was first predicted by Gurevitch and Lebedinski as early as 1947 (see [440])[7].

A detailed investigation of 13 ThF in evolutionary calculations of an AGB star of $1M_{\odot}$ with initial composition $x_{\mathrm{H}} = 0.9$, $x_{\mathrm{He}} = 0.099$, $x_{\mathrm{Z}} = 0.001$, the Ledoux criterion for convection, $l = H_{\mathrm{p}}$ in the envelope and the adiabatic convection approximation in the region of nuclear burning has been made in [885]. More than 25,000 equilibrium models have been constructed to cover 4×10^6 yr of the evolution of the star. The first flash occurred when the middle of the helium shell was at $m_{\mathrm{He}} = 0.465M_{\odot}$, and the last of all calculated flashes at $m_{\mathrm{He}} = 0.539M_{\odot}$. The respective masses for the middle M_{c} of the hydrogen shell have been 0.55 and $0.60M_{\odot}$. The subsequent evolution may comprise a significantly larger number of flashes.

The major role in luminosity at quiet evolutionary phases belongs to hydrogen burning because of its high caloricity. The energy release in flashes due to helium burning may be five orders of magnitude higher than the hydrogen burning. The overall characteristics of thermal flashes are as follows. The central temperatures at quiet evolutionary phases are $(1.92\text{–}2.14) \times 10^8$ K, in flash peaks $(1.84\text{–}2.04) \times 10^8$ K; the central densities are $(4.35\text{–}7.08) \times 10^5\,\mathrm{g\,cm^{-3}}$ and $(4.11\text{–}6.64) \times 10^5\,\mathrm{g\,cm^{-3}}$ respectively, so the core expands and cools slightly during ThF. In the middle of the helium shell, the temperatures are $(1.26\text{–}1.29) \times 10^8$ K during quiet phases, and $(1.69\text{–}2.54) \times 10^8$ K during flashes, the respective densities are $(1.38\text{–}3.46) \times 10^4\,\mathrm{g\,cm^{-3}}$ and $(4.06\text{–}4.88) \times 10^3\,\mathrm{g\,cm^{-3}}$. Variations of helium-burning rate L_{He} during the first nine flashes are shown in Fig. 9.32; L_{He}, L_{H} and stellar luminosity L during the ninth flash are given in Fig. 9.33 from [885].

Each flash comprises two or three peaks with time intervals between them varying from 30,000 yr for the first flash to 1,000 yr for the 13th flash. The time width of the main peak decreases from 300 to 1 yr, and the height of the peak, L_{He}, varies from $\sim 10^5$ to $\sim 10^7 L_{\odot}$. It should be noted that the intensity of flashes increases with time, the convection in the burning zone grows stronger and, from the ninth flash on, the convection reaches hydrogen-rich regions, leading to substantial changes in chemical composition that result from participation of free neutrons produced in the reaction $^{12}\mathrm{C}(\mathrm{p},\gamma)^{13}\mathrm{N}(\beta^+\nu)^{13}\mathrm{C}(\alpha,\mathrm{n})^{16}\mathrm{O}$.

---

[7] In some papers (e.g., [885]), the term "flash" denotes what we call here "flash peak", while the "relaxation cycle" is identical to our "flash". Other terms may also be encountered in texts.

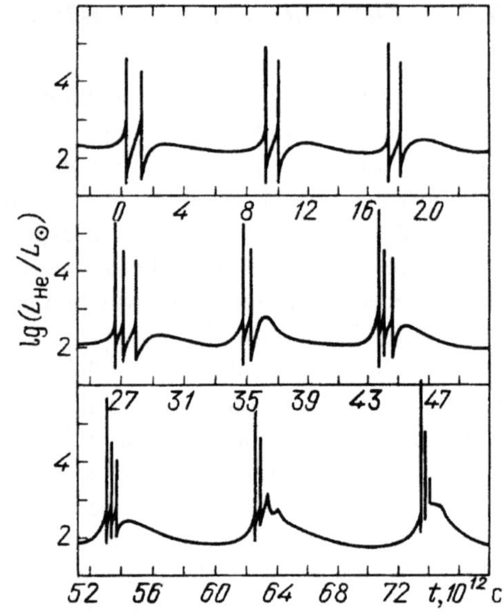

Fig. 9.32 Helium-burning rate L_{He} as a function of time during the first nine flashes caused by the thermal instability of helium-burning shell in a population II star of $M = M_{\odot}$ ($x_H = 0.9$, $x_{He} = 0.099$, $x_Z = 0.001$). The mass of the core with exhausted hydrogen varies from the first to the 13th flash from 0.554 to 0.6 $M_{\odot}$, from [885]

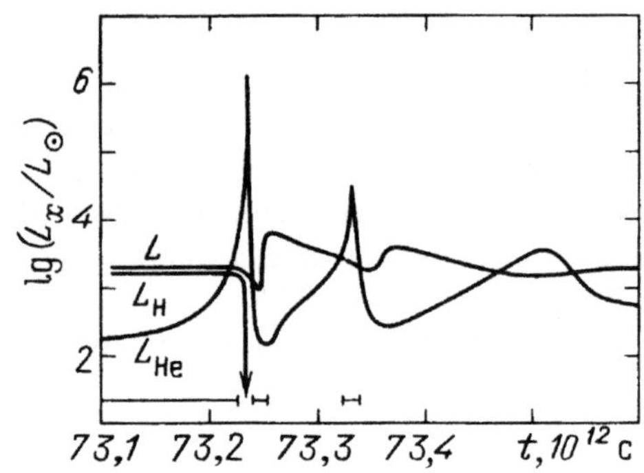

Fig. 9.33 The helium, L_{He}, and hydrogen, L_H, burning rates and surface luminosity of star during the ninth flash, see Fig. 9.32. The instability regions are labeled by |—|, from [570]

The evolution of population II stars with initial $x_H = 0.732$, $x_{He} = 0.266$, $x_Z = 0.001$ and masses $M = 0.6$ and $M = 0.8 M_{\odot}$ has been calculated in [954] up to the AGB, where few ThF have been studied in detail. The results for variations of L, L_H and L_{He} coincide qualitatively with [885]. It was first pointed out that the effective temperature did not vary significantly over the flash: $\Delta \lg T_e \leq 10^{-2}$, so that the star moves nearly vertically in the HR diagram (see Fig. 9.35 from [500]). ThF occurs only in stars with $M \geq 0.52$–$0.53\,M_{\odot}$, and the total number of flashes increases with increasing stellar mass. The evolution of a model with $M = 0.6 M_{\odot}$ has been calculated in [500]. Following the AGB, this model has experienced only ten flashes, shown in Fig. 9.34, before leaving the AGB and moving towards the white dwarf region (for more detail, see Sect. 9.3.6). Between the first and tenth flashes, for flash maxima, the core mass changes from 0.5272 to 0.5901 $M_{\odot}$, and the helium-burning rate L_{He} increases from 3.3×10^4 to $2.9 \times 10^7 L_{\odot}$. The corresponding luminosity

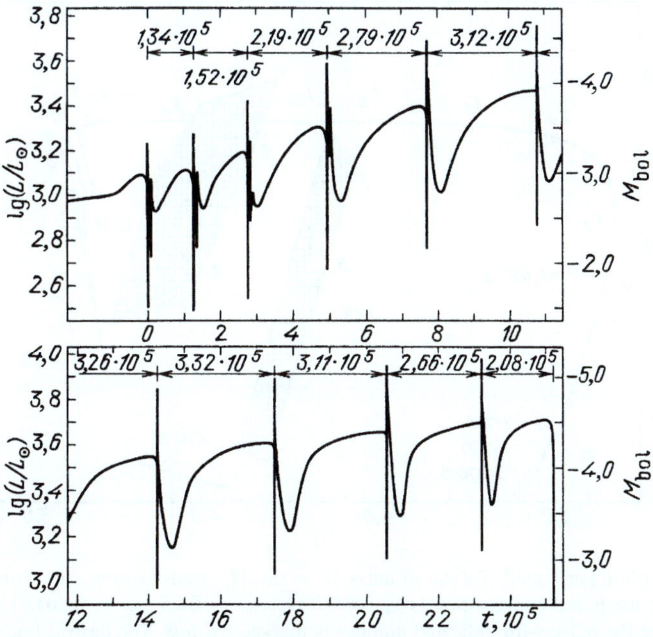

Fig. 9.34 Time dependence of luminosity and bolometric stellar magnitude during the thermally flashing phase for a model of $M = 0.6\,M_\odot$ and initial composition $x_{\rm H} = 0.749$, $x_{\rm He} = 0.25$, $x_{\rm Z} = 0.001$. The numbers above the horizontal arrows indicate the time that elapses between flashes in years, the mixing length is taken to be $l = 1.5\,H_{\rm p}$, $M_{\rm bol} = -2.5\,{\rm lg}(L/L_\odot) + 4.77$, from [500]

changes are significantly less steep (see Fig. 9.34). The time structure of flashes for a model with $M = 0.6 M_\odot$ is qualitatively the same as for $1\,M_\odot$ (compare Figs. 9.33 and 9.34), but from the sixth flash onwards, the subsequent flashes become single-peaked, which might be due to the low mass of the envelope surrounding $M_{\rm c}$.
E – AGB is the early asymptotic giant branch (without flashes) where the star arrives after the core helium has been exhausted. Hatched are the regions MS, the MS, and CHB, core helium burning, where are given approximate evolutionary tracks for stars with $M = 3, 5, 7\,M_\odot$ and $x_{\rm H} = 0.719$, $x_{\rm He} = 0.28$, $x_{\rm Z} = 0.001$. ZAHB is the zero-age horizontal branch corresponding to the onset of the helium burning in static core. The *dashed line* on the left represents a star of the constant radius $R = 0.0285\,R_\odot$ (hot white dwarf), from [500, 502].

The dependence of the time interval between flashes $\Delta t_{\rm ThF}$ on the core mass $M_{\rm c}$ has been studied in [779, 780]. The following approximate dependence is obtained in [780] (for $x_{\rm H} = 0.7$, $x_{\rm He} = 0.27$, $x_{\rm Z} = 0.03$):

$$\lg \Delta t_{\rm ThF} \approx 3.06 - 4.5\left(\frac{M_{\rm c}}{M_\odot} - 1\right). \qquad (9.105)$$

Here, $t_{\rm ThF}$ is measured in years.

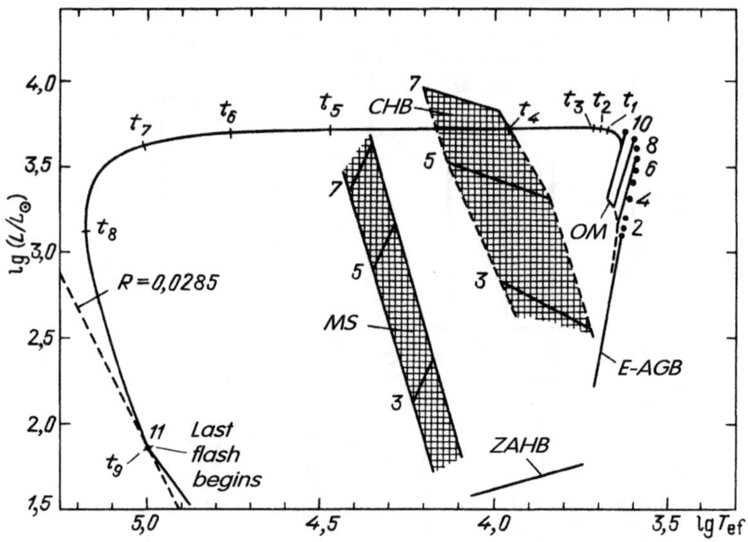

Fig. 9.35 Evolutionary track of a star of mass $M = 0.6\,M_\odot$ transforming into white dwarf, from E – AGB on; the initial composition is $x_\mathrm{H} = 0.749$, $x_\mathrm{He} = 0.25$, $x_Z = 0.001$. The location of the star before the pulse with indicated number is marked by *dots*. The dashed line OM contours the extended dips in luminosity during flashes (Fig. 9.34); shown are the tracks for flashes no. 7, 9 and 10. Evolution times t_1 at tips and corresponding masses in hydrogen-rich envelopes are in Table 9.10

Table 9.10 Table to Fig. 9.35

i	1	2	3	4	5	6	7	8	9
t_i (10^4 yr)	−3.0	−2.0	−1.0	0.5	0	0.5	1.0	1.5	1.86
$M_{e,\mathrm{H}}$ ($10^{-3}\,M_\odot$)	3.15	2.53	1.84	1.47	1.13	0.80	0.49	0.27	0.27

The dependence of the stellar luminosity in flash peak L_m on the core mass M_c obtained from numerical calculations is given in [1048]:

$$\frac{L_\mathrm{m}}{L_\odot} = 97,000\left(\frac{M_\mathrm{c}}{M_\odot} - 0.52\right). \tag{9.106}$$

9.3.5 *The Mass Loss in AGB Stars*

An overwhelming evidence for mass losses in low- and intermediate-mass (LI) stars in the evolutionary process is provided by observations of clusters that comprise stars of nearly equal age. The turning point off the MS for the Pleiades cluster corresponds to $M = M_\mathrm{t} > 6M_\odot$. On the other hand, this cluster has been found to contain single white dwarfs which clearly originate from stars with initial mass

$M > M_t$ and, consequently, lose most of their mass during the evolution [1038]. All single LI-stars with $M < \sim 8 M_\odot$ thus transform into white dwarfs losing up to $7 M_\odot$ during the evolution. The major loss occurs during the AGB-phase. When the envelope is ejected from the star, a core remains in the form of a hot star which heats the escaping matter so as to make it observable in the form of a PN. After cooling, the core transforms into a white dwarf. The evolutionary scheme red giant – PN – white dwarf was first suggested by Shklovski in 1956 [904] on the basis of observational data studies.

Several mechanisms for mass loss by LI-stars have been proposed so far, but the problem is not completely resolved as yet. The possibility for an extended red-giant envelope to be ejected owing to a release of hydrogen ionization energy that makes it possible for the envelope to escape to infinity has been considered in [652, 786]. The role of this mechanism cannot be estimated rigorously without dynamical calculations of such an ejection.

Nevertheless, investigation with the help of the conventional static code led to important results [1027]. It was etablished, probably for the first time in evolutionary calculations, that ThF on AGB leads to the development of an instability resulting in a thermal run-away and loss of the hydrogen envelope. This instability is acting in stars whose initial masses are in the range 0.8–5 $M_\odot$, which coincides with the results of the stability analysis made in [786]. This instability is attributed to the energy of the hydrogen recombination, which makes positive the total energy of the envelope

$$\Delta W = \int_{M_c}^{M} \left(-\frac{GM}{r} + E \right) \mathrm{d}m \geq 0 \tag{9.107}$$

and makes it energetically preferable for a loss of the hydrogen envelope to infinity. In [453], the relation (9.107) with $\Delta W = 0$ was used for estimation of the mass of the white dwarf remaining after the loss of its envelope. In [1027] this instability, leading to the mass loss, was obtained in the calculations. It is interesting to mention possible reasons, why this instability had not been realized in previous numerous evolutionary calculations of AGB stars [1027].

1. Use of the old opacities.
2. Neglecting the energy production in the outermost layers, leading to loss of the energy of the hydrogen recombination in these layers.
3. Crude resolution in time, which smears the released recombination energy and makes it ineffective.
4. Attributing of such an instability to the instability of the numerical scheme, instead of having a physical basis, which is connected with numerical difficulties at the evolutionary calculations of the AGB stars.

The matter outflow under the action of light pressure at a large opacity in the envelope may represent another mass-loss mechanism [385, 599, 600]. This mechanism is similar to that examined in [172] for the case of giants (see Sect. 9.2.4, also [171]). Note that equations in [599] remain valid even when the critical point is at an optical thickness $\tau < 1$.

A mechanism for mass loss arising from pulsation instability and shock generation which leads to an occasional ejection of matter is now believed to be very plausible. Numerical calculations of such an ejection have been made in [368, 981, 1046, 1047] on the basis of one-dimensional non-stationary equations of hydrodynamics. Calculations carried out for fairly realistic conditions have revealed a sufficient efficiency of this mechanism. An additional mass loss produced by the pressure on dust grains is included in calculations in [368, 981]. Observational arguments for the scheme proposed by Shklovski [904] are given in [601].

The above mechanisms of mass loss have not been included so far as a component in the evolutionary calculation scheme. Contrary to the first group of quasistationary mechanisms for which a scheme similar to Sect. 9.2.4, or code based on the relaxation Henyey-type method with inclusion of the dynamical terms (see Chap. 6, Vol. 1) could be used, in the case of the third mechanism the principles of its self-consistent inclusion in the evolutionary scheme are obscure. Lamers [610] has obtained the empirical dependence of $\dot{M}$ on stellar parameters:

$$\begin{aligned} \lg F_{\mathrm{m}} &= -5.23(\pm 0.06) + 4.60(\pm 0.45)\lg(T_{\mathrm{ef}}/3\times 10^4\,\mathrm{K}) \\ &\quad - 0.48(\pm 0.11)\lg(g_{\mathrm{ef}}/10^3)\,\mathrm{g\,cm^{-2}\,s^{-1}}, \end{aligned} \tag{9.108}$$

$$\begin{aligned} \lg \dot{M} &= -4.83(\pm 0.28) + 1.42(\pm 0.40)\lg(L/(10^6 L_\odot)) \\ &\quad - 0.99(\pm 0.47)\lg(M/30M_\odot) \\ &\quad + 0.61(\pm 0.13)\lg(R/30R_\odot) M_\odot/\mathrm{yr}. \end{aligned} \tag{9.109}$$

Another relation is given by Reimers (see [505], for observational data see in [826]):

$$\dot{M} = -4\times 10^{-13}\eta\left(\frac{L}{L_\odot}\frac{g_\odot}{g}\frac{R_\odot}{R}\right)\frac{M_\odot}{\mathrm{yr}}, \quad \eta\sim 1. \tag{9.110}$$

9.3.6 Evolution with Mass Loss: From AGB to White Dwarf State

The evolutionary track of a star with mass $M = 0.6M_\odot$ evolving from AGB to white dwarf state is represented in Fig. 9.35 from [500, 502]. A low mass has allowed accurate calculation of the evolution, and calculations being carried through for ten helium-shell flashes. The 11th and last flash occurs when the star is already moving at $M_{\mathrm{c}} = 0.5997\,M_\odot$ towards the white dwarf region. As the position of a star on the AGB is determined mainly by the mass M_{c}, this track, with the exception of the number and some properties of flashes, may serve to describe the evolution of a star with a larger initial mass.

Analyzing observational data of AGB stars, planetary nebulae and their cores gave rise to the conception of two types of mass loss on the AGB: a quiescent stellar wind with a rate given by the empirical relations (9.109) or (9.110), and the rapid ejection phase (superwind) occurring immediately before the departure of the star from the AGB [505]. Observations have yielded the formula (see [505]) relating the mass of a PN M_{PN} with that of its nucleus (PNN) M_{PNN}:

$$M_{PN} \approx b(1.69 - 8.09\,M_{PNN} + 11.69\,M_{PNN}^2 - 4.34\,M_{PNN}^3). \tag{9.111}$$

Here, $b = 0.5$–1 is an empirical coefficient. During the evolution of a star along the AGB with quiescent mass loss, M_c increases while the hydrogen envelope mass M_e decreases. A rapid envelope ejection is assumed to take place when the equality $M_e = M_{PN}$ becomes valid; this envelope, after being illuminated by a hot central star with mass M_{PNN}, will transform later into PN. Using (9.110) for finding $\dot{M}$ and applying (9.111) enables us to obtain the dependence of the maximum value of $\dot{M}$ during the evolution on initial stellar mass M_i for the case of quiescent outflow. This dependence is given in Fig. 9.36 from [505]. For masses M_i preceding the peak in Fig. 9.36, when $M_i < M_W$, the quiescent outflow ends with a superwind starting at $M_e = M_{PN}$ from (9.111). At a peak value $M_i = M_W$, the core has time at the point of rapid ejection to attain mass $M_{Ch} = 1.39\,M_\odot$ corresponding to the onset of explosive carbon burning. For larger masses, $M_i > M_W$, the equality $M_c = M_{Ch}$ is reached earlier than $M_e = M_{PN}$. In this case, the quiescent outflow phase is followed by the carbon ignition, the maximum quiescent outflow rate $\dot{M}$ decreases with increasing M_i by reason of gR growth in (9.110) at the same L.

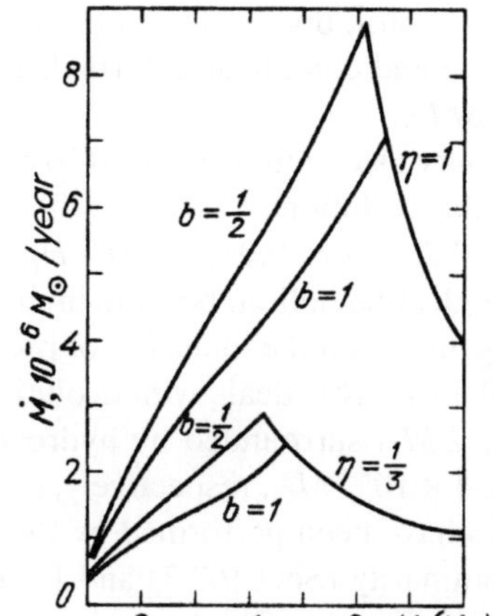

Fig. 9.36 The maximum quiescent mass loss rate $\dot{M}$ for a model with initial composition $x_H = 0.7$, $x_{He} = 0.28$, $x_Z = 0.02$ on the AGB, reached before the carbon ignition and explosion at $Mc = M_{Ch}$ for $M_i > M_W$ (M_W is the mass corresponding to peak), or before the rapid envelope ejection and white dwarf formation for $M_i < M_W$; $\dot{M}$ has been evaluated from (9.110), the rapid envelope ejection occurs when the envelope mass falls down to $M_e = M_{PN}$ from (9.111). The quantity M_W depends on initial composition, uncertainties in the relation (9.110) and (9.111) for $\dot{M}$ and M_{PN} and other factors pointed out in Sect. 9.1, from [505]

Mass losses taken into account, and the track in Fig. 9.35 corresponds to a model of mass $\sim 2M_\odot$. The star loses $\sim 1.2M_\odot$ during the quiet phase, whereupon on reaching $M_e = M_{PN}$, it ejects rapidly $\sim 0.2\,M_\odot$. The lifetime of the star from the onset of rapid ejection to its appearance on the horizontal portion of the track may be well below the time indicated in Fig. 9.35, which is determined by the hydrogen-burning rate.

The evolution of a star with $M_i > M_W$ is likely to end with the thermal instability development and an explosive process (probably, supernova explosion); white dwarfs arise most probably for $M_i < M_W$. The dependence of the mass of the white dwarf m_f on initial stellar mass M_i and parameter η from (9.110) for $x_{He} = 0.28$, $x_Z = 0.02$ is given in [505]:

$$m_f \approx 0.53\,\eta^{-0.082} + 0.15\,\eta^{-0.35}(M_i - 1). \tag{9.112}$$

The value M_W equals the value of M_i from (9.112) at $m_f = M_{Ch} = 1.39\,M_\odot$ and is determined (with the inclusion of the dependence on b) by [505]

$$M_W = 1.0 + 9.33\,\eta^{0.35} - 3.53\,\eta^{0.27} + 0.8(b - 1.0)\,.$$

It then follows that $M_W/M_\odot = 4.7, 4.3, 8$ for $(\eta, b) = (1/3, 1), (1/3, 1/2), (2, 1)$. The value $M_W = 8M_\odot$ (see Sect. 9.3.5) fits observations better, but at $\eta = 2$ giants are not present for $M_i \leq 1.15\,M_\odot$; therefore, this value is not consistent with observations. Such a discrepancy seems to originate from insufficient accuracy of the relation (9.110) with $\eta =$ const. during the evolution. Including variability of η or using other, more accurate and complicated formulae for $\dot{M}$ is likely to yield a consistent value of $M_W = 8M_\odot$.

One of the first evolutionary calculations from AGB to the white dwarf state has been made in [838], where the evolution of a model with constant mass $M = M_i = 0.85\,M_\odot$; $x_H = 0.7$, $x_{He} = 0.29$, $x_Z = 0.01$, $l = 0.7\,H_p$ has been studied. The ThF have been suppressed by neglecting helium burning in the shell.

A more realistic investigation with the inclusion of precedent mass loss has been made by Paczynski in [770, 774]. This deals with evolutionary calculations of stars with $M_c = 0.6$, 0.8 and $1.2\,M_\odot$ surrounded by hydrogen envelopes with $M_e = 1.2 \times 10^{-3}$, 1.4×10^{-4} and $4 \times 10^{-6}\,M_\odot$, respectively, that is, of almost completely stripped cores. Calculations have been performed by the ordinary Henyey method without requirement of stationarity (Sect. 9.3.3) and for initial composition $x_H = 0.7$, $x_{He} = 0.27$ $x_Z = 0.03$ at $l = H_p$. The logarithms of effective temperature of initial models have been arbitrarily taken to be 4.191, 4.202 and 4.728, respectively. According to [786], where the role of ionization energy in the envelope ejection has been studied, the masses of initial models are assumed to be 0.8, 1.5, and 3.0, respectively, but this relationship is uncertain. The calculated evolutionary tracks are shown in Fig. 9.37 from [774]. The selected envelope masses are so low that the models start moving immediately towards the white dwarf region with, however, one helium-shell flash experienced by all of them during this period. This final flash discovered in [770, 774] is accompanied by a quick loop-like motion of the model

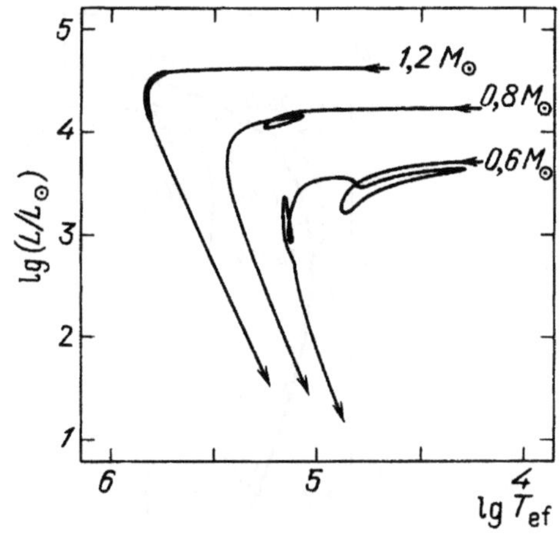

Fig. 9.37 Evolutionary tracks for models with masses of 0.6, 0.8 and 1.2 $M_\odot$ in PN phase, from [774]

in the HR diagram, particularly marked on the 0.6 $M_\odot$ model track. The phase of expansion lasts several hundred years from the left to the right end of the loop, while returning takes $\sim 10^3$ yr. The quick changes (decreasing T_{ef}) observed in FG Sagittae have been associated in [770, 774] with the effect of this flash.

Following Paczinski's calculations, the final flash on the model path from AGB to the white dwarf sequence has been studied by many authors [504, 874, 876, 955]. An example of a track with indication of time intervals is given in Fig. 9.38 from [504]. The dependence of the final flash shape on the point on the evolutionary track (ejection phase) corresponding to the onset of the rapid envelope ejection (superwind) generating a PN, and on the mass of the hydrogen envelope left behind by the superwind has been studied in detail in [501, 1048]. As pointed out also in [874], these tracks are very different from each other. The expected uncertainty in the distribution of models over ejection phases and residual envelope masses makes plausible all evolutionary tracks obtained in [501, 1048] for the final flash.

Evolutionary calculations including an empirical dependence for $\dot{M}$, similar to analogous calculations for massive stars (see Sect. 9.2.4), was first performed in [457]. This investigation deals with the effect produced on the evolution by a rapid mass loss $\sim 10^{-3} M_\odot/$ yr starting at a peak of one ThF. Over a short period of time a star with $1 M_\odot$ leaves behind a core of mass 0.65 $M_\odot$. During the subsequent evolution, this core remains very bright ($\geq 10^3 L_\odot$) over a time substantially exceeding the lifetime of the known planetary nebulae. Evolutionary calculations for more realistic laws of mass loss occurring in the transition from AGB to the white dwarf state have been made in [874, 876]. Calculations have been compared in [875, 877] with observations of PN nuclei. The evolution of stars with $M = 0.8$ and $1 M_\odot$ and initial composition $x_H = 0.739$, $x_{He} = 0.240$, $x_Z = 0.021$ from AGB to the white dwarf configuration has been calculated in [876] with the inclusion of the quiescent and rapid phases of mass loss. To obtain remnants with mass $M < 0.6 M_\odot$, the onset of a rapid mass loss has been specified at sufficiently early time: before the onset of ThF at $L = 1400 L_\odot$ for $M = 0.8 M_\odot$ and after the

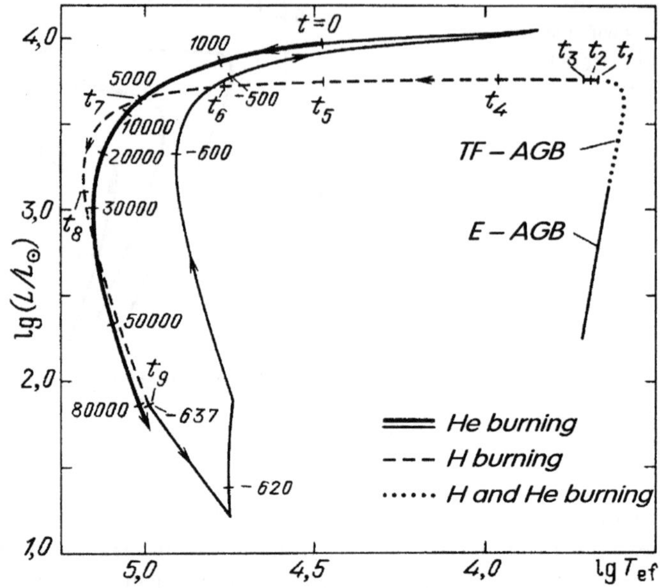

Fig. 9.38 The final helium flash in a star of mass $M = 0.6\,M_{\odot}$ represented by the loop in the HR diagram, the onset of the flash corresponds to the time $t = -637$ yr. Shown also is a track before the flash, starting from E – AGB (Fig. 9.35). Times t_i on this track are the same as in Fig. 9.35, from [504]

Fig. 9.39 Mass loss rate $|\dot{M}|$ adopted in calculations [876] versus $T_{\rm ef}$ (*solid lines*) the increase of core mass, $\dot{M}_{\rm c}$, due to hydrogen burning, is given for comparison (*dashed lines*)

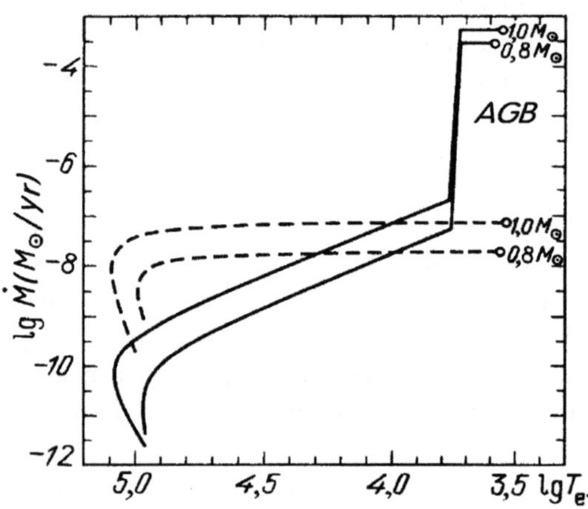

fifth ThF at $L = 4500\,L_{\odot}$ for $M = 1 M_{\odot}$. Figure 9.39 shows the dependence $\dot{M}(T_{\rm ef})$ adopted in the computations and the rate of increase of the core mass $\dot{M}_{\rm c}$. The value $\dot{M} = (2\text{–}4) \times 10^{-4} M_{\odot}/$ yr has been adopted for $T_{\rm ef} < 10^{3.7}$ K, while for larger $T_{\rm ef}$ Reimers' law (9.110) with $\eta = 1$ has been used. The results of calculations are presented in Fig. 9.40. The rapid mass loss causes the models to move rapidly to the left in the HR diagram. Small departures from thermal equilibrium

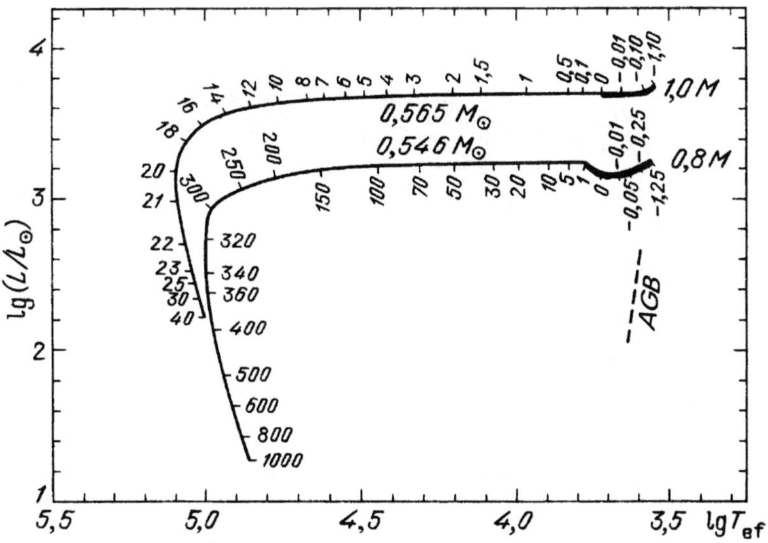

Fig. 9.40 Evolutionary tracks for stars moving from the AGB to a white dwarf state, from [876]. The numbers indicate the model age in 10^3 yr, zero age is assumed to be at $T = 5{,}000$ K. The thick portions of the tracks represent phases of rapid mass loss. Indicated also are the initial (on the right) and final model masses

give rise to a moderate fall in luminosity (~20% for 0.8 $M_\odot$). The rapid mass loss lasts ~10^3 yr, whereupon the models of 0.8 and 1 $M_\odot$ leave stars with masses 0.546 and 0.565 $M_\odot$, respectively, that is, almost completely stripped cores evolving to a white dwarf configuration. Comparing times of cooling for two models in Fig. 9.40 yields a very steep dependence on final mass. Increasing the mass by 0.019 $M_\odot$ leads to a decrease in the cooling time by ~20 times (see also [774, 874], Figs. 9.35 and 9.38).

The linear relations (9.99) or (9.101) between L and M_c are not relevant for $M_c < 0.6\, M_\odot$ because the contribution of helium burning to luminosity is substantial for this range: ~30% for 0.546 $M_\odot$. In addition, the relation (9.105) for Δt_{ThF} does not hold for small cores. The time interval between flashes is $\sim 7 \times 10^4$ yr for $M_c = 0.56\, M_\odot$ and increases to ~10^5 yr for $M_c = 0.57 \sim 0.58 M_\odot$. Only for larger M_c does the time Δt_{ThF} decrease with increasing M_c in accordance with (9.105).

Besides the tracks shown in Fig. 9.40, a track with a final ThF providing a loop in the HR diagram has been obtained in [876] by means of a slight change in initial conditions. This track, given in Fig. 9.41, arises when a 0.553 $M_\odot$ remnant results from the superwind action during the evolution of a star with $M = 1 M_\odot$. The appearance of the loop in the HR diagram depends on the phase of the time interval between two successive ThF when the star starts moving away from the AGB. If the onset of this motion occurs soon after the last ThF, then the star has a sufficient time up to the next ThF for the hydrogen envelope to pass through the shell into the core so that the next flash does not occur. If, on the contrary, the departure from the

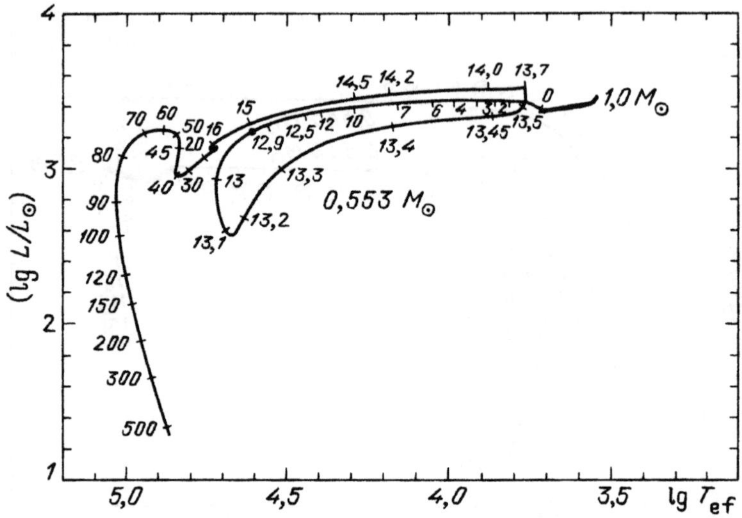

Fig. 9.41 The same as in Fig. 9.40 but now for a final mass 0.553 $M_\odot$, when the final helium flash occurs far from the AGB (see Fig. 9.38). The peak helium-burning rate with $L_{He} = 6.3 \times 10^4\, L_\odot$ occurs at *filled circle*. Visible also is a very small loop at $T_{ef} = 5 \times 10^4$ K caused by a second, minor helium flash with $L_{He} \leq 1.8 \times 10^3\, L_\odot$, from [876]

AGB is not far from the next ThF, then the hydrogen envelope has no time to pass quiescently into the core, and the phenomenon of final flash arises. The probability for a post-AGB model to experience a final flash is estimated to be some 20% [876]. Besides a possible explanation provided by the final flash for phenomena observed in FG Sge, it may give rise to additional mixing due to convection development and enrichment of surface layers with helium, which was observed in nuclei of the planetary nebulae A30 and A78 [501,504,876], and may also take off the star in the region of low T_{ef} and cause an ejection of the second envelope of lower mass in the regime of superwind [501].

9.3.7 On Mixing on the AGB and in Neighbourhoods

The problem of convection, and the mixing it leads to, is one of the most difficult and obscure items in the theory of stellar evolution. The reason for this is the absence of a reliable theory of convection and far-reaching simplifications in the description of convection in evolutionary problems. The approximation of adiabatic convection is usually used for stellar interiors including shells, and the effects of non-adiabaticity, in addition to those of non-stationarity and overshooting, are not considered. Conclusions based on mixing and convection development are usually tested against observations and taken as satisfactory only if corroborated by some observational fact referring mostly to chemical composition of stellar atmospheres.

The results below are satisfactory from this standpoint but one should bear in mind that the theory underlying them is not really reliable, and attempts to explain the same observational data from other theoretical models are not rare.

The transfer of convection from the envelope into deep regions of star is the main mechanism of mixing leading to enrichment of outer shells with heavy elements. The first mixing occurs when hydrogen has been exhausted in the central core and the star has evolved into the red-giant region, and involves stars with initial masses M_i [505]

$$M_i < M_i^{\max} = 8.95 + 69.4(x_Z - 0.02) - 31.3(x_{\mathrm{He}} - 0.28)(M_\odot). \tag{9.113}$$

The second mixing takes place after helium has exhausted in the centre, when the star arrives on the AGB, and involves stars with initial masses M_i [505]

$$M_i > M_i^{\min} = 4.59 + 82.5(x_Z - 0.02) - 6.88(x_{\mathrm{He}} - 0.28)(M_\odot). \tag{9.114}$$

For stars with $M_i < M_i^{\min}$ arriving on the AGB, the second mixing does not occur. According to calculations in [84], the core mass M_c before the second mixing is (for $x_{\mathrm{He}} = 0.28$, $x_Z = 0.02$)

$$M_c^{(1)} = 0.2954\, M_i - 0.5\,(M_\odot) \tag{9.115}$$

and after it

$$M_c^{(2)} = 0.0526\, M_i + 0.59\,(M_\odot). \tag{9.116}$$

For $M_i = 5M_\odot$ and the other chemical compositions

$$\begin{aligned}(x_{\mathrm{He}}; x_Z) = (0.2; 0.02);\ (0.36; 0.02);\ (0.2; 0.001)\\ (0.28; 0.01);\ (0.28; 0.001);\ (0.28; 0.02)\end{aligned} \tag{9.117}$$

these masses are, respectively, (in $M_\odot$)

$$\begin{aligned}(M_c^{(1)}; M_c^{(2)}) = (0.86; 0.855);\ (1.176; 0.900);\ (1.129; 0.906)\\ (1.18; 0.967);\ (1.283; 0.930);\ (0.977; 0.853)\,.\end{aligned} \tag{9.118}$$

The last values in (9.118) are obtained from (9.115) and (9.116). Thermal flashes on the AGB start soon after the second mixing. They lead, in turn, to the third mixing.

During ThF there occurs a formation in a helium shell of a convective layer with mass ΔM_{csh} and maximum temperature $T_{\mathrm{csh}}^{\max}$ such that [505]

$$\begin{aligned}\lg \Delta M_{\mathrm{csh}} &= -1.835 + 1.73\, M_c - 2.67\, M_c^2\,,\\ T_{\mathrm{csh}}^{\max} &= [3.1 + 2.85(M_c - 0.96)] \times 10^8\,\mathrm{K},\\ &\qquad \text{for } M_c > 0.9, \text{ all } M \text{ are in } M_\odot.\end{aligned} \tag{9.119}$$

The mixing during ThF may be due to both the penetration of convection from the helium shell into the hydrogen envelope, as is the case after the ninth flash in calculations [885], and its penetration from the envelope into deep regions. The peculiarity of this kind of mixing is that the convective envelope and convective helium-burning shell never come into contact. During the evolution, they penetrate the same shell of matter alternately, thus giving rise to substantial changes in composition. This may be seen from Fig. 9.42 from [497] for $7M_\odot$, where the convective envelope penetrates the region occupied previously by the convective shell to a depth $\Delta_d = 3.9 \times 10^{-4} M_\odot$. Variations of parameters during a ThF with mixing of this kind are given in Fig. 9.43 from [495] for a star of $M = 7M_\odot$.

It should be noted that at various evolutionary phases, the occurrence of mixing and its quantitative characteristics are strongly dependent on the adopted parameters, such as mixing length l, opacity and reaction rates. The inclusion of non-adiabatic convection inside the shell, non-locality and overshooting of convection into the stable region may have a significant effect on the results. Discrepancies arise also in describing the convective zone passing through the discontinuity in chemical composition [866]. The variant of mixing we give here is thought to be satisfactory since its consequences fit observations well [505].

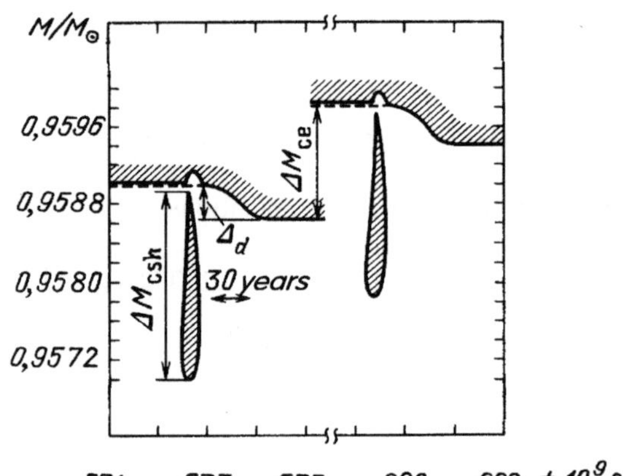

Fig. 9.42 Convective regions (hatched) during the 15th and 16th flashes on the AGB in a model of $M = 7\,M_\odot$ and initial composition $x_H = 0.7$, $x_{He} = 0.28$, $x_Z = 0.02$. The convective shell in the helium-burning shell has a maximum mass $\Delta M_{csh} = 1.98 \times 10^{-3}\,M_\odot$, its outer boundary does not touch the inner boundary of the convective envelope, but the latter does penetrate the region occupied before by the convective shell to a depth $\Delta_d = 3.9 \times 10^{-4}\,M_\odot$, thus causing an enrichment of the surface with heavy elements (dredge-up), $\Delta M_{ce} = 1.13 \times 10^{-3}\,M_\odot$ is the mass shell penetrated by the hydrogen-burning shell during the time interval between two flashes, the dashed lines indicate the boundary of the core containing no hydrogen before "dredge-up" begins, from [418,423]

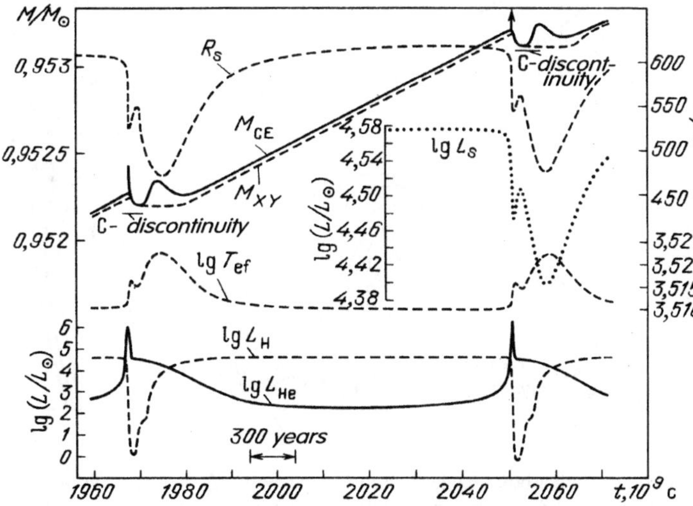

Fig. 9.43 Time variations of several characteristics in a model with $M = 7\,M_{\odot}$ and initial composition $x_{\mathrm{H}} = 0.7$, $x_{\mathrm{He}} = 0.28$, $x_{\mathrm{Z}} = 0.02$ between its seventh and eighth thermal flashes as in as AGB star (TF–AGB); L_{H} and L_{He} are the helium- and hydrogen-burning rates, L_{s} and R_{S} are the model luminosity and radius, T_{ef} is the effective temperature, M_{XY} is the mass in the core containing no hydrogen, M_{CE} is the mass inside the convective envelope, C-discontinuity determined the mass in the core containing no helium, from [495, 505]

9.3.8 Thermal Instability in Degenerate Carbon Core

If the carbon core of an AGB star reaches the Chandrasekhar limit $\sim 1.39 M_{\odot}$ (see Sect. 11.2), then the equilibrium is no longer supported by the degenerate electron pressure. The core starts contracting, the temperature rises to the carbon ignition point, and degeneracy gives rise to a thermal flash that may release, under certain conditions, an energy of the order of supernova energy. The time of carbon ignition is then determined by the counterwork of the heat release in the $^{12}\mathrm{C}(^{12}\mathrm{C},\gamma)^{24}\mathrm{Mg}$ reaction and energy losses due to neutrino, mainly plasma neutrino, emission. Evolutionary changes in models with no mass loss before $^{12}\mathrm{C}$ ignition are given in Figs. 9.44–9.47 from [773] for masses $M = 3, 5, 7 M_{\odot}$. For all these models, the line $\epsilon_{\nu} = \epsilon_{\mathrm{CC}}$ with neutrino losses by URCA processes, plasma neutrino, etc. (see Sects. 4.3, 5.2, Vol. 1), which is natural to be taken for the onset of thermal instability, is first crossed in the centre of the star, though for the $7 M_{\odot}$ model the regions far from the centre have been close to this line (Fig. 9.47). The core structure at the onset of instability is nearly the same for all the models and is given in Table 9.11 from [773].

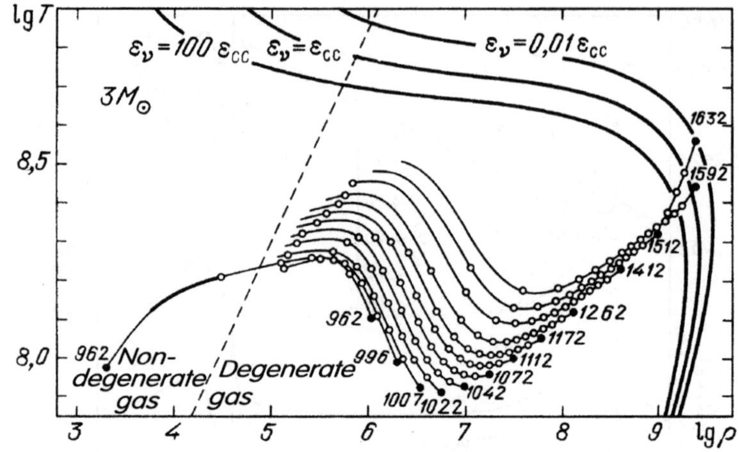

Fig. 9.44 Evolution of the carbon–oxygen core of a star with $M = 3\,M_{\odot}$ on the ($\lg T - \lg \rho$) plane. Each model is represented by a curve labeled by its number, the circles separate mass shells with $0.1\,M_{\odot}$. The location of the helium-burning shell is shown as a heavy portion of the curve. Shown also are the lines of a constant ratio $\epsilon_{\nu}/\epsilon_{CC}$, ϵ_{ν} is the rate of neutrino losses, ϵ_{CC} is the ^{12}C burning rate for $x_{^{12}C} = 0.5$. Carbon ignites when the model line crosses the line $\epsilon_{\nu} = \epsilon_{CC}$, from [773]

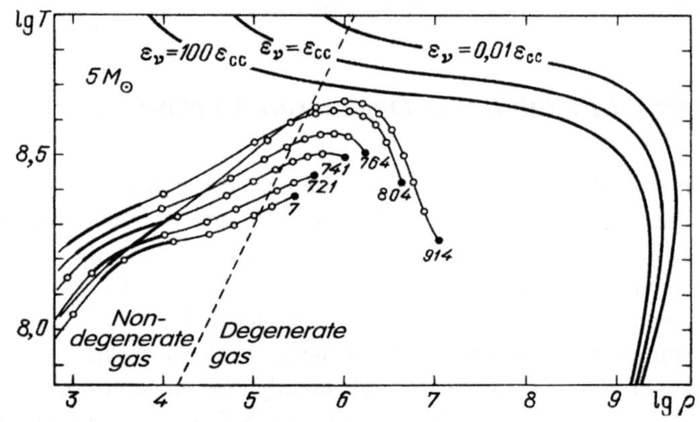

Fig. 9.45 The same as in Fig. 9.44 for $M = 5\,M_{\odot}$, from [773]

9.3.9 Convective URCA Shells

The increase in neutrino luminosity due to URCA shells may significantly change the character of instability development. Supernova explosions of a type Ia are believed to happen due to the carbon ignition in a highly degenerate matter of a medium mass star, leading to a rapid carbon burning, and total disruption of the star [47]. It was suggested in [775] that cooling of the matter in a convective region may

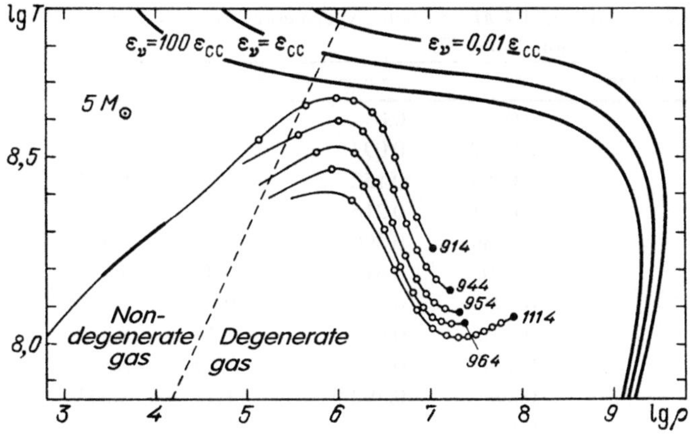

Fig. 9.46 The same as in Fig. 9.44 for $M = 5\,M_{\odot}$, from [773]

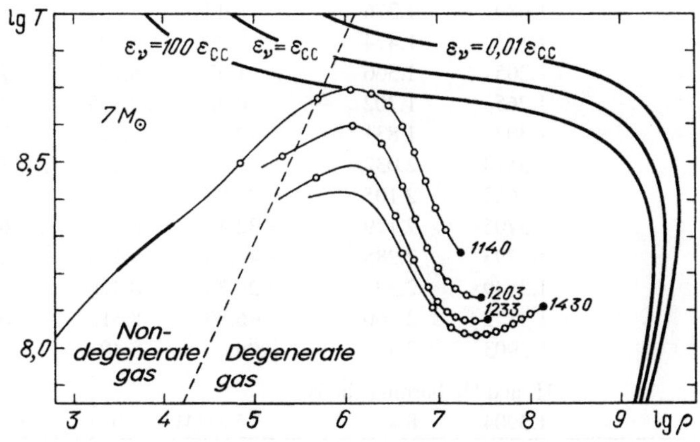

Fig. 9.47 The same as in Fig. 9.44 for $M = 7\,M_{\odot}$, from [773]

be enhanced in the presence of URCA shells, first considered in [980]. The URCA shell appears when the matter contains an isotope with a threshold Fermi energy for an electron capture, corresponding to a density less than the central one.

In the presence of convection or oscillatory motion of matter in the star, e^--capture occurs at densities higher than at the boundary $\epsilon_{\rm Fe} = \Delta_{\rm Z-1,Z} - m_{\rm e}c^2$, and e^--decay at lower densities. Hence, the neutrino losses result from beta reactions on both sides of the boundary $\epsilon_{\rm Fe} = \Delta_{\rm Z-1,Z} - m_{\rm e}c^2$. In [980], the shell of the star at $\epsilon_{\rm Fe} = \Delta_{\rm Z-1,Z} - m_{\rm e}c^2$ has been called the URCA shell. The URCA shells consisting of elements with odd A and a low threshold for electron capture have been examined in [980]: ^{35}Cl$\leftrightarrow^{35}$S with $\Delta = m_{\rm e}c^2 + 0.168$ MeV, ^{31}P$\leftrightarrow^{31}$Si with $\Delta = m_{\rm e}c^2 + 1.48$ MeV. The elements ^{21}Ne, ^{23}Na, ^{25}Mg, and ^{55}Mn have a thresholds for

Table 9.11 Models of 1.4 $M_\odot$ carbon–oxygen cores for stars with 3, 5, 7 $M_\odot$ at the time of carbon ignition, from [773]

Number of zone	$\frac{M_r}{M_\odot}$	$\frac{r}{0.001R_\odot}$	$\frac{L_r}{L_\odot}$	lg T	lg ρ
Center	0.0	0.0	0.0	8.443	9.428
2	0.0017	0.097	1.05	8.438	9.420
11	0.0065	0.152	1.86	8.431	9.408
18	0.0181	0.216	0.23	8.418	9.388
25	0.0489	0.306	0.26	8.405	9.348
31	0.110	0.414	0.47	8.390	9.285
36	0.208	0.531	0.80	8.372	9.199
39	0.295	0.617	1.13	8.357	9.125
42	0.408	0.717	1.49	8.338	9.031
44	0.498	0.792	1.72	8.323	8.953
46	0.598	0.876	2.04	8.306	8.863
48	0.706	0.968	2.20	8.287	8.759
50	0.818	1.070	2.51	8.265	8.638
52	0.923	1.175	2.51	8.243	8.507
54	1.014	1.277	2.55	8.222	8.377
56	1.089	1.376	2.47	8.204	8.247
57	1.123	1.424	2.29	8.195	8.182
60	1.205	1.566	1.67	8.175	7.987
63	1.265	1.702	0.32	8.165	7.792
66	1.307	1.831	−2.26	8.172	7.597
72	1.3513	2.032	−13.72	8.240	7.269
76	1.3620	2.105	−21.97	8.288	7.140
81	1.3705	2.179	−32.40	8.348	6.999
86	1.3793	2.285	−42.16	8.436	6.777
94	1.3849	2.393	−29.35	8.496	6.517
106	1.3872	2.460	−6.80	8.511	6.327
	1.3903	3.4	200	8.40	4.95
	H- and He-burning shells				
	1.3904	8.4	5.21(4)	7.0	−3.5

the electron capture $\Delta = m_e c^2 + (5.19, 3.89, 3.25, 1.99)$ MeV, which correspond to densities $(3.44, 1.68, 1.17, 0.42)$ in the units of 10^9 g/cm^3 relatively, for a number of electrons per one electron $\mu_e = 2$. In conditions when the central density in the pre-supernova may be close to 10^{10} g/cm^3 (see [499]) the presence of such isotope leads to the existence of a jump in the composition at a density, corresponding to a threshold energy.

The physical processes accompanying the eddies motion around the URCA shell are rather complicated. Different interpretations had been suggested, with different conclusions about the direction of influence: stabilizing or destabilizing the carbon burning in the convective degenerate core [67, 233, 310, 358, 498, 499, 626, 706, 777, 935]. It was concluded in [67] that non-equilibrium heating is balanced by the change in the convective flow, leading to the net cooling due to the convective URCA shell. Nine years later, the same authors (Stein, Barkat, Wheeler, 1999) changed their mind, concluding that "convective URCA process can reduce the rate of heating by nuclear reactions but cannot result in a net decrease in entropy, and hence in temperature, for a constant or increasing density." This conclusion, as well as the opposite one, made using thermodynamic relations only, seems to be unconvincing. Following this line, let us present two plausible scenarios, leading to two opposite conclusions.

A. Due to an action of a nonlinear bulk viscosity [114], the convection is damping in the vicinity of the URCA shell, decreasing the convective heat flux from the central part of the star[8]. In the general heat balance of the star it means that cooling become less effective, and nuclear reactions become thermally unstable and lead to a nuclear explosion earlier, than without the presence of the URCA shell. Non-equilibrium heating gives additional heating, supporting the earlier nuclear explosion.

B. Due to an action of a nonlinear bulk viscosity [114], the convection is damping in the vicinity of the URCA shell, decreasing the convective heat flux from the central part of the star. Due to a local decrease in the heat flux from the core, the average temperature gradient increases, leading finally to increase in the convective flux soon after entering the URCA shell into the convective zone. If the increase in the convective flux prevails the non-equilibrium heating in the URCA shell, the general heat balance would be shifted to a larger temperature with more effective cooling, and the boundary of the thermal explosion would be postponed in time, if not eliminated.

In such a highly nonlinear system, as a star with nuclear reactions, neutrino losses, degeneracy, convection and many feedback influences the numerical simulations are needed, because it seems to be impossible to make a conclusion about the direction of the process under the action of an additional URCA shell, basing only on the thermodynamical ground.

The main difference between the dissipation of pulsations due to nonlinear bulk viscosity and dissipation of a convective motion is connected with the role of the excitation of the sound waves. Such excitation is of the most importance for dissipation of plane parallel oscillations of the slab, which are analogous to radial oscillations of a star, both belonging to the p-mode family of oscillations [317]. In contrast to them, the convective modes belong to another family of perturbations, related to stellar g-modes, in which the local pressure perturbations are much smaller and could be neglected when the convective velocity is much less than the velocity of the sound. In this situation, the sound wave dissipation of the convective modes imposed by the URCA shell is much less important than in the case of p-modes. The convective URCA shell may only change the convective velocity, and dissipation of the convection connected with the wave excitation by the convective motion, which is neglected in the standard mixing length model. Therefore, the dynamical dissipation of the convective motion in the model of the convection, where the URCA shell influence is taken into account, is not taken into account in [114].

[8] Nonlinear character of the bulk viscosity in the highly degenerate matter is due to nonlinear dependence of the reaction rate of the electron capture on the matter density, see Sect. 5.1.4, Vol.1.

The equations of stellar evolution in the presence of the URCA shell should take into account the following physical processes:

1. A loss of the energy due to neutrino emission in the URCA shell
2. Heating of the matter in the convective region around the URCA shell due to the non-equilibrium beta processes
3. Decrease in the convective velocity in the layer around the URCA shell due to the energy dissipation connected with the non-equilibrium beta processes. The kinetic energy of the convection is the source of the energy for neutrino losses and non-equilibrium heating of the matter

9.3.9.1 Energy Equation in Presence of the Convective URCA Shell

In the condition of a static equilibrium, only the energy and heat transfer equations should be modified. In the energy equation

$$T\frac{\mathrm{d}S}{\mathrm{d}t} = \frac{\mathrm{d}E}{\mathrm{d}t} - \frac{P}{\rho^2}\frac{\mathrm{d}\rho}{\mathrm{d}t} = \epsilon_n - \epsilon_\nu + \epsilon_\nu^{CU} - \frac{1}{4\pi\rho r^2}\frac{\mathrm{d}L_\mathrm{r}}{\mathrm{d}r}, \tag{9.120}$$

in addition to the conventional neutrino cooling processes ϵ_ν, the new term $\epsilon_\nu^{\mathrm{CU}}$ is connected with heating due to the non-equilibrium beta processes around the URCA shell, having in mind the strong degeneracy of electrons in this region. The neutrino emission in the non-equilibrium URCA processes is accompanied by heating at high degeneracy, because the positive term $\sum \mu_i dn$ exceeds the energy carried away by the neutrinos [189]. The convective motion, consisting of convective vortexes around the URCA shell, is a source of additional neutrino energy losses and of heating of the matter. This dissipation of the convective energy may be described in the same way, as corresponding dissipation and heating during stellar pulsations. The formulae for a description of these processes are obtained in [114].

$$\dot{Q}_\nu = \tilde{R}\frac{M_\mathrm{e}c^2}{Am_\mathrm{p}}\frac{\delta^8\alpha_\mathrm{p}^5}{g\,2^{10}\times 75\pi}, \tag{9.121}$$

where

$$\tilde{R} = \ln 2\frac{(\delta^2-1)^{1/2}\delta}{12(Ft_{1/2})_{Z-1}}\frac{M_\mathrm{e}^4c^5}{12\pi^2\hbar^3}\left(1+\frac{g_{Z-1}}{g_Z}\right). \tag{9.122}$$

The equation (9.121) is related to energy losses averaged over the whole convective motion layer around the URCA shell radius r_*

$$r_* + l_\mathrm{conv} < r < r_* - l_\mathrm{conv}, \quad l_\mathrm{conv} = \alpha_\mathrm{p}\frac{P}{\nabla P}, \quad \delta = \frac{\Delta}{M_\mathrm{e}c^2}, \tag{9.123}$$

here l_conv is taken from the mean free path model. The local rate is obtained from (9.121), if we take into account that the whole heating is concentrated inside the layer (9.123). We then get

$$\epsilon_\nu^{CU} = \frac{\dot{Q}_\nu}{2\rho l_{\text{conv}}}. \tag{9.124}$$

9.3.9.2 Convective Flux

The heat flux consists of the radiative and convective parts $L_\text{r} = L_\text{r}^\text{rad} + L_\text{r}^\text{conv}$. The convective part L_r^conv should be modified with the account of damping of the convective motion in the layer around the URCA shell. The convective heat flux density $F_\text{conv} = \frac{L_\text{r}^\text{conv}}{4\pi r^2}$ is written in the mean free path model of the convection as (see (3.1.13), Vol. 1)

$$F_\text{conv} = (\Delta\nabla T)\,\text{d}r\; c_\text{p}\,\rho\, v_\text{conv}. \tag{9.125}$$

Here, $\Delta\nabla T = \frac{\text{d}T}{\text{d}r}\big|_\text{ad} - \frac{\text{d}T}{\text{d}r}$ is the difference between the adiabatic temperature gradient and the actual one, c_p is the heat capacity at a constant pressure, v_conv is the convective velocity and $\text{d}r$ is the length which the convective element crosses before its merging with the background. In the conventional mixing length theory, the convective velocity is a result of an action of the buoyancy force on the length $\text{d}r$. While neutrino energy losses and non-equilibrium beta heating have the kinetic energy of a convective motion as a source, we should also add explicitly the corresponding terms to the equation determining the convective velocity. With account of this, we write the equation for the convective velocity in a form

$$\frac{1}{2}\rho v_\text{conv}^2 = \frac{1}{2}(\Delta\nabla\rho)\text{d}r^2\, g - \frac{\dot{E}_\text{conv}^{(\beta)}}{v_\text{conv}}. \tag{9.126}$$

Here the difference between the actual density gradient and the adiabatic one $\Delta\nabla\rho$ is written as

$$\Delta\nabla\rho = \frac{d\rho}{\text{d}r} - \frac{\text{d}\rho}{\text{d}r}\bigg|_\text{ad} = -\Delta\nabla T\left(\frac{\partial\rho}{\partial T}\right)\bigg|_P. \tag{9.127}$$

The average value of the convective element path $\text{d}r$ is connected with the mixing length l as $\bar{\text{d}r} = \frac{l}{2}$, in the approximation of small density variations, accepted here we have the relation (see Sect. 5.3.1, Vol. 1) $\dot{E}_\text{conv}^{(\beta)} = 4\dot{Q}_\nu$. We get a more complicated relations for the convective energy flux density, than in the standard mixing length model, because the convective velocity is not expressed explicitly, but should be found from the third-order algebraic equation. Finally, the temperature gradient $\frac{\text{d}T}{\text{d}r}$ and the total radial heat flux L_r are connected by the following system of algebraic equations in the region (9.123) around the URCA shell

$$L_\text{r} = 4\pi r^2\left(\frac{4acT^3}{3\kappa\rho}\frac{\text{d}T}{\text{d}r} + F_\text{conv}\right), \tag{9.128}$$

$$F_\text{conv} = \frac{1}{2}(\Delta\nabla T)\, l\, c_\text{p}\,\rho\, v_\text{conv}, \tag{9.129}$$

$$\frac{1}{2}\rho v_{\rm conv}^2 = -\frac{1}{8}(\Delta\nabla T)l^2\left(\frac{\partial\rho}{\partial T}\right)\bigg|_P g - \frac{\dot{E}_{\rm conv}^{(\beta)}}{v_{\rm conv}}, \tag{9.130}$$

$$\delta\nabla T = \left(\frac{\partial T}{\partial P}\right)\bigg|_S \frac{dP}{{\rm d}r} - \frac{{\rm d}T}{{\rm d}r}, \tag{9.131}$$

where $\dot{E}_{\rm conv}^{(\beta)} = 4\,\dot{Q}_\nu$ is found from (9.121), κ is the Rosseland matter opacity and the pressure gradient is determined by the equation of static equilibrium of the star. These relations may be applied for a description of the URCA shell convection only for sufficiently rapid convective motion, when the first term in (9.126) exceeds

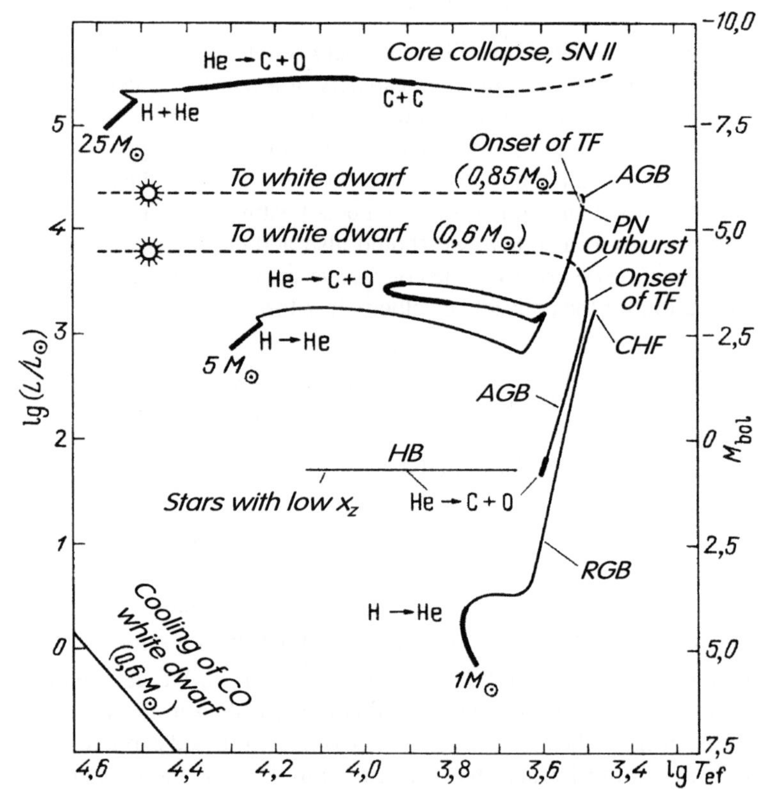

Fig. 9.48 Coarse evolutionary tracks for stars with $M_i = 1, 5, 25\,M_\odot$. Heavy portions represent principal burning phases in the core. For $M_i < 2.3\,M_\odot$, CHF occurs after which quiescent ^{4}He burning begins. After ^{4}He has been exhausted in the core, the star arrives on the AGB. When the core with no helium in it reaches the mass $\sim$0.53 $M_\odot$, TF (thermal flashes in ^{4}He shell) start. An AGB star loses mass, and this process terminates with a rapid ejection of the residual hydrogen envelope in the form of PN. C–O core with $m_{\rm f} \approx 0.6\,M_\odot$ transforms into a white dwarf. More massive AGB and post-AGB stars with $M_i \leq 9\,M_\odot$ evolve in a similar way, $m_{\rm f}$ increases with increasing M_i and equals 1.08 $M_\odot$ at $M_i = 8.8\,M_\odot$. The symbol indicates the onset of PN luminescence, when $T_{\rm ef}$ of the star reaches 3×10^4 K and the gas ionization in PN begins, from [502]

considerably the second one. For a slow convective motion, this model cannot be applied, because it was suggested that the change in the initial composition by beta processes in the URCA shell is small during the period of damping of oscillations. For the case of a convective URCA shell where the convection is excited by the external source, this condition may be softened and reduced to the demand that the changes of composition are small during one rotation of the convective vortex. The equation (9.130) has roots only when

$$\dot{E}_{\text{conv}}^{(\beta)} < \frac{1}{8\sqrt{\rho}} \left[-\frac{1}{3}(\Delta \nabla T) l^2 \left(\frac{\partial \rho}{\partial T} \right) \bigg|_P g \right]^{3/2} . \tag{9.132}$$

The violation of this inequality may result in an abrupt termination of the convection in the layer (9.123) around the URCA shell.

When analyzing the URCA shell convection in a star, it would be premature to predict the results of the evolutionary calculations with the account of a convective URCA shell according to (9.128)–(9.131), before such calculations are done. But there may be expected two possibilities. One possibility is connected with obtaining of a definite result which has a little sensitivity to the input parameters of the problem, such as α_p, $Ft_{1/2}$, accepted rates of nuclear reactions and neutrino losses. Another possibility could be a great sensitivity of the result to the same input parameters. If the second possibility would be realized, we could still remain in the situation of an ambiguity, because the set of the input parameters for pre-supernova model cannot be established with a sufficient precision.

Numerical two-dimensional simulations of the URCA shell convection could be useful for further clarifying the problem. Nevertheless, numerical problems connected with a simulation of the convective motion seems to remain severe enough; therefore, we should first obtain some understanding from the simplified one-dimensional model of the convective URCA shell, generalizing the mixing length model, the variant of which is presented in this chapter.

To conclude this chapter, we reproduce Fig. 9.48 from [502] sketching out the evolution of stars of diverse masses from the MS to the white dwarf formation or supernova explosion.

Chapter 10
Collapse and Supernovae

Supernova explosions are the most spectacular events in the universe of stars. The major part of the energy releases in a time interval (<1 s) is relatively small not only to stellar but also to human lifetime, whereas its quantity exceeds by an order of magnitude and more the energy emitted by the star over its total lifetime reaching $\sim 10^{10}$ yr for Sun-like stars. During the neutron-star formation, most of the energy releases are in the form of scarcely observable neutrinos.

A supernova explosion is the end of the life of most massive stars with $M > 8M_{\odot}$. The flash itself results either from the thermal instability development in the degenerate core or from gravitational and partly nuclear energy release during collapse which leads to neutron-star formation. The rotation and magnetic field may play an important role in conversion of gravitational energy into the energy of the observable flash. A small number of stars (the most massive ones) seem to end their life with collapse and black hole formation. The collapse in this case may be "silent" and not lead to supernova explosion.

A wealth of texts that we are unable to include in this book are concerned with both observational and theoretical studies of supernovae. From an observational standpoint, they fall into Type I supernovae (SN I) with no hydrogen lines in their emission spectrum and numerous absorption lines of various heavy elements instead, and Type II supernovae (SN II) with much hydrogen and nearly normal chemical composition. Type I SN are divided into three groups, two of which, SN Ib, c are substantially different from SN Ia. SN II have been established to be a result of the evolution of massive stars, while SN I have less massive stars for progenitors; many SN I are likely to occur in binaries.

The pulsars are thought to be produced by SN II explosions. The common feature between SN II and SN Ib, c is a non-thermal radio emission visible for several years after the explosion. This is a real indication that SN Ib, c also give birth to pulsars. On the contrary, SN Ia never show such a radio emission; hence, it is suggested that they are produced by nuclear explosions, leading to total disruption of the star, while other types of SN result from hydrodynamical collapse, leading to formation of a neutron star. In SN II, the collapsing core is surrounded by a large hydrogen envelope, while in SN Ib, c the collapsing core is almost naked, presumably originated from less massive stars, which have lost their hydrogen-rich

G.S. Bisnovatyi-Kogan et al., *Stellar Physics: 2: Stellar Evolution and Stability*, Astronomy and Astrophysics Library, DOI 10.1007/978-3-642-14734-0_4,

envelope in their preceding evolution, may be in a binary. There are some spectral differences between the three types of SN I, while they all do not show hydrogen lines. The main difference appears after ~250 days from the explosion. In SN Ia the strongest emission lines are represented by FeII and FeIII ions, and in SN Ib, c the main emission is determined by O I lines. The difference between SN Ib and SN Ic is the absence of helium lines in SN Ic, which may be called helium-poor SN Ib [207, 384, 719, 748].

The presence of hydrogen in the SN II spectra provides evidence that the explosion occurred before the star has lost its hydrogen envelope. The shape of the light curve shows that before the explosion, this envelope could be extremely extended $(10^3–10^4)R_\odot$ [525]. Observations in the Large Magellanic Cloud of SN1987A, the first supernova visible to the naked eye for more than 300 years, provide evidence for a SN II explosion in a fairly compact blue supergiant. The SN1987A light curve may be reproduced only in the case of a prolonged energy release after the main explosion. The variety of SN II light curves is substantially larger than those of SN I [950].

The absence of hydrogen in the SN I spectra leads to the conclusion that the star loses its hydrogen envelope during its pre-explosion evolution, and that the presupernova is compact. The shape of a SN I light curve can be explained if the emitted energy is accounted for by the radioactive decay ${}^{56}_{28}\mathrm{Ni} \rightarrow {}^{56}_{27}\mathrm{Co} \rightarrow {}^{56}_{26}\mathrm{Fe}$; to maintain the luminescence energetically, it must be in this case $M_{^{56}\mathrm{Ni}} = 0.3–1.1\,M_\odot$ [1053]. The slow energy release during 2–20 days makes the maximum light phase prolonged (10–20 days) for a compact stellar model [525]. The same radioactive decay accounts for the character of the SN1987A light curve at $t \geq 120$ days after the explosion; here it is necessary to have initially $0.078 M_\odot$ of ^{56}Ni [526].

The exact relationship between SN I and SN II, on the one hand, and the initial stellar masses, mechanisms of explosion and resulting remnants, on the otherhand, is not reliably established. New observational data may change some of our present model conceptions. Various aspects of the supernova problem arising from observations and their interpretation are given in the book [905].

The items of this chapter are concerned with the author's special interests, and their selection is therefore somewhat subjective.

The physical processes treated in Vol. 1: nuclear reactions, neutrino processes and so on, proceed during supernova explosions and collapse in the same way as in quiet evolutionary phases. The supernova theory differs from the theory of evolution using non-stationary hydrodynamic equations instead of hydrostatic equations, by involving regions of higher temperature and density, and the predominant role played by neutrino processes. Despite much effort, the supernova theory is far from complete even in the spherically symmetrical approximation by reason of serious numeric and fundamental difficulties related to non-stationary convection, neutrino transport and the equation of state for matter of a density above the nuclear.

10.1 Presupernova Models

As pointed out above, a supernova is a result of hydrodynamical or thermal instability development.

10.1.1 Stellar Cores at Threshold of Hydrodynamical Stability: Energetic Method

Consider polytropes with $P = K\rho^\gamma$. Combining the equilibrium and continuity equations,

$$\frac{\mathrm{d}P}{\mathrm{d}r} = -\rho\frac{Gm}{r^2}, \tag{10.1}$$

and (33.13a), we have the equation for $\rho(r)$

$$\frac{\mathrm{d}}{\mathrm{d}r}\left[K\gamma r^2\rho^{\gamma-2}\frac{\mathrm{d}\rho}{\mathrm{d}r}\right] = -4\pi G\rho r^2. \tag{10.2}$$

Transforming to dimensionless quantities θ and ξ, defined as

$$\rho = \rho_c\theta^n, \quad r = \alpha\xi, \quad \alpha = \left[\frac{(n+1)K}{4\pi G}\rho_c^{\frac{1}{n}-1}\right]^{1/2}, \quad \gamma = 1 + \frac{1}{n}, \tag{10.3}$$

gives the equilibrium Lane–Emden equation (see [269])

$$\frac{1}{\xi^2}\frac{\mathrm{d}}{\mathrm{d}\xi}\left(\xi^2\frac{\mathrm{d}\theta}{\mathrm{d}\xi}\right) = -\theta^n \tag{10.4}$$

with the boundary conditions

$$\theta = 1, \quad \frac{\mathrm{d}\theta}{\mathrm{d}\xi} = 0 \ \text{at}\ \xi = 0. \tag{10.5}$$

The stellar boundary corresponds to $\xi = \xi_1$ so that $\theta(\xi_1) = 0$. The stellar mass M expressed in terms of variables (10.3) becomes

$$\begin{aligned} M &= 4\pi\int_0^R \rho r^2 \mathrm{d}r = 4\pi\rho_c\alpha^3\int_0^{\xi_1}\theta^n\xi^2\mathrm{d}\xi \\ &= 4\pi\left[\frac{(n+1)K}{4\pi G}\right]^{3/2}\rho_c^{\frac{3}{2n}-\frac{1}{2}}\int_0^{\xi_1}\theta^n\xi^2\mathrm{d}\xi, \end{aligned} \tag{10.6}$$

and hence from (10.4),

$$\int_0^\xi \theta^n \xi^2 \mathrm{d}\xi = -\xi^2 \frac{\mathrm{d}\theta}{\mathrm{d}\xi}, \quad -\xi^2 \frac{\mathrm{d}\theta}{\mathrm{d}\xi}\bigg|_{\xi=\xi_1} = M_n. \tag{10.7}$$

Obviously, at $n = 3$, $\gamma = 4/3$, the stellar mass is independent of ρ_c and is exactly determined by the constant K in the equation of state. The stellar mass increases at $\gamma > 4/3$ with increasing ρ_c, and at $\gamma < 4/3$ decreases. If the polytropic power coincides with the adiabatic power, $\gamma = \gamma_{ad}$, then the star is stable at $\gamma > 4/3$ and unstable at $\gamma < 4/3$, and $\gamma = 4/3$ corresponds to the boundary case and represents the indifferent equilibrium.

Real stars are not polytropes, but the condition $\gamma = 4/3$ is approximately valid at the boundary of stability if γ is treated as an adiabatic power properly averaged over the star.

A strict derivation of stability conditions is made in Chap. 12 using the variational method. Equating the first variation to zero yields the equilibrium equation, while the stability condition requires the second variation to be positive. For an isentropic polytrope with $\gamma = 4/3$, ρ_c is arbitrary, whereas the density distribution $\theta(\xi)$ is invariant against homologous contraction or expansion. Let us treat these properties as valid also for the case where $\gamma = 4/3$ only on average. We then derive the equilibrium and stability conditions using the simplified variational method based on the assumption of homology and conservation of stellar structure at density variations [1080], usually called the energetic method. We write down the total energy of an instantaneously static star analogous to the potential energy of conservative mechanical system:

$$\epsilon = \int_0^M E(\rho, T)\,\mathrm{d}m - \int_0^M \frac{Gm\,\mathrm{d}m}{r} - 5.06\frac{G^2M^3}{R^2c^2},$$

$$\mathrm{d}m = 4\pi\rho r^2 \mathrm{d}r. \tag{10.8}$$

The first term here represents the internal energy ϵ_i, the second the Newtonian gravitational energy ϵ_G and the third, ϵ_{GR}, a small general-relativity correction $\left(r_g/r = 2Gm/c^2r \ll 1 \text{ is the small parameter}\right)$ evaluated in [1080] for the matter distribution over a $n = 3$ polytrope (see Sect. 12.2).

The term containing the Newtonian gravitational energy of an equilibrium star may be explicitly evaluated for an arbitrary polytropic equation of state [617]. The equilibrium equation (10.1) gives

$$\begin{aligned}\epsilon_G &= -G\int_0^M \frac{m\,\mathrm{d}m}{r} = 4\pi\int_{P_c}^0 r^3 \mathrm{d}P \\ &= -12\pi\int_0^R Pr^2\mathrm{d}r = -3\int_0^M \frac{P}{\rho}\,\mathrm{d}m.\end{aligned} \tag{10.9}$$

A formal integration of the equilibrium equation

$$\frac{1}{\rho}\frac{\mathrm{d}P}{\mathrm{d}r} + \frac{\mathrm{d}\varphi_G}{\mathrm{d}r} = 0 \tag{10.10}$$

yields the integral relation

$$(n+1)\frac{P}{\rho} + \varphi_G = -\frac{GM}{R}. \tag{10.11}$$

The constant on the right follows from $P/\rho = 0$ on the boundary of the star and the normalization condition $\varphi = 0$ at $r = \infty$. We evaluate ϵ_G in another way, using (10.9–10.11):

$$\begin{aligned}\epsilon_G &= \frac{1}{2}\int_0^M \varphi_G\,\mathrm{d}m = -\frac{GM^2}{2R} - \frac{n+1}{2}\int_0^M \frac{P}{\rho}\,\mathrm{d}m \\ &= -\frac{GM^2}{2R} + \frac{n+1}{6}\epsilon_G,\end{aligned} \tag{10.12}$$

and so from (10.12)

$$\epsilon_G = -\frac{3}{5-n}\frac{GM^2}{R}. \tag{10.13}$$

We then have $E = nP/\rho$ for an adiabat with $\gamma = \gamma_{ad}$, and (10.9–10.13) give, for a star in equilibrium,

$$\begin{aligned}\epsilon_i &= -\frac{n}{3}\epsilon_G = \frac{n}{5-n}\frac{GM^2}{R}, \\ \epsilon_N &= \epsilon_i + \epsilon_G = \frac{n-3}{5-n}\frac{GM^2}{3},\end{aligned} \tag{10.14}$$

where ϵ_N is the total energy of a Newtonian star. The total energy of a stable star is negative, and therefore the stability requires that $n < 3$, $\gamma > 4/3$. The radius of a polytrope is, using (10.3), (10.6) and (10.7),

$$\begin{aligned}R = \alpha\xi_1 &= \left[\frac{(n+1)}{4\pi G}K\right]^{1/2} \rho_c^{(1-n)/2n}\xi_1 \\ &= \left(\frac{\xi_1^3}{4\pi M_n}\right)^{1/3} M^{1/3}\rho_c^{-1/3} = \frac{M^{1/3}\rho_c^{-1/3}}{0.426}.\end{aligned} \tag{10.15}$$

Here, the values for a polytrope of $n = 3$ are used (see [269] and the Problem in this section): $\xi_1 = 6.89685$, $M_3 = 2.01824$. The ratio of ρ_c to the average density $\overline{\rho}\,(M = 4\pi\overline{\rho}R^3/3)$ is

$$\frac{\rho_c}{\overline{\rho}} = \frac{4\pi}{3}\frac{1}{(0.426)^3} = 54.18\,.$$

From (10.15) and (10.13), we have [1080]

$$\epsilon_{\mathrm{G}} = -0.639\, GM^{5/3}\, \rho_{\mathrm{c}}^{1/3}, \tag{10.16}$$

and from (10.8), we have

$$\epsilon_{\mathrm{GR}} = -0.918\, \frac{GM^{7/3}}{c^2}\, \rho_{\mathrm{c}}^{2/3}. \tag{10.17}$$

Only one parameter, $\rho_{\mathrm{c}}^{1/3}$ or R, varies with homologous variations:

$$\rho = \rho_{\mathrm{c}}\varphi\left(\frac{m}{M}\right), \quad \varphi\left(\frac{m}{M}\right) \tag{10.18}$$

is an invariant function and hence, the energy variations reduce to ordinary derivatives. Using (10.16–10.18) and taking the entropy to be constant at variations, we obtain from (10.8) the equilibrium condition

$$\begin{aligned} \frac{\partial \epsilon}{\partial \rho_{\mathrm{c}}^{1/3}} &= 3\rho_{\mathrm{c}}^{4/3} \int_0^M P\, \frac{\mathrm{d}m}{\varphi\left(\frac{m}{M}\right)} - 0.639\, GM^{5/3} \\ &\quad - 1.84\, \frac{G^2 M^{7/3}}{c^2}\, \rho_{\mathrm{c}}^{1/3} = 0. \end{aligned} \tag{10.19}$$

The second derivative of the energy at $S = \text{const.}$ becomes zero on the boundary of stability:

$$\begin{aligned} \frac{\partial^2 \epsilon}{\partial \left(\rho_{\mathrm{c}}^{1/3}\right)^2} &= 9\rho_{\mathrm{c}}^{-5/3} \int_0^M \left(\gamma - \frac{4}{3}\right) P\, \frac{\mathrm{d}m}{\varphi(m/M)} \\ &\quad - 1.84\, \frac{G^2 M^{7/3}}{c^2} = 0. \end{aligned} \tag{10.20}$$

We use here the thermodynamic relations

$$\begin{aligned} \frac{\partial E}{\partial \rho_{\mathrm{c}}^{1/3}} &= \left(\frac{\partial E}{\partial \rho}\right)_S \frac{\partial \rho}{\partial \rho_{\mathrm{c}}} \cdot 3\rho_{\mathrm{c}}^{2/3} = \frac{P}{\rho^2}\, \varphi(m/M) \cdot 3\rho_{\mathrm{c}}^{2/3} \\ &= 3\, \frac{P}{\varphi\left(\frac{m}{M}\right)}\, \rho_{\mathrm{c}}^{-4/3}, \\ \frac{\partial P}{\partial \rho_{\mathrm{c}}^{1/3}} &= \left(\frac{\partial P}{\partial \rho}\right)_S \varphi\left(\frac{m}{M}\right) 3\rho_{\mathrm{c}}^{2/3} = 3\gamma P \rho_{\mathrm{c}}^{-1/3}, \\ \gamma \equiv \gamma_1 &= \left(\frac{\partial \ln P}{\partial \ln \rho}\right)_S. \end{aligned} \tag{10.21}$$

Equations (10.19) and (10.20) have been obtained in [148] and used for determining the boundary of stability for various stellar models.

The critical mass reaches its minimum for cold stars. At low temperatures, the electrons in a star near the critical point are close to degeneracy and ultrarelativism. The nuclei may be taken as non-degenerate. The temperature is nearly constant over the star by virtue of the large thermal conductivity of degenerate electrons (see Chap. 2, Vol. 1). Such stars are called white dwarfs (see Sect. 11.2). The equation of state of matter in white dwarfs with a mass close to the critical one, at finite temperature, should include the pressure of cold almost ultrarelativistic degenerate (URD) electrons, taking into account, to first order, the deviation from the pressure of UR electrons, the thermal correction to the pressure of URD electrons and the pressure of non-relativistic, non-degenerate nuclei (see Chap. 1, Vol. 1). Ignoring Coulomb corrections, this equation of state may be written in the form

$$P = \frac{m_e^4 c^5}{12\pi^2\hbar^3} y^4 \left(1 - \frac{1}{y^2} + \frac{2\pi^2}{3\alpha^2 y^2} + \frac{4}{Z\alpha y}\right),$$

$$y = \left(\frac{3\pi^2\rho}{\mu_Z m_u}\right)^{1/3} \frac{\hbar}{m_e c}, \quad \alpha = \frac{m_e c^2}{kT}. \tag{10.22}$$

This relation includes the pressure of non-degenerate nuclei. If the star consists of iron ^{56}Fe at $\rho_c = 1.15 \times 10^9$ g cm^{-3} (1.24×10^9 g cm^{-3} for a capture on the excited state of ^{56}Mn), the neutronization (capture of electrons by nuclei) breaks out, because the Fermi energy of electrons starts to exceed the difference of energies between nuclei of ^{56}Fe and ^{56}Mn (Chap. 1, Vol. 1). Neutronization proceeds initially at $A =$ const., leading to an increase in the number of nucleons per electron

$$\mu_Z = \left(\sum_i \frac{Z_i x_i}{A_i}\right)^{-1}. \tag{10.23}$$

Taking into account successive electron captures during the equilibrium neutronization of iron ^{56}Fe, we obtain approximately [188, 1014]

$$\mu_Z = \frac{56}{26}(1 + \nu y), \quad \nu = 6 \times 10^{-3}. \tag{10.24}$$

Evaluating γ ($\equiv \gamma_1$ from (10.21)) gives [148, 188, 1014]

$$P\left(\gamma - \frac{4}{3}\right) = \frac{m_e^4 c^5}{9\pi^2\hbar^3}\left[\frac{y^2}{2} - \frac{\nu y^5}{3} + \frac{y^3}{Z\alpha} - \frac{\pi^2 y^2}{2\alpha^2}\left(\frac{3}{2} + \frac{\pi^2 Z}{\alpha y}\right)^{-1}\right]. \tag{10.25}$$

Only the first, leading term of the pressure in (10.22) allows for μ_Z changes here. Since the leading term of the pressure at $\mu_Z = \text{const.}$ corresponds to $\gamma = 4/3$, finding the critical parameters from (10.19) and (10.20) is in this case an asymptotically exact procedure. The evolutionary sequence $T(\rho_c)$ for radiating contracting isothermal stars with a given mass M may be found from (10.19) using (10.22), while obtaining the critical parameters requires additional use of (10.20) and (10.25). A solution to these equations without neutronization ($\nu = 0$) is obtained in [148]. The relevant evolutionary curves and critical curve are shown in Fig. 10.1. The neutronization is included in calculations in [188, 1014]. The equilibrium neutronization causes the star to lose its stability at densities less by an order of magnitude with respect to the GR case, and is therefore the principal reason for the loss of stability. In reality, the electron capture rate is fairly low, while the course of neutronization is non-equilibrium, and the Fermi energy of electrons may considerably exceed the energy difference between subsequent nuclei (see Chap. 5, Vol. 1). The timescale of neutronization is well above the hydrodynamical time scale near the boundary of stability and determines the contraction rate (see Sect. 10.3 for more detail). The instability due to GR effects leads to hydrodynamical collapse. For sufficiently cold stars $\pi^2 Z/6\alpha y \ll 1$, (10.19) and (10.20) with $\nu = 0$ have the approximate analytical solution to ~1% of accuracy [148, 189]

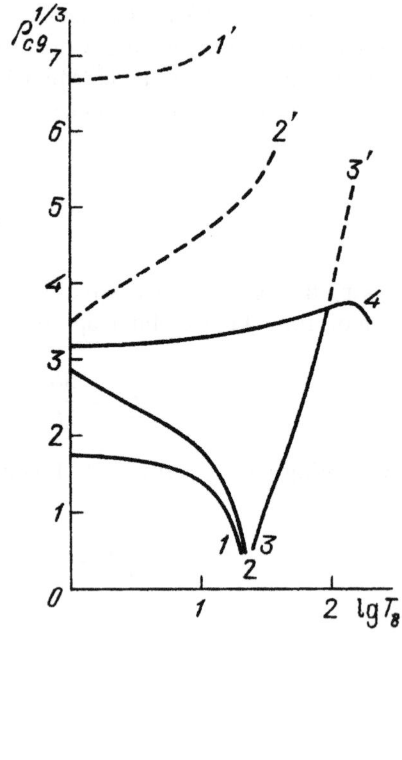

Fig. 10.1 Evolutionary curves and critical states on the ($\rho_{c0}^{1/3}$, lg T_8) plane with GR effects, without neutronization, from [148]; 1, 2 and 3 are the evolutionary curves for $M = 1.19, 1.20$ and $1.36\,M_\odot$, respectively. The dashed lines 1′, 2′, 3′ represent the unstable equilibrium states of these stars, 4 represents the critical states, its endpoint corresponds to $M = 1.7\,M_\odot$

Table 10.1 The densities ρ_n corresponding to the onset of neutronization of various nuclei, and critical densities $\rho_{c,GR}$ of cold stars due to GR effects [1081]

Element	$\epsilon_\beta = \epsilon_{Fe}$, MeV	ρ_n, 10^9 g cm^{-3}	$\rho_{c,GR}$, 10^9 g cm^{-3}
$^{56}Fe \to {}^{56}Mn$	3.7 (3.81)	1.15 (1.24)	31
$^{32}S \to {}^{32}P$	1.7	0.15	27
$^{28}Si \to {}^{28}Al$	4.64	1.97	27
$^{24}Mg \to {}^{24}Na$	5.5	3.2	27
$^{20}Ne \to {}^{20}F$	7.03	6.2	27
$^{16}O \to {}^{16}N$	10.4	19	27
$^{12}C \to {}^{12}B$	13.4	39	27
$^{4}He \to {}^{3}H + n$	20.6	137	27

Given in parentheses for ^{56}Fe are values of ϵ_β and ρ_n corresponding to the capture onto the first excitation level 109 KeV of the final nucleus, since the capture onto the ground level is strongly forbidden by selection rules

$$\rho_{c,cr} = 6.7 \times 10^9 \mu_Z^2 \left(1 + 17\,\frac{\mu_Z^{4/3}}{\alpha A}\right) \text{ g cm}^{-3},$$

$$M_{cr}/M_\odot = \frac{5.83}{\mu_Z^2}\left(1 - \frac{0.046}{\mu_Z^{2/3}} + \frac{0.53\,\mu_Z^{2/3}}{\alpha A}\right), \tag{10.26}$$

$$\alpha \gg 0.087\, Z \mu_Z^{1/3}.$$

The critical mass of iron stars at $T = 0$ due to GR effects is somewhat higher ($1.222\, M_\odot$) than in the case of neutronization ($1.16\, M_\odot$).[1] Comparing densities corresponding to the onset of neutronization of various nuclei with the density (10.26) shows (Table 10.1) that carbon and helium white dwarfs lose their stability owing to GR effects. The GR influence on the white dwarf stability was first investigated by Kaplan [554].

Increasing the temperature causes M_{cr} and $\rho_{c,cr}$ to increase because of the stabilizing effect of non-relativistic nuclei. At $T = 1.4 \times 10^{10}$ K we have $M_{cr} = 1.4\, M_\odot$ for ^{56}Fe in GR, the quantity $\rho_{c,cr}$ reaches the maximum $\sim 5.2 \times 10^{10}$ g cm^{-3} and decreases with further increase in the critical mass (Fig. 10.1) [148].

When a star becomes non-degenerate, we may take an adiabatic star rather than an isothermal one for our considerations, which is possible because of the convection effect. With increasing mass the critical entropy of the star S_{cr} increases, but the central temperature falls. The numerical computations of critical parameters of isentropic stars are made in [174] by solving (10.19) and (10.20). For the ranges

[1] This value is obtained from (10.57) by the energetic method at $T = 0$, $\rho_c = 1.24 \times 10^9$ g cm^{-3}. An exact calculation in [189] without Coulomb corrections yields for a cold white dwarf of iron ^{56}Fe a critical mass of $1.181\, M_\odot$ for the onset of instability at $\rho_c = 1.15 \times 10^9$ g cm^{-3}. The new phase core has a finite mass $m_c = 1.4 \times 10^{-3} M$. Increasing the mass of the new phase core from zero to m_c leads to an increase in stellar mass $\Delta M = 0.15\, m_c$ (see (11.12))

$10^5 \le \rho \le 10^{10}$ g cm^{-3}, $10^9 \le T \le 2 \times 10^{10}$ K, thermodynamic functions are taken from [523], where the nuclear equilibrium is examined, taking into account of ^{56}Fe, α, n, p and the approximation $\mu_{\nu_e} = 0$ for the chemical potential of the electronic neutrino. To the left and downwards from this region the thermodynamic functions are evaluated in [174] by numerical computation of integrals, determining thermodynamic functions of semidegenerate, semirelativistic electrons in iron ^{56}Fe (see Chap. 1, Vol. 1). Approximate parameters of isentropic stars of various masses in the critical state are given in Table 10.2 and in Fig. 10.2 from [174]. The total energy of an equilibrium star, according to (10.16–10.19), is

$$\epsilon_{\mathrm{eq}} = \int_0^M \left(E - 3\frac{P}{\rho}\right) \mathrm{d}m + 0.918\,\frac{G^2 M^{7/3}}{c^2}\,\rho_c^{2/3} \tag{10.27}$$

Table 10.2 Core parameters of stars in critical state

$\frac{M}{M_\odot}$	ρ_c, g cm^{-3}	T_c, K	S, erg g^{-1} K^{-1}	ϵ, erg	$\frac{\epsilon}{M}$, erg s^{-1}
5	1.0(8)	6.7(9)	2.1(8)	−1.3(51)	−1.3(17)
10	4.2(7)	6.4(9)	3.2(8)	−1.9(51)	−9.4(16)
50	1.0(7)	6.0(9)	6.8(8)	–	–
100	9.4(6)	6.4(9)	9.9(8)	–	–
500	3.1(6)	6.0(9)	2.1(9)	–	–
1000	6.3(3)	1.1(9)	2.8(9)	−3.7(51)	−1.8(15)

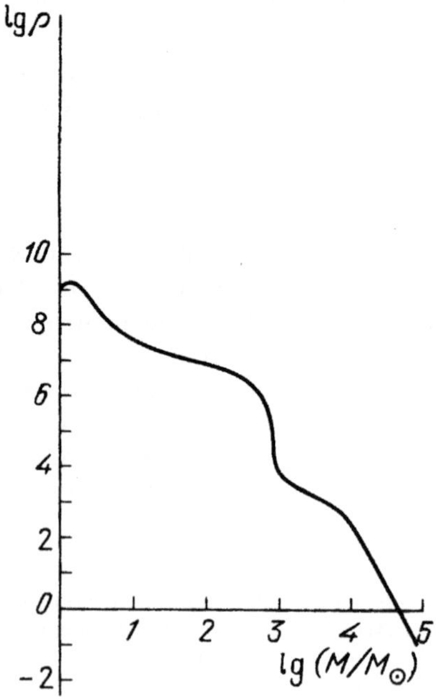

Fig. 10.2 Central density as a function of stellar mass for the critical state of a star with $M \ge 5\,M_\odot$, from [174], with the inclusion of neutronization for low masses, from [188]

and is calculated in [174]. The specific binding energies of stars from Table 10.2 in the critical state are substantially less than the energy released by burning of 1 g of hydrogen ($\sim 6 \times 10^{18}$ erg g^{-1}), helium ($\sim 5.8 \times 10^{17}$ erg g^{-1}) and carbon ($\sim 5.6 \times 10^{17}$ erg g^{-1}).

The critical parameters for $M = 1.5\,M_\odot$ calculated from (10.19–10.25) exhibit a satisfactory agreement with the corresponding parameters at the starting point of collapse for stars with $M = 15$ and $25M_\odot$ from Table 9.6. The iron core masses for these stars in the critical state are $1.56\,M_\odot$ for $15M_\odot$ and $1.61\,M_\odot$ for $25M_\odot$ (see Figs. 9.11 and 9.13).

Iron cores of different masses lose their stability for different reasons. For $M > 1.2M_\odot$, neutronization as a reason for instability is gradually substituted by dissociation of iron which remains the main reason of instability up to $\sim 500M_\odot$. For $M = 500 - 10^4 M_\odot$, the instability is due to pair creation, while for $M > 10^4 M_\odot$ the predominant role belongs to GR effects. The entropy of such supermassive stars in a critical state is so large that the pressure is determined mainly by the radiation with a small admixture of plasma, important for stability. According to [1080], the dependence $\rho_{c,cr}(M)$ due to GR effects is determined by

$$\rho_{c,cr}(M) = 2.4 \times 10^{17} \frac{1}{\mu^3} \left(\frac{M_\odot}{M} \right)^{7/2} \text{ g cm}^{-3}, \tag{10.28}$$

where μ is the molecular weight.

Evolutionary calculations performed to date for single stars with constant masses [1033, 1040, 1053] give the iron core masses at the stability threshold $M_{\text{Fe,cr}}$ not exceeding $2.45\,M_\odot$. By reason of uncertainties in these calculations[2] and the necessity to include the mass loss and various types of mixing (convection overlap, meridional circulation), their results are not quite reliable, and it is not to be ruled out that in reality the values of $M_{\text{Fe,cr}}$ may be significantly larger.

10.1.2 *Stellar Cores at Thermal Instability Threshold*

Hoyle and Fowler [480] first assumed that the thermal instability development in a degenerate carbon core of mass $1.4\,M_\odot$ would lead to a supernova explosion, and Arnett [47] performed model computations for such an explosion. The presupernova model was obtained from a crude evolutionary calculation giving the C–O core parameters for the boundary of stability: $M_{\text{core}} = 1.37\,M_\odot$, $\rho_c = 1.7 \times 10^9$ g cm^{-3}.

[2] Cf. Figs. 9.11–9.14 from [1033] and the results of subsequent calculations [1040, 1053]. A more accurate inclusion of neutrino processes on nuclei [402–405], revision of the $^{12}C(\alpha,\gamma)^{16}O$ reaction rate giving a value for physical conditions in massive stars approximately three times larger compared to the rate from [393], other corrections have resulted in changes in iron core masses before stability loss from 1.56 to $1.33\,M_\odot$ for $15M_\odot$ and from 1.61 to $2.22\,M_\odot$ for $25M_\odot$. Also, the central entropy has decreased for $15M_\odot$ and increased for $25M_\odot$ models.

The initial temperature for hydrodynamical calculations $T_c = 3.5 \times 10^8$ K was only slightly greater than the threshold temperature for the onset of thermal instability development, when $\epsilon_\nu = \epsilon_{CC}$ (see Sect. 9.3) but, by reason of a very steep dependence of the rate of the nuclear reaction on T, following an exponential law, the perturbation turned out to be important, leading immediately to detonation and complete run-away. For further calculations of supernova models based on this mechanism, see Sect. 10.2. More accurate evolutionary calculations in [773] have given the following critical parameters for the C–O core: $M_{core} = 1.39\, M_\odot$, $\rho_c = 2.68 \times 10^9$ g cm^{-3}, $T_c = 2.77 \times 10^8$ K (Table 9.11). A universal character of the critical model for a wide range of masses (3–8 $M_\odot$ in [773]) is a characteristic feature of this type of presupernova. Only the most massive stars of this range may produce supernova explosions, since, for smaller masses, the C–O core never attains the mass threshold because of mass loss and remains in the form of a white dwarf.

Evolution in binaries may give rise to models at thermal stability threshold, which differ from the model in Table 9.11. Simplified evolutionary calculations for C–O core growth due to accretion in binaries have been made in [359]. Besides $\dot{M}_c$, the critical model is substantially dependent on the shape of the curve $\epsilon_\nu = \epsilon_{CC}$ which is determined in the region of high densities by screening effects and the neutrino-loss function. It has been pointed out in [946] that the inclusion of neutral currents into the cooling function [324, 720] changes it slightly compared to the function [82] for charged currents (see Chap. 5, Vol. 1). Changes due to various screening modifications are substantially more important. Figure 10.3 from [946] shows evolutionary tracks for C–O cores on the AGB and for the case of accretion in a binary

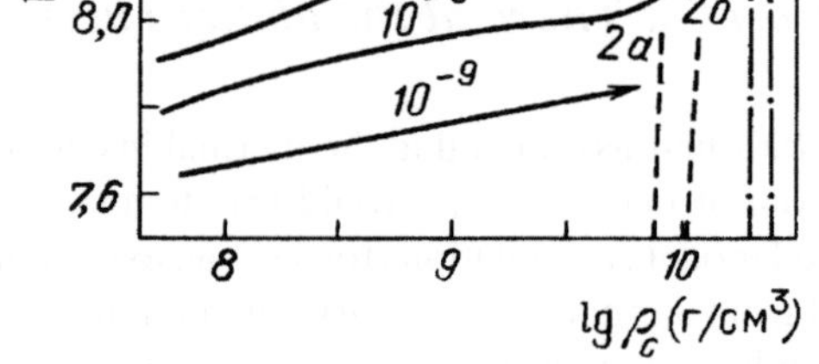

Fig. 10.3 Evolution of degenerate C–O cores on the ($\lg T_c - \lg \rho_c$) plane, RG is the evolution of the AGB giant core with $\dot{M}_c = 6 \times 10^{-7}\, M_\odot$/yr, the numbers on evolutionary tracks indicate $\dot{M}_c$ in $M_\odot$/yr; 1 are the curves $\epsilon_\nu = \epsilon_{CC}$; 1a for the screening according to [537, 538]; 1b according to [859]; 2 are the curves $\epsilon_\nu = \epsilon_{CC}$ for the pycnonuclear regime; 2a for the static approximation; 2b for the approximation of complete lattice relaxation from [859], see Sect. 4.5.4, Vol. 1); 3 are the critical states for dynamical stability loss due to oxygen neutronization; 4 is the same due to GR effects, from [946]

according to [359]. Shown are critical curves $\epsilon_\nu = \epsilon_{CC}$ for screening according to [859] and [537, 538]. As may be seen from Fig. 10.3, the critical density decreases with increasing $\dot{M}_c$ and reaches a maximum in the pycnonuclear limit with $\rho_{c,max} = (0.7 - 1) \times 10^{10}$ g cm^{-3} for $\dot{M}_c \leq 10^{-9} M_\odot$/yr. So, in the case of accretion in binaries, the critical density for low $\dot{M}_c$ could thus be three to four times larger than in the single star. It has been shown in [746] that the equality $\epsilon_\nu = \epsilon_{CC}$ may be first achieved out of the centre thereby leading to a non-central explosion. Evolution of a close binary with two C–O white dwarf may be ended by merging of this binary because of a loss of the orbital angular momentum due to gravitational radiation. After merging, a thermally unstable core with a mass exceeding the chandrasekhar limit is formed, giving a strong thermonuclear explosion. In the "scenario machine" calculations, the frequency of such explosive events may be higher than the frequency of explosions, resulting from the accretion [551].

The carbon ignition in stars with $M > 8M_\odot$ takes place in the absence of degeneracy, but the resulting core of ^{16}O, ^{20}Ne and ^{24}Mg turns out to be completely or partially degenerate for $M < 13M_\odot$ [740]. For stars with $8 - 10M_\odot$ a degenerate ^{16}O + ^{20}Ne + ^{24}Mg core forms with a temperature insufficient for ^{20}Ne ignition. When the core mass reaches 1.37 $M_\odot$, the neutronization begins (see Table 10.1), determining the initial contraction rate [189, 473]. The oxygen O + O flash occurs in the contraction phase, when ρ_c approaches $\sim 2 \times 10^{10}$ g cm^{-3} [473, 946].

Burning in stars of mass 10–13$M_\odot$ leads to formation of an ^{16}O+^{20}Ne core with mass $1.37 - 1.5\, M_\odot$. As the rate of plasma neutrino losses increases with increasing density due to the increase of the plasmon energy $\hbar\omega_p$ with $\omega_p^2 = 4\pi n_e e^2/m_e$ (see Chap. 5, Vol. 1), the central regions cool to lower temperatures, and the temperature attains a maximum outside the centre, at a mass coordinate $m < 0.8\, M_\odot$ decreasing with increasing mass. The temperature inversion vanishes at $M = 13M_\odot$. The inversion leads to a non-central neon-oxygen flash.[3] The burning zone extends inward and outward and reaches the centre. The flash results in the formation of a non-degenerate core which evolves quietly until iron peak elements form. The dynamical effect of the flash is essential only for $M =$ 10–11$M_\odot$, when the helium shell ejection is possible. For $M = 11M_\odot$, we have $M_{He} = 2.8\, M_\odot$, $M_{ONe} = 1.42\, M_\odot$, where M_{He} is the mass of the core inside the helium shell.

Evolutionary calculations for helium stars with masses $M_{He} = 8, 3.3, 3.0, 2.8, 2.2\, M_\odot$ are presented in Fig. 10.4 from [747]. The initial stellar masses were $25, 13, 12, 11, 9M_\odot$, respectively. Figure 10.4 shows that a star with $M_{He} = 2.2M_\odot$ traverses the density of ^{24}Mg and ^{20}Ne neutronization before the neon and oxygen flashes, stars with $M_{He} = 3M_\odot$ ignite ^{16}O and ^{20}Ne on the boundary of strong degeneration, giving a weak flash removing the degeneracy, while for $M_{He} = 3.3$ and $8M_\odot$ neon and oxygen ignite at a weak degeneracy.

Note that the area with $\gamma < 4/3$ in Fig. 10.4 is different from the calculations made in [174, 523] and presented in Fig. 1.6, Vol. 1, where the region with $\gamma < 4/3$

[3] The neon burning begins by photodetachment of an α-particle which is subsequently captured. The neon burning increases the concentration of ^{16}O, and the formation of ^{24}Mg and ^{28}Si takes place.

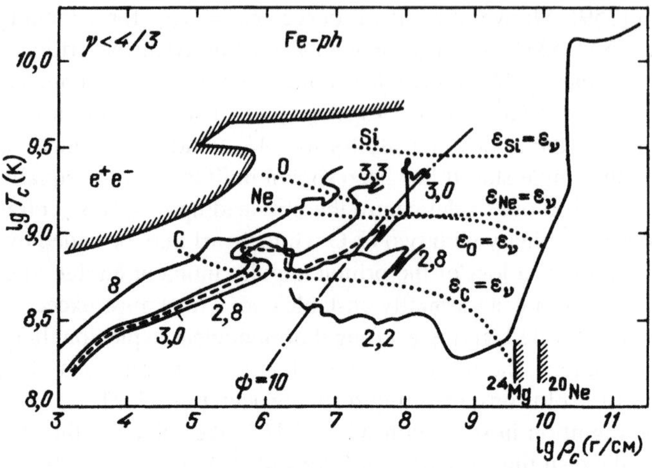

Fig. 10.4 Evolutionary curves for helium stars with $M_\alpha = 8, 3.3, 3.0, 2.8, 2.2\,M_\odot$ in the $(\lg T_c, \lg \rho_c)$ diagram. The masses equal 25, 13, 12, 11, $9\,M_\odot$, respectively. Given are approximate lines of ^{12}C, ^{20}Ne, ^{16}O, ^{28}Si ignition, for which burning rates are equal to rates of neutrino losses. To the right and underneath are marked the densities of the onset of ^{24}Mg and ^{20}Ne neutronization, while in the upper left corner is the area marked with $\gamma_1 < 4/3$ (γ_1 from (8.26)) resulting from the pair creation (e^+e^-) and photodissociation of iron peak elements (Fe–ph). The *dot–dashed line* $\psi = \mu_{te}/kT = 10$ (μ_{te} is the electron chemical potential) indicates a strong-degeneration region (on the right), from [747]

was found for iron dissociation with the formation of α-particles, protons and neutrons, taking into account electron degeneracy, relativism, pair formation and zero neutrino chemical potential. This might be due to the difference in the choice of iron photodissociation products in thermodynamic calculations.

Problem 1. For a polytrope $n = 3$, evaluate integrals of the form

$$I_{lk} = \frac{1}{M}\int_0^M \varphi^l(m/M)\xi^k(m/M)\,\mathrm{d}m \tag{10.29}$$

resulting from the substitution of (10.22–10.25) into (10.19) and (10.20).

Solution. Using (10.7), we have

$$\mathrm{d}m/M = \theta^3\xi^2\mathrm{d}\xi/M_3. \tag{10.30}$$

Substituting (2) into (1) and, bearing in mind that $\varphi\,(m/M) \equiv \theta^3(\xi)$ in conformity with (10.18) and (10.3), we obtain

$$I_{lk} = \frac{1}{M_3}\int_0^{\xi_1} \theta^{3l+3}\xi^{k+2}\mathrm{d}\xi = \frac{J_{pr}}{M_3} = \frac{J_{pr}}{J_{32}},$$

$$p = 3l + 3, \quad r = k + 2. \tag{10.31}$$

Table 10.3 Integrals J_{pr}

r \ p	1	2	3	4	5	6	7	8
1	4.848	2.132	1.293	0.9144	0.7043	0.5718	0.4809	0.4148
2	14.19	4.327	2.018	1.181	0.7881	0.5709	0.4372	0.3482
3	52.47	11.67	4.224	2.037	1.170	0.7517	0.5216	0.3823
4	222.8	37.94	10.85	4.318	2.127	1.207	0.7559	0.5080

The integrals

$$J_{pr} = \int_0^{\xi_1} \theta^p \xi^r \mathrm{d}\xi, \quad \xi_1 = 6.897 \tag{10.32}$$

are evaluated by integrating (10.4) numerically and finding $\theta(\xi)$. The results of the numerical evaluation of the integrals are given in Table 10.3.

Addendum.

The energetic method for investigation of stability of rotating stars in GR implies double integrals [138] of the type

$$I_{mn,pq} = \int_0^{\xi_1} \theta^m \xi^n \mathrm{d}\xi \int_0^{\xi} \theta^p \xi^q \mathrm{d}\xi,$$

$$I_{mn,(pq)^2} = \int_0^{\xi_1} \theta^m \xi^n \mathrm{d}\xi \left(\int_0^{\xi} \theta^p \xi^q \mathrm{d}\xi \right)^2,$$

equal for values occurring in [138] to

$$I_{3-1,12} = 0.4745 \qquad I_{31,74} = 0.4117$$

$$I_{31,34} = 2.096 \qquad I_{4-4,(12)^2} = 0.08089$$

$$I_{31,42} = 0.6609 \qquad I_{5-1,12} = 0.2139$$

$$I_{31,52} = 0.5161 \qquad I_{71,52} = 0.09474$$

Problem 2. Find analytical solutions of the Lane–Emden equation (10.4).

Solution [269]. Analytical solutions exist for $n = 0, 1, 5$. The case of $n = 0$ corresponds to an incompressible fluid, and the solution with boundary conditions (10.5) is

$$\theta_0 = 1 - \frac{\xi^2}{6}, \qquad \rho \sim \theta^n = \text{const.}, \qquad P \sim \theta^{1+n} \sim \theta. \tag{10.33}$$

Solution (1) represents the first two terms in the expansion of the solution near the centre for arbitrary n.

For $n = 1$, (10.4) becomes linear. Substitution of $y = \theta\xi$ leads to the oscillator equation $y'' + y = 0$. The solution satisfying the boundary conditions (10.5) reads

$$\theta_1 = \frac{\sin\xi}{\xi}. \tag{10.34}$$

Looking for algebraic solutions of (10.4) in the form

$$\theta_n = (1 + a\,\xi^l)^{-k} \tag{10.35}$$

we obtain, using expansion (1), that $l = 2$, $a = 1/6k$, so $\theta_n = \left(1 + (\xi^2/6k)\right)^{-k}$. Substituting into (10.4) gives

$$\left(1 + \frac{\xi^2}{6k}\right)^{-k-2}\left(1 + \frac{1-2k}{18k}\xi^2\right) = \left(1 + \frac{\xi^2}{6k}\right)^{-kn}. \tag{10.36}$$

This equality may hold only at $k = 1/2$, $n = 5$, so that

$$\theta_5 = \frac{1}{\sqrt{1 + (\xi^2/3)}}. \tag{10.37}$$

10.2 Explosions Resulting from the Thermal Instability Development in Degenerate Carbon Cores

10.2.1 Basic Equations

In the case of spherical symmetry the equations of stellar hydrodynamics in Lagrangian coordinates, including thermal processes, have the form (see Sect. 7.2.2)

$$\frac{\partial r}{\partial t} = v, \quad \frac{\partial v}{\partial t} = -4\pi r^2 \frac{\partial P}{\partial m} - \frac{Gm}{r^2}, \quad \frac{\partial r^3}{\partial m} = \frac{3}{4\pi\rho}, \tag{10.38}$$

$$\frac{\partial E}{\partial t} + P\frac{\partial}{\partial t}(1/\rho) = \epsilon_n - \epsilon_\nu + \epsilon_{\nu d} - \frac{\partial L_{\text{conv}}}{\partial m}, \tag{10.39}$$

$$\frac{\partial x_i}{\partial t} = -\sum_j \frac{\epsilon_{n,ij}}{E_{ij}}, \quad E_{ij} = \frac{Q_j}{a_{ij} A_i m_{\text{u}}}. \tag{10.40}$$

Here, in the energy equation (10.39) ϵ_n is the nuclear energy release rate, ϵ_ν is the rate of neutrino losses and $\epsilon_{\nu\text{d}}$ is the rate of heating due to interactions with neutrinos propagating from stellar interiors. In primary calculations of hydrodynamical collapse of an iron core, the latter phenomenon was called deposition [302]. In calculations of thermal explosions [277], this heating, promoting the nuclear fuel ignition, was called ignition. Only the convective heat flux L_{conv} (erg s^{-1}) is included in (10.39) since the heat transfer caused by heat conduction is usually negligible. In (10.40), i stands for elements involved in burning; the sum is taken over j reactions involving these elements; Q_j is the energy release in the j-th reaction; a_{ij} is the number of nuclei of the i-th element involved in the j-th reaction, $a_{ij} < 0$ for resulting nuclei.

Equations (10.38–10.40) have been solved in all studies of thermal flash, from [47]. In addition to differences in computational schemes, these studies vary in the adopted initial central densities, temperature profiles and methods for calculations of convective heat transfer. Let us now discuss briefly the results and group them by burning regimes.

10.2.2 Detonation

It has been assumed in [47] that after having begun in the centre, the burning will propagate over the star in the form of a detonation wave satisfying the Chapmen–Jouguet condition, i.e., travelling through the hot gas at the sound velocity of this gas. Subsequent calculations (see review in [946]) have shown that the detonation regime is self-maintaining. The explosion with detonation of the star on the boundary of thermal stability results in its complete run-away. Such a result follows from simple energetic estimations. As the density distribution over the star before explosion is nearly polytropic with $n = 3$ since degenerate electrons are the main contributors to the pressure, the binding energy $\overline{\epsilon}_b = 5 \times 10^{50}$ erg is therefore well below the gravitational stellar energy $\overline{\epsilon}_g = 3.1 \times 10^{51}$ erg. The carbon completely burnt out up to the iron peak elements with $Q = 7.7 \times 10^{17}$ erg g^{-1} releases for $M = 1.4\,M_{\odot}$ the energy $\overline{\epsilon}_n = 2.2 \times 10^{51}$ erg by far exceeding $\overline{\epsilon}_b$ and the neutrino losses $\overline{\epsilon}_\nu = 6 \times 10^{49}$ erg. Detailed numerical investigation of the propagation of a detonation wave in the degenerate core has related an oscillatory instability of the front, leading to its decay into the hydrodynamical shock wave, and slow burning (deflagration) front [520, 579].

10.2.3 Deflagration

The thermal instability development in a degenerate carbon core has been studied numerically in [541]. It has been found that the detonation does not set in, and the burning front travels at subsonic velocity in the deflagration regime. As the burning front propagates outward, pulsations develop and eventually bring about the run-away of the entire star. Subsequent calculations by the same group [277] have included the neutrino ignition caused by heating due to neutrino scattering off electrons. The calculations have been performed for several initial central densities and are reviewed in [525]. At $\rho_{c0} = 5.03 \times 10^9$ g cm^{-3}, the small pulsation is followed by strong expansion and contraction resulting in the formation of a powerful shock and a complete run-away of the star. At the very end, carbon burns in the detonation regime. Diverse cases of such a "delayed" detonation have been investigated later in more detail in [562,563]. The kinetic energy of the run-away is sufficient to account for the supernova energetics. In the variant with $\rho_{c0} = 9.22 \times 10^9$ g cm^{-3} the burning wave, before entering the envelope, becomes detonational as well, but results in a weak ejection only: $\sim 10^{-2} M_\odot$ with $\overline{\epsilon_{\rm kin}} = 1.4 \times 10^{49}$ erg. The major part, $\sim 1.4\, M_\odot$, collapses to form a neutron star. To account for the observed energetics of supernova explosions with remnants in the form of a neutron star, it is assumed that a magnetorotational mechanism acts subsequently [155], see Sect. 10.5. The central density $\rho_{c0} = \rho_c^* = 9 \times 10^9$ g cm^{-3} is assumed in [277] to be the boundary between models with complete run-away and a large energy release ($\rho_{c0} < \rho_c^*$), and collapsing models with a weak ejection and low energy release. As is visible from Fig. 10.3, the variant with $\rho_{c0} > \rho_c^*$ may occur only at a low accretion rate in the binary if the pycnonuclear reaction can be described using the approximation of complete relaxation (see Chap. 5, Vol. 1).

The convection is ignored in the above calculations, and all dissipative processes arise from the scheme viscosity due to the crudeness of the calculation grid. Calculations based on the mixing-length theory and using diverse convection parameters exhibit a high sensitivity of the results to the mixing length (see reviews [207,946]). The thermonuclear explosion theory cannot be regarded as complete because of the crudeness of calculation grids arising from limited capacities of even the most powerful computers, and by reason also of the absence of a rigorous convection theory. Various restrictions to the thermonuclear explosion models may be derived from comparing these explosions to observations of Type I supernovae (SN I) they are associated with, and from calculations of nucleosynthesis in degenerate matter [207]:

– The "radioactive" model of the light curve due to the decay chain ^{56}Ni $\to$ ^{56}Co $\to$ ^{56}Fe requires production of a large amount of ^{56}Ni in the ejected matter. Such an amount is produced in both the deflagration and detonation models, though the latter fits the light curve somewhat worse than the former. The half-lives in this chain are 6.1 and 77 days, and positronic β-decay energies are $\epsilon_{\beta+} = 2.133$ and 4.568 MeV, respectively [590]. The energy $\epsilon_{\beta+}$ produced in positronic β-decay includes the energy of subsequent annihilation and equals

$$\begin{aligned}\epsilon_{\beta+}(A,Z) &= (m_{A,Z} - m_{A,Z-1})\,c^2 + m_e c^2 \\ &= B_{A,Z-1} - B_{A,Z} - (m_n - m_p)c^2 + m_e c^2 = -\epsilon_\beta(A,Z).\end{aligned}$$

Here, the nucleus $(A, Z-1)$ has a larger binding energy than the unstable nucleus (A, Z), leading to positronic beta decay of the latter. Stable nuclei (A, Z) are bound more strongly than nuclei $(A, Z-1)$ with $B_{A,Z} > B_{A,Z-1}$, like ^{56}Fe and ^{56}Mn, and electron capture occurs when $\epsilon_{\rm Fe} > \epsilon_\beta(A, Z) > 0$, from the above definition.

– In the deflagration model, the outer regions of a star have the time to expand considerably before the burning front approaches them. This causes the nuclear static equilibrium to have no time to set in, and the ejected matter to contain considerable amounts, $\sim 0.3\, M_\odot$, of the intermediate elements ^{28}Si, ^{32}S, ^{36}Ar, ^{40}Ca observed in SN I near the light maximum. The detonation model would supply intermediate elements only under conditions of a lowered density, as is the case at a rapid rotation, for instance.

– In the deflagration model, substantial discrepancies with observations arise for an isotropic composition of iron peak elements. Because of a slow burning in the deflagration regime, a large number of electron captures have time to occur at increased temperature, thus raising the degree of neutronization, or the N_n/N_p ratio. The ^{54}Fe/^{56}Fe ratio in the ejected matter turns out to be three to five times the observed ratio equal to 0.061 for the solar system, the Ni/Fe ratio over all stable isotopes is approximately five times larger as well, and the ^{58}Fe/^{56}Fe ratio for $\rho_{c0} = 2.5 \times 10^9$ g cm^{-3} proves to be $\sim$40 times the solar ratio.[4] Such a strong discrepancy makes doubtful the deflagration model arising in most calculations of carbon explosion.

10.2.4 Spontaneous Burning and Detonation

In the absence of satisfactorily reliable calculations, a natural initial temperature profile is adiabatic. The possibility for such a profile to be adopted is provided by the convection action at the beginning of carbon burning at $T \geq 3 \times 10^8$ K. The temperature distribution near the centre then has the form

$$T = T_c \left[1 - \left(\frac{2}{3} \pi G \rho_c^2 P_c^{-1} \gamma_{2c} \right) r^2 \right]. \tag{10.41}$$

Here, $\gamma_2 = (\partial \ln T / \partial \ln P)_S$ (see (8.22)). The expansion (10.41) follows from (10.1) and (9.97) which, for a region near the centre, give

$$m = \frac{4\pi}{3} \rho_c r^3, \quad P = P_c - \frac{2\pi G}{3} \rho_c^2 r^2, \tag{10.42}$$

and from the expansion of T for the adiabatic case

$$T = T_c + \frac{1}{2} \frac{d^2 T}{dr^2} r^2 = T_c + \frac{1}{2} \frac{T_c}{P_c} \gamma_{2c} \frac{d^2 P}{dr^2} r^2. \tag{10.43}$$

In the case of smooth variations of T, the burning rate depends weakly on radius, so the burning front travels with a velocity determined by the initial temperature profile, and it is not excluded that $v_f \gg v_s$. The velocity $v_f \to \infty$ at $T \to$ const., i.e., the exhaustion may occur simultaneously throughout the region. When $v_f > v_s$, the regime of front propagation is called spontaneous [1076]. The possibility for

[4] Note that ^{56}Fe results mainly from the ^{56}Ni $\to$ ^{56}Co $\to$ ^{56}Fe decay so that $(N/Z)_{^{56}\mathrm{Ni}} = 1 < (N/Z)_{^{54}\mathrm{Fe}} = 14/13$; stable isotopes of Ni have $A = 58, 60, 61, 62$ and 64.

such a regime to set in during carbon flashes is pointed out in [213], where a model is examined with $\rho_{c0} = 3 \times 10^9$ g cm^{-3} and containing equal amounts of ^{12}C and ^{16}O. The time τ_e for ^{12}C to be exhausted is determined for $0.6 < T_9 < 0.9$ by

$$\tau_e = 10^{-3} T_9^{-20} \text{ s.} \tag{10.44}$$

Evaluating the coefficient in (10.41), we have

$$T = T_c(1 - 2.0 \times 10^{-16} r^2). \tag{10.45}$$

The velocity at which the front of spontaneous burning propagates is given by

$$D_{sp} = \left(\frac{d\tau_e}{dr}\right)^{-1} = 1.2 \times 10^{17}\, T_{c9}^{20}\, r^{-1} \times (1 - 2.0 \times 10^{-16} r^2)^{21} \text{ cm s}^{-1}. \tag{10.46}$$

At

$$r > r_s = \frac{10^5}{\tau_e} \text{ cm}, \tag{10.47}$$

where $\tau_e = 10, 1, 0.1$ for $T_{c9} = 0.63, 0.71, 0.80$, respectively, the velocity D_{sp} becomes lower than the sound velocity v_s, and the regime of spontaneous burning comes to an end. All the above estimates refer to a constant density. The numerical calculations for the propagation of a burning wave, performed in [213] by the method of characteristics, show that after the end of the spontaneous burning regime a shock forms and the detonation regime may set in. As the detonation front enters the regions of sufficiently low densities, the burning front broadens and the detonation breaking becomes possible with subsequent transition to the deflagration regime. It is pointed out in [213] that numerous calculations of thermal explosions involving the artificial viscosity method are too crude and can barely resolve the picture of hydrodynamic flow during carbon explosion. The emergence of the deflagration or detonation in these calculations is due to the numerical scheme properties and may not fit the reality. Further progress in the field seems to be possible as a result of applying other numerical schemes, say, the method of characteristics without artificial broadening of the shock front.

10.2.5 Instabilities of Nuclear Flames

To fit observational constraints, the speed of the flame in the deflagration model must be larger than in the simple laminar model. Such acceleration occurs if the surface of the flame front is unstable, and its wrinkling leads to an increase in the effective burning surface, which in turn increases the effective speed of the flame propagation.

Two types of instabilities, Rayleigh–Taylor (RT) and Landau–Darrieus (LD), are considered in this relation. RT instability is developed due to accelerating propagation of the flame front and has a hydrodynamical origin. In the non-linear regime, the turbulent scale cascade is formed with a Kolmogorov spectrum of the turbulent velocities $v(l)$ and energy per unit of the wave vector $\varepsilon(k)$ of fluctuations [614]

$$v(l) = v(L)\left(\frac{l}{L}\right)^{1/3}, \quad \varepsilon \sim k^{-5/3}, \quad l = \frac{2\pi}{k}. \tag{10.48}$$

In [744], the cut-off scale of hydrodynamic instability of the flame front was taken at the level where the turbulent velocity (10.48) is equal to the laminar flame velocity $v_{\rm lf}$, which is highly subsonic. This minimal value $l_{\rm min}$, called the Gibson-scale $\lambda_{\rm Gibs}$, is defined by the relation

$$v(\lambda_{\rm Gibs}) = v_{\rm lf}. \tag{10.49}$$

LD instability [614] is connected with the flame propagation process and has a universal character. Its minimal length is defined by dissipative effects (like thermal conductivity) which smear out small-scale perturbations with $l < \lambda_{\rm Mark}$, where $\lambda_{\rm Mark}$ is a so-called Markstein scale. In the numerical experiment [743], the LD instability was detected on scales of the order of several $\lambda_{\rm Mark}$. The Markstein scale is about an order of magnitude larger than the flame front thickness $\lambda_{\rm fl} \simeq 10^{-5}$–$10^{-2}$ cm [744], so that

$$\lambda_{\rm Mark} \approx 10\,\lambda_{\rm lf}. \tag{10.50}$$

In the absence of the flame, the cascade of the hydrodynamical turbulence is developed on the Reynolds scale

$$\lambda_{\rm Re} = L\,Re^{-3/4}, \tag{10.51}$$

where Reynolds number $Re \approx 10^{14}$ in a typical white dwarf with $L \simeq 10^8$ cm, so $\lambda_{\rm Re} \simeq 10^{-3}$ cm. This is of the order of the laminar front width $\lambda_{\rm lf}$. The influence of LD instability may change the spectrum (10.48) in the region $l < \lambda_{\rm Gibs}$, where burning effects could be more important than the hydrodynamical ones. If we take into account that [203] the speed of the laminar flame is about a few per cent of the sound speed $v_{\rm s}$

$$v_{\rm lf} \simeq 0.02\,v_{\rm s} \tag{10.52}$$

and the maximal turbulent velocity $v(L)$ is of the order of $v_{\rm s}$, we arrive using (10.48), (10.49) and (10.52) to the value $\lambda_{\rm Gibs} = L\,(v_{\rm lf}/v_{\rm s})^3 \approx 10^{-5}L = 10^3$ cm.

Another scale that has a pure numerical origin is a length of the grid Δ. The number of grids usually does not exceed a few tens of thousands, so in 2D simulations $\Delta \geq 5 \times 10^{-3}\,L \simeq 5 \times 10^5$ cm. We see that Δ in all simulations strongly exceeds all the above values of the scales connected with the instabilities.

In the calculations of a SN Ia explosion, instabilities on the scales of several Δ are suppressed due to damping, having a pure numerical origin. For the case of RT instability, this means only an account of modes with wavelengths of a few $\Delta < \lambda < L$, and LD instability operating on smaller scales cannot be detected in such a grid.

Understanding that small-scale instabilities are very important for increasing the effective burning surface and the effective burning speed leads to their account in the frame of a statistical model, called subgrid (SG), because it deals with scales that are less than the size of the grid Δ. In [744], the SG model was used for description of a turbulent flame propagation, and it was suggested that the energy spectrum of perturbations is the Kolmogorov one at all scales. The input of the turbulence was considered by adding terms into equations of motion and energy, representing the turbulent stresses and dissipation on SG scales. The energy generation rate was given indirectly by prescription of the speed of the flame propagation v_t, using the power dependence

$$v_t(\Delta) = v_{\mathrm{lf}} \left(\frac{\Delta}{\lambda_{\mathrm{Gibs}}} \right)^{D_F - 2}, \tag{10.53}$$

which may be connected with a fractal structure of the flame due to RT instability. A preliminary set of calculations made for D_F in the interval 2.17–2.33 and $\lambda_{\mathrm{Gibs}} = 0.123\,\Delta$ and $0.051\,\Delta$ shows that the model is insufficient to explain the flame acceleration needed to reproduce observational data.

In [203], the fractal behaviour of the effective surface of burning S_{ef} was considered, which depends on the radius R as

$$S_{\mathrm{ef}} \sim R^{D_F} \tag{10.54}$$

with the fractal dimension $D_F > 2$ due to development of LD instabilities in the region $l_{\min} < l < l_{\max}$. Here, $l_{\min}$ is of the order of λ_{Mark}. With (10.54), the effective speed of the flame propagation would be similar to (10.53)

$$v_t = v_{\mathrm{lf}} \left(\frac{l_{\max}}{\lambda_{\mathrm{Mark}}} \right)^{D_F - 2}. \tag{10.55}$$

To estimate D_F for LD instability, an integro-differential equation, describing nonlinear cascades was solved. The value of D_F depends on $\gamma = 1 - (\rho_b/\rho_u)$, where ρ_b and ρ_u are densities before and after the density jump on the flame front. The dependences which were used for modeling of the flame propagation in SN Ia, accounting for LD instability, are

$$D_F(\gamma) = 2 + 0.6\gamma^2, \quad \text{and} \quad D_F(\gamma) = 2 + \gamma^2. \tag{10.56}$$

In the laboratory flames there is $\gamma \simeq 0.8$, and in SN explosions the density jump is suppressed due to degeneracy of the matter, and $\gamma \leq 0.4$. Model calculations have been performed in [203] for a SN Ia explosion with account of both RT and LD instabilities. RT instability was supposed to be acting on scales larger than a few tens of λ_{Gibs}, having fractal dimension 2.50. Values smaller than about 2.4 do not give acceptable explosions, even with account of LD instability. LD formalism is applied to all shorter wavelengths down to ten times the flame thickness λ_{lf}. Calculations have shown that account of LD instability helps, making robust explosions out of what would have been weak ones. However, the complete solution depends heavily on what one does outside the LD domain, where RT instability prevails.

The problems of turbulent flame acceleration and the possibility of transition from the deflagration to the detonation regime have been considered in [745]. Additional diffficulties in the realization of this transition have been revealed and analyzed.

10.3 Collapse of Low-Mass Stellar Cores

After quiescent ^{12}C burning in stars with initial mass $8 - 10M_{\odot}$, a degenerate O+Ne+Mg core forms and then loses its stability owing to ^{24}Mg and ^{24}Ne neutronization. According to calculations, the nuclear fuel effect consists in slowing to some extent the contraction relative to iron core collapse, but no ejection occurs [747].

The neutronization may be the reason for the onset of collapse of accreting an iron white dwarf in a binary when its mass exceeds $M_{\rm cn} = 1.18\ M_{\odot}$ (see Sect. 10.1).

If a single star with initial mass $M > 10M_{\odot}$ rapidly loses its envelope after the iron core formation, and the remnant mass slightly exceeds the Chandrasekhar limit (see Sect. 11.1) $M_{\rm ch}$, then after cooling, when $\rho_{\rm c}$ approaches $1.15 \times 10^9 (1.24 \times 10^9)$ g cm^{-3}, the star will start to lose its stability owing to ^{56}Fe neutronization. For $M - M_{\rm ch} \ll M_{\rm ch}$, the cooling time may be sufficiently large, thus being consistent with the observed Type I supernova explosions in elliptic galaxies, when massive stars are rare or completely absent [188]. A slow accretion in a binary may cause a similar loss. Note that the same statement remains qualitatively valid for cooling or accreting white dwarfs that consist of ^{24}Mg, ^{40}Ca or another heavy element. The energy released by the ^{56}Fe formation after the onset of collapse is even less efficient here than during the O+Ne+Mg core collapse. Neglecting the terms $\sim T^2$ and GR effects, the dependence $M(\rho_{\rm c}, T)$ becomes, on substituting (10.22) without the third term in parentheses into (10.19) without the last term [189],

$$M = \frac{5.83}{\mu_{\rm Z}^2}\left(1 - 5.4 \times 10^4 \frac{\mu_{\rm Z}^{2/3}}{\rho_{\rm c}^{2/3}} + 1.7 \frac{10^{-7}}{A} \mu_{\rm Z}^{4/3} \frac{T}{\rho_{\rm c}^{1/3}}\right). \tag{10.57}$$

At temperature T and the same central density, the stellar mass exceeds the mass of the cold star by ΔM:

$$\frac{\Delta M}{M} = \frac{1.7 \times 10^{-3}}{A} \mu_{\rm Z}^{4/3} \frac{T_7}{\rho_{c9}^{1/3}}, \quad T_7 = \frac{T}{10^7\ {\rm K}},$$
$$\rho_9 = \rho/10^9\ {\rm g\ cm}^{-3}. \tag{10.58}$$

The stability loss will occur at the stage of the finite size core of the new phase, when $\rho_{\rm c}$ exceeds the neutronization density $\rho_{\rm n}$ by a small but finite value [189], see Sects. 11.1 and 12.4. To obtain an estimate for ΔM, it is sufficient to substitute the initial neutronization density from Table 10.1 into (10.58). Combining (10.58) with the cooling curve for white dwarfs (Sect. 11.1) gives the dependence $\tau(\Delta M)$, i.e., the delay of collapse as a function of mass excess ΔM[5]

$$\tau = 2.18 \times 10^8 \frac{(\mu_Z/2)^{16/3}\left(M/10^3 \Delta M\right)}{(\mu/2)(A/12)^{7/2}\rho_{c9}^{5/6}}\ \text{yr}$$

$$\text{for} \quad x_Z = 1, \quad \mu_Z = A/Z, \quad \mu = A/(Z+1). \tag{10.59}$$

The Debye temperature θ is defined as

$$\theta = 0.775\frac{\hbar\omega_\mathrm{i}}{k} = \frac{3.5 \times 10^3 \sqrt{\rho}}{\mu_Z}\ \text{K}, \qquad \omega_i^2 = \frac{4\pi e^2 n_\mathrm{e}}{\mu_Z m_\mathrm{u}}, \tag{10.60}$$

(see also Chap. 1, Vol. 1).

A qualitative picture of the instability development due to neutronization is given in [189]. When the central density exceeds the initial neutronization density ρ_n, the star becomes unstable for an equilibrium composition, that is, for a very high neutronization rate. However, it remains stable with respect to a frozen composition, that is, in the absence of neutronization at a constant μ_Z. At a small $\rho_\mathrm{c} - \rho_\mathrm{n}$, the neutronization ($\beta$-process) rate is low, and it is only this rate that determines the initial rate of star contraction. With increasing density

- The β-process rate increases
- GR effects become more important
- Heating of the star occurs owing to adiabatic contraction and non-equilibrium β-reactions. The last type of heating occurs when the Fermi energy of electrons in strongly degenerate matter considerably exceeds the value of the binding energy difference between successive nuclei equal to the Fermi energy of electrons at which neutronization starts. Under such conditions, neutronization is accompanied by simultaneous energy losses due to outflowing neutrinos, and heating due to entropy increase (see Chap. 5, Vol. 1). The heating results eventually in the dissociation of iron and decreases the value of $\gamma < 4/3$ in frozen β-processes.

All these phenomena lead to hydrodynamical collapse at a rate close to the free-fall velocity. Calculations of contraction caused by neutronization require simultaneous solution of the hydrodynamical equations (10.38–10.40) and equations of neutronization kinetics, describing time variations of chemical composition due to beta capture and decay reactions in nuclei (see Chap. 5, Vol. 1), which determine the contraction rate at this stage. The non-equilibrium heating leads to an appreciable temperature increase and makes the "cold" collapse impossible.

---

[5] The relation (10.59) is valid for $\Delta M/M > 6 \times 10^{-5}(\rho_\mathrm{c}^{1/6}\mu_z^{1/3}/A)$, when the temperature at the point of stability loss T_cr is above $0.1\,\theta < T_\mathrm{cr}$, where θ is the Debye temperature of crystal degeneration, at which the thermal energy of the ion is of the order of the lowest oscillation energy of the Coulomb crystal lattice $\hbar\omega_i$.

Calculations of the collapse of low-mass stellar cores in the approximation of homologous contraction at a given density profile [150] have been performed in [72, 73]. In [210] the gasodynamic equations (10.38–10.40) have been solved simultaneously with kinetic equations of neutronization for the case of collapse of a low-mass iron core. A hydrostatically equilibrium star of ^{56}Fe with $M = 1.2\,M_{\odot}$, $\rho_c = 1.78\times10^9$ g cm^{-3}, $T_c = 10^8$ K and temperature distribution $T = T_c(\rho/\rho_c)^{0.1}$ has been taken for the initial model. The equation of state allows for the degeneracy and relativism of electrons and for a perfect nuclear gas. The stellar mass exceeds the mass limit (Sect. 10.1), and accordingly, the initial central density exceeds the neutronization density $\rho_n = 1.15\times10^9$ g cm^{-3} by a finite magnitude. The difference $\rho_c - \rho_n$ determines the initial neutronization and star contraction rates. An approach based on solving the time differential equations yielded by dividing the star into Lagrangian zones by an implicit method has been worked out in [210] for thorough calculation of slow initial and rapid hydrodynamic stages. The artificial viscosity has been used, neutrino cooling completely taken into account (i.e. electron and positron captures and decays (URCA processes), plasma neutrinos, etc.), while convection and heat conduction have been ignored. The conditions for velocity, $v = 0$ at the centre, and pressure, $P = 0$ at the outer boundary, have been taken for boundary conditions. Equations of neutronization kinetics are solved at $T < 3\times10^9$ K for all the reactions of the chain $^{56}\mathrm{Fe} \to {}^{56}\mathrm{Mn} \to {}^{56}\mathrm{Cr} \to {}^{56}\mathrm{V} \to {}^{56}\mathrm{Ti}$. At $T > 3\times10^9$ K, the transition to the nuclear statistic equilibrium was switched on. To obtain a smooth description of this transition, a kinetic equation is introduced for quantity f, the weight fraction of matter in statistic equilibrium. The variation rate for f is chosen according to the (γ, p) reaction on iron, which depends exponentially on T. The fraction $(1-f)$ is still described by a set of equations of neutronization kinetics. At $f > 0.99$, the equilibrium concentration of nuclei with only one kinetic equation, describing the change of the total number of protons and neutrons (free and bound ones) with time is adopted (Chap. 1, Vol. 1).

The distribution of parameters in the initial model is given in Fig. 10.5. Figure 10.6 shows temperature T, entropy S and μ_Z in the centre of the star

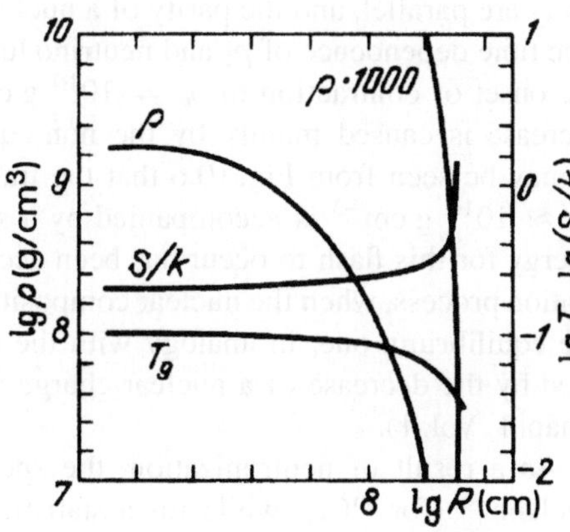

Fig. 10.5 The distributions of density ρ, temperature T and entropy S over the initial hydrostatic model ($T_9 = T/10^9$ K), from [210]

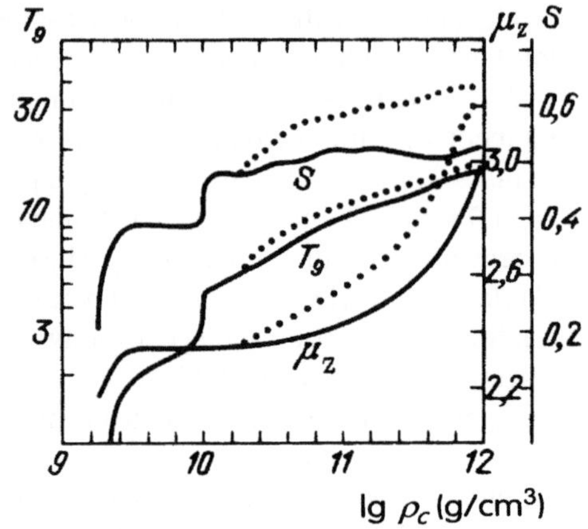

Fig. 10.6 T_9, S (in units of k/m_u), and $\mu_Z = \left[\sum_i (Z_i x_1/A_i)\right]^{-1}$, the number of nucleons per electron, versus central density ρ_c. Solid lines do not take into account of the Gamov–Teller resonance, dotted lines do it. For $10 \le \lg \rho_c \le 10.3$, curves are shown smoothed over pulsations (from [210])

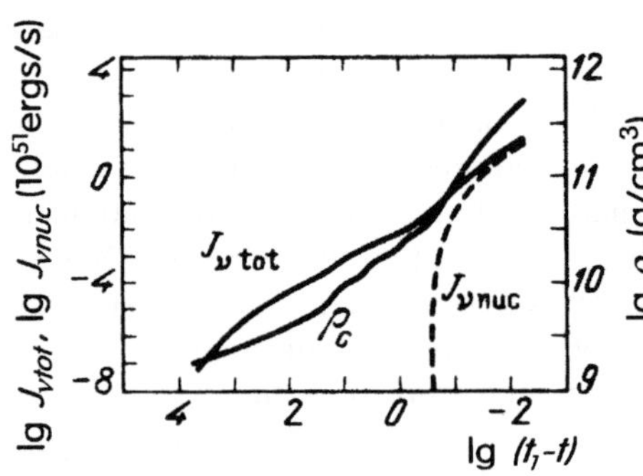

Fig. 10.7 Total neutrino luminosity $J_{\nu,\mathrm{tot}}$, luminosity due to β-processes on nucleons $J_{\nu,\mathrm{nuc}}$, and ρ_c versus time t. The time $t = t_1 = 5098$ s corresponds to $\lg \rho_c = 11.96$ (from [210])

versus central density ρ_c. The dotted line represents the results of more accurate calculations of neutrino emission including electron captures through the giant Gamov–Teller resonance, from [402–405]. The efficiency of non-equilibrium heating increases here, being partially due to the high excitation energy of nuclei involved in Gamov–Teller transitions in which the spins of the leptons ($e\nu$) flying away are parallel, and the parity of a nucleus does not change (see Chap. 5, Vol. 1). The time dependence of ρ_c and neutrino luminosity J_ν are given in Fig. 10.7. From the onset of contraction to $\rho_c = 10^{10}$ g cm^{-3}, $T = 3 \times 10^9$ K, the temperature increase is caused mainly by the non-equilibrium heating through β-processes. It may be seen from Fig. 10.6 that the transition to static equilibrium at $T_9 > 3$, $\rho_c \approx 10^{10}$ g cm^{-3} is accompanied by a strong heating, like a thermal flash. The energy for this flash to occur has been accumulated during the preceding neutronization process, when the nuclear composition has been increasingly departing from the equilibrium one, in analogy with the cool neutronization, which is accompanied by the decrease of a nuclear charge at constant nuclear mass [122] (see also Chap. 1, Vol. 1).

As a result of neutronization, the specific binding energy equals 8.72 MeV nucleon^{-1} (for ^{56}Cr), while for a statistically equilibrium state at the same density $\rho_c = 10^{10}$ g cm^{-3}, $E_b = 8.74$ MeV nucleon^{-1}. An excess in binding energy

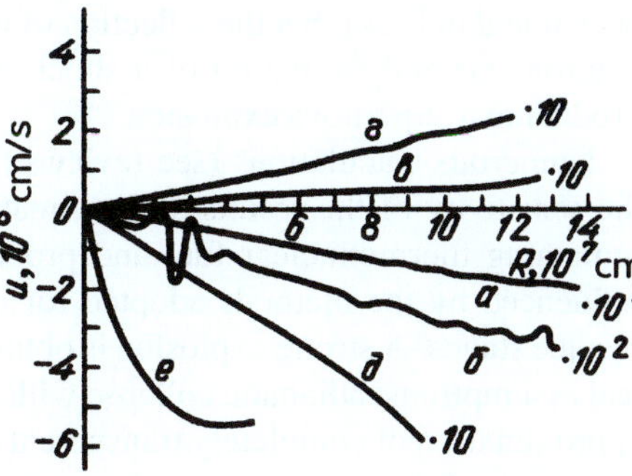

Fig. 10.8 Profiles of velocity u versus radius R at $\lg \rho_c = 9.96$ (**a**), 10.00 (**b**), 10.07 (**c**), 10.18 (**d**), 10.52 (**e**), 11.96 (**f**), from [210]. The real velocities are multiplied by numbers that mark the curves

of 0.02 MeV nucleon^{-1} converts into heat during the transition to equilibrium and increases the central temperature T_c from 3×10^9 to 4.7×10^9 K. The velocity distribution along a radius for different central densities is presented in Fig. 10.8. The thermal flash gives rise to a sound wave (curve (*b*) in Fig. 10.8) which turns into almost homologous pulsation of the star as a whole. After several tens of pulsations, they decay at $\rho_c \approx 2 \times 10^{10}$ g cm^{-3} by being damped by neutrino radiation flow. At $\rho_c \geq 10^{12}$ g cm^{-3}, the collapse becomes non-homologous, and the central stellar regions become non-transparent for neutrino radiation. Further calculations in [210] have been performed with the adiabatic approximation and resulted in the formation of a homologous core with mass 1.1 $M_\odot$, reflecting the shock wave (bounce) (see also [473]). A correct inclusion of neutrino processes may substantially decrease the power of the bounce, as in the case of collapse of more massive stars discussed in Sect. 10.4.

10.4 Hydrodynamical Collapse of Stellar Cores

Iron cores with mass $M_{\rm Fe} \geq 1.4\,M_\odot$ lose their stability through the iron dissociation which directly leads to a rapid collapse. The neutronization role and time interval in which the contraction proceeds with β-decay rate increase with decreasing mass. As noted in Sect. 10.1.2, the iron core forms in stars with initial mass $M_i > 10M_\odot$, while for $M > 13M_\odot$ all stages of nuclear burning proceed smoothly. As all evolutionary calculations yield significant uncertainty in the relationship $M_{\rm Fe}(M_i)$, the stability loss due to the iron dissociation is certain for single stars with $M_i > 13 \pm 3M_\odot$.

Hydrodynamical calculations of iron core collapse were first performed in [302], and soon after in [46, 540]. A solution has been searched for the hydrodynamical equations (10.38–10.40), and massive stellar cores ($M \geq 2M_\odot$) on the boundary of hydrodynamical stability have been taken for initial conditions. These studies concern the effect of electron and muon neutrinos on collapse, the role of neutrino deposition leading to the envelope heating and probable ejection, and the effect of burning of thermonuclear fuel ^{12}C and ^{16}O remaining around the iron core. It has been noted in [136] that the reflection of infalling matter from the surface of a stable neutron star and formation of a shock wave (bounce) may also be important for producing a supernova explosion.

Numerous calculations (see reviews [207, 472, 525, 1040, 1053]) have revealed the sensitivity of the results to the equation of state of nuclear matter, quantity of remaining thermonuclear fuel and probable convection. The results are strongly influenced by the methods adopted for including neutrinos at the transparent and opaque stages. A strong explosion is obtained by many authors under opposite physical assumptions: adiabatic collapse with a density-dependent adiabatic index γ and approximation of completely transparent neutrinos. The former yields the explosion as a result of the bounce of incident matter from the surface of the forming neutron star and formation of a powerful shock wave propagating outward. In the approximation of completely transparent neutrinos the mean neutrino energy, 30–50 MeV, proves sufficient for a strong deposition leading to explosion.

Calculations including neutrino processes in a self-consistent way over the entire star were first performed by Nadyozhin [711, 729]. These calculations yield the formation of a core opaque for neutrinos. For such a core, the energy equation (10.39) was completed by the neutrino heat transfer term, and the equation for diffusion of a lepton charge in the neutrino opaque core was added [728] (see also Chap. 5, Vol. 1). These calculations take into account both the emission of neutrinos outside the core which is opaque for neutrinos (neutrinosphere) and absorption of neutrinos outgoing of this core with heating of matter by them (deposition). The absorption is calculated from the diluted Fermi spectrum of neutrinos propagating from the neutrinosphere [525, 728]. The calculations [729] show that promoting the conversion of kinetic energy into heat and eliminating the bounce almost completely are the major effects of neutrinos. The maximum velocity of fall obtained in [729] for collapse of a $2M_\odot$ core took place for 0.12 s after the onset of collapse and was equal to 3700 km s^{-1}. This velocity is 1.5 times less than the free-fall velocity. The temperature in the post-shock region increases to $(4-5) \times 10^{10}$ K and the matter there becomes opaque for neutrinos. The shock boundary almost coincides with the neutrinosphere and neutrino core boundary. The growth of the neutrino core is caused by conversion of the kinetic energy of matter into heat energy when this matter traverses the shock front. The mass of the hot neutron core grows rapidly (in ~0.04 s) to $\sim 0.8\,M = 1.6\,M_\odot$, the radius to ~80 km. After that, there occurs a long stage of residual envelope accretion onto a neutrino core which loses energy and gradually contracts. Four seconds later, the neutrino star turns into a hydrostatically equilibrium hot neutron star with $R \approx 15.5$ km that keeps cooling. Most of the energy radiated in the form of neutrinos, $\sim 5 \times 10^{53}$ erg $\approx 0.15\,\mathrm{Mc}^2 \approx 0.3\,\mathrm{M}_\odot \mathrm{c}^2$, is lost during cooling of the hot neutron star, over a time $t \approx 20$ s. The mean energies of neutrinos arising in collapse are ~10 MeV,[6] and their deposition is insufficient to reverse envelope collapse and to obtain a powerful explosion. Inclusion of thermonuclear burning of oxygen

[6] In [679], the mean energies of electron neutrinos equals 14 MeV, electron antineutrinos, 15 MeV, other neutrino species (ν_μ, $\tilde{\nu}_\mu$, ν_τ, $\tilde{\nu}_\tau$), 32 MeV. The total energy of emitted neutrinos $\sim 6 \times 10^{53}$ erg is distributed almost equally between these six types of particles.

in the envelope, muon- and tau-neutrinos, momentum transfer from neutrinos to nuclei caused by coherent scattering due to neutral currents in the weak interaction Lagrangian does not significantly alter the results. A number of studies published after [711, 729] still use the bounce mechanism to obtain a powerful explosion in a crude description of neutrino processes. The inclusion of neutrinos becomes more accurate in later studies, so that obtaining an explosion becomes difficult.

Studies reviewed in [207, 472, 1040, 1053] give more exact descriptions of neutronization kinetics, describe neutrinos with use of a transfer equation and include non-ideal effects in the equation of state at near-nuclear densities. Under conditions of high neutrino concentration, heavy nuclei exist up to almost nuclear densities because of suppression of electron captures. This makes leptons with $\gamma = 4/3$ predominate in pressure at the stage of an opaque neutrino core. At $\gamma = 4/3$, the homologous contraction of matter with $\tau_\nu > 1$ accounts for the name of homologous core given to central regions with $\tau_\nu > 1$. On approaching the density $\rho = 2.5 \times 10^{14}$ g cm^{-3}, the nuclei dissociate, and the contribution of non-relativistic nucleons to pressure grows drastically, resulting in $\gamma > 4/3$ and the non-homologeneity of the collapse. A strong shock with specific entropy increased by six or seven times arises on the boundary of the homologous core but well inside the neutrinosphere. This shock loses an important part of its energy via the dissociation of heavy nuclei and, on reaching the neutrinosphere boundary, emits the residual energy in the form of a neutrino pulse with $L_\nu \approx 10^{54}$ erg s^{-1}. Consequently, the shock decays and outburst does not occur during the collapse of a star with $M_{\rm core} \geq 1.5\, M_\odot$.

Another attempt to obtain a powerful outburst during iron core collapse has been made in [95]. This study deals with the cooling stage of a neutrino star with luminosity $L_{\nu_e} = L_{\tilde{\nu}_e} = 4 \times 10^{52}$ erg s^{-1}, the mean energy of emitted neutrinos $\sim$5 MeV, neutrinosphere radius $R_\nu \approx 30$ km. The mass of the collapsing core is $1.64 - 1.69\, M_\odot$. A small part of the neutrino flux ($\sim$0.1%) is absorbed in outer layers, causing their appreciable heating. According to calculations [95] for $t = 0.1 - 0.8$ s, the neutrino heating results in an increase in pressure and the formation of a shock which propagates outward to give an outburst of $\sim 4 \times 10^{50}$ erg, insufficient to account for the energy of the most powerful supernovae. The fall velocity of the material before the shock is $(2 - 3) \times 10^9$ cm s^{-1}, and the velocity of the shock front motion outward is $\sim 5 \times 10^8$ cm s^{-1} for a time $t > 0.5$ s from the onset of collapse. Calculations in [723] use a description of neutrino transfer which is more accurate compared to [95] and give no explosion or outburst at all. Another possibility for an explosion has been considered in [436].

According to estimations, the reaction of the neutrino–antineutrino annihilation

$$\nu + \tilde{\nu} \to e^+ + e^-, \tag{10.61}$$

opposite to the reaction of neutrino creation due to pair annihilation is likely to be important for the shock formation and supernova explosion. The results [436] based on neutrino fluxes and spectra from [679] show that more than 10^{51} erg can be converted through the $(\nu\tilde{\nu})$ annihilation into (e^+e^-)-pair energy and subsequently into heat above the neutrinosphere. The contribution of all the three species of neutrinos to the pair formation is equal. For the same energy, the cross-section for $(\nu_e \tilde{\nu}_e)$ annihilation caused by charged and neutral currents is larger than $(\nu_\mu \tilde{\nu}_\mu)$ and $(\nu_\tau \tilde{\nu}_\tau)$ annihilation cross-section caused only by neutral currents, but the average μ- and τ-neutrino energy is more than twice the e-neutrino energy, so that their contributions become nearly equal.

The neutrino fireball is enhanced by increasing neutrino flux outward due to convection, and neutrino production in the process of run-away accretion of the residual matter into the neuron star [183, 1079]. The possibilities for obtaining a robust supernova explosion due to neutrino energy deposition from this fireball were analyzed in [298, 301]. Next, we discuss some interesting physical effects arising from hydrodynamical collapse.

10.4.1 Low-Energy Window for Neutrinos

As the cross-section for interaction of neutrinos with matter increases with energy approximately as $\sim E_\nu^2$, the interaction depth decreases with decreasing neutrino energy, and they escape more freely. The escape of such neutrinos becomes still easier because their mean free path is further increased by electron and neutrino degeneracy, due to the occupation of a larger number of quantum states, and only a few free states remain for particles which appear as a result of interactions. It has been assumed in [684] that a low-energy window would increase the outward neutrino energy flux and promotes the supernova explosion during collapse.

The calculations [604] have shown that the neutrino degeneracy reduces only the $(\nu\, e)$-scattering, but does not noticeably alter the scattering from nucleons and coherent scattering from nuclei. The reason for this is that the energy transfer is very small for collisions of neutrinos with heavy nuclei and nucleons, and changes occur only in momentum, while collisions with electrons lead to comparable magnitudes of energy and momentum transfer. It follows from the collision integral for fermions that when the energy exchange is negligible and $\epsilon = \epsilon'$, the terms representing degeneracy cancel out. Very much in the same way, the induced processes acting for bosons (instead of the degeneracy of fermions) give no contribution to the collosion integral in the case of Thomson scattering with negligible energy transfer by the photon (see Chaps. 2 and 5, Vol. 1). The scattering on nucleons and nuclei is the major source of neutrino opacity. Calculations of neutrino scattering due to neutral currents [324, 604] have given for the mean free path $\lambda_{\rm el}$ with respect to scattering from nucleons and coherent scattering from nuclei

$$\lambda_{\rm el} \approx 1.0 \times 10^6 \left(\frac{\rho}{10^{12}\ {\rm g\ cm^{-3}}}\right)^{-1} \left(\frac{x_Z \overline{A}}{12} + x_{\rm n}\right)^{-1}, \qquad (10.62)$$

where x_Z and $x_{\rm n}$ are the weight concentrations of heavy nuclei and free neutrons. An analogous calculation of the mean free path $\lambda_{\rm e}$ with respect to neutrino scattering from electrons has given [604]

$$\frac{\lambda_e}{\lambda_{el}} \approx 20\left(\frac{\mu_Z}{2}\right)\left(\frac{x_Z\overline{A}}{12}+x_n\right)\left(\frac{\epsilon_{Fe}}{30\text{ MeV}}\right)^2\left(\frac{kT}{3\text{ MeV}}\right)^{-2}, \tag{10.63}$$

where the kinetic Fermi energy of electrons is

$$\epsilon_{Fe} = (m_e c^4 + p_{Fe}^2 c^2)^{1/2} - m_e c^2,$$
$$p_{Fe} = \hbar\left(\frac{3\pi^2\rho}{\mu_Z m_u}\right)^{1/3} = \left(\frac{1.027\rho}{10^6\mu_Z}\right)^{1/3} m_e c. \tag{10.64}$$

Here, T is the temperature in the scattering region, and μ_Z is the number of nucleons per electron from (10.23). We may see from (10.63) that for the characteristic values $\mu_Z = 2$, $x_Z\overline{A} \geq 12$, $\epsilon_{Fe} \geq 30$ MeV and $kT \leq 3$ MeV, the inequality $\lambda_e \gg \lambda_{el}$ holds at any degree of neutrino degeneracy, and from (10.62), that the path λ_{el} is less than the neutrinosphere radius R_ν. The neutrino degeneracy thus has no effect on the low-energy window which has a small aperture and does not insure any considerable increase in neutrino flux density that would remove difficulties encountered in obtaining a supernova explosion during collapse.

10.4.2 Asymmetric Neutrino Emission During Collapse of a Star with a Strong Magnetic Field

In a strong magnetic field, the electron energy depends on its spin orientation [616]

$$E = \left(n + \frac{1}{2} + \sigma_e\right)\frac{|e|\hbar}{m_e c} B + \frac{p_z^2}{2}. \tag{10.65}$$

Here, $n = 0, 1\ldots$ is the number of the Landau level and $\sigma_e = \pm 1/2$ is the electron spin component along the **B** direction. The state with $\sigma_e = -1/2$, $n = 0$ has a minimum energy so that electrons in a strong magnetic field are polarized. The state with $\sigma_e = 1/2$, $n = 0$ is energetically preferable for positrons owing to the charge of opposite sign. The degree of the electron polarization depends on the parameter p determined by relations [294, 329]:

$$\begin{aligned}
&\text{(a) } p = \frac{\hbar\omega_B}{kT} \text{ for non-degenerate, non-relativistic gas,}\\
&\text{(b) } p = \frac{\hbar\omega_B}{\epsilon_{Fe}} \text{ for degenerate, non-relativistic gas,}\\
&\text{(c) } p = \frac{\sqrt{\hbar\omega_B m_e c^2}}{\epsilon_{Fe}} \text{ for degenerate, relativistic gas,}\\
&\text{(d) } p = \frac{\sqrt{\hbar\omega_B m_e c^2}}{kT} \text{ for non-degenerate, relativistic gas.}
\end{aligned} \tag{10.66}$$

Here, $\epsilon_{\rm Fe}$ is given in (10.64), and

$$\omega_{\rm B} = \frac{eB}{m_{\rm e}c} \tag{10.67}$$

is the Larmor frequency.

The degree of polarization is close to unity at $p > 1$. The existence of strong magnetic fields, $B = 10^{12} - 10^{13}$ G, at the surface of neutron stars follows from pulsar observations [670]. Fields in stellar interiors may be even stronger than this. As neutrinos have left-handed helicity, they have a preferred escape orientation after being produced in reactions with polarized $e^{\pm}$. Owing to magnetic anisotropy, the angular dependence for the intensity of neutrino emission is determined by the function

$$g(\theta) = \frac{1}{4\pi}(1 + a\cos\theta), \tag{10.68}$$

where θ is the angle at which neutrinos escape with respect to the magnetic field. Evaluating the energy density E_ν and the radial neutrino energy flux F_ν averaged over the solid angle,

$$E_\nu = \frac{1}{c}\int_\Omega I_\nu \, d\Omega, \quad F_\nu = \int_\Omega I_\nu \cos\theta \, d\Omega, \tag{10.69}$$

and using the intensity of neutrino emission $I_\nu \sim g(\theta)$, we obtain for the distribution (10.68)

$$F_{\nu B} = \frac{acE_\nu}{3} \text{ erg cm}^{-2}\text{ s}^{-1}, \tag{10.70}$$

where $F_{\nu B}$ is that part of the neutrino flux due to magnetic anisotropy. If the total momentum of directed neutrino emission of a neutron star is

$$P_\nu = f\frac{Q_\nu}{c}, \tag{10.71}$$

where $Q_\nu = 0.1\,Mc^2$ is the total emitted energy which is nearly equal to the binding energy of the neutron star, and the neutron star acquires the velocity

$$v_n = 0.1\,cf. \tag{10.72}$$

The asymmetry factor f is obtained by averaging (10.70) over the neutrinosphere surface, and the value of a is proportional to the factor p from (10.66). If the poloidal component of the neutron star field has the dipole (case D) form

$$B_z = d\,(3\cos^2\theta - 1), \quad B = d\sqrt{3\cos^2\theta + 1} \tag{10.73}$$

and $a \sim B_z$, from (10.66a,b), then for $E_\nu = $ const. integrating over the neutrinosphere gives $P_\nu = 0$. For $a \sim B_z/\sqrt{B}$ from (10.4.4c, d), we obtain

$$f = \frac{4a_0}{3\sqrt{2}} \int_0^1 \frac{3x^2 - 1}{(3x^2 + 1)^{1/4}} \, \mathrm{d}x \approx -0.0776\, a_0, \tag{10.74}$$

where a_0 is the value of a for $\theta = 0$ at the pole, and

$$\int_0^1 \frac{3x^2 - 1}{(3x^2 + 1)^{1/4}} \, \mathrm{d}x = -0.0823.$$

In this case, the mean neutrino momentum is opposite to its directions at the poles.

If the field inside the star is uniform and directed along the z axis (case P), then $a =$ const. and

$$P_\nu = \frac{4a}{3c} Q_\nu, \text{ i.e., } f = \frac{4a}{3}. \tag{10.75}$$

Note that the main flux on the neutrinosphere is $F_\nu = cE_\nu/4$. The URCA process on nucleons

$$\begin{aligned} &\text{(a) } \mathrm{p} + \mathrm{e}^- \to \mathrm{n} + \nu, \\ &\text{(b) } \mathrm{n} + \mathrm{e}^+ \to \mathrm{p} + \tilde{\nu}, \end{aligned} \tag{10.76}$$

is the major source of opacity for neutrinos in neutron stars in the phase of emission of the main neutrino pulse. Nucleons are usually in a non-polarized state for their large masses. The probability of the reaction (10.76) in a strong magnetic field with non-polarized nucleons has been calculated in [329]. An asymmetry in neutrino escape arises only when electrons are captured from the low Landau level ($\sigma_\mathrm{e} = -1/2, \mathrm{n} = 0$), the maximum asymmetry corresponding to

$$a_{\max} = \frac{1 - \alpha^2}{1 + 3\alpha^2}, \quad \alpha = G_A/G_V \approx 1.25. \tag{10.77}$$

This asymmetry is solely due to a difference in vector and axial coupling constants. For small values of B, we have approximately

$$a = a_{\max} p^2 = 0.099\, p^2, \quad p \ll 1. \tag{10.78}$$

Substituting f from (10.74) and (10.75) into (10.72) and using (10.78), and (10.4.4c, d) gives

$$\begin{aligned} v_\mathrm{n} &= \frac{4c}{A} \frac{B}{B_\mathrm{c}} \left(\frac{m_\mathrm{e} c^2}{\epsilon_\mathrm{Fe}} \right)^2, \; \epsilon_\mathrm{Fe} \gg kT \\ &= \frac{4c}{A} \frac{B}{B_\mathrm{c}} \left(\frac{m_\mathrm{e} c^2}{kT} \right)^2, \quad \epsilon_\mathrm{Fe} \ll kT \\ B_\mathrm{c} &= \frac{m_\mathrm{e}^2 c^3}{e\hbar} = 4.414 \times 10^{13}\,\mathrm{G} \\ A &= 303\ (P); \quad A = -5.21 \times 10^3\ (D). \end{aligned} \tag{10.79}$$

A negative value of v_n implies a neutron star velocity opposite to the magnetic field at the poles. For strongly degenerate electrons with $\epsilon_{Fe} = 60\, m_e c^2$, we have the neutron star velocity due to neutrino emission in the reaction (10.4.13a)

$$v_n^{(\nu)} \approx 1.2(B/B_c)\,\mathrm{km\,s^{-1}} \qquad (P) \qquad (10.80)$$
$$\approx -0.064(B/B_c)\,\mathrm{km\,s^{-1}} \qquad (D).$$

When electrons are strongly degenerate, positrons have a small density and are weakly degenerate with a mean energy $\sim kT$. Under these conditions, the parameter p is determined from (10.4.4d), and the degree of asymmetry of antineutrinos resulting from the reaction (10.4.13b) is $(\epsilon_{Fe}/kT)^2$ times the degree of asymmetry of neutrinos in the reaction (10.4.13a). For a steady-state composition, the numbers of escaping neutrinos and antineutrinos are equal: $\overline{\epsilon}_\nu = \epsilon_{Fe}$, $\overline{\epsilon}_{\tilde{\nu}} \approx kT$. As noted by Voloshin (private communication), the momentum transferred from antineutrinos to the neutron star is $\sim\epsilon_{Fe}/kT$ times the momentum transferred from neutrinos under the same conditions and has the same direction. At $\epsilon_{Fe} = 10\,kT$, the velocity acquired by the neutron star is

$$v_n = v_n^{(\nu)} + v_n^{(\tilde{\nu})} \approx 13.2(B/B_c)\,\mathrm{km\,s^{-1}} \qquad (P)$$
$$\approx -0.072(B/B_c)\,\mathrm{km\,s^{-1}} \qquad (D). \qquad (10.81)$$

These velocities are too low to break a close binary by explosion or to account for the large velocities of radio pulsars.

10.4.3 Neutrino Oscillations in Matter

Differences in interactions of various neutrino species with matter change possible neutrino oscillations in matter in comparison with vacuum [697]. The reason for these differences is that the interactions of electron neutrinos with matter are due to charged and neutral currents, while μ- and τ- neutrino interactions are almost solely due to neutral currents because of large muon and tau-lepton masses. The effect of the medium reduces to altering the oscillation depth (the mixing angle θ_m) and oscillation length l_m, in which case the interactions may both damp ($\theta_m < \theta$) and enhance ($\theta_m > \theta$) oscillations (θ, l refer to vacuum). The enhancement of oscillations may occur either for neutrinos (then the antineutrino oscillations are damped) or for antineutrinos. A medium of varying density may produce a resonant enhancement of oscillations when neutrinos of a given energy become mixed in a thin layer at [810]: $\rho = \rho_0 = (\Delta(m^2)/\epsilon_\nu)\,(\mu_Z/0.65 \times 10^{-7}) \cos 2\theta$ g cm^{-3}, $\sin 2\theta_m = \sin 2\theta/A$, $l_m = l/A$, $A = \left[\cos^2 2\theta\ (1 - \rho/\rho_0)^2 + \sin^2 2\theta\right]^{1/2}$, $l = 2.5\,\epsilon_\nu/\Delta(m^2)$(m); the resonance width is $\Delta\rho/\rho_0 = \tan 2\theta$; here, ϵ_ν is the neutrino energy (MeV), $\Delta\, m^2)$ is the difference in squares of neutrino masses (eμ) or (eτ) in eV^2.

Calculations show that the resonant matter-induced oscillations $(\nu_e \nu_\mu)$ and $(\nu_e \nu_\tau)$ on the Sun may considerably reduce the neutrino flux from it [93, 697]. This may account for the very small value of the measured fluxes of neutrino ν_e radiation from the Sun [61].

The same mechanism is applied in [698] to neutrinos emitted during collapse. As the sign of this effect is different for interactions with electrons and neutrons, the oscillations in the core and envelope may be partially compensated. In the absence of oscillations, the mean energies of emitted neutrinos are lower, and their neutrinospheres are larger for ν_e and $\tilde{\nu}_e$ than for ν_μ, $\tilde{\nu}_\mu$, ν_τ, $\tilde{\nu}_\tau$, so that equipartition is nearly achieved between energies of all the six neutrino species emitted during collapse [679]. The oscillation action outside the neutrinospheres may give rise to significant departures from this equipartition. Either neutrinos or antineutrinos are subject to oscillations, so one of these species goes out almost without oscillational distortions, and the other changes significantly. Removal of equipartition between different neutrino species arises because of the large difference between energies of (e) and (μ, τ) neutrino, and energy dependence of the resonant mixing. Since detection methods are substantially different for ν_e and $\tilde{\nu}_e$, for observational predictions to be correct it is important to know for what particles, ν_e or $\tilde{\nu}_e$, oscillations become enhanced in matter. Experimental hints to the existence of neutrino oscillations (also from solar neutrino experiment) emerge from anomalous cosmic ray data [880], and from Los Alamos experiments with a neutrino beam [881].

10.4.4 Convective Instability in Collapsing Stellar Cores

To maintain the neutrino energy flux out of the core, the concentration of leptons in the region of the neutrinosphere falls abruptly outward, following the drop in pressure. This leads to the convective instability development [357] since the element motion inward is related to an excess of the element density over that of the surroundings for the pressure to be compensated, and vice versa. Convective motions in the neutrinosphere might bring hot material outward and increase the mean energy and flux of escaping neutrinos whose deposition could initiate an explosion.

The possibility for such an instability to develop and its consequences have been analyzed in [623, 625]. An equilibrium equation of state of a hot dense matter [605] with neutrinos in thermodynamical equilibrium has been used. The convective instability investigations have led to the conclusion that within the neutrinosphere, at $\mathrm{d}x_l/\mathrm{d}r < 0$ (x_l is the fractional concentration of leptons), the convection development is possible only at $\gamma_l = (\partial S/\partial x_l)_{\rho,P} < 0$. Studying the shock motion inside the neutrinosphere reveals that the entropy in the post-shock region really increases with decreasing density. Calculating γ_l from the equation of state in [605] shows that at high T and ρ insuring a purely nucleon composition, the value $\gamma_l > 0$, so the convection development is hardly possible at $\rho > 10^{14}$ g cm^{-3}. In the region transparent to neutrinos at $\rho < 10^{12}$ g cm^{-3}, the lepton fractional concentration increases outside due to increases in electron fraction, and lepton gradient will tend to stabilize these regions against convective overturn. Thus, lepton-driven overturn is possible only at densities $10^{12} < \rho < 10^{14}$ g cm^{-3}, where neutrino trapping occurs, and $\gamma_l < 0$. Account of convection in the local approximation in hydrodynamic calculations has shown that the degree of enhancement of neutrino flux is ambiguous and the possibilities for obtaining an explosion remain doubtful [623].

If neutronization and dissociation of nuclei occur simultaneously in the region of the shock dissipation near $1 M_{\odot}$ (enclosed mass), the entropy gradient becomes negative and entropy-driven convection is developed. The evolution of a new-born neutron star with account of this convection was studied in [246] on the basis of general relativistic "Henyey-like" code [245] with inclusion of a mixing-length scheme for convective energy and lepton transport. It was found that such convection can enhance the neutrino luminosity in the quasistatic post-collapse stage of the core of a massive star by at times an order of magnitude, and neutrinospheric temperature and the energy of the emitted neutrinos are increased by up to ~50%. This helps in obtaining the supernova explosion, but is not enough to solve all the problems [247].

It was suggested in [299] that in the presence of sufficiently large destabilizing lepton gradients, the core may become unstable to a large-scale overturn that dredges up neutrinos from the very centre of the core. This can increase dramatically the neutrinos flux through the infalling envelope, which together with the large kinetic energy of the convective plumes may lead to explosion. 3D calculations of large-scale overturn in protoneutron stars have been performed in [288, 1003].

10.4.5 Two-Dimensional and Three-Dimensional Calculations of Neutrino Convection

An unstable lepton and entropy profiles formed after ~10 ms of the creation of a shock wave and bounce of the core, can drive a violent Rayleigh–Taylor-like overturn as studied in [247]. Combination of 2D hydrodynamics treated by the piecewise-parabolic method (PPM), with independent 1D neutrino transport were used. The explosion was obtained in this model, while without neutrino transport, or with account of convection in the 1D hydrodynamical model, the explosion did not happen. Extended calculations in a similar model with 2D neutrino transport have been presented in [248]. Here, convection becomes so violent that spherical and even plane symmetry of the core are strongly broken, neutrino emission and mass ejection proceed anisotropically, inducing the explosion with ejection of a few tens of high entropy clumps, and giving a kick to a neutron star, which by estimations can reach a speed of ~500 km/s.

3D simulations of convection in the shocked matter of the supernova core have been performed in [908], assuming that the neutrino radiation from the protoneutron star is radial, but axisymmetric. The asphericity of the neutrino flux was connected with rapid rotation of the protoneutron star. The formation of high-entropy hot bubbles and a jet-like explosion was obtained as a result, but the explosion energy problem was not considered.

The PPM method was used in 2D calculations [546, 547] of neutrino-driven supernova with convective overturn and accretion. The effects of convection obtained here are less pronounced than in [248], while a powerful explosion is obtained in a certain, although rather narrow, window of core ν fluxes in which 1D models do not explode. The maximum attainable velocities of the kick are estimated to be around 200 km/s.

Extensive 2D studies of a supernova explosion following the collapse of cores of two massive stars (15 and $25 M_{\odot}$) have been performed in [467]. The calculations begin at the onset of core collapse and stop several hundred milliseconds after the bounce, at which time successful explosions of appropriate magnitudes have been obtained. The explosion is powered by the heating of the envelope due to neutrinos emitted by the protoneutron star. This heating generates strong convection outside the neutrinosphere which was demonstrated to be critical to the explosion. Convection leads to violation of the radiative equilibrium between neutrino emission and absorption. Thus, explosions become quite insensitive to the physical input parameters, such as neutrino cross-section or the nuclear equation of state parameter.

A smooth particle hydrodynamics (SPH) code was used for 2D calculations with spherically symmetric gravity and a realistic equation of state. A 2D explicit code for neutrino transport was developed with account of the most important processes of neutrino emission, absorption and scattering. A peculiar characteristic of neutrino processes in supernova is that the dominant process which leads to neutrino trapping does not affect the neutrino spectrum, because elastic scattering between nucleons and neutrinos happens almost without the energy exchange. This leads to the situation where the optical depth can be large without thermalizing the fields. As a result, local thermodynamic equilibrium cannot be assumed.

The main features common to all 2D simulations made in [467] are the following. After an initial period of dynamical infall lasting a few hundred milliseconds, the central density becomes supernuclear, the core hardens and a bounce shock is launched. Within a few milliseconds, this shock stalls due to energy losses at a radius $\sim$150 km. At this point, 2D calculations begin to differ greatly from 1D computations because of the onset of hydrodynamical instabilities. Most important to the supernova is the neutrino-driven convection that lasts for over 100 ms until a successful explosion is achieved.

We can distinguish between lepton- and entropy-driven mechanisms of convective instability. The lepton instability develops over a time scale of 15 ms after the bounce (Fig. 10.9) in the region around the neutrinosphere of the protoneutron star, situated at $r \sim 40$ km. Maximum turnover velocities are about 4000 km s^{-1}, while at other times the velocity can decrease below 1000 km s^{-1}. The width of the unstable region is about 15 km.

The entropy instability develops farther out ($r \sim 50$–150 km) and is driven by a negative entropy gradient. At first, this negative gradient is due to the stall of the bounce shock and becomes smaller as the shock weakens. Subsequently, this unstable entropy gradient is maintained by neutrinos, emerging from the hot protoneutron star heating the matter preferentially at smaller radii. The entropy instability is fully developed 20–25 ms after the bounce. Figure 10.9 illustrates the convective patterns

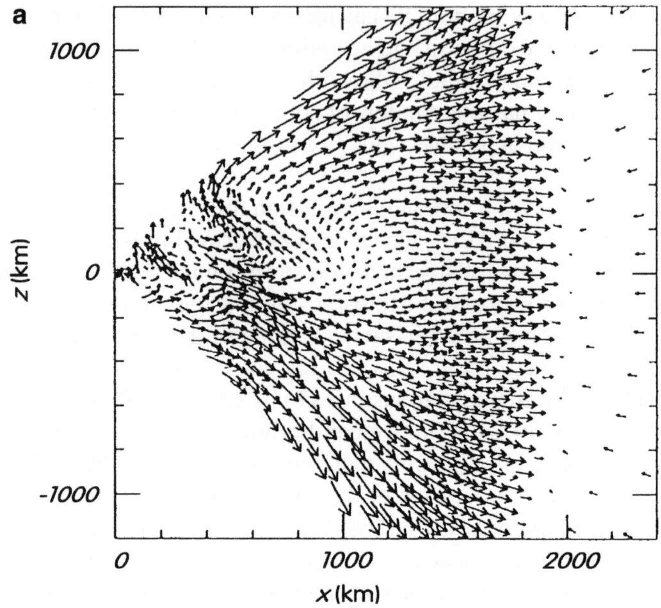

Fig. 10.9 (**a**) Lepton-driven convection in the 25 $M_\odot$ progenitor 15 ms after the bounce

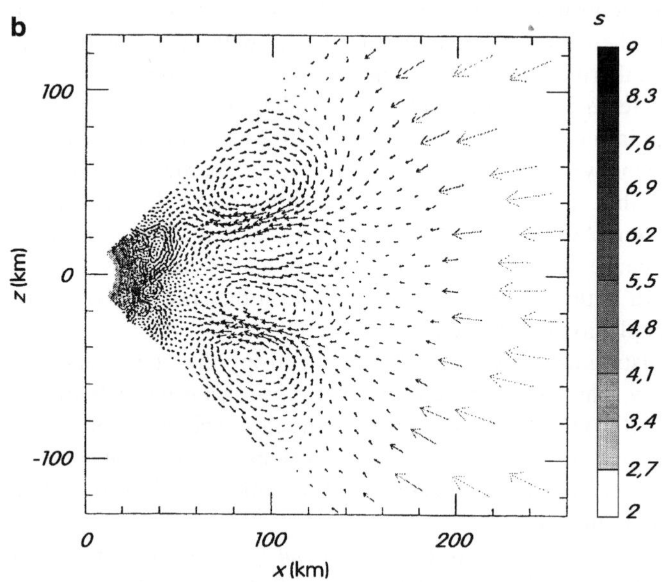

Fig. 10.9 (**b**) Entropy-driven convection in the 25 $M_\odot$ progenitor 25 ms after the bounce

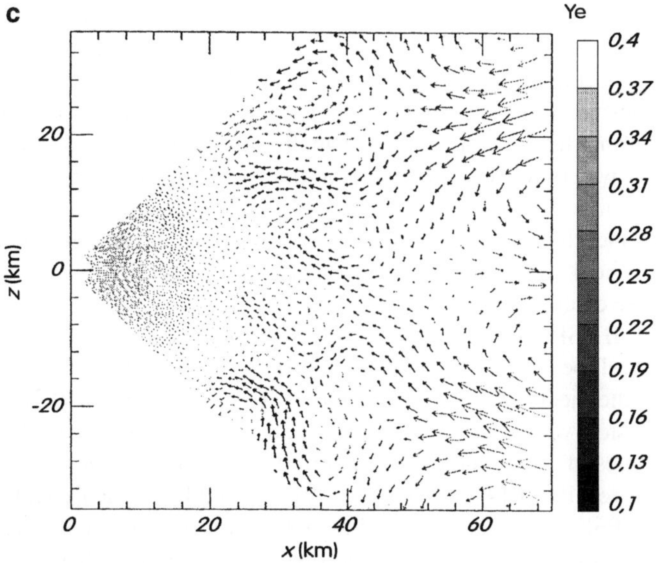

Fig. 10.9 (**c**) Exploding 25 $M_\odot$ progenitor 200 ms after the bounce. Postshock velocities are $\sim$15,000 km s^{-1}

present in the supernova at 25 ms after the bounce. Note the two distinct regions of instability, the lepton-driven instability at 40 km and the entropy instability at 100 km. The convective velocities in the entropy instability region are much larger than in the lepton-driven case and sometimes exceed 10,000 km/s^{-1}.

After an amount of time, which varies depending on the progenitors and the details of the physics, but which is always of the order of 50–100 ms after the bounce, the energy build up in the convective region and the thinning of the infalling envelope allow the shock to move forward decisively. Adiabatic expansion cools the material behind the shock and allows the free nucleons to recombine. In particular, the reconstruction of alpha particles liberates $\sim$7 MeV/nucleon of energy.

By 100 ms after the bounce, the shock is located about 1,000 km above the neutron star and is able to impact significant velocities (comparable to or larger than the escape velocity) to the infalling matter. At a time $\sim$200 ms after the bounce, the entropy of the region above the protoneutron star reaches about $\sim$20 k/nucleon due to heating by the electron neutrinos which is faster than its replacement by an infall of cold matter, and the density is only $\sim 3 \times 10^9$ g/cm^3. By that time, the energy of the explosion is larger than one foe (10^{51} ergs), producing a supernova, independently of future events in the neighbourhood of the protoneutron star. The beginning of the propagation of the successful explosion shock through the envelope of the progenitor is shown in Fig. 10.9.

Investigations of collapse and explosion of rotating cores have shown that the explosion proceeds in the same manner as in the non-rotating case, except the rotation has a strong influence on the shape of the convective patterns that develop above the protoneutron star. As was pointed in [467], the most wanting aspect of the calculations remains in neutrino physics, because of the obvious difficulties of radiation transport in multidimensions. The basic flux-limited approximation leaves much to be desired in nearly optically thin regions, and also inelastic neutrino nucleus interactions have been ignored, which could play an important role in the explosion. As was indicated in [467], at a time $\sim$200 ms the SPH model ceased to adequately resolve the atmosphere above the neutron star and simulations had to be stopped. The fact that this time is close to the time of the formation of the successful explosion shock makes it desirable to check more carefully the role of numerical effects in these calculations.

The same 2D SPH code was used in [466] for investigation of mixing in the expanding envelope of SN 1987A due to hydrodynamical instabilities. The necessity of mixing in the ejecta of SN 1987A follows from observations of X-rays at early stages, expansion velocities seen in line widths of different infrared lines and in gamma-ray lines of ^{56}Co, and in the form of the optical light curve.

Numerical results of [467] have been reproduced by an analytical description, presented in [96]. There, preference has been given to these results leading to a powerful supernova explosion, but not to results obtained by other authors [704] in 2D simulations with similar input physics, but different numerical scheme and initial conditions, where no explosion was obtained in the presence of the neutrino-driven convection.

10.4.6 Explosion of Rapidly Rotating Star

The model of the core-collapse supernova produced by rapidly rotating star was considered in [518]. A scenario is advanced of the formation of a binary system of neutron stars as a result of fragmentation of a neutron protostar almost in static equilibrium, formed after the stage of collapse. In the formation of the binary system, the redistribution of mass and angular momentum between its components is presumed, suggesting the orbit of the binary system to be circular and the axial rotation of the components to be negligible. The subsequent fate of the binary system follows the way described in [208, 209] where the idea is developed of the explosion of a low-mass neutron star as a component of a binary system of neutron stars. This alternative scenario can be treated basically as a supplement to those papers, considering the formation of such a binary system during the collapse of a massive stellar core that initially was rotating fairly rapidly at the end of the evolution in static equilibrium of a massive enough star. Such stars are often considered to be the progenitors of type II supernovae, including the remarkable explosion of SN 1987A in the LMC. As shown in [518], the lifetime of a binary system of neutron stars deep inside the envelope of a progenitor is brief, because of energy and angular momentum losses due to gravitational radiation, especially if the masses of the components are comparable to each other. Limiting that time to about 1 h, compatible with data on SN 1987A, it was found that the ratio of the masses $0.3 - 0.4 \lesssim (M_2/M_1) \lesssim 1$, where M_2 is related to the low-mass component, and M_l is a mass of the more massive component. After the components converge, the low-mass neutron star fills its Roche lobe, and its mass is then transferred rapidly to the more massive neutron star. When its mass has decreased to its critical value $M_{2cr} \simeq 0.1 M_\odot$, an explosion happens with an energy release $\sim 10^{51}$ erg, typical of supernova explosions, according to [208, 209]. It is undoubtedly that the convergence of neutron stars in a close binary system does not lead to a trivial finale – their coalescence – since it is forestalled by the explosive instability of the low-mass neutron star. In the case of coalescence, everything would end in silent collapse due to the transformation of angular momentum into gravitational radiation, but an explosion would not ensue. The key problems with the present scenario – the fragmentation of a rotating stellar core and mass transfer in a close binary system of neutron stars – need further study and resolution, by numerical modeling, in particular. Further development of this model is presented in [519].

10.4.7 Standing Accretion Induced Instability

The mechanism of instability in shocked converging flows, considered in [388], is based on the cycle of entropic and acoustic waves in the subsonic region of the flow between the shock and the sonic surface surrounding the accretor. The advection of entropy perturbations towards the accretor produces outgoing acoustic waves which propagate towards the shock. Perturbed by these acoustic waves, the shock produces new entropy perturbations, thus closing the entropic–acoustic cycle. This cycle is unstable (i.e. the amplitude of the new entropy perturbation exceeds the initial one) if the sound speed at the sonic point is much larger than the sound speed at the shock. This mechanism might play a destabilizing role in various astrophysical environments such as wind accretion or disc accretion.

Standing accretion induced instability (SASI) is widely considered as one of the candidates to solve the problem of the explosion of core-collapse supernova, without including magnetic field action. In [761], this phenomenon was studied by 2D simulations with including neutrino heating and realistic EOS. It was found that SASI may enhance neutrino heating. Although g-mode of proto-neutron star may enhance the SASI growth, the simulations just including the pressure perturbation as a mimic of g-mode induced sound wave reveal no significant effect on the shock dynamics. Similar results have been obtained in 3D simulations of this problem in [542].

By performing axisymmetric hydrodynamic simulations of core-collapse supernovae with spectral neutrino transport based on the isotropic diffusion source approximation scheme, support of the assumption was obtained in [952] that the neutrino-heating mechanism aided by the standing accretion shock instability and convection can initiate an explosion of a 13 $M_\odot$ star. These results show that bipolar and unipolar explosions are likely to be associated with models that do or do not include rotation. Models that include rotation form a north-south symmetric bipolar explosion that leads to a larger enclosed mass behind the shock than in the corresponding unipolar explosions. For the relatively low mass progenitor chosen in this study, the results suggest that rotation significantly aids the occurrence of a neutrino driven explosion. The explosion energies at the termination of the simulation were less than 10^{50} erg for all the models, which seems to be not enough for these type of supernovae.

10.4.8 Acoustic Explosion Model

A mechanism for core-collapse supernova explosions was considered in [249] which relies upon acoustic power generated in the inner core as the driver. In the simulation using an $11 M_{\odot}$ progenitor, an SASI-type oscillation with a period of $\sim 25 - 30$ milliseconds (ms) arises $\sim$200 ms after bounce. However, this instability is not the primary agent of explosion. Rather, it is the acoustic power generated early on in the inner turbulent region, and most importantly, but later on, by the excitation and sonic damping of core g-mode oscillations, see Sect. 14.1.3. An $l = 1$ mode with a period of $\sim$3 ms grows at late times to be prominent around $\sim$500 ms after bounce. The accreting protoneutron star is a self-excited oscillator, "tuned" to the most easily excited core g-mode. It was claimed in [249] that the acoustic power seen in 11 $M_{\odot}$ simulation is sufficient to drive the explosion >550 milliseconds after bounce.

In simulations [707], the acoustic oscillations also have been generated in the neutron core, and in Fig. 10.10, the kinetic energy of these oscillations is represented. This plot was made by subtracting the kinetic energy of the matter expanding by the supernova shock from the total kinetic energy of the star. The amplitude of the kinetic energy of the core oscillations is $1.3 \cdot 10^{48}$ erg. The simulations in [707]

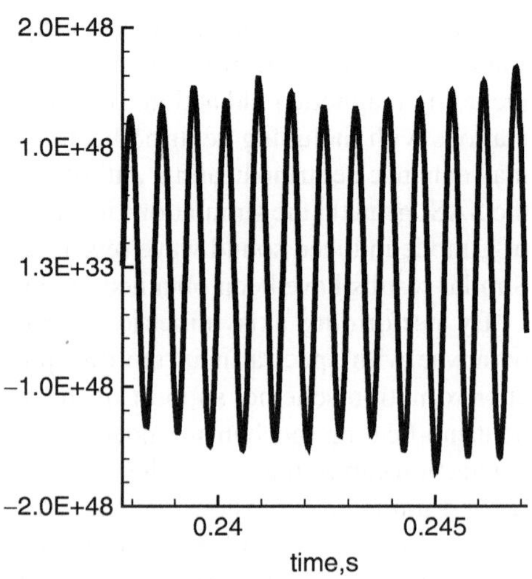

Fig. 10.10 The time evolution of the kinetic energy of the core only during the MR explosion, from [707]

have shown that the amount of acoustic energy generated in the collapsing core is much less than the expected energy of core-collapse supernova, contrary to the results obtained in [249]. Note that the explosion in [249] happens only after about 200 oscillation periods, and the numerical scheme should be checked for its stability during a large number of time steps.

10.5 Magnetorotational Model of Supernova Explosion

The 1D spherically symmetrical simulations of the core collapsed supernova do not lead to the explosion, see [243, 248]. The 2D and 3D simulations of the neutrino driven supernova mechanism do not give supernova explosions with the sufficient level of confidence. The recent improved models of the core collapse where the neutrino transport was simulated by solving the Bolzmann equation do not explode [243]. When all the above mechanisms of explosion prove inefficient, the magnetic field may convert the rotational energy of a neutron star resulting from collapse into kinetic energy of the envelope and thus ensure a supernova explosion. A magnetorotational model of explosion has been suggested in [155].[7] First numerical calculations for this model have been made to the cylindrical approximation in [34, 136], the spherically symmetric approximation in [718] and in a simplified 2D formulation in [760]. The results of these calculations are in qualitative agreement and give a conversion of $\sim 3\%$ of the rotational energy into kinetic energy of the outburst. For $E_{\rm rot} = 10^{53}$ erg, we have the energy $E_{\rm kin} = 3 \times 10^{51}$ erg sufficient to account for a supernova explosion.

10.5.1 Mechanism of Magnetorotational Explosion

When a rapidly rotating presupernova collapses, it leads to formation of a rapidly rotating neutron star surrounded by a differentially rotating envelope in which the centrifugal forces are comparable with gravitational ones. The differential rotation twists the lines of magnetic force, thereby causing the magnetic field with initial energy $\epsilon_{\rm M} \ll \epsilon_{\rm G}$ to increase linearly in time. When the energy of the field in the envelope approaches $\epsilon_{\rm M} \sim \epsilon_{\rm G}$, the magnetic pressure pushes the material outward. The arising wave of compression propagates over the medium with falling density, becomes enhanced and transforms into a shock to result in a powerful explosion. As the compression wave and subsequent shock move outward, their energy keeps

[7] Observations of radio emission at a relatively early period of the flash, roughly 1 year after the light peak, provide indirect evidence for an essential role played by the magnetic field in a supernova explosion [1022]. A radio emission has been detected in SN II, where a pulsar birth is assumed, and in those types of SN I (SN Ib, c) where a neutron star birth is suspected as well [929].

increasing, maintained by rotational energy supplied by magnetic field. The magnetic field serves also to transfer to outer layers an essential part of the total angular momentum. A qualitative picture of explosion is given in [104, 136, 155, 598], and some analytical estimations are given in [693]. Twisting of magnetic lines has been used earlier [558] to account for the energy transfer from the neutron star to the Crab nebula and maintaining its luminescence. Mention also should be made of the paper [629] that provides a numerical calculation of collapse for a rotating star with a strong initial magnetic field see also [957]. The obtained picture of an explosion in the form of outbursts lined up along the dipole axis (see also [300]) differs from the magnetorotational explosion where the major part of the outburst occurs in the equatorial plane. Such geometry of the outburst in [629] was the result of a specific, rather unrealistic choice of the magnetic field configuration, which was produced by a current ring, at the equator, out of the stellar centre at a radius where the matter density was an order of magnitude less than the central one. The magnetic field of this ring had a zero radial component in the equatorial plane, and a magnetic pressure gradient was formed in the Z-direction due to the differential rotation, having zero toroidal component in the equatorial plane, due to the choice of the initial configuration. Such a magnetic field structure had led to a matter stream pattern appearing preferentially along the symmetry axis of the magnetic field.

10.5.2 Basic Equations

The equations of magnetohydrodynamics (MHD) with gravitation in the cylindrical Eulerian system (r, φ, z), at an infinite conductivity and axial symmetry $\partial/\partial\varphi = 0$ read [618]

$$\frac{\partial v_{\mathrm{r}}}{\partial t} + v_{\mathrm{r}} \frac{\partial v_{\mathrm{r}}}{\partial r} + v_{\mathrm{z}} \frac{\partial v_{\mathrm{r}}}{\partial z} - \frac{v_{\varphi}^2}{r} = -\frac{1}{\rho}\frac{\partial P}{\partial r} - \frac{\partial \varphi_{\mathrm{G}}}{\partial r} + \frac{1}{\rho c}\left(j_{\varphi} B_z - j_z B_{\varphi}\right), \quad (10.82)$$

$$\frac{\partial v_{\varphi}}{\partial t} + v_{\mathrm{r}} \frac{\partial v_{\varphi}}{\partial r} + v_{\mathrm{z}} \frac{\partial v_{\varphi}}{\partial z} + \frac{v_{\mathrm{r}} v_{\varphi}}{r} = \frac{1}{\rho c}\left(j_z B_r - j_r B_z\right), \quad (10.83)$$

$$\frac{\partial v_{\mathrm{z}}}{\partial t} + v_{\mathrm{r}} \frac{\partial v_{\mathrm{z}}}{\partial r} + v_{\mathrm{z}} \frac{\partial v_{\mathrm{z}}}{\partial z} = -\frac{1}{\rho}\frac{\partial P}{\partial z} - \frac{\partial \varphi_{\mathrm{G}}}{\partial z} + \frac{1}{\rho c}\left(j_r B_{\varphi} - j_{\varphi} B_r\right), \quad (10.84)$$

$$\frac{\partial \rho}{\partial t} + v_{\mathrm{r}} \frac{\partial \rho}{\partial r} + v_{\mathrm{z}} \frac{\partial \rho}{\partial z} + \rho\left[\frac{1}{r}\frac{\partial (r v_{\mathrm{r}})}{\partial r} + \frac{\partial v_{\mathrm{z}}}{\partial z}\right] = 0, \quad (10.85)$$

$$\frac{\partial B_r}{\partial t} = -\frac{\partial}{\partial z}\left(v_{\mathrm{z}} B_r - v_{\mathrm{r}} B_z\right), \quad (10.86)$$

$$\frac{\partial B_{\varphi}}{\partial t} = \frac{\partial}{\partial z}\left(v_{\varphi} B_z - v_{\mathrm{z}} B_{\varphi}\right) - \frac{\partial}{\partial r}\left(v_{\mathrm{r}} B_{\varphi} - v_{\varphi} B_r\right), \quad (10.87)$$

$$\frac{\partial B_z}{\partial t} = \frac{1}{r}\frac{\partial}{\partial r}\left[r\left(v_z B_r - v_r B_z\right)\right], \tag{10.88}$$

$$\frac{1}{r}\frac{\partial}{\partial r}(rB_r) + \frac{\partial B_z}{\partial z} = 0, \tag{10.89}$$

$$j_r = -\frac{c}{4\pi}\frac{\partial B_\varphi}{\partial z}, \tag{10.90}$$

$$j_\varphi = \frac{c}{4\pi}\left(\frac{\partial B_r}{\partial z} - \frac{\partial B_z}{\partial r}\right), \tag{10.91}$$

$$j_z = \frac{c}{4\pi r}\frac{\partial}{\partial r}(rB_\varphi), \tag{10.92}$$

$$\frac{1}{r}\frac{\partial}{\partial r}\left(r\frac{\partial \varphi_G}{\partial r}\right) + \frac{\partial^2 \varphi_G}{\partial z^2} = 4\pi G\rho, \tag{10.93}$$

$$\frac{\partial E}{\partial t} + v_r\frac{\partial E}{\partial r} + v_z\frac{\partial E}{\partial z} + \frac{P}{\rho}\left[\frac{1}{r}\frac{\partial (rv_r)}{\partial r} + \frac{\partial v_z}{\partial z}\right] = -f_\nu, \tag{10.94}$$

$$P = P(\rho, T), \quad E = E(\rho, T), \quad f_\nu = f_\nu(\rho, T), \tag{10.95}$$

(see Chap. 1 and Sect. 5.2, Vol. 1).

Here, (10.82–10.84) are the equations of motion with magnetic fields, (10.85) is the continuity equation, (10.86–10.88) are the "frozen-in" field equations ($\partial\mathbf{B}/\partial t = \mathrm{rot}(\mathbf{v}\times\mathbf{B})$), (10.89) is the equation for a field with no divergence ($\mathrm{div}\mathbf{B} = 0$), (10.90–10.92) are the equations for field generation by electric currents (with no term for displacement currents, $\mathrm{rot}\mathbf{B} = ((4\pi/c)\mathbf{j})$, (10.93) is Poisson's equation, (10.94) is the energy equation, $\mathbf{v}(v_r, v_\varphi, v_z)$ is the velocity, $\mathbf{B}(B_r, B_\varphi, B_z)$ is the magnetic field strength, $\mathbf{j}(j_r, j_\varphi, j_z)$ is the current density, c is the velocity of light and φ_G is the gravitational potential.

Under the assumption of a plane symmetry, the set of equations (10.82–10.85) is solved for a star of mass M with the following boundary conditions:

(a) $P = \rho = T = B_\varphi = 0$ on the outer boundary,

(b) $v_r = j_r = B_r = 0$ at r = 0,

(c) $v_\varphi = j_\varphi = B_\varphi = 0$ at r = 0,

(d) $v_z = 0$, $\frac{\partial j_z}{\partial z}$. or $j_z = 0$, $\frac{\partial B_z}{\partial z}$ or $B_z = 0$ at z = 0. (10.96)

Conditions $\partial B_z/\partial z = 0$ and $B_z = 0$ at $z = 0$ correspond to dipole-like and quadrupole-like magnetic fields, respectively. Condition $\partial j_z/\partial z = 0$ at $z = 0$

corresponds to a plane symmetric azimuthal magnetic field B_φ, while the condition $j_z = 0$ with $\partial j_z/\partial z \neq 0$ at $z = 0$ corresponds to plane antisymmetric magnetic field B_φ. The dissipative processes are neglected in calculations, the neutrino emission is allowed for by f_ν, and an artificial viscosity is used for shock calculations. The introduction of artificial viscosity in [35], where a Lagrangian coordinate system is used, implies replacing P in the equations of motion (10.82–10.84) and energy (10.94) by

$$P + \omega_1 = P - \nu \mathrm{div}\mathbf{v} = P - \nu \left[\frac{1}{r} \frac{\partial (r v_\mathrm{r})}{\partial r} + \frac{\partial v_\mathrm{z}}{\partial z} \right], \tag{10.97}$$

where ν is the viscosity coefficient. The distributions $\rho(\mathbf{r})$, $T(\mathbf{r})$ and $\mathbf{B}(\mathbf{r})$ are specified at some initial time, and the last of them should satisfy the condition of the absence of magnetic charges (10.89) and yield finite values of $\mathbf{j}(\mathbf{r})$ throughout the star, in accordance with (10.90–10.92). Surface and linear currents arising from singularities in $\mathbf{j}(\mathbf{r})$ are usually not considered in calculations. If the equality (10.89) does hold at the beginning, it will remain valid with time provided that only (10.86–10.88) are used to determine the field.

10.5.3 Cylindrical Approximation

A cylinder uniform along the z-axis with $v_\mathrm{z} = B_z = j_r = j_\varphi = 0$ is considered in a 1D formulation. This means that in a real star, the motion along the z-axis is neglected. The basic equations with the Lagrangian independent variable

$$s = \int_0^r \rho' r' \mathrm{d}r' \tag{10.98}$$

become [34, 136]

$$\frac{\partial}{\partial t}\left(\frac{1}{\rho}\right) = \frac{\partial}{\partial s}(r v_\mathrm{r}), \quad \frac{\partial r}{\partial t} = v_\mathrm{r}, \quad \frac{\partial \varphi}{\partial t} = \frac{v_\varphi}{r}, \tag{10.99}$$

$$\frac{\partial v_\mathrm{r}}{\partial t} - \frac{v_\varphi^2}{r} = -r \frac{\partial P}{\partial s} - \frac{1}{8\pi r}\frac{\partial}{\partial s}(r B_\varphi)^2 + g. \tag{10.100}$$

$$\frac{\partial}{\partial t}(r v_\varphi) = \frac{r B_r}{4\pi}\frac{\partial}{\partial s}(r B_\varphi), \tag{10.101}$$

$$\frac{\partial}{\partial t}\left(\frac{B_\varphi}{\rho r}\right) = r B_r \frac{\partial}{\partial s}\left(\frac{v_\varphi}{r}\right), \quad r B_r = A = \mathrm{const.}, \tag{10.102}$$

$$\frac{\partial E}{\partial t} = -P \frac{\partial}{\partial s}(r v_\mathrm{r}) - f_\nu, \tag{10.103}$$

$$g = -2G\,\frac{2\pi(M_0 + s)}{r}. \tag{10.104}$$

Here, M_0 is the mass per radian of a unit length of a core with a uniform rigid-body rotation and g is a gravitational acceleration. An approximate equation of state in the form

$$P = 3.09 \times 10^{12} \rho^{5/3} \left(1 + 1.59 \times 10^{-3} \rho^{1/3}\right) / \left(1 + 3.18 \times 10^{-3} \rho^{1/3}\right)^2$$

$$+ 6.5 \times 10^4 \rho^2 + aT^4/3 + \rho \mathscr{R} T \quad \text{for } \rho < 3 \times 10^9 \text{ g cm}^{-3} \tag{10.105}$$

$$P = 2.04 \times 10^{27} + 6.5 \times 10^4 \rho^2 + aT^4/3 + \rho \mathscr{R} T \qquad \text{for } \rho \geq 3 \times 10^9 \text{ g cm}^{-3}$$

$$E = 4.635 \times 10^{12} \rho^{2/3} / (1 + 3.18 \times 10^{-3} \rho^{1/3}) + 6.5 \times 10^4 \rho$$

$$+ \frac{aT^4}{\rho} + \frac{3}{2} \mathscr{R} T \qquad \text{for } \rho < 3 \times 10^9 \text{ g cm}^{-3}$$

$$5.19 \times (10^{27}/\rho) + 2.41 \times 10^{18} \left(\rho - 3 \times 10^9\right) / \rho$$

$$+ 6.5 \times 10^4 \rho + (aT^4/\rho) + (3/2) \mathscr{R} T \qquad \text{for } \rho \geq 3 \times 10^9 \text{ g cm}^{-3} \tag{10.106}$$

has been used in [34, 136]. The equation allows, approximately, for the transition from non-relativistic to relativistic electrons occurring at strong degeneracy. The electron pressure has been taken as constant after the onset of neutronization. The function of neutrino losses due to URCA processes has been taken from [540] in the form

$$Q_{\text{URCA}} = 1.3 \times 10^9 \rho \eta(T) \Phi \left(\rho / T_9^3\right) T_9^6 \quad (\text{erg cm}^{-3} \text{s}^{-1}), \tag{10.107}$$

where

$$\eta(T) = 1 \quad \text{at} \quad T_9 < 7,$$

$$\eta(T) = 664.31 + 51.024(T_9 - 20) \quad at \quad 7 < T_9 < 20,$$

$$\eta(T) = 664.31 \quad at \quad T_9 > 20,$$

$$\Phi \left(\rho / T_9^3\right) = \left[1 + \left(7.1 \times 10^{-5} \rho / T_9^3\right)^{2/5}\right]^{-1}. \tag{10.108}$$

In addition to the artificial viscosity (10.97), another type of viscosity has been examined in [34, 136] for describing rotational discontinuities:

$$\omega_2 = \mu \rho r^2 \frac{\partial}{\partial s} (v_\varphi / r), \tag{10.109}$$

so that the term $\partial\omega_2/\partial s$ is added to the right side of (10.101), and the term $\omega_2(\partial/\partial s)\,(v_\varphi/r)$ to the right side of the energy equation (10.103); μ is an artificial viscosity coefficient for ω_2.

At initial time $t = 0$, $T = 0$, has been adopted, and the density distribution has been specified in the form [34]

$$\rho(s,0) = a\exp\left[-b(r-R_0)^2\right]; \quad a,b = \text{const.} \tag{10.110}$$

Here, $R_0 = R(0,t)$ is the core radius; M is the envelope mass per unit length per radian. It is adopted that $v_r(s,0) = 0$, $B_\varphi(s,0) = 0$, and the boundary conditions 10.96a. Also specified are the constant A from (10.102) and the initial distribution $v_\varphi(s,0)$ from the radial equilibrium equation (10.100) with $\partial v_r/\partial t = 0$

$$\frac{v_\varphi^2(s,0)}{r(s,0)} - r(s,0)\,\frac{\partial}{\partial s}\,P(s,0) + g(s) = 0. \tag{10.111}$$

The angular momentum of the system core+envelope is assumed to be conserved throughout the calculations; the relation for this conservation, owing to the continuity of v_φ on the core boundary, is written in the form of the boundary condition

$$\frac{M_0}{2}\,\frac{\partial h}{\partial s} - h = 0 \ \text{ at } \ s = 0, \qquad h = rB_\varphi. \tag{10.112}$$

10.5.4 Calculational Results

The problem was calculated numerically in the region

$$t > 0, \quad 0 < s < M \quad (R_0 < r < R(t)). \tag{10.113}$$

As rotational discontinuities are absent in [34], it has been possible to use only the artificial viscosity ω_1 from (10.97), while the coefficient ν is chosen in such a way as to make the effective broadening of the shock equal to a few intervals of the mass grid. The basic dimensionless parameters of the problem are

$$\alpha = \frac{A^2}{4\pi M V_0^2}, \qquad \left(V_0 = \sqrt{2\pi G M_0,}\right), \quad \beta = \frac{M_0}{M}. \tag{10.114}$$

The solution in [34] was obtained for $\beta = 1$, $\alpha = 10^{-2}, 10^{-4}, 10^{-8}$. To introduce dimensionless quantities, all the variables are taken in the form $F = F_0\tilde{F}$ with the following scale variables F_0:

$$v_0 = V_0, \quad r_0 = R_0, \quad t_0 = R_0/V_0, \quad h_0 = A, \quad \rho_0 = M/R_0^2,$$

$$P_0 = MV_0^2/R_0^2, \quad E_0 = V_0^2, \quad \Omega_0 = V_0/R_0 \quad (\Omega = v_\varphi/r),$$

$$s_0 = M, \quad T_0 = V_0^2/10^3\mathscr{R}, \quad f_{\nu 0} = V_0^3/R_0;$$

$$\text{with } R_0 = 10^6 \text{ cm}, \quad 2\pi M_0 = 0.5 \times 10^{-6} M_\odot. \tag{10.115}$$

Decreasing the parameter α causes the time scales of processes to increase as $\alpha^{-1/2}$. As $\alpha \to 0$, it is convenient to introduce the dimensionless functions

$$t_\alpha = t\alpha^{1/2}, \quad v_{r\alpha} = v_r\alpha^{-1/2}, \quad h_\alpha = h\alpha^{1/2}, \quad f_{\nu\alpha} = f_\nu\alpha^{-1/2}, \tag{10.116}$$

having the same relationships between them for all small α. For other functions $F_\alpha = F$. The results of numerical calculations are presented in Figs. 10.11–10.15 from [34]. Propagation of a slow ($v < v_A = B\sqrt{4\pi\rho}$) MHD shock over the envelope may be seen in Figs. 10.11 and 10.12. The region of the sharp temperature peak beyond the discontinuity front is the major source of neutrino emission (Fig. 10.14). It is clear from Figs. 10.11, 10.14 and 10.15 that the relationships between the variables (10.116) are only slightly sensitive to decreasing α. The characteristic time scale proportionality to $\alpha^{-1/2}$ is caused by an increase $\sim\alpha^{-1/2}$ in the number of turns of magnetic lines required for achieving the condition for the onset of

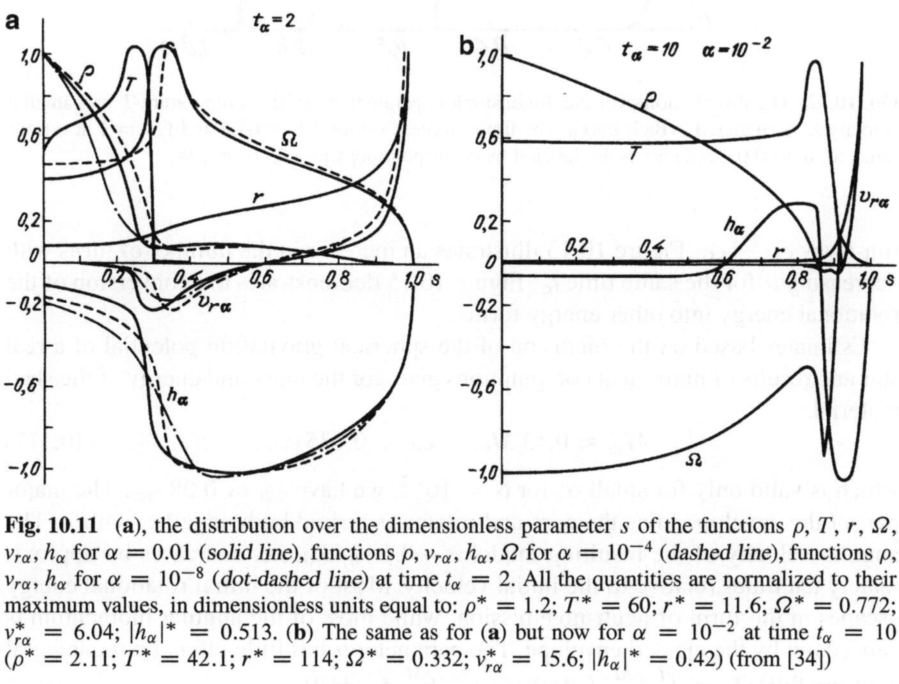

Fig. 10.11 (**a**), the distribution over the dimensionless parameter s of the functions ρ, T, r, Ω, $v_{r\alpha}$, h_α for $\alpha = 0.01$ (*solid line*), functions ρ, $v_{r\alpha}$, h_α, Ω for $\alpha = 10^{-4}$ (*dashed line*), functions ρ, $v_{r\alpha}$, h_α for $\alpha = 10^{-8}$ (*dot-dashed line*) at time $t_\alpha = 2$. All the quantities are normalized to their maximum values, in dimensionless units equal to: $\rho^* = 1.2$; $T^* = 60$; $r^* = 11.6$; $\Omega^* = 0.772$; $v_{r\alpha}^* = 6.04$; $|h_\alpha|^* = 0.513$. (**b**) The same as for (**a**) but now for $\alpha = 10^{-2}$ at time $t_\alpha = 10$ ($\rho^* = 2.11$; $T^* = 42.1$; $r^* = 114$; $\Omega^* = 0.332$; $v_{r\alpha}^* = 15.6$; $|h_\alpha|^* = 0.42$) (from [34])

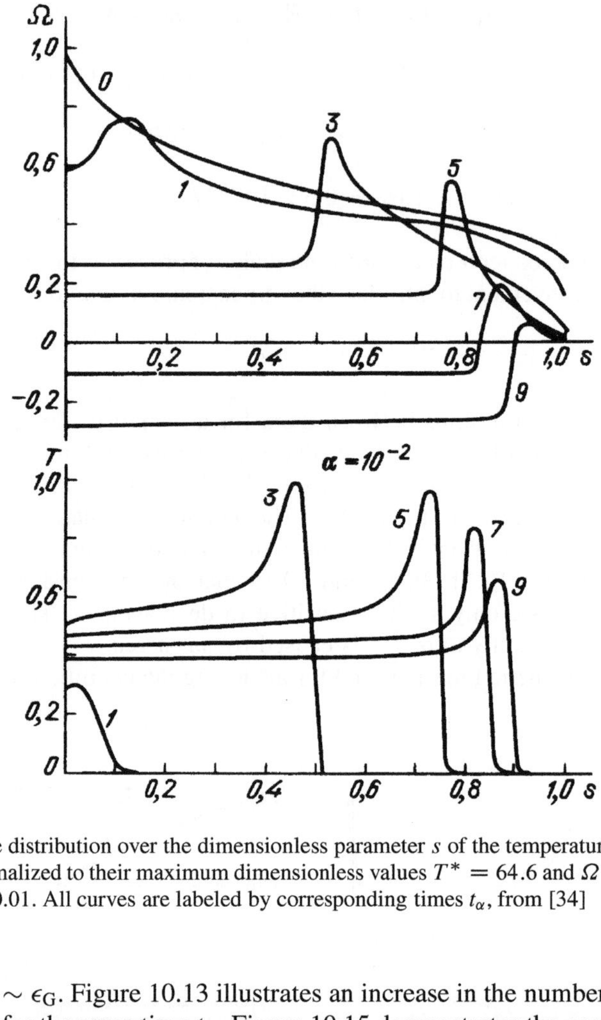

Fig. 10.12 The distribution over the dimensionless parameter s of the temperature T and angular velocity Ω normalized to their maximum dimensionless values $T^* = 64.6$ and $\Omega^* = 1$ at various times for $\alpha = 0.01$. All curves are labeled by corresponding times t_α, from [34]

run-away $\epsilon_M \sim \epsilon_G$. Figure 10.13 illustrates an increase in the number of turns with decreasing α for the same time t_α. Figure 10.15 demonstrates the conversion of the rotational energy into other energy forms.

Estimates based on the inclusion of the spherical gravitation potential of a real star and results of numerical computations give, for the mass and energy of the shed material,

$$M_{sh} \approx 0.13\, M_\odot, \quad \epsilon_{sh} \approx 0.035\, \epsilon_{rot}, \qquad (10.117)$$

which is valid only for small α; for $\alpha = 10^{-2}$, we have $\epsilon_{sh} \approx 0.08\, \epsilon_{rot}$. The major part of the envelope joins the core and rotates as a rigid body together with it. The angular velocity of the resulting model is $\sim 0.1\, V_0/R_0$, i.e. decreases by approximately ten times relative to the initial velocity. Most of the initial rotational energy escapes in the form of neutrino emission, while most of the angular momentum is carried out by the ejected envelope. The parameter α has little effect on the integral neutrino flux $Q_\nu = \int_0^t \int_0^M f_\nu \mathrm{d}s\, \mathrm{d}t = \int_0^{t_\alpha} \int_0^M f_{\nu\alpha} \mathrm{d}s\, \mathrm{d}t_\alpha$.

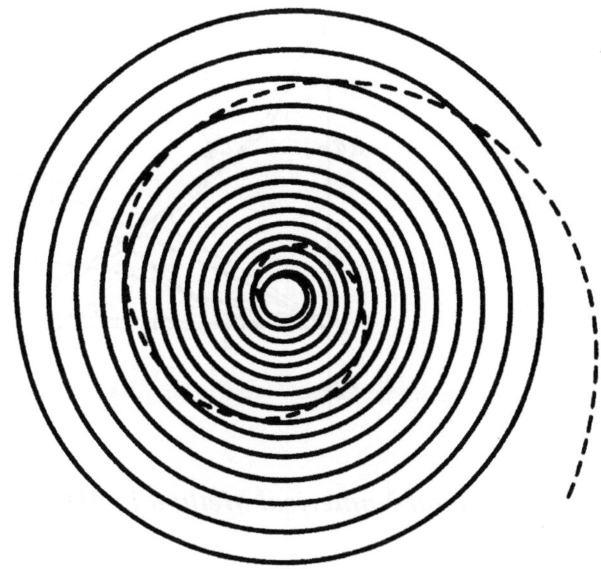

Fig. 10.13 The shape of lines of magnetic force in a region nearby the core at time $t_\alpha = 7$ for $\alpha = 0.01$ (*dashed line*) and $\alpha = 10^{-4}$ (*solid line*) (from [34])

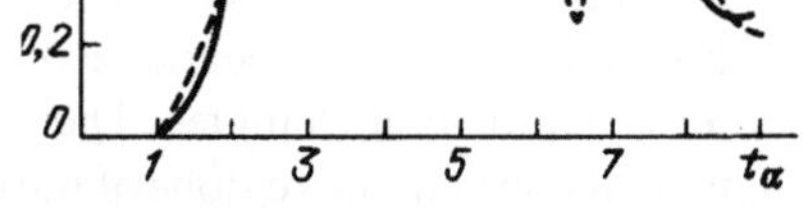

Fig. 10.14 Neutrino luminosity $q_\alpha = \int_0^1 f_{\nu\alpha}\, ds$ as a function of time t_α, normalized to the maximum dimensionless value $q_\alpha^* = 1.4$ for $\alpha = 0.01$ (*dashed line*) and $\alpha = 10^{-4}$ (*solid line*) (from [34])

An interesting result of calculations is a possible stage of magnetorotational oscillations of the core-envelope system, during which the angular velocity changes its sign. The angular velocity of the resulting core may be opposite to the initial angular velocity.

The envelope ejection leaves behind a young pulsar that keeps supporting the envelope acceleration by the pressure of its radiation in the form of high-energy particles and dipole electromagnetic radiation [217, 766]. The action of the magnetohydrodynamical rotational mechanism of envelope ejection and subsequent activity of the young pulsar could affect the explosion and light curve formation of the supernova SN 1987A [950].

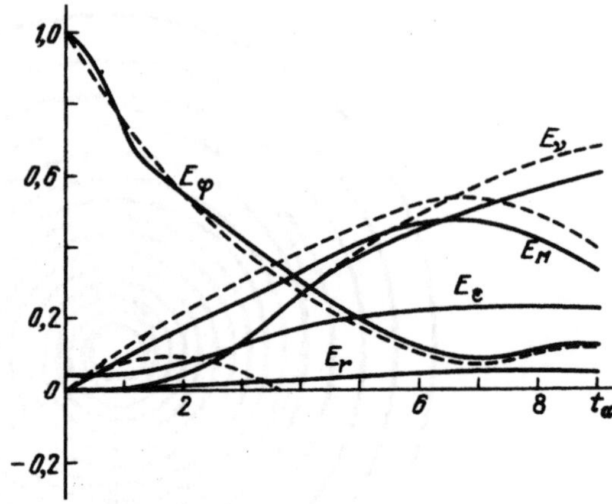

Fig. 10.15 Variations with t_α of the rotational E_ϕ, radial kinetic E_r, internal thermal E_ϵ, magnetic E_M energies, of the total neutrino energy losses E_ν for $\alpha = 0.01$ (*solid line*) and $\alpha = 10^{-4}$ (*dashed line*). All the quantities are normalized to the maximum rotational energy $E^*_\phi = 10^{47}$ erg/cm per unit cylinder length, from [34]

10.5.5 Two-Dimensional Numerical Method in MHD

The implicit operator-difference completely conservative scheme on the triangular grid of variable structure is applied for the numerical modelling of the problem of the magnetorotational explosion. The scheme was suggested and investigated in [32, 33, 40, 41].

The solution of the problem is a sequence of time-steps. Calculation of every time-step can be divided into two parts. The first part is the calculation of the values of the functions on the next time level using implicit completely conservative operator-difference scheme on the triangular grid in lagrangian variables [40, 41]. The coordinates of grid knots are changing at this stage.

The second part is an analysis of the quality of the grid, its improvement and adaptation (grid reconstruction). The improvement of the quality of the grid is necessary because of the appearance of "poor" cells, i.e. triangles that strongly deviate from the equilateral triangles. The dynamical adaptation of the grid allows to concentrate the grid in the regions of the computational domain where spatial resolution need to be increased and rarefy the grid in the regions where the flow is smooth. It allows to reduce significantly the dimensionality of the grid and hence strongly reduce the computation time.

The grid reconstruction procedure itself consists of the following two stages. The first stage is a local correction of the structure of the grid. The second stage is a calculation of the values of the functions defined in cells and knots in the regions of corrected structure.

The local correction of the structure of the grid can be done using the following three local operations (see Fig. 10.16, and details in [42]):

1. Replacement of the diagonal of quadrangle formed by two triangles by another diagonal
2. Joining up of two neighboring grid knots
3. Addition of the knot at the middle of the cell which connects two knots

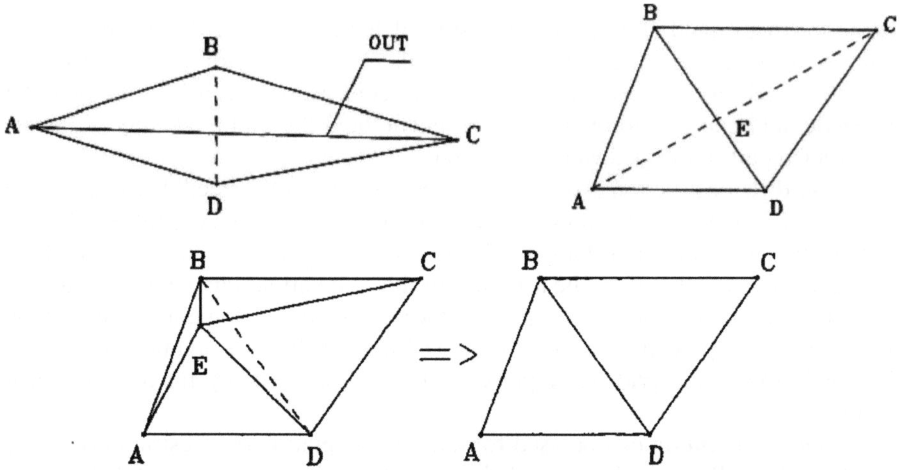

Fig. 10.16 Elementary reconstructions: (*Left up*) BD connection is introduced instead of AC connection. The total number of knots and cells in the grid is not changed. (*Right up*) Adding a knot at the middle of the connection: the knot E is added to the existing knots ABCD in the middle of BD connection, two connections AE and EC appear, and the total number of cells is increased by two cells. (*Down*) Removing a knot: the knot E is removed from the grid, and the total number of cells is decreased by two cells, from [42]

The improvement of the grid structure is made using the first two operations. The grid adaptation is made by application of the local operations (II) and (III).

The values of functions defined in cells and knots involved in grid structure modification are calculated at the second stage of the grid reconstruction.

The application of simple interpolation for the calculations of the new values of grid function leads to the violation of the conservation laws and adds significant errors in the regions with high gradients (e.g. at shock waves). The goal of the calculation of the new values of functions is to minimize numerical errors introduced by this procedure. To achieve the goal the error not only in the values of the functions but also on their gradients has to be minimized. It is important to fulfill conservation laws (mass, momentum, energy, magnetic flux) in the vicinity of the local grid reconstruction. The method for the calculation of these new values of grid functions is based on minimization of the functionals containing the values of the functions, its gradients and grid analogs of the conservation laws [40].

The method of the conditional minimization of the functionals is used for the solution, guaranteeing exact fulfilment of the conservation laws. It is important to fulfill conservation laws for the solution of the collapse and magnetorotational explosions problems, because a large number of the time-steps need to be made. In such a situation, even slight violation of the conservation laws at a time-step could lead to a significant growth in the errors and hence to qualitative distortion of the results.

The method of calculating hydrodynamical values ρ and P during grid reconstruction is taken from [40]. In [44] for the calculations of the new values of the magnetic field components, the sum of the products of the magnetic field components values and the volumes of the cells in the vicinity of the reconstruction part of the grid were conserved. To obtain a better precision, in [45] the poloidal magnetic field energy and the toroidal magnetic field flux in the vicinity of the reconstructed part of the grid were conserved.

The grid reconstruction procedure not only allows to "correct" the Lagrangian grid, but also dynamically adapt it using different criteria for the grid in different parts of the computational domain. The grid can be refined in the regions where it is necessary, and therefore the accuracy of the calculations can be increased. It is possible to rarefy the grid in those parts of the computational domain where the flow is "smooth". The procedure of the rarefying of the grid allows to significantly reduce the dimension of the grid while preserving the same accuracy for the numerical solution.

Geometrical criteria can be used for the grid adaptation (i.e. restrictions on the length of the cell size, which are defined by the cell coordinates only), but such adaptation criteria are suitable for the flows of simple or easily predictable structures only.

For grid adaptation in the case of complicated flows or flows with unknown structures, it is better to apply dynamical criteria which are defined by the solution behavior. The dynamical criteria for the grid adaptation applied here were suggested in [42].

The characteristic length of the cell side l_k was chosen as a local criterion for the grid reconstruction. As an example, consider the criterion where l_k is a function of ρ and gradρ. Let us introduce the function

$$f(\rho, \mathrm{grad}\rho) = \frac{\alpha}{(\rho+\varepsilon)^{\gamma_1}} + \frac{\beta}{(|\mathrm{grad}\rho|+\varepsilon)^{\gamma_2}}, \tag{10.118}$$

where $0 < \varepsilon << 1$, $\alpha \geq 0$, $\beta \geq 0$, $\alpha + \beta = 1$. γ_1, γ_2 are power indexes and gradρ is the grid analogue of the density gradient. In limited cases, $\alpha = 1$, $\beta = 0$, f depends on the density only, and $\alpha = 0$, $\beta = 1$, f depends on the gradient of the density only.

Let N be the total number of the grid cells. The characteristic length of the side l_k of the cell with the number k will be calculated as a function of the density, the gradient of the density and coordinates r, z (implicitly) by the following formula

$$l_k = 2\sqrt{\frac{s_k}{3}}, \quad s_k = \frac{f(\rho_k, \mathrm{grad}\rho_k)}{\sum\limits_{n=1}^{N} f(\rho_n, \mathrm{grad}\rho_n)} S. \tag{10.119}$$

Here, S is a square of the computational domain, which consists of the triangular cells. The summation in the denominator is made for all grid cells. Note that $S = \sum\limits_{k=1}^{N} s_k$, where s_k is equal to the square of the equilateral triangle with the length of the side l_k. The function f in equation (10.118) was used in the following form $f(\rho_k) = 1/(\rho_k + \varepsilon)^{0.5}$.

The criterion described was applied in the following way. When the length of the side of the cell i is larger than $2l_k$, a new knot is added to the middle of this side of the cell. When the length of the side of the cell i is less than $0.7l_k$, the operation of the joining up of these knots is applied. The application of the dynamical adaptation criterion allowed us not only to adapt the grid to the specialities of the solution, but also to provide an acceptable accuracy of the calculations with a small fluctuation of the total number of grid knots and cells. In the calculations [45, 133, 707], the total number of knots (5,000) and cells (10,000) was violated not more than 5%.

At the moment of the maximal compression (for the collapse problem) the minimal size of the cell side is so small that application of the uniform grid with the same spatial resolution would require a grid with the dimensions $\sim$1,000 $\times$ 1,000 (!) cells.

The calculation of the gravitational potential, in the frames of the applied numerical method, is made on the basis of the finite element method of higher order [1070]. This procedure allowed to increase the accuracy of the calculations and eliminate the loss of approximation near z axis. The linear artificial viscosity (e.g. [862]) was introduced into the numerical scheme for the shock capturing .

10.5.6 Magnetorotational Explosion of the Initially Uniform Cloud

The attempt to obtain a magnetorotational explosion (MRE) in a realistic 2D scheme has been carried out in [43]. The simplified problem was solved for an initially uniform and rigidly rotating gas cloud with the equation of state (7.43), initial density and size as in (7.46), and initial values of internal, rotational and magnetic energies taken as

$$\frac{E_{\text{in0}}}{|E_{\text{gr0}}|} = 0.1, \qquad \frac{E_{\text{rot0}}}{|E_{\text{gr0}}|} = 0.04, \qquad \frac{E_{\text{mag1}}}{|E_{\text{in1}}|} = 0.05\,. \tag{10.120}$$

Here, the index "0" is related to the initial state of the collapse, index "1" is related to the quasistationary state, which the rotating cloud without magnetic field reaches in the process of collapse. The magnetic field was included in calculations in the point "1". This simplifies the calculations of the collapse and may be justified for a realistic case of the neutron star with $E_{\text{mag1}}/E_{\text{in1}} \ll 1$, where it does not influence the process of the collapse. The dynamical action of the magnetic field begins to be important only after its considerable amplification in the process of twisting, which takes a much longer time than the time of the collapse and establishing of the quasistationary state. The quadrupole component of the magnetic field is expected to be the most important for MRE, because it has a large radial component near the equator, which is amplified by the field twisting. To avoid a central singularity, a regular field of similar topology was chosen instead of the quadrupole in the form (in non-dimensional variables)

$$\begin{aligned} B_{r1} &= F_r(0.5\,r,\ 0.5\,z - 2.5) - F_r(0.5\,r,\ 0.5\,z + 2.5), \\ B_{\phi 1} &= 0, \\ B_{z1} &= F_z(0.5\,r,\ 0.5\,z - 2.5) - F_z(0.5\,r,\ 0.5\,z + 2.5), \\ F_r(r,\,z) &= k\left(\frac{2rz}{(z^2+1)^3} - \frac{2r^3z}{(z^2+1)^5}\right), \\ F_z(r,\,z) &= k\left(\frac{1}{(z^2+1)^2} - \frac{r^2}{(z2+1)^4}\right). \end{aligned} \tag{10.121}$$

In the transition to non-dimensional variables in (10.82–10.95), the scaling (7.39) was used for hydrodynamical variables, and the factor

$$B_0 = \sqrt{P_0} = \rho_0^{1/2} r_0/t_0 \tag{10.122}$$

was used for the magnetic field components. The value $k = 0.43$ adjusts the energy relation (10.120). The numerical method is based on the generalization of the implicit Lagrangian code described above. On the outer boundary, a non-zero value of the pressure $P_{\rm out} = 10^{-3} P_0$ was kept, which did not influence the MRE process, but solved some numerical problems. At $r = 0$, it was assumed that

$$v_{\rm r} = v_\phi = B_r = B_\phi = (\nabla \times \mathbf{B})_r = (\nabla \times \mathbf{B})_\phi = 0, \tag{10.123}$$

and at $z = 0$

$$v_{\rm z} = B_z = \frac{\partial(\nabla \times \mathbf{B})_z}{\partial z} = 0 \tag{10.124}$$

was taken. The results of computations are presented in Figs. 10.17a–i. The quasistationary state in Figs. 10.17a,b is presented at the moment very close to $t_1 = 23.920$, when the magnetic field (10.121) was included in the computations. The magnetic field configuration (practically initial) at almost the same time is presented in Fig. 10.17c. The azimuthal component of the magnetic field increases until it becomes important for a dynamical influence. Magnetic pressure pushes out the matter, mainly in the equatorial plane, which expands and part of it (about 2.4%) flys away to infinity, carrying away about 0.5% of the rotational energy of the configuration, formed after the collapse. Growth of the toroidal magnetic field during twisting is seen in Fig. 10.17d, and its decrease in the process of the matter outburst is shown in Fig. 10.17e. Density contours and velocity fields in subsequent moments of time showing the development of the outburst are given in Figs. 10.17f–i. Calculations of MRE have been performed in [44] on the refined grid with 2200 nodes (4400 cells), instead of 1000 nodes (2000 cells) in [43]. More realistic choices of the initial magnetic field have been made, which were produced by a toroidal electrical current in the central part of the star at $\sqrt{r^2 + z^2} < 1/3$ (in non-dimensional variables (7.39))

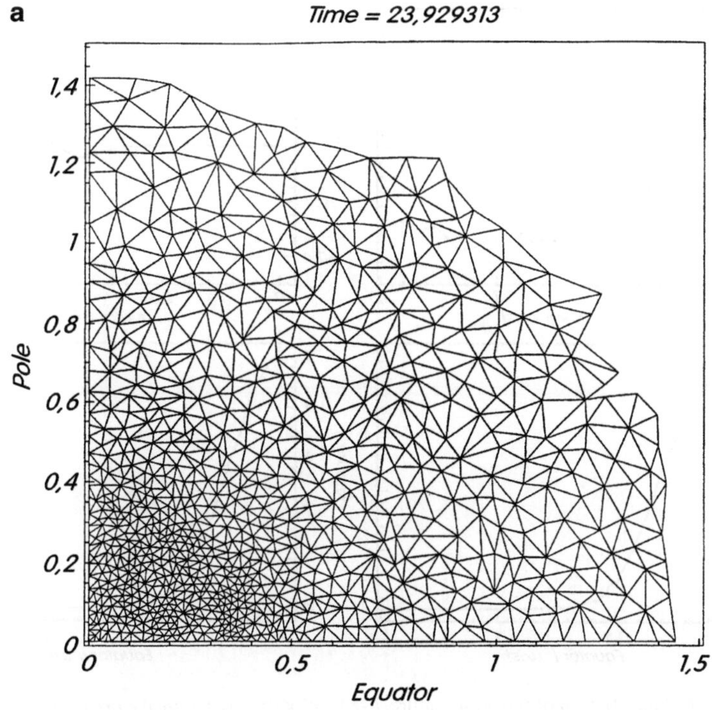

Fig. 10.17 (**a**) Triangular grid at $t = 23.929313$. For (**b**)–(**i**)

The current was taken as plane anti-symmetric to obtain the quadrupole-like field with large equatorial $B_r\,(B_z = 0)$. Calculations in [44] have been performed for the same initial value of rotational energy, and ten times smaller internal energy than in (10.120). The initial values of the magnetic energy have been considered to be 5, 500 and 5×10^4 times smaller than in (10.120). It was shown that the scaling (10.116) obtained in 1D calculations is valid also in the 2D case. The qualitative picture of MRE in [44] is similar to that in [43], but the explosion is much stronger. In the refined calculations [44], the amount of the ejected matter was about 7% of the total mass, carrying away approximately 14% of the rotational energy of the collapsed configuration.

10.5.7 Magnetorotational Supernova: Quadruple and Dipole Magnetic Configurations

Calculations of the magnetorotational processes in the strongly magnetized collapsed core were performed in [584] and [1060]. They assumed that the initial poloidal magnetic field is uniform and parallel to the rotation axis. The initial strong

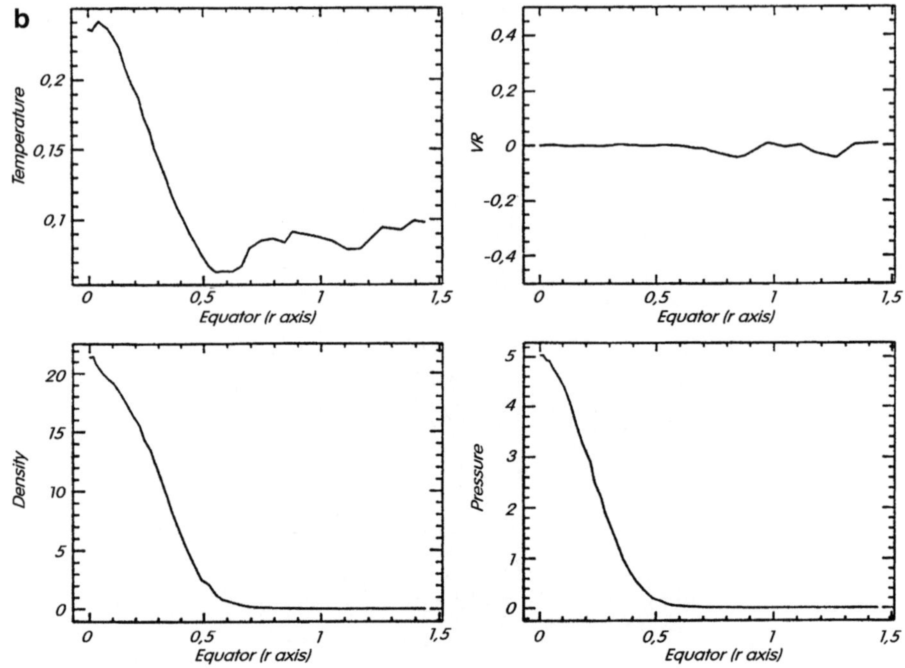

Fig. 10.17 (**b**) Distribution of the temperature $T(r)$, velocity $v_r(r)$, density $\rho(r)$ and pressure $P(r)$ along r-axis at the same time, as in Fig. 10.14a

toroidal magnetic field was also used as the initial condition. The values of the magnetic fields used by them are usual for "magnetars" (see Sect. 11.40), but too large for the ordinary core collapsed supernova, see review [583]. Calculations of magnetorotational supernova with realistic values of magnetic fields of different topology have been made in [45, 133, 707]. Results of these calculations are described in this section. Two-dimensional magnetorotational core-collapse simulations of massive Pop III stars, leading to the black hole formation, have been performed in [951]. Relativistic effects in magnetorotational supernova explosion have been investigated in [757]. They do not introduce big changes in the model of explosion. Simulations of supernova core collapse, bounce, and explosion by method of multi-group, radiation magnetohydrodynamics were done in [250]. All calculations show that the presence of realistic magnetic field in the new-born differentially rotating neutron star produces a strong explosion with a kinetic energy output comparable with values, observed in the core-collapse supernovae.

One of the main problems for the numerical simulation of the magnetorotational supernova is the smallness of the initial magnetic fields. The ratio of the initial magnetic and gravitational energies is in the range of $10^{-6} - 10^{-10}$. This value characterizes the stiffness of the set of MHD equations. There is a problem with strongly varying characteristic time-scales. A small time-scale is defined by a high sound speed in the center of the core and a large time-scale is a characteristic time of

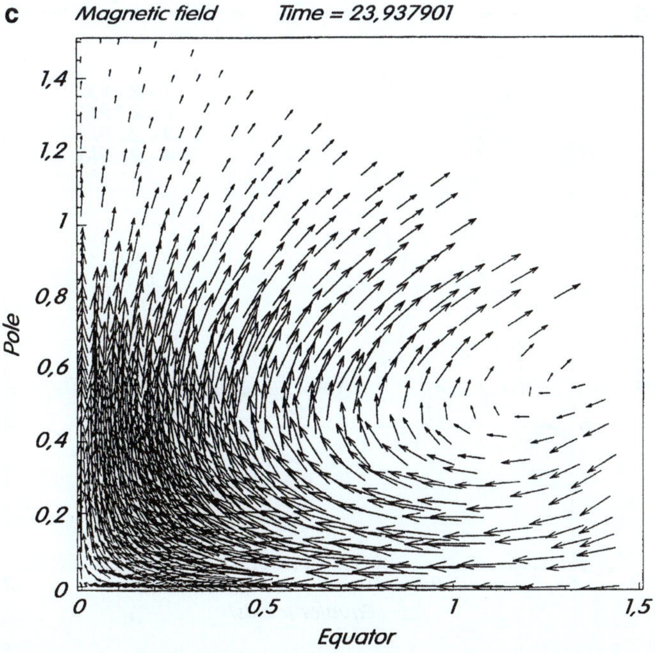

Fig. 10.17 (**c**) Magnetic field patterns at $t = 23.937901$

the evolution of the magnetic field. In such situation, the explicit numerical methods that are widely used in astrophysical hydro-simulations require an enormously large number of time-steps and a possible loss of accuracy due to the large numerical errors. The implicit approach should be used in that case. It is well known that implicit schemes are free from the Courant time-step limitation, which is too strong for such problems.

At the initial stages of the process, the toroidal magnetic field is produced by a twisting of the radial component due to differential rotation and is growing linearly.

$$B_\phi \sim \frac{\partial \omega}{\partial r} r t. \tag{10.125}$$

The linear growth is terminated when the toroidal component becomes so strong that magnetohydrodynamic instability starts to develop. This instability leads to poloidal motion of the matter that increases the radial component, which in turn strongly amplifies the growth of the toroidal field. As a result, both components, toroidal and poloidal, start to grow almost exponentially, shortening greatly the time between the collapse and magnetorotational explosion. The equation of state used in the calculations [37, 45, 133, 707] is

$$P \equiv P(\rho, T) = P_0(\rho) + \rho \Re T + \frac{\sigma T^4}{3}, \tag{10.126}$$

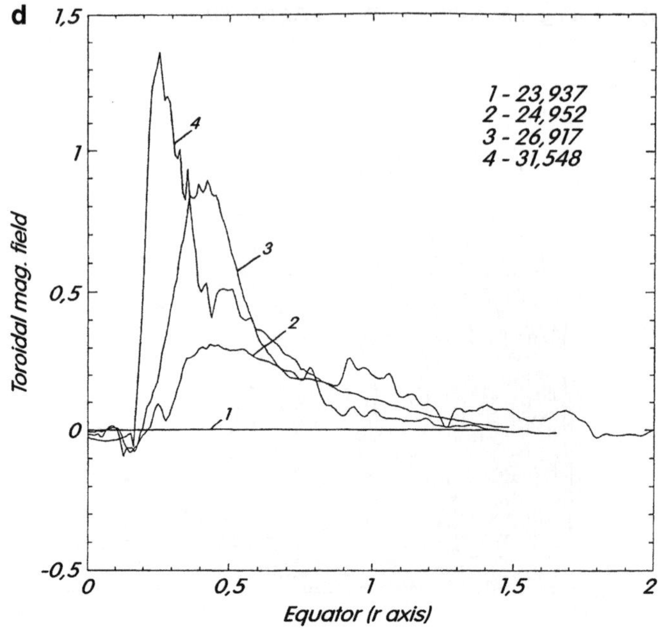

Fig. 10.17 (**d**) Distribution of the toroidal magnetic field B_ϕ at different time moments during its amplification. 1 - $t = 23.937$, 2 - $t = 24.952$, 3 - $t = 26.917$, 4 - $t = 31.548$

where $P_0(\rho)$ was taken from the problem to the Sect. 1.4.7, Vol. 1. The specific energy (per mass unit) was defined thermodynamically as

$$\varepsilon = \varepsilon_0(\rho) + \frac{3}{2}\Re T + \frac{\sigma T^4}{\rho} + \varepsilon_{\mathrm{Fe}}(\rho, T). \tag{10.127}$$

The value $\varepsilon_0(\rho)$ is defined by the relation

$$\varepsilon_0(\rho) = \int_0^\rho \frac{P_0(\tilde{\rho})}{\tilde{\rho}^2} \mathrm{d}\tilde{\rho} \tag{10.128}$$

The term $\varepsilon_{\mathrm{Fe}}$ is responsible for iron dissociation. It was used in the following form:

$$\varepsilon_{\mathrm{Fe}}(\rho, T) = \frac{E_{\mathrm{b,Fe}}}{A\, m_{\mathrm{p}}} \left(\frac{T - T_{0\mathrm{Fe}}}{T_{1\mathrm{Fe}} - T_{0\mathrm{Fe}}} \right). \tag{10.129}$$

It was supposed that in the region of the iron dissociation the iron amount was about 50% of the mass, $E_{\mathrm{b,Fe}} = 8 \cdot 10^{-4}$erg is the iron-binding energy, $A = 56$ is the iron atomic weight, m_{p} is the proton mass, $T_{0\mathrm{Fe}} = 0.9 \cdot 10^{10}$K and $T_{1\mathrm{Fe}} = 1.1 \cdot 10^{10}$ K. For the numerical calculations, (10.129) has been slightly modified (smoothed):

$$\varepsilon_{\mathrm{Fe}}(\rho, T) = \frac{1}{2} \frac{E_{\mathrm{b,Fe}}}{A\, m_{\mathrm{p}}} \left\{ 1 + \sin\left[\pi \left(\frac{T - T_{0\mathrm{Fe}}}{T_{1\mathrm{Fe}} - T_{0\mathrm{Fe}}} \right) - \frac{\pi}{2} \right] \right\}. \tag{10.130}$$

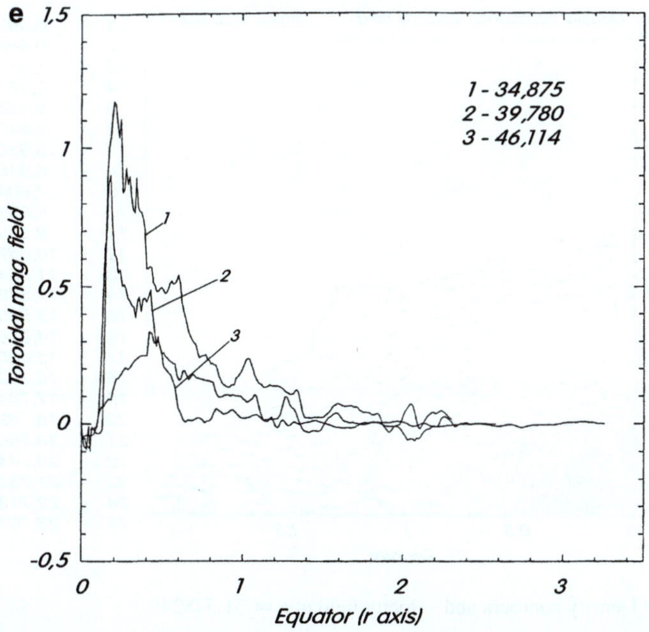

Fig. 10.17 (**e**) Same as in Fig. 102.3 during the matter ejection. 1 - t = 34.875, 2 - t = 39.78, 3 - t = 46.114

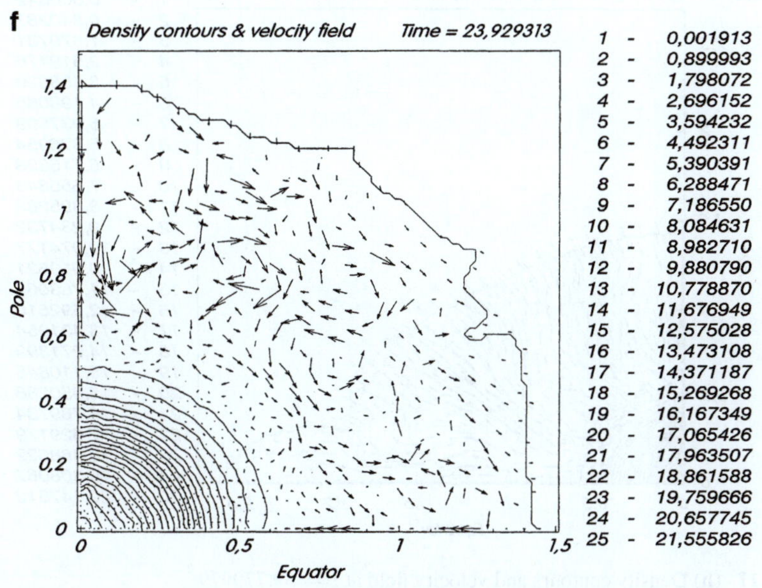

Fig. 10.17 (**f**) Density contours and velocity field at t = 23.929313

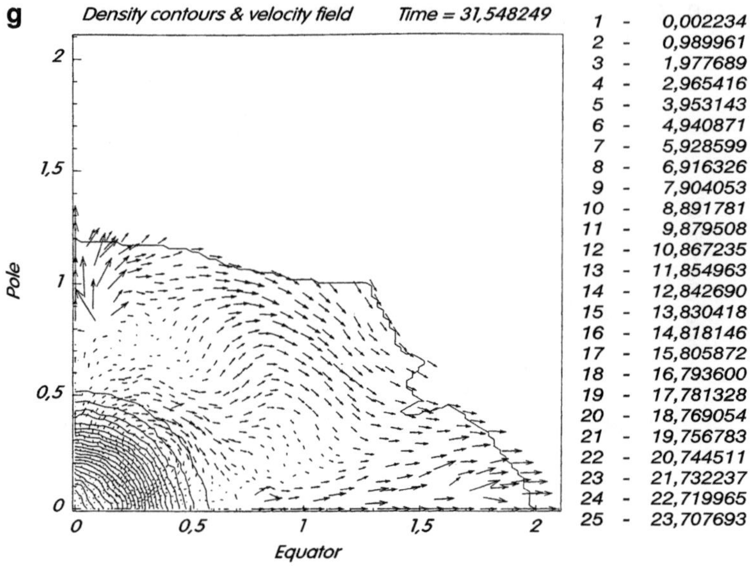

Fig. 10.17 (**g**) Density contours and velocity field at $t = 31.548249$

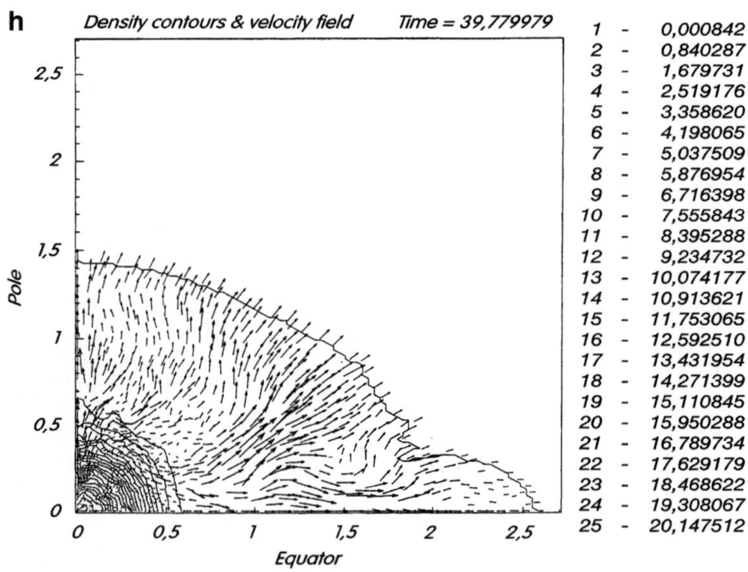

Fig. 10.17 (**h**) Density contours and velocity field at $t = 39.779979$

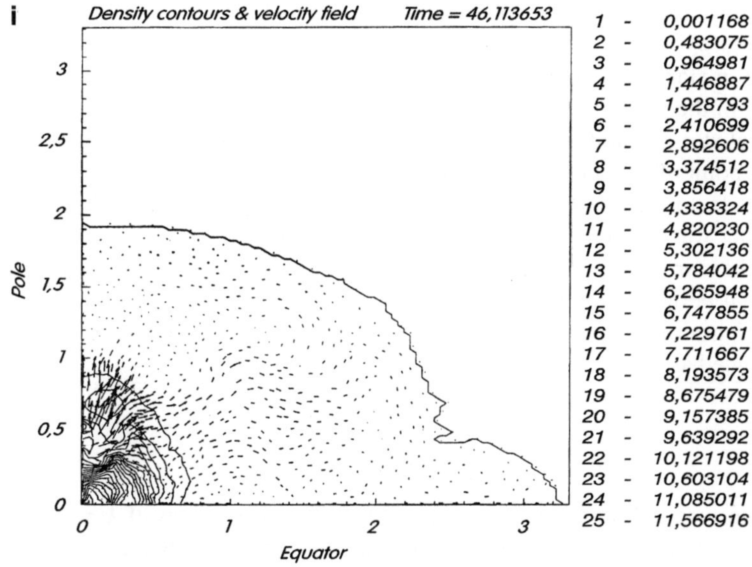

Fig. 10.17 (**i**) Density contours and velocity field at $t = 46.113653$

Neutrino losses were taken into account, which included Urca processes (Q_{URCA}), according to (10.107), (10.108); losses due to pair, photo and plasma neutrino production (Q_{tot}), described in Sect. 5.2.4, Vol. 1. A dense core of the neutron star becomes opaque to neutrino, decreasing neutrino losses. It was taken into account approximately by multiplying the losses from a transparent medium by the factor $\exp(-\frac{\tau_\nu}{10})$, where the neutrino optical depth is connected with a number density of weakly interacting particles n, neutrino cross-section σ_ν (see Sects. 5.1 and 5.2, Vol. 1) and characteristic length l_ν. The formula for neutrino losses $F(\rho, T)$ was used as

$$F(\rho, T) = (Q_{\text{URCA}} + Q_{\text{tot}})\mathrm{e}^{-\frac{\tau_\nu}{10}},$$
$$\tau_\nu = \sigma_\nu n l_\nu, \quad \sigma_\nu = \frac{10^{-44}T^2}{(0.5965 \cdot 10^{10})^2}, \quad n = \frac{\rho}{m_{\text{p}}}, \tag{10.131}$$

where

$$l_\nu = \frac{\rho}{|\nabla \rho|} = \frac{\rho}{((\partial\rho/\partial r)^2 + (\partial\rho/\partial z)^2)^{1/2}}. \tag{10.132}$$

Here, l_ν monotonically decreases towards the surface, having its maximum at the center; it is approximately equal to the thickness of the neutrino absorbing matter. The factor 1/10 in the expression $\exp(-\frac{\tau_\nu}{10})$ is used because only about 1/10 of the nucleons with energies near the Fermi energy interact with neutrinos in the degenerate matter of a hot neutron star.

The modeling of an MR supernova explosion consists of two subsequent stages. The first one is simulating the collapse of the iron core [37]. As the initial conditions for collapse calculations, a spherically symmetrical model of a star with a central density of $\rho_c = 4.5 \cdot 10^9$ g/cm^3 was used. This corresponds to the maximum mass of an iron star on the curve $M(\rho_c)$, for T = 0. For the approximate equation of state from the problem to the Sect. 1.4.7, Vol.1 used here, the maximum mass of a spherical star is equal to $1.0042 M_\odot$. At the beginning of the collapse, the density at each point of the star (and consequently the mass of the star) was increased by a constant factor λ, which specifies the mass of the collapsing core. The temperature of the star was taken as is $T = \delta\rho^{2/3}$, with $\delta = 1$ K$\cdot$cm$^2\cdot$ g$^{-2/3}$. At the initial time t = 0, the star undergoes rigid-body rotation with the angular velocity $\omega = \omega_0$ (the rotational period of the star is $\tau = 2\pi/\omega$). It was assumed also that there are no poloidal motions in the star at time t = 0 ($v_r = v_z = 0$), and rotational energy in [37] was defined as

$$\frac{E_{\rm rot}}{E_{\rm grav}} = 0.0057, \; \frac{E_{\rm int}}{E_{\rm grav}} = 0.727, \; \omega_0 = 2.52 {\rm s}^{-1},$$

which corresponds to the rotational period 2.496 s. The values of λ = 1.2, 1.3, 1.4, 1.5, and 1.7 have been used in the calculations. Since the selected initial model is not in equilibrium, a rapid collapse occurs at the beginning of the calculations.After the contraction of the star, at time t = 0.1377 s, a recoil shock originates at a distance of $6 \cdot 10^5$ cm. Directly behind the shock front, the temperature of the matter rapidly increases and the radiation of neutrinos and dissociation of iron are switched on. The matter of the star outside the shock continues to collapse towards the centre of the star. The collapse is finished by the formation of a stable rotating neutron star, in which the poloidal motion is damped rapidly, without producing any substantial explosion. The collapsed core rotates differentially. The central portion, with a radius of 10 km, rotates almost rigidly with a rotation period of 0.0015 s. With increasing distance from the centre the angular velocity falls off rapidly. Particles of matter located at the outer boundary near the equator rotate with a period of 0.3 s. The distributions of the rotation period and angular velocity in the equatorial plane are shown in Fig. 10.18.

At the second stage, the poloidal magnetic field is switched on, which is specified by the shape of the toroidal current j_ϕ in the form

$$j_\varphi = \begin{cases} j_\varphi^u \text{ for } z \geq 0, \; r^2 + z^2 \leq 0.025^2, \\ j_\varphi^d \text{ for } z \leq 0, \; r^2 + z^2 \leq 0.025^2, \\ 0 \quad \text{for} \qquad r^2 + z^2 \geq 0.025^2, \end{cases} \tag{10.133}$$

where

$$j_\varphi^u = A_j\left[\sin\left(\pi\frac{r}{0.025} - \frac{\pi}{2}\right) + 1\right]\left[\sin\left(\pi\frac{z}{0.025} - \frac{\pi}{2}\right) + 1\right]$$
$$\times\left[1 - \left(\frac{r}{0.025}\right)^2 - \left(\frac{z}{0.025}\right)^2\right],$$
$$j_\varphi^d = -j_\varphi^u \text{ (quadrupole symmetry)} \quad \text{or,} \quad j_\varphi^d = j_\varphi^u \text{ (dipole symmetry)}.$$

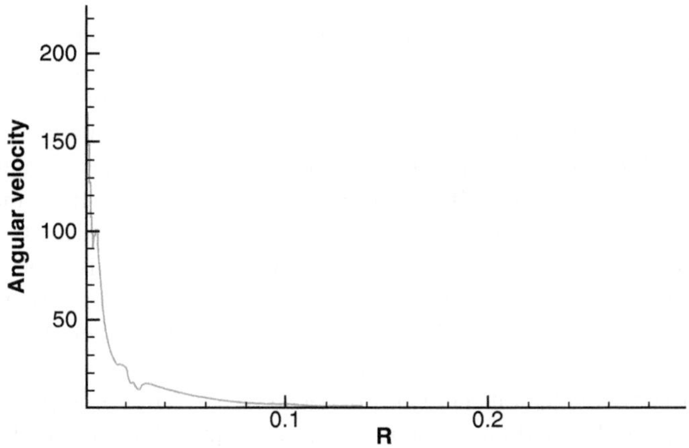

Fig. 10.18 The angular velocity in s^{-1} as a function of the coordinate r in the equatorial plane

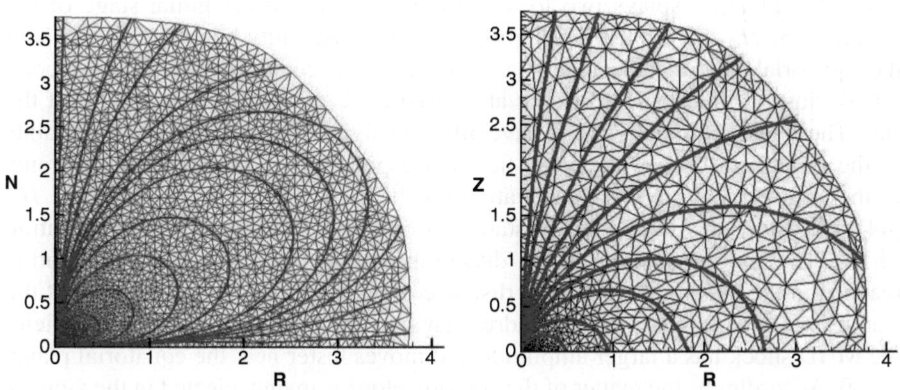

Fig. 10.19 Initial poloidal ($B_{\mathrm{pol,in}}r$) quadrupole (*left*) and dipole (*right*) magnetic field configurations

Here, A_j is a coefficient that is used to calibrate the initial toroidal current and, consequently, the initial magnetic field. The indices "u" and "d" refer to the upper and lower hemispheres of the star. The initial configurations of the magnetic field are presented in Fig. 10.19.

The second stage is finished by the MR supernova explosion. The length of this stage is determined by the initial magnetic field. The weaker the initial poloidal magnetic field, the longer the stage of the strengthening of the toroidal field component before the explosion. In the 1D model, the time for the development of the MR explosion from the beginning of the magnetic field evolution is proportional to $\sim\frac{1}{\sqrt{\alpha}} \sim B_{\mathrm{pol,in}}^{-1}$, see (10.114), and growing inversely proportional to the initial magnetic field strength. In 2D calculations, the time of the growth of the magnetic field increases much slower (logarithmically, $\sim|\log\alpha|$) than in 1D case. due to development of magnetorotational instability (MRI) [10] and rapid (exponential) increase of the strength of the magnetic field, both toroidal and poloidal components.

The differential rotation and amplification of the toroidal component of the magnetic field, and also of the poloidal component during the development of MRI, lead to the formation of the MHD compression wave moving from the centre of the star. This compression wave goes along steeply decreasing density profile and soon transforms into the MHD shock wave. This MHD shock is supported by the flux of the matter from the central parts of the star. This matter flux works like a piston increasing the velocity and the strength of the shock. When the MHD shock reaches the outer boundary of the collapsed iron core, the kinetic energy of the radial motion (supernova explosion energy) is between $(0.5-2.6)\cdot 10^{51}$ ergs, according to the calculations [133], and the amount of the mass ejected by the shock is between $(0.09-0.22)\cdot M_\odot$, for the total masses of the core between $(1.2-1.7)\cdot M_\odot$, and specific rotation energy changing by the coefficient 2, see Table 10.4.

In the case of quadrupolar symmetry of the poloidal magnetic field, the generated toroidal field H_ϕ displays two local extremum points, at the initial stage of the evolution of H_ϕ, before the development of MRI instability begins. The first is in the equatorial plane at a distance of $\approx$ 15 km from the center of the star. The second one is close to the rotational axis z at a distance $\approx$ 10 km from the centre of the star. These two extremum points have different signs of H_ϕ. The extremum points of the generated toroidal magnetic field correspond approximately to the extrema of the term $r\mathbf{H}\,\nabla(V_\phi/r)$ in the equation for the evolution of the toroidal field H_ϕ [44], since the star is in a steady state and only this term specifies the evolution of H_ϕ. The maximum value of H_ϕ during the calculations is $8.9\cdot 10^{16}$ G and is reached in the equatorial plane at a distance of $1.3\cdot 10^6$ cm from the centre of the star at time $=0.02$ s. Due to the quadrupolar symmetry of the initial magnetic field, the MHD shock has a larger amplitude and moves faster near the equatorial plane, $z=0$. Accordingly, the matter of the stars envelope is mainly ejected in the vicinity of the equatorial plane in an MR explosion.

In contrast to the explosion with quadrupolar symmetry, where matter was ejected mainly near the equatorial plane, the supernova explosion for an initial dipolar field develops mainly along the rotational axis. Estimates show that the

Table 10.4 Results of MR-supernova simulations, from [133]

Core mass $M/M_\odot$	Specific rotational energy when magnetic field is switched on E_{rot}/M_{core}, 10^{19} erg/g	Energy of explosion E_{expl}, 10^{51} erg	Ejected mass, $M_\odot$	Final central density $\rho_c\cdot(10^{14})$g/cm^3
1.3	0.19	0.5	0.087	3.7
1.5	0.21	0.7	0.091	3.8
1.7	0.23	1.1	0.095	3.9
1.2	0.39	1.8	0.160	2.4
1.4	0.40	2.6	0.220	2.5

characteristic time for reconnection of the magnetic field substantially exceeds the time for the development of MRI. The energy and ejected mass for an MR supernova with an initial dipolar field differ only slightly from the corresponding values for the quadrupolar field. The basic qualitative distinguishing feature of the dipolar case, in comparison with the quadrupolar one, is that the compression wave, and later the MHD shock, propagates mainly along the rotation axis. Further propagation of the shock in the supernova results in the formation of a weakly collimated directed jet.

10.5.8 Development of the Magnetorotational Instability in 2D Simulations

MRI in the magnetized star with differential rotation was analyzed in [927], where dynamo action accompanied the development of such instability. In our axially symmetric 2D simulations, dynamo action is prohibited [314]. The possibility of the appearance of 2D instability was mentioned in [300], in application to the SN1987a. This type of MPI was investigated earlier in [965, 1009], without any connection with a supernova explosion. The qualitative picture of the MRI in 2D is the following. At the initial stages, the differential rotation leads to a linear growth of the toroidal field described qualitatively as

$$\frac{\mathrm{d}H_\varphi}{\mathrm{d}t} = H_\mathrm{r}\left(r\frac{\mathrm{d}\Omega}{\mathrm{d}r}\right). \tag{10.134}$$

The right-hand side is almost constant at the initial stage of the process. After about 100 rotational periods of the central core, the MRI instability starts to develop. The appearance of MRI is characterized by the formation of multiple poloidal differentially rotating vortexes, which twist the initial poloidal field, leading to its amplification according to

$$\frac{\mathrm{d}H_\mathrm{r}}{\mathrm{d}t} = H_{r0}\left(\frac{\mathrm{d}\omega_v}{\mathrm{d}l}l\right), \tag{10.135}$$

where l is the coordinate, directed along the vortex radius, and ω_v is the angular velocity of the poloidal vortex. Qualitatively the poloidal field amplification due to the vortexes induced by MRI is shown in Fig. 10.20.

The enhanced poloidal field immediately starts to take part in the toroidal field amplification according to equation (10.134). With further growth of H_ϕ., the poloidal vortex speed increases. The calculations [45] give the values of $\omega_v = 0.0132\ \mathrm{s}^{-1}$ at $|H_\phi| = 2.46 \cdot 10^{15}$ G corresponding to $t = 0.041$ s, and $\omega = 0.052$ s^{-1} at $|H_\phi| = 4.25 \cdot 10^{15}$ G corresponding to $t = 0.052\ \mathrm{s}^{-1}$ for the same Lagrangian particle. In general, we may approximate the value in brackets in equation (10.135) by a linear function on the value $(H_\phi - H_\phi^*)$ as

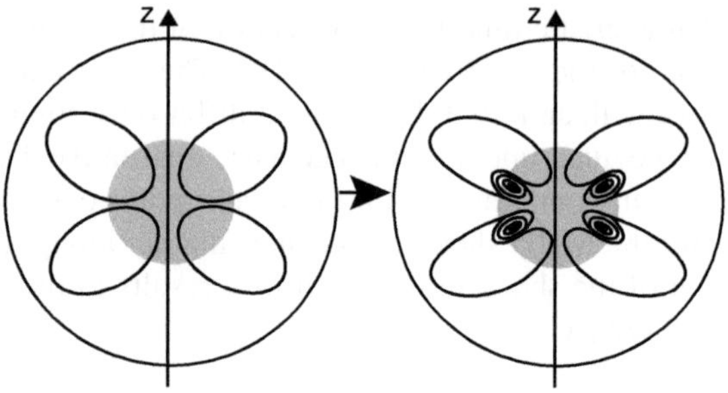

Fig. 10.20 Qualitative picture of the development of the magnetorotational instability, from [45]

$$\left(\frac{d\omega_v}{dl}l\right) = \alpha(H_\varphi - H_\varphi^*). \qquad (10.136)$$

H_r(or $H_{poloidal}$ in general case) in equation (10.134) is no longer constant after the onset of MRI, but described by equations (10.135) and (10.136). Assuming, for simplicity, that $[r(d\Omega/dr)] = A$ is constant during the first stages of MRI, and taking H_ϕ^* as constant, we obtain to the following equation:

$$\frac{d^2}{dt^2}(H_\varphi - H_\varphi^*) = AH_{r0}\alpha(H_\varphi - H_\varphi^*), \qquad (10.137)$$

This gives the exponential growth of magnetic field components

$$H_\varphi = H_\varphi^* + H_{r0}e^{\sqrt{A\alpha H_{r0}}(t-t^*)},$$
$$H_r = H_{r0} + \frac{H_{r0}^{3/2}\alpha^{1/2}}{\sqrt{A}}\left(e^{\sqrt{A\alpha H_{r0}}(t-t^*)} - 1\right),$$

taking H_{r0} as the seed field for the development of MRI. The above model shows the way of development of MRI instability in the 2D case. In this case, there is no direct influence of toroidal field on to a poloidal field, and therefore there is no dynamo action. Nevertheless, both components grow exponentially chaotically for the poloidal field and both chaotically and regularly for the toroidal one. Magnetorotational explosion is produced by both regular and chaotic magnetic fields. After rapid neutrino cooling and damping of the hydrodynamic motion, the chaotic component remains frozen into the matter due to high electric conductivity, and the chaotic magnetic field could be the source of magnetar activity in soft gamma repeaters. Exponential growth of the magnetic field in MRI determines logarithmical growth of the explosion time, because the explosion takes place at approximately same density of the magnetic energy, independently on the initial magnetic field strength.

It is known from the observations that the shapes of core-collapse supernovae are different. From our simulations, it follows that the MR core-collapse explosion model arises after the development of the MRI. The development of the MRI is a stochastic process, and hence the resulting shape of the supernova can vary. We may conclude that the MRE mechanism can lead to a different shape of the supernova. It is important to point out that the MR mechanism of the supernova explosion leads always to asymmetrical outbursts.

10.5.9 Symmetry Breaking Of the Magnetic Field, Anisotropic Neutrino Emission and High Velocity Neutron Star Formation

Observations of the pulsars moving at velocities up to $500\,\mathrm{km\,s^{-1}}$ [659] is a challenge to the theory of neutron-star formation, and the large neutron-star velocities should be taken into account for the determination of pulsar radiation properties from observational data [137]. The plausible explanation for the birth of rapidly moving pulsars seems to be the suggestion of the kick at the birth from the asymmetric explosion. We now make estimations for the strength of the kick, produced by the asymmetric neutrino emission during the collapse. The asymmetry of the neutrino pulse is produced by the asymmetry of the magnetic field distribution, formed during the collapse and differential rotation.

Consider a rapidly and differentially rotating new-born neutron star with the dipole poloidal and symmetric toroidal fields. A field amplification during the differential rotation leads to the formation of an additional toroidal field from the poloidal one. This field, made from the dipole poloidal one by twisting, is antisymmetric with respect to the symmetry plane. The sum of the initial symmetric with the induced antisymmetric toroidal fields has no plane symmetry.

In the absence of dissipative processes, the neutron star returns to the state of rigid rotation, losing the induced toroidal field and restoring mirror symmetry of the matter distribution. Formation of the asymmetric toroidal field distribution is followed by a magnetorotational explosion, which is asymmetric, leading to neutron star recoil and star acceleration [184]. The neutron star acceleration occurs also [110] due to the dependence of the cross-section of weak interactions on the magnetic field [758] (see also Chap. 5, Vol. 1).

After a collapse of a rapidly rotating star, the neutron star rotates at a period P about 1 ms. Differential rotation leads to the linear amplification of the toroidal field

$$B_\phi = B_{\phi 0} + B_\mathrm{p}(t/P). \tag{10.138}$$

The time of the neutrino emission is several tens of seconds [730]. After 20 s, the induced toroidal magnetic field will be about $2 \times 10^4 B_\mathrm{p}$, corresponding to 10^{15}–10^{17} Gs for $B_\mathrm{p} = 10^{11}$–10^{13} Gs, observed in the pulsars. Adopting the initial toroidal field $B_{\phi 0} = (10–10^3)B_\mathrm{p} = 10^{12}–10^{16}$, we may estimate an asymmetry of the neutrino pulse. For symmetric $B_{\phi,0}$ and dipole poloidal field, the difference ΔB_ϕ between the magnetic field's absolute values in two hemispheres increases, until it reaches the value $2B_{\phi 0}$. It remains constant later, while the relative difference

$$\delta_B = \frac{\Delta B_\phi}{B_{\phi+} + B_{\phi-}} \tag{10.139}$$

decreases. The main neutrino flux is formed in the region where the mean free path of the neutrino is smaller than the stellar radius. The neutrino luminosity is $L_\nu = 4\pi\, r^2\, H_\nu$, where H_ν is connected with a temperature gradient as

$$H_\nu = -\frac{7}{8}\frac{4acT^3}{3} 4\pi r^2 \rho l_{\rm T} \frac{\partial T}{\partial m}. \tag{10.140}$$

Here, the part of H_ν connected with a gradient of a lepton charge is omitted (see Sect. 5.4, Vol. 1). The quantity $l_{\rm T}$, having the meaning of the neutrino mean free path, is connected with the neutrino opacity κ_ν by

$$\kappa_\nu = 1/(l_{\rm T}\rho). \tag{10.141}$$

Calculations of the spherically symmetrical collapse [730] have shown that during the phase of the main neutrino emission, a hot neutron star consists of a quasiuniform, quasi-isothermal core with temperature T_i, whose mass increases with time, and the region between the neutrinosphere and the isothermal core, where the temperature decreases smoothly by about ten times while the density, which finally drops about six times decreases non-monotonically. A neutrino flux is forming in this region, containing about one half of the neutron star mass. We suggest, for simplicity, power–law dependences for the temperature and $l_{\rm T}$:

$$T = T_i \left(\frac{r_i}{r}\right)^m, \qquad l_{\rm T} = \frac{1}{\kappa\rho} = l_{Ti}\left(\frac{r}{r_i}\right)^n. \tag{10.142}$$

The neutrinosphere with the radius r_ν is determined approximately by the relation

$$\int_{r_\nu}^{\infty} \kappa_\nu \rho \mathrm{d}r = \int_{r_\nu}^{\infty} \frac{\mathrm{d}r}{l_{\rm T}} = 1. \tag{10.143}$$

Using (10.142) outside the neutrinosphere, we obtain from (10.143) the relation

$$r_\nu = r_i \left(\frac{r_i}{(n-1) l_{Ti}}\right)^{1/(n-1)}. \tag{10.144}$$

Finally, we obtain the temperature of the neutrinosphere T_ν and the neutrino luminosity L_ν

$$T_\nu = T_i \left(\frac{(n-1) l_{Ti}}{r_i}\right)^{m/(n-1)}, \tag{10.145}$$

$$L_\nu = 4\pi r_\nu^2 B_\nu = \frac{7}{8} m \frac{16\pi ac T_i^4}{3} (n-1)^{(4m-n+1)/(n-1)} r_i^2 \left(\frac{l_{Ti}}{r_i}\right)^{(4m-2)/(n-1)}. \tag{10.146}$$

To estimate the anisotropy of the neutrino flux, we compare two stars with the same radius and temperature of the core r_i and T_i and different opacities. Assume l_{Ti} is different and constant in two hemispheres, and each one is radiating according to (10.146). The anisotropy of the flux is

$$\delta_L = \frac{L_+ - L_-}{L_+ + L_-}, \tag{10.147}$$

where L_+ and L_- are luminosities in the different hemispheres, calculated using (10.146). For a small difference between the hemispheres

$$\delta_L = \frac{\Delta L}{L} = \frac{4m-2}{n-1}\frac{\Delta l_{Ti}}{l_{Ti}}. \tag{10.148}$$

Here, $n > 1$, when $m = 1/2$ the neutrino fluxes in both hemispheres are equal because the smaller opacity and larger neutrinosphere temperature T_ν from (10.145) are compensated by a smaller neutrinosphere radius r_ν from (10.144). The equation of motion of the neutron star with mass M_n

$$M_n \frac{\mathrm{d}v_n}{\mathrm{d}t} = \frac{L_+ - L_-}{c}, \qquad L_+ + L_- = \frac{2}{\pi} L_\nu(t). \tag{10.149}$$

For the power distributions, it follows from (10.146) that

$$L_\pm = A l_{Ti\pm}^{(4m-2)/(n-1)}. \tag{10.150}$$

In general, l_{Ti} is determined by various neutrino processes and depends on B. As an example, consider the dependence on B in the form

$$W_{\mathrm{n}B} = W_{\mathrm{n}} \left[1 + 0.17 \left(\frac{B}{B_{\mathrm{cr}}} \right)^2 + \ldots \right] \quad \text{for } B \ll B_{\mathrm{cr}}, \tag{10.151}$$

and

$$W_{\mathrm{n}B} = W_{\mathrm{n}}\, 0.77 \left(\frac{B}{B_{\mathrm{cr}}} \right) \quad \text{for } B > 2.7\, B_{\mathrm{cr}}. \tag{10.152}$$

In a strongly relativistic ($E_{\mathrm{Fe}} >> m_{\mathrm{e}}c^2$) or very hot ($kT >> m_{\mathrm{e}}c^2$) plasma the value

$$B_{\mathrm{c}} = \lambda B_{\mathrm{cr}}, \quad \lambda = \frac{E_{\mathrm{Fe}}}{m_{\mathrm{e}}c^2}, \text{ or } \quad \lambda = \frac{\kappa T}{m_{\mathrm{e}}c^2}, \tag{10.153}$$

with maximal value of λ should be used in (10.151 and 10.152) instead of B_{cr}. At large values of the energy of beta decay $\Delta >> m_{\mathrm{e}}c^2$, also $\lambda = \Delta/m_{\mathrm{e}}c^2$ should be added to (10.153). Making the interpolation between two asymptotic forms, we obtain

$$l_{Ti\pm} \sim \frac{1}{W} = l_{T0} \frac{1+(B/B_c)^3}{1+0.17(B/B_c)^2+0.77\left(\frac{B}{B_c}\right)^4} \equiv l_{T0} F^{(n-1)/(4m-2)}(B) \tag{10.154}$$

The time dependence of the average value of B in each hemisphere can be found from (10.138) using

$$B_{p+} = -B_{p-}, \quad B_{\phi 0+} = B_{\phi 0-}. \tag{10.155}$$

By B_p, we mean the average radial component of the poloidal field taking part in the amplification of B_ϕ. The time dependence of L_ν is taken from the spherically symmetric calculations of the collapse.

For $(4m-2)/(n-1) = 1$ and in the condition when the neutron star is accelerated at $B \gg B_c$, we have $F_\pm = B_c/0.77B_\pm$. The equation of motion (10.149) may be written as

$$M_n \frac{dv_n}{dt} = \frac{2}{\pi} \frac{L_\nu}{c} \frac{|B_+| - |B_-|}{|B_+| + |B_-|}, \tag{10.156}$$

with the linear functions for $B_\pm$. Take the constant $L_\nu = 0.1 M_n c^2/20\,\mathrm{s}$. With these simplifications, the final velocity of the neutron star v_{nf} follows as a result of the solution of (10.156) in the form

$$v_{nf} = \frac{2}{\pi} \frac{L_\nu}{M_n c} \frac{PB_{\phi 0}}{|B_p|} \left[0.5 + \ln\left(\frac{20\,s}{P} \frac{|B_p|}{B_{\phi 0}} \right) \right]. \tag{10.157}$$

For $P = 10^{-3}\,s$ we obtain

$$v_{nf} = \frac{2}{\pi} \frac{c}{10} \frac{P}{20\,\mathrm{s}} x \left[0.5 + \ln\left(\frac{20\,\mathrm{s}}{P} \frac{1}{x} \right) \right] \approx 1 \frac{\mathrm{km}}{\mathrm{s}} x \left[0.5 + \ln\left(\frac{2 \times 10^4}{x} \right) \right]. \tag{10.158}$$

For the value $x = B_{\phi 0}/|B_p|$ ranging between 20 and 10^3, we have v_{nf} between 140 and 3,000 km/s^{-1}, which can explain the nature of the most rapidly moving pulsars. Formula (10.158) can be applied when $B_{\phi 0} \gg B_c$ and $x \gg 1$.

The analysis of the spatial structure of pulsar emission sources using interstellar scintillation had shown [919] that the velocity vector of the pulsar motion in space is close to the orientation of the rotation axis of the pulsar, in accordance with prediction made in [110].

The observational analysis permitted [474] to make a catalogue of 233 pulsars with proper motion measurements. The sample contains a wide variety of pulsars including recycled objects and those associated with globular clusters or supernova remnants. After taking the most precise proper motions for those pulsars for which multiple measurements are available, the majority of the proper motions (58%) are derived from pulsar timing methods, 41% using interferometers and the remaining 1% using optical telescopes. Many of the 1D and 2D speeds (referring to speeds measured in one coordinate only and the magnitudes of the transverse velocities respectively) derived from these measurements are somewhat lower than earlier estimates because of the use of the most recent electron density model in determining pulsar distances. The mean 1D speeds for the normal and recycled pulsars are 152 ± 10 and 54 ± 6 km/s respectively. The corresponding mean 2D speeds are 246 ± 22 and 87 ± 13 km/s. The observed 1D and 2D speeds clearly demonstrate that the 3D velocity vector is isotropic. PSRs B2011 + 38 and B2224 + 64 have the highest inferred 2D speeds of 1,600 km/s. Applying a novel deconvolution technique to the sample of 73 pulsars with characteristic ages less than 3Myr, the mean 1D and 2D speeds for pulsars with characteristic ages less than 3Myr are 192 ± 20 and 307 ± 47 km/s, indicating to the mean 3D pulsar birth velocity to be 400 ± 40 km/s. The distribution of velocities is well described by a Maxwellian distribution with 1D rms = 265 km/s. There is no evidence for a bimodal velocity distribution. Most pulsars within supernova remnants are found to have lower velocities than other pulsars. This is most likely due to selection effects where fast moving pulsars leave the supernova shell within a relatively short period of time. The proper motions for PSRs B1830-08 and B2334 + 61 are consistent with their proposed associations with the supernova remnants W41 and G114.3 + 0.3 respectively. Large values of the proper motion velocities are $\sim$1,600 km/s in 2D, which corresponds to $\sim$2,100 km/s for 3D velocity, assuming equipartition of the kinetic energy of the pulsar in 3 axis (isotropic distribution). It looks out that the largest values of the proper motion velocities could be gained only in the presence of anisotropic neutrino emission, which carries away more than 15% of the rest mass energy of the neutron star.

The simulations of the MR core-collapse explosion model for the initial quadrupole-like magnetic field described in [45], and the MR core-collapse explosion model with the dipole field [707] are restricted by the symmetry to the equatorial plane. While in reality this symmetry can be violated due to the MRI, the simultaneous presence of the initial dipole and quadrupole-like magnetic field [1031] and the initial toroidal magnetic field [184]. The violation of the mirror symmetry of the magnetic field in rotating stars can lead to the kick effect and formation of rapidly moving radio pulsars. When rotational and magnetic axes do not coincide, the whole picture of the explosion process is 3D. Nevertheless, the magnetic field twisting happens always around the rotational axis; so, we may expect the kick velocity of the neutron star to be strongly correlated with its spin direction, also due to the anisotropy of the neutrino flux [110]. Simultaneously, because of the stochastic nature of the MRI, the level of the anisotropy should be strongly variable, leading to a large spreading in the neutron star velocities. This prediction of the MRE differs from the models with powerful neutrino convection, where the arbitrary direction of the kick velocity is expected [248].

Three-dimensional magnetohydrodynamical simulations of a core-collapse supernova in which the progenitor has magnetic fields inclined to the rotation axis have been made in [703]. The simulations used a simple empirical equation of state in which the pressure of degenerate gas is approximated by piecewise polytropes for simplicity. Energy loss due to neutrinos is not taken into account for simplicity as well. The simulations start from the stage of dynamical collapse of an iron core. The dynamical collapse halts at t =189 ms by the pressure of high-density gas, and a proto-neutron star (PNS) forms. The evolution of the PNS was followed for about 40 ms in typical models. When the initial rotation is mildly fast and the initial magnetic fields are mildly strong, bipolar jets are launched from the upper atmosphere (r 60 km) of the PNS. The jets are accelerated to $\sim 3 \cdot 10^4$ km/s, which is comparable to the escape velocity at the footpoint. The jets are parallel to the initial rotation axis. Before the launch of the jets, magnetic fields are twisted by rotation of the PNS. The twisted magnetic fields form torus-shaped multilayers in which the azimuthal component changes alternately. The formation of magnetic multilayers is due to the initial condition in which the magnetic fields are inclined with respect to the rotation axis. The energy of the jet depends only weakly on the initial magnetic field assumed. When the initial magnetic fields are weaker, the time lag is longer between the PNS formation and jet ejection. It is also shown that the time lag is related to the Alfvén transit time. The morphology of twisted magnetic field and associate jet ejection is not affected by the simplification of the equation of state and neutrino treatment. We may expect from the simulations [703], that the jet formation along the rotational axis will be accompanied by the same type of anisotropy of the neutrino emission due to the mirror symmetry violation relative to the plane perpendicular to the rotational axis.

Recent analysis of observations of pulsars [549, 550] had confirmed the result of [919] that rotation and velocity vectors of pulsars are aligned, as is predicted by the MR supernova mechanism. Strong observational evidence for a relationship between the direction of a pulsar's motion and its rotation axis have been presented in [549]. Carefully calibrated polarization data for 25 pulsars have been shown, 20 of which display linearly polarized emission from the pulse longitude at closest approach to the magnetic pole. Such data allow determination of the position angle of the linear polarization which in turn reflects the position angle of the rotation axis. Of these 20 pulsars, 10 show an offset between the velocity vector and the polarization position angle which is either less than 10° or more than 80°, a fraction which is very unlikely by random chance. It was concluded that the bimodal nature of the distribution arises from the presence of orthogonal polarization modes in the pulsar radio emission. In some cases, this orthogonal ambiguity is resolved by observations at other wavelengths so that the velocity vector and the rotation axis are aligned at birth. Strengthening the case is the fact that four of the five youngest pulsars in the sample with ages less than 3 Myr, which suffer from the smallest unknown errors, show the best alignment including the Vela pulsar. The observer sees a projection of the direction of the rotation axis, PA_r, and the velocity vector, PA_v, onto the plane-of-the sky, and it is concluded in [549] that PA_v and PA_r are parallel in all cases. It is pointed out that if the kicks impart both linear and angular momentum to the new-born neutron star, the information about the stellar axis before the supernova explosion is lost. Observations of double neutron star systems cannot therefore rule out aligned kicks, contrary to claims in the literature.

The results of observations of 22 pulsars at frequencies of 0.7, 1.4 and 3.1 GHz and their polarization profiles are presented in [550]. The orientation of the spin and velocity vectors have been compared to verify the proposed alignment of these vectors in [549]. It was found that for the 14 pulsars for which it was possible to determine both vectors, 7 are plausibly aligned, a fraction which is lower than, but consistent with, earlier measurements. There appears to be no obvious correlation between a pulsar's velocity and its alignment angle. The correlation between the velocity and rotation vectors therefore appears less strong than in [549]. However, the pulsars in the current sample are both older and more distant than the earlier sample. Both these parameters make it more likely that the gravitational pull of the Galaxy has altered the pulsar's proper motion vector from its birth direction. Although the numbers are small and the uncertainties large, the results do not obviously follow the prediction [741] where it is postulated that only the fast moving pulsars should show a correlation between their velocity and rotation vectors.

The acceleration of the collapsing star by anisotropic neutrino emission can happen even when the star collapses into a black hole, and the efficiency of acceleration decreases with increasing mass. We may expect black holes of stellar origin to move rapidly, like radio pulsars, and they may be found well above the galactic disk. This is observed among the soft X-ray novae – the most probable candidates for black holes in the Galaxy.

10.5.10 A Kick Due to Hydrodynamic Instabilities

Hydrodynamic instabilities during the first seconds of core-collapse supernovae has been studied in [871] by means of 2D simulations with approximative neutrino transport and boundary conditions that parameterize the effects of the contracting neutron star and allow to obtain sufficiently strong neutrino heating and, hence, neutrino-driven explosions. Confirming more idealized studies, as well as supernova simulations with spectral transport, it was found that random seed perturbations can grow by hydrodynamic instabilities to a globally asymmetric mass distribution in the region between the nascent neutron star and the accretion shock, leading to a dominance of dipole ($l = 1$) and quadrupole ($l = 2$) modes in the explosion ejecta, provided the onset of the supernova explosion is sufficiently slower than the growth time scale of the low-mode instability. It was obtained that by gravitational and hydrodynamic forces, the anisotropic mass distribution causes an acceleration of the nascent neutron star, which lasts for several seconds and can propel the neutron star to velocities of more than 1,000 km/s. Because the explosion anisotropies develop chaotically and change by small differences in the fluid flow, the magnitude of the kick varies stochastically. No systematic dependence of the average neutron star velocity on the explosion energy or the properties of the considered progenitors is found. Instead, the anisotropy of the mass ejection, and hence of the kick, seems to increase when the nascent neutron star contracts more quickly, and thus low-mode instabilities can grow more rapidly. More than 70 models have been constructed, which separate into two groups, one with high and the other with low neutron star velocities and accelerations after 1 s of post-bounce evolution, depending on whether the $l = 1$ mode is dominant in the ejecta or not. This leads to a bimodality of the distribution when the neutron star velocities are extrapolated to their terminal values. Establishing a link to the measured distribution of pulsar velocities, however, requires a much larger set of calculations and ultimately 3D modelling. Note that in the observational analysis of the pulsar velocities in [474], no evidence for a bimodal velocity distribution was found.

The growth of the instabilities proceeds extremely nonlinearly and chaotically such that the final ejecta anisotropy turns out to be sensitive to the initial random pattern of the seed perturbations, as well as to small differences between numerical runs (connected, e.g., to small changes in the grid zoning, machine roundoff errors, or small differences of the input physics). Despite the different ejecta geometry, however, integral parameters of the models, such as the neutron star mass, explosion time scale, or explosion energy, show little variability.

The proposed hydrodynamic kick mechanism, however, leads to an unambiguous prediction, which might be tested by future detailed observations of supernova remnants: the measured neutron star velocity should be directed against the momentum of the gaseous supernova ejecta. It is claimed in [871] that this is different from theories that explain pulsar kicks by anisotropic neutrino emission from the nascent neutron star, because in that case the direction of the acceleration can be independent of ejecta asymmetries. Actually this statement is not consistent with the anisotropic neutrino emission in the magnetorotational explosion, where both neutrino and hydrodynamic anisotropies are elongated by the same rotational axis.

The deficiency in the analysis of [871] is that it is based on simulations that assume axial symmetry with the polar axis being a coordinate singularity that is impenetrable for the fluid flow. Currently, it is neither clear to what degree pronounced $l = 1$ modes of the ejecta distribution and long-lasting downflows of matter to the neutron star can develop in the 3D environment, and how common they are, although first 3D simulations with the setup and input physics described here are promising, see [548], nor is it clear what the distribution of neutron star recoil velocities from 3D models will be. The large number of long-time simulations required by the stochastic nature and long duration of the proposed hydrodynamic kick mechanism, is currently out of reach due to its enormous demand in computer time. The results in [871] must therefore be considered as indicative, but they are far from providing definitive answers.

After a magnetorotational explosion, we may expect periods of the order of tens of ms for a neutron star rotation. When a neutron star is formed in a binary system, it either accelerates its rotation in low mass X-ray binaries (LMXB) or decelerates it in high mass XB. The former are transformed into recycled (ms, binary) pulsars, and the latter (after explosion of the massive component and disruption of the binary) may form a group of very slowly rotating neutron stars, one of which was observed in the strongest gamma ray burst of 5 March 1979, now related to soft gamma repeater family, and containing about ten objects.

Chapter 11
Final Stages of Stellar Evolution

During the final evolutionary stages, the nuclear fuel is consumed, and the star emits radiation owing to cooling. It is relatively cold and has a very high density, while the pressure arises mainly from the matter degeneracy. Chandrasekhar [264] in 1931 obtained the fundamental result that a star where the pressure is due to degenerate electrons has a maximum mass. At $\rho > 1.15 \times 10^9$ g cm^{-3} (for ^{56}Fe, for other nuclei see Table 10.1) the neutronization begins, and stars become unstable. Stable (neutron) stars reappear only at $\rho_c \approx 1.5 \times 10^{14}$ g cm^{-3} and exist to densities $\rho_c \approx 1.15 \times 10^{15}$ g cm^{-3}, where instability arises from general relativity (GR) effects. Oppenheimer and Volkov established in 1939 the existence of a mass limit for neutron stars [765], but its value has been recalculated many times for various equations of state. Solving the equilibrium equations for cold stars in Newtonian theory (9.97) and (10.1) and in GR (11.62) and (11.63) at a given equation of state, $P(\rho)$ has allowed derivation of the curve $M(\rho_c)$ demonstrating the existence of two maximum masses and an instability region (falling $M(\rho_c)$). Figure 11.1 represents the curve $M(\rho_c)$ from [80] for the equation of state corresponding to a minimum energy of matter in full thermodynamic equilibrium.

White dwarfs preceding the first maximum in $M(\rho_c)$ originate from stars with initial masses $M_i \leq 8M_\odot$, neutron stars (between the minimum and second maximum) from stars with initial masses $M_i = 8$–$M_{\rm lim}$, where $M_{\rm lim} = (20$–$50)\ M_\odot$. If $M_i > M_{\rm lim}$, the collapse may result in black hole formation.

The evolutionary path determines the real masses of white dwarfs and neutron stars. From observations, the masses of single white dwarfs are $\geq 0.6\ M_\odot$ [656, 877] and may be substantially less in binaries (see, for example, [160]). Observational estimates for neutron star masses are $\sim 1.4\ M_\odot$ [969] and some of them are known with a high accuracy [966, 968].

G.S. Bisnovatyi-Kogan et al., *Stellar Physics: 2: Stellar Evolution and Stability*, Astronomy and Astrophysics Library, DOI 10.1007/978-3-642-14734-0_5,

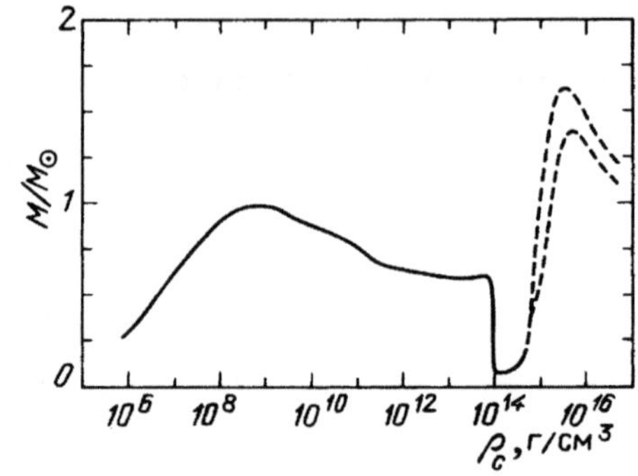

Fig. 11.1 Mass versus central density for zero-temperature non-rotating stars in complete nuclear equilibrium. Stars to the left of the maximum (Chandrasekhar limit) at $\rho_c = 1.4 \times 10^9\ \mathrm{g\,cm^{-3}}$ are stable white dwarfs, while stars to the right of the minimum at $\rho_c = 1.55 \times 10^{14}\ \mathrm{g\,cm^{-3}}$ are neutron stars. The *dashed curves* are constructed from Pandharipande's hyperonic equation of state (*lower curve*) and Pandharipande's equation of state for pure neutrons (*upper curve*). Neutron stars beyond the second maximum are unstable

11.1 White Dwarfs

11.1.1 Case T = 0

When the pressure is determined by the fully degenerate electrons with constant μ_Z from (10.23), the equilibrium equations (9.97) and (10.1) reduce to one equation [269]

$$\frac{1}{r^2}\frac{\mathrm{d}}{\mathrm{d}r}\left(r^2\frac{\mathrm{d}\sqrt{y^2+1}}{\mathrm{d}r}\right) = -\frac{y^3}{l_1^2},$$

$$l_1 = \frac{1}{2\mu_Z m_u m_e}\left(\frac{3\pi\hbar^3}{cG}\right)^{1/2} = \frac{7.77\times 10^8}{\mu_Z}\ \mathrm{cm}, \tag{11.1}$$

where

$$y = \frac{p_{\mathrm{Fe}}}{m_e c}. \tag{11.2}$$

Using (10.64) we obtain

$$\rho = By^3, \qquad B = \frac{m_e^3 c^3 \mu_Z m_u}{3\pi^2\hbar^3} = 9.74\times 10^5 \mu_Z. \tag{11.3}$$

The solution of (11.1) enables us to find the functions $M(\rho_c)$, $R(\rho_c)$ for cold white dwarfs [269]. For high ($y \gg 1$) and small ($y \ll 1$) densities, the equation of state for fully degenerate electrons in the ultrarelativistic and non-relativistic limits reduces to polytropes. From the exact formulae and expansions (see Chap. 1, Vol. 1), we obtain

$$P = \frac{m_e^4 c^5}{15\pi^2\hbar^3} y^5 = \frac{(3\pi^2)^{2/3}}{5} \frac{\hbar^2}{m_e(\mu_Z m_u)^{5/3}} \rho^{5/3}$$

$$= \frac{1.0036 \times 10^{13}}{\mu_Z^{5/3}} \rho^{5/3} \quad \text{for } y \ll 1, \tag{11.4}$$

$$P = \frac{m_e^4 c^5}{12\pi^2\hbar^3} y^4 = \frac{(3\pi^2)^{1/3}}{4} \frac{\hbar c}{(\mu_Z m_u)^{4/3}} \rho^{4/3}$$

$$= \frac{1.2435 \times 10^{15}}{\mu_Z^{4/3}} \rho^{4/3} \quad \text{for } y \gg 1. \tag{11.5}$$

From (11.4), (10.3) and (10.6), we obtain the relations for non-relativistic white dwarfs

$$M = \frac{3\pi^{3/2}}{2(\mu_Z m_u)^{5/2}} \left(\frac{\hbar^2}{2Gm_e}\right)^{3/2} M_{1.5}\rho_c^{1/2} = \frac{2.81 M_\odot}{\mu_Z^2} \left(\frac{\rho_{c6}}{\mu_Z}\right)^{1/2}, \tag{11.6}$$

where [269] $M_n = 2.71406$ for $n = 1.5$, $\rho_{c6} = \rho_c \times 10^{-6}$ g cm^{-3} $< 0.3\,\mu_Z$,

$$R = \frac{(9\pi)^{1/6}}{2\sqrt{2}} \frac{\hbar \xi_1 \rho_c^{-1/6}}{(Gm_e)^{1/2}(\mu_Z m_u)^{5/6}} = \frac{2.00 \times 10^9}{\mu_Z} \left(\frac{\mu_Z}{\rho_{c6}}\right)^{1/6} \text{ cm}, \tag{11.7}$$

where [269] $\xi_1 = 3.65375$ for $n = 1.5$. From (11.6) and (11.7), the dependence $R(M)$ becomes

$$R = \frac{2.82 \times 10^9}{\mu_Z^{5/3}} \left(\frac{M_\odot}{M}\right)^{1/3} \text{ cm}. \tag{11.8}$$

For ultrarelativistic electrons with $\gamma = 4/3$, $n = 3$, there is only a single value of equilibrium mass from (11.5) and (10.7)

$$M_{Ch} = \frac{\sqrt{3}\pi}{2} \left(\frac{\hbar c}{G}\right)^{3/2} \frac{1}{(\mu_Z m_u)^2} M_3 = \frac{5.83}{\mu_Z^2} M_\odot, \tag{11.9}$$

(see M_3 in (10.15)) called the Chandrasekhar limit and obtained in [264], [265]. A simple derivation of the formula (11.9), similar to [265], was performed independently by Landau [613]. The radius of a $n = 3$ polytrope may be arbitrary since the polytrope is in equilibrium at any central density (see (10.6)). As M approaches M_{Ch} and $n \to 3$, the white dwarf radius decreases rapidly and depends only on ρ_c. In this limit

$$R = 5.31 \times 10^8 \rho_{c9}^{-1/3} \mu_Z^{-2/3} \text{ cm} \quad \text{for } \rho_{c9} = \frac{\rho_c}{10^9 \text{ g cm}^{-3}} \gg 0.3\mu_Z. \tag{11.10}$$

The inclusion of a Coulomb interaction decreases the pressure at the same density and, correspondingly, decreases M in comparison with [269]. In the ultrarelativistic limit, the pressure is reduced by a constant multiplier, independently of the density. The limiting mass is reduced proportionally to this multiplier to the power (3/2). Using pressure calculations from [857] (see also Chap. 1, Vol. 1) in (10.6) and (11.9), the Chandrasekhar limit $M_{\mathrm{Ch,q}}$, with the Coulomb interaction taken into account, is

$$M_{\mathrm{Ch,q}} = \frac{5.83}{\mu_Z^2}\left(1 - 4.56\times 10^{-3} Z^{2/3} - 1.78\times 10^{-5} Z^{4/3} + 1.16\times 10^{-3}\right)^{3/2} M_{\odot}. \tag{11.11}$$

At $Z = 2, 6, 12, 26$, we have

$$M_{\mathrm{Ch,q}}/M_{\mathrm{Ch}} = 0.991, 0.979, 0.965, 0.940,$$
$$M_{\mathrm{Ch,q}} = 1.44, 1.43, 1.41, 1.18$$
$$\text{for } {}^{4}\mathrm{He}, {}^{12}\mathrm{C}, {}^{24}\mathrm{Mg}, {}^{56}\mathrm{Fe}, \text{ respectively.}$$

Decreasing the density leads to more important Coulomb corrections. The neutronization slows down the pressure growth and brings about white dwarf instability which occurs at a peak on the curve $M(\rho_c)$ at a density slightly above the neutronization density given in Table 10.1 (see Fig. 11.1 and Chap. 12). Models for white dwarfs including Coulomb corrections and neutronization are investigated in [452] using the equation of state of cold matter with account of Coulomb interaction in Thomas–Fermi approximation (see Sect. 1.4, Vol. 1), and are given in Table 11.1 Analyzing the stability of a star with a phase transition (see Sect. 12.4) shows that for equilibrium neutronization the loss of stability occurs at the finite new-phase core.[1] For the chain ${}^{56}\mathrm{Fe} \rightarrow {}^{56}\mathrm{Mn} \rightarrow {}^{56}\mathrm{Cr}$, we have at the point of stability loss [189]

$$\frac{M_{\mathrm{nc}}}{M} = 1.4\times 10^{-3}, \quad \frac{\Delta M}{M} = 2.1\times 10^{-4}, \quad \frac{\Delta\rho_c}{\rho_{c1}} = 0.022, \tag{11.12}$$

where M_{nc} is the mass of the new-phase core, ΔM is the increase in mass from the onset of neutronization to the loss of stability, and $\Delta\rho_c$ is the corresponding increase in central density. In [189], the onset of neutronization was assumed to occur at $\rho_{c0} = 1.15\times 10^9$ g cm^{-3} (the capture onto the ground level, see Table 10.1), Coulomb corrections have been ignored. On the boundary of the new-phase core, the density makes a jump of 26/24 times and attains $\rho_{c1} = 1.25\times 10^9$ g cm^{-3}. Small changes in ρ and M will result in changes in the relative quantities (11.12) which are small as well. The structure and stability of rotating white dwarfs will be discussed in Sect. 12.3. Decreasing the mass leads to lower stellar densities making

[1] Stable models for white dwarfs with finite new-phase core have been omitted in [452].

Table 11.1 Masses M and radii R of white dwarfs of ^{4}He, ^{12}C, ^{24}Mg and ^{54}Fe. Given also are M_{Ch} and R_{Ch} for $\mu_Z = 2; 56/26$ without Coulomb interaction and neutronization. The values of $(M_{Ch}\mu_Z^2)$ and $(R_{Ch}\mu_Z)$ are the same for all μ_Z at the same (ρ_c/μ_Z) (from [384])

	ρ_c/cm^3	3.16(3)	7.08(3)	8.71(3)	1.23(4)
$\mu_Z = 2$	R_{Ch}, cm				
	M_{Ch}, $M_\odot$				
	R_{He}, cm				
	M_{He}, $M_\odot$				
	R_C, cm				
	M_C, $M_\odot$				
	R_{Mg}, cm				
	M_{Mg}, $M_\odot$				
$\mu_Z = \frac{56}{26}$	R_{Ch}, cm				
	M_{Ch}, $M_\odot$				
	R_{Fe}, cm	1.402(9)	1.463(9)	1.465(9)	1.459(9)
	M_{Fe}, $M_\odot$	7.0(−3)	1.5(−2)	1.8(−2)	2.4(−2)
	ρ_c/cm^3	1.95(6)	3.58(6)	6.95(6)	1.56(7)
$\mu_Z = 2$	R_{Ch}, cm	9.702(8)	8.658(8)	7.614(8)	6.47(8)
	M_{Ch}, $M_\odot$	0.512	0.622	0.748	0.899
	R_{He}, cm	9.521(8)	8.519(8)	7.510(8)	6.40(8)
	M_{He}, $M_\odot$	0.499	0.609	0.734	0.885
	R_C, cm	9.382(8)	8.408(8)	7.426(8)	6.33(8)
	M_C, $M_\odot$	0.488	0.597	0.722	0.872
	R_{Mg}, cm	9.222(8)	8.289(8)	7.336(8)	6.26(8)
	M_{Mg}, $M_\odot$	0.476	0.584	0.708	0.857
$\mu_Z = \frac{56}{26}$	R_{Ch}, cm	9.138(8)	8.150(8)	7.176(8)	6.10(8)
	M_{Ch}, $M_\odot$	0.430	0.525	0.633	0.765
	R_{Fe}, cm	8.394(8)	7.579(8)	6.74(8)	5.78(8)
	M_{Fe}, $M_\odot$	0.380	0.471	0.576	0.703
	ρ_c/cm^3	1.95(9)	3.16(9)	6.00(9)	7.67(9)
$\mu_Z = 2$	R_{Ch}, cm		1.82(8)	1.52(8)	
	M_{Ch}, $M_\odot$		1.411	1.426	
	R_{He}, cm				
	M_{He}, $M_\odot$				
	R_C, cm		1.80(8)	1.50(8)	
	M_C, $M_\odot$		1.381	1.396	
	R_{Mg}, cm		1.79(8)	1.69(8)	
	M_{Mg}, $M_\odot$		1.363	1.282	
$\mu_Z = \frac{56}{26}$	R_{Ch}, cm				
	M_{Ch}, $M_\odot$				
	R_{Fe}, cm	1.99(8)	1.37(8)		
	M_{Fe}, $M_\odot$	1.093	1.028		

(continued)

Table 11.1 Masses M and radii R of white dwarfs of ^{4}He, ^{12}C, ^{24}Mg and ^{54}Fe. Given also are M_{Ch} and R_{Ch} for $\mu_Z = 2; 56/26$ without Coulomb interaction and neutronization. The values of $(M_{\mathrm{Ch}}\mu_Z^2)$ and $(R_{\mathrm{Ch}}\mu_Z)$ are the same for all μ_Z at the same (ρ_c/μ_Z) (from [384])

3.16(4)	1.22(5)	2.44(5)	5.47(5)	1.06(6)
	1.582(9)	1.405(9)	1.220(9)	1.084(9)
	0.164	0.204	0.316	0.411
	1.518(9)	1.359(9)	1.189(9)	1.061(9)
	0.154	0.213	0.305	0.399
	1.469(9)	...	...	...
	0.147	...	...	...
	1.414(9)	1.283(9)	1.136(9)	1.022(9)
	0.139	0.196	0.286	0.378
	1.488(9)	1.322(9)	1.148(9)	1.020(9)
	0.136	0.187	0.265	0.345
1.395(9)	1.233(9)	1.136(9)	1.019(9)	9.243(8)
4.6(−2)	0.103	0.149	0.222	0.298
	1.61(8)	6.68(8)	1.15(9)	1.92(9)
5.00(8)	3.87(8)	2.74(8)		2.08(8)
1.097	1.235	1.347		1.396
4.91(8)	3.81(8)	2.70(8)		2.06(8)
1.070	1.206	1.318		1.366
4.86(8)	3.79(8)	2.69(8)		2.05(8)
1.053	1.190	1.300		1.348
4.51(8)	3.52(8)	2.50(8)	2.19(8)	
0.872	0.991	1.088	1.112	
	2.45(10)	2.51(10)	3.61(10)	
1.32(8)		1.01(8)		
1.434		1.444		
1.45(8)		1.18(8)		
1.349		1.174		
1.50(8)				
1.205				
9.60(7)		8.84(7)		
1.014		0.990		

the Coulmb corrections that render the equation of state stiffer (reduce n, increase γ) more important. This causes departures from (11.8) such that the radius reaches its maximum value and falls with further decrease in M (Table 11.1). Equilibrium states of cold low-mass stars are constructed in [1069] using a more precise equation of state which allows large deviations from an ideal one. The results of calculations are approximated by

Table 11.2 Maximum R_{cr} and corresponding masses M_{cr} for various compositions (from [1069])

Element	$\frac{M_{cr}}{M_\odot}$	R_{cr}, cm	Element	$\frac{M_{cr}}{M_\odot}$	R_{cr}, cm
H	3.16(−3)	8.15(9)	^{12}C	2.24(−3)	2.74(9)
$x_H = 0.75$	2.63(−3)	6.99(9)	^{24}Mg	3.89(−3)	2.28(9)
$x_{He} = 0.25$			^{56}Fe	5.89(−3)	1.70(9)
^{4}He	1.12(−3)	3.57(9)			

$$0.424\left(\frac{M}{M_0}\right)^{1/3}\frac{R}{R_0} = \left(1 - \frac{R^3}{R_0^3}\frac{M_0}{M}\phi\right)^5, \tag{11.13}$$

where

$$M_0 = \left(\frac{P_0}{G}\right)^{3/2}\rho_0^{-2} = \left(\frac{Z}{A}\right)^2 Z \times 3.58 \times 10^{30}\ \text{g},$$

$$R_0 = \left(\frac{P_0}{G}\right)^{1/2}\rho_0^{-1} = \frac{Z}{A}Z^{-1/3} \times 9.73 \times 10^9\ \text{cm}.$$

Here,

$$\rho_0 = 3.88ZA\ \text{g cm}^{-3}, \quad P_0 = 9.52 \times 10^{13}Z^{10/3}\text{dyn cm}^{-2},$$

$$\varphi = \frac{1}{20}3^{1/3} + \frac{1}{8}\left(\frac{3}{4}\pi^{-2}Z^{-2}\right)^{1/3}, \tag{11.14}$$

see also Sect. 1.4, Vol. 1. At $R^3M_0\phi/R_0^3M \ll 1$ the relation, (11.13) reduces to (11.8). The former implies the existence of a peak on the curve $R(M)$. The maximum radii R_{cr} and corresponding M_{cr} obtained from numerical calculations are given in Table 11.2 from [1069] for various chemical compositions. The values of R_{cr} and M_{cr} following from (11.13) have uncertanties of ∼1% for R_{cr} and ∼25% for M_{cr}. The errors yielded by (11.13) increase with decreasing M. Stars with $M \le M_{cr}$ have their matter in the liquid state and are true Jupiter-like planets. The relation $R(M)$ from Table 11.1 is in good agreement with observed values for three white dwarfs in binaries [656].

11.1.2 Account for a Finite Value of T and Cooling

An approximate theory for the cooling of a white dwarf has been created by Kaplan [555] and Mestel [688] independently and is given in [878]. A white dwarf is divided into a non-degenerate radiative envelope and degenerate isothermal core. The pressures of degenerate (P_d) and non-degenerate ($P_{nd} = k\rho T/\mu_Z m_u$) electrons are assumed to be equal on the boundary between these regions, resulting, for the non-relativistic case of P_d from (11.4), in the relation between the boundary values ρ_* and T_*

$$\rho_* = \left(\frac{5m_e}{\hbar^2}\right)^{3/2} \frac{\mu_Z m_u}{3\pi^2} (kT_*)^{3/2} = 2.38 \times 10^{-8} \mu_Z T_*^{3/2} \text{ g cm}^{-3}. \tag{11.15}$$

Solving, in the radiative envelope, the equation of heat conductivity

$$\frac{dT}{dr} = -\frac{3}{4ac} \frac{\kappa\rho}{T^3} \frac{L}{4\pi r^2} \tag{11.16}$$

and the equilibrium equation (10.1) at $m = M = \text{const.}$

$$\frac{dP}{dr} = -\frac{GM\rho}{r^2}, \tag{11.17}$$

for Krammers' law of opacity due to free-bound transitions at $t/\overline{g}_{\text{bf}} = 10$ (see Chap. 1, Vol. 1)

$$\kappa = \kappa_0 \rho T^{-3.5}, \quad \kappa_0 = 4.34 \times 10^{24} x_Z(1 + x_H), \tag{11.18}$$

we obtain at $L = \text{const.}$ the equation

$$\frac{dP}{dT} = \frac{4ac}{3} \frac{4\pi GM}{\kappa_0 L} \frac{T^{6.5}}{\rho}, \qquad \rho = \frac{\mu m_u P}{kT}, \tag{11.19}$$

with a solution in the form

$$\rho = \left(\frac{8ac}{3 \cdot 8.5} \frac{4\pi GM}{\kappa_0 L} \frac{\mu m_u}{k}\right)^{1/2} T^{3.25}. \tag{11.20}$$

Assuming (11.20) to be valid on the boundary between the core and envelope and using (11.15), we obtain finally

$$L = \frac{8\pi^6}{8.5 \times 625} \frac{4\pi c GM}{\kappa_0} \frac{\hbar^3}{m_e^3 m_u c^3} \frac{\mu T_*^{3.5}}{\mu_Z^2}$$

$$= 5.75 \times 10^5 \frac{\mu T_*^{3.5}}{\mu_Z^2} \frac{M/M_\odot}{x_Z(1 + x_H)} \text{ erg s}^{-1}. \tag{11.21}$$

The temperature in the degenerate core is almost constant owing to the high thermal conductivity of degenerate electrons at high densities (see also Chap. 2, Vol. 1) and is equal to T_*, hence (11.21) gives the white dwarf luminosity as a function of the core temperature.

The available thermal energy Q_T of an isothermal white dwarf with temperature T is provided mainly by non-degenerate nuclei:

$$Q_T = \frac{3}{2} \frac{kT}{Am_u}. \tag{11.22}$$

Solving the energy equation

$$\dot{Q} = -L, \tag{11.23}$$

where $Q = Q_T$, we obtain the time τ of cooling from the core temperature T_0 to T

$$\tau = \frac{3 \times 8.5 \times 625}{16\pi^6 \times 2.5} \frac{k}{A} \frac{\kappa_0}{4\pi c G} \left(\frac{m_e c}{\hbar}\right)^3 \frac{\mu_Z^2}{\mu} \left(T^{-2.5} - T_0^{-2.5}\right)$$

$$= 1.72 \times 10^{35} \frac{x_Z(1 + x_H)}{A} \frac{\mu_Z^2}{\mu} \left(T^{-2.5} - T_0^{-2.5}\right) \text{ s}$$

$$= 1.72 \times 10^{10} \frac{x_Z(1 + x_H)}{A} \frac{\mu_Z^2}{\mu} \left(T_7^{-2.5} - T_{7,0}^{-2.5}\right) \text{ yr}$$

$$T_7 = T \times 10^{-7} \text{ K}. \tag{11.24}$$

With (11.21) and (11.24), the dependence $L(t)$ for $T_0 \gg T$ is given by

$$L = \frac{1.0 \times 10^{-3}}{t_{10}^{7/5}} \frac{M}{M_\odot} \frac{x_Z^{2/5}(1 + x_H)^{2/5}}{A^{7/5}} \frac{\mu_Z^{4/5}}{\mu^{2/5}} L_\odot, \tag{11.25}$$

$$t_{10} = t \times 10^{-10} \text{ yr}$$

For white dwarf envelopes, it is usual to take [878, 899] $x_Z = 0.1$ and $x_H = 0$, and hence, $\mu_Z = 2$, $\mu = 1.38$, $\mu/\left[\mu_Z^2 x_Z(1 + x_H)\right] = 3.45$. As the cooling proceeds and the white dwarf temperature falls below

$$T = T_g \approx 150 T_m \approx 2 \times 10^7 Z^{5/3} [\rho/(\mu_Z 10^6)]^{1/3} \text{K}, \tag{11.26}$$

the perfect gas approximation is no longer valid and we cannot use (11.19). Further cooling causes the matter to crystallize at

$$T = T_m = 1.3 \times 10^5 Z^{5/3} \left(\frac{\rho}{\mu_Z 10^6}\right)^{1/3} \text{K}, \tag{11.27}$$

while at $T < \theta$ from (10.60), the crystal becomes quantum-like. The phase transition into the crystalline state is likely to be accompanied by heat release

$$\delta U_{\text{Coul}} \approx -\frac{3}{4} k T_m. \tag{11.28}$$

The thermal energy of the crystal is given by the Debye formula

$$E_{iT} = \frac{3kT}{A m_u} \mathscr{D}\left(\frac{\theta}{T}\right),$$

where $\mathscr{D}(x)$ is the Debye function

$$\mathscr{D}(x) = \frac{3}{x^3}\int_0^x \frac{z^3 \mathrm{d}z}{\mathrm{e}^z - 1} = (\pi^4/5x^3) - 3\mathrm{e}^{-x} \quad \text{for} \quad x \gg 1$$
$$= 1 - (3/8)x + (x^2/20) \quad \text{for} \quad x \ll 1. \qquad (11.29)$$

For $T_\mathrm{m} < T < T_\mathrm{g} = 150\,T_\mathrm{m}$, we may use an interpolation between the perfect gas

$$E_i = \frac{3kT}{2Am_\mathrm{u}}, \qquad (11.30)$$

and a classical crystal from (11.29) at $x \ll 1$, see Chap. 1, Vol. 1. The inclusion of non-ideality effects first decelerates cooling owing to the increased heat capacity of a classical crystal and a possible heat release during phase transition, and after the quantum crystal has formed, the cooling accelerates because of a reduction in heat capacity $\sim(T/\theta)^3$ from (11.29). These factors, together with Coulomb corrections to the equation of state, neutrino cooling, partial ionization, electron heat conductivity in non-isothermal core and convection have been taken into account in [606], where the cooling of a carbon white dwarf of mass $1M_\odot$ has been numerically explored on the basis of stellar evolution equations (see Chap. 6, Vol. 1). The calculation results [606] are given in Figs. 11.2, 11.3 and Table 11.3. Figure 11.2 shows a slight deceleration of cooling between models 6 and 10 from Table 11.3 caused by the increased heat capacity of a classical crystal and crystallization heat and subsequent rapid cooling of the degenerate crystalline core. The crystallization is assumed in calculations to occur at $T = (15/16)T_\mathrm{m}$ from (11.27). The Debye

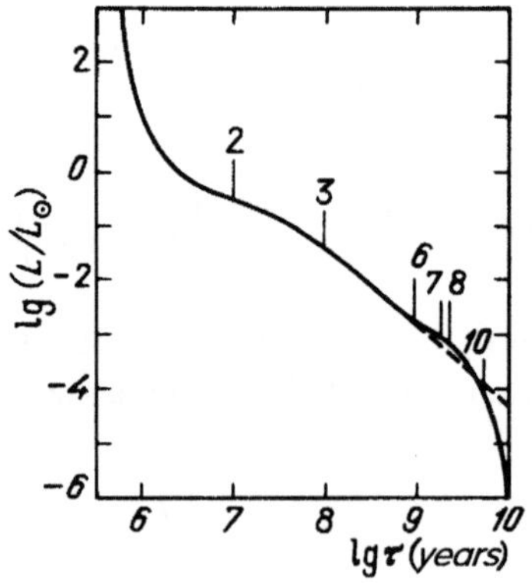

Fig. 11.2 The cooling curve, luminosity L versus age τ, for a 1 $M_\odot$ white dwarf. The numbers correspond to models from Table 11.3: 2 is the model with $L\nu = L$, 3 is the model where electron heat capacity $c_{ve,1/2}$ is equal to ionic $c_{vi,1/2}$, 6 is the onset of crystallization, 7 is the model where $\theta_{1/2} = T_{1/2}$, 8 is the model of the contact of the convective region with the degenerate core, 10 is the model where the crystalline core reaches $q = 0.99$. The dashed line shows the cooling curve obeying the power law (from [606])

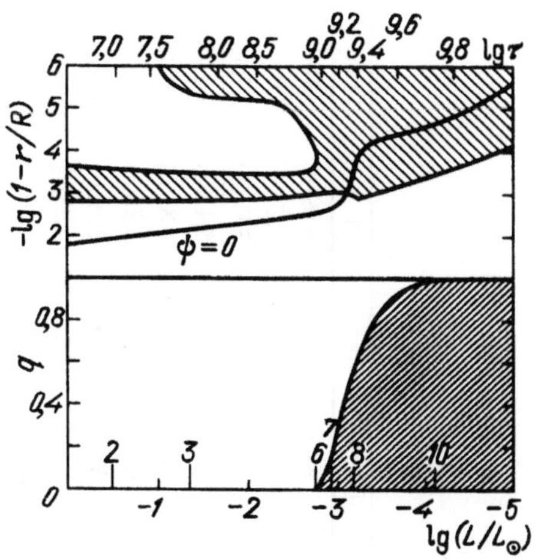

Fig. 11.3 Crystalline core growth (*thick hatch*) and behaviour of the outer convective zone (*thin hatch*) during white-dwarf cooling, shown is also the degeneracy curve $\psi = 0$. Epochs 2–10 correspond to models listed in Table 11.3 (see caption to Fig. 11.2); $q = m/M$ is the interior mass fraction (from [606])

temperature θ is taken in [606], analogously to (10.60), to be $\theta = 0.444\hbar\omega_{pi}/k$, and the thermodynamic functions are taken from numerical calculations. As may be seen from Fig. 11.3 and Table 11.3, a crystalline core arises originally at luminosity $L = 1.6 \times 10^{-3} L_\odot$ and age $\tau = 9.2 \times 10^8$ yr. The boundary of crystal degeneracy $T = \theta$ reaches the middle of the star $q = m/M = 1/2$ at $L = 1.1 \times 10^{-3} L_\odot$ and $\tau = 1.3 \times 10^9$ yr. Figure 11.3 represents the crystalline core growth, the convection development in the outer envelope of the star and the motion of the degenerate core boundary $\psi = 0$, where $\psi = \left(\mu_{et} - m_e c^2\right)/kT$ (see Sect. 1.2, Vol. 1). A possible hydrogen or helium burning in the envelope during cooling of a carbon–oxygen $0.6\, M_\odot$ white dwarf is calculated in [508]. The contribution of hydrogen burning is important up to the age of $\tau = 2 \times 10^9$ yr, and the subsequent cooling proceeds analogously to the calculations [606] for $1 M_\odot$ (see Table 11.3).

11.1.3 Cooling of White Dwarfs Near the Stability Limit with the Inclusion of Heating by Non-Equilibrium β-Processes [34]

The cooling of a star with mass close to M_{Ch} leads to the formation of a new-phase core and heating by the non-equilibrium two-step beta capture. This happens because the energy of the electron which may be captured by even–even nuclei is substantially larger than the corresponding energy for odd–odd nuclei [189] (see also Chap. 5, Vol. 1). An amount of $q = 200 - 500$ KeV per nucleus is converted into heat in this reaction. Consider the cooling of an iron white dwarf with $q = 476$ KeV/nucleus.[2] If T_f is the temperature of the star by the onset of the phase

[2] Including the excitation energy, equal to 109 keV, of the excited nucleus ^{56}Mn produced during the first electron capture.

Table 11.3 Properties of a 1 $M_{\odot}$ carbon white dwarf during cooling

No.	1	2	3	4	5	6	7	8	9	10	11	12
τ, yr	3.02(6)	9.46(6)	8.97(7)	3.13(8)	8.18(8)	9.16(8)	1.29(9)	2.09(9)	3.69(9)	5.52(9)	8.09(9)	9.10(9)
$L/L_{\odot}$	6.81(−1)	3.13(−1)	4.29(−2)	7.96(−3)	1.92(−3)	1.63(−3)	1.12(−3)	6.19(−4)	2.36(−4)	7.53(−5)	1.29(−5)	3.04(−6)
T_{ef}, K	5.81(4)	4.82(4)	2.96(4)	1.95(4)	1.37(4)	1.32(4)	1.20(4)	1.04(4)	8.15(3)	6.12(3)	3.94(3)	2.75(3)
R, cm	5.71(8)	5.65(8)	5.52(8)	5.47(8)	5.46(8)	5.45(8)	5.43(8)	5.42(8)	5.41(8)	5.41(8)	5.41(8)	5.41(8)
T_c, K	6.98(7)	5.31(7)	2.48(7)	1.32(7)	7.64(6)	7.11(6)	6.05(6)	4.49(6)	2.59(6)	1.42(6)	4.81(5)	1.67(5)
ρ_c, g cm^{-3}	3.15(7)	3.21(7)	3.30(7)	3.34(7)	3.36(7)	3.37(7)	3.37(7)	3.37(7)	3.38(7)	3.38(7)	3.38(7)	3.38(7)
ψ_c	146.0	193.5	419.3	795.3	1374	1474	1737	2346	4059	7428	21880	63130
Γ_c	16.2	21.4	46.2	87.6	151	162	191	257	446	815	2400	6940
θ_c/T_c	0.140	0.185	0.402	0.764	1.32	1.42	1.67	2.25	3.42	7.13	21.0	60.7
$R_{1/2}$, cm	2.50(8)	2.49(8)	2.46(8)	2.45(8)	2.45(8)	2.45(8)	2.45(8)	2.44(8)	2.44(8)	2.44(8)	2.44(8)	2.44(8)
$\psi_{1/2}$	81.31	106.6	229.0	433.6	748.7	803.5	946.6	1278	2211	4047	11920	34390
$\Gamma_{1/2}$	10.9	14.2	30.4	57.3	98.9	106	125	169	292	534	1570	4540
$\theta_{1/2}/T_{1/2}$	0.0759	0.0995	0.214	0.404	0.698	0.750	0.884	1.19	2.06	3.78	10.9	32.1
T_e, K	1.23(7)	1.11(7)	8.45(6)	6.24(6)	4.55(6)	4.38(6)	3.96(6)	3.12(6)	1.92(6)	1.15(6)	4.22(5)	1.51(5)
ρ_e, g cm^{-3}	350	402	546	689	807	820	847	906	971	1000	1030	1040
R_e, cm	5.64(8)	5.58(8)	5.48(8)	5.43(8)	5.42(8)	5.41(8)	5.41(8)	5.41(8)	5.40(8)	5.40(8)	5.40(8)	5.40(8)
ψ_e	−0.504	−0.161	0.804	1.95	3.354	3.556	4.074	5.528	9.557	16.46	45.87	129.3
Γ_e	2.05	2.37	3.46	5.04	7.32	7.65	8.54	11.1	18.4	31.1	85.7	240
$\lg(1-q)_{conv}$	−10.7	−10.5	−8.98	−8.81	−8.85	−8.86	−8.50	−7.58	−7.53	−8.22	−9.11	−9.24

$q = m/M$ is the interior mass fraction, the subscripts denote: "c" the centre, "1/2" the point $q = 0.5$, "e" the point $q = 1 - 10^{-6}$, "conv" the inner boundary of the convective zone (from [606])

transition at $\rho_c = \rho_{c0}$, then its equilibrium mass exceeds the cold star mass M_0 at this density by a quantity δM related to T_f by (see (10.58))

$$T_f = \frac{A}{\mu_Z^{4/3}} \frac{\rho_{c0}^{1/3}}{1.7 \times 10^{-7}} \frac{\delta M}{M_0} = 1.24 \times 10^{11} \frac{\delta M}{M_0} = \beta \frac{\delta M}{M_0}. \tag{11.31}$$

Near the mass maximum on the curve $M_T(\rho_c)$ and temperature minimum on $T_M(\rho_c)$ the following square dependences are valid:

$$M \equiv M_T(\rho_c) = M_{max} - \alpha(\rho_c - \rho_{cm})^2,$$
$$T \equiv T_M(\rho_c) = T_{min} + \gamma(\rho_c - \rho_{cm})^2. \tag{11.32}$$

Take, approximately, the parameters α, γ and $\rho_{cm} = \rho_{c1} + \Delta\rho_c$ to be constant, $\Delta\rho_c$ is given in (11.12). For the phase transition point with $T = T_f$, $\rho_c = \rho_{c1}$, we have from (11.32)[3]

$$T_f = T_{min} + \gamma(\rho_{c1} - \rho_{cm})^2. \tag{11.33}$$

From (11.31) to (11.33), we obtain for a star of mass M

$$T = T_f + \gamma\left[(\rho_c - \rho_{cm})^2 - (\rho_{c1} - \rho_{cm})^2\right]$$
$$= \beta\frac{\delta M}{M_0} + \gamma\left[(\Delta\rho_c - \delta\rho_c)^2 - \Delta\rho_c^2\right], \tag{11.34}$$

$$\delta\rho_c = \rho_c - \rho_{c1}.$$

The constant γ may be obtained from the condition that at $\delta M = \Delta M$ and $T = 0$, the equality $\delta\rho_c = \Delta\rho_c$ from (11.12) is valid. We then have

$$T = \beta\frac{\Delta M}{M_0}\left[\frac{\delta M}{\Delta M} + \left(1 - \frac{\delta\rho_c}{\Delta\rho_c}\right)^2 - 1\right]. \tag{11.35}$$

The central density $\rho_{c1} + \delta\rho_c$, for the growing stable branch of white dwarfs to the left of the maximum (see Fig. 11.1), is determined by

$$\delta\rho_c = \Delta\rho_c\left[1 - \left(1 + \frac{T}{\beta}\frac{M_0}{\Delta M} - \frac{\delta M}{\Delta M}\right)^{1/2}\right]. \tag{11.36}$$

The density near the centre varies as a square radius. For a polytrope, this is easily seen from the expansion (10.4) near the centre

$$\theta = 1 - \frac{\xi^2}{6}, \quad \rho = \rho_c\theta^n = \rho_c\left(1 - \frac{n}{6}\xi^2\right). \tag{11.37}$$

[3] For stars with mass below the mass limit, we may have formally in (11.32) and (11.33) $T_{min} < 0$. Of course, only ρ_c for which $T > 0$ in (11.34) have a physical sense.

The mass of the chromium core $m_{\rm nc}$ is related here with $\rho_{\rm c} = \rho_{\rm c1} + \delta\rho_{\rm c}$ by

$$\frac{m_{\rm nc}}{M_0} = \frac{M_{\rm nc}}{M_0}\left(\frac{\delta\rho_{\rm c}}{\Delta\rho_{\rm c}}\right)^{3/2} \approx 1.1\times 10^{-14}(\delta\rho_{\rm c})^{3/2}. \qquad (11.38)$$

Due to the increase in the mass of the chromium core, the heat is produced due to non-equilibrium beta capture, which is emitted at the rate

$$\dot{Q}_\nu = -(476\ \text{KeV})\,\frac{\dot{m}_{\rm nc}}{Am_{\rm u}}. \qquad (11.39)$$

From (11.22) and (11.39), using (11.36) and (11.38), we have for the total losses

$$\dot{Q} = \dot{Q}_T + \dot{Q}_\nu = \frac{3}{2}\frac{kM}{Am_{\rm u}}$$
$$\times\left\{1 + 3.9\times 10^6\,\frac{\left[1-(1+(TM_0/\beta\Delta M)-(\delta M/\Delta M))^{1/2}\right]^{1/2}}{(1+(TM_0/\beta\Delta M)-(\delta M/\Delta M))^{1/2}}\,\frac{M_0}{\beta\Delta M}\right\}\dot{T}. \qquad (11.40)$$

For ^{56}Fe and $(\mu/[\mu_Z^2 x_Z(1+x_{\rm H})] = 3.45)$, (11.23) combined with (11.21) and (11.40) becomes

$$0.7\times 10^{16}\left\{1 + 0.15\,\frac{\left[1-(1+(T_7/2.6)-(\delta M/\Delta M))^{1/2}\right]^{1/2}}{(1+T_7/2.6-\delta M/\Delta M)^{1/2}}\right\}\frac{{\rm d}T_7}{{\rm d}t} = -T_7^{3.5}. \qquad (11.41)$$

The equality $\beta\Delta M/M_0 = 2.6\times 10^7$ K is taken into account here according to (11.31) and (11.12). For the critical mass $M = M_0 + \Delta M$, $\delta M = \Delta M$ we have from (11.41) the equation

$$0.7\times 10^{16}\left[1+\frac{0.24}{T_7^{1/2}}\left(1-\sqrt{T_7/2.6}\right)^{1/2}\right]\frac{{\rm d}T_7}{{\rm d}t} = -T_7^{3.5}. \qquad (11.42)$$

Solving (11.42) for $T_7 \ll 2.6$, we obtain the cooling time in the form[4]

$$\tau = 2.8\times 10^{15}\left[T_7^{-2.5} - T_{7,0}^{-2.5} + 0.2\left(T_7^{-3} - T_{7,0}^{-3}\right)\right]\ \text{s}. \qquad (11.43)$$

[4] Analogous estimates for the cooling time have been made in [894].

When there is $T \leq 0.1\,\theta_c$ (see (10.60)), the heat capacity of a degenerate crystal

$$Q_D = \frac{4\pi^4}{5}\frac{kMT}{Am_u}\left(\frac{T}{\theta_c}\right)^3, \tag{11.44}$$

should be used instead of (11.22). The coefficient in (11.44) is taken from [899] and is $4/3$ times larger than the value following from the expansion of the Debye function for $x \gg 1$ in (11.29), Substituting (11.44) for (11.22) gives, in lieu of (11.43)

$$\tau = 2.8 \times 10^{15}\left\{\left(\frac{T_0}{0.1\,\theta_c}\right)^3 T_{7,0}^{-2.5} - \left(\frac{T_0}{0.1\,\theta_c}\right)^3 T_7^{-2.5} + 0.2\left(T_7^{-3} - T_{7,0}^{-3}\right)\right\}\ \text{s}. \tag{11.45}$$

It follows from (11.43) that the inclusion of non-equilibrium heating increases the time of cooling to $T = 0.1\,\theta_c = 5.5 \times 10^6$ K for ^{56}Fe by ~27%. According to (11.45), the non-equilibrium heating becomes very important for $T < 0.1\,\theta_c$. Without it, an almost complete cooling of the white dwarf, according to (11.45), requires 8×10^8 yr, (4×10^8 yr from $T_0 = 5.5 \times 10^6$ K), while in the presence of non-equilibrium heating a white dwarf of critical mass cools to $\sim 10^6$ K over a cosmological time of 2×10^{10} yr. An accurate calculation of a $1 M_\odot$ carbon white dwarf cooling with a gradual growth of a degenerate crystal core, but without a new-phase core gives after 9×10^9 yr a core temperature of 1.7×10^5 K (see Table 11.3). It may be seen from comparison of (11.45) with Fig. 11.2 and Table 11.3 that for $T < 3.6 \times 10^6$ K, the non-equilibrium heating will become predominant.

11.1.4 On the Evolution of Magnetic Fields in White Dwarfs

Over 10% of the known white dwarfs have strong magnetic fields ranging from 10^6 to 10^8 G, detected by polarization of optical radiation in single stars and binaries. The latter are coupled with red dwarfs and belong to cataclysmic variables, where flashes with $\Delta m_V = 4$–5^m are observed several times a year. They are also X-ray sources, and the degree of polarization of optical radiation is a few tens of per cent, giving a reason for calling them polars. So far, it is unclear whether magnetic fields are left behind by a prior phase of stellar evolution or generated by the dynamo mechanism in late evolutionary phases characterized by a strong convection (see Sect. 9.3.7).

The problem of the magnetic field decay in white dwarfs may have a more reliable solution. Observations of strong magnetic fields in relatively cold white dwarfs with $T_{ef} \sim 6{,}000$ K suggest that the decay timescale is comparable to or exceeds the cosmological time (see (11.24) and Table 11.3). In [1036], calculations have been performed for the decay of poloidal magnetic fields of various shapes with the inclusion of material motion. A solution is sought for the equation reading in the vector form

$$\frac{\partial \mathbf{B}}{\partial t} = -\nabla \times \left(\frac{c^2}{4\pi\sigma} \nabla \times \mathbf{B} - \mathbf{v} \times \mathbf{B} \right). \tag{11.46}$$

Equation (11.46) with conductivity $\sigma \to \infty$ and axial symmetry $\partial/\partial\varphi = 0$ yields (10.86)–(10.88). For poloidal fields, the vector equation (11.46) reduces to a scalar equation on defining the vector potential

$$\mathbf{A} = \left(0, 0, A_\varphi(r, \theta, t)\right), \quad \mathbf{B} = \nabla \times \mathbf{A}, \quad S = -r \sin\theta A_\varphi,$$

$$\frac{\partial S}{\partial t} + \mathbf{v} \cdot \nabla S = \frac{c^2}{4\pi\sigma} \left[\frac{\partial^2 S}{\partial r^2} + \frac{\sin\theta}{r^2} \frac{\partial}{\partial\theta} \left(\frac{1}{\sin\theta} \frac{\partial S}{\partial\theta} \right) \right]. \tag{11.47}$$

The solution for the function S is looked for at a spherically symmetric distribution of density and radial velocity $v = (v_r(r), 0, 0)$, in the case of validity of the expansion

$$S(r, \theta, t) = \sum_{l \geq 1} R_l(r, t) \sin\theta \, P_l^1(\cos\theta), \tag{11.48}$$

where $P_l^1(\cos\theta)$ are the associated first-order Legendre functions [428], while for R_l we have the equation

$$\frac{\partial^2 R_l}{\partial x^2} - \frac{l(l+1)}{x^2} R_l = \frac{4\pi R_*^2 \sigma}{c^2} \left(\frac{\partial R_l}{\partial t} + \frac{v_r}{R_*} \frac{\partial R_l}{\partial x} \right), \tag{11.49}$$

where $x = r/R_*$, $R_*(t)$ is the stellar radius. For the expansion (11.48), the magnetic field $\mathbf{B}$, according to (11.47), has the form

$$\mathbf{B}(x, \theta, t) = \frac{1}{R_*^2} \left[-\sum_{l \geq 1} \frac{l(l+1)}{x^2} P_l(\cos\theta) \, R_l(x, t) \, , \right.$$

$$\left. \sum_{l \geq 1} P_l^1(\cos\theta) \, \frac{1}{x} \frac{\partial R_l}{\partial x}, 0 \right], \tag{11.50}$$

where $P_l(\cos\theta)$ are the Legendre polynomials, $l = 1$ corresponds to dipole, $l = 2$ to quadrupole and so forth. Solutions to (11.49) looked for in [1036] are limited in the centre and continuous on the boundary with vacuum. The electronic current j_i is determined in the form

$$j_i = -e n_e \langle v_i \rangle, \tag{11.51}$$

where $e > 0$, and $\langle v_i \rangle$ is the electron diffusive velocity in the presence of electric and magnetic fields. Large differences in ion and electron accelerations $\mathbf{a}_{\mathrm{e,i}}$ under the action of the electromagnetic force

$$\mathbf{a}_{\mathrm{e,i}} = \frac{\mathbf{F}_{\mathrm{e,i}}}{m_{\mathrm{e,i}}} = \frac{e}{m_{\mathrm{e,i}}} \left(\mathbf{E} + \frac{1}{c} \mathbf{v} \times \mathbf{B} \right) \tag{11.52}$$

lead to a strong enhancement of the diffusion, thereby producing the electrical current j_i which is mainly due to the motion of the electrons. The conductivity σ is defined as the coefficient of proportionality between j_i and E_i at $\mathbf{B} = 0$:

$$j_i = \sigma E_i . \tag{11.53}$$

The calculations of the electrical conductivity of electrons σ_{e} are performed in the same way as those of the heat conductivity of electrons λ_{e}. One should also take into account the electromagnetic acceleration (11.52) in the original Boltzmann equation. The same Coulomb logarithm

$$\Lambda = \ln \left(\frac{b_{\max} \bar{v^2} m_{\mathrm{e}}}{Z e^2} \right), \tag{11.54}$$

appears in the resulting expression. Calculations of λ_{e}, $b_{\max}$ (related to Debye screening radii) and $\bar{v}^2$ for the electrons based on a solution of the Boltzmann equation are given in Chap. 2, Vol. 1, for various cases. The conductivity σ is given in [1036] for a wide range of parameters in the form

$$\begin{aligned} \sigma &= \gamma_E \frac{2(2kT)^{3/2}}{\pi^{3/2} m_{\mathrm{e}}^{1/2} Z e^2 \Lambda} \approx 3.31 \times 10^7 \frac{T^{3/2}}{\Lambda} \ \mathrm{s}^{-1} \quad (\text{for } ^{12}\mathrm{C}) \\ &\text{for } \ \lg \rho < 4 \lg T - 29.825, \\ &= \frac{\lambda_{\mathrm{e}}}{T} \frac{3}{\pi^2} \left(\frac{\mathrm{e}}{k} \right)^2 \text{ for higher } \rho \text{ with} \end{aligned} \tag{11.55}$$

$$\begin{aligned} \lambda_e &= \frac{16\sqrt{2}}{\pi^{3/2} \lambda} k \frac{n_e}{n_N} \left(\frac{kT}{e^2 Z} \right)^2 \left(\frac{kT}{m_e} \right)^{1/2} \quad \text{(ND)} \\ &= \frac{1}{32\lambda} \frac{k^2 T n_e^2 h^3}{m_e^2 n_N Z^2 e^4} \quad \text{(D)} \end{aligned} \tag{11.56}$$

for non-reativistic electrons, and

$$\begin{aligned} \lambda_e &= \frac{4.11 \times 10^{15} (\rho_6/\mu_{\mathrm{Z}}) T_6}{[1 + (\rho_6/\mu_{\mathrm{Z}})^{2/3}]^{1/2}} \left(\frac{10^{16}\,\mathrm{s}^{-1}}{\upsilon_{\mathrm{e}}} \right) \mathrm{ergs\,cm^{-1}\,K^{-1}}, \\ \rho_6 &= \rho/10^6\,\mathrm{g\,cm^{-3}}, \qquad \mathrm{T_6 = T/10^6\,K}, \end{aligned} \tag{11.57}$$

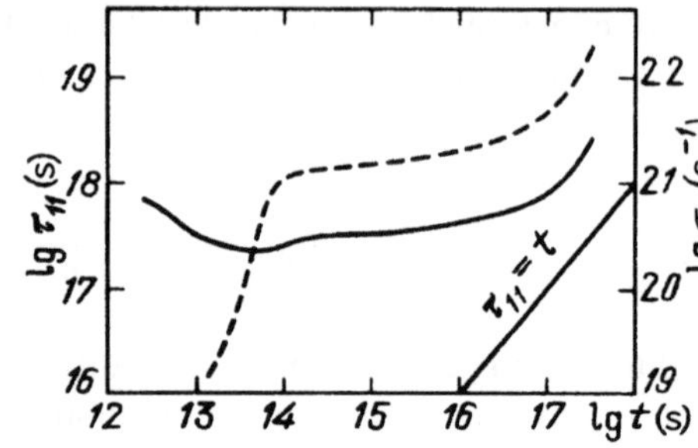

Fig. 11.4 The time scale for decay of a large-scale magnetic field τ_{11} versus white-dwarf age t; obviously, $\tau_{11} > t$. The *dashed line* represents the conductivity δ_e in the centre of the star (from [1036])

for relativistic degenerate electrons. Here υ_e is the total frequency of electron collisions in the medium (see Sect. 2.4.4, Vol. 1), the coefficient $\gamma_E = 0.582, 0.683$ *and* 0.838 for H, ^{4}He and ^{12}C, respectively. The second equality in (11.55) is determined from the Wiedemann–Franz relation. The relations (11.55) have been fitted to one another using a smooth interpolation. The time scale for field decay is obtained from (11.49) for $\sigma = \text{const.}$, $v_r = 0$, $l = 1$ in the form

$$\tau_{ln} \approx \frac{4\pi R_*^2 \sigma}{c^2 \pi^2 n^2}, \quad \text{integer } n \geq 1; \quad B_{ln} \sim \exp(-t/\tau_{ln}). \tag{11.58}$$

A numerical solution to an eigenvalue problem for (11.49) with σ from (11.55) is found in [1036] for a 0.6 $M_\odot$ carbon white dwarf along its evolutionary track. The calculation results for a maximum-scale field ($l = 1$, $n = 1$) are given in Fig. 11.4, from which it is clear that the decay time scale τ_{11} is always larger than the cooling time τ, so the large-scale field of a white dwarf experiences almost no decay with time.

11.1.5 Nova Outbursts

Single white dwarfs are quiet objects, but exhibit a high activity when they are in close binaries. These violent activities are caused by the accretion of matter from the neighbour star, a low-mass hydrogen dwarf, leading to instability development. During accretion, hydrogen-rich material is accumulated in the envelope of the white dwarf. After the envelope mass has reached 10^{-6}–$10^{-4} M_\odot$ and the hydrogen electrons have become degenerate, a thermal instability develops resulting in a thermonuclear explosion. The explosion power depends on the mass of the exploding envelope which, in turn, depends on the accretion rate $\dot{M}$ and the white dwarf mass M_{WD}. This dependence is inverse: the higher the accretion rate, the higher the matter temperature in the envelope, the easier the ignition, the lower the envelope mass before the explosion and, ultimately, the lower the explosion power. Any further increase in $\dot{M}$ makes the degeneracy insufficient, and the explosion does not occur.

The critical values of $\dot{M}_{cr}$ and corresponding L_{cr} have been found theoretically in [661] as functions of white dwarf mass M_{WD} and chemical composition:

$$\text{for } x_Z = 0.02 :$$

$$\lg \dot{M}_{cr} = -8.775 - 15.088\,(M_{WD}/M_\odot - 1.459)^2,$$
$$\lg(L_{cr}/L_\odot) = -0.629 - 5.923 \times (M_{WD}/M_\odot - 1.766)^2;$$
$$\text{for } \quad x_Z = 0.51 :$$
$$\lg \dot{M}_{cr} = -8.632 - 4.596\,(M_{WD}/M_\odot - 1.334)^2, \tag{11.59}$$
$$\lg(L_{cr}/L_\odot) = -1.375 - 7.027 \times (M_{WD}/M_\odot - 1.308)^2.$$

Here, $\dot{M}$ is in $M_\odot$/yr. Envelope ejections are absent for $\dot{M} > M_{cr}$, and the hydrogen burning becomes stationary. Observations of novae give substantially higher values for $\dot{M}$: up to $\times 10^{-8} - 10^{-7}$ $M_\odot$/yr after the explosion. This might be due to relaxation processes, and later the accretion rate may become lower than $\dot{M}_{cr}$ from (11.59), but it is not excluded that the theoretical estimates are not sufficiently accurate. The increase in luminosity during nova outburst may reach $\Delta m_V = 10$–15^m and even 19^m for nova Cygni 1975 [813]. The larger the value of $\dot{M}$ and the weaker the outburst, the more often it occurs. In reality, all the novae are recurrent but the brightest of them have a time interval between outbursts Δt_B which is too large for astronomical observations to be possible. The outburst amplitude A_m versus Δt_B is given in Fig. 11.5 from [813]. The nova outbursts are calculated from the hydrodynamical equations (10.38)–(10.40), analogously to supernova calculations, but the region of Lagrangian calculations is spatially restricted by the envelope. The relevant studies are reviewed in [363].

The nova outbursts occur in close binaries, where the companion – a hydrogen star – fills the Roche lobe, therefore, the matter accreting onto the white dwarf has the shape of a disk (see Sect. 11.3). A stationary disk accretion in low-mass close binaries is possible only in the case of a turbulent accretion disk. In a laminar case, the material remains in the disk because of a low viscosity and does not accrete onto

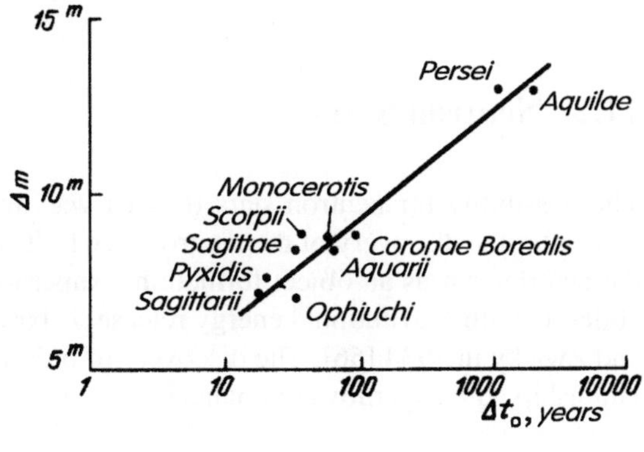

Fig. 11.5 Outburst amplitude Δm versus time interval between outbursts Δt_O for recurrent novae, from [813]

the white dwarf. When the accretion rate is small ($\dot{M} < 10^{-9} M_{\odot}/\text{yr}$), the disk turbulization is likely to be impeded, and additional forms of activity emerge owing to the instability of the disk accretion leading to sporadic disk turbulization. Just these forms of activity account for the dwarf nova outbursts, also referred to as nova-like or cataclysmic variables. To interpret observational properties of the light curve of dwarf novae, one usually presumes the existence of a hot spot originating from a shock of accretion flow against the disk. Another interpretation, based on 3-D numerical simulations [145], which does not include the hot spot was considered in [146]. A small accretion rate thus causes the disk to remain laminar for a part of the time, the viscosity to be low and the material to accumulate in the outer parts of the disk. When the disk mass begins to exceed a certain critical value, an instability develops in it to result in turbulization, viscosity growth and a steep increase in the mass flow onto the white dwarf. An outburst then occurs. After the disk mass has reduced, it becomes laminar again until the next outburst. Such a model has been developed in [784, 915, 916]. The time interval between two successive outbursts ranges for various stars from days to several hundred days, with corresponding outburst amplitudes ranging from 2^m to $5 - 6^m$ [813]. For some stars, the intervals between outbursts and their shapes vary appreciably, while for others the occurrence of outbursts is more or less regular. In [916], the relatively regular outbursts are incorporated in type A outbursts with the onset of instability in the outer parts of the disk, while the irregular outbursts belong to the type B, where the instability starts to develop in the innermost parts of the disk. It is noted that the type A is characterized by higher values of $\dot{M}$ at the same stellar masses M_1 and M_2. Type A outbursts occur in U Gem and SS Aur stars, whereas type B outbursts are detected in SS Cyg and AH Her stars. Observations and theoretical models for cataclysmic variables are reviewed in [799].

As the accretion rate is very low in cataclysmic variables, the hydrogen accumulation rate in the white dwarf envelope is also very low, but the amount accumulated over a period of $10^4 - 10^5$ yr is sufficient for the instability of thermonuclear burning to develop and a nova to outburst. It follows from the empirical dependence in Fig. 11.5 that these stars are characterized by the most powerful and rare nova outbursts. The search of cataclysmic phenomena is likely to have a sense in outburst remnants of the most bright novae Cygni 1975, Puppis 1942 and Cygni 1920.

11.2 Neutron Stars

The possibility for neutron stars to exist was first suggested by Landau in 1932, soon after the discovery of the neutron (see [836]). The modern conception treating the neutron star as an object formed in a supernova explosion during gravitational collapse, with gravitational energy release, is based on the idea proposed by Baade and Zwicky in 1934 [56]. The discovery of pulsars and X-ray sources, and their relationship with supernova remnants have conclusively corroborated this prediction.

The first neutron star model was calculated by Oppenheimer and Volkov [765]. They used the equation of state of a perfect degenerate neutron gas written in a way similar to the degenerate electron gas (see Chap. 1, Vol. 1) in the form

$$E_{\mathrm{n}} = \frac{6.860 \times 10^{35}}{\rho_0} g(y_{\mathrm{n}}) \quad P_{\mathrm{n}} = 6.860 \times 10^{35} f(y_{\mathrm{n}})$$
$$f(y) = y(2y^2 - 3)\sqrt{y^2+1} + 3\sinh^{-1} y,$$
$$g(y) = 3y(2y^2 + 1)\sqrt{y^2+1} - 3\sinh^{-1} y,$$
$$y_{\mathrm{n}} = \frac{p_{\mathrm{n,Fe}}}{m_{\mathrm{n}}c}, \quad p_{\mathrm{n,Fe}} = \hbar\left(\frac{3\pi^2\rho_0}{m_{\mathrm{n}}}\right)^{1/3} \hbar = \left(\frac{1.638\rho_0}{10^{16}}\right)^{1/3} m_{\mathrm{n}}c. \quad (11.60)$$

Studies of such compact objects as neutron stars must be based on general relativity. The ratio of gravitational radius

$$R_{\mathrm{g}} = \frac{2GM}{c^2} \approx 2.95 \times 10^5 \frac{M}{M_\odot} \text{ cm} \quad (11.61)$$

to stellar radius is the relativistic parameter in gravitation theory [615]. For a white dwarf of ^{56}Fe, at the point of stability loss due to neutronization, according to (11.10), (11.12) and (11.61), the ratio $R_{\mathrm{g}}/R \approx 1.00 \times 10^{-3}$. For neutron stars $R_{\mathrm{g}}/R > 0.1$ (see [80] and Fig. 11.1), and the inclusion of general relativity is customarily accurate. Applications of general relativity are usually much more complicated than Newtonian gravitation theory, but the equations obtained in [765] for spherically symmetric stars do not exceed much in complexity the Newtonian equations (9.97) and (10.1):

$$\frac{\mathrm{d}P}{\mathrm{d}r} = -\frac{G(\rho + P/c^2)(m + 4\pi r^3 P/c^2)}{r^2(1 - 2Gm/rc^2)}, \quad (11.62)$$

$$\frac{\mathrm{d}m}{\mathrm{d}r} = 4\pi\rho r^2. \quad (11.63)$$

Contrary to (9.97) and (10.1), where ρ has been taken to equal the rest-mass density ρ_0 because of $E \ll \rho_0 c^2$, the density in (11.62) and (11.63) is

$$\rho = \rho_0\left(1 + \frac{E}{c^2}\right). \quad (11.64)$$

Besides the total stellar mass $M = m(R)$ in (11.62) and (11.63), determining the gravitational field, general relativity also treats the rest mass M_0 which for a spherically symmetric neutron star will be written as

$$m_0 = 4\pi \int_0^r \frac{\rho_0 r'^2 \mathrm{d}r'}{(1 - 2Gm'/r'c^2)^{1/2}}, \quad M_0 = m_0(R), \quad N_{\mathrm{b}} = \frac{M_0}{m_{\mathrm{n}}}, \quad (11.65)$$

where N_b is the total number of baryons in the neutron star. The quantity M includes the positive contribution of the energy weight and negative gravitational energy. For stable stars $M < M_0$ usually, but for neutron stars with a mass close to the minimum (Fig. 11.1), the inequality changes sign [24, 1081].

11.2.1 Cold Neutron Stars

The perfect gas does not fit well the equation of state of a neutron gas, for which at densities above the nuclear density a predominant role belongs to interactions, so that we should speak of a neutron liquid rather than a neutron gas (see Chap. 1, Vol. 1). Models for cold neutron stars are usually obtained by solving (11.62) and (11.63) for equilibrium matter. The dependences $M(\rho_c)$ for the equation of state for different models of nuclear interaction are given in Fig. 11.6 [668]. Properties of various models for maximum masses are given in Table 11.4a [668]. Table 11.4a gives masses and radii of white dwarfs and neutron stars from [80] versus total stellar density for the lower curve in Fig. 11.1. The matter in a neutron star has a large variety of properties arising from the diversity in chemical compositions, varying state of aggregation, possible presence of nucleon superfluidity and proton superconductivity.

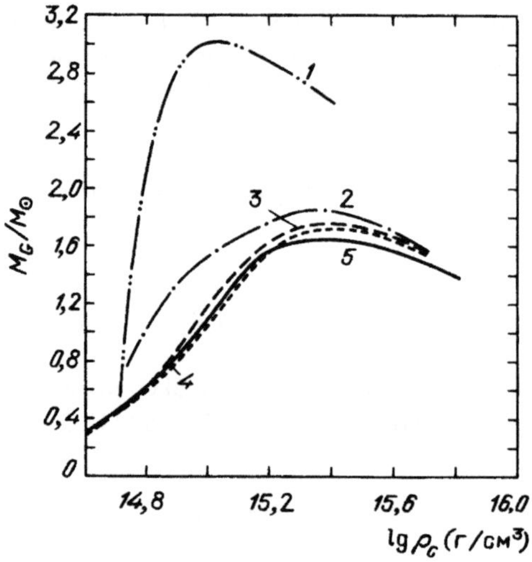

Fig. 11.6 Total mass of neutron star M_G versus rest-mass central density ρ_{c0} for various nuclear interaction models [94]: *double-dot–dashed line* (1) represents the extremely stiff equation of state $P = P^* + (\epsilon - \epsilon^*)$ which can be fitted at $\rho_0 = 5.02 \times 10^{14}$ g cm^{-3} to models (3) and (5), coinciding at this density (from [668]); (2) is the model based on the Reid potential equal for all baryons; *long dashes* (3) represents a model for nucleon interaction based on the experimental data on the ω-meson formation at high energies; *short dashed* (4) the same as (2) but with a more plausible potential for (np)-interaction; the *solid line* (5) the same as (3) but including the hyperon generation (see Sect. 1.4, Vol. 1)

Table 11.4 a. Properties of neutron stars in maximum mass for different nuclear models

No. in Fig. 11.6	$\frac{M}{M_\odot}$	$\frac{M_0}{M_\odot}$	ρ_{c0}, g cm^{-3}	P_c, dyn cm^{-2}	R, km	$\frac{M_0-M}{M_\odot}$
2	1.85	2.15	2.28(15)	1.26(36)	9.73	0.294
4	1.73	2.02	2.69(15)	1.58(36)	8.88	0.292
3	1.76	2.06	2.49(15)	1.33(36)	9.18	0.301
5	1.65	1.91	2.47(15)	1.00(36)	9.38	0.256
1	3.02	3.92	1.13(15)	1.00(36)	12.9	0.901

Table 11.4 b. Masses and radii of cold neutron stars as functions of total central density $\rho_c = \rho_{0c}\left(\frac{1+E_c}{c^2}\right)$, from [80]

ρ_c, g cm^{-3}	$M/M_\odot$	R (km)
4.33 (6)	0.502	7380
2.97 (7)	0.759	5110
1.38 (9)	1.00	2140
3.01 (10) u	0.845	944
1.00 (14) u	0.612	2620
1.55 (14)	0.0925	164
2.00 (14)	0.100	56.5
2.30 (14)	0.105	40.9
2.60 (14)	0.111	32.9
2.90 (14)	0.117	27.7
3.20 (14)	0.124	24.2
3.60 (14)	0.132	20.9
4.00 (14)	0.142	18.6
4.50 (14)	0.154	16.5
5.00 (14)	0.167	15.0
6.00 (14)	0.244	11.5
8.00 (14)	0.424	10.1
1.00 (15)	0.516	9.8
3.00 (15)	1.25	8.1
6.20 (15)	1.41	7.0
1.00 (16) u	1.37	6.4

Letter "u" denotes instability of the star to collapse

The envelope may undergo departures from equilibrium composition, connected with non-equilibrium neutronization and the absence of nuclear reactions between charged particles at lower temperatures due to high Coulomb barriers.

A neutron star with equilibrium composition comprises the following layers [899],

1. The surface layer ($\rho \leq 10^6$ g cm^{-3}), a region in which the temperature and magnetic field can affect the equation of state and envelope structure, equilibrium composition is the iron ^{56}Fe.

2. The outer crust ($10^6 \le \rho \le 4.3 \times 10^{11}$ g cm^{-3}), a region in which a Coulomb crystal of nuclei exists and where the A/Z ratio increases with increasing ρ.
3. The inner crust ($4.3 \times 10^{11} \le \rho \le (2\text{–}2.4) \times 10^{14}$ g cm^{-3}), a region in which a crystalline lattice of nuclei near the neutron drip coexists with free neutrons that may be superfluid.
4. The neutron liquid ($(2\text{–}2.4) \times 10^{14}$ g cm$^{-3} < \rho < \rho_b$), which contains superfluid neutrons and protons and normal electrons.
5. A core region ($\rho > \rho_b$), a hypothetical region where pion condensation may occur, or neutron solid matter, quark generation or some other extraordinary phase physically distinct from the neutron liquid. The value of $\rho_b \approx 6 \times 10^{14}$ g cm^{-3}.

The neutron star mass is measured with the best accuracy in binary pulsars PSR 1913+16 with orbital period $P_{\rm orb} = 7.75$ h, and eccentricity $e = 0.617$ [670]. In the case of an elliptical orbit, relativistic effects cause the apsid line, or major axis of the orbit, to move with angular velocity $\omega = 4.2$ degree/yr, and the orbital period to reduce owing to emission of gravitational waves so that $\dot{P}_{\rm orb} = -2.3 \times 10^{-12}$, according to observations. The Doppler shift of the pulsar period $P_p = 0.059$ s is used to determine the amplitude of the neutron star orbital velocity v_p and the mass function

$$f(M_p, M_2, i) = \frac{P_p v_p^3}{2\pi G} = \frac{(M_2 \sin i)^3}{(M_p + M_2)^2} \approx 0.1312\, M_\odot. \qquad (11.66)$$

The known values $\dot{P}_{\rm orb}$, ω and f allow determination of unambiguous $M_p = 1.4411 \pm 0.0007\, M_\odot$, $M_2 = 1.3874 \pm 0.0007\, M_\odot$ and the angle between the normal to the plane of the orbit and the direction to the observer $i \approx 46°$ [966, 969, 970]. Stellar masses in another four binary radiopulsars with neutron star or white dwarf companions have also been obtained from observations of GR effects. The best data in this group are obtained for PSR 1543 + 16, giving $M_p = 1.34 \pm 0.07 M_\odot$ and $M_2 = 1.34 \pm 0.07 M_\odot$ [966, 1045]. It is obvious that all measured masses of neutron stars are lower than the mass limit of neutron stars for any equation of state from Table 11.4. The error in measurement of neutron star masses in binary X-ray sources and X-ray bursters is more than 100 times larger than for PSR 1913 + 16 (see [899]).

The most accurately measured parameters of neutron stars are obtained from observations of the most close binary, consisting of two pulsars, one of which is a "normal pulsar", similar to single objects, and another is a "recycled" pulsar, going through the accretion phase, when it increases is rotational frequency and decreases the magnetic field. This unique system PSR J0737 – 3039a,b discovered in 2003–2004 [244, 660] is the best laboratory for checking the gravitational theory, which is represented by GR within 0.1% of precision. This unique system has the following parameters: orbital period $P_{\rm orb} = 2.45$ h, low eccentricity $e = 0.088$, periods of the pulsars $P_p(\mathrm{a}) = 22.7$ ms (recycled pulsar, age $\approx 210 \cdot 10^6$ years) and $P_p(\mathrm{b}) = 2.77$ s ("normal" pulsar, age $\approx 50 \cdot 10^6$ years). The masses of the neutron stars in this system are less than in other known systems, and are equal to

$M_a \approx 1.338 \pm 0.001 M_\odot$ and $M_b \approx 1.249 \pm 0.001 M_\odot$. The error boxes of these numbers are decreasing with time due to accumulation of observational data, and the indicated numbers are related to year 2005, see review [161].

The rotation increases the maximum mass of the neutron star, reducing the central density and increasing the radius for the same mass. For neutron stars with the equation of state (determining the curve 2) from Fig. 11.6, uniformly rotating at a limiting velocity that corresponds to the onset of outflow from the equator, the maximum in mass occurs at [399] $\Omega = 1.11 \times 10^4\ \mathrm{s}^{-1}$, $M = 2.16\, M_\odot$, $M_0 = 2.47\, M_\odot$, $R_e = 13.0$ km (equator), $T/|W| = 0.11$, where T and W are the rotational and the gravitational energies. For most equations of state treated in [399], the quantity $T/|W|$ does not exceed 0.12 for uniform rotation, thereby giving evidence for stability to transformation into triaxial figures. Calculation of exact models for rotating stars in the framework of general relativity performed in [399] represents a much more serious problem than in Newtonian theory. The maximum mass of a uniformly rotating neutron star for all the equations of state in [399, 856] is not more than 20% larger than the same value for the non-rotating star.

11.2.2 Hot Neutron Stars

When a neutron star results from a collapse, the matter there achieves enormous temperatures of 10^{11}–10^{12} K. The mass limit of a hot star is higher than that of a cold star and increases with increasing temperature. Hot neutron star models have been calculated and their stability studied in [152] by solving equations (11.62) and (11.63) and using the stability criterion

$$I = \delta^2\epsilon \Big/ \left[4\pi\alpha^2 \exp\left(-\int_0^R \frac{\tilde{E}n + P}{1 - 2Gm/c^2 r} \frac{4\pi G r}{c^4}\, dr \right) \right]$$

$$= \int_0^R \exp\left(\int_0^r \frac{\tilde{E}n + P}{1 - 2Gm/c^2 r} \frac{4\pi G r}{c^4}\, dr \right)$$

$$\times \left\{ \gamma P r^2 \left[3 - \frac{Gm}{c^2 r} \frac{1 + 4\pi r^3 P/mc^2}{1 - 2Gm/c^2 r} \right]^2 \right.$$

$$- \frac{(P + \tilde{E}n)(1 + 4\pi r^3 P/mc^2)^2}{(1 - 2Gm/c^2 r)^2} \left(\frac{Gm}{c^2 r} \right)^2 r^2$$

$$\left. - 4 \frac{En + P}{1 - 2Gm/c^2 r} \left(1 + \frac{1}{2} \frac{4\pi r^3 P}{mc^2} \right) \frac{Gm}{c^2 r} r^2 \right\} dr > 0. \qquad (11.67)$$

Here, $\tilde{E}n$ includes the rest energy $\rho_0 c^2$.[5] The criterion (11.67) follows from (10.47) and from the general relation (10.48) for the quantity $\delta^2\epsilon$ of spherically symmetric stars in general relativity with a linear trial function $\delta r = \alpha r$. In the Newtonian limit, $\tilde{E}n = \rho_0 c^2$, $P \ll \tilde{E}n$, $c \to \infty$, and the criterion (11.67) gives the stability condition $\gamma > 4/3$. The post-Newtonian approximation with the inclusion of terms $\sim P/\rho_0 c^2$, $(\tilde{E}n/\rho_0 c^2 - 1)$ and terms $\sim G^2$ for the density distribution over the polytrope $n = 3$, applied to (11.67), leads to the stability condition (10.20). For the Oppenheimer–Volkoff curve [765], the criterion (11.67) yields the stability loss for the density differing from the maximum-mass density by $\Delta\rho < 10^{-5}\rho$, thus proving fairly accurate.

The equation of state allows for the presence of non-degenerate protons p, neutrons n and nuclei of iron ^{56}Fe in nuclear equilibrium with respect to each other. Neutrinos are assumed to escape freely with $\mu_\nu = 0$; hence, the equilibrium composition is found from equations for chemical potentials

$$\begin{aligned} \mu_{56,26} &= 26\mu_{\rm p} + 30\mu_{\rm n}, \\ \mu_{\rm n} &= \mu_{\rm p} + \mu_{\rm e^-}. \end{aligned} \tag{11.68}$$

The radiation and $e^\pm$-pairs in the ultrarelativistic approximation are taken into account. The entropy is assumed to be constant over the star. Thermodynamic functions in the presence of nuclear and pair equilibrium are described in Chap. 1, Vol. 1. The nuclear interaction has been included using a model with extremely stiff equations of state [1073] with the coefficient from [552]

$$P_{ni} = E_{ni}\rho_0 = 6\pi \frac{\hbar^3}{m_{\rm p}^4 c}\rho_0^2. \tag{11.69}$$

The integration of the system (11.62) and (11.63) has been performed with the aid of the Runge–Kutta scheme. For the entropy to hold constant along the star, the temperature at the subsequent step T_n is expressed in terms of T_{n-1} and densities ρ_n and ρ_{n-1}:

$$T_n = T_{n-1}\left(\frac{\rho_n}{\rho_n - 1}\right)^{\gamma_3} \tag{11.70}$$

with γ_3 defined as

$$\gamma_3 = \left(\frac{\partial \ln T}{\partial \ln \rho}\right)_S = -\left(\frac{\partial S}{\partial \ln \rho}\right)_T \Big/ \left(\frac{\partial S}{\partial \ln T}\right)_\rho. \tag{11.71}$$

This ensures the constancy of S with an uncertainty within 2%. Also, the stellar rest mass $M_0 = m_0(R)$ has been evaluated using the relation (11.65). The results of calculations are given in Fig. 11.7 and Table 11.5 from [152]. Models for cold

[5] $\tilde{E}$ is the energy per baryon, n is the concentration of baryons.

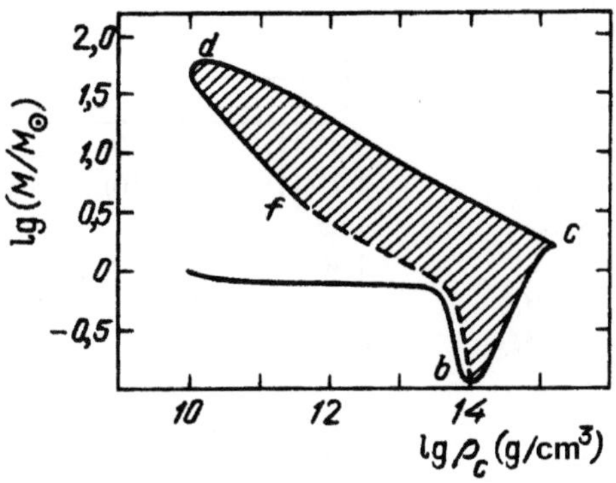

Fig. 11.7 Equilibrium stable configurations on the mass M, central rest-mass density ρ_c diagram for superdense isentropic stars (hatched). The curve of secondary maxima cd and part of minima curve df are plotted according to Table 11.5, fb is an approximate interpolation. The curve for cold neutron stars is taken from [552] (from [152])

Table 11.5 Characteristics of hot neutron stars [152]

Characteristic	Model 1	2	3	4	5
ρ_c, g cm^{-3}	2.93(14)	2.93(14)	4.40(13)	4.40(13)	1.47(13)
T_c, K	9.31(11)	9.65(11)	6.23(11)	6.52(11)	4.80(11)
S/S_0	4.82	4.99	7.12	7.47	8.54
γ_{1c}	1.60	1.59	1.49	1.48	1.46
$M/M_\odot$	2.91	3.03	5.68	6.18	8.49
$M_0/M_\odot$	2.97	3.09	5.75	6.24	8.57
R, km	38.6	40.2	120.7	103	215
I/ϵ_0	7.7(−5)	−5.2(−6)	4.6(−5)	−9.0(−5)	2.0(−5)
$\tilde{n}_{e,c}$	0.226	0.239	0.331	0.358	0.399

Characteristic	Model 6	7	8	9	10
ρ_c, g cm^{-3}	1.47(13)	2.93(12)	2.93(12)	2.93(11)	2.93(11)
T_c, K	4.86(11)	3.20(11)	3.26(11)	7.89(10)	1.48(11)
S/S_0	8.66	10.8	11.0	7.62	11.9
γ_{1c}	1.45	1.43	1.43	1.52	1.43
$M/M_\odot$	8.72	15.0	15.8	4.76	19.5
$M_0/M_\odot$	8.80	15.1	15.9	4.87	19.6
R, km	193	408	364	1070	1090
I/ϵ_0	−2.1(−5)	1.3(−5)	−4.1(−5)	1.0(−6)	3.1(−4)
$\tilde{n}_{e,c}$	0.407	0.508	0.525	0.184	0.512

(continued)

neutron stars with the equation of state (11.69) have been calculated in [552], giving $M_{max} = 1.60\, M_\odot$ and $M_{0,max} = 1.71\, M_\odot$ as a result.

With increasing stellar entropy, the curves $M_S(\rho_c)$ are located above one another, and the minimum approaches the second maximum until they merge at the point (ρ_{cc}, M_{cc}) at $S = S_{cc}$. This point corresponds to the mass limit

Table 11.5 (continued)

	Model					
Characteristic	11	12	13	14	15	16
ρ_c, g cm^{-3}	2.93(11)	2.93(10)	2.93(10)	2.93(10)	1.47(10)	1.47(10)
T_c, K	1.75(11)	6.61(10)	8.60(10)	9.13(10)	6.02(10)	6.82(10)
S/S_0	14.5	12.6	16.9	18.4	14.8	17.3
γ_{1c}	1.40	1.43	1.39	1.39	1.41	1.39
$M/M_\odot$	33.2	22.2	51.6	64.8	35.7	54.8
$M_0/M_\odot$	33.4	22.3	51.7	65.0	35.8	55.0
R, km	1250	3070	4820	3550	3910	4190
I/ϵ_0	1.4(−5)	6.0(−6)	2.0(−4)	1.4(−5)	1.5(−5)	1.1(−4)
$\tilde{n}_{e,c}$	0.705	0.491	0.807	0.918	0.636	0.812

	Model				
Characteristic	17	18	19	20	21
ρ_c, g cm^{-3}	1.47(10)	8.8(9)	8.8(9)	5.9(9)	5.9(9)
T_c, K	7.23(10)	5.93(10)	5.98(10)	4.74(10)	5.17(10)
S/S_0	18.7	18.3	18.5	16.5	18.4
γ_{1c}	1.39	1.39	1.39	1.40	1.39
$M/M_\odot$	68.7	64.5	66.7	48.6	66.7
$M_0/M_\odot$	68.9	64.7	66.9	48.7	66.9
R, km	4610	5090	5100	4860	5690
I/ϵ_0	4.3(−5)	1.1(−5)	2.3(−5)	−1.0(−4)	−3.1(−5)
$\tilde{n}_{e,c}$	0.920	0.871	0.887	0.735	0.873

Here $S_0 = k/m_p = 0.831\times10^8$ erg g^{-1} K^{-1}, $\epsilon_0 = \sqrt{\pi/4}\,(\hbar^3c^7/G^3(100m_e)^4)^{1/2} \approx 556\,M_\odot c^2$, I from (11.67); $\tilde{n}_e = n_e m_u/\rho_0$.

of a "neutron star"[6] because the minimum and second maximum are not present for $S > S_{cc}$, the curve $M(\rho_c)$ falls off monotonically after the first maximum, and stable states do not emerge. The maximum mass of a hot "neutron star" is closest to model 17 from Table 11.8 with

$$\rho_{cc} = 1.5\times10^{10}\ \text{g cm}^{-3} \quad T_{cc} = 7.2\times10^{10}\ \text{K},$$
$$M_{cc} = 68.7\,M_\odot \quad M_{0,cc} = 68.9\,M_\odot. \tag{11.72}$$

There is no iron in the centre of all the models from Table 11.5. while $x_{n,c}$ changes from 0.794 for model #1 to 0.553 for model #21; accordingly, $x_{p,c}$ changes from 0.206 to 0.467.

The maximum mass of a cold neutron star is determined by general relativity effects. The closer to the point d, the greater the role of the iron dissociation in

[6] Stars near the mergence point contain nearly equal numbers of neutrons and protons and are referred to as "neutron" conventionally for their topological similarity to cold neutron stars presented in Fig. 11.7.

stability loss, and appearance of $\gamma < 4/3$ in the outer layers of the star. The return of stability on the line of minima *bfd* is related to the formation of a dense neutron core with $\gamma > 4/3$. General relativity effects are of little importance for low-mass cold neutron stars, and their role nevertheless increases when approaching the point d (see Fig. 11.7).

Comparing binding energies $(M_0 - M)c^2$ from Table 11.5 with the binding energy in critical states from (10.27) given in Table 10.2, we see that the binding energy is larger in a superdense state than in the critical state (curve of first maxima) for $M_0 < 15M_\odot$. This means that the star can (but need not) stop on the curve of second maxima during collapse only if $M_0 \leq 15M_\odot$.

11.2.3 Cooling of Neutron Stars

Neutrino losses represent the main mechanism for cooling of neutron stars at $T_c \geq 4 \times 10^8$ K. As the cross-section for neutrino interaction with matter increases with energy approximately $\sim E_\nu^2$, hot neutron stars are opaque for neutrinos. Estimates in [152] show that this happens for all models from Table 11.5. Using the neutrino losses due to URCA processes (10.107) and (10.108), with $\Phi = 1$ and $\eta = 664.31$, and the available energy of non-degenerate nucleons, gives the cooling time $\tau_{\nu f}$ for the case of freely escaping neutrinos[7]

$$t_{\nu f} = \frac{E_T}{Q_{\mathrm{URCA}}} \approx 3 \times 10^5 T_9^{-5}\ \mathrm{s}. \tag{11.73}$$

To estimate neutrino thickness τ_{ν_e} we use the cross-section for interactions averaged over the energy

$$\overline{\sigma}_{\nu_e} = 2 \times 10^{-44} \left(\frac{\overline{\epsilon_{\nu_e}}}{m_e c^2}\right)^2 \approx 2 \times 10^{-44} \left(\frac{5kT}{m_e c^2}\right)^2\ \mathrm{cm}^2, \tag{11.74}$$

and then

$$\tau_{\nu_e} \approx \frac{\rho}{m_u} \overline{\sigma}_{\nu_e} R_{\mathrm{ef}} \approx 4 \times 10^{-15} \rho R_{\mathrm{ef}} \left(\frac{kT}{100 m_e c^2}\right)^2. \tag{11.75}$$

[7] The cross-section for neutrino capture $\sigma_{\nu_e}^{\mathrm{cap}}$ is twice σ_e^{cap} (see Problem 2, Chap. 5, Vol. 1) if ϵ_e is substituted for ϵ_{ν_e} in the expression for u_e: $u_{\nu_e} = \epsilon_{\nu_e}/m_e c^2$. A factor 2 is due to the summation over electron spins for ν_e-capture instead of averaging for e-capture. The pressure of equilibrium neutrinos P_{ν_e} is equal to one half of the pressure of ultrarelativistic $e^\pm$ pairs and does not exceed $7/26 \approx 27\%$ of the total pressure of pairs and radiation. The inclusion of P_{ν_e} in models from Table 11.5 does not change them by more than 27%. The effect of ν_μ and ν_τ on these models is substantially less because of a smaller cross-section for their interactions. Some models from Table 11.5, like model 1, are also opaque for ν_μ and ν_τ owing to interactions of neutral currents. This removes thresholds and makes possible interactions of ν_μ and ν_τ with matter without creating heavy leptons.

For all models in Table 11.5, the cooling time under opaque conditions (for not very large t_ν),

$$t_\nu = e^{\tau_{\nu e}} t_{\nu f},$$

exceeds a few seconds for $R_{\rm ef} = 1/10\,R$, whereas τ_{ν_e} varies from $\tau_{\nu_e} = 10^6$ for model #1 to $\tau_{\nu_e} = 7$ for model #21 from Table 11.5.

Numerical calculations for the early phases of neutron-star cooling have been made simultaneously with collapse calculations in the framework of Newtonian theory in [730], and with evolutionary calculations in the framework of general relativity in [245] using an equation of state analogous to [152], with the inclusion of neutrino pressure. In both cases, the time of cooling to the formation of a neutron star with $T_c \le 10^{10}$ K is $\sim$20 s. Figures 11.8–11.10 from [245] depict temperature, density and radius evolution over the first 20 s of cooling of a neutron star with a baryon rest mass $1.4\,M_\odot$. The temperature maximum of the initial model lies far from the centre at $m_0 \approx 1.15\,M_\odot$ and is related to heating and entropy increase occurring at the break of contraction. The adiabatic contraction brings about an increase in the neutron star temperature at early evolutionary phases (Fig. 11.8). The heating of central regions is stronger because of heat conductivity and the non-equilibrium character of β-processes (see also Chap. 5, Vol. 1). We may see from Figs. 11.8–11.10 that the radius and density distributions in the neutron star become steady state in 20 s because of decreasing temperature and a strong degeneration of matter, while the temperature distribution acquires a monotonic character.

Decreasing the temperature renders the neutron star transparent with respect to neutrino absorption and scattering, so $\tau_{\nu_e} < 1$. Evolutionary calculations for these conditions in the framework of general relativity have been carried out in [752]. All available mechanisms of neutrino cooling have been taken into account.

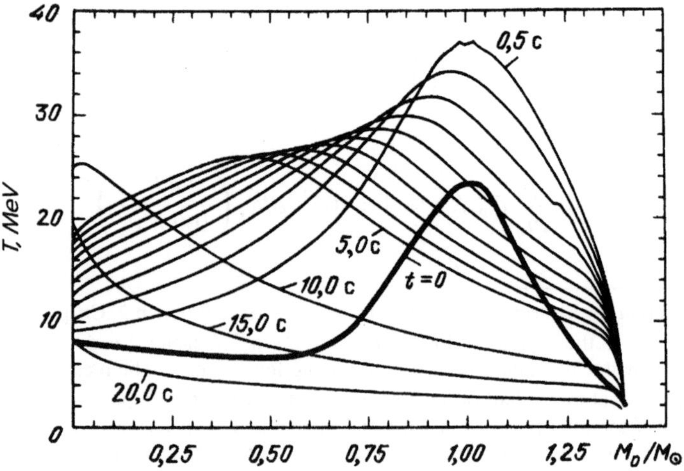

Fig. 11.8 Evolution of neutron star temperature distributions over enclosed baryon rest mass. Distribution are taken every 0.5 s for $t > 5$ s and every 5 s for $t > 5$ s (from [245])

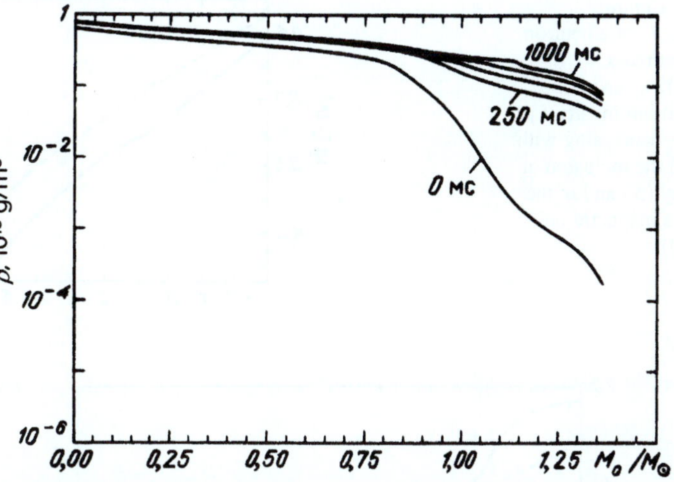

Fig. 11.9 Evolution of rest-mass density distribution in neutron star. Distributions are taken every 250 ms for $t \leq 1$ s (from [245])

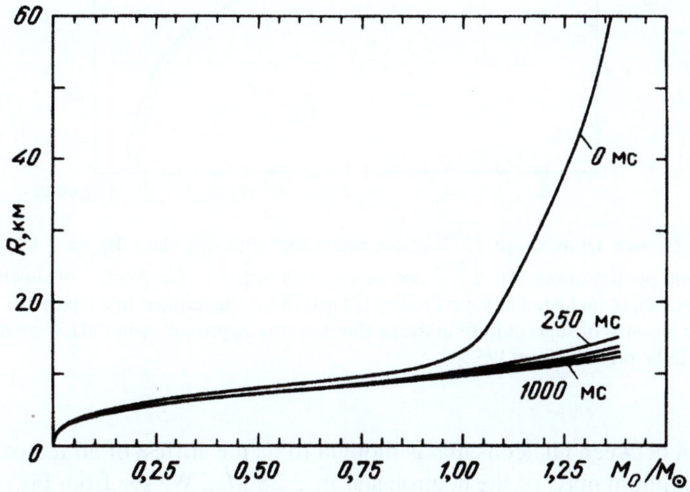

Fig. 11.10 Evolution of neutron-star radius distribution over enclosed baryon rest mass, distributions are taken every 250 ms for $t < 1$ s (from [245])

As in [245], the neutron star rest mass has been taken to be 1.4 $M_\odot$, with, however, other equations of state. Heat transfer in a neutron star has been considered with the inclusion of radiative and electron heat conductivity. The effect of superfluidity on the heat capacity of matter has been included as well. The results of calculations for the equation of state from [398] are given in Figs. 11.11 and 11.12. The effect of magnetic field on the cooling of a neutron star of the same mass has been studied in [751]. This study uses the equation of state from [791] with tensor forces of

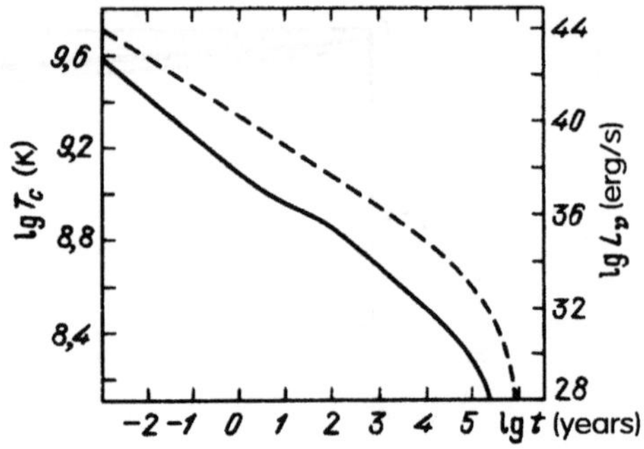

Fig. 11.11 Central temperature T_c of a neutron star with rest mass $M_0 = 1.4\, M_\odot$, *solid line*, and neutrino luminosity L_ν, *dashed line*, decreasing with time t, with the inclusion of superfluidity (S) and in the absence of a magnetic field (from [752])

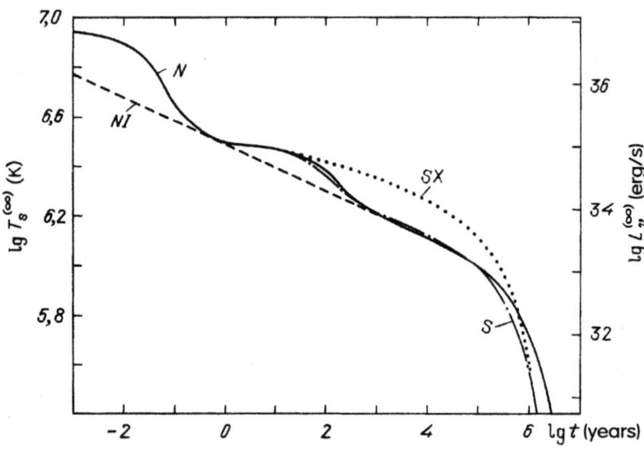

Fig. 11.12 Surface temperature $T_s^{(\infty)}$ of a neutron star with rest mass $M_0 = 1.4\, M_\odot$ observed at infinity and photon luminosity $L_\gamma^{(\infty)}$ decreasing with time t in the absence of superfluidity (N) and for two cases of inclusion of superfluidity (S) and (SX). The *dashed line* represents the cooling curve in the absence of superfluidity in the isothermal core approximation (NI). No magnetic field is assumed to be present (from [752])

interaction between nucleons that is thought to be the stiffest of all realistic models with a maximum mass of the neutron star of $2.28\, M_\odot$. We see from Fig. 11.13 that eventually the magnetic field **B** enhances the neutron-star cooling. This is caused by decreasing the Krammers mean opacity of matter in a magnetic field due to a fall $\sim 1/B^2$ of the cross-section for scattering of quanta with $\omega < \omega_B = eB/m_e c$ and circular polarization corresponding to counterrotation of the photon field vector to the electron rotation in magnetic field.

Account of pion condensation in the neutron star cores implies a rapid drop in luminosity within $10^{1.5}$–10^3 yr, depending on the physical conditions used in calculations [990].

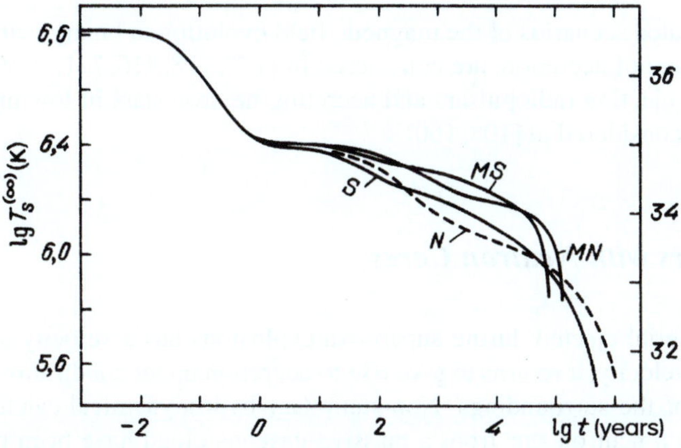

Fig. 11.13 Surface temperature $T_s^{(\infty)}$ of a neutron star with rest mass $M_0 = 1.4\,M_\odot$ observed at infinity and photon luminosity $L_\gamma^{(\infty)}$ as functions of time t. Letters stand for: (N) no superfluidity, zero magnetic field, (S) superfluidity included, without magnetic field, (MN) and (MS) the same as (N) and (S) but in the presence of a uniform magnetic field of strength $B = 4.4 \times 10^{12}$ g (from [751])

11.2.4 Magnetic Field Decay in Neutron Stars

Theoretical studies of magnetic field decay in neutron stars are impeded by incomplete knowledge as to the origins of these fields, mechanisms of their generation and the physical properties of matter. The fields are usually generated by ohmic currents which decay owing to a finite conductivity and do not decay at all under conditions of superconductivity. It is shown in [891] that the neutron magnetic alignment may generate a magnetic momentum of a star of $10^{27} - 10^{30}$ G cm^{-3}, while the field must not decay in this case.

Statistical analysis of observational data on pulsars suggests that the mean time for field decay is $\tau_m \approx 2\times10^6$ yr [819], but the same data could be compatible with no field decay hypothesis [100], see review [99]. Optical observations of pulsars in the binaries PSR 0655 + 64 ($P = 0.196$ s) and PSR 0820 + 02 ($P = 0.865$ s) led to the discovery of the optical companions of these pulsars which turned out to be white dwarfs. Lower estimates of their ages, as obtained from the cooling curve of Sect. 11.1, are $\tau_e = 2 \times 10^9$ and 10^7 yr, respectively. Evidently the same estimates are applicable to the pulsars themselves. The magnetic fields of these pulsars estimated from the time of deceleration $P/\dot{P}$ and the relation for a dipole emission [615] are $B = 10^{10}$ G for PSR 0655 + 64 and $B = 3 \times 10^{11}$ G for PSR 0820 + 02. We thus see from observations that even in 2×10^9 yr the magnetic field does not decay completely. It is assumed in [596], where the results of optical observations are listed, that the magnetic field of a neutron star may comprise two components, one of them being large and having a short lifetime $\sim \tau_m$, while the other is small and does not decay appreciably. The component might be due to currents in the superconductive component of the matter and/or to baryon magnetic alignment.

More complicated scenarios of the magnetic field evolution in binary neutron stars, passing the stage of accretion, are considered in [177, 178, 416, 721, 1066]. Transitions between ejecting radiopulsars and accreting neutron stars in low mass X-ray binaries were considered in [108, 160].

11.2.5 Stars with Neutron Cores

When the material ejected during supernova explosions has a velocity lower than the parabolic velocity, it returns to give rise to accretion under conditions of a fairly high density of the surroundings. Non-stationary hydrodynamical calculations of accretion onto a neutron star from a massive gaseous cloud have been performed in [1079]. The accretion rate turns out to be extremely high, corresponding to $\sim 10^7 L_{cr}$, where L_{cr} is the critical Eddington luminosity from (9.27). The accretion is nevertheless almost stationary, because most of the energy flux is carried away by neutrinos (only ν_e, $\tilde{\nu}_e$ have been included), and the photon luminosity does not exceed L_{cr}.

Another approach to this problem has been used in [974], where an equilibrium stellar model with a neutron core inside it has been calculated. The adopted conditions have allowed a low neutrino luminosity stationary model (LNLSM) with critical photon luminosity: $L_{ph} \simeq L_{cr}$, $L_\nu \ll L_{ph}$ to be obtained. The formation of stars with neutron cores could occur in close binaries, where the dragging of the neutron star orbital motion in a dense envelope of a giant or supergiant leads to the neutron star spiraling towards the centre of its companion [191]. The low neutrino luminosity model can live for a long time, $\sim 5 \times 10^7$ yr, and have a very high luminosity, $\sim 4 \times 10^4 L_\odot$.

An attempt to construct LNLSM without the restrictions used in [974] has been undertaken in [183], where the inner boundary of the model is shifted towards a deeper layer of the neutron star. This attempt has failed, so the LNLSM has been supposed to be thermally unstable and, even if appearing, to transform into models with high neutrino luminosity similar to those calculated in [1079] (neutrino run-away model). The lifetime of such a model does not exceed a few tens of years, and the existence of long-lived very luminous stars is impossible. Arguments in favour of LNLSM have been discussed in [253, 349]. Neutrino run-away phenomena were considered in [808].

11.2.6 Quark stars

The inner cores of neutron stars are the only known sites where one could expect degenerate quark matter in nature [539]. Figure 11.14 shows the cross section of a neutron star interior. The crust is typically ~ 1 km thick, and consists of a lattice of bare nuclei immersed in a sea of degenerate electrons, as in a normal metal.

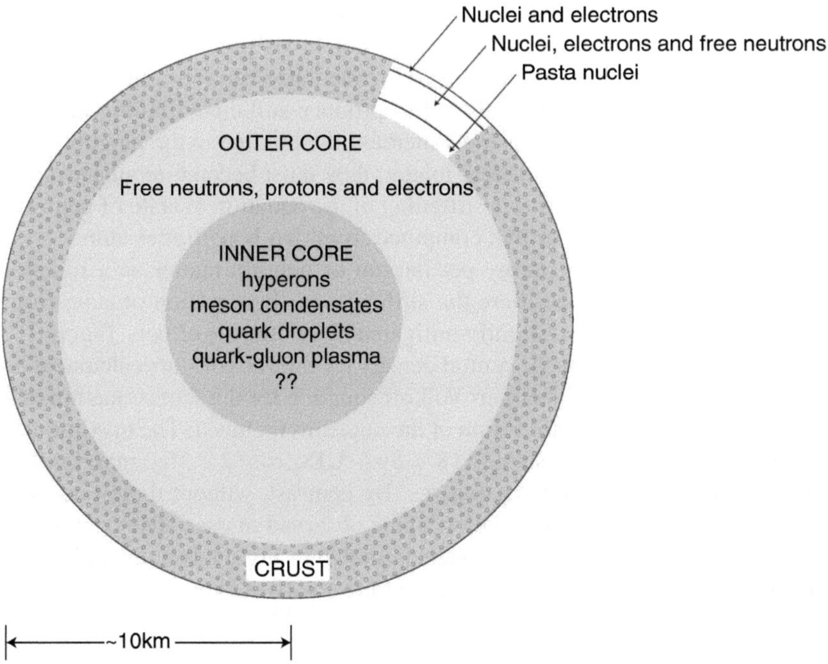

Fig. 11.14 Schematic cross section of a neutron star (from [77]). Pasta nuclei denotes the state of the matter, where drops of a lower density are surrounded by a higher density nuclear see [516]

The matter becomes more neutron rich with increasing density, a result of the increasing electron Fermi energy favoring electron capture on protons, $e^- + p \to n + \nu_e$. Beyond the *neutron drip* point, $\rho_{drip} \simeq 3 \cdot 10^{11}$g/cm^3($\simeq 2 \times 10^{-4}$ fm^{-3}), the matter becomes so neutron rich that the continuum neutron states begin to be filled, and the still solid matter becomes permeated by a sea of free neutrons in addition to the electron sea. At a density of order of nuclear matter density, $\rho_0 \simeq 2.7 \cdot 10^{14}$ g/cm^3 ($n_0 \simeq 0.16\,\mathrm{fm}^{-3}$), the matter dissolves into a uniform liquid composed primarily of neutrons, plus $\sim$5% protons and electrons, and a sprinkle of muons [77].

The nature of the extremely dense matter in the cores of neutron stars, while determining the gross structure of neutron stars, e.g., density profiles $\rho(r)$, radii R, moments of inertia, and the maximum neutron star mass, M_{max}, remains uncertain. Scenarios, from nuclear and hadronic matter, to exotic states involving pionic [700] or kaonic [556] Bose–Einstein condensation, to bulk quark matter and quark matter in droplets, including superconducting states, as well as strange quark matter, have been proposed. The uncertainies in the properties of matter at densities much greater than ρ_0 are reflected in uncertainties in $\mathrm{M}_{\mathrm{max}}$, important in distinguishing possible black holes from a neutron stars by measurement of their masses, and in inferring whether an independent family of denser quark stars, composed essentially of quark matter, can exist.

The properties of the liquid near ρ_0 can be readily determined by extrapolation from laboratory nuclear physics. Two-body potentials predict a reasonable binding energy of nuclear matter; however, the calculated equilibrium density is too high. Similarly, two-body potentials fail to produce sufficient binding of light nuclei [802]. The three-body forces must increase the binding in the neighborhood of n_0, but, to avoid overbinding nuclear matter, they must become repulsive at higher densities. This repulsion leads to a stiffening of the equation of state of neutron star matter at higher densities over that computed from two-body forces alone.

Figure 11.15 shows the energy per baryon of neutron matter as a function of baryon density [11]. One sees here the stiffening of the equation of state from inclusion of three-body forces, slightly mitigated by relativistic effects. Figure 11.16a shows the gravitational mass vs. central density for families of stars calculated by integrating the Tolman–Oppenheimer–Volkoff equation for the same equation of state as in Fig. 11.15, with beta equilibrium of the nucleons included. The maximum mass for the nucleonic equation of state, A18 + δv + UIX, is $\simeq 2.2\ M_\odot$, marginally consistent with observed neutron star masses. By contrast, without three-body forces, the maximum mass is $\sim 1.6 M_\odot$, below some observed masses. The corresponding mass vs. radius of the families of models is shown in Fig. 11.16b; the radii of these models vary little with mass, and are in the range of 10–12 km.

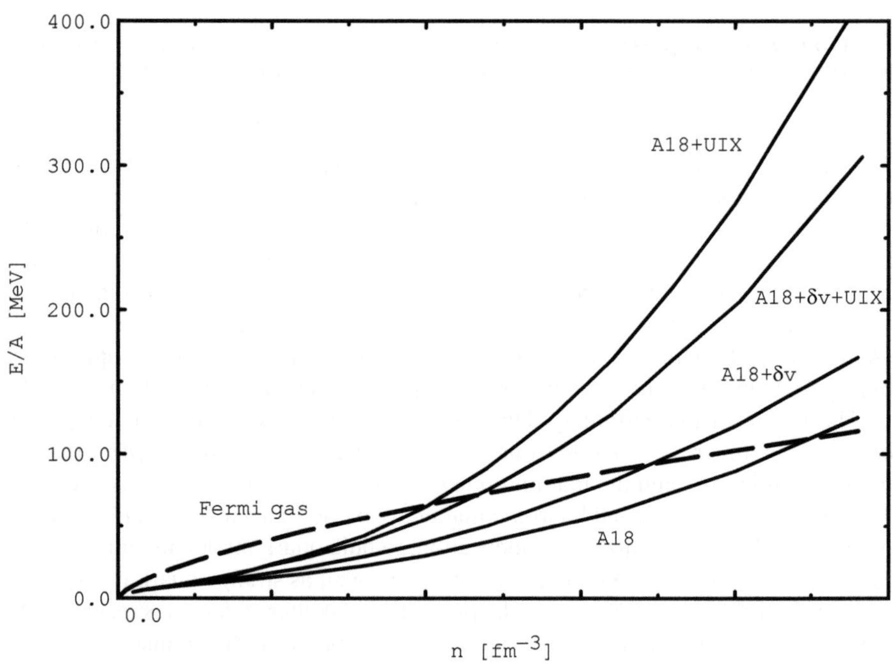

Fig. 11.15 Energy per baryon of pure neutron matter as a function of baryon density, n, calculated with the A18 two-body potential with and without the Urbana IX (UIX) three-body potential, and lowest order relativistic corrections, δv. From [77]

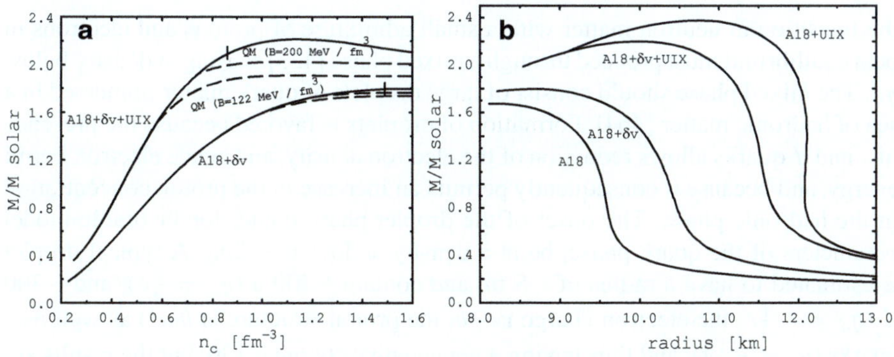

Fig. 11.16 (**a**) Neutron star mass vs. central baryon density for the equations of state shown in Fig. 11.15, including beta equilibrium. The curves labelled QM show the effect of allowing for a transition to quark matter described in the simple MIT bag model, with bag constants $B = 122$ and $200\,\text{MeV/fm}^3$. (**b**) Mass vs. radius of neutron stars for the same models. From [77]

An equation of state based on nucleon interactions alone, while accurately describing neutron star matter in the neighborhood of ρ_0, has several fundamental limitations. One should not expect beyond a few times ρ_0 that the forces between particles can be described in terms of static few-body potentials. The nucleonic equation of state furthermore does not take into account the rich variety of hadronic (Δ, hyperonic, mesonic, etc.) and quark degrees of freedom in the nuclear system which become important with increasing density. Nor can one continue to assume at higher densities that the system can even be described in terms of well-defined "asymptotic" laboratory particles. As one sees in Fig. 11.16a, the density in the central cores rises well above ρ_0; equations of state and neutron star models based on consideration of nuclear matter alone should not be regarded as definitive.

Nuclear matter is expected to turn into a quark-gluon plasma at sufficiently high baryon density. Figure 11.16a shows effects of including quark matter cores, naively calculated in the simple MIT bag model, with bag parameter $B = 122$ and $200\,\text{MeV/fm}^3$. Because of the well-known technical problems in implementing lattice gauge theory calculations at non-zero baryon density, to date there is no reliable estimate of the transition density at zero temperature or even a compelling answer as to whether there is a sharp phase transition or a crossover. Although estimates of the density range of the transition, $\sim 5-10n_0$, are possibly above the central density found in neutron stars models based on nuclear equations of state, the question of whether the dominant degrees of freedom of the matter in the deep cores of neutron stars are quark-like remains open. In the absence of information about the equation of state at very high densities, the issue of whether a distinct family of quark stars with higher central densities than neutron stars can exist also remains open.

If pure neutron matter and quark matter are distinct phases with a first order transition between them, the transition occurs at nucleonic density n_q where the energy per baryon of quark matter crosses below that in neutron matter. However, the transition in neutron matter with a small admixture of protons and electrons in beta equilibrium must proceed through a mixed phase [423] starting at density below n_q. The mixed phase should consist of large droplets of quark matter immersed in a sea of hadronic matter [464]. Formation of droplets is favored because the presence of s and d quarks allows reduction of the electron density, and hence electron Fermi energy, and because it consequently permits an increase in the proton concentration in the hadronic phase. The onset of the droplet phase could, for favorable model parameters of the quark phase, be at a density as low as $\sim 2n_0$. A typical droplet is estimated to have a radius of $\sim$5 fm and contain $\sim$100 u ($q_u = \frac{2}{3}e$), and $\sim$300 d ($q_d = -\frac{1}{3}e$, the electron charge is $-e$, the proton structure is uud) as well as s quarks ($q_s = -\frac{1}{3}e$), and thus having a net negative charge $\sim$130, but the results are very model dependent [77].

Observations of neutron star masses constrain the equation of state in the cores of neutron stars. The general rule obeyed by families of neutron stars generated at various central densities from a given equation of state is that the stiffer the equation of state, the higher is the maximum mass that a neutron star can have, but the lower is the central mass density, ρ_c at the maximum mass. Lower central density means that there is less room for exotic matter in the interior including π and K meson condensates, as well as quark matter. Observations of millisecond binary radio pulsars, consisting of two orbiting neutron stars, have permitted accurate determinations of their neutron star masses, as well as confirmed the existence of gravitational radiation; the well-measured masses lie in the interval, $\sim 1.35 \pm 0.1 M_\odot$ [161]. Were the maximum neutron star mass of order $1.4 M_\odot$, the central densities could be sufficiently large to allow substantial exotica in the interior. Even though the range of measured masses of neutron stars in binary neutron-star systems is tightly constricted, not all neutron stars must have such small masses. Nice et al. [742] have reported mass determination of the neutron star in the 3.4 ms pulsar PSR J0751 + 1807 in a close circular (6 h) binary orbit about a helium white dwarf. The measured neutron star mass is $2.1 \pm 0.2 M_\odot$, almost at the limit of compatibility with the nucleonic equation of state. The companion mass is $0.191 \pm 0.015 M_\odot$.

Observations of masses close to the maximum, $\simeq 2.2 M_\odot$, predicted by the nucleonic equation of state challenge our knowledge of the physics of neutron star interiors. The existence of high mass neutron stars immediately indicates that the equation of state must be very stiff, whether produced by interacting nucleons or other physics. Nonetheless, quarks degrees of freedom – not accounted for by interacting nucleons interacting via static potential – are expected to play a role in neutron stars. As nucleons begin to overlap, quark degrees of freedom should become more important. Indeed, once nucleons overlap considerably the matter should percolate, opening the possibility of their quark constituents propagating throughout the system. A firm assessment of the role of quarks in neutron stars must await a better understanding of mechanisms of quark deconfinement with increasing baryon density [77].

11.2.6.1 Strange Quark Matter

Strange quark matter (SQM) is an infinite deconfined mixture of u, d and s quarks in a colour singlet state (colourless), together with an appropriate number of electrons to guarantee electrical neutrality (see review [221]). According to the so-called strange matter hypothesis [1043], SQM is the true ground state of matter. The energy per baryon of SQM, at the baryon density where the pressure is equal to zero, is supposed to be less than the lowest energy per baryon found in atomic nuclei, which, with account of the rest energy, is about 930.4 MeV for the most bound nuclei (^{62}Ni, ^{58}Fe, ^{56}Fe).

If the strange matter hypothesis is true, then a nucleus with A nucleons could in principle lower its energy by converting to a *strangelet*. However, this process requires the simultaneous weak decay of about a number A of u and d quarks of the nucleus into strange quarks. The probability for such a process is extremely low, and the mean life time for an atomic nucleus to decay to a strangelet is much higher than the age of the Universe. In addition, finite size effects (surface, coulomb and shell effects) place a lower limit ($A_{\rm min} \sim 10 - 10^3$, depending on the values of the model parameters) on the baryon number of a stable strangelet even if in bulk SQM is stable.

The quarks in a nucleus are confined within protons and neutrons. Thus, the energy per baryon for a droplet of u, d quark matter (the so-called nonstrange quark matter in the bulk limit) must be higher than the energy per baryon of a nucleus with the same baryon number. These conditions may be used to constrain the values of the parameters entering in models for the equation of state of SQM and to find the region in the parameter space where the strange matter hypothesis is fulfilled and nonstrange quark matter is unstable. Our present understanding of the properties of ultra-dense hadronic matter does not allow us to exclude or to accept a priori the validity of the strange matter hypothesis.

A schematic model [81] for the equation of state of SQM is inspired to the MIT bag model for hadron. Despite this model is applicable at asymptotic densities (where perturbative QCD is valid), it has become very popular in the study of SQM in astrophysics [425]. The basic idea of the model is to suppose that quarks are confined within a spherical region (the *bag*) of the QCD vacuum. Inside the bag, quarks interact very weakly (perturbatively) with each other. The vacuum inside the bag (perturbative vacuum) is considered as an excited state of the true QCD vacuum outside the bag. Perturbative vacuum is characterized by a constant energy density B, the *bag constant*, which accounts in a phenomenological way of nonperturbative aspects of QCD. This gives rise to an inward pressure $P_B = -B$ on the surface of the bag, which balances the outward pressure originating from the Fermi motion of quarks and from their perturbative interactions. Thus, in the MIT bag model for SQM, the essential phenomenological features of QCD, i.e. quark confinement and asymptotic freedom, are postulated from the beginning. The short range qq interaction can be introduced in terms of a perturbative expansion in powers of the QCD structure constant α_c. The *up* and *down* quarks are assumed to be massless

($m_u = m_d = 0$), and the *strange* quark to have a finite mass, m_s, which is taken as a free parameter.[8]

Let us consider [221] for simplicity the limiting case of massless non-interacting quarks ($m_s = 0$, $\alpha_c = 0$). In this case, SQM is composed of an equal number of u,d, s quarks with no electrons (i.e. $n_e = 0$, $n_u = n_d = n_s$,) and the EOS takes the following simple form:

$$\varepsilon = Kn^{4/3} + B, \qquad P = \frac{1}{3}Kn^{4/3} - B, \qquad K = \frac{9}{4}\pi^{2/3}\hbar c. \tag{11.76}$$

Eliminating the total baryon number density, one has:

$$P = \frac{1}{3}(\varepsilon - 4B). \tag{11.77}$$

The properties of SQM, within this model, depend on the values of the three parameters B, m_s and α_c. The EOS, in the form $P = P(\rho)$, is essentially determined by the value of the bag constant B. The net fraction of leptons (e^-, e^+), which neutralize the electric charge of the quark component of SQM, will mainly depend on the values of m_s and α_c. For the most plausible values of the model parameters[369], the quark component has a positive electric charge, thus, electrons will be present in SQM to neutralize it. [9]

11.2.6.2 Strange Stars

One of the most important and fascinating consequences of the strange matter hypothesis is the possible existence of *strange stars*, that is, compact stars which are completely (or almost completely) made of SQM.[10] These stars have bulk properties (mass and radius) very similar to those of neutron stars (hadronic stars). Thus, pulsars could be strange stars.

The structural properties of non-rotating compact stars are obtained integrating numerically the Tolman–Oppenheimer-Volkoff (TOV) equations 11.62,11.63. The basic input to solve these equations is the stellar matter EOS. In the case of strange stars, one has to use one of the various models for SQM.

The properties of the maximum mass configuration for strange star sequences obtained using a few models for the SQM equation of state are reported in Table 11.6 from [221]. In this table, the EOS labeled as $B60_0$ refer to the bag model EOS

---

[8] The value of the *current quark mass*, as reported by the Particle Data Group (http://pdg.lbl.gov/), are the following: $m_u = 1$–3 MeV, $m_d = 3$–7 MeV, $m_s = 95 \pm 25$ MeV.

[9] For small values of m_s and large values of α_c, the quark component of SQM has a negative charge[369], and thus positrons will be present to guarantee global charge neutrality.

[10] When the values of the EOS parameters are such that the strange matter hypothesis is not fulfilled, the possible compact stars containing deconfined quark matter are the hybrid stars.

Table 11.6 Properties of the maximum mass configuration for strange stars obtained from different equations of state of SQM (see text for details). M is the gravitational stellar (maximum) mass in unit of the solar mass $M_\odot$, R is the corresponding radius, ρ_c the central density, n_c the central number density in unit of the saturation density ($n_0 = 0.16\,\text{fm}^{-3}$) of nuclear matter, P_c is the central pressure

EOS	$M/M_\odot$	R (km)	ρ_c (g/cm^3)	n_c/n_0	P_c (dyne/cm^2)
$B60_0$	1.964	10.71	2.06×10^{15}	6.94	0.49×10^{36}
$B60_{200}$	1.751	9.83	2.44×10^{15}	7.63	0.54×10^{36}
$B90_0$	1.603	8.75	3.09×10^{15}	9.41	0.73×10^{36}
$SS1$	1.438	7.09	4.65×10^{15}	14.49	1.40×10^{36}
$SS2$	1.324	6.53	5.60×10^{15}	16.34	1.64×10^{36}

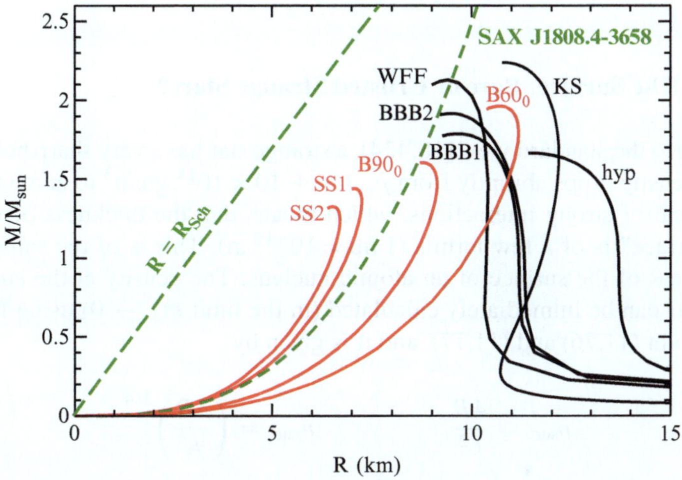

Fig. 11.17 The mass–radius relation for different theoretical models of compact stars. Black curves refer to hadronic stars and red curves to strange stars. The dashed curve gives the upper limit, extracted from observational data in [633], for the radius of the compact star in SAX J1808.4-3658. The dashed straight line labeled R_{Sch} gives the Schwarzschild radius as a function of the stellar mass

with $B = 60\,\text{MeV/fm}^3$ and $m_s = 0$; model $B60_{200}$ is for the same value of the bag constant but taking $m_s = 200\,\text{MeV}$; model $B90_0$ refer to the case $B = 90\,\text{MeV/fm}^3$ and $m_s = 0$ ($\alpha_c = 0$ in all the three cases). The models denoted as $SS1$ and $SS2$ refer to stellar sequences obtained with the equation of state for SQM from [323]

In Fig. 11.17 from [221], the mass–radius (MR) relation is plotted for strange stars (left bundle of curves) obtained using different model for SQM. For comparison, in the same figure, the MR relations for hadronic stars are plotted (right bundle of curves). The curves labeled with $BBB1$, $BBB2$ [64], WFF [1042] and KS [589] are for neutron stars having a β-stable nuclear matter core, the curve labeled *hyp* refers to an *hyperon star* [425]. As we can see, there is a striking qualitative difference between the mass–radius relation of strange stars with respect to that of neutron stars. For strange stars with "small" (i.e. M not to close to $M_{\max}$) mass,

M is proportional to R^3. In contrast, neutron stars (hadronic stars) have radii that decrease with increasing mass. This difference in the MR relation is a consequence of the differences in the underlying interactions between the stellar constituents for the two types of compact stars. In fact, "low" mass strange stars are bound by the strong interaction, contrary to the case of neutron stars, which are bound by gravity.

As it is well known, there is a minimum mass for a neutron star ($M_{\rm min} \sim 0.1 M_{\odot}$). In the case of a strange star, there is essentially non-minimum mass. As the stellar central density $\rho_{\rm c} \to \rho_{\rm surf}$ (surface density) a strange star (or better to say a lump of SQM for low value of the baryon number) is a self-bound system, until the baryon number becomes so low that finite size effects destabilize it. Review of the properties of strange stars is given in [1015].

11.2.6.3 The Surface: Bare or Crusted Strange Stars?

According to the standard view [18,424], a strange star has a very sharp boundary. In fact, the density drops abruptly from $\rho_{\rm surf} \sim$ 4–10 $\times 10^{14}$ g/cm^3 to zero on a length scale typical of strong interactions, which means that the thickness of the stellar "quark surface" is of a few fermis (1 fm $= 10^{-15}$ m). This is of the same order of the thickness of the surface of an atomic nucleus. The density at the surface of a strange star can be immediately calculated (in the limit $m_s \to 0$) using the simple EOS given in (11.76) and (11.77), and it is given by

$$\rho_{\rm surf} = \frac{4B}{c^2}, \qquad n_{\rm surf} = \left(\frac{3B}{K}\right)^{3/4}. \tag{11.78}$$

Strange stars with this sort of exposed quark matter surface are known as *bare* strange stars.

Electrons are bound to the star by the electromagnetic force; thus, they can extend for several hundreds of fermis above the "quark surface". This thin layer is usually referred to as the *electrosphere*. As a consequence of this charge distribution, a very strong electric field is established at the stellar surface. This field has been estimated to be of about 10^{17} V cm^{-1} and directed outward [18]. Such a huge electric field is expected to produce an intense emission of e^+e^- pairs [999] and a subsequent hard X-ray spectrum, at luminosities well above the Eddington limit,[11] as long as the stellar surface temperature is above [1000] $\sim 5 \times 10^8$ K. Thus, a bare strange star should produce a striking characteristic signal, which differs both qualitatively and quantitatively from the thermal emission from a neutron star (hadronic star), and which could provide an observational signature for their existence [13,544,999, 1000]. These special conditions (i.e. bare quark surface and high temperature) are realized in the case of "young" strange stars [787, 1001].

[11] A super-Eddington luminosity ($L > L_{\rm Edd}$) is allowed for a bare strange star, because at its surface SQM is bound by strong interaction rather than gravity [18, 999].

Older and "cold" ($T < 5 \times 10^8$ K) strange stars will likely form a crust of "normal" (nuclei and electrons) matter via mass accretion onto the star from the interstellar medium or from a companion star, or from the matter left in the supernova explosion which could have formed the strange star. In fact, the strong electric field at the stellar surface will produce a huge outward-directed force on any single positive ion (nucleus) of normal matter which is accreted onto the star. This force greatly overwhelms the force of gravity acting on the incoming positive ion. This accreted material will start to accumulate on the top of the electrosphere. Thus a strange star will form a crust of normal matter [18], which is suspended above the quark surface by the tiny ($\sim 10^2$–10^3 fm) electrostatic *gap*, and will completely obscure the "quark surface". This crust is similar in composition to the outer crust of a neutron star (i.e. a Coulomb lattice of heavy neutron rich nuclei plus an electron gas) In fact, when the density of matter at the base of this crust will reach the neutron drip density ($\rho_n^{drip} \simeq 4.3\ 10^{11} \times \mathrm{g/cm^3}$), the neutrons could freely enter in the SQM stellar core, and there they will be dissolved (their constituent quarks will be deconfined) due to the assumed absolute stability of SQM. Thus, the density at the base of the crust cannot exceed the value of the neutron drip density. This condition sets an upper limit for the mass of the stellar crust. For a strange star with a mass of $1.4\ M_\odot$, the mass of its crust is of the order of $M_{\rm crust} \sim 10^{-5} M_\odot$ and its thickness of the order of a few hundreds of meters [18, 424].

11.3 Black Holes and Accretion

If the mass of a collapsing core exceeds the mass limit of the neutron star, a black hole emerges, that is, an object with a very strong gravitational field $\varphi_G \sim c^2$, the existence of which is connected to general relativity. The black hole properties are described in a series of monographs (see, for instance, [272, 756, 1081]). The most important observational property of a black hole is that it does not let any light escape, so that in vacuum it can be detected only by making some known light source invisible. The space between stars and galaxies is, however, filled with a gas that falls into the black hole, becomes heated and emits radiation, making, in principle, the black hole observable. The most powerful accretion and best observational conditions occur when the black hole is incorporated in a binary, and the material flows from the neighbouring star onto the black hole. Such are X-ray sources in binaries—candidates for black holes are: CygX-1, LMCX-1, A0620-00, 4U1658-48(GX339-4) and others [283–285, 667].

The temperature of Hawking radiation $T_M \simeq 10^{26}/M(g) \sim 10^{-8}$K for $M = 5M_\odot$. This radiation is completely negligible for stellar mass black holes.

We discuss below various models for accretion onto black holes. All of them are hydrodynamical, i.e., the mean free path of particles is assumed to be less than the binary size. For strong ionization in the interstellar medium, this is accounted for by entanglement of electron and ion trajectories by the magnetic field, while in binaries, where the density is high, the mean free path is determined by Coulomb collisions and is also short.

11.3.1 Spherically Symmetric Accretion

In the case of a spherically symmetric accretion, the flow passes through the sound-velocity point at a saddle-like singularity of the system of hydrodynamical equations [224]. The condition for the flow to pass through the sound-velocity point determines unambiguously the mass flux $\dot{M}$ and all the flow properties for given T_∞ and ρ_∞. An adiabatic flow in the point-mass gravitational field can pass through the sound-velocity point only if the adiabatic index $\gamma < 5/3$. The theory of adiabatic accretion is given in [899, 1081], and is similar to the stellar wind theory (see Sect. 8.3.1). We discuss hereafter the flows passing through the sound-velocity point and supersonic near the centre of gravity.

Studies of spherically symmetric accretion of interstellar gas onto black holes have shown that in the presence of a magnetic field frozen in plasma, the effectivity of the kinetic energy conversion into heat approaches $\eta \approx 10\%$ [913], while in the absence of a field the bremsstrahlung causes $\eta \approx 10^{-8}$. When the gas falls radially, the lines of magnetic force stretch along a radius, $B_r \sim r^{-2}$, and the magnetic energy per unit volume $E_M \sim B^2 \sim r^{-4}$ increases more rapidly than the kinetic energy $E_{\rm kin} \sim \rho v^2 \sim \dot{M}v/r^2 \sim r^{-5/2}$ ($\dot{M} = 4\pi\rho v r^2$ is the stationary mass flux, free-fall velocity $v \sim r^{-1/2}$). Since the energy E_M cannot physically exceed $E_{\rm kin}$, it is assumed in [913] that an equipartition of energy $E_M \approx E_{\rm kin}$ is supported by the dissipation of magnetic energy, the excess of which is consumed by plasma heating. This heating is taken into account in [139] and leads to an increase in effectivity η to 30% that may be considered under these assumptions as a realistic estimate.

If $E_M \sim r^{-4}$ is the energy of a magnetic field with no dissipation, and $E'_M = E_{\rm kin} \sim r^{-5/2}$ is the energy of the magnetic field in the flow, then the increase in entropy per unit volume along a radius due to the field annihilation in stationary flow is given by

$$Q_M = \left(\rho T \frac{\mathrm{d}S}{\mathrm{d}r}\right)_M = \left(\frac{\mathrm{d}E_M}{\mathrm{d}r} - \frac{\mathrm{d}E'_M}{\mathrm{d}r}\right)_{E_M=E'_M}$$

$$= -4\frac{E_M}{r} + \frac{5}{2}\frac{E_M}{r} = -\frac{3}{2r}\frac{B^2}{8\pi}. \tag{11.79}$$

Consider separately the regions with non-relativistic electrons, having

$$kT \ll m_{\rm e}c^2, \quad \gamma_1 = 5/3$$

and relativistic electrons with[12]

$$kT \gg m_{\rm e}c^2,$$

[12] Protons are always non-relativistic.

$$\rho E = \frac{1}{2}\left(3P + \frac{3}{2}P\right) = \frac{9}{4}P = nP,$$

$$\gamma_1 = 1 + \frac{1}{n} = 13/9. \tag{11.80}$$

Here, $P_e = P_p = P/2$, γ_1 is given in (8.26), n is the adiabatic index, and for simplicity we treat a hydrogen plasma. From the energy balance equation

$$\frac{\mathrm{d}E}{\mathrm{d}r} - \frac{P}{\rho^2}\frac{\mathrm{d}\rho}{\mathrm{d}r} = \frac{Q_M}{\rho} - \frac{\epsilon_B}{v_r}, \tag{11.81}$$

where ϵ_B (erg g^{-1} s^{-1}) is the rate of magneto-bremsstrahlung losses of the Maxwellian plasma, with

$$\frac{B^2}{8\pi} = \frac{1}{2}\rho v_r^2, \quad v_r = \alpha v_{ff} = \alpha\sqrt{\frac{2GM}{r}}, \quad \overline{B_\perp^2} = \frac{2}{3}\overline{B^2}, \tag{11.82}$$

and

$$\epsilon_B = 2\frac{e^2}{m_p c}\left(\frac{eB_\perp}{m_e c}\right)^2$$

$$\times \frac{kT}{m_e c^2} \approx 0.46\,T B_\perp^2 \text{ erg g}^{-1}\text{ s}^{-1} \quad \text{for} \quad kT \ll m_e c^2 \quad \text{(NR)}, \tag{11.83}$$

$$\epsilon_B = 8\frac{e^2}{m_p c}\left(\frac{eB_\perp}{m_e c}\right)^2$$

$$\times \left(\frac{kT}{m_e c^2}\right)^2 \approx 3.2\times10^{-10}\,T^2 B_\perp^2 \text{ erg g}^{-1}\text{ s}^{-1} \quad \text{for} \quad kT \gg m_e c^2 \quad \text{(R)}. \tag{11.84}$$

We obtain the equations for $T(r)$

$$\frac{3}{2}\frac{\mathrm{d}T}{\mathrm{d}r} + \frac{3}{2}\frac{T}{r} + \frac{3}{4}\frac{\alpha^2 2GM}{\mathcal{R}_g r^2} - 1.5\frac{T\dot{M}}{\mathcal{R}_g r^2} = 0 \quad \text{(NR)} \tag{11.85}$$

$$\frac{9}{4}\frac{\mathrm{d}T}{\mathrm{d}r} + \frac{3}{2}\frac{T}{r} + \frac{3}{4}\frac{\alpha^2 2GM}{\mathcal{R}_g r^2} - 2.2\times10^{-10}\frac{T^2\dot{M}}{\mathcal{R}_g r^2} = 0. \quad \text{(R)} \tag{11.86}$$

Here, $\mathcal{R}_g = 2k/m_p$ is the gas constant for ionized hydrogen. For specified values of ρ_∞, T_∞, M, the mass flux is determined by

$$\dot{M} = 4\pi\rho v_r r^2 = \frac{10^{32}}{c^2}\left(\frac{M}{M_\odot}\right)^2\left(\frac{\rho_\infty}{10^{-24}\ \mathrm{g\ cm^{-3}}}\right)\left(\frac{T_\infty}{10^4\ \mathrm{K}}\right)^{-3/2} \mathrm{g\ s^{-1}}. \quad (11.87)$$

Neglecting the radiation in (11.85) and adiabatic heating in (11.86), we find the solution in the form

$$T = 2\times 10^{12}x + 2.7\times 10^{12}x\alpha^2 \ln\left(\frac{10^8 x}{T_\infty/10^4\ \mathrm{K}}\right) \quad \text{(NR)} \quad (11.88)$$

$$T = \frac{m_e c^2}{k} + T_1\frac{e^{a(x-x_0)}-1}{e^{a(x-x_0)}+1}, \quad T_1 \gg \frac{m_e c^2}{k} \quad \text{(R)}. \quad (11.89)$$

Here,

$$x = r_g/r = \frac{2GM}{rc^2} < 1,$$

$$T_1 = 2.8\times 10^{12}\alpha\left(\frac{T_\infty}{10^4}\right)^{3/4}(M/M_\odot)^{-1/2}\left(\frac{\rho_\infty}{10^{-24}}\right)^{-1/2},$$

$$a = 1.3\,\alpha\left(\frac{M}{M_\odot}\right)^{1/2}\left(\frac{T_\infty}{10^4\ \mathrm{K}}\right)^{-3/4}\left(\frac{\rho_\infty}{10^{-24}\ \mathrm{g\ cm^{-3}}}\right). \quad (11.90)$$

The quantity x_0 depends slightly on T_∞, ρ_∞, and for various values of α is

α^2	1	1/3	1/10
x_0	2×10^{-4}	5×10^{-4}	1.2×10^{-3}

When $x = x_0$, we have $T = m_e c^2/k$, and the solutions (11.88) and (11.89) become fitted to one another. The luminosity and spectrum for this model have been calculated in [139]. The luminosity due to magneto-bremsstrahlung is determined mainly by relativistic electrons from (11.89) and equals

$$L_B = 2.7\times 10^{31}\alpha^4(M/M_\odot)^3(\rho_\infty/10^{-24}\ \mathrm{g\ cm})^{-3}(T_\infty/10^4\mathrm{K})^{-3}, \quad a < 1$$

$$L_B = 9\times 10^{31}\alpha^2(M/M_\odot)^2(\rho_\infty/10^{-24}\ \mathrm{g\ cm})^{-3}(T_\infty/10^4\mathrm{K})^{-3/2}, \quad a \gg 1 \quad (11.91)$$

Comparing this expression with (11.87), we find that at a realistic value $\alpha^2 = 1/3$ the quantity $\eta = L_B/\dot{M}c^2 \leq 30\%$. An approximate emission spectrum L_ω ($L_B = \int_0^\infty L_\omega\, \mathrm{d}\omega$) of a black hole with mass $M = 10M_\odot$ is given in Fig. 11.18 from [157]. The range with $L_\omega \sim \omega^{3/5}$ is related to the emission of non-relativistic electrons; at $kT \gg m_e c^2$, $\hbar\omega \ll kT$, $L_\omega \sim \omega^{1/3}$, while at $\hbar\omega \approx \hbar\omega_{B,\max}(kT_1/m_e c^2)^2 \sim$

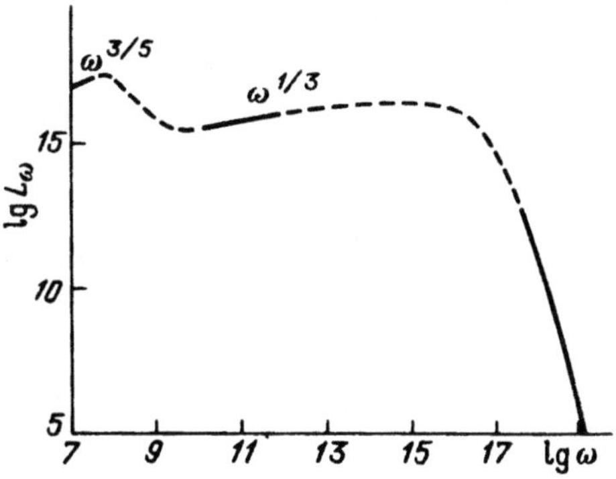

Fig. 11.18 Magneto-bremsstrahlung spectrum of a black hole of $M = 10\,M_\odot$ for a spherically symmetric accretion and random magnetic field at $\rho_\infty = 10^{-24}$ g$\cdot$cm^{-3}, at $T_\infty = 10^4$ K, $\alpha^2 = 1/3$. The *solid lines* represent asymptotic dependencies; *dashed lines* show extrapolation

10 eV for $\rho_\infty = 10^{-24}$ g cm^{-3}, $T_\infty = 10^4$ K, $B_{\max} \sim 10^5$ G an exponential cut-off occurs.[13]

The visible magnitude m_V for such a black hole under the assumption of a plane spectrum is [139]

$$m_V = 4.8 - 2.5\lg L/L_\odot + 5\ln(R/10\text{ pc}) \approx 14.1 - 7.5\lg M/M_\odot$$

$$-2.5\lg\left[\left(\frac{\rho_\infty}{10^{-24}\text{ g cm}^{-3}}\right)^{3/2}\left(\frac{T_\infty}{10^4\text{ K}}\right)^{-9/4}\right] + 5\lg\left(\frac{R}{10\text{ pc}}\right). \quad (11.92)$$

At high luminosity and high density, the role of bremsstrahlung increases and the interaction of the accretion flow with outgoing flow of radiation becomes essential, thereby changing (11.85)–(11.92). Calculations of accretion onto a black hole including the reciprocal effect of radiation have been performed in [795,796]. For the case of a neutron star, this effect has been included in [121].

The above discussion assumes the equipartition to set in at any point in the flow. In reality, the field annihilation may occur discontinuously and leads to the shock formation [191]. This is why the emission spectrum may differ even at low luminosities from Fig. 11.18, which should be treated as a rough estimate.

---

[13] The synchrotron radiation spectrum of a unit volume of relativistic Maxwellian plasma is [157]

$$I_\omega = (\sqrt{2}\rho e^2/6\mu_Z m_p c)\,\omega_B z \exp\left[-(9/2)^{1/3} z^{1/3}\right], \quad z \gg 1$$

$$= 3^{1/6}/\pi\Gamma(4/3)\Gamma(5/3)\,(\rho e^2/\mu_Z m_p c)\,\omega_B z^{1/3}, \quad z \ll 1$$

$$z = \frac{\omega}{\omega_B}\left(\frac{m_e c^2}{kT}\right)^2, \quad I\ (\text{erg cm}^{-3}\,\text{s}^{-1}) = \int_0^\infty I_\omega\,\mathrm{d}\omega.$$

11.3.2 Accretion at an Ordered Magnetic Field

If the characteristic scale of non-uniformity of magnetic field greatly exceeds the accretion radius

$$r_a = \frac{GM}{v_s^2}, \quad \text{where } v_s \text{ is the sound velocity in the gas,} \tag{11.93}$$

the flow is no longer spherically symmetric. In the case of a uniform magnetic field the accretion symmetry is cylindrical. For a black hole at rest, a stationary pattern of magnetic lines sets in, the material flows along them and forms a disk in the plane of symmetry. A qualitative picture of the flow is shown in Fig. 11.19. At a large but finite conductivity the disk material infiltrates slowly through lines of magnetic force towards the black hole. The formation process has been studied, the structure of a disk supported by a magnetic field and its radiation have been calculated in [103, 143].

Consider, approximately, the stationary disk structure. Equilibrium of a non-rotating disk is determined by the balance between magnetic forces and gravity:

$$\frac{GM\Sigma}{r^2} = \frac{1}{c} B_\theta I_\varphi \approx \frac{2\pi}{c^2} I_\varphi^2. \tag{11.94}$$

Here, $\Sigma = 2h\rho$ is the surface density, ρ is the mean matter density, I_φ is the circular current propagating through an element of disk area with unit length along r and height $2h$. We have roughly [118]

$$B_\theta \approx B_r \approx \frac{2\pi}{c} I_\varphi. \tag{11.95}$$

Equilibrium along the vertical is supported by the pressure gradient

$$\frac{dP}{dz} = -\frac{\rho GM}{r^2}\frac{z}{r}, \quad h \approx \left(\frac{r^3}{GM}\frac{P}{\rho}\right)^{1/2}. \tag{11.96}$$

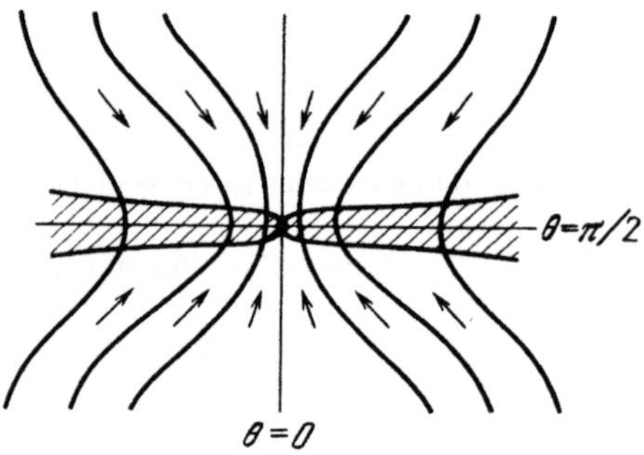

Fig. 11.19 Schematic pattern of magnetic field lines in the matter around a black hole for a field uniform at infinity, with the inclusion of distortions due to disk currents. The *arrows* indicate the direction of the gas flux, hatched is the region filled in by the dense disc (from [139, 143])

Taking into account the conversion into heat of the energy of material inflowing to the disk at a velocity close to the free-fall velocity, we have for the energy flux from unit disk surface

$$F = \frac{GM\dot{M}}{4\pi r^3}\left[1 + \frac{1}{2}\left(\frac{r}{R}\right)^{3/2}\right]. \tag{11.97}$$

From the mass conservation, we have for the radial velocity

$$v_{\rm r} = -\frac{\dot{M}}{2\pi r\Sigma}\left[1 - \left(\frac{r}{R}\right)^{3/2}\right]. \tag{11.98}$$

Here $\dot{M}$ is determined by the values at infinity in (11.87), $R \approx r_{\rm a}$ from (11.93). The disk undergoes an ohmic dissipation that causes the material to infiltrate through the magnetic field and determines most of the released energy

$$\frac{GM\dot{M}}{4\pi r^3}\left[1 - \left(\frac{r}{R}\right)^{3/2}\right] = \frac{I_\varphi^2}{4\pi\sigma}, \tag{11.99}$$

where σ is the conductivity. If the disk is opaque to radiation, the energy is carried to its surface by radiative heat conduction, so that we have roughly

$$acT^4 = \kappa\Sigma F, \tag{11.100}$$

where $T(r)$ is the mean disk temperature. For a transparent disk, we have

$$2F = \Sigma(\epsilon_{\rm ff} + \epsilon_B). \tag{11.101}$$

This relation includes plasma bremsstrahlung $\epsilon_{\rm ff}$ and magneto-bremsstrahlung ϵ_B. Equations (11.94–11.3.22) with known functions P, κ, σ, $\epsilon_{\rm ff}$ and ϵ_B give the structure of a non-rotating disk with magnetic field around the black hole.

In a laminar approximation, the disk is always optically thick, electrons are non-degenerate and non-relativistic, and the pressure is determined mainly by the ideal ionized gas $P = \rho\mathscr{R}T$. The conductivity due to Coulomb collisions is given in (11.55). Two regions may be discerned in such a disk: an outer region, where bremsstrahlung and photo-ionization processes with

$$\kappa_{\rm ff} + \kappa_{bf} \approx 2 \times 10^{24}\rho T^{-7/2} \tag{11.102}$$

predominate in the opacity, and an inner one, with predominance of the opacity due to magneto-bremsstrahlung absorption of non-relativistic electrons with

$$\kappa_B \approx 40\, B^2/T^3, \tag{11.103}$$

see [143]. As the Coulomb conductivity is small, the material infiltrates slowly through lines of magnetic force, and in a stationary case the disk mass turns out to be large: for a black hole of $M = 100M_{\odot}$, the mass of the stationary disk is $M_d \approx 0.2\,M_{\odot}$. Most of the disk mass is accumulated in its outer layers. The disk interiors are the main source of radiation, the temperature in them attains $10^8 - 10^9$ K, the magnetic field $10^{10} - 10^{12}$ G. The disk thickness nowhere exceeds 0.01 of the radius.

In the turbulent disk approximation, the dissipation proceeds far more rapidly owing to entanglement of magnetic lines, and [143]

$$\sigma \approx \sigma_T \approx \frac{c^2}{\tilde{\alpha} 4\pi h \sqrt{P/\rho}}, \quad \tilde{\alpha} = 0.1 - 0.01. \tag{11.104}$$

The outer regions of such a disk are transparent for radiation, electrons are non-relativistic there, the gas pressure predominates, and the contributions of bremsstrahlung and recombination radiation with

$$\epsilon_{\rm ff} + \epsilon_{fb} \approx 2 \times 10^{22} \rho T^{1/2}, \tag{11.105}$$

and magnetobremsstrahlung with ϵ_B from (11.3.5) losses are of the same order. At $(Mr_{\rm g}/M_{\odot} r) = mx \approx 100\alpha^2$ the electrons become relativistic and ϵ_B from (11.84) greatly exceeds

$$\epsilon_{\rm ff}^{\rm rel} \approx 2 \times 10^{16} \rho T \ln \frac{kT}{m_{\rm e} c^2}, \tag{11.106}$$

see [143]. Note the connection between opacity and emissivity $\epsilon = A\sigma T^4 \kappa$, following from Kirchhoff's law. The constant $A = 170$ was chosen in estimations of κ_B, and the relativistic case of $\kappa_{\rm ff}$, in accordance with the corresponding constant for non-relativistic Rosseland $\kappa_{\rm ff}$ and $\epsilon_{\rm ff}$, calculated in textbooks (see, e.g., Chap. 2, Vol. 1). The zone of relativistic electrons is narrow in radius because the optical thickness increases rapidly and the disk becomes opaque with decreasing radius. In disk interiors, the gas pressure and electron scattering due to opacity $\kappa_{\rm es} r \equiv \sigma_T$ from (9.41) predominate for $10 < mx < 1000$, while for $mx > 1000$ the predominant role in pressure belongs to radiation. The mass of a turbulent disk always remains small because of a large dissipation and rapid infiltration of material through the magnetic field. Note again the non-monotonical character of the function $T(r)$ arising from the transformation of a transparent disk into an opaque one with decreasing r. Relations describing the distribution of disk parameters are given in [143] for various disk regions.

As the gravitational energy always has time to convert into heat during stationary disk accretion, the black-hole luminosity at a minimum radius of $1.5\,r_{\rm g}$ is

$$L = \frac{1}{3} \dot{M} c^2 \tag{11.107}$$

with $\dot{M}$ from (11.87).

The emission spectrum of an opaque disk is related to its effective temperature determined locally by

$$\frac{ac}{4}T_{\mathrm{ef}}^4 = F \tag{11.108}$$

at any point on the disk surface provided that the shock energy has been completely thermalized. Hence, using (11.97), we have for the disk interiors with $r \ll R$, where most of the energy is released,

$$T_{\mathrm{ef}} \sim r^{-3/4}. \tag{11.109}$$

The distribution (11.109), upon being integrated over the disk surface, yields the spectrum

$$L(\omega) \sim \omega^{1/3} \tag{11.110}$$

with an exponential cut-off at $\hbar\omega \sim kT_{\max}$, $T_{\max} \approx 7 \times 10^5$ K for $M = 10M_{\odot}$. So, despite substantial differences in flow patterns for random and ordered magnetic fields, for the case of a laminar disk their emission spectra are similar, and the relation (11.92) for m_V is valid in such a case as well. A turbulent disk has a large transparent region with magneto-bremsstrahlung in the infrared range that may be comparable in power with the ultraviolet and soft X-ray emission of opaque disk interiors.

If the accreting material has an intrinsic angular momentum, the magnetized disk generates electric fields giving rise to the formation of relativistic particles [162]. Such a mechanism is likely to act in Cyg X-1, galactic nuclei [201,643], and is analogous to the unipolar mechanism proposed for explaining the pulsar radiation [431].

11.3.3 Conical Accretion on to a Rapidly Moving Black Hole

A rapidly moving black hole is streamlined by gas, thus causing formation of a conical shock behind it. Passing through the shock, the material decelerates and falls onto the black hole inside the cone. A qualitative picture of conical accretion, first considered in [225], is given by Salpeter [858]. For a primarily cold gas falling onto the gravitating centre along parabolic trajectories, a self-similar solution has been obtained in [175] with properties substantially dependent on the adiabatic power γ and density distribution in the falling gas. A system of steady-state hydrodynamical equations for gas motion in the field of central mass M reads, in a spherical coordinate system with $v_\varphi = \partial/\partial_\varphi = 0$, as

$$v_{\mathrm{r}}\frac{\partial v_{\mathrm{r}}}{\partial r} + \frac{v_\theta}{r}\frac{\partial v_{\mathrm{r}}}{\partial \theta} - \frac{v_\theta^2}{r} + \frac{1}{\rho}\frac{\partial P}{\partial r} + \frac{GM}{r^2} = 0,$$

$$v_{\mathrm{r}}\frac{\partial v_\theta}{\partial r} + \frac{v_\theta}{r}\frac{\partial v_\theta}{\partial \theta} + \frac{v_{\mathrm{r}} v_\theta}{r} + \frac{1}{\rho r}\frac{\partial P}{\partial \theta} = 0,$$

$$\frac{1}{r^2}\frac{\partial}{\partial r}\left(\rho r^2 v_r\right) + \frac{1}{r\sin\theta}\frac{\partial}{\partial\theta}\left(\rho v_\theta \sin\theta\right) = 0,$$
$$v_r\frac{\partial\tilde{S}}{\partial r} + \frac{v_\theta}{r}\frac{\partial\tilde{S}}{\partial\theta} = 0,$$
$$P = \tilde{S}\rho^\gamma, \tag{11.111}$$

where $\tilde{S}$ is a function of entropy S.

A self-similar solution to the system (11.111) is looked for in the form

$$v_r = \sqrt{\frac{2GM}{r}}\,U(\theta), \quad v_\theta = \sqrt{\frac{2GM}{r}}\,V(\theta),$$
$$\rho = \rho_0\sqrt{2GM}r^{-\alpha}R(\theta), \quad P = \rho_0(2GM)^{3/2}r^{-\alpha}r^{-(1+\alpha)}\mathscr{P}(\theta). \tag{11.112}$$

For the pre-shock region, we have the solution

$$\rho = \rho_0 r^{-\alpha}\left(\cos\frac{\theta}{2}\right)^{-2\alpha}\cot\frac{\theta}{2}\sqrt{2GM}, \quad \mathscr{P} = 0,$$
$$v_r = -\sqrt{\frac{2GM}{r}}\sin\frac{\theta}{2}, \quad v_\theta = -\sqrt{\frac{2GM}{r}}\cos\frac{\theta}{2}. \tag{11.113}$$

A uniform density at infinity corresponds to $\alpha = 1/2$. The shock coincides with a straight cone with apex angle $\theta = \theta_s$, θ_s being an eigenvalue of this problem. The solution to the system of ordinary equations for $U(\theta)$, $V(\theta)$, $R(\theta)$, $\mathscr{P}(\theta)$ is obtained numerically in [175] and has the following properties.

The solution, corresponding to the cone beyond the black hole filled in completely with material, exists for $\alpha = 1/2$ only at $1.31 \leq \gamma \leq 5/3$, and the apex angle of the shock cone increases with γ (Fig. 11.20 and Table 11.7 [175]). For $\alpha > 1/2$ the filled cone exists only at a single value of γ for each α (Table 11.8 [175]). For α and γ corresponding to the solution with a filled cone and for all other

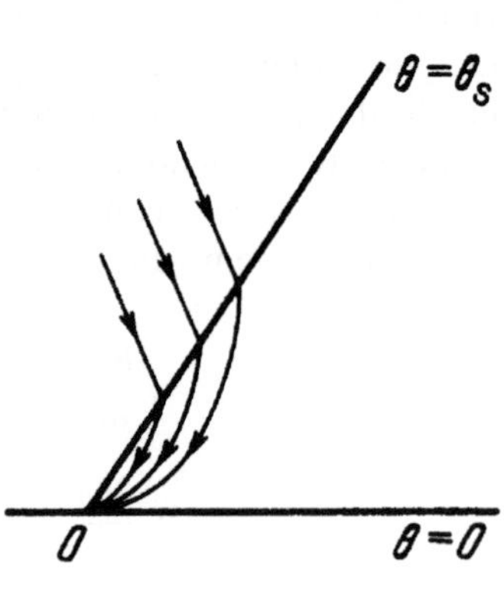

Fig. 11.20 Flow pattern for a gas filling in all the space, θ_s is the slope angle of the shock, from [175]

Table 11.7 The dependence of the slope angle of the sock θ_s on adiabatic power γ for flows filling in all the space at $\alpha = 1/2$

Parameter	Value						
γ	1.31	1.34	1.4	1.45	1.5	1.55	1.6
θ_s	1.298	1.426	1.679	1.905	2.131	2.375	2.651

Table 11.8 The dependence of the adiabatic power γ corresponding to the flow with a filled in cone and corresponding slope angle of the shock θ_s on index α

Parameter	Value							
α	0.6	0.8	1.0	1.2	1.3	1.34	1.4	1.45
γ	1.008	1.02	1.08	1.18	1.26	1.24	1.15	1.05
θ_s	0.04	0.09	0.22	0.51	0.7	0.61	0.33	0.11

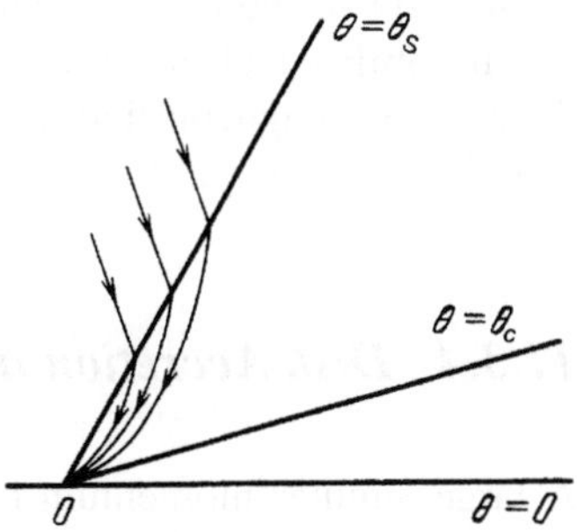

Fig. 11.21 Flow pattern in the presence of an empty cone with slope angle $\theta = \theta_c$ beyond the moving center, θ_s is the slope angle of the shock, from [175]

α and γ as well, there exist solutions with density going to infinity and v_θ to zero at a finite apex angle $\theta = \theta_c$, and the empty region inside θ_c (Fig. 11.21 [175]). For given α and γ, a θ_s corresponds to each θ_c so that $\theta_1(\gamma, \alpha) < \theta_s < \theta_2(\gamma, \alpha)$. For $\alpha = 1/2$, $\theta_1(\gamma, 1/2) = \theta_s$ from Table 11.7, that is, when a solution with a filled cone exists, the slope angle of the corresponding shock is minimal. Complicated properties of self-similar solutions arise from the presence of a singular line in a self-similar set of ordinary equations and a varying number of intersections of the integral curve with this line.

If the flow properties are such that the radiative losses in the shock are essential, then the apex angle of the cone beyond the black hole may be substantially less than in Table 11.7 at the given γ. This corresponds effectively to γ approaching 1.

The numerical solution of the problem of accretion onto a moving gravitating centre requires knowledge of boundary conditions at a finite radius r that affect the solution. For modeling a black hole, the condition of complete absorption of material at a small finite radius r_0 has been used in [483, 907]. 2D calculations have been made for Mach numbers of inflowing material at infinity $M = 0.6$, 1.4, 2.4, 5.0 and adiabatic powers $\gamma = 1.1$, 4/3, 5/3. It has been found in [907] that at $\gamma = 5/3$ and $4/3$, the bow shock is formed in front of the centre, while the density beyond the shock is well above the density on the cone axis beyond the black hole.

At $\gamma = 1.1$, the shock becomes adjacent to the absorbing sphere surface, and the density contrast in the cone beyond the shock is not large.

This problem was computed with higher resolution in 2D and in 3D simulations. It was found in 2D simulations [581] that the flow is not completely in the steady state, but a dome-like shock is formed quasiperiodically in front of the compact object for higher Mach cases. 2D and 3D simulations in [677] have shown that at higher resolution the "flip-flop" instability is developed in the flow, which appears to be stable at low resolution. This instability results in shock cone oscillations from side to side, accompaned by brief periods of disk formation. It was suggested that accretion flow instabilities are intrinsic to all accretion flows. Calculations of the same problem made in [806] using another numerical method have led to construction of stationary solutions in the range of parameters, where instabilities have been obtained in previous calculations. In the complicated numerical situation with a singular point in the centre, additional analysis is needed for studying the nature of this instability and influence of the numerical effects. The luminosity in this case depends on magnetic field and should be estimated analogously to the case of a spherical accretion in Sect. 11.79.

11.3.4 Disk Accretion in Binaries

A large angular momentum in close binaries causes the accreting matter to take the form of a disk, though at a high velocity of the stellar wind from the normal star it is possible under certain conditions that the disk will not form and a conical accretion will take place instead [517]. A theory of disk accretion has been proposed in [657, 816, 897] and developed in [755, 898] under the assumption of a stationary thin disk and in the presence of turbulent viscosity. General relativity equations have been treated in [657, 755, 833], other studies use Newtonian theory. General relativity can be included in a semi-quantitative way by adopting [785]

$$\varphi_G = \frac{GM}{r - r_g}, \quad r_g = \frac{2GM}{c^2}. \tag{11.114}$$

To model the boundary conditions on the inner edge of the disk, the advective terms, representing the radial velocity terms, as well as derivatives of pressure and entropy along the disk radius should be taken into account. For a low radial velocity v_r, the equations of stationary disk accretion will be written as ([783], see also [158])

$$-2\pi r \Sigma v_r = \dot{M}, \quad \text{for conservation of mass,} \tag{11.115}$$

$$\dot{M}(l - l_{in}) = 4\pi r^2 \alpha P h, \quad \text{for conservation of angular momentum,} \tag{11.116}$$

$$\dot{M}\left[-(l - l_{in})\frac{d\Omega}{dr} + B_1 T \frac{dS}{dr}\right] = B_3 4\pi r \frac{acT^4}{3\kappa\rho h},$$

$$\text{for conservation of energy,} \tag{11.117}$$

$$h^2 \Omega_K^2 = B_4 \frac{P}{\rho} \quad \text{for equilibrium along the disk depth,} \tag{11.118}$$

$$\frac{1}{\rho}\frac{\mathrm{d}P}{\mathrm{d}r} = r\left(\Omega^2 - \Omega_K^2\right) \quad \text{for equilibrium along the radius.} \tag{11.119}$$

Here, $\rho, T, P, h, \Sigma, \dot{M}$ are defined above, and $\Omega, l = \Omega r^2$ are the current angular velocity and angular momentum, $\alpha < 1$ is the turbulent viscosity parameter,

$$\Omega_K = \left(-\frac{1}{r}\frac{\mathrm{d}\varphi_{\mathrm{G}}}{\mathrm{d}r}\right)^{1/2} = \left(\frac{GM}{r^3}\right)^{1/2}\frac{r}{r - r_{\mathrm{g}}}$$

is the Keplerian angular velocity, (11.120)

$$l_{\mathrm{in}} = \left(\frac{1}{3}r^4\frac{\mathrm{d}^2\varphi_{\mathrm{G}}}{\mathrm{d}r^2}|_{r=r_{\mathrm{in}}}\right)^{1/2} = \left(\frac{2}{3}GMr_{\mathrm{in}}\right)^{1/2}\left(\frac{r_{\mathrm{in}}}{r_{\mathrm{in}} - r_{\mathrm{g}}}\right)^{3/2}. \tag{11.121}$$

The value l_{in} is the angular momentum at the innermost stable orbit, $r_{\mathrm{in}} = 3r_{\mathrm{g}}$ for the potential (11.114). This point simultaneously happens to be an extremum and bending point of the total (gravitational plus centrifugal) potential. In the α-disk approximation [897], the stress-tensor component $t_{r\varphi}$ is supposed to be related to pressure by (see also Problem 3)

$$t_{r\varphi} = \mu r\frac{\mathrm{d}\Omega}{\mathrm{d}r} = -\alpha P. \tag{11.122}$$

For a fully ionized opaque gas including radiation pressure, we have pressure, entropy and molecular weight as

$$P = \frac{\rho k T}{\mu m_{\mathrm{u}}} + \frac{1}{3}\alpha T^4, \quad S = \frac{k}{\mu m_{\mathrm{u}}}\ln(T^{3/2}/\rho) + \frac{4}{3}\frac{aT^3}{\rho} + \text{const.}$$

$$\mu \simeq \left[2x_{\mathrm{H}} + \frac{3}{4}x_{\mathrm{He}} + \frac{1}{2}x_A\right]^{-1}, \qquad x_A = \sum_{i\geq 6} x_i, \tag{11.123}$$

where x_k are weight fractions of different elements. The opacity is determined by the free–free and free–bound transitions, and electron scattering, and are given in (11.102) and (9.41). The dimensionless constants B_i are determined by the density distribution over the disk thickness. The following were obtained in [717, 783]

$$B_1 = 0.5, \quad B_3 = B_4 = 6. \tag{11.124}$$

Far from the inner edge of the thin disk, the terms with dS/dr and dP/dr in (11.117) and (11.119) are small and a standard theory of disk accretion [755, 897, 898] corrected for a potential implying general relativity properties follows from (11.115)–(11.124), with the inner radius $r_{\mathrm{in}} = 3r_{\mathrm{g}}$, as in the Schwarzschild metric [1081].

When dP/dr is taken into account for equilibrium, we must add the term $\int dP/\rho$ to the total potential. The inner edge corresponding to the last stable orbit neglecting the radial motion is determined by

$$r_{\rm in} = 3r_{\rm g} - 3\,\frac{(r_{\rm in} - r_{\rm g})^3}{GM}\left[\left(\frac{1}{\rho}\frac{{\rm d}P}{{\rm d}r}\right) + r\frac{\rm d}{{\rm d}r}\left(\frac{1}{\rho}\frac{{\rm d}P}{{\rm d}r}\right)\right]_{\rm in}. \tag{11.125}$$

The choice of a pressure on the inner edge of the disk is somewhat arbitrary in the given formulation of the problem, since with decreasing radius and approaching the last stable orbit the velocity $v_{\rm r}$ tends to the sound velocity $v_{\rm s}$, and the system of equations becomes more complicated [476, 477, 716, 717]. The condition $v_{\rm r} = v_{\rm s}$ is adopted for the inner edge $r = r_{\rm in}$, $r_{\rm in}$ may then be obtained as an eigenvalue for a set of differential equation with respect to S, P and Ω. For $M = 10M_\odot$, $\dot M = \dot M_{\rm cr} = 64\pi GM/c\kappa_{es}$, $x_{\rm H} = 0.7$ and $\alpha = 0.001$, it has been found in [783] that the ratio $r_{\rm in}/r_{\rm g} = 2.885$.

When a full system of equations is solved, including radial velocity terms, the point where $v_{\rm r} = v_{\rm s}$ is a critical point of the equations of a saddle type, and the eigenvalue for $r_{\rm in}$ is determined by the condition for the solution to pass this point.

As shown in [897,898], there are three characteristic zones in a thin accretion disk without advection (Fig. 11.22). When $\dot M > (1/20 - 1/50)\dot M_{\rm cr}$ (1/20 corresponds to convective, 1/50 to radiative energy transfer along z [120]), there exists a radiative zone 1 with $P = P_{\rm r} \gg P_g$ and $\kappa = \kappa_{es}$. Zone 2 with $P = P_g \gg P_{\rm r}$, $\kappa = \kappa_{es}$ and zone 3 with $P = P_g \gg P_{\rm r}$, $\kappa = \kappa_{\rm ff} + \kappa_{\rm fb}$ are located at larger radii. With decreasing $\dot M$ the radiative zone vanishes, and the boundary between the radiative zones 2 and 3 shifts towards the centre. If the term with dS/dr is neglected, all the gravitational energy released in the disk is emitted from the surface. Taking the spectrum to be locally black-body with effective temperature $T_{\rm ef}$, we obtain

$$\pi racT_{\rm ef}^4 = -\dot M\,(l - l_{\rm in})\,\frac{{\rm d}\Omega}{{\rm d}r}. \tag{11.126}$$

Using (11.120) and (11.121), we have, putting $\Omega = \Omega_k$ [158]

$$T_{\rm ef}(r) = 2.05 \times 10^7 \left(\frac{\dot M}{\dot M_{\rm cr}}\frac{M_\odot}{M}\right)^{1/4}\left(\frac{3r_{\rm g}}{r}\right)^{3/4}\varphi^{1/4}_{r,r_{\rm in}}, \tag{11.127}$$

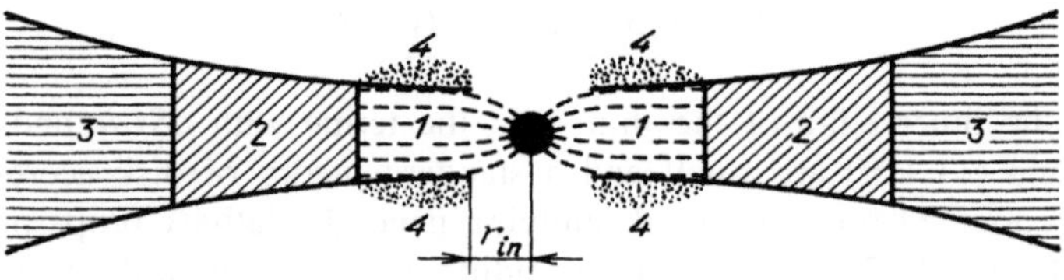

Fig. 11.22 Disk accretion onto black hole. 1 is the radiative zone with $P = P_{\rm r}$, $\kappa = \kappa_{\rm es}$, 2 is the gas zone with scattering, $P = P_g$, $\kappa = \kappa_{\rm es}$, 3 is the gas zone with bremsstrahlung processes, $P = P_g$, $\kappa = \kappa_{\rm ff} + \kappa_{\rm fb}$, 4 is the corona with $T = 10^8$–10^9 K (from [158])

where

$$\varphi_{r,r_{\rm in}} = 1 - \sqrt{3r_{\rm g}/r} \quad \text{for the Newtonian potential}$$
$$= \left(1 - \sqrt{3r_{\rm g}/r}\right)^2 (1 - r_{\rm g}/3r)\left(1 + \frac{1}{2}\sqrt{3r_{\rm g}/r}\right)(1 - (r_{\rm g}/r))^{-3}$$
$$\text{for the potential (11.114).} \qquad (11.128)$$

The total disk spectrum has the power form $F_\omega \sim \omega^{1/3}$, similar to (11.110), with an exponential cut-off at

$$\omega = \frac{kT_{\rm ef}(\beta r_{\rm in})}{\hbar} = 2.7 \times 10^{18}\left(\frac{\dot{M}}{M_{\rm cr}}\frac{M_\odot}{M}\right)^{1/4} \beta^{-3/4}\varphi^{1/4}_{\beta r_{\rm in},r_{\rm in}}. \qquad (11.129)$$

Here, $r = \beta r_{\rm in}$ corresponds to the maximum $T_{\rm ef}$, $\beta = 49/36$, $\varphi_{\beta r_{\rm in},r_{\rm in}} = 1/7$, $\beta^{-3/4}\varphi^{1/4} = 0.49$ for a Newtonian potential; $\beta = 1.59$, $\varphi_{\beta r_{\rm in},r_{\rm in}} = 0.113$, $\beta^{-3/4}$ $\varphi^{1/4} = 0.41$ for the potential (11.114).

As shown in [120, 162], the innermost region of the disk (zone 1 in Fig. 11.22) is unstable to convection. The mechanical transfer of the energy generated by convection leads to the formation above the radiative zone of a hot ($T = 10^8 - 10^9$ K) corona with $n \sim 10^{15}$ cm^{-3}, see also [634]. The inverse Compton interaction between thermal radiation of the disk and hot electrons of the corona leads to a power form of dependence in the hard end of the spectrum [341,949] and may thus account for the observable hard (up to ~ 150 KeV) portion of the CygX-1 spectrum.

It was shown in [900] that in addition to the solution for optically thick disk, presented in the Problem 2, there is another solution, where disk is optically thin and has much higher temperature. The viscous dissipation heats mainly protons, and radiation is produced mainly by electrons. In the optically thin case, electrons gain their energy from collisions with protons. At low density, the energy exchange between them by binary collisions is slow, so protons could become much hotter than electrons and carry into a black hole almost all the energy. This could decrease the efficiency of accretion up to $\sim 10^{-4}$, which is why this regime was called [734] advection-dominated accretion flow (ADAF). The presence of a magnetic field leading to both the heating of matter (electrons and protons) during its dissipation (11.79) and the enhanced energy exchange between electrons and protons in a turbulent plasma should substantially increase the efficiency of accretion and does not let it become less than $1/4$ of the corresponding value for an optically thick disk [128, 129]. Far from the stellar surface or from the Alfven surface of radius r_A, where $B^2/8\pi = (1/2)\rho v^2$, the accretion onto a neutron star or white dwarf is not distinct from that onto a black hole. Near the magnetosphere at the radius r_A, or at the stellar surface when the field is low, the accretion pattern becomes markedly complicated because of the interaction of the accretion flow with the magnetosphere [644] and/or the surface. Studies of accretion onto neutron stars are reviewed in [158, 899], and onto white dwarfs in [17].

When the magnetic field of a neutron star is strong, and $r_A > R_{\rm ns}$ (neutron star radius), the matter halts at the Alfven surface and flows along magnetic lines towards the magnetic poles. The emission of such a star is anisotropic, and its rotation gives the observed pattern of an X-ray pulsar. When the field is weak and $r_A < R_{\rm ns}$, no hot spot forms at the poles and no X-ray pulsar emerges. The accretion disk approaches the neutron star surface and makes it rotate rapidly, giving rise to a millisecond radiopulsar after the accretion has ceased [21].

During the disk accretion into a slowly rotating star, a boundary layer (BL) is formed with rapid transition from Keplerian to stellar angular velocity, where a substantial part of the energy is released. When the star accelerates its rotation the energy release inside the BL decreases, and finally it disappears for critically rotating stars. The accretion disk around a critically rotating star increases its luminosity several times, compared to the case of BL existence [110].

11.3.5 Accretion Disc Structure with Optically Thin/Thick Transition

Equations describing accretion disks with continuous description of optically thick/thin transition have been derived in [50]. To find a description valid in both limiting cases and also between them, the Eddington approximation was used for obtaining a formula for the heat flux which may be used instead of (11.256) which is valid only for the optically thick case.

Suppose that the disc is geometrically thin and has a constant density along the z-axis. Defining S as the energy density of the radiation, $F_{\rm rad}$ as the radiation flux in the z-direction and $P_{\rm rad}$ as the radiation pressure, we write first two momentum equations for radiative transfer for frequency-independent α_a and σ_T in the form

$$\frac{dF_{\rm rad}}{dz} = -\rho c \alpha_a a T^4 \left(\frac{S}{aT^4} - 1\right), \tag{11.130}$$

$$c\frac{dP_{\rm rad}}{dz} = -\sigma_T \rho F_{\rm rad}. \tag{11.131}$$

Consider the case when the scattering opacity σ_T is much larger than the absorption opacity α_a, and suggest that the heat production rate is proportional to the mass density ρ. Then, neglecting the flux in the radial direction, we obtain

$$F_{\rm rad} = 2\frac{F_0}{\Sigma_0}\rho z, \tag{11.132}$$

where F_0 is the flux from the unit surface of the disc at $z = h$, and surface density

$$\Sigma_0 = 2\rho h. \tag{11.133}$$

Substituting (11.132) into (11.130), we obtain

$$S = 3P_{\text{rad}} = aT^4\left(1 - \frac{2F_0}{c\alpha_a aT^4\Sigma_0}\right). \tag{11.134}$$

Using (11.132) in (11.131), we obtain

$$c\frac{\mathrm{d}P_{\text{rad}}}{\mathrm{d}z} = -2\sigma_T\frac{F_0}{\Sigma_0}\rho^2 z. \tag{11.135}$$

Introducing the scattering optical depth

$$\tau = \int_z^\infty \sigma_T\rho\mathrm{d}z = \sigma_T\rho(h-z) = \tau_0 - \sigma_T\rho z,$$

$$\tau_0 = \sigma_T\rho h = \frac{1}{2}\sigma_T\Sigma_0, \tag{11.136}$$

we rewrite (11.135) in the form

$$c\frac{\mathrm{d}P_{\text{rad}}}{\mathrm{d}\tau} = 2\frac{F_0}{\Sigma_0}\frac{\tau_0-\tau}{\sigma_T}. \tag{11.137}$$

Solving (11.137) with the following boundary condition

$$F_{\text{rad}}|_{\tau=0} = F_0 = \frac{cS|_{\tau=0}}{2} = \frac{3cP_{\text{rad}}|_{\tau=0}}{2}, \tag{11.138}$$

results in

$$cP_{\text{rad}} = F_0\left(\frac{2}{3} + \tau - \frac{\tau^2}{2\tau_0}\right). \tag{11.139}$$

In the symmetry plane of the disc at $\tau = \tau_0$, we have

$$cP_{\text{rad,c}} = F_0\left(\frac{2}{3} + \frac{\tau_0}{2}\right). \tag{11.140}$$

Using (11.140) in (11.134), we obtain in the symmetry plane, where $T = T_c$

$$F_0 = caT_c^4\left(2 + \frac{3\tau_0}{2} + \frac{1}{\tau_{\alpha 0}}\right)^{-1}, \tag{11.141}$$

where the absorption optical depth

$$\tau_\alpha = \int_z^\infty \alpha_a\rho\mathrm{d}z = \alpha_a\rho(h-z) = \tau_{\alpha 0} - \alpha_a\rho z,$$

$$\tau_{\alpha 0} = \alpha_a\rho h = \frac{1}{2}\alpha_a\Sigma_0. \tag{11.142}$$

Introducing the effective optical depth

$$\tau_* = (\tau_0 \tau_{\alpha 0})^{1/2}, \tag{11.143}$$

we finally obtain the expressions for the vertical energy flux from the disc F_0 (analogous to Q_- in (11.256)) and the radiation pressure in the symmetry plane as

$$F_0 = \frac{2acT_c^4}{3\tau_0}\left(1 + \frac{4}{3\tau_0} + \frac{2}{3\tau_*^2}\right)^{-1}, \tag{11.144}$$

$$P_{\text{rad,c}} = \frac{aT_c^4}{3}\frac{1 + (4/3\tau_0)}{1 + (4/3\tau_0) + (2/3\tau_*^2)}. \tag{11.145}$$

At $\tau_0 \gg \tau_* \gg 1$, we have (11.256) from (11.144). In the optically thin limit $\tau_* \ll \tau_0 \ll 1$, we obtain

$$F_0 = acT_c^4 \tau_{\alpha 0}, \tag{11.146}$$

$$P_{\text{rad,c}} = \frac{2}{3} acT_c^4 \tau_{\alpha 0}. \tag{11.147}$$

Using F_0 instead of Q_- and the equation of state $P = \rho \mathcal{R} T + P_{\text{rad,c}}$, the algebraic equations (11.253), (11.255), (11.257) and (11.268), together with equation

$$Q_+ = F_0 \tag{11.148}$$

with Q_+ from (11.254), have been solved numerically in [50].[14]

It occurs that two solutions, optically thick and optically thin, exist separately when luminosity is not very large. Two solutions intersect at $\dot{m} = \dot{m}_{\text{b}}$ and there is no global solution for accretion disc at $\dot{m} > \dot{m}_{\text{b}}$ (see Fig. 11.23). It is concluded in [50] that to obtain a global physically meaningful solution at $\dot{m} > \dot{m}_{\text{b}}$, an advective term in (11.148), as in (11.117), needs to be accounted for.

11.3.6 Black Hole Advective Accretion Disks with Optical Depth Transition

In the widely used standard model of accretion disks around black holes [755, 897, 898] it is assumed that an accretion flow has a small radial velocity, and the disk has a small geometrical, large optical thicknesses, and rotates with a nearly Keplerian angular velocity. The solution of the structure of the "standard" model accretion disk is presented in Problem 2. For the optically thick and low accretion rate disks

[14] Viscosity in α presentation was used in [50] in the form (11.284) instead of the original (11.122) from [897].

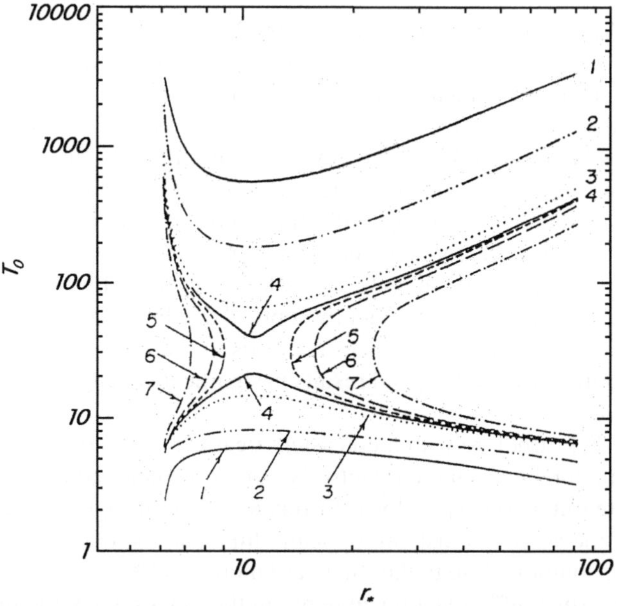

Fig. 11.23 The dependence of the optical depth τ_0 on the radius r_* for the case of the black hole mass $M_{BH} = 10^8 M_\odot$, $\alpha = 1$ and various values of $\dot{m}$. The curves are numbered in increasing order $\dot{m} = 1.0, 3.0, 8.0, 9.35, 10.0, 11.0$ and 15.0. The upper curves correspond to the optically thick family, the lower curves correspond to the optically thin family. The non-dimensional radius r_* and mass flux $\dot{m}$ are defined in (11.267), $r_* \equiv x$, and τ_0 is defined in (11.136)

with $\dot{M} \ll \dot{M}_{\rm Edd}$, where $\dot{M}_{\rm Edd} = L_{\rm Edd}/c^2$ is the Eddington accretion rate, these assumptions are generally considered as being reasonable. At higher accretion rate flows at $\dot{M} \gtrsim \dot{M}_{\rm Edd}$, the advection should crucially modify the properties of the innermost parts of the standard disks – the disks become hotter and thicker, and their rotation laws deviate from the Keplerian law. The system of equations (11.115)–(11.119) considered in [783] included only the effects of heat advection and the radial pressure gradient. The importance of the transonic nature of accretion flows was emphasized in [635]. It was shown in [678] that for $\alpha \gtrsim 0.2$, and viscosity prescription $t_{r\phi} = \alpha P$, the singular point, where the radial velocity approximately equals the sound velocity, is changed from a saddle to a nodal type, with increase in the accretion rate. Extensive investigation of accretion disk models with advection for a wide range of the parameters $\dot{M}$ and α was conducted in [4]. The advection problem was investigated in [263] involving shock waves near the innermost disk region by considering accretion through saddle points.

As follows from the solution in the previous section, neglecting the advective heat transport in the high- $\dot{M}$ optically thick disks leads to qualitatively wrong conclusions about the topology of the family of solutions of the disk structure. Without the advection terms, these equations give rise to two branches of solutions, optically thick and optically thin solutions, which do not intersect if $\dot{M} < \dot{M}_{\rm cr} \approx$

$(0.6 - 0.9)\dot{M}_{\rm Edd}$ for $\alpha = 1$ and $M_{BH} = 10^8 M_\odot$ [50], see Fig. 11.23. For larger accretion rates, there are no solutions for these equations extending continuously from large to small radii and having the Keplerian boundary conditions at the outer boundary of the disk (see also [636, 655, 1030]). It was argued in [50] that for the accretion rates larger than $\dot{M}_{\rm cr}$, advection becomes critically important and would allow solutions that are extended all the way towards the inner disk edge to exist. The transonic solutions for the optically thick advective accretion disks was constructed in [51]. However, for some choices of $\dot{M}$ and α, these solutions were not consistent in the respect that in spite of the large vertical Thomson (or scattering) optical depths, the corresponding effective optical depths (11.143) become less than unity in the inner disk regions. This effect leads to violation of the optically thick disk approximation used in [51] and results in significant changes in the structure of the inner regions.

Advective disk structure with a consistent treatment that correctly describes the intermediate region between the effectively optically thick and thin disk from [50] was constructed numerically in [52]. In this solution, at high accretion rates, the inner parts of the accretion disk are optically thin, and the outer ones are optically thick, with the solution close to the "standard model" [898].

In spite of a small effective optical depth in the inner region, the radiation pressure (11.145), being considerably less than the thermodynamic equilibrium one, nevertheless dominates over the gas pressure (Fig. 11.29). The solution with the optically thick/thin transition is the only equilibrium solution for the high-luminosity one-temperature advective disk. Therefore, it is stable to the global transitions, contrary to the stationary solution from [900], which is unstable, and exists at small and moderate luminosities simultaneously with a stable optically thick model. The considered models always have large Thomson optical depths (11.136), and all the discussion [52] regarding the optical depth transitions deals with the effective optical depths. In the study of transonic models, the radial pressure gradient and radial drift velocity were included, and a non-Keplerian rotation was allowed.

The geometrically thin disk approximation is used correctly, because the relative thickness of the disk in the considered models is always less than unity. Using of the one-temperature approximation for the accreting plasma is consistent with the estimates of the short electron-ion Coulomb coupling time in comparison with the long accretion timescale. Note that the real coupling time could be even shorter if the relaxation due to plasma instabilities is more efficient than that due to Coulomb collisions. Global solutions extending from large radii to the inner edge of the disk are constructed for a wide range of accretion rates considerably larger than $\dot{M}_{\rm cr}$, and the smooth transition between the optically thick and thin regions appears when the mass accretion rate is close to $\dot{M}_{\rm cr}$. The temperature in the optically thin region can be increased by up to a factor of 30 in comparison with the solutions obtained using the optically thick approximation. This high-temperature inner disk region can explain the emission of the hard X-ray tails observed in spectra of bright X-ray sources with black hole components such as CygX-1, LMCX-3 and LMCX-1 [284]. These tails are interpreted in different models, such as the model of Comptonization in the hot disk corona [162] or the (Thomson) optically thin disk [900]. Hot optically thin regions appear in the central parts of highly luminous accretion disks, and the hard X-ray tails in such objects are an inevitable consequence of this disk equilibrium state.

11.3.6.1 Basic Equations

The one-dimensional, height-averaged stationary accretion disk model, with account of the effect of advection, and arbitrary vertical optical depth that includes optically thin, intermediate and optically thick accretion regimes, is described by the following equations [52]:
The mass conservation equation is

$$\dot{M} = 4\pi r h \rho v, \tag{11.149}$$

the radial and angular momentum equations are

$$v\frac{\mathrm{d}v}{\mathrm{d}r} = -\frac{1}{\rho}\frac{\mathrm{d}P_{\mathrm{tot}}}{\mathrm{d}r} + (\Omega^2 - \Omega_K^2)r, \tag{11.150}$$

$$\frac{\dot{M}}{2\pi}(l - l_{\mathrm{in}}) = 2hr^2\alpha P_{\mathrm{tot}}, \tag{11.151}$$

and the internal energy equation is

$$Q_{\mathrm{adv}} = Q^+ - Q^-, \tag{11.152}$$

where

$$Q_{\mathrm{adv}} = -\frac{\dot{M}}{4\pi r}\left[\frac{\mathrm{d}E}{\mathrm{d}r} + P\frac{\mathrm{d}}{\mathrm{d}r}\left(\frac{1}{\rho}\right)\right], \tag{11.153}$$

$$Q^+ = -\frac{\dot{M}}{4\pi}r\Omega\frac{\mathrm{d}\Omega}{\mathrm{d}r}\left(1 - \frac{l_{\mathrm{in}}}{l}\right), \tag{11.154}$$

$$Q^- = \frac{2aT^4c}{3\kappa\rho h}\left(1 + \frac{4}{3\tau_0} + \frac{2}{3\tau_*^2}\right)^{-1}, \tag{11.155}$$

are the heat advection, viscous dissipation and radiative cooling rates per disk unit surface, respectively. In equations (11.149)–(11.155), it is assumed that ρ, T, v, Ω, P_{tot} and E are the midplane density, temperature, radial velocity, angular velocity, total (gas plus radiation) pressure, and specific internal energy, respectively. Other quantities have the following notations: κ is the Thomson opacity, a is the radiation constant, $\Omega_K = \sqrt{GM/r(r-2r_{\mathrm{g}})^2}$ is the Keplerian angular velocity in the framework of the pseudo-Newtonian approach (11.114), M is the black hole mass, $2r_{\mathrm{g}} = 2GM/c^2$ is the gravitational radius, $\ell = \Omega r^2$ is the specific angular momentum, ℓ_{in} is a constant defined by the global solution, $h = c_{\mathrm{s}}/\Omega_K$ is the disk half-thickness and $c_s = \sqrt{P_{\rm tot}/\rho}$ is the midplane 'isothermal' sound speed. The accretion rate $\dot{M}$ and velocity v are assumed to be positive in these equations. The value of $\ell_{\rm in}$ has an obvious physical meaning: it is an asymptotic angular momentum of accretion matter at the black hole horizon [515]. In equation (3), following [897], the viscosity prescription is used in which the main $(r\phi)$ component of the viscous stress tensor $\mathbf{t}_{ik}$ is proportional to the total pressure,

$$t_{r\phi} = \alpha P_{\rm tot}. \tag{11.156}$$

where $\alpha \lesssim 1$ is a parameter. The radiative cooling term (11.155) can be applied to optically thin and thick regimes, where $\tau_0 = \kappa\rho h$ is the total Thomson scattering depth of the disk, and

$$\tau_* = [(\tau_0 + \tau_\alpha)\tau_\alpha]^{1/2} \tag{11.157}$$

is the effective optical depth. Here, τ_α is the optical depth with respect to bremsstrahlung absorption,

$$\tau_\alpha \simeq 5.2 \cdot 10^{21} \frac{\rho^2 T^{1/2} h}{acT^4}. \tag{11.158}$$

In the disk models considered below, the inequality $\tau_0 \gg \tau_\alpha$ is valid. The equation of state of the accretion matter consisting of a gas and radiation mixture reads

$$P_{\rm tot} = P_{\rm gas} + P_{\rm rad}, \tag{11.159}$$

where the gas pressure $P_{\rm gas}$ is as for an ideal gas with the gas constant $\mathscr{R}$,

$$P_{\rm gas} = \rho\mathscr{R}T, \tag{11.160}$$

and the radiation pressure $P_{\rm rad}$ is represented by the bridging formula for the optically thick and thin regimes [50],

$$P_{\rm rad} = \frac{aT^4}{3}\left(1 + \frac{4}{3\tau_0}\right)\left(1 + \frac{4}{3\tau_0} + \frac{2}{3\tau_*^2}\right)^{-1}. \tag{11.161}$$

The internal energy density of the mixture takes the form,

$$\rho E = \frac{3}{2}P_{\rm gas} + 3P_{\rm rad}. \tag{11.162}$$

11.3.6.2 Singular Points and Uniqueness of Solutions

The system of equations (11.149)–(11.152) can be reduced to the following system of two ordinary differential equations,

$$x\frac{\tilde{v}'}{\tilde{v}} = \frac{N}{D}, \tag{11.163}$$

$$x\frac{\tilde{c}_s{}'}{\tilde{c}_s} = 1 + \frac{x^2}{c_s^2}\left[\tilde{\Omega}^2 - \frac{1}{x(x-2)^2}\right] + \frac{3x-2}{2(x-2)} - \left(\frac{\tilde{v}^2}{\tilde{c}_s^2} - 1\right)\frac{N}{D}, \tag{11.164}$$

where the notations N and D are used to denote algebraical expressions, which depend only on r, v, c_s and ℓ_{in}. Explicit forms of N and D are given in [52]. All quantities in equations (11.163) and (11.164) have been converted to the dimensionless form, with variables

$$x = \frac{r}{r_g}, \quad \tilde{v} = \frac{v}{c}, \quad \tilde{c}_s = \frac{c_s}{c}, \quad \tilde{\Omega} = \frac{\Omega r_g}{c}, \quad \tilde{\ell} = \frac{\ell}{r_g c}, \quad m = \frac{M}{M_\odot}. \tag{11.165}$$

Singular points of equations (11.163)-(11.163) are defined by the condition

$$D\,|_{r=r_s} = 0, \tag{11.166}$$

where r_s is the location of the singular point. Continuous solutions, which smoothly go through the singular points, also have to satisfy the regularity condition at these points,

$$N\,|_{r=r_s} = 0. \tag{11.167}$$

The type of the singular points must be consistent with a transonic nature of the global steady-state solutions. Typically, there is at least one 'inner' singular point in the global accretion disk solutions, which is associated with the sonic point near the black hole horizon at $r_s \approx 6r_g$ (e.g., see [783]). This point can be a saddle or nodal-type point [678]. Any global solution, which is subsonic at large radii and supersonic near the black hole horizon, must go through the inner singular point. The solutions, which have a saddle inner singular point, are obviously unique. Uniqueness of the solutions with a nodal inner singular point requires an additional study. Some formulations of the equations for the accretion disk structure, including our formulation, result in appearance of the second 'outer' singular point located in the subsonic region of the accretion disk solution at $r \gg 6r_g$. It was found in [51] that the outer singular points are always saddle points. Therefore, these global solutions with two singular points are unique independently of the type, saddle or nodal, of the inner singular point. For a given black hole mass, the global solution is determined by the two parameters, $\dot{M}$ and α. The parameter ℓ_{in}, implicitly present in equations (11.163) and (11.164) via the expression for Ω (see (11.151)

$$\Omega = \frac{\ell_{in}}{r^2} + \alpha\frac{c_s^2}{vr}, \tag{11.168}$$

is not an independent parameter, but it is an eigenvalue of the problem, i.e. ℓ_{in} is determined by the global solution.

11.3.6.3 Method of Solution

A finite-difference method was used to solve equations (11.163) and (11.164). The basic features of this method were outlined in [515] and [51]. The method uses a standard numerical discretization technique to reduce the system of differential equations to a system of nonlinear algebraic equations, which are then solved using the Newton–Raphson iteration scheme. Important ingredients of the method, introduced in [51], are that it involves an expansion of a solution around the inner singular point and uses $l_{\rm in}$ as an independent variable in the iteration scheme.

The solutions do not explicitly depend on the outer and inner boundary conditions. Formally, the numerical method requires specification of the outer boundary values for v and $c_{\rm s}$. However, it was found that the specific values of v and $c_{\rm s}$ are not important, because the solution quickly settles down on the separatrix, which behavior does not depend on the outer boundary values (see [51]). The behavior of the solutions in the inner region is defined by the regularity conditions (11.166) and (11.167) at the inner singular point, and therefore no explicit inner boundary conditions are required. Note that the analytical forms of of the partial derivatives of the numerator N and denominator D have been used to accurately treat the inner singular point in the iteration procedure [51]. The Mathematica system [1044] was used to perform the needed extended algebraic calculations.

11.3.6.4 Numerical Results and Physical Effects

Solutions of three types of accretion disk models have been computed : (1) non-advective models with optical depth transition, (2) advective models with optical depth transition and (3) advective models using the optically thick approximation only. In the case of the small accretion rates, $\dot{m} \lesssim 0.1$, where $\dot{m} = \dot{M}c^2/L_{\rm Edd}$, the models of all types are very similar at all radial distances. When increasing the accretion rates, the models of different types deviate significantly from each other at the small radial distances, whereas they remain similar at the larger radial distances. A comparative analysis of the obtained models is given below, focusing on the models with large accretion rates, $\dot{m} \gtrsim 10$.

Solutions of equations (11.149)–(11.152) without the advection terms were obtained in [50] (Sect. 11.3.5) using the Newtonian potential. Similar solutions are reproduced below using the pseudo-Newtonian potential. The solutions of this type are global, i.e. they exist for the all radial range, only if the accretion rates $\dot{m} < \dot{m}_{\rm cr}$, where the critical accretion rate $\dot{m}_{\rm cr}$ depends on α and the black hole mass. In particular, $\dot{m}_{\rm cr} = 36$ in the case of $\alpha = 0.5$ and $M_{BH} = 10M_{\odot}$, and $\dot{m}_{\rm cr} = 9$ in the case of $\alpha = 1$ and $M_{BH} = 10^8 M_{\odot}$. Figure 11.24 shows the radial dependence of the Thomson scattering depth (left) and the effective optical depth (right) for models with $\alpha = 0.5$, $M_{BH} = 10M_{\odot}$, and different $\dot{m}$. Dashed lines represent two families of the global models with $\dot{m} < \dot{m}_{\rm cr}$, which will be called as the 'optically thick' and 'optically thin' families. The three upper lines on the left panel correspond to the optically thick family with $\dot{m} = 10, 20, 30$ (*from top to bottom*).

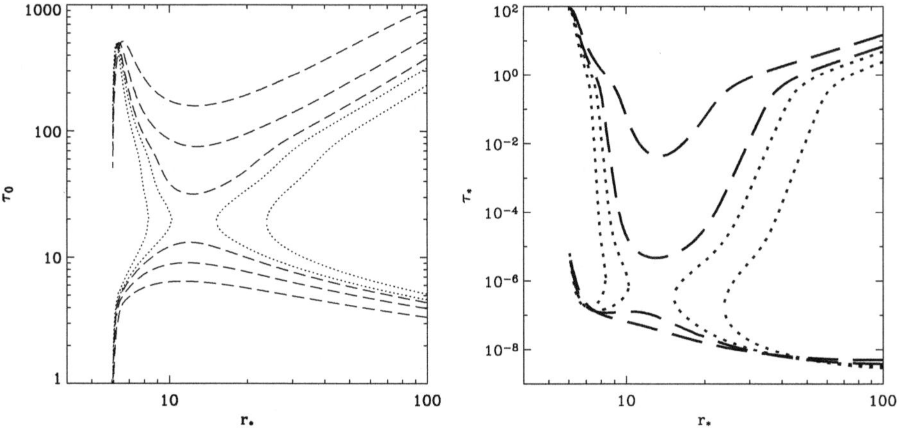

Fig. 11.24 Radial dependence of the Thomson scattering depth τ_0 (left panel), and the effective optical depth τ_* (right panel) for the models with $\alpha = 0.5$ and $M_{BH} = 10M_{\odot}$ constructed neglecting the radial heat advection. Dashed lines correspond to the solutions with $\dot{m} < \dot{m}_{cr} = 36$. The three upper lines on the left panel correspond to the optically thick family of solutions with $\dot{m} = 10, 20$ *and* 30 (from top to bottom). The three lower lines correspond to the optically thin family of solutions with the same $\dot{m} = 10, 20$ and 30 (from bottom to top). On the right panel, the two upper dashed lines correspond to the optically thick solutions with $\dot{m} = 20$ *and* 30 (from top to bottom) and the two lower lines correspond to the optically thin solutions with $\dot{m} = 20$ *and* 30 (from bottom to top). Dotted lines on the left and right panels correspond to the unphysical solutions with $\dot{m} = 36$ (left and right inner lines) and $\dot{m} = 50$ (left and right outer lines), from [52].

The three lower dashed lines correspond to the optically thin family and have the same $\dot{m} = 10, 20, 30$ (*from bottom to top*). On the right panel, the two upper dashed lines correspond to the optically thick family with $\dot{m} = 20, 30$ (*from top to bottom*), and the two lower lines correspond to the optically thin family with $\dot{m} = 20, 30$ (*from bottom to top*). Note that the effective optical depth of the optically thin family of the solutions is always very small, whereas the optical depth of the optically thick family is usually large and becomes less than unity only in the limited ranges of radii $\simeq 10r_g$–$40r_g$ and accretion rates $\dot{m} \simeq 10$–30. When $\dot{m} > \dot{m}_{cr}$, there are no global solutions. Dotted lines in Figure 11.24 (*left and right panels*) show solutions for $\dot{m} = 36$ (*two inner left and right lines*) and $\dot{m} = 50$ (*two outer left and right lines*). These solutions are clearly unphysical.

Solutions of equations (11.149)–(11.152) with the advection terms were obtained only in the case of the optically thick family. These solutions coincide closely with the non-advective solutions of the optically thick family at large radii, $r \gtrsim 10^2 r_g$. At smaller radii, however, there are significant deviations between the advective and non-advective solutions. In Figures 11.25 and 11.26, one can compare the radial profiles τ_0 and τ_* for the advective and non-advective solutions in the case when $\dot{m}$ is close to $\dot{m}_{cr}$. Figure 11.25 (*left and right panels*) shows by dashed lines the same non-advective global solutions with $\dot{m} = 30 < \dot{m}_{cr}$ that are shown in Fig. 11.24,

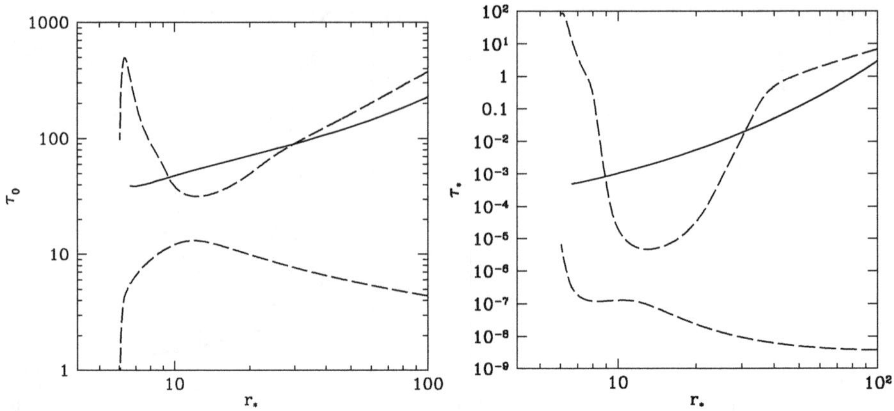

Fig. 11.25 Radial dependence of the Thomson scattering depth τ_0 (*left panel*) and the effective optical depth τ_* (*right panel*) for the model with $\alpha = 0.5$, $M_{BH} = 10M_\odot$, and $\dot{m} = 30$. The *dashed lines* correspond to the solutions without advection (see the caption of Fig. 1 for details of notations). The *solid lines* correspond to the solution with advection and optical depth transition, from [52]

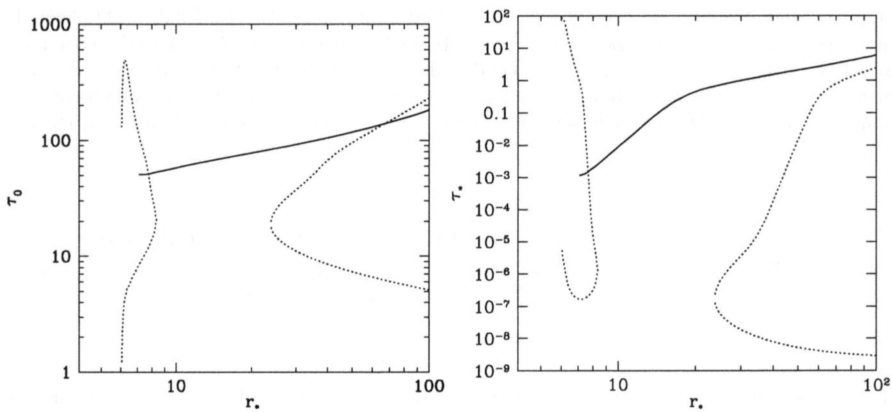

Fig. 11.26 Radial dependence of the Thomson scattering depth τ_0 (*left panel*) and the effective optical depth τ_* (*right panel*) for the model with $\alpha = 0.5$, $M_{BH} = 10M_\odot$, and $\dot{m} = 50$. The *dotted lines* correspond to the unphysical solutions without advection. The *solid lines* correspond to the global solution with advection and optical depth transition, from [52]

and by solid lines the corresponding global advective solution. Note a significant quantitative difference in the values of the effective optical depths inside $40r_g$ in the two solutions.

Figure 11.26 represents the case $\dot{m} = 50 > \dot{m}_{cr}$. The dotted lines on the left and right panels show the non-advective solutions (Fig. 1) with the gap at $r \approx 10r_g$–$20r_g$ where the solutions do not exist, and the solid lines are the corresponding global advective solution. It was discussed above that there are no global and, therefore, physical solutions without advection for the large $\dot{m}$. The reason is that the local

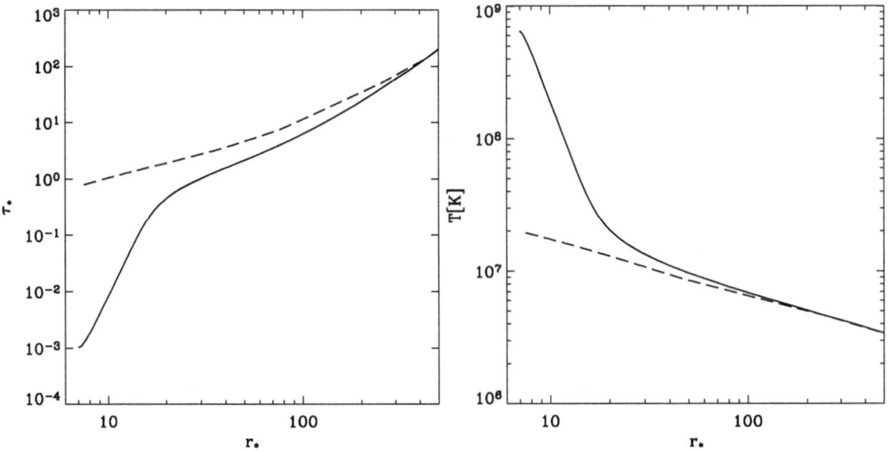

Fig. 11.27 Radial dependence of the effective optical depth τ_* (*left panel*) and the midplane temperature T (*right panel*) for the models with $M_{BH} = 10M_\odot$, $\alpha = 0.5$, and $\dot{m} = 48$. The *dashed lines* correspond to the advective optically thick solution and the *solid lines* correspond to the advective solution with the optical depth transition, from [52]

thermal equilibrium without the advection term, $Q^+ = Q^-$ (see (11.152)), cannot be satisfied anymore for $\dot{m} > \dot{m}_{cr}$ in the case of the effectively optically thin plasma that radiates inefficiently. Therefore, in the global solutions with the optical depth transition, the advection cooling term Q_{adv} is an essential and important ingredient in equation (11.152) to balance the viscous energy release term Q^+.

The advective models constructed using the optically thick approximation can significantly overestimate the effect of radiative cooling in the effectively optically thin regions of accretion disks. This results in the cooler interior of the disks, which then produce a softer outgoing radiation. Figure 11.27 shows an example of the radial dependence of the effective optical depth (*left*) and midplane temperature (*right*) for the advective models with and without the optical depth transition (*solid and dashed lines, respectively*). These models are characterized by $\alpha = 0.5$, $M_{BH} = 10M_\odot$ and $\dot{m} = 48$. Note the significant increase of the temperature up to 6×10^8K and corresponing drop of the effective optical depth down to 10^{-3} inside the radius $\sim 20r_g$ in the model with optical depth transition. This is an illustrative example of the super-Eddington accretion, which can be discriminated in observations due to production of hard-X-rays excess.

In contrast, the models with the accretion rates smaller than the critical one, $\dot{m} < \dot{m}_{cr}$, do not show the prominent effectively optically thin regions and relative temperature increase. As an example of the latter case, we show the model characterized by $\alpha = 0.1$, $M_{BH} = 10M_\odot$ and $\dot{m} = 50 < \dot{m}_{cr}$ with the optical depth transition (Fig. 11.28, *solid lines*), in which the central temperature is increased to $T = 4 \times 10^7 K$ – only a factor of 2 larger than the maximum temperature in the optically thick counterpart of this model (Fig. 11.28, *dashed lines*).

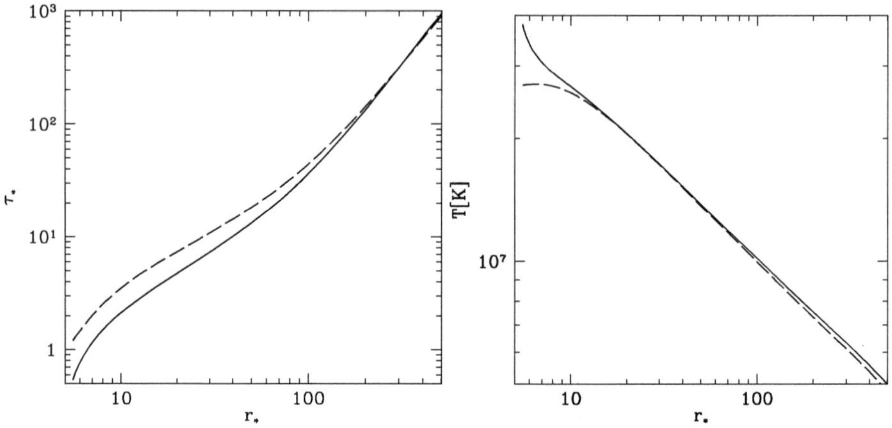

Fig. 11.28 Radial dependence of the effective optical depth τ_* (*left panel*) and the midplane temperature T (*right panel*) for the models with $M_{BH} = 10M_\odot$, $\alpha = 0.1$, and $\dot{m} = 50$. The *dashed lines* correspond to the advective optically thick solution and the *solid lines* correspond to the advective solution with the optical depth transition

Using the analysis of singular points discussed in [51], the types of singular points have been determined in the numerical solutions [52]. With the viscosity prescription (11.156), the advective models with the optical depth transition have two singular points, similar to optically thick models in [51]. The inner singular points, $(r_s)_{\rm in}$ can be saddles or nodes depending on the values of α and $\dot{m}$. In the case of the models with $\alpha = 0.5$ and $M_{BH} = 10M_\odot$, the inner singular points are nodes if $\dot{m} > 1$ and saddles otherwise. Note that the change of the type from the saddle to nodal one does not introduce any features in the solutions. The outer singular points, $(r_s)_{\rm out}$, are always saddle points in the studied models.

The difference between the advective models with and without the optical depth transition can be investigated considering the ratio of the gas to total pressure at this point, $\beta_s = P_g/P_{\rm tot}$, as functions of $\dot{m}$ in Fig. 11.29 for the case of the models with $\alpha = 0.5$ and $M_{BH} = 10M_\odot$. The solid line corresponds to the models with the optical depth transition, and the dashed lines correspond to the optically thick models. In spite of some quantitative changes, there are no changes in the qualitative behaviour of the value β_s with change of $\dot{m}$ in the considered advective models with and without the optical depth transition, as well as in the behaviour of $\ell_{\rm in}$ and r_s with change of $\dot{m}$ [52].

As was found in [52], the advective accretion disk models constructed using the optically thick approximation for the vertical radiative transport significantly overestimate the effect of radiative cooling in the effectively optically thin regions. This results in the cooler interior of the disks. The models with the correct account of the optical depth transition show significant temperature increase (more than one order of magnitude) in the disk interior at the inner radii. This increase result in harder emitting spectra from the disks, and therefore such disks can be discriminated in observations. The high temperatures in the inner parts of the accretion disk may

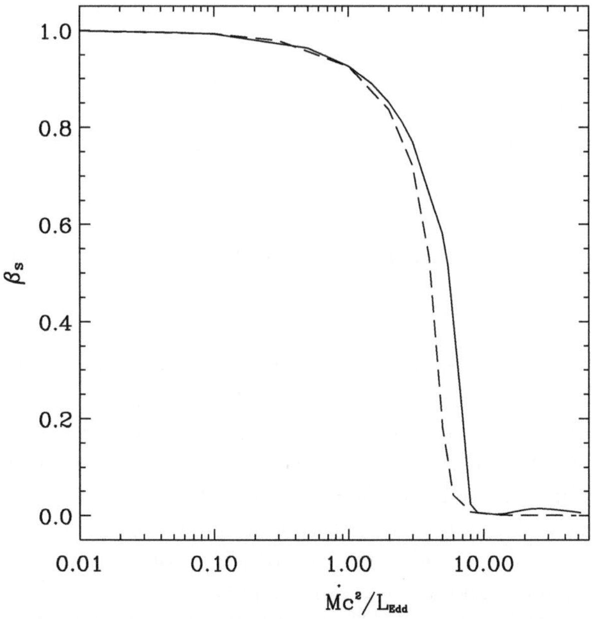

Fig. 11.29 Ratio of the gas pressure to total pressure β_s at the inner singular point, $(r_s)_{\rm in}$, as a function of the mass accretion rate $\dot{M}$. The models with $\alpha = 0.5$ and $M_{BH} = 10 M_\odot$ are shown. See the caption of Fig. 8 for other details of notations, from [52]

lead to electron–positron pair creation and modification of the energy spectrum of the source in the high-energy region over 500 keV. The radiation energy density in the inner regions is at least 4 orders of a magnitude less than the thermodynamic equilibrium radiation energy. It follows from the comparison with optically thick models that when increasing the temperature by 30 times and decreasing the density, the ratio of the gas pressure to the radiation pressure increases (Fig. 11.29). Therefore, bremsstrahlung radiation is more efficient in this situation.

Accretion disks around rotating black holes can be situated closer to the centre, until $r = r_{\rm g}$ for the extreme Kerr metric. Therefore, the temperature in the inner parts of such disks with optically thin/thick transition increases, leading to higher importance of electron–positron pairs. It the temperature in the inner optically thin disk parts exceeds the limiting temperature of about $kT \approx 20$ MeV, the unlimited birth pairs start [173] producing a separate optically thick region around the inner boundary. Another important changes in the structure of the advective disks with the optically thin/thick transition should be introduced by the presence of the magnetic field, regular as well as chaotic. Magnetodipole radiation and absorption in the presence of magnetic field may be more important than the bremstrahlung, increasing the effective optical depth. Both effects should change considerably the values of $\dot{m}_{\rm cr}$ and quantitive characteristics of the advective disks with the optically thin/thick transition.

A unified model of the advective accretion disks around black holes, using the bridging formula for the optical depth transition from [1030], was constructed in [279]. The models in [279] have $\dot{m} < \dot{m}_{\rm cr}$ and therefore did not differ much from the pure optically thick models, with the interior temperature not exceeding $T \approx 6 \times 10^7 K$, similar to Fig. 11.28.

11.3.7 Large-Scale Magnetic Fields Dragging in Accretion Disks

Early work on disk accretion to a black hole argued that a large-scale magnetic field of, for example, the interstellar medium would be dragged inward and greatly compressed by the accreting plasma [139, 143, 643]. Subsequently, analytic models of the field advection and diffusion in a turbulent rotating disk suggested that the large-scale field diffuses outward rapidly [646, 651] and prevents a significant amplification of the external poloidal field by electrical current in the accretion disk.

This has led to the suggestion that special conditions (non-axisymmetry) are required for the field to be advected inward [928]. In accretion with nonuniform vertical structure the high electrical conductivity of the surface layers of the disk, where the turbulence is suppressed by the radiation flux and the relatively high magnetic field, prevents outward diffusion of the magnetic field. This leads in general to a strong magnetic field in the inner parts of accretion disks around black holes [131, 647].

Consider first the fully turbulent disks. There are two limiting accretion disk models which have analytic solutions for a large-scale magnetic field structure. The first was constructed in [143] for a stationary non-rotating accretion disk, (see Sect. 11.3.2). A stationary state in this disk (with a constant mass flux onto a black hole) is maintained by the balance between magnetic and gravitational forces, and thermal balance (local) is maintained by Ohmic heating and radiative radiative conductivity for an optically thick conditions. The mass flux to the black hole in the accretion disk is determined by the finite conductivity of the disk matter and the diffusion of matter across the large-scale magnetic field as sketched in Fig. 11.19. The value of the large-scale magnetic field in stationary conditions is determined by the accretion disk mass, which in turn is determined by the magnetic diffusivity of the matter. For a laminar disk with Coulomb conductivity (which is very large), the mass of the stationary disk is also very large. Correspondingly, the magnetic field needed to support a mechanical equilibrium is also very large.

It is widely accepted that the laminar disk is unstable to different hydrodynamic, magnetohydrodynamic and plasma instabilities, which implies that the disk is turbulent. In X-ray binary systems, the assumption about turbulent accretion disk is necessary for construction of a realistic models [898]. Therefore, the turbulent accretion disks had been constructed also for non-rotating models with a large-scale magnetic field. A turbulent magnetic diffusivity was considered in [797] and [143], where the turbulent diffusivity was scaled by the parameters of the turbulent motion (11.104), similar to the scaling of the shear α-viscosity in turbulent accretion disk in binaries. In a consistent representation (first relation in (11.122)), the coefficient of turbulent kinematic viscosity $\nu = \mu/\rho$ in the Navier–Stokes equation is taken in the form $\nu = \frac{2}{3}\alpha v_s h$, where $v_s = \sqrt{P/\rho}$ is the isothermal sound speed and ρ is the midplane density of the disk.

The evolution of a large-scale magnetic field threading a turbulent Keplerian disk can be estimated easily. This field arises from two sources: external electrical currents and currents in the accretion disk. Evidently, the field generated by the currents in the disk can be much larger than that due to the external currents. The magnetic field may become dynamically important, influencing the accretion disk structure and leading to powerful jet formation, only if it is strongly amplified during the radial inflow of the disk matter. This amplification is possible only when the radial accretion speed of matter in the disk is larger than the outward diffusion speed of the poloidal magnetic field due to the turbulent diffusivity $\eta_t = c^2/(4\pi\sigma_t)$. Estimates in [651] have shown that for a turbulent conductivity (11.104), the outward diffusion speed is larger than the accretion speed. Thus, it appears that there is no large-scale magnetic field amplification during Keplerian disk accretion. This conclusion is discouraging because the most plausible models of many phenomena observed in the the systems with black holes inside the galaxy, as well as in the extragalactic supermassive black holes, are connected with a large values of a large-scale magnetic fields. Nevertheless, this result directly follows from the equations of the standard disk structure, with the turbulent electric conductivity (11.104), and was analytically obtained in [131]. The standard accretion disk structure (see Sects. 11.3.4, 11.3.6) can be written as

$$\dot{M} = 4\pi\rho v_r r h, \quad h = \frac{v_s}{\Omega_K}, \quad v_s = \sqrt{\frac{P}{\rho}}, \quad 4\pi r^2 h\alpha P = \dot{M}(j - j_{in}),$$
$$\frac{3}{2}\Omega_K \alpha P h = \frac{2aT^4 c}{3\kappa\rho h}, \tag{11.169}$$

where $v_K = r\Omega_K$ is the Keplerian velocity (see e.g. [130]). For regions far from the inner disk boundary, the specific angular momentum $j \gg j_{in}$. The characteristic time t_{visc} of the accretion disk matter advection due to the shear viscosity is $t_{visc} = r/v_r$. From the first three relations in equation (11.169), we obtain

$$t_{visc} = \frac{r}{v_r} = \frac{j}{\alpha v_s^2}. \tag{11.170}$$

To estimate the time of the magnetic field diffusion, we use

$$t_{diff} = \frac{r^2}{\eta}\frac{h}{r}\frac{B_z}{B_r}, \tag{11.171}$$

[651], where B_r and B_z are the large-scale field components evaluated at the top surface of the disk. Here the coefficient of the magnetic turbulent diffusivity η is obtained from (11.104)

$$\eta = \frac{c^2}{4\pi\sigma_t} = \tilde{\alpha} h v_s. \tag{11.172}$$

For stationary conditions, the large-scale magnetic field in the accretion disk is determined by the equality $t_{\text{visc}} = t_{\text{diff}}$, which implies

$$\frac{B_r}{B_z} = \frac{\alpha}{\tilde{\alpha}} \frac{v_s}{v_K} = \frac{\alpha}{\tilde{\alpha}} \frac{h}{r} \ll 1. \tag{11.173}$$

In contrast, the coronal poloidal field solution typically have $B_{\text{zd}} \sim B_{\text{rd}} \sim 1$ at the disk surface [118,1002] which implies that $t_{\text{visc}} \gg t_{\text{diff}}$. This inequality indicates that the magnetic field is not amplified during accretion for these physical conditions.

11.3.7.1 Turbulent Disk with Radiative Outer Zones

Near the surface of the disk, in the region of low optical depth, the turbulent motion is suppressed by the radiative flux, similar to the suppression of the convection over the photospheres of stars with outer convective zones. The presence of the outer radiative layer does not affect the estimate of the characteristic time t_{visc} of the matter advection in the accretion disk because it is determined by the main turbulent part of the disk. The time of the field diffusion, on the contrary, is significantly changed, because the electrical current is concentrated in the radiative highly conductive regions, which generate the main part of the magnetic field. The structure of the magnetic field with outer radiative layers is shown schematically in Fig. 11.30.

Inside the turbulent disk, the electrical current is negligibly small so that the magnetic field there is almost fully vertical, with $B_r \ll B_z$, according to (11.173). In the outer radiative layer, the field diffusion is very small, so that matter advection is leading to strong magnetic field amplification. The field amplification will last until the magnetic forces in the region over the photosphere become of the order of the gravitational ones and start to participate in the equilibrium balance. In such conditions, the MHD and plasma instabilities are developed, decreasing the effective electrical conductivity. In the stationary state, the magnetic forces could support the

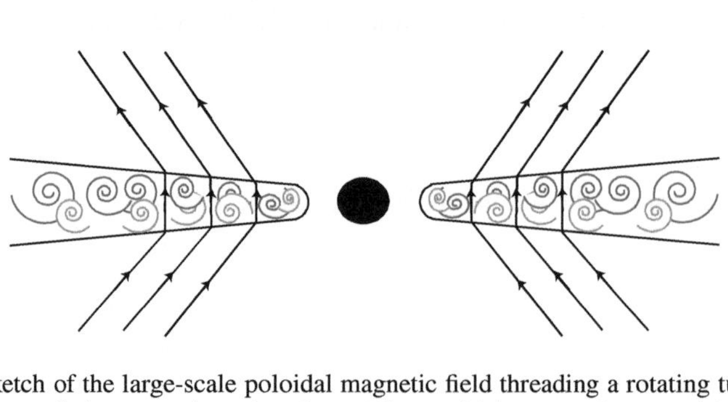

Fig. 11.30 Sketch of the large-scale poloidal magnetic field threading a rotating turbulent accretion disk with a radiative outer boundary layer. The toroidal current flows mainly in the highly conductive radiative layers. The large-scale (average) field in the turbulent region is almost vertical

optically thin regions against gravity. In the nonrotating magnetized disk, magnetic forces support the whole disk against the gravity; so they should be much higher. When the magnetic force balances the gravitational force on the outer optically thin part of the disk of surface density $\Sigma_{\rm ph}$, one finds that the following relation takes place [143]

$$\frac{GM\Sigma_{\rm ph}}{r^2} \simeq \frac{B_z I_\phi}{2c} \simeq \frac{B_z^2}{4\pi}, \tag{11.174}$$

The surface density over the photosphere corresponds to a layer with effective optical depth close to $2/3$ (e.g. [724]). The lower limit of the magnetic field strength is obtained taking $\kappa_{\rm es}$ (instead of the effective opacity $\kappa_{\rm eff} = \sqrt{\kappa_{\rm es}\kappa_a}$). Writing

$$\kappa_{\rm es}\Sigma_{\rm ph} = 2/3, \tag{11.175}$$

it follows $\Sigma_{\rm ph} = 5/3$ (g/cm^2) for the opacity of the Thomson scattering, $\kappa_{\rm es} = 0.4$ cm^2/g. The absorption opacity κ_a is much less than $\kappa_{\rm es}$ in the inner regions of a luminous accretion disk. Thus, using $\Sigma_{\rm ph}$ from equation (11.175) in equation (11.174), the lower bound on the large-scale magnetic field of a Keplerian accretion disk is estimated as

$$B_z = \sqrt{\frac{5\pi}{3}}\frac{c^2}{\sqrt{GM_\odot}}\frac{1}{x\sqrt{m}} \simeq 10^8 \mathrm{G}\frac{1}{x\sqrt{m}}, \tag{11.176}$$

where $x = \frac{r}{r_g}$ and $m = \frac{M}{M_\odot}$. For comparison, the surface density Σ_d of the disk in the inner radiation dominated region, where we may expect the largest values of the magnetic fields, is (see Problem. 2, Sect. 11.3):

$$\Sigma_d = \frac{80\sqrt{2}}{9\alpha}\frac{x^{3/2}}{\dot{m}}\left(1 - \sqrt{\frac{3}{x}}\right)^{-1}, \tag{11.177}$$

$$\dot{m} = \frac{\dot{M}c^2}{L_c}, \qquad L_c = \frac{4\pi c GM}{\kappa_{\rm es}},$$

The maximum magnetic field is reached when the outward magnetic force balances the gravitational force on the disk of surface mass density $\Sigma_{\rm ph}$. In equilibrium, $B_z \sim \sqrt{\Sigma_{\rm ph}}$. We find that B_z in a Keplerian accretion disk is about 20 times less than its maximum possible value for $x = 10$, $\alpha = 0.1$ and $\dot{m} = 10$.

An important question is the energy density of the large-scale stationary magnetic field in comparison with the rotational or gravitational energy density of the disk. For a non-rotating magnetized accretion disk, the energy density of the field is of the order of the gravitational one, so that the magnetic field strength is very large in the vicinity of a black hole [143], and may be many order of magnitude larger than the external seed field. In the case of the fully turbulent Keplerian disk, the poloidal magnetic field tends to drift outward [646, 651] so that its value cannot significantly exceed the strength of the large-scale seed magnetic field.

The suggestion of a fully turbulent accretion disk with a small turbulent conductivity is violated in the outer surface layer of the accretion disk where the optical depth is small and the turbulence is suppressed by the strong radiative flux. This is similar to the radiative layer above the main body of a convective star (similar to the Sun; e.g., [878]). In the radiative layer, the magnetic field diffusion is much slower than in the region of fully developed turbulence. In the radiative layer, the diffusion is determined by the classical Coulomb conductivity which is very large. The diffusion in this layer is practically negligible. The electric current is concentrated in the radiative layer, and the main body of the turbulent disk is almost current-free and thus force-free. The magnetic field lines in this region are almost straight as shown in Fig. 11.30. Because of the negligible diffusion in the radiative layer, the large-scale field drifts inward until the dynamical action of the magnetic field on the photosphere becomes comparable with that of centrifugal and gravitational forces. At this point, the inward drift of the field will be halted and the stationary state formed where the magnetic, centrifugal and gravitational forces on the optically thin region will be comparable, and deviations from the Keplerian angular velocity will be significant. The strength of the magnetic field for such conditions is smaller than in the nonrotating disk of [143], but still it is very large in the vicinity of the black hole. In this situation, we may, expect a nonuniform distribution of the angular velocity over the disk thickness: The main body of the turbulent disk rotates with the velocity close to the Keplerian one, and outer optically thin layers rotate substantially slower. The strong shear in the disk surface layers may be a source of turbulence development in the whole body of the accretion disk.

11.3.8 Battery Effect in Accretion Disks

The classical battery mechanism of magnetic field generation is connected with a non-coincidence of surfaces of constant pressure and constant density, where forces connected with pressure gradients become nonpotential. In this situation no static equilibrium in the gravitational field is possible. When considering separately the motion of electrons and ions, there is always a difference in the velocities of electrons and ions which creates electric currents and an associated magnetic field. Self-induction is very important in the battery mechanism, determining the rate of increase of the magnetic field.

Along with ion and electron pressure gradients, a nonpotential force field may arise due to the radiation force which acts predominantly on the electrons. In a spherically symmetric star, the radiation force has a potential so that no magnetic field is generated: equilibrium in the two-fluid plasma results from a distribution of an electric charge and a static radial electric field. It was shown in [120] that in *geometrically thin, optically thick* accretion disks, the radiation force above the disk has a nonpotential or vortex-like component. Under this condition, no electric charge distribution can give a static equilibrium. Instead, electric currents and a corresponding magnetic field are generated. The radiation forces above a thin disk gives rise to poloidal electrical current flow and a toroidal magnetic field. Self-consistent solution for *optically thin, geometrically thick* accretion disk flow was obtained in

[900]. At very low mass accretion rates, suggesting two-temperature plasma with $T_i \neq T_e$, there is a solution where the ion temperature is close to the virial temperature, and $T_e \ll T_i$. In the absence of a magnetic field, and neglecting relaxation processes between electrons and ions except for binary collisions, these flows are referred to as advection dominated accretion flows (ADAF) [509, 734]. In the ADAF regime the radiative efficiency of accretion may be very low, $\sim 10^3$ times less then the standard value for a geometrically thin, optically thick accretion disk. Account of processes connected with the presence of a magnetic field increases the efficiency up to at least $1/3$ of the standard value [128, 129].

It was proposed in [305] that a cosmic battery may operate in ADAF accretion flows owing to the Poynting–Robertson (PR) effect. The PR effect acts to generate a toroidal electrical current and poloidal magnetic field. The authors claimed that the magnetic field may be amplified up to $\sim 10^7$G in the vicinity of a black hole of stellar mass. The PR mechanism of magnetic field generation is similar to the mechanism in [120] based on the nonpotential radiative force, but the magnetic field reached there only values of ≤ 10 G for a stellar mass black hole. In an optically thin disk both mechanisms act together leading to the generation of toroidal and poloidal components of the magnetic field. The radiative force initially causes the magnetic field to grow linearly with time. However, this linear growth holds for only a *restricted* time interval which is of the order of the accretion time of the matter [120], while in [305] the interval of linear growth is unrestricted.

11.3.8.1 Radiatively Induced Current and Toroidal Magnetic Field Production in Accretion Disks

Above a geometrically thin accretion disk around a black hole, the electrons are acted on by a nonpotential radiation force $\mathbf{F}_{\mathrm{L}}$ due to Thomson scattering which is found in [120]

$$\mathbf{F}_{\mathrm{L}} = -R\cos\theta\,\nabla\phi_L = (F_{\mathrm{LR}},\, F_{L\theta},\, 0), \tag{11.178}$$

where a spherical coordinate system (R, θ, ϕ) is used, and ϕ_L is the "pseudopotential" of the radiation force, which may be expressed as

$$\phi_L = \frac{\sigma_T}{c}\int_{r_{\mathrm{in}}}^{\infty} \frac{H(r)\,r\,\mathrm{d}r}{(R^4 + r^4 + 2R^2r^2\cos 2\theta)^{1/2}}. \tag{11.179}$$

Here, the disk thickness is neglected and the cylindrical radius is $r = R\cos\theta$. The function $H(r)$ is the radiative flux emitted per unit area from one side of the disk. In the standard local accretion model,

$$H(r) = \frac{3}{8\pi}\frac{GM\dot{M}}{r^3}\mathscr{J}, \tag{11.180}$$

where $\mathcal{J} \equiv 1 - (r_{\text{in}}/r)^{1/2}$, and $r_{\text{in}} = 3r_S = 6GM/c^2$ is the inner radius of the disk for a non-rotating black hole of Schwarzschild radius r_S. In the disk plane, $\theta = \pi/2$, the radiative force is perpendicular to the disk,

$$F_{L\theta} = -\frac{\sigma_T}{c} H(r) = \frac{3GMm_p}{r^2} \frac{r_{\text{in}}}{r} \frac{L}{L_{\text{Edd}}} \mathcal{J}, \tag{11.181}$$

where

$$L = \frac{GM\dot{M}}{2r_{\text{in}}}, \qquad L_{\text{Edd}} = \frac{4\pi c GMm_p}{\sigma_T}, \tag{11.182}$$

where σ_T is the Thomson cross-section. Due to the interaction of the radiation flux mainly with the electrons, the accretion disk becomes positively charged up to a value where the electrostatic attraction of the electrons balances the radiation force. The vertical component of the electrical field strength E_θ in the disk plane is written as

$$E_\theta(r) = -\frac{\sigma_T}{c|e|} H(r) = -E_0 \frac{L}{L_{\text{Edd}}} \left(\frac{r_{\text{in}}}{r}\right)^3 \mathcal{J}, \tag{11.183}$$

where

$$E_0(M) \equiv \frac{m_p c^4}{12|e|GM} \approx 1.76 \frac{M_\odot}{M} \text{ cgs} \approx 528 \frac{M_\odot}{M} \frac{\text{V}}{\text{cm}} \tag{11.184}$$

Thus, the surface charge-density of the disk is

$$\sigma_{\text{e}}(r) = \frac{E_\theta(r)}{2\pi}.$$

The influence of this charge on the structure and stability of the accretion disk is negligible.

Both the gravitational and electrical forces have a potential, so that they cannot balance the nonpotential radiation force. Due to the radiation and electric forces, electrons move with respect to protons, which to a first approximation do not acquire the poloidal motion. Thus, a poloidal electrical current is generated with an associated toroidal magnetic field. The finite disk thickness may create poloidal motion of all of the matter of the accretion disk, similar to meridional circulation in rotating stars [568]. To estimate the magnetic field strength, we write the electromotive force (EMF) as

$$\mathcal{E} = \frac{1}{e} \oint \mathbf{F}_L \times d\mathbf{l} = \frac{1}{e} \int\int d\mathbf{S} \cdot \nabla \times \mathbf{F}_L \sim E_\theta h, \tag{11.185}$$

where h is the half-thickness of the disk. Thus the stationary current-density is

$$J_{\text{st}} \sim \frac{\sigma_{\text{e}} \mathcal{E}}{r} \sim \frac{\sigma_{\text{e}} E_\theta h}{r}, \tag{11.186}$$

where σ_e is the conductivity of the disk plasma. The stationary toroidal magnetic field is [120]

$$B_{\phi 0} \sim \frac{4\pi}{c} J_{st} h \sim \frac{4\pi\sigma_e}{c} \frac{E_\theta h^2}{r}. \quad (11.187)$$

In the radiation-dominated inner region of the standard α-disk model, h can be written as [898]

$$h = 3\frac{L}{L_{Edd}} \mathscr{J} r_{in}, \quad (11.188)$$

Finally, the stationary value of toroidal magnetic field in the disk is equal to

$$B_{\phi 0} \sim \frac{36\pi\sigma_e}{c} E_0 \left(\frac{L}{L_{Edd}}\right)^3 \left(\frac{r_{in}}{r}\right)^4 r_n \mathscr{J}^3,$$

$$= \frac{12\pi\sigma_e h}{c} E_0 \left(\frac{L}{L_{Edd}}\right)^2 \left(\frac{r_{in}}{r}\right)^4 \mathscr{J}^2. \quad (11.189)$$

For the plasma conductivity σ_{Coul} owing to binary collisions,

$$\sigma_e = \sigma_{Coul} \approx 3 \times 10^6 T^{3/2} s^{-1}, \quad (11.190)$$

and the conductivity σ_T in the presence of well-developed turbulence (11.104), the stationary field from (11.189) is equal to $B_{\phi 0} \sim 10^{13}$G, and $B_{\phi 0} \sim 10$ G (for $\alpha = 0.1$), respectively. The time-scale τ_m for reaching the stationary field given by (11.189) is determined by the self-induction of the disk. This is equivalent to the "$\mathscr{L}$ over $\mathscr{R}$ time" of circuit with inductance $\mathscr{L}$ and resistance $\mathscr{R}$. For the electrical conductivity σ_e, this time-scale is equal to

$$\tau_m \simeq \frac{4\pi\sigma_e h r}{c^2}. \quad (11.191)$$

The crossing time of matter passing through the radiation dominated region of the disk is

$$t_c \approx \frac{r}{v_r}. \quad (11.192)$$

During the time t_c, there is linear growth of the magnetic field after which the matter falls into the black hole. The battery value of the large-scale toroidal magnetic field is equal to

$$B_\phi \approx B_{\phi 0}\frac{t_c}{\tau_m} = 3\frac{c}{v_r} E_0 \left(\frac{L}{L_{Edd}}\right)^2 \left(\frac{r_{in}}{r}\right)^4 \mathscr{J}^2. \quad (11.193)$$

For a turbulent conductivity (11.104), the growth time-scale of the magnetic field, using equations (11.104) and (11.191), is equal to

$$\tau_m^{turb} \approx \frac{r}{\alpha c_s}. \quad (11.194)$$

Taking into account, from (11.266), that $v_r = \alpha c_s (L/L_{\rm Edd})(r_{\rm in}/r) < \alpha c_s$ and $t_m^{\rm turb} < t_c$, we find

$$B_\phi = B_{\phi 0}^{\rm turb} \approx 3 \frac{c}{\alpha c_s} E_0 \left(\frac{L}{L_{\rm Edd}}\right) \left(\frac{r_{\rm in}}{r}\right)^3 \mathcal{J}^2, \tag{11.195}$$

where

$$c_s = 7 \times 10^9 \,\frac{\rm cm}{\rm s} \left(\frac{L}{L_{\rm Edd}}\right) \left(\frac{r_{\rm in}}{r}\right)^{3/2} \mathcal{J}.$$

The strength of the stationary toroidal magnetic field produced by the battery effect in the radiation-dominated region of an accretion disk with the turbulent or higher conductivity from (11.190) is equal to

$$B_\phi \approx \frac{22}{\alpha} \left(\frac{M_\odot}{M}\right) \left(\frac{r_{\rm in}}{r}\right)^{3/2} \mathcal{J}. \tag{11.196}$$

At a distance $r = 3r_{\rm in}$, we have

$$B_\phi \approx \frac{2}{\alpha} \left(\frac{M_\odot}{M}\right) \text{G}, \tag{11.197}$$

in accordance with [120]. The corresponding magnetic energy-density is much less than the energy-density associated with the turbulent motion in the disk $\rho v_t^2/2$.

11.3.8.2 Production of a Poloidal Magnetic Field in Optically Thin Accretion Flows by Poynting–Robertson Effect

In optically thin accretion flows, the radiation flux interacts with the inspiraling matter by the Poynting–Robertson (PR) effect [834]. The PR effect was studied in [292, 305, 306] as a mechanism for generating poloidal magnetic field in an optically thin accretion flow, where it was obtained that dynamically important magnetic field strengths could result from this effect. The linear growth of the magnetic field due to the radiative force on the electrons found in [305] is similar to that analyzed in [120], but the PR effect implies an additional (small) numerator, (v_ϕ/c). Also, for a quasi-spherical optically thin accretion flow the characteristic scale is r instead of h, and the quasi-spherical luminosity is $L/(4\pi r^2)$ instead of H in (11.180). Then, using equations (11.187), (11.191), (11.193), we obtain the rate of growth the poloidal magnetic field due to the PR effect, as [305],

$$B_z \approx B_{z0} \frac{t}{\tau_m} = \frac{E_0}{3\alpha} \left(\frac{L}{L_{\rm Edd}}\right) \left(\frac{r_{\rm in}}{r}\right)^2 \left(\frac{t v_r}{r}\right). \tag{11.198}$$

Here, E_0 is defined in equation (11.183), and $r_{\rm in}$ in equation (11.180).

It is essential to take into account that an element of matter with the induced magnetic field reaches the black hole in time $t_c \approx r/v_r$. This means that the magnetic field grows *only* during the time t_c. Consequently, the magnetic field reaches a value

$$B_z \approx \frac{E_0}{3\alpha}\frac{L}{L_{\mathrm{Edd}}}\left(\frac{r_{\mathrm{in}}}{r}\right)^2 \approx \frac{0.7}{\alpha}\frac{L}{L_{\mathrm{Edd}}}\frac{M_\odot}{M}\left(\frac{r_{\mathrm{in}}}{r}\right)^2 \qquad (11.199)$$

Taking into account that the luminosity is $\le 10^{-3}L_{\mathrm{Edd}}$ for optically thin accretion, and taking $\alpha = 0.1$, we get a maximum value of the magnetic field created by the PR effect in an ADAF to a black hole [132],

$$B_z \approx 7\times 10^{-3}\left(\frac{M_\odot}{M}\right)\ \mathrm{G}. \qquad (11.200)$$

This estimate of the field is about 10 orders of magnitude less that the value obtained in [305]. The difference in the estimates results from the fact that in [305] it was assumed that magnetic flux accumulates continuously near the black hole during a long time, reaching the equipartition with the kinetic energy. The accumulation actually occurs only during the time the plasma (which carries the field or current loops) takes to move inward to the black hole horizon. The current loops created by the PR effect disappear as the matter approaches the horizon. In the case of accretion onto a neutron star or a white dwarf, matter containing the current loops merges with the stellar matter, which is typically much more strongly magnetized. Analysis of the magnetic field generation in the vicinity of the horizon of the schwarzschild black hole, made in [132], had shown that the magnetic field generated by the current loops, for the distant observer, is tending to zero at approaching r_g.

More complicated model with a specific inner boundary conditions was constructed in [292, 306], where it was claimed about a possibility of a strong PR battery in the accretion disk. This boundary condition is equivalent to the suggestion of non-penetration and stopping of the accreting matter at radius larger than the horizon, which seems to be nonrealistic and artificial.

11.3.9 Screening of the Magnetic Field of Disk Accreting Stars

The first evidence for accretion-induced magnetic field decay was obtained from observations of the first binary pulsar PSR 1913 + 16 [481]. This pulsar rapidly rotates with the period 0.059 s in an orbit with the period 7^h45^m and large eccentricity $\varepsilon = 0:6171$. Immediately after the discovery, this pulsar was interpreted to be an old recycled pulsar with a weak magnetic field [156]. The subsequent measurements [967] of the period derivative $\dot{P}$ allowed estimating its age as $\tau = P/\dot{P} \approx 10^8$ years and magnetic field as $B \approx 2\cdot 10^{10}$ G, which proved the proposed interpretation. In more than 100 recycled pulsars, the companion to the neutron star is usually a low-mass degenerate dwarf. Binaries consisting of two neutron stars, like PSR 1913 + 16, are much less numerous, with about ten such systems known. All recycled pulsars, irrespective of the nature of their companions, have magnetic fields

10^8-10^{10} G, which fully confirms the hypothesis of the accretion-induced magnetic field decay [177]. The discovery of the first binary system with two radio pulsars [660] provided one more firm proof of this prediction. According to simple estimates, we expect screening of the neutron star magnetic field by accretion. In [103], it was shown that if instabilities causing penetration of the accreting plasma inside the pulsar magnetosphere can be neglected, the pressure of the accreting plasma overcomes the magnetic pressure of a dipole with the strength $B = 10^{12}$ G in a time interval as short as 1 day, assuming the subcritical accretion rate $10^{-9}\,M_{\odot}$/year. The original magnetic dipole then turns out to be buried under a plasma layer in which currents prevent the magnetic field of the star from emerging to the surface. In reality, instabilities lead to the penetration of gas through the magnetosphere and strongly slow down the process of magnetic field screening. During accretion, the magnetic field value is determined by the balance between the screening action of the accreting gas and turbulent field diffusion. As the accretion mass increases, the thickness of the layer through which the field should diffuse increases and, correspondingly, the external magnetic field decreases. Many papers (see the references in [290]) considered the neutron star magnetic field decay due to the heating of the crystal crust of the neutron star during accretion that decreases the matter conductivity and accelerates the Ohmic dissipation of the magnetic field. This mechanism works, however, only if electric currents forming the pulsar magnetic field flow in the layers of the neutron star that can be heated by accretion. Clearly, the strongly degenerate neutron star core remains insensible to such a heating and preserves a very high conductivity. Therefore, the magnetic field formed by deep electric currents can be decreased only by the accreting plasma screening [177]. Several models of the accretion screening of the neutron star magnetic field have been studied in [280, 281, 290, 291]. It is usually assumed that the infalling matter is canalized by the magnetic field and flows along the field lines to the magnetic pole region of the dipole [25, 193]. As the magnetic field decays due to the screening, the canalization of the accretion flux decreases and the polar cap area increases. It is easy to estimate the polar cap angular width θ_{P} as a function of the neutron star surface magnetic field B_s [899]. The magnetic force line that reaches the Alfvenic radius r_A starts at the angle θ_P on the surface of a neutron star of radius r_s and is the last closed line of the dipole field. It is easy to show that

$$\sin\theta_{\mathrm{P}} = \left(\frac{r_{\mathrm{s}}}{r_{\mathrm{A}}}\right)^{1/2}. \tag{11.201}$$

Setting the dynamic pressure of freely falling gas equal to the magnetic pressure at the Alfvenic surface, we find the Alfvenic radius value

$$r_{\mathrm{A}} = (2GM)^{-1/7} r_{\mathrm{s}}^{12/7} B_{\mathrm{s}}^{4/7} \dot{M}^{-2/7}, \tag{11.202}$$

where M is the neutron star mass and $\dot{M}$ is the accretion rate. Equations (11.201) and (11.202) imply that

$$\sin\theta_{\mathrm{P}} \propto B_{\mathrm{s}}^{-2/7}. \tag{11.203}$$

Therefore, the polar cap size increases as the magnetic field decreases until θ_P becomes equal to 90° and (11.203) cannot be applied any more. For $M = 3 \cdot 10^{33}$ g, $\dot{M} = 10^{-8} M_\odot$/year, $r_s = 10$ km and $B_s = 10^{12}$ G, we obtain from (11.202) that $r_A \approx 260$ km, and from (11.201), we find the initial angular size of the polar cap of the order of 10°. The accreting matter falls onto the polar caps and then spreads over the stellar surface towards the equator. In the model considered in [291,582], the matter that flows from both poles meet near the equator and go inside the star. In a kinematical model calculated in [291], the matter that flows cover the star magnetic field and effectively decrease its strength; the sizes of the polar caps increase accordingly. When the polar cap size attains 90°, the accretion flow becomes spherical and radial and, as argued in [291], the magnetic field decay stops due to the assumed low efficiency of the magnetic field screening by the radial flow. If $\theta_{P,i}$ is the initial size of the polar cap, then according to (11.203) the magnetic field decreases by $\sin 90° / \sin \theta_{P,i})^{7/2}$ times during the accretion stage. Assuming $\theta_{P,i}$ to be in the range $5° - 10°$, we obtain the possibility of the accretion-induced magnetic field decay by $10^3 - 10^4$ times, which is exactly the difference between the magnetic fields of ordinary and recycled pulsars. Assuming that the magnetic field during accretion reaches some stationary value when the Alfvenic radius equals the neutron star radius, we obtain

$$B_{\text{asymp}} = (2GM)^{1/4} \dot{M}^{1/2} r_s^{-5/4}. \tag{11.204}$$

from (11.202) with $r_A = r_s$. Using standard values of the parameters as discussed above, we find $B_{\text{asymp}} \approx 10^8$ G. After the accretion has stopped, a recycled millisecond radio pulsar emerges with a magnetic field close to 10^8 G. The development of magneto-hydrodynamic instabilities near the Alfvenic surface facilitates plasma penetrating the magnetosphere and slows down the process of the magnetic field screening by accretion, considered in [645]. It was noted in [178] that "neutron star magnetic fields screened by intensive accretion can percolate outwards after the accretion has stopped." In this picture, the neutron star magnetic field value should increase with time, which can increase the rotational energy loss rate $\dot{E}$ with time, in contrast to the usual decrease $\dot{E}$ with time. By writing the field growth as $B \sim \Omega^{-k}$, we obtain $\dot{E} \sim \Omega^{4-2k}$. Here the rotational energy losses are determined by the magneto-dipole radiation and and pulsar wind [669, 670]

$$\dot{E} = -\frac{r_s^6 B_p^2 \Omega^4}{6c^3}, \tag{11.205}$$

where r_s is the neutron star radius, B_p is the magnetic pole field strength and Ω is the neutron star angular velocity. For $k > 2$, the rotational energy losses increase as the pulsar rotation decelerates. Another characteristic of the pulsar parameters is the braking index $n = \Omega\ddot{\Omega}/\dot{\Omega}$ (for the above relation, $n = 3-2k$). If observations yield $n < 3$, this may be due to the neutron star magnetic field increase with time [699]. The estimates of the magnetic field percolation outward were obtained in [722]. The authors considered the case of a very rapid percolation and the field growth over

$10^3 - 10^4$ years until the original value before the accretion is recovered. Clearly, this conclusion contradicts observations because all recycled pulsars, including very old ones with the age up to 10^{10} years [161, 641, 642], have equally small magnetic fields. Such a discrepancy is due to the artificial model considered in [722]. The percolation in that model started from depths smaller than 260 m, which corresponded to the accreted mass smaller than $5 \cdot 10^{-5}\,M_\odot$. The conductivity is not very high in the outer layers of the neutron star, which allows rapid percolation of the field outward. Actually, the accretion stage can last up to 10^7 to 10^8 years or more, and the mass of the accreted matter can be several orders of magnitude higher. The magnetic field is then buried in much deeper layers of the star with much higher conductivity. Accordingly, in recycled pulsars, the magnetic field percolation outward becomes much less effective. As noted in [699], the magnetic field percolation outward can occur in young neutron stars, which accreted an insignificant amount of screening matter after the supernova explosion. In these stars, a sufficiently rapid percolation of the magnetic field and its growth on the surface are possible.

11.3.10 Jet Confinement by Magneto-Torsional Oscillations

Objects of very different sizes and natures, from young or very old stars to galactic nuclei, are associated with the existence of collimated outbursts jets. The geometrical size of these jets varies from parsecs to megaparsecs. The origin of jets is not well understood: several qualitative mechanisms have been proposed, but these are not justified by calculations. A theory of jets must give answers to three main questions. How are jets formed? How are they stabilized? How do they radiate? The last question is related to the problem of the origin of relativistic particles in outbursts from active galactic nuclei (AGN), where synchrotron emission is observed. Relativistic particles ejected from the central machine rapidly lose their energy, so the problem of particle acceleration inside the jet arises (see the review [111]). It is sometimes convenient to investigate jets by means of a simple model of an infinitely long circular cylinder [266]. The magnetic field in the collimated jet determines the direction of the jet, and the axial current may stabilize the elongated form of the jet at large distances from the source (e.g. in AGNs) [92, 176, 533]. When observed with high angular resolution, these jets show a structure of bright knots separated by relatively dark regions [973]. High percentages of polarization, sometimes exceeding 50%, indicate the non-thermal nature of the radiation, which is well explained as synchrotron emission of the relativistic electrons in an ordered magnetic field. Estimates of the lifetime of these electrons, based on the observed luminosities and spectra, often give values much lower than the kinematic ages $t_k = d/c$, where d is the distance of the emitting point from the central source. An explanation of these observations requires continuous re-acceleration of the electrons in the jets. The acceleration mechanism for electrons in extragalactic jets proposed in [127] assumes that intense long-wavelength electromagnetic oscillations accompany a relativistic jet, and the electromagnetic wave amplitudes envisioned are sufficient to give in situ acceleration of electrons to the very high energies of $>10^{13}$ eV that are observed. The model assumes that jets are formed by a sequence of outbursts from the nucleus with considerable charge separation at the moment of the outburst. The direction of motion of the outbursts is determined by the large-scale magnetic field. When the emitted wave is strong enough it washes out the surrounding medium and the density can become very small, with the medium containing only the accelerated particles. The action of the oscillating knot is similar to the action of a pulsar, considered as an inclined magnetic rotator. Both emit strong electromagnetic waves, which could accelerate particles effectively. The stabilization of a jet against its expansion, by a pure magnetohydrodynamic mechanism associated with torsional oscillations, was considered in [117]. Oscillations of this type occurring in neutron stars have been considered in [74]. It is suggested in [117] that the matter in the jet is rotating, and that different parts of the jet rotate in different directions. Such a distribution of the rotational velocity produces an azimuthal magnetic field, which prevents a disruption of the jet. The jet is representing a periodical, or quasi-periodical structure along the axis, and its radius oscillates with time all along the axis. The space and time periods of oscillations depend on the conditions at jet formation: the length-scale, the amplitude of the rotational velocity, and the strength of the magnetic field. The time period of oscillations should be obtained during the construction of the dynamical model, and the model should also show at what input parameters a long jet stabilized by torsional oscillations could exist. Two-dimensional non-stationary MHD calculations are needed to solve the problem numerically. A very simplified model of this phenomenon was constructed in [117], which, nonetheless, permited to confirm the reality of such stabilization, to estimate the range of parameters at which it takes place and to establish the connection between the time and space scales, the magnetic field strength and the amplitude of the rotational velocity.

Axially symmetric MHD equations at $\frac{\partial}{\partial \phi}$, for the perfect gas with an infinite conductivity, in cylindric coordinates (r, ϕ, z), are given in (10.82)–(10.95). Let us consider a long cylinder with a magnetic field directed along its axis. This cylinder will expand without limit under the action of pressure and magnetic forces; so no confinement will be reached. It is possible, however, that a limiting value of the radius of the cylinder could be reached in a dynamic state, in which the whole cylinder undergoes magnetotorsional oscillations. Such oscillations produce a toroidal field, which prevents radial expansion. There is therefore competition between the induced toroidal field, compressing the cylinder in the radial direction, and gas pressure, together with the field along the cylinder axis (poloidal), tending to increase its radius. During magnetotorsional oscillations, there are phases when either the compression or expansion force prevails, and, depending on the input parameters, there are three possible kinds of behaviour of such a cylinder that has negligible self-gravity.

1. The oscillation amplitude is low, so the cylinder suffers unlimited expansion (no confinement).
2. The oscillation amplitude is too high, so the pinch action of the toroidal field destroys the cylinder, and leads to formation of separated blobs.

3. The oscillation amplitude is moderate, so the cylinder survives for a long time, determined by electromagnetic losses, and its parameters (radius, density, magnetic field, etc.) change periodically or quasi-periodically in time.

Solution of MHD equations (10.82)–(10.95) could give, in principle, the answer about a correctness of the above scenario. It is reasonable nevertheless to try to find a simple approximate way for obtaining a qualitative answer, and to make a rough estimation of parameters leading to different regimes.

11.3.10.1 Profiling in Axially Symmetric MHD Equations

Let us simplify the system of equations in such away that the resulting system contains the most important property of the dynamical competition between the different forces, to check for the possibility of dynamical confinement. For this purpose, we use a profiling procedure. Let us neglect the gravity in the direction of the cylinder axis (z), and approximate the density by a function $\rho(t,z)$, suggesting a uniform density along the radius. The components of the velocity and magnetic field are approximated as

$$v_{\rm r} = r\,a(t,z), \quad v_\varphi = r\,\Omega(t,z), \quad v_z = 0; \tag{11.206}$$

$$B_r = r\,h_r(t,z), \quad B_\varphi = r\,h_\varphi(t,z), B_z = B_z(t,z). \tag{11.207}$$

In this case, the current components (10.90)–(10.92) are written as

$$j_r = -\frac{cr}{4\pi}\frac{\partial h_\varphi}{\partial z}, \quad j_\varphi = \frac{cr}{4\pi}\frac{\partial h_r}{\partial z}, \quad j_z = \frac{ch_\varphi}{2\pi}. \tag{11.208}$$

After neglecting velocity v_z along the axis, we should omit the corresponding Euler equation (10.84), and the radial pressure gradient is approximated by the linear function

$$\frac{\partial P}{\partial r} = \lambda\frac{P}{R^2}r, \tag{11.209}$$

where the constant $\lambda \sim 1$ is determined by the pressure dependence on the radius, which is connected with the equation of state, $P(t,z)$ is the pressure, and $R(t,z)$ is the radius of the cylinder. From here on, we consider the adiabatic case, in which the polytropic equation of state $P = K\rho^\gamma$ is considered instead of the energy equation (10.94). Neglecting the z derivatives in the Poisson equation (10.93), we obtain $\varphi_{\rm G} = \pi G\rho r^2$. Substituting (11.206)–(11.209) into the original system of equations, we obtain for the profiling functions the following equations

$$\frac{\partial a}{\partial t} + a^2 - \Omega^2 = \lambda\frac{P}{\rho R^2} - 2\pi G\rho + \frac{1}{4\pi\rho}\left(B_z\frac{\partial h_r}{\partial z} - 2h_\varphi^2\right), \tag{11.210}$$

$$\frac{\partial \Omega}{\partial t} + 2a\Omega = \frac{1}{4\pi\rho}\left(B_z \frac{\partial h_\varphi}{\partial z} + 2h_r h_\varphi\right), \tag{11.211}$$

$$\frac{\partial h_r}{\partial t} = \frac{\partial (a B_z)}{\partial z}, \tag{11.212}$$

$$\frac{\partial h_\varphi}{\partial t} = \frac{\partial (\Omega B_z)}{\partial z} - 2(a h_\varphi - \Omega h_r), \tag{11.213}$$

$$\frac{\partial B_z}{\partial t} = -2a B_z, \tag{11.214}$$

$$\frac{\partial \rho}{\partial t} = -2a\rho, \tag{11.215}$$

$$\frac{\partial R}{\partial t} = aR. \tag{11.216}$$

It follows from relationships (11.214)–(11.216), representing conservation of mass, and the conservation of magnetic flux, equivalent to freezing condition, that

$$\rho R^2 = C_m(z), \quad B_z R^2 = C_b(z), \quad B_z = \frac{C_b(z)}{C_m(z)}\rho. \tag{11.217}$$

In the subsequent consideration, the arbitrary functions will be taken as constants: $C_m(z) = C_m$, $C_b(z) = C_b$. The algebraic relationships (11.217) can be used instead of any two equations from (11.214) to (11.216).

11.3.10.2 Further Simplification: Reducing the Problem to an Ordinary Differential Equation

While in the relativistic jet the self-gravitating force is expected to be much lower than the magnetic and pressure forces, let us neglect gravity in the subsequent consideration. Without gravity, the equilibrium static state of the cylinder does not exist. It is necessary to solve numerically the system of non-linear equations (11.210) – (11.213), (11.216), (11.217) to check for the existence of a cylinder whose radius remains finite as a result of torsional oscillations (dynamic confinement). Let us try instead to reduce the system to ordinary differential equations by making additional simplifications. Let us consider an axially symmetric jet moving along the z-axis with a constant bulk motion velocity in the comoving coordinate frame. In a jet in which confinement is reached as a result of standing magnetotorsional oscillations, there are points along the z-axis where the rotational velocity always remains zero in this frame. Let us take $\Omega = 0$ in the plane $z = 0$. Let us consider standing-wave torsional oscillations with the space period along the z-axis equal to z_0. Thus nodes with $\Omega = 0$ are situated at $z = \pm n z_0/2$, $n = 0, 1, 2, \ldots$. Let us write the equations describing the behaviour of the cylinder in the plane $z = 0$, where $\Omega = 0$. Denote all values in this plane by a tilde. Let us take, for simplicity, $\lambda = 11$. Then, equations in the plane $z = 0$ are written as

$$\frac{\mathrm{d}\tilde{a}}{\mathrm{d}t} + \tilde{a}^2 = \frac{K}{\tilde{R}^2} + \frac{C_b}{4\pi C_m}\left(\frac{\partial h_r}{\partial z}\right)_{z=0} - \frac{\tilde{h}_\varphi^2 \tilde{R}^2}{2\pi C_m}, \tag{11.218}$$

from equation (11.210);

$$\frac{C_b}{4\pi C_m}\left(\frac{\partial h_\varphi}{\partial z}\right)_{z=0} + \frac{\tilde{h}_\varphi \tilde{h}_r \tilde{R}^2}{2\pi C_m} = 0, \tag{11.219}$$

from equation (11.211);

$$\frac{d\tilde{h}_r}{dt} = C_b\left(\frac{\partial(a/R^2)}{\partial z}\right)_{z=0}, \tag{11.220}$$

from equation (11.212);

$$\frac{d\tilde{h}_\varphi}{dt} = C_b\left(\frac{\partial(\Omega/R^2)}{\partial z}\right)_{z=0} - 2\tilde{a}\tilde{h}_\varphi, \tag{11.221}$$

from equation (11.213);

$$\frac{d\tilde{R}}{dt} = \tilde{a}\tilde{R}. \tag{11.222}$$

from equation (11.216). The integrals of motion (11.217) in the plane $z = 0$ are written as

$$\tilde{\rho}\tilde{R}^2 = C_m, \quad \tilde{B}_z \tilde{R}^2 = C_b, \quad \tilde{B}_z = \frac{C_b}{C_m}\tilde{\rho}. \tag{11.223}$$

Initial conditions for the system (11.218)–(11.223) are

$$\tilde{R} = R_0, \ \tilde{\rho} = \rho_0 = \frac{C_m}{R_0^2}, \ \tilde{B}_z = \frac{C_b}{R_0^2}, \ \tilde{a} = \tilde{h}_r = \tilde{h}_\varphi = 0 \text{ at } t = 0. \tag{11.224}$$

In (11.218)–(11.223) were used relationships

$$\tilde{\rho} = \rho_0\frac{R_0^2}{\tilde{R}^2}, \ \tilde{B}_z = \rho_0\frac{C_b}{C_m}\frac{R_0^2}{\tilde{R}^2}, \tag{11.225}$$

valid for any time. If the rotational velocity of the cylinder is antisymmetrically relative to the plane $z = 0$, $\Omega = 0$, and the density distribution of the cylinder is symmetric relative to this plane, then we have an extremum (maximum) of the azimuthal magnetic field h_ϕ, with $\left(\frac{\partial h_\varphi}{\partial z}\right)_{z=0} = 0$, and $\tilde{h}_r = 0$ reaches an extremum (minimum) in this plane with $\left(\frac{\partial h_r}{\partial z}\right)_{z=0} = 0$. The product $a\rho$ also reaches an extremum in the plane $z = 0$, so that $\left(\frac{\partial(a/R^2)}{\partial z}\right)_{z=0} = 0$. The term with the z derivative in equation (11.221) is not equal to zero, and changes periodically during the torsional oscillations. We approximate the derivative d/dz by the ratio $1/z_0$, where z_0 is the space period of the torsional oscillations along the z-axis. While $\Omega = 0$ in the plane $z = 0$, its derivative along z changes periodically with an amplitude Ω_0 and frequency ω, which should be found from the solution of the problem. We therefore approximate

$$\left(\frac{\partial(\Omega/R^2)}{\partial z}\right)_{z=0} = \frac{\Omega_0}{z_0\tilde{R}^2}\cos\omega t. \tag{11.226}$$

Finally, we have from (11.218),(11.221) and (11.222) the following approximate system of equations, describing the non-linear torsional oscillations of the cylinder at given z_0 and Ω_0.

$$\frac{d\tilde{a}}{dt} + \tilde{a}^2 = \frac{K}{\tilde{R}^2} - \frac{\tilde{h}_\varphi^2\tilde{R}^2}{2\pi C_m},$$

$$\frac{d\tilde{h}_\varphi}{dt} = C_b\frac{\Omega_0}{z_0\tilde{R}^2}\cos\omega t - 2\tilde{a}\tilde{h}_\varphi,$$

$$\frac{d\tilde{R}}{dt} = \tilde{a}\tilde{R}. \tag{11.227}$$

Combining the last two equations gives

$$\frac{d(\tilde{h}_\varphi R^2)}{dt} = \frac{C_b\Omega_0}{z_0}\cos\omega t, \tag{11.228}$$

with the solution, satisfying initial condition (11.224), in the form

$$\tilde{h}_\varphi R^2 = \frac{C_b\Omega_0}{z_0\omega}\sin\omega t. \tag{11.229}$$

Taking account of (11.229), the first and third equations in (11.227) give the equation

$$\tilde{R}\frac{d(\tilde{a}R)}{dt} = K - \left(\frac{C_b\Omega_0}{z_0\omega}\right)^2\frac{\sin^2\omega t}{2\pi C_m}, \tag{11.230}$$

Two differential equations, namely (11.230) and the third equation (11.227), determine the behaviour of the cylinder during magnetotorsional oscillations. Solutions in which the radius does not go to infinity with time determine a dynamically confined cylinder. The formation of blobs occurs, when the radius tends to zero. A long dynamically confined cylinder exists when its radius changes with time between two finite values.

11.3.10.3 Numerical Solution

Let us introduce non-dimensional variables

$$\tau = \omega t, \; y = \frac{\tilde{R}}{R_0}, \; z = \frac{a\tilde{R}}{a_0 R_0}, \; a_0 = \frac{K}{\omega R_0^2} = \omega, \; R_0 = \frac{\sqrt{K}}{\omega}, \qquad (11.231)$$

in which the differential equations have the form

$$\frac{dy}{d\tau} = z, \quad \frac{dz}{d\tau} = \frac{1}{y}(1 - D\sin^2\tau); \quad y(0) = 1, \; z = 0 \text{ at } \tau = 0. \qquad (11.232)$$

Therefore, the problem is reduced to a system (11.232) with only two non-dimensional parameters $D = \frac{1}{2\pi K C_m}\left(\frac{C_b \Omega_0}{z_0 \omega}\right)^2$, and $y(0)$, and the second one is taken equal to unity in further consideration. The solution of this nonlinear system changes qualitatively with changes in the parameter D.

The solution of this system was obtained numerically in [117] for $D = 2, 2.1, 2.11, 2.15, 2.2, 2.25, 2.28, 2.4, 2.5, 2.6, 2.9, 3.0$. The results of the numerical solution are represented in Figs. 11.31–11.42. Roughly the solutions may be divided into three groups.

(1) At $D \le 2$ there is no confinement, and radius grows to infinity after several low-amplitude oscillations (Fig. 11.31).
(2) With increasing D the amplitude of oscillations increase, and at $D = 2.1$ radius is not growing to infinity, but is oscillating around some average value, forming rather complicated curve (Figs. 11.32, and 11.33).
(3) At $D = 2.28$ and larger the radius finally goes to zero with time, but in a manner that depends on the value of D. For D between 2.28 and 2.9, the dependence of the radius y with time may be very complicated, consisting of low-amplitude and large-amplitude oscillations that finally go to zero. The time at which the radius becomes zero depends on D in a somewhat peculiar way: it may occur at $\tau \le 100$, as it does for $D = 2.4, 2.6$ (Figs. 11.36 and 11.39); or it may go through a very large radius and return to zero at a very large time, for example, $\tau \sim 10^7$ for $D = 2.5$ (Figs. 11.37, and 11.38). For $D = 3$ and larger, the solution becomes very simple, and the radius goes to zero at $\tau < 2.5$ (Fig. 11.42), before the right-hand side of the second equation (11.232) returns to a positive value.

11.3.10.4 Restrictions of the Model

Let us consider a jet that has an equation of state of the form $P = K\rho = v_s^2\rho$, $v_s^2 \le c^2/3$, v_s is the sound speed in the jet's matter. For ultrarelativistic pair-plasma, we have $P = c^2/3$. The non-dimensional parameter D, as a function of the characteristic radius R_0, periodic length z_0 along z-axis, initial density ρ_0 and magnetic field B_{z0}, Ω_0 and ω is written in the form

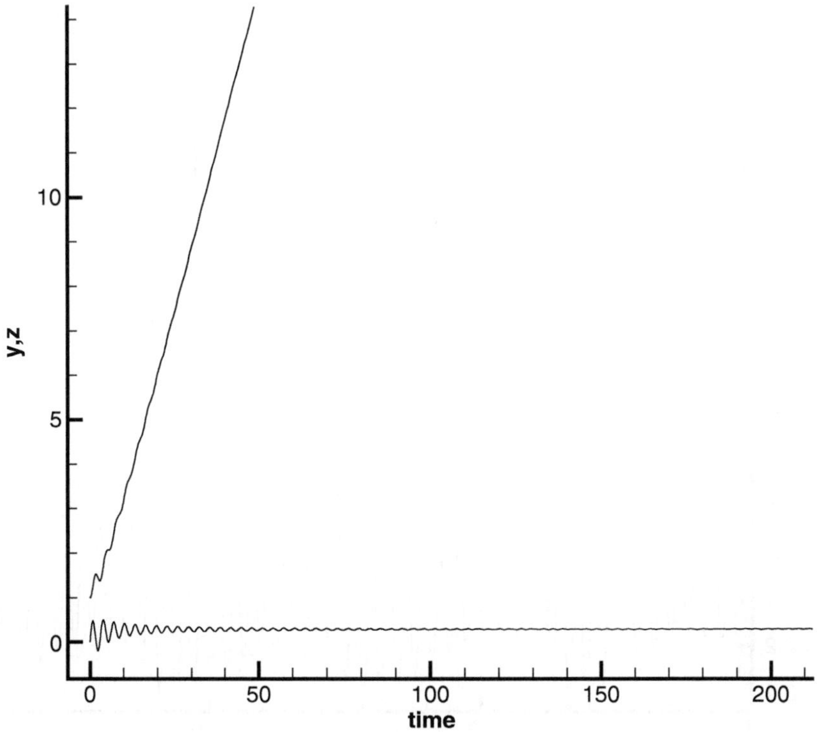

Fig. 11.31 Time dependence of non-dimensional radius y (*upper curve*) and non-dimensional velocity z (*lower curve*) for $D = 2.0$ from [117]

$$D = \frac{1}{2\pi\rho_0} \frac{B_{z0}^2 R_0^2 \Omega_0^2}{z_0^2 \omega^2 v_s^2}. \tag{11.233}$$

The amplitude of oscillations Ω_0, and ω should be found from the solution of the non-linear system (11.210)–(11.216), together with determination of the interval of values of D at which confinement occurs. To find approximately a self-consistent model with $\Omega_0, \omega(D)$, we may use the solution of linearized equations (11.210)–(11.216) with $\omega = k V_A$, with alfvenic velocity $V_A = B_{z0}/\sqrt{4\pi\rho}$. The frequency of non-linear oscillations is smaller, and we can write

$$\omega = \alpha_n\, k\, V_A,\ \alpha_n < 1,\ k = \frac{2\pi}{z_0}, \tag{11.234}$$

so that

$$\omega^2 = \alpha_n^2 k^2\, V_A^2 = \alpha_n^2 \frac{\pi B_{z0}^2}{\rho_0 z_0^2}. \tag{11.235}$$

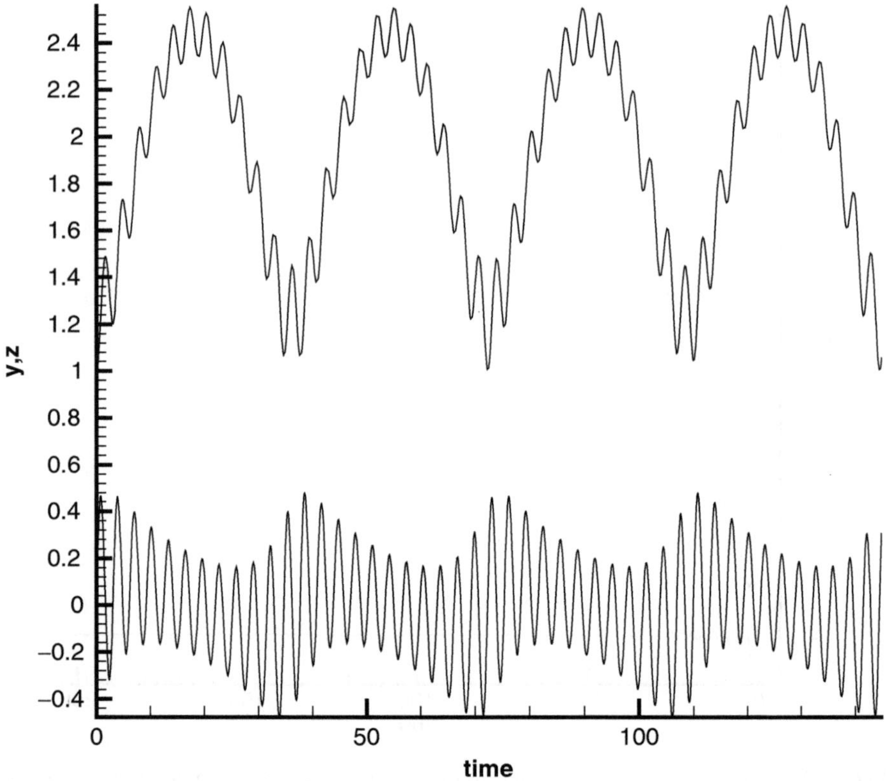

Fig. 11.32 Time dependence of non-dimensional radius y (*upper curve*) and non-dimensional velocity z (*lower curve*), for $D = 2.1$ from [117]

Using it in (11.233), and taking into account of (11.231), we obtain

$$\Omega^2 R_0^2 = 2\pi^2 D\alpha_n^2 v_s^2 < c^2, \;\; R_0^2 = \frac{K}{\omega^2} = z_0^2 \frac{\rho_0 v_s^2}{\alpha_n^2 \pi B_{z0}^2}. \tag{11.236}$$

On the edge of the cylinder, the rotational velocity cannot exceed the light velocity, so the solution with initial conditions in (11.232), corresponding to $y_0 = 1$, has a physical sense only at $v_s^2 < \frac{c^2}{2\pi^2 D\alpha_n^2} \approx \frac{c^2}{40\alpha_n^2}$. Taking $\alpha_n^2 = 0.1$ for a strongly non-linear oscillations, we obtain a very moderate restriction $v_{s0}^2 < \frac{c^2}{4}$. While in the intermediate collimation regime, the amplitude of the outer tangential velocity is not changing significantly; this restriction would hold also for the whole period of the time. To have the sound velocity not exceeding $c/2$, the jet should contain baryons. The density of the barions ρ_0 must be above a certain value and must contribute more than about 30% to the total density of the jet.

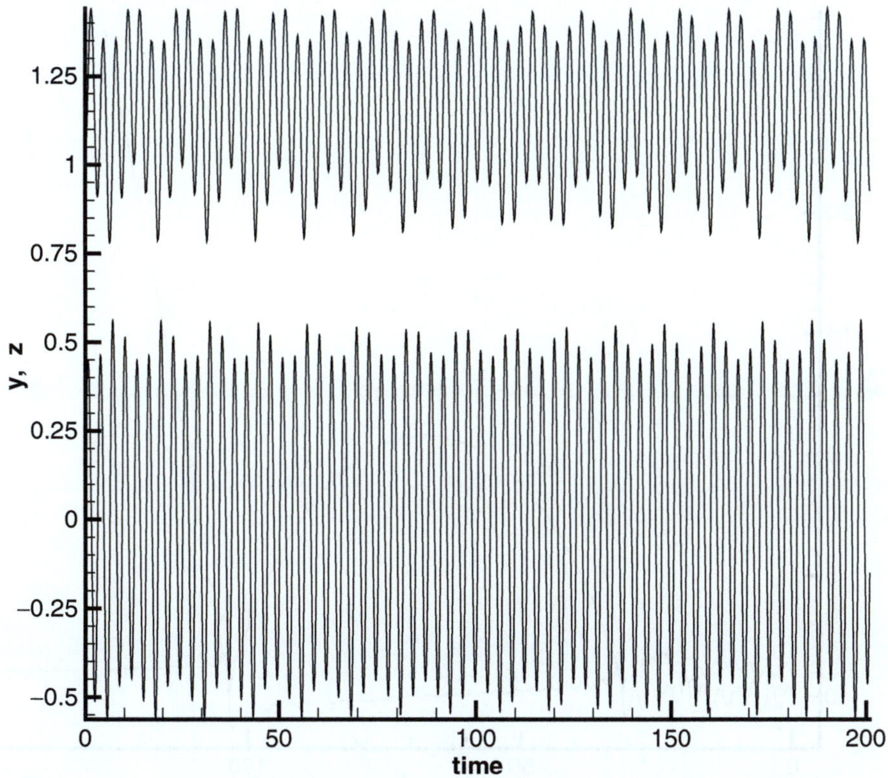

Fig. 11.33 Time dependence of non-dimensional radius y (*upper curve*), and non-dimensional velocity z (*lower curve*), for $D = 2.25$, during a long time period from [117]

Confinement by torsional oscillations starts at $D = 2.1$, and at $D \geq 2.28$ the jet is divided into separate blobs according to Figs. 11.34–11.42. Confinement by magnetotorsional oscillations is therefore physically realizable [117]. Extending the choice of parameters outside the point $y_0 = 1$, permits to construct all possible kinds of restricted solutions of the system (11.232). There are chaotic and regular solutions, depending on the combination of parameters D and y_0. The regular solutions exist for D in the interval 2.1 – 2.28, where jets with $y_0 = 1$ remain finite with non-zero radius, but also outside this interval for different y_0. Varying the initial y_0 for the same value of D, it is possible to find a pure periodic solution for a particular value of y_0. It is interesting to follow the evolution of this periodic solution with increasing D. The behaviour of this solution resembles the behaviour of solutions of many different equations, where change of a parameter is accompanied by the transition from the periodic solution into the chaotic one, via period doubling [382]. The sequence of values of corresponding period doubling D_i may follow the low

$$\lim |_{i\to\infty} \frac{D_i - D_{i-1}}{D_{i+1} - D_i} = \text{Const.}$$

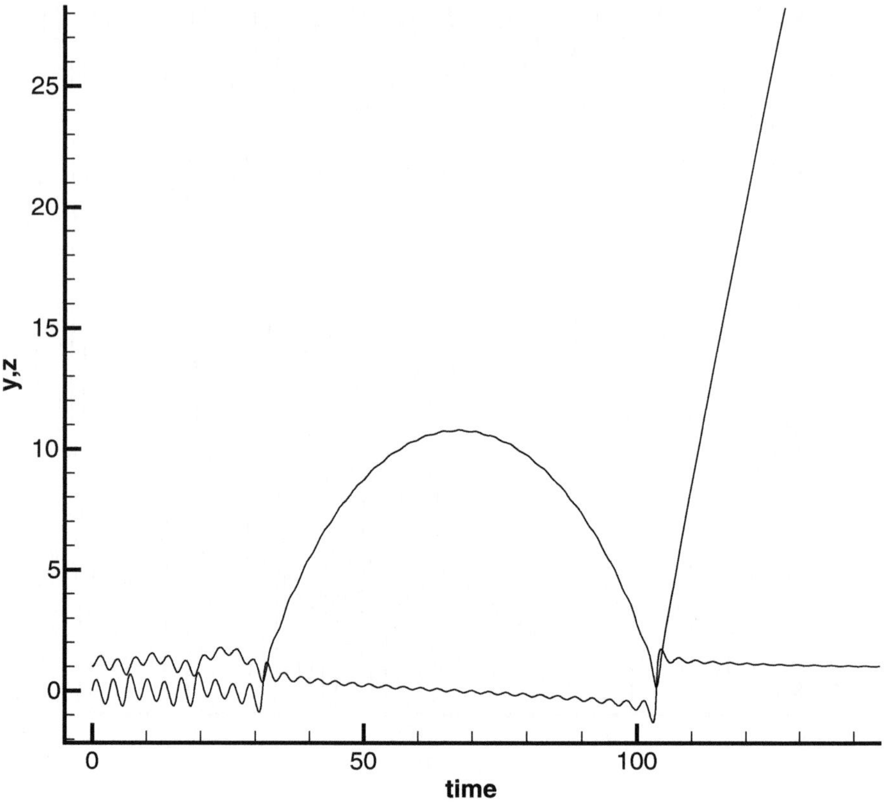

Fig. 11.34 Time dependence of non-dimensional radius y (*upper curve*), and non-dimensional velocity z (*lower curve*), for $D = 2.28$ from [117]

There are several (at least two) different constants in this limit, which appear in a wide family of different equations, describing different physical processes. One of them is related to systems with dissipation [382] (Const ≈ 2.412), and another one appears in the conservative systems [441], (Const ≈ 8.721). It is not yet clear which value of the constant governs the approach of the system (11.232) to chaos; the work is in progress [135].

11.4 Cosmic Gamma Ray Bursts: Observations and Modeling

Nowadays, it is commonly accepted that cosmic gamma-ray bursts (GRBs), the discovery of which was reported in 1973 [574], are of cosmological origin. The first cosmological model based on explosions in active galaxy nuclei (AGN) was proposed in [815]. GRB formation near a collapsing object due to neutrino antineutrino annihilation was investigated in [90]. Previously, the model of GRB formation

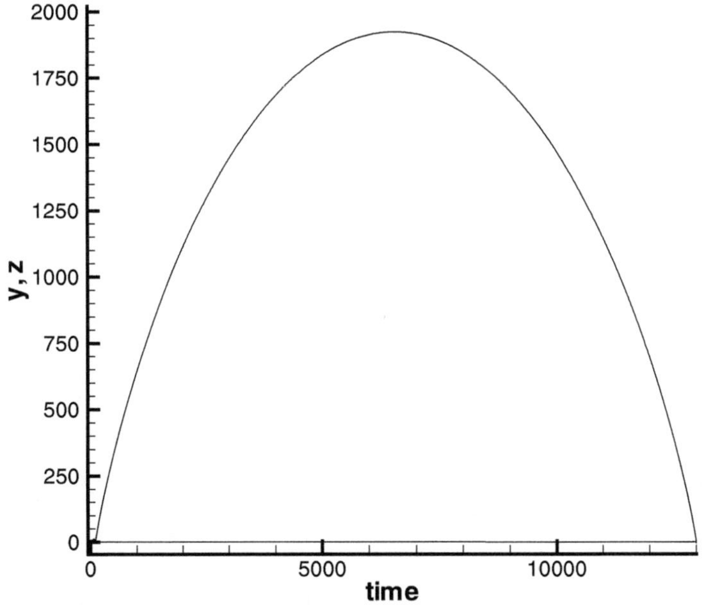

Fig. 11.35 Time dependence of non-dimensional radius y (*upper curve*), and non-dimensional velocity z (*lower curve*), for $D = 2.28$, during a long time period from [117]

during collapses and explosions of supernovae was considered in [126], where the following chain of reactions resulting in gamma-photon production was considered:

$$\tilde{\nu} + p \rightarrow n + e^{+}, \tag{11.237}$$

$$e^{+} + e^{-} \rightarrow 2\gamma \ (0.5 \text{ MeV}), \tag{11.238}$$

$$n + p \rightarrow d + \gamma \ (2.3 \text{ MeV}), \tag{11.239}$$

$$d + p \rightarrow \text{He}^3 + \gamma \ (5.5 \text{ MeV}), \tag{11.240}$$

$$\text{He}^3 + \text{He}^3 \rightarrow \text{Be}^7 + \gamma \qquad (1.6 \text{ MeV}). \tag{11.241}$$

In brackets, we give photon energies. The reactions with heavy nuclei were also considered in [126]:

$$\nu + (A, Z) \rightarrow (A, Z + 1) + e^{-}, \tag{11.242}$$

$$\tilde{\nu} + (A, Z) \rightarrow (A, Z - 1) + e^{+}. \tag{11.243}$$

The starquake, the subsequent explosion and the ejection of matter from a nonequilibrium layer in the neutron-star crust discovered in [122, 194] are accompanied by the gamma rays induced by the fission of superheavy nuclei. This scenario was

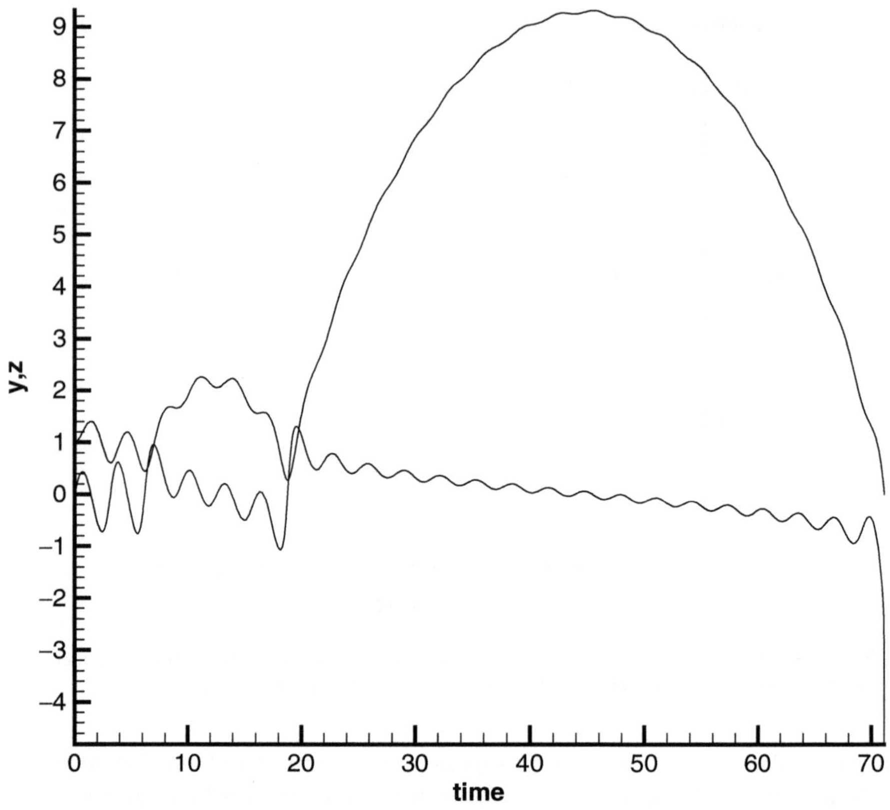

Fig. 11.36 Time dependence of non-dimensional radius y (*upper curve*), and non-dimensional velocity z (*lower curve*), for $D = 2.4$ from [117]

proposed in [126] as an alternative model of the galactic GRB origin. The total number of various galactic models exceeds 100. Even now, after the discovery of redshifts in optical afterglows, which became possible due to the X-ray observations of GRB with the Beppo-SAX satellite and their subsequent optical identification, the galactic models are of more than simply historical interest because the origin of the activity of soft gamma repeaters (SGR) – neutron stars inside the Galaxy – is not quite certain. In [90], it was obtained that the efficiency of transformation of neutrino-flux energy $W_\nu \sim 6 \cdot 10^{53}$ erg into the energy of the X and γ -ray burst amounts to a fraction $\alpha \sim 6 \cdot 10^{-6}$ with a total GRB energy release $W_{X,\gamma} \sim 3 \cdot 10^{48}$ erg. The 3D numerical simulation of two colliding neutron stars [848] and a hot torus around a black hole [849] showed higher efficiency of X- and γ-ray production, reaching 0.5% in the first case and 1% in the second case. In part, the distinction with [90] can be associated with the geometry more preferential for the neutrino-flux outflow when the annihilation rate increases in comparison with that estimated in the spherical geometry [90]. Nevertheless, even in such an optimistic

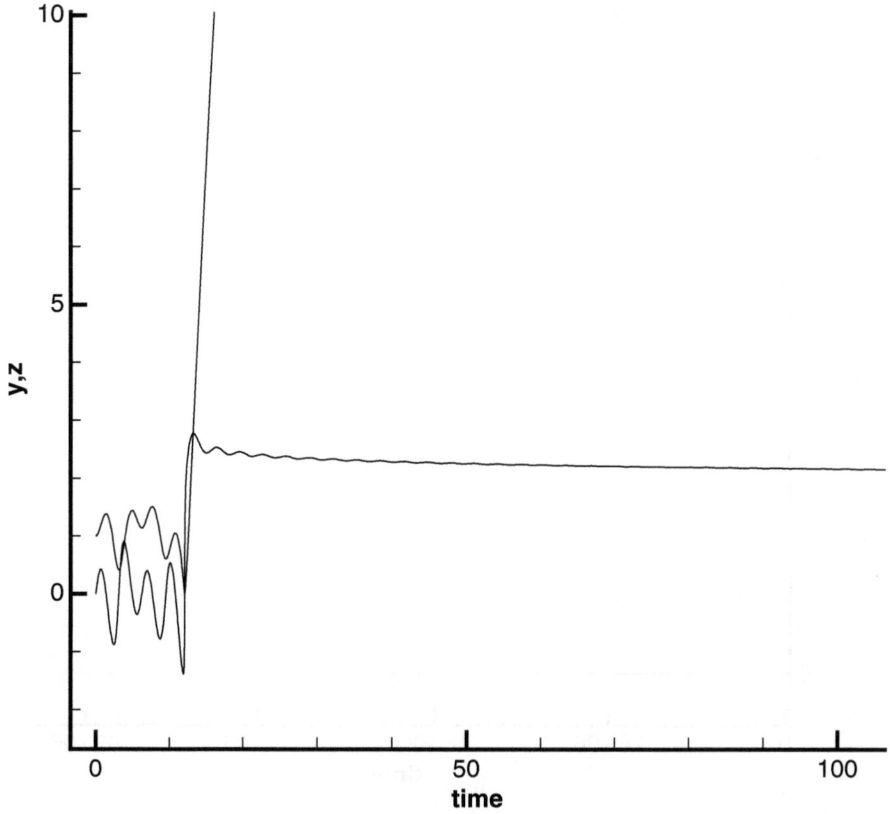

Fig. 11.37 Time dependence of non-dimensional radius y (*upper curve*), and non-dimensional velocity z (*lower curve*), for $D = 2.5$ from [117]

variant, the GRB formation with a total energy yield in the X- and γ-ray regions is not exceeding $5 \cdot 10^{50}$. This energy is insufficient for explaining the energy resource of many GRBs because only the direct optical GRB radiation can attain 10^{51} erg, and the isotropic flux in the gamma region reaches $2.3 \cdot 10^{54}$ erg for GRB 990123 with the redshift $z \sim 1.6$ [8, 597]. To explain such a high observable energy release, it is necessary to have a strong collimation.

11.4.1 Central Engine of Cosmic Gamma-Ray Bursts

The following models of central engine as a source of the energy are worth to be mentioned.

1. *The coalescence of two neutron stars or a neutron star and a black hole of a stellar mass.* This mechanism was numerically investigated in [848, 849]. The γ

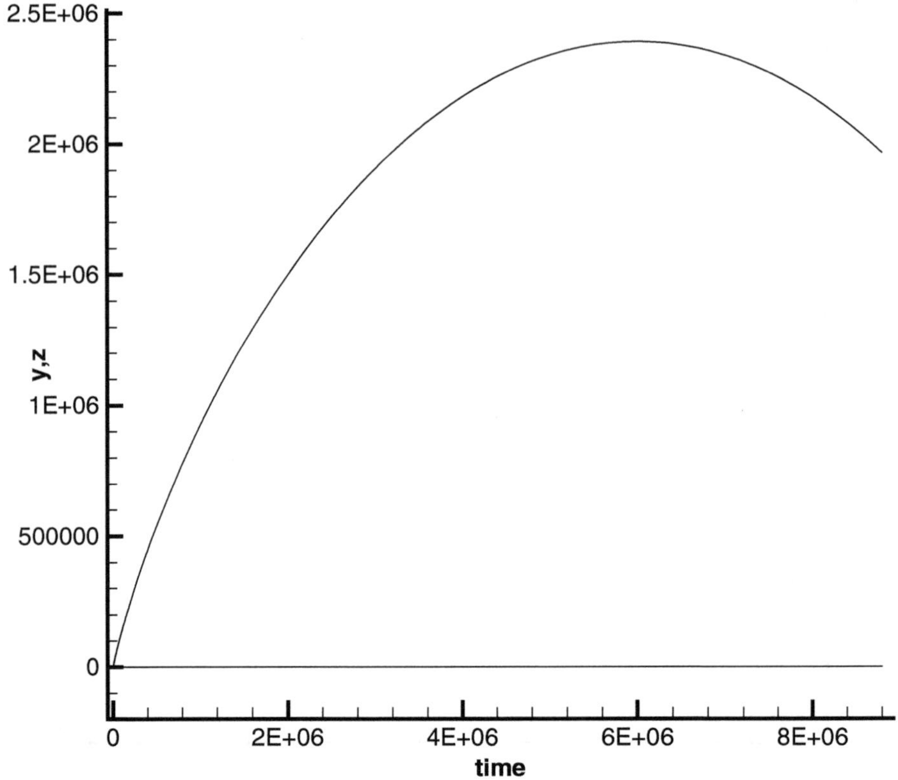

Fig. 11.38 Time dependence of non-dimensional radius y (*upper curve*), and non-dimensional velocity z (*lower curve*), for $D = 2.5$, during a long time period from [117]

rays are produced here due to the $(\nu, \tilde{\nu})$ annihilation, and the energy yield proves to be insufficient for explaining the most powerful GRBs even assuming a strong collimation of the GRB radiation. The energy emitted only in the GRB 990123 optical afterglow exceeds the total radiation energy yield in this model approximately by an order of magnitude.

2. *Magnetorotational explosion.* The 2D numerical calculations of a magnetorotational supernova explosion in [45, 707] showed that the efficiency of transformation of the rotation energy into explosive energy amounts to about 10%. The released energy is sufficient for explaining the explosion of a collapsing supernova but proves to be insufficient for cosmological GRB.
3. *Hypernova (very powerful supernova).* A very extensive radiation from a new born pulsar, very rapid, with very high magnetic field, was suggested in [998]. W-R stars as GRB predecessors have been considered in [414, 982]. The model, based on a collapse of a very massive star, formation of a black hole (BH) with a massive disk, with a rapid accretion into BH have been considered in [662].

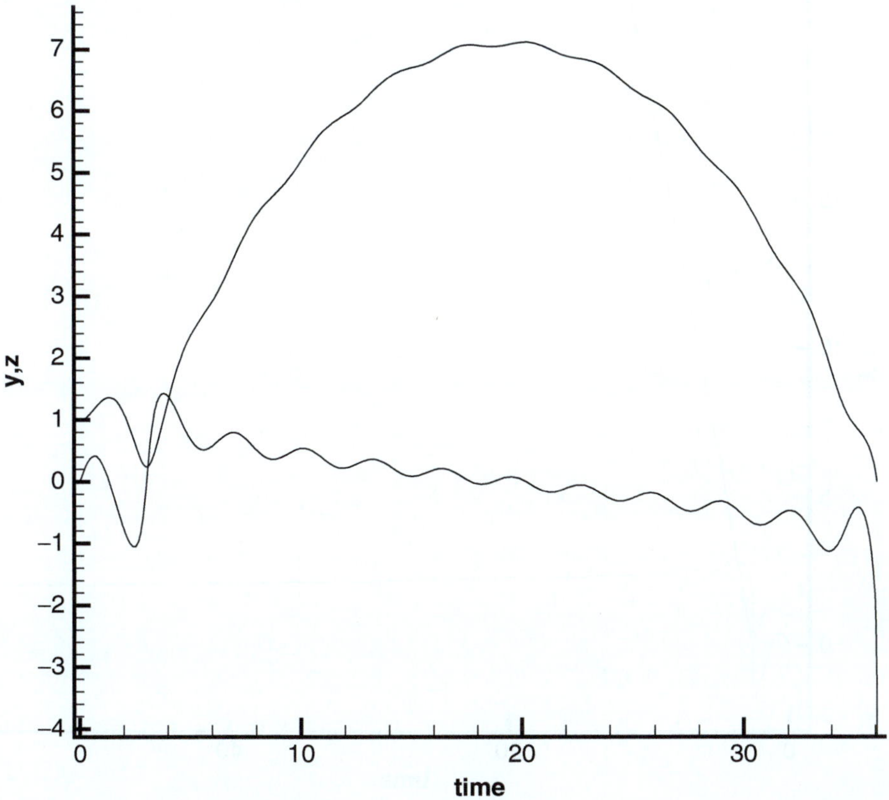

Fig. 11.39 Time dependence of non-dimensional radius y (*upper curve*), and non-dimensional velocity z (*lower curve*), for $D = 2.6$ from [117]

Accretion into BH from a magnetized disk was calculated in [68]. The models of this type are the most popular now. SNe spectra are found in optical afterglows of several GRB [923,934].

4. *Magnetized disk around a (Kerr) rotating black hole (RBH).* This model is based on extracting the RBH rotation energy for the GRB production due to magnetic coupling of the RBH and the surrounding accretion disk or torus [1024]. Magnetic extraction of energy from the rotating black hole (Blandford–Znajek mechanism) is claimed to work also after the collapse of the magnetized disk [68,69,971].
5. *Dyadosphere.* In the model [850], the GRB originates from an explosive formation of electron–positron pairs and the electromagnetic pulse from a strongly electrically charged black hole. The key problem here consists in the possibility of formation of such a strongly charged BH.

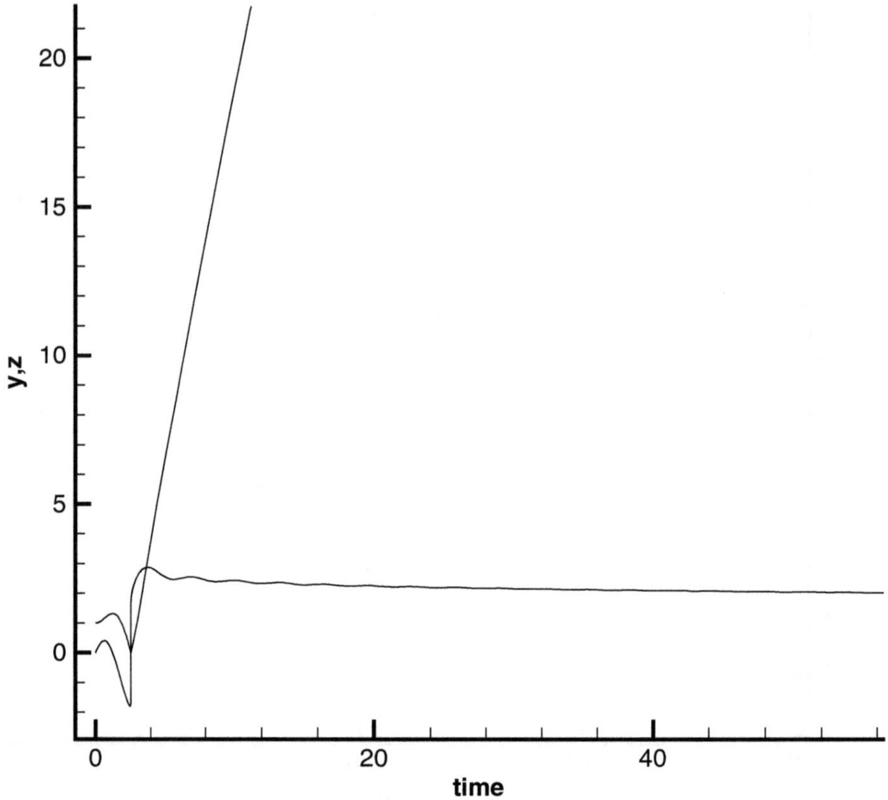

Fig. 11.40 Time dependence of non-dimensional radius y (*upper curve*), and non-dimensional velocity z (*lower curve*), for $D = 2.9$ from [117]

11.4.2 Optical Afterglows

The X-ray afterglow discovered by the Italian Beppo-SAX satellite made it possible to identify GRBs optically and to obtain their spectra. These spectra showed the presence of a large redshift z, reaching ~ 8 [961], which points to the cosmological origin of GRBs and an enormous energy release. The comparison of redshifts and total energy fluxes shows no correlation [116] between the distance and observable flux (Fig. 11.43). This fact is usually explained by a strong collimation leading to a wide spread in observed fluxes due to observations of a narrow beam under different angles of view.

As was noted in [803], the GRB-afterglow properties are better explained under the assumption that GRB sources are located in the star-formation regions with a high gas and dust density. The interaction of a powerful GRB pulse with the surrounding gas of a density $n = 10^4 - 10^5\ \mathrm{cm}^{-3}$ results in the occurrence of a special-form optical afterglow, the duration of which can reach up to 10 years. The

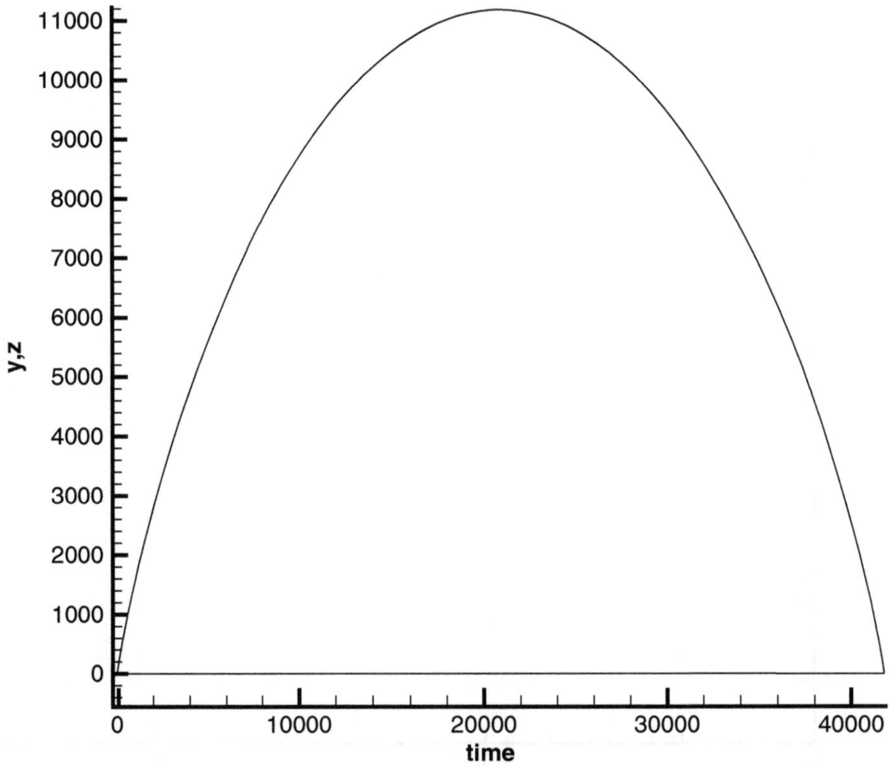

Fig. 11.41 Time dependence of non-dimensional radius y (*upper curve*), and non-dimensional velocity z (*lower curve*), for $D = 2.9$, during a long time period from [117]

light curve and the spectrum of such an afterglow in the uniform cloud were calculated in [192] (Fig. 11.44). It was shown in [70] that the duration and spectrum of the GRB optical afterglow depend on the GRB location with respect to the dense cloud. In the case of an anisotropic GRB, they depend also on the orientation of the observer and the cloud with respect to the GRB axis. The light curves presented in Fig. 11.45 correspond to the anisotropic GRB beam in the anisotropic cloud, with the explosion at the distance $R_0 = 1$ pc from the center of the molecular cloud. The angular distribution of the energy in the GRB impulse is specified as $E = E_0\, e^{(-\theta/\theta_0)^2}$, where $E_0 = 10^{52}$ erg is the isotropic energy of the GRB, and the density distribution as $n = 10^5\, e^{-(r/r_0)^2}$ cm^{-3}, where r is the distance from the center of the molecular cloud, and r_0 is the radius of the quasi-homogeneous central region of the molecular cloud. The GRB axis is directed from the cloud center.

Prompt optical emission was first observed in GRB990123. Its total duration is about 100 s, T(50%) = 30 s, T(90%) = 63 s. Optical observations were started by ROTSE, at $t = 22.18$ s after the beginning of GRB, and reached 8.95^m at maximum. Observations on Keck revealed the redshift z = 1.61, which corresponds to

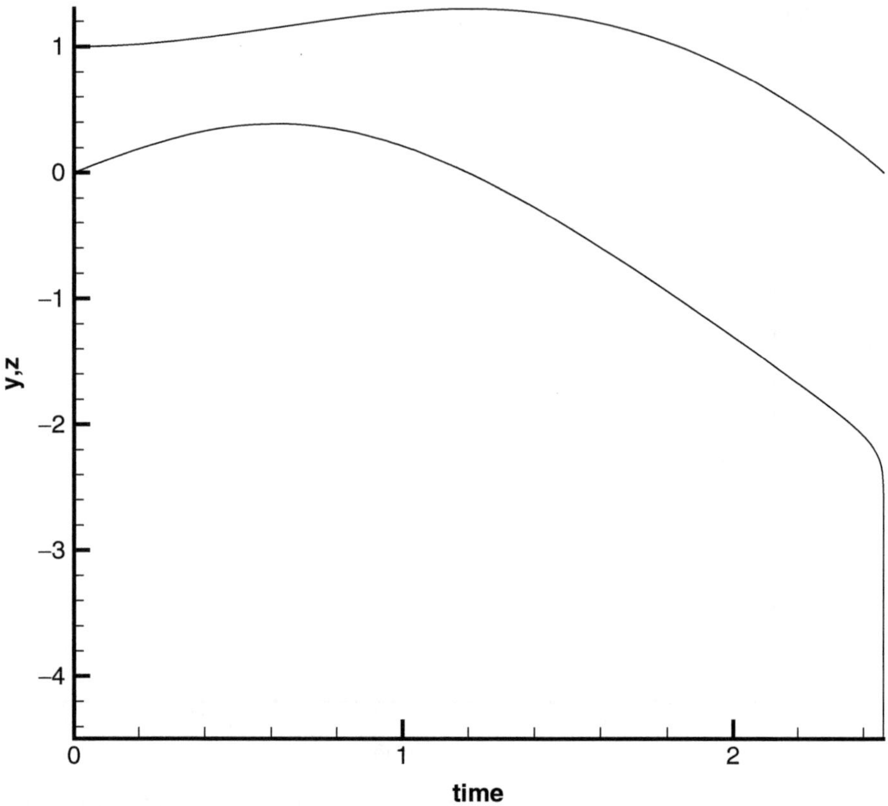

Fig. 11.42 Time dependence of non-dimensional radius y (*upper curve*) and non-dimensional velocity z (*lower curve*), for $D = 3.0$, from [117]

the energy release $Q_\gamma > 2.3 \cdot 10^{54}$ erg, and $L_{opt} > 2 \cdot 10^{16} L_\odot = 8 \cdot 10^{49}$ erg/s. Light curve of the prompt optical emission together with γ is represented in Fig. 11.46, due to courtesy of A. Pozanenko (2003, Private communication). The strategy of the afterglow observations, based on continuous sky monitoring by wide-field cameraes, suggested in [811], led to discovery of the brightest optical afterglow with peak brightness 5.3^m, available to the naked eye. The first wild-field camera FAVOR (Fig. 11.47) for observations of rapid transients was established in SAO, Caucasus, Russia, in the year 2003. In the year 2006, working in combination with the Italian telescope REM, the similar wide-field camera TORTORA, where the discovery was done, was installed in La Silla, Chile (Fig. 11.48). In Fig. 11.49, the gamma-ray light curve from KONUS, and in Fig. 11.50 the one from SWIFT, together with the simultaneous TORTORA optical light curve are presented [433], [818]. It is clear from Figs. 11.49,11.50 that prompt optical emission is strongly correlated with gamma ray pulse. May be it has the same collimation, as gamma radiation. Optical afterglows at longer times should not be collimated in the same way,

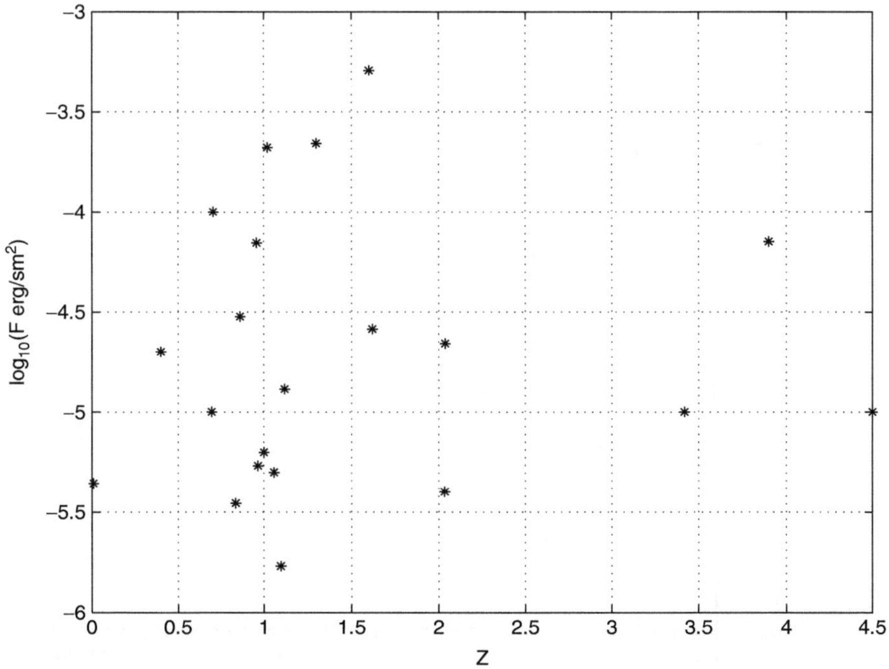

Fig. 11.43 Total observed energy fluence F as a function of the redshift z, from [116]

or be not collimated at all. If prompt optical and gamma ray radiation have the same collimation angle, then the number of bright orphan optical flashes should be very limited, or absent at all.

Infrared afterglow without any optics was observed in GRB041219 [708]. The explanation of this unusual afterglow was done [70], based on the model of reradiation by dust in the dense dust cloud, which is not evaporated by the GRB flash.

11.4.3 Short GRB and Giant SGR Bursts

Possibility to observe SGR giant bursts in the nearby galaxies was discussed in [681] and [115]. First identification of the giant burst 1 February 2007 in Andromeda galaxy M31 was done in [682]. It was observed as a short GRB070201, but the detailed analysis of gamma-ray emission in the tail have indicated its origin as SGR 0044 + 42 (M31). The energy of the burst in M31 is about $1.5 \cdot 10^{45}$ erg, which is smaller than the energy of the giant burst of SGR1806-20, estimated as $2.3 \cdot 10^{46}$ erg [377]. The giant burst in SGR1806-20 was the strongest among those observed in other SGR and was the only one where reliable data about total energy

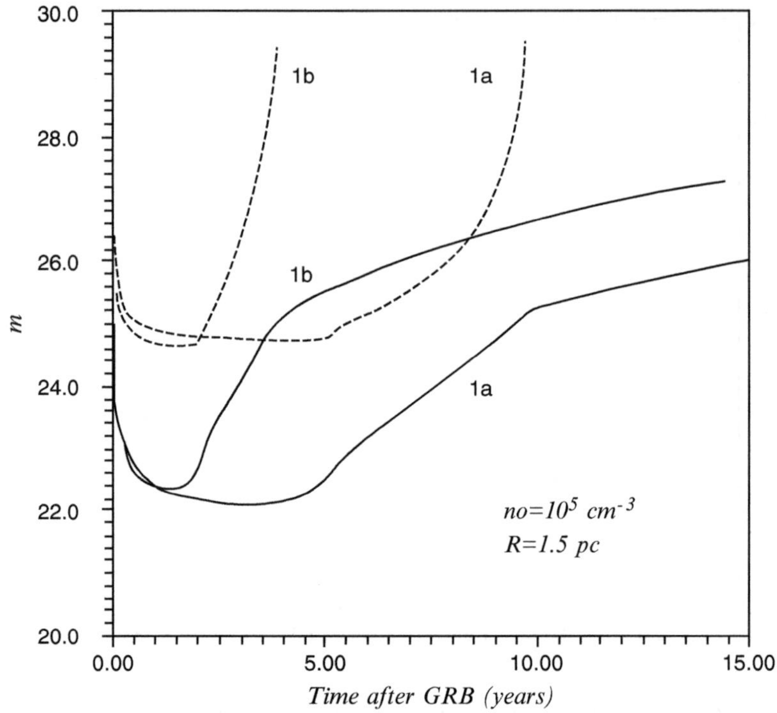

Fig. 11.44 Light curve of the optical afterglow in stellar magnitudes. The *solid line* gives the upper limit and the *dashed line* the lower limit. The origin of coordinates corresponds to the moment of the GRB flare with an energy E and a total fluence near the Earth $F_{\rm GRB} = 10^{-4}$ erg cm^{-2}: curves 1a and 1b correspond to the cases of $E = 10^{52}$ erg, $n_0 = 10^5$ cm^{-3}; and $E = 10^{51}$ erg, $n_0 = 10^5$ cm^{-3}, respectively, from [192]

release have been obtained, due to observation of the flux, reflected from the Moon. Note [682], that we have only lower limits: $7 \cdot 10^{44}$ erg in SGR 0526-66 (LMC) and $4.3 \cdot 10^{44}$ erg in SGR 1900 + 14 because of the absence of data about maximum brightness in these 2 SGR. It is necessary also to have in mind that three above estimations are based on the identification of SGR with SN remnants. If these, rather questionable, identifications are not valid, then the energy estimations become not valid too, because of very poor possibilities to estimate the distance to these SGR by any other method. The strong burst in SGR 1627-41 was considerably lower energetic ($1 \cdot 10^{43}$ erg). Short GRB 051103, interpreted as giant bursts of SGR, was also observed on 3 November 2005 in M81 group of galaxies [378]. Its energy release $7 \cdot 10^{46}$ erg is the largest among SGR giant bursts. Only the form of the short burst and its spectrum indicate to its possible origin as SGR 0952 + 69 (M81).

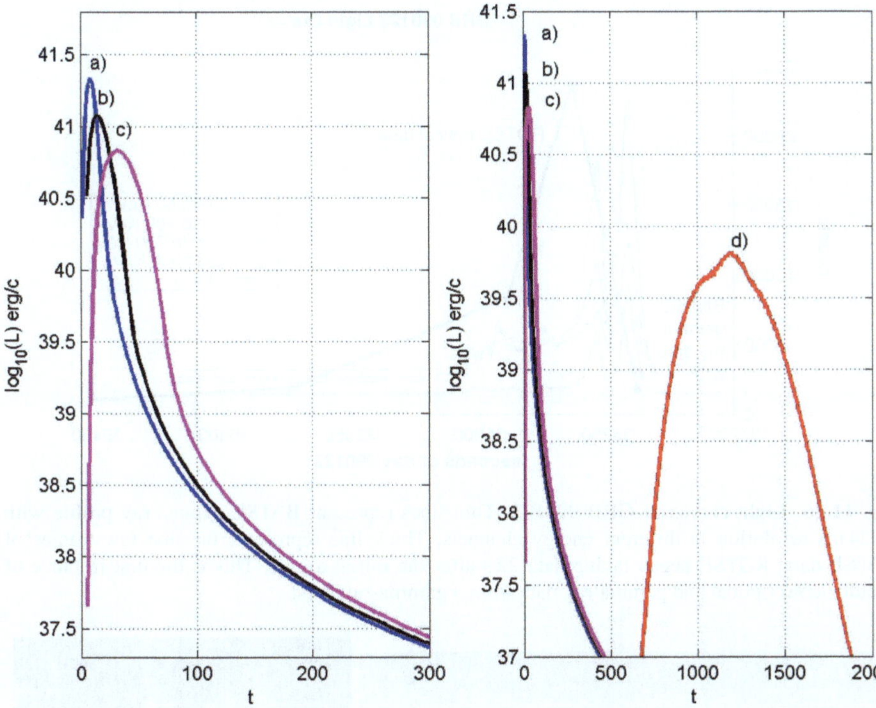

Fig. 11.45 Light curves for the parameters of the GRB and molecular cloud, indicated in the text, for an observer located on the GRBcloud axis (curve A), at an angle of $\alpha = 0.1$ rad (curve B), at an angle of $\alpha = 0.2$ rad (curve C), and at an angle of $\alpha = \pi/2$ rad (curve D). The time is measured in days after the GRB, from [70]

11.4.4 High Energy Afterglows (30–10,000 MeV)

Observations in the EGRET experiment onboard of the orbital Compton γ-ray observatory (CGRO) show that the GRBs radiate also in hard gamma regions up to an energy of 20 GeV [386]. Hard γ-rays were detected approximately for 10 GRBs by EGRET, and this number is increasing due to observations from the LAT experiment on the Fermi γ-ray satellite. As a rule, the hard γ rays last longer than the basic soft GRB by up to 1.5 h in GRB 940217 (Fig. 11.51). The comparison of the angular aperture in the experiments EGRET and BATSE, and the duration of their operation at the orbit, results in the conclusion that the hard γ rays should be observed in (1/3 – 1/2) of all GRB. If a GRB is associated with the supernova explosion and the neutron-star birth, the neutron-star vibrations after the explosion can be responsible for a afterglow in the hard gamma region [113].

Evidently, it is necessary to obtain optical GRB spectra at the earliest stages as close as possible to the time of the γ-ray burst itself. This problem is formulated for wide-angle automatic telescopes [811]. The best would be to have a system of continuously operating wide-angle telescopes covering the entire sky, or a specialized

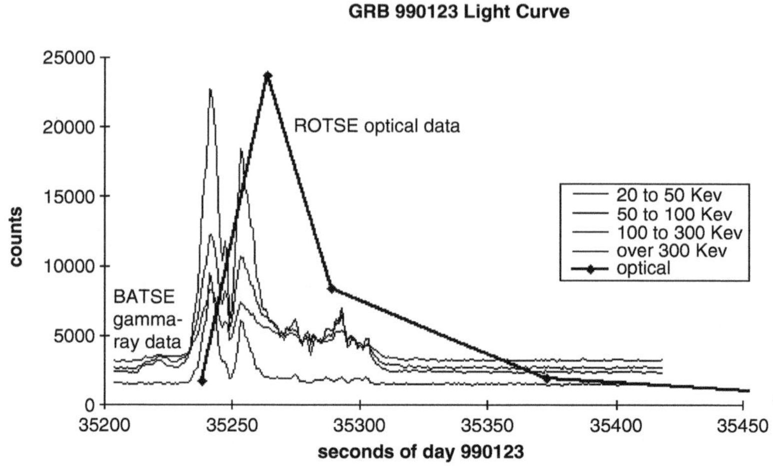

Fig. 11.46 Light curves of GRB 990123. Thin lines represent BATSE gamma-ray profile with 1,024 ms resolution in different energy channels. Thick line represents the first few frames of ROTSE data. ROTSE began taking data 22 s after the initial trigger. This is the first instance of simultaneous optical and gamma-ray data from a gamma-ray burst

Fig. 11.47 FAVOR (FAst Variability Optical Registrator) camera - SAO RAS, operating since 2003. Built in collaboration with IPI and IKI (Moscow), from [91]

Fig. 11.48 Wide field camera for the search of GRB optical afterglows TORTORA in La Silla, Chile, from [91], Credit ESO

satellite with an optical all-sky monitor. Simultaneous polarization measurements of the prompt optical afterglow would be very important for the determination of emission mechanism.

Problem 1. Find the stationary solution for spherically symmetric accretion of a polytropic fluid into a gravitating centre.

Solution. [224] The equations describing the problem are

$$u\frac{\mathrm{d}u}{\mathrm{d}r} + \frac{1}{\rho}\frac{\mathrm{d}P}{\mathrm{d}r} + \frac{GM}{r^2} = 0,$$

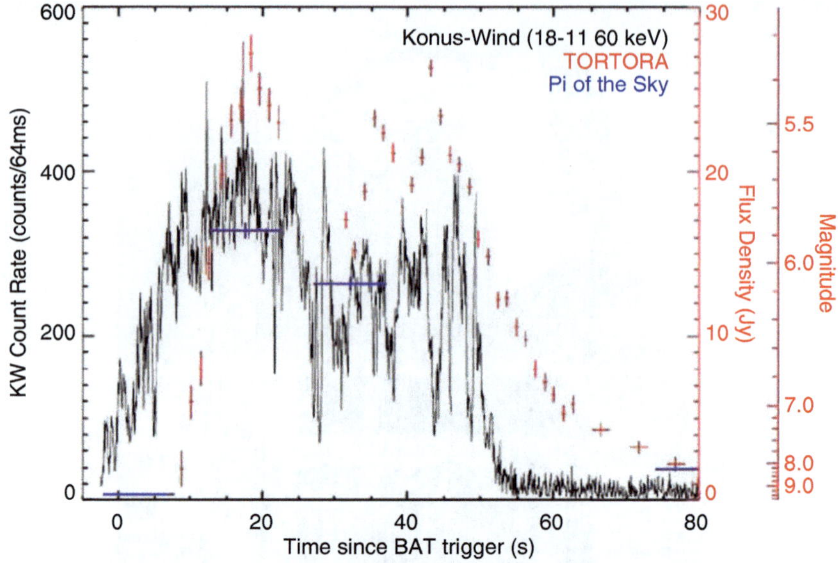

Fig. 11.49 Prompt Emission Light Curve. The Konus-Wind background-subtracted γ- ray lightcurve (*black*), shown relative to the *Swift* BAT trigger time, T_0. Optical data from Pi of the sky (Polish instrument, long horizontal blue lines) and TORTORA (*red crosses*) are superimposed for comparison. The optical emission begins within seconds of the onset of the burst. The TORTORA data have a gap during the slew of the REM telescope to this field, but show 3 sub-peaks in the optical brightness, reaching a peak brightness of 5.3 magnitudes (*white*). The optical flash begins and ends at approximately the same times as the prompt γ-ray emission, providing strong evidence that both originate at the same site, from [818]

$$\dot{M} = -4\pi\rho r^2 u, \quad P = K\rho^\gamma, \quad \gamma = 1 + \frac{1}{n}. \tag{11.244}$$

Excluding ρ from the first equation, we obtain

$$\frac{\mathrm{d}u}{\mathrm{d}r} = \frac{u}{r}\frac{(GM/r) - 2a^2}{u^2 - a^2}, \quad a^2 = \frac{\mathrm{d}P}{\mathrm{d}\rho} = \gamma K\rho^{\gamma-1}. \tag{11.245}$$

A singular point in (11.245) is of a saddle-type and solution, describing the flow from a region far outside must go through this point, where

$$a_c^2 = \frac{GM}{2R_C} = u_c^2, \quad \rho_c = \left(\frac{n}{n+1}\frac{GM}{2KR_C}\right)^n. \tag{11.246}$$

The system (11.244) has a Bernoulli integral (8.34)

$$\frac{u^2}{2} + (n+1)K\rho^{1/n} - \frac{GM}{r} = H. \tag{11.247}$$

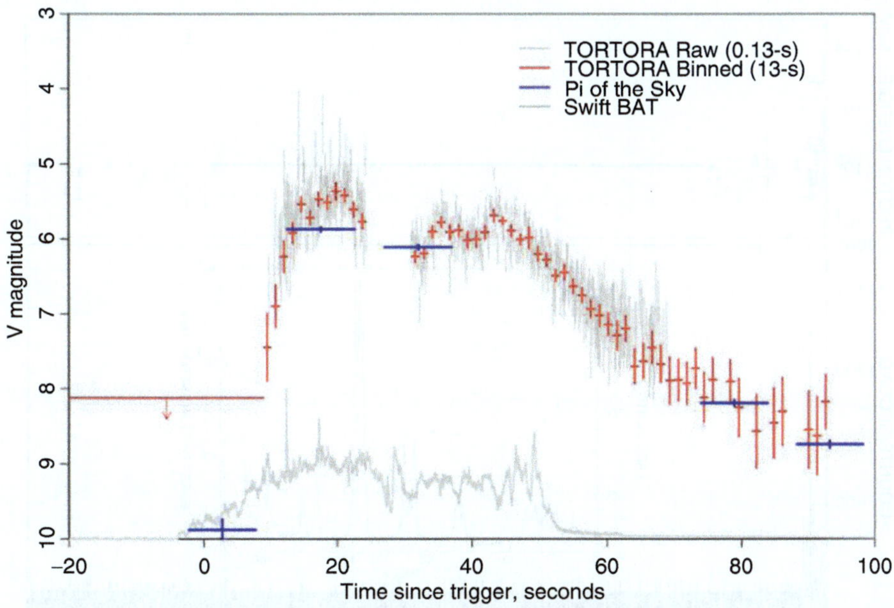

Fig. 11.50 Light curves of GRB080319B obtained in optics by TORTORA with 0.13 s (*vertical grey lines*), and 1.3 s. (*red crosses*) resolution. Also represented are optical data from Pi of the sky (*long horizontal blue lines*), and Swift – BAT prompt γ-ray emission light curve, from [91]

Using (11.246), we can express H as

$$H = \frac{2n-3}{4}\frac{GM}{R_C}. \tag{11.248}$$

Taking the density at a faraway radius r_{out}, where $u_{out} = 0$ as ρ_{out} and assuming the gaseous equation of state $P = \rho \mathscr{R} T = K\rho^{\gamma}$, we obtain

$$H = (n+1)K\rho_{out}^{1/n} - \frac{GM}{r_{out}} = (n+1)\mathscr{R}T_{out} - \frac{GM}{r_{out}}, \tag{11.249}$$

$$R_C = \frac{2n-3}{4}\frac{GM}{(n+1)\mathscr{R}T_{out} - (GM/r_{out})}, \quad K = \rho_{out}^{-1/n}\mathscr{R}T_{out}. \tag{11.250}$$

It is clear from (11.250) that stationary continuous supersonic accretion occurs only at $H > 0$, $R_C < r_{out}$, $n > 3/2$. We obtain also

$$\begin{aligned}\dot{M} &= 2^{3/2-n}\pi\left(\frac{n}{n+1}\right)^{n}\frac{(GM)^{n+(1/2)}}{K^n}R_C^{(3/2)-n}\\ &= \pi\left(\frac{n}{n+1}\right)^{n}\left(\frac{2n-3}{2}\right)^{(3/2)-n}\frac{(GM)^2\rho_{out}}{(\mathscr{R}T_{out})^n[(n+1)\mathscr{R}T_{out} - (GM/r_{out})]^{(3/2)-n}}.\end{aligned} \tag{11.251}$$

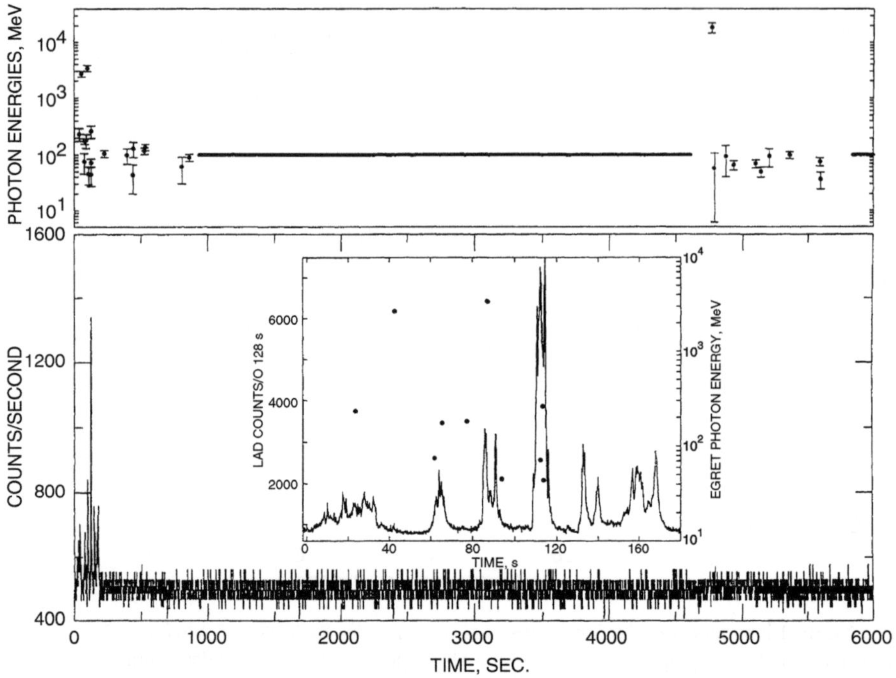

Fig. 11.51 GRB flare of February 17, 1994, the gamma rays from which were observed in GeV-energy region in the experiment EGRET within 1.5 h after the basic flare. The composite figure includes the EGRET data and the data of observations of the basic flare in the BATSE experiment and at the satellite Ulysses

When outer boundary conditions are given at infinity, we obtain

$$\dot{M} = \pi \left(\frac{n}{n+1}\right)^n \left(\frac{2n-3}{2n+2}\right)^{(3/2)-n} \frac{(GM)^2 \rho_{\text{out}}}{(\mathcal{R}T_{\text{out}})^{3/2}}. \tag{11.252}$$

Here, $(n/(n+1))^n\ ((2n-3)/(2n+2))^{3/2-n} = 1.089, 1.509, 1.831, 2.651, e^{3/2} = 4.482$ for $n = 2, 2.5, 3, 5, \infty$. The approximate relation (11.87) follows from (11.252).

Problem 2. Find the solution for an optically thick accretion disc structure in Newtonian gravity with a α-prescription for viscosity $t_{r\phi} = -\alpha P$.

Solution. [755, 897, 898] Without terms with radial derivatives (advection), the angular velocity is equal to the Keplerian one

$$\Omega = \Omega_K = (GM/r^3)^{1/2}, \tag{11.253}$$

and the rate of heat production due to viscosity Q_+ per unit area of the disc, and the equation for angular momentum conservation (11.116) remain the same:

$$Q_+ = \frac{1}{2} t_{r\phi} r \frac{d\Omega}{dr} = \frac{3}{8\pi} \dot{M} \frac{GM}{r^3} \left(1 - \frac{l_{in}}{l}\right), \tag{11.254}$$

$$\dot{M}\Omega \left(1 - \frac{l_{in}}{l}\right) = 4\pi\alpha P h. \tag{11.255}$$

Suggesting that heat in the vertical direction is carried out mainly by radiation, we use a connection between radiation flux Q_- and temperature T in the plane of the disc in the form

$$Q_- = \frac{4acT^4}{3\kappa\Sigma} \tag{11.256}$$

with a surface density

$$\Sigma = 2\rho h. \tag{11.257}$$

Without advection, we have $Q_+ = Q_-$

$$\frac{3}{2}\dot{M}\Omega^2 \left(1 - \frac{l_{in}}{l}\right) = \frac{16\pi acT^4}{3\kappa\Sigma}. \tag{11.258}$$

This corresponds to $B_3 = 2$ in (11.117). Equation (11.115) is used for obtaining the radial velocity v_r. To close the system of equations, we need an additional relation between Σ, ρ and h following from the equilibrium along the z-axis, like (11.118).

In the region I with $P = aT^4/3$, $\kappa = \sigma_T = 0.4$ cm^2/g, the equations of equilibrium and heat transfer along the z-axis reduce to

$$\frac{dP}{dz} = -\rho g = -\frac{\sigma_T \rho}{c} Q, \qquad g = \frac{GM}{r^3} z, \tag{11.259}$$

which corresponds to the local balance between radiation force and gravity. Using (11.259) near the surface of the disc where $z = h$, $Q = Q_+$ we obtain the expression for the disc thickness (all values here are given in CGS units)

$$h = \frac{\sigma_T}{c} \frac{Q_+ r^3}{GM} = \frac{3}{8\pi} \frac{\sigma_T}{c} \dot{M} \left(1 - \frac{l_{in}}{l}\right)$$
$$= \frac{3GM_\odot}{2c^2} m\dot{m} \left(1 - \frac{l_{in}}{l}\right) = 7.41 \times 10^4 m\dot{m} \left(1 - \frac{l_{in}}{l}\right). \tag{11.260}$$

Substituting (11.260) into (11.255), we obtain for the pressure in the plane of the disc

$$P = \frac{2c\Omega}{3\alpha\sigma_T} = \frac{2}{3} \frac{c^4}{\alpha\sigma_T GM_\odot} \frac{1}{mx^{3/2}} = \frac{1.01 \times 10^{16}}{\alpha m x^{3/2}}. \tag{11.261}$$

Substituting (11.260) and (11.261) into (11.258) gives

$$\Sigma = \frac{64\pi}{9\alpha}\frac{c^2}{\sigma_T^2}\frac{1}{\Omega\dot{M}}\left(1-\frac{l_{\rm in}}{l}\right)^{-1}$$
$$= \frac{16}{9\alpha\sigma_T}\frac{x^{3/2}}{\dot{m}}\left(1-\frac{l_{\rm in}}{l}\right)^{-1} = \frac{40}{9\alpha}\frac{x^{3/2}}{\dot{m}}\left(1-\frac{l_{\rm in}}{l}\right)^{-1}. \quad (11.262)$$

The density in the plane of the disc follows from (11.257), (11.260) and (11.262)

$$\rho = \frac{\Sigma}{2h} = \frac{256\pi^2}{27\alpha}\frac{c^3}{\sigma_T^3}\frac{1}{\Omega\dot{M}^2}\left(1-\frac{l_{\rm in}}{l}\right)^{-2}$$
$$= \frac{16}{27\alpha}\frac{c^2}{\sigma_T G M_\odot}\frac{x^{3/2}}{\dot{m}^2 m}\left(1-\frac{l_{\rm in}}{l}\right)^{-2} = \frac{10^{-5}}{\alpha}\frac{x^{3/2}}{\dot{m}^2 m}\left(1-\frac{l_{\rm in}}{l}\right)^{-2}. \quad (11.263)$$

From (11.115), we have

$$v_{\rm r} = -\frac{\dot{M}}{2\pi r\Sigma} = -\frac{9\alpha}{128\pi^2}\frac{\sigma_T^2}{c^2}\frac{\Omega}{r}\dot{M}^2\left(1-\frac{l_{\rm in}}{l}\right)$$
$$= -\frac{9\alpha}{8}c\frac{\dot{m}^2}{x^{5/2}}\left(1-\frac{l_{\rm in}}{l}\right). \quad (11.264)$$

Write also

$$T = \left(\frac{3P}{a}\right)^{1/4} = \left(\frac{2}{\alpha\sigma_T a G M_\odot}\right)^{1/4}\frac{c}{m^{1/4}x^{3/8}} = \frac{4.48\times 10^7}{m^{1/4}x^{3/8}}\,\mathrm{K}, \quad (11.265)$$

$$\left|\frac{v_{\rm r}}{v_{\rm s}}\right| = \frac{3\sqrt{2}\alpha}{16\pi}\frac{\sigma_T\dot{M}}{cr} = \frac{3\sqrt{2}\alpha}{4}\frac{\dot{m}}{x}. \quad (11.266)$$

Here

$$m = \frac{M}{M_\odot}, \quad \dot{m} = \frac{\dot{M}c^2}{L_c}, \quad L_c = \frac{4\pi c G M}{\sigma_T}, \quad x = \frac{rc^2}{GM}. \quad (11.267)$$

Also

$$v_{\rm s}^2 = \frac{P}{\rho}, \quad \Omega = \frac{c^3}{GM}x^{-3/2} = \frac{2.02\times 10^5}{m x^{3/2}}.$$

It follows from (11.260), (11.264) and (11.266) that

$$h = \sqrt{2}\frac{v_{\rm s}}{\Omega} = \frac{\sqrt{2}}{\Omega}\sqrt{\frac{P}{\rho}}. \quad (11.268)$$

Relation (11.268) is used in other regions instead of (11.260). In region II with $P = \rho \mathscr{R} T$, $\kappa = \sigma_T$ with $\mathscr{R} = 10^8$, we obtain, using (11.244), (11.255), (11.257), (11.258) and (11.268),

$$T = \left(\frac{9\sigma_T}{64\pi^2\alpha}\frac{\dot{M}^2\Omega^3}{ac\mathscr{R}}\right)^{1/5}\left(1-\frac{l_{\rm in}}{l}\right)^{2/5}$$

$$= \left(\frac{9c}{4\sigma_T}\frac{1}{a\mathscr{R}GM_\odot}\right)^{1/5}\frac{c\dot{m}^{2/5}}{m^{1/5}x^{9/10}\alpha^{1/5}}\left(1-\frac{l_{\rm in}}{l}\right)^{2/5}$$

$$= 5.22\times10^8\frac{\dot{m}^{2/5}}{m^{1/5}x^{9/10}\alpha^{1/5}}\left(1-\frac{l_{\rm in}}{l}\right)^{2/5}\ \mathrm{K}, \tag{11.269}$$

$$\rho = \frac{1}{2\pi\alpha}\left(\frac{2\pi^2\alpha ac}{9\sigma_T}\right)^{3/10}\frac{\dot{M}^{2/5}\Omega^{11/10}}{\mathscr{R}^{6/5}}\left(1-\frac{l_{\rm in}}{l}\right)^{2/5}$$

$$= \frac{2^{1/10}}{3^{3/5}}\frac{a^{3/10}c^{16/5}}{\mathscr{R}^{6/5}\sigma_T^{7/10}(GM_\odot)^{7/10}}\frac{\dot{m}^{2/5}}{(\alpha m)^{7/10}x^{33/20}}\left(1-\frac{l_{\rm in}}{l}\right)^{2/5}$$

$$= 26.6\frac{\dot{m}^{2/5}}{(\alpha m)^{7/10}x^{33/20}}\left(1-\frac{l_{\rm in}}{l}\right)^{2/5}, \tag{11.270}$$

$$h = \sqrt{2}\left(\frac{9\sigma_T}{64\pi^2\alpha}\frac{\dot{M}^2\mathscr{R}^4}{ac\Omega^7}\right)^{1/10}\left(1-\frac{l_{\rm in}}{l}\right)^{1/5}$$

$$= \sqrt{2}\left(\frac{3}{2}\right)^{1/5}\frac{(GM_\odot)^{9/10}\mathscr{R}^{2/5}}{c^{12/5}(a\alpha\sigma_T)^{1/10}}\dot{m}^{1/5}m^{9/10}x^{21/20}\left(1-\frac{l_{\rm in}}{l}\right)^{1/5}$$

$$= 1.60\times10^3\frac{\dot{m}^{1/5}m^{9/10}x^{21/20}}{\alpha^{1/10}}\left(1-\frac{l_{\rm in}}{l}\right)^{1/5}, \tag{11.271}$$

$$\Sigma = \left(\frac{2}{9\pi^3}\right)^{1/5}\left(\frac{ac}{\sigma_T}\right)^{1/5}\frac{\dot{M}^{3/5}\Omega^{2/5}}{(\alpha\mathscr{R})^{4/5}}\left(1-\frac{l_{\rm in}}{l}\right)^{3/5}$$

$$= \frac{2^{7/5}}{3^{2/5}}\left(\frac{c}{\sigma_T\mathscr{R}}\right)^{4/5}\frac{(aGM_\odot)^{1/5}}{\alpha^{4/5}}\frac{\dot{m}^{3/5}m^{1/5}}{x^{3/5}}\left(1-\frac{l_{\rm in}}{l}\right)^{3/5}$$

$$= 8.54\times10^4\frac{\dot{m}^{3/5}m^{1/5}}{\alpha^{4/5}x^{3/5}}\left(1-\frac{l_{\rm in}}{l}\right)^{3/5}, \tag{11.272}$$

$$v_{\rm r} = -\left(\frac{9}{64\pi^2}\right)^{1/5}\left(\frac{\sigma_T}{ac}\right)^{1/5}\left(\frac{\dot{M}}{\Omega}\right)^{2/5}\frac{(\alpha\mathscr{R})^{4/5}}{r}\left(1-\frac{l_{\rm in}}{l}\right)^{-3/5}$$

$$= -\left(\frac{3}{2}\right)^{2/5}\left(\frac{c\mathscr{R}^4}{a\sigma_T GM_\odot}\right)^{1/5}\frac{\alpha^{4/5}\dot{m}^{2/5}}{m^{1/5}x^{2/5}}\left(1-\frac{l_{\text{in}}}{l}\right)^{-3/5}$$

$$= -1.76\times 10^6\frac{\alpha^{4/5}\dot{m}^{2/5}}{m^{1/5}x^{2/5}}\left(1-\frac{l_{\text{in}}}{l}\right)^{-3/5}. \qquad (11.273)$$

In region III there is $P=\rho\mathscr{R}T$, and $\kappa \approx 2\times 10^{24}\,\rho T^{-7/2} = \kappa_1\rho T^{-7/2}$ cm^2/g is defined by free–free and free-bound processes. Here, we obtain from (11.253), (11.255), (11.257), (11.258) and (11.268)

$$T = \frac{3^{1/5}}{2^{9/10}\pi^{3/10}}\frac{\kappa_1^{1/10}\dot{M}^{3/10}\Omega^{1/2}}{\alpha^{1/5}\mathscr{R}^{1/4}(ac)^{1/10}}\left(1-\frac{l_{\text{in}}}{l}\right)^{3/10}$$

$$= \frac{3^{1/5}}{2^{3/10}}\frac{\kappa_1^{1/10}c^{11/10}}{\sigma_T^{3/10}\mathscr{R}^{1/4}a^{1/10}(GM_\odot)^{1/5}}\frac{\dot{m}^{3/10}}{(\alpha m)^{1/5}x^{3/4}}\left(1-\frac{l_{\text{in}}}{l}\right)^{3/10}$$

$$= 1.85\times 10^8\frac{\dot{m}^{3/10}}{(\alpha m)^{1/5}x^{3/4}}\left(1-\frac{l_{\text{in}}}{l}\right)^{3/10}\ \text{K}, \qquad (11.274)$$

$$\rho = \frac{1}{2^{23/20}3^{3/10}\pi^{11/20}}\frac{\Omega^{5/4}\dot{M}^{11/20}(ac)^{3/20}}{\alpha^{7/10}\kappa_1^{3/20}\mathscr{R}^{9/8}}\left(1-\frac{l_{\text{in}}}{l}\right)^{11/20}$$

$$= \frac{1}{2^{1/20}3^{3/10}}\frac{c^{67/20}a^{3/20}}{(GM_\odot)^{7/10}\sigma_T^{11/20}\kappa_1^{3/20}\mathscr{R}^{9/8}}\frac{\dot{m}^{11/20}}{(\alpha m)^{7/10}x^{15/8}}\left(1-\frac{l_{\text{in}}}{l}\right)^{11/20}$$

$$= 1.28\times 10^2\frac{\dot{m}^{11/20}}{(\alpha m)^{7/10}x^{15/8}}\left(1-\frac{l_{\text{in}}}{l}\right)^{11/20}, \qquad (11.275)$$

$$h = \left(\frac{18}{\pi^3}\right)^{1/20}\frac{\kappa_1^{1/20}\mathscr{R}^{3/8}\dot{M}^{3/20}}{\Omega^{3/4}\alpha^{1/10}(ac)^{1/20}}\left(1-\frac{l_{\text{in}}}{l}\right)^{3/20}$$

$$= 2^{7/20}3^{1/10}\frac{\kappa_1^{1/20}\mathscr{R}^{3/8}(GM_\odot)^{9/10}}{\sigma_T^{3/20}a^{1/20}c^{49/20}}\frac{\dot{m}^{3/20}m^{9/20}x^{9/8}}{\alpha^{1/10}}\left(1-\frac{l_{\text{in}}}{l}\right)^{3/20}$$

$$= 1.21\times 10^3\frac{\dot{m}^{3/20}m^{9/20}x^{9/8}}{\alpha^{1/10}}\left(1-\frac{l_{\text{in}}}{l}\right)^{3/20}, \qquad (11.276)$$

$$\Sigma = \frac{1}{2^{1/10}3^{1/5}\pi^{7/10}}\frac{\Omega^{1/2}\dot{M}^{7/10}(ac)^{1/10}}{\alpha^{4/5}\kappa_1^{1/10}\mathscr{R}^{3/4}}\left(1-\frac{l_{\text{in}}}{l}\right)^{7/10}$$

$$= \frac{2^{3/10}}{3^{1/5}}\frac{(GM_\odot)^{1/5}c^{9/10}a^{1/10}}{\sigma_T^{7/10}\kappa_1^{1/10}\mathscr{R}^{3/4}}\frac{\dot{m}^{7/10}m^{1/5}}{\alpha^{4/5}x^{3/4}}\left(1-\frac{l_{\text{in}}}{l}\right)^{7/10}$$

$$= 1.48 \times 10^4 \frac{\dot{m}^{7/10} m^{1/5}}{\alpha^{4/5} x^{3/4}} \left(1 - \frac{l_{\rm in}}{l}\right)^{7/10}, \tag{11.277}$$

$$v_{\rm r} = -\frac{3^{1/5}}{2^{9/10}\pi^{3/10}} \frac{\alpha^{4/5}\kappa_1^{1/10}\mathscr{R}^{3/4}\dot{M}^{3/10}}{r\Omega^{1/2}(ac)^{1/10}} \left(1 - \frac{l_{\rm in}}{l}\right)^{-7/10}$$

$$= -\frac{3^{1/5}}{2^{3/10}} \frac{\kappa_1^{1/10}\mathscr{R}^{3/4}}{\sigma_T^{3/10}(GM_\odot)^{1/5}} \left(\frac{c}{a}\right)^{1/10} \frac{\alpha^{4/5}\dot{m}^{3/10}}{x^{1/4}m^{1/5}} \left(1 - \frac{l_{\rm in}}{l}\right)^{-7/10}$$

$$= -6.16 \times 10^5 \frac{\alpha^{4/5}\dot{m}^{3/10}}{x^{1/4}m^{1/5}} \left(1 - \frac{l_{\rm in}}{l}\right)^{-7/10}. \tag{11.278}$$

The integration constant $l_{\rm in}$ multiplied by $\dot{M}$ represents the total flux of angular momentum (advective plus viscous) flowing through the accretion disc. It may be scaled by the angular momentum of matter on the inner Keplerian orbit

$$l_{\rm in} = \xi r_{\rm in}^2 \Omega_{K\rm in} = \xi \sqrt{GM r_{\rm in}}. \tag{11.279}$$

In the case of accretion into a black hole or slowly rotating neutron star, where $d\Omega/dr = 0$ on the inner edge of the disc, one usually takes $\xi = 1$.

Problem 3. Find the solution for an accretion disc structure with polytropic equation of state.

Solution. The hydrodynamical equations, describing the structure of a stationary thin accretion disk in the usual notations, have the form (in cylindrical coordinates r, ϕ, z with $\partial/\partial\phi = 0$)

$$-\frac{\mathrm{d}(\dot{M}u_r)}{\mathrm{d}r} - 2\pi\Sigma r\left(\Omega^2 r - \frac{GM}{r^2}\right) + 2\pi r\frac{\mathrm{d}\mathscr{P}}{\mathrm{d}r} = 0, \tag{11.280}$$

$$\frac{\dot{M}}{2\pi}\frac{\mathrm{d}j}{\mathrm{d}r} = -\alpha\frac{d}{dr}\left(\Sigma u_{s0} z_0 r^3 \frac{\mathrm{d}\Omega}{\mathrm{d}r}\right), \tag{11.281}$$

$$\rho\frac{GMz}{r^3} = -\frac{\partial P}{\partial z}, \tag{11.282}$$

$$\frac{\mathrm{d}\dot{M}}{\mathrm{d}r} = 0. \tag{11.283}$$

Following [897], we write the viscosity coefficient μ in the α approximation as

$$\mu = \alpha\rho u_{s0} z_0, \tag{11.284}$$

where u_{s0} is the sound velocity in the equatorial plane $z = 0$ and z_0 is the semi-thickness of the disk. The average values, surface density Σ, mass flux over the disk $\dot{M}$, integrated pressure $\mathscr{P}$ and the values of angular velocity Ω and specific angular momentum j are determined as follows

$$\Sigma = \int_{-z_0}^{z_0} \rho dz, \qquad \dot{M} = -2\pi r \int_{-z_0}^{z_0} \rho u_r dz = -2\pi r \Sigma u_r,$$
$$\mathscr{P} = \int_{-z_0}^{z_0} P dz, \quad \Omega = \frac{u_\phi}{r}, \quad j = r u_\phi. \tag{11.285}$$

The equations (11.280)–(11.283) determine the variables $(\rho, \Sigma, \dot{M}, j)$, where (u_r, Ω) are obtained from (11.285) with

$$u_r < 0, \qquad \dot{M} > 0. \tag{11.286}$$

The solution of the equation of vertical balance (11.282) gives the variables

$$u_{s0} = \left(\frac{\mathrm{d}P_0}{\mathrm{d}\rho_0}\right)^{1/2}, \quad z_0, \quad \mathscr{P}, \ \Sigma \tag{11.287}$$

as functions of (r, ρ_0), when the equation of the state is given in the form $P(\rho)$ (e.g., a polytropic). Here, ρ_0 and P_0 are the density and the pressure in the equatorial plane, respectively.

For a polytropic equation of state

$$P = K\rho^{1+(1/n)} \tag{11.288}$$

we obtain [476] solving (11.282)

$$\rho = \rho_0 \left(1 - \frac{z^2}{z_0^2}\right)^n, \tag{11.289}$$

where the density in the equatorial plane ρ_0 is connected with r and z_0 by the relation

$$\rho_0 = \left[\frac{GM}{2K(n+1)}\right]^n \frac{z_0^{2n}}{r^{3n}}. \tag{11.290}$$

Some physical quantities from (11.285) to (11.287) can be expressed in terms of r and ρ_0.

$$z_0 = \left[\frac{2K(n+1)}{GM}\right]^{1/2} \rho_0^{1/2n} r^{3/2} = \beta_{z0}\rho_0^{1/2n} r^{3/2},$$

$$\Sigma = \sqrt{\pi}\frac{\Gamma(n+1)}{\Gamma\,(n+(3/2))}\rho_0 z_0 = \beta_\Sigma \rho_0^{(2n+1)/2n} r^{3/2},$$

$$\mathscr{P} = \sqrt{\pi}\frac{\Gamma(n+2)}{\Gamma\left(n+(5/2)\right)}K\rho_0^{1+(1/n)}z_0 = \beta_{\mathscr{P}}\rho_0^{(2n+3)/2n}r^{3/2},$$

$$u_{s0} = \left(\frac{n+1}{n}K\right)^{1/2}\rho_0^{1/2n} = \beta_{s0}\rho_0^{1/2n}, \tag{11.291}$$

where

$$\beta_\Sigma = \sqrt{\pi}\frac{\Gamma(n+1)}{\Gamma\left(n+(3/2)\right)}\left[\frac{2K(n+1)}{GM}\right]^{1/2},$$

$$\beta_{\mathscr{P}} = \sqrt{\pi}\frac{\Gamma(n+2)}{\Gamma\left(n+(5/2)\right)}K\left[\frac{2K(n+1)}{GM}\right]^{1/2}. \tag{11.292}$$

For the isothermal case, corresponding to $n = \infty$, $P = K\rho$, we have instead of (11.289)

$$\rho = \rho_0\exp\left(-\frac{GMz^2}{2Kr^3}\right) = \rho_0\exp\left(-\frac{z^2}{z_i^2}\right), \quad z_i = \left(\frac{2Kr^3}{GM}\right)^{1/2}. \tag{11.293}$$

Formally an isothermal disk has infinite thickness z_0, but the density falls exponentially with the characteristic height z_i from (11.293). Instead of (11.291), we have

$$\Sigma = \left(\frac{2\pi K}{GM}\right)^{1/2}\rho_0 r^{3/2}, \quad \mathscr{P} = K\left(\frac{2\pi K}{GM}\right)^{1/2}\rho_0 r^{3/2}, \quad u_{s0} = \sqrt{K}. \tag{11.294}$$

Equation (11.281) can be integrated, giving [897]

$$\frac{\dot{M}}{2\pi}(j - j_0) = -\alpha\Sigma u_{s0}z_0 r^3\frac{d\Omega}{dr}. \tag{11.295}$$

The integration constant j_0 after multiplication by $\dot{M}$ gives the total (advective plus viscous) flux of angular momentum within the accretion disk. A positive value of j_0 corresponds to a total flux directed to the centre, when the central body accretes angular momentum. For a thin disk neglecting u_r and $\mathscr{P}$, we obtain from (11.280)

$$\Omega^{(0)} = \left(\frac{GM}{r^3}\right)^{1/2} = \Omega_K, \quad j^{(0)} = j_K = (GMr)^{1/2}. \tag{11.296}$$

The solution (11.296) is used up to the inner boundary of the accretion disk at $r = r_{in}$. The integration constant j_0 can be scaled by the Keplerian angular momentum $j_K(r_{in})$, so that

$$j_0 = \xi j_K(r_{in}) = \xi\sqrt{GMr_{in}}. \tag{11.297}$$

Substituting (11.297) into (11.295), we obtain with account of (11.292) and (11.296) the relation for the density in the equatorial plane

$$\rho_0^{(0)} = b_{\rho 0}\left[\frac{\dot{M}}{r^3}\left(1-\xi\sqrt{\frac{r_{\rm in}}{r}}\right)\right]^{2n/(2n+3)},$$

$$b_{\rho 0} = \left[\frac{1}{6\pi^{3/2}\alpha}\frac{\sqrt{n}}{(n+1)^{3/2}}\frac{\Gamma(n+(3/2))}{\Gamma(n+1)}\frac{GM}{K^{3/2}}\right]^{2n/(2n+3)}. \qquad (11.298)$$

Using (11.298) in (11.291), we obtain

$$z_0^{(0)} = b_{z0}\left[\dot{M}\left(1-\xi\sqrt{\frac{r_{\rm in}}{r}}\right)\right]^{1/(2n+3)} r^{3/2(2n+1)/(2n+3)},$$

$$\Sigma^{(0)} = b_{\Sigma}\left[\dot{M}\left(1-\xi\sqrt{\frac{r_{\rm in}}{r}}\right)\right]^{(2n+1)/(2n+3)} r^{-3(2n-1)/2(2n+3)},$$

$$\mathscr{P}^{(0)} = b_{\mathscr{P}}\frac{\dot{M}}{r^{3/2}}\left(1-\xi\sqrt{\frac{r_{\rm in}}{r}}\right),$$

$$u_{s0}^{(0)} = b_{s0}\left[\frac{\dot{M}}{r^3}\left(1-\xi\sqrt{\frac{r_{\rm in}}{r}}\right)\right]^{1/(2n+3)}, \qquad (11.299)$$

where

$$b_{z0} = [2(n+1)]^{1/2}\left[\frac{1}{6\pi^{3/2}\alpha}\frac{\sqrt{n}}{(n+1)^{3/2}}\frac{\Gamma(n+(3/2))}{\Gamma(n+1)}\right]^{1/(2n+3)}$$
$$\times K^{n/(2n+3)}(GM)^{-(2n+1)/2(2n+3)},$$

$$b_{\Sigma} = [2\pi(n+1)]^{1/2}\frac{\Gamma(n+1)}{\Gamma(n+(3/2))}$$
$$\times\left[\frac{1}{6\pi^{3/2}\alpha}\frac{\sqrt{n}}{(n+1)^{3/2}}\frac{\Gamma(n+(3/2))}{\Gamma(n+1)}\right]^{(2n+1)/(2n+3)}$$
$$\times K^{-2n/(2n+3)}(GM)^{(2n-1)/2(2n+3)},$$

$$b_{\mathscr{P}} = \frac{\sqrt{nGM}}{\alpha 3\pi\sqrt{2}\,(n+3/2)}, \qquad b_{s0} = b_{\rho 0}^{1/2n}. \qquad (11.300)$$

In the formula for viscosity coefficient (11.284), z_0 must be replaced by z_i from (11.293) in the isothermal case,

$$\mu = \alpha\rho u_{s0} z_i. \qquad (11.301)$$

Then, instead of (11.298)–(11.300), we obtain a solution

$$\rho_0^{(0)} = \frac{GM}{6\pi^{3/2}\alpha K^{3/2}}\frac{\dot{M}}{r^3}\left(1-\xi\sqrt{\frac{r_{\rm in}}{r}}\right),$$

$$\Sigma^{(0)} = \frac{\sqrt{GM}}{\alpha 3\pi\sqrt{2}K}\frac{\dot{M}}{r^{3/2}}\left(1-\xi\sqrt{\frac{r_{\rm in}}{r}}\right),$$

$$\mathcal{P}^{(0)} = \frac{\sqrt{GM}}{\alpha 3\pi\sqrt{2}}\frac{\dot{M}}{r^{3/2}}\left(1-\xi\sqrt{\frac{r_{\rm in}}{r}}\right). \tag{11.302}$$

For accretion onto a slowly rotating star with angular velocity smaller than the Keplerian velocity on the equator, the drop of the angular velocity from Keplerian in the disk to stellar equatorial occurs in a thin boundary layer, which must be considered separately with proper account of the pressure term in (11.280).

Problem 4. Find the structure of the boundary layer between an accretion disc and a slowly rotating star with a polytropic equation of state.

Solution. [108, 825] Inside the BL, variables change considerably over the small thickness of the layer $H_b \ll r_{\rm in}$. The radial velocity term in (11.280) is negligible for small α, but the pressure term is comparable with the gravitational and centrifugal forces [825]. The thickness H_b of the BL is smaller than its vertical size z_0, which also remains small. The adopted inequalities for the boundary layer parameters

$$H_b \ll z_0 \ll r_* \tag{11.303}$$

will be confirmed by the results. The radius of the star r_* differs from the radius, at which $d\Omega/dr = 0$, by the very small value H_b. In the asymptotic consideration of the BL, we use r_* as an inner boundary for the disk solution $r_{\rm in} = r_*$ The variable x

$$r = r_* + \delta x, \quad \delta = \frac{H_b}{r_*} \ll 1 \tag{11.304}$$

is used inside BL instead of r. The inner solution within the BL is looked for in the region $0 < x < \infty$, while the outer solution (11.296)–(11.299) of Problem 3 is valid in $r_* < r < \infty$. According to the method of matched asymptotic expansion (MAE) (see [735]) the inner and outer solutions are matched so that values for the outer solution at $r = r_*$ are equal to the corresponding values of the inner solution at $x = \infty$. This condition is valid asymptotically at $\delta \to 0$.

Consider a stationary accretion disk BL where equations (11.280), (11.283) and (11.295) are valid with the integration constant ξ_b instead of ξ in (11.297). The thickness of the disk remains small in the boundary layer, so relations (11.288)–(11.292) for the polytropic case and (11.293)–(11.294) for the isothermal case are valid. Taking account of only the main terms in the asymptotic expansion inside the BL, we obtain the equations from (11.280) and (11.295) [825]

$$\frac{d\mathcal{P}}{dx} = -\Omega_{K*}^2 H_b \Sigma(1-\omega^2), \tag{11.305}$$

$$\frac{d\omega}{dx} = \frac{\dot{M}H_b}{2\pi\Sigma\nu_b r_*^2}(\xi_b - \omega). \tag{11.306}$$

Here,

$$\omega = \frac{\Omega}{\Omega_{K*}}, \quad \Omega_{K*}^2 = \frac{GM}{r_*^3}. \tag{11.307}$$

The viscosity coefficient μ is expressed in BL through the coefficient of kinematical viscosity ν_b, so that $\mu = \rho\nu_b$. While the radial extent of the BL H_b is much less than its vertical size z_0, the formula (11.284) cannot be used for a viscosity coefficient approximation. Inside the BL, we thus use the α approximation in the form

$$\nu_b = \alpha_b u_{s0} H_b. \tag{11.308}$$

For the polytropic case, it is convenient to use the variables (Σ, ω) in (11.305) and (11.306). We obtain from (11.291)

$$\mathscr{P} = d_{\mathscr{P}} \Sigma^{(2n+3)/(2n+1)}, \quad u_{s0} = d_{s0} \Sigma^{1/(2n+1)}, \tag{11.309}$$

$$d_{s0} = \left(\frac{n+1}{n}K\right)^{1/2} \left[\frac{1}{\sqrt{2\pi}} \frac{\Gamma(n+(3/2))}{\Gamma(n+1)}\right]^{1/(2n+1)} \left[\frac{GM}{r_*^3 K(n+1)}\right]^{1/2(2n+1)},$$

$$d_{\mathscr{P}} = \frac{n+1}{n+3/2} K \left[\frac{1}{\sqrt{2\pi}} \frac{\Gamma(n+3/2)}{\Gamma(n+1)}\right]^{2/(2n+1)}$$

$$\times \left[\frac{GM}{r_*^3 K(n+1)}\right]^{1/(2n+1)} = \frac{n}{n+3/2} d_{s0}^2. \tag{11.310}$$

From the matching conditions in MAE, we need to obtain Keplerian angular velocity $\Omega = \Omega_{K*}$, $\omega = 1$ from the inner solution at $x \to \infty$ to fit the outer solution (11.296) at the inner boundary $r = r_*$. This implies $\xi_b = 1$ for the constant in (11.306). We obtain after transition to the variable Σ in (11.305) and (11.306)

$$\frac{\mathrm{d}\Sigma}{\mathrm{d}x} = -\frac{1}{\mathrm{d}_{\mathscr{P}}} \frac{2n+1}{2n+3} \Omega_{K*}^2 H_b (1-\omega^2) \Sigma^{(2n-1)/(2n+1)}, \tag{11.311}$$

$$\frac{\mathrm{d}\omega}{\mathrm{d}x} = \frac{\dot{M}}{2\pi\alpha_b r_*^2 d_{s0}} (1-\omega) \Sigma^{-2(n+1)/(2n+1)}. \tag{11.312}$$

Dividing (11.311) by (11.312), we obtain

$$D_b \frac{\mathrm{d}\Sigma}{\mathrm{d}\omega} = -(1+\omega)\Sigma^{(4n+1)/(2n+1)}, \quad D_b = \frac{2n+3}{2n+1} \frac{d_{\mathscr{P}}}{\mathrm{d}_{s0}} \frac{\dot{M}}{2\pi\alpha_b H_b} \frac{1}{r_*^2 \Omega_{K*}^2}. \tag{11.313}$$

The solution of (11.313) must fit the boundary condition on the stellar surface $\Sigma = \Sigma_*$ at $\omega = \omega_*$. Taking into account that the surface density rapidly grows into the star, we may put, with sufficient accuracy, $\Sigma_* = \infty$ and obtain the solution, using (11.309) and (11.313), in the form

$$\Sigma = d_\Sigma \frac{\sqrt{K\Omega_{K*}^{1/n}}}{(\Omega_{K*}r_*)^{(2n+1)/n}} \left(\frac{\dot{M}}{\alpha_b H_b}\right)^{(2n+1)/2n}$$

$$\times \left[(\omega - \omega_*)\left(1 + \frac{\omega + \omega_*}{2}\right)\right]^{-(2n+1)/2n},$$

$$d_\Sigma = \sqrt{n+1}\,n^{-(2n+1)/4n}(2\pi)^{-(4n+3)/4n}\left[\frac{\Gamma\,(n+3/2)}{\Gamma(n+1)}\right]^{1/2n}. \qquad (11.314)$$

For fitting the inner BL solution and the outer solution for an accretion disk, we must make the surface densities (11.314) at $\omega = 1$ and (11.299) at $r = r_{\text{in}}$ equal. This fitting uniquely determines the outer integration constant ξ. After some algebraic calculations, we obtain

$$1 - \xi = d_n \alpha \alpha_b^{-(2n+3)/2n}\left(\frac{r_*}{H_b}\right)^{(2n+3)/2n}\left[\frac{\dot{M}K^n}{r_*^2(\Omega_{K*}r_*)^{2n+1}}\right]^{3/2n}$$

$$\times \left[(1-\omega_*)(3+\omega_*)\right]^{-(2n+3)/2n},$$

$$d_n = 2^{(2n+3)/2n}3\sqrt{\pi}(2\pi)^{-(2n+9)/4n}n^{-(4n+3)/4n}(n+1)^{3/2}\left[\frac{\Gamma\,(n+3/2)}{\Gamma(n+1)}\right]^{3/2n}.$$

(11.315)

The value of H_b is still not determined. It must be found from (11.312), where Σ is substituted from the solution (11.314). By order of magnitude, we have in the accretion disc (Problem 3)

$$u_K = \Omega_K r, \quad \frac{u_{s0}}{u_K} \sim \frac{z_0}{r}, \quad \frac{u_r}{u_K} \sim \alpha\left(\frac{z_0}{r}\right)^2\left(1 - \xi\sqrt{\frac{r_{\text{in}}}{r}}\right)^{-1}. \qquad (11.316)$$

Equation (11.312) contains a non-physical logarithmic divergency because of using the MAE method. For an approximate estimation of the value of H_b, we use a characteristic thickness over which the ω variation occurs, and for this, using the definition (11.304), the following relation may be written

$$\frac{2\pi\alpha_b r_* d_{s0}}{\dot{M}}\Sigma^{2(n+1)/(2n+1)}|_{\omega=1} = 1. \qquad (11.317)$$

With the help of (11.309) and (11.314), we obtain from (11.317)

$$\frac{H_b}{r_*} \approx d_H \alpha_b^{-1/(n+1)} \left[\frac{\dot{M} K^n}{r_*^2 (\Omega_{K*} r_*)^{2n+1}}\right]^{1/(n+1)} \left[(1-\omega_*)(3+\omega_*)\right]^{-1},$$

$$d_H = 2(2\pi)^{-3/2(n+1)} n^{-(2n+1)/2(n+1)} (n+1)^{n/(n+1)} \left[\frac{\Gamma\,(n+3/2)}{\Gamma(n+1)}\right]^{1/(n+1)}. \tag{11.318}$$

Using (11.318) in (11.315), we finally obtain the expression for $(1-\xi)$

$$1-\xi \approx d_\beta \alpha \alpha_b^{-(2n+3)/2(n+1)} \left[\frac{\dot{M} K^n}{r_*^2 (\Omega_{K*} r_*)^{2n+1}}\right]^{1/2(n+1)},$$

$$d_\beta = d_n d_H^{-(2n+3)/2n} 3\sqrt{\pi} (2\pi)^{-(2n+5)/4(n+1)} n^{1/4(n+1)} (n+1)^{n/2(n+1)}$$

$$= \left[\frac{\Gamma\,(n+3/2)}{\Gamma(n+1)}\right]^{1/2(n+1)}. \tag{11.319}$$

By order of magnitude, taking into account (11.316), we obtain, combining (11.318) and (11.319),

$$(1-\xi) \sim \frac{\alpha}{\alpha_b}\frac{z_0}{r_*}, \quad \frac{H_b}{r_*} \sim \frac{1}{1-\omega_*}\left(\frac{z_0}{r_*}\right)^2. \tag{11.320}$$

Thus, we obtain a complete analytical solution of the problem. For an isothermal disk with the same viscosity as in (11.308), we obtain, using (11.294), (11.305) and (11.306),

$$\mathscr{P} = K\Sigma, \quad \frac{d\Sigma}{dx} = -\frac{\Omega_{K*}^2}{K} H_b \Sigma (1-\omega^2), \quad \frac{d\omega}{dx} = \frac{\dot{M}(1-\omega)}{2\pi\alpha_b \sqrt{K}\,\Sigma r_*^2}. \tag{11.321}$$

Dividing the last two equations and using the approximate boundary condition $\Sigma_* = \infty$, we obtain the solution

$$\Sigma = \left(\frac{\dot{M}\sqrt{K}}{2\pi\alpha_b H_b}\right) \frac{1}{r_*^2 \Omega_{K*}^2} \left[(\omega-\omega_*)\left(1+\frac{\omega+\omega_*}{2}\right)\right]^{-1}. \tag{11.322}$$

From (11.321) and (11.322), we obtain the characteristic scale, which we identify with H_b

$$\frac{H_b}{r_*} \approx \frac{2K}{r_*^2 \Omega_{K*}^2} \left[(1-\omega_*)(3+\omega_*)\right]^{-1}. \tag{11.323}$$

Matching (11.302) and (11.322) with account of (11.323), we obtain for the integration constant

$$1 - \xi = \frac{3}{\sqrt{2}} \frac{\sqrt{K}}{r_* \Omega_{K*}} \frac{\alpha}{\alpha_b}. \tag{11.324}$$

Comparing (11.323) and (11.324) with (11.320), we see that the order of magnitude estimates for the polytropic case become exact for the isothermal case apart from numerical coefficients close to unity. In the isothermal case, there is a simple solution of (11.321) when (11.322) is substituted for Σ,

$$(1-\omega)^{-2}(\omega-\omega_*)^{(3+\omega_*)/(1+\omega_*)}(2+\omega+\omega_*)^{-(1-\omega_*)/(1+\omega_*)}$$
$$= \exp\left[2\frac{H_b}{K}\Omega_{K*}^2 x(1-\omega_*)(3+\omega_*)\right]. \tag{11.325}$$

We can see from (11.325) that the scale (11.323) appears in the exponent. It is clear that this solution has a physical sense only over several characteristic scales H_b. It is proved [735] that the region of applicability of the inner and outer solutions, obtained by the MAE method, do overlap, and the formula, constructed from inner (i) and outer (e) solutions, gives good interpolation also for the intermediate region. The formula, describing the function f in the whole region, has a structure

$$f = f_i + f_e - (f_i)_e. \tag{11.326}$$

Here, $(f_i)_e$ is the value of the inner solution on the outer edge (equal to the value of the outer solution on the inner edge $(f_e)_i$) with $f = \Sigma,\ \Omega,\ \mathcal{P}$. Using the relation $\dot{M} = -2\pi r_* \Sigma u_r$ for the estimation of the radial velocity u_r in the boundary layer, we obtain $u_r/u_{s0} \sim \alpha_b$. This means that the solution obtained, where radial velocity was neglected, is valid only for sufficiently small viscosity with $\alpha_b \ll 1$.

Problem 5. Find the solution for an accretion disc structure around a rapidly rotating star.

Solution. [110] Investigations of low-mass X-ray binaries (LMXB) containing a neutron star have led to the conclusion that the accreting object can rotate rapidly with a surface angular velocity close to the Keplerian value. The question of accretion onto a rapidly rotating star also arises in cataclysmic variables, in which the primary is a white dwarf. The evolution of the star under mass accretion may be characterized by the variations of two values: the mass M and the total angular momentum J and by one function, $j(r)$, determining the angular momentum distribution within the star; $j(r)$ is determined by the viscosity law inside the star. For a given mass flux $\dot{M}$, determined by conditions far away from the star, the star gradually increase its mass and angular momentum. When the stellar rotation rate is smaller than the Keplerian limit, the specific angular momentum of matter accreting onto the star is approximately equal to $\dot{M} v_{\mathrm{Ke}} R_{\mathrm{se}}$, ($v_{\mathrm{K}} = (GM/r^3)^{1/2}$ is the Keplerian velocity, R_{se}- the stellar equatorial radius, index "e" referes to values at the stellar equator). In this case, the angular velocity of the disk has a maximum near R_{se}; at this point, the viscous flux is zero and the momentum flux is determined only by the advective term $\dot{M} j_{\mathrm{e}}$, $j = v_\phi r$. The maximum of the rotational velocity disappears when the star rotates with the critical velocity and the (negative) viscous momentum flux becomes important.

The demand of self-consistency during accretion onto a rapidly rotating star may be formulated as the condition that the star absorb the accreted matter with a specific angular momentum j_0 such that the star remains in the state of critical rotation $j_0 = j_a = (dJ/dM)_c$. If viscosity inside the star is high (turbulent), the star rotates rigidly with a high precision. When the star-disc system can be described by a polytropic equation of state, there is a self-similar solution for such a system which does not depend on mass. In this case the value of the specific angular momentum of the accreted matter is obtained from the structure of the polytropic star. It is convenient to solve the problem in non-dimensional polytropic variables [268]

$$\tilde{\xi} = r/r_*, \quad \Theta^n = \rho/\rho_c, \quad \omega = \Omega/\Omega_*, \tag{11.327}$$

where ρ_c is the central density of the star, and r_* and Ω_* are given by

$$r_* = \left(\frac{(n+1)K\rho_c^{(1/n)-1}}{4\pi G}\right)^{1/2}, \quad \Omega_* = \sqrt{2\pi G\rho_c}. \tag{11.328}$$

In the non-dimensional variables (11.327) and (11.328), the equation of stellar structure, obtained from equilibrium and the Poisson equation for $\Omega = \text{const.}$ (see Chap. 6, Vol. 1), can be written in the form

$$\frac{1}{\tilde{\xi}^2}\frac{\partial}{\partial\tilde{\xi}}\left(\tilde{\xi}^2\frac{\partial\Theta}{\partial\tilde{\xi}}\right) + \frac{1}{\tilde{\xi}^2}\frac{\partial}{\partial\mu}\left((1-\mu^2)\frac{\partial\Theta}{\partial\mu}\right) + \Theta^n - \omega^2 = 0, \tag{11.329}$$

with $\mu = \cos\theta$. The solution of (11.329) is completed by the boundary conditions and fixed value of ω.

For compressible matter with $n > 0$, a solution exists only if $\omega < \omega_c$; for $\omega = \omega_c$ the centrifugal force on the equatorial boundary of the star exactly balances gravity (see the proof in [165]). Numerical calculations have been carried out which gave the structure of rotating polytropic stars for different ω up to the critical value ω_c [204, 532, 545], see also [964].

For a given n, the structure of a critically rotating star does not depend on its mass. By definition, the mass M and the total angular momentum J are given by

$$M = 2\pi\rho_c r_*^3 \int_{-1}^{1} d\mu \int_0^{\tilde{\xi}_{\text{out}}(\mu)} \Theta^n(\tilde{\xi},\mu)\tilde{\xi}^2 d\tilde{\xi}, \tag{11.330}$$

$$J = 2\pi\rho_c r_*^5 \omega\sqrt{2\pi G\rho_c} \int_{-1}^{1} (1-\mu^2)d\mu \int_0^{\tilde{\xi}_{\text{out}}(\mu)} \Theta^n(\tilde{\xi},\mu)\tilde{\xi}^4 d\tilde{\xi}. \tag{11.331}$$

Introducing non-dimensional values of the mass $\mathscr{M}_n$ and of the total angular momentum $\mathscr{J}_n$, defined as

$$\mathscr{M}_n = \frac{1}{2}\int_{-1}^{1} d\mu \int_0^{\tilde{\xi}_{\rm out}(\mu)} \Theta^n(\tilde{\xi},\mu)\tilde{\xi}^2 d\tilde{\xi}, \tag{11.332}$$

$$\mathscr{I}_n = \frac{\omega}{2}\int_{-1}^{1} (1-\mu^2) d\mu \int_0^{\tilde{\xi}_{\rm out}(\mu)} \Theta^n(\tilde{\xi},\mu)\tilde{\xi}^4 d\tilde{\xi}, \tag{11.333}$$

M and J may be written as

$$M = 4\pi\rho_c r_*^3 \mathscr{M}_n, \tag{11.334}$$

$$J = 4\pi\rho_c r_*^5 \sqrt{2\pi G\rho_c}\,\mathscr{I}_n. \tag{11.335}$$

The average specific angular momentum of the star j_s and the derivative along the critical states $j_a = (dJ/dM)_c$ are written, using (11.328) and (11.334), as

$$j_s = J/M = r_*^2\sqrt{2\pi G\rho_c}\,\mathscr{I}_n/\mathscr{M}_n \quad j_a = \frac{5-2n}{3-n} j_s. \tag{11.336}$$

Using (11.334), one may write ρ_c and r_*, which appear in (11.328), as a function of M

$$\rho_c = \left(\frac{4\pi G}{(n+1)K}\right)^{3n/(3-n)} \left(\frac{M}{4\pi\mathscr{M}_n}\right)^{2n/(3-n)},$$

$$r_* = \left(\frac{4\pi G}{(n+1)K}\right)^{n/(n-3)} \left(\frac{M}{4\pi\mathscr{M}_n}\right)^{(1-n)/(3-n)}. \tag{11.337}$$

The specific angular momentum of matter j_e at the stellar equator r_e may be written as

$$j_e = \Omega r_e^2 = r_*^2\sqrt{2\pi G\rho_c}\,\tilde{\xi}_e^2\omega, \quad \tilde{\xi}_e = r_e/r_* = \tilde{\xi}_{\rm out}(\pi/2). \tag{11.338}$$

From a comparison of (11.336) and (11.338), it follows that the ratio

$$\frac{j_a}{j_e} = \frac{5-2n}{3-n}\frac{j_s}{j_e} = \frac{5-2n}{3-n}\frac{\mathscr{I}_n}{\mathscr{M}_n\tilde{\xi}_e^2\omega} \tag{11.339}$$

does not depend on the stellar mass M and is a function of n and ω only. The non-dimensional values of the angular velocity ω_c, the equatorial radius $\tilde{\xi}_{\rm out}(\theta = \pi/2) = \tilde{\xi}_e$, the mass $\mathscr{M}_n$, the momentum of inertia around the rotational axis $\mathscr{I}_n = \mathscr{I}_n/\omega_c$, the average specific angular momentum of the star ζ_s , the non-dimensional derivative ζ_a

$$\zeta_s = j_s/(r_*^2\sqrt{2\pi G\rho_c}) = \mathscr{I}_n/\mathscr{M}_n, \quad \zeta_a = \frac{5-2n}{3-n}\zeta_s, \tag{11.340}$$

and the angular momentum of matter on the stellar equator

$$\zeta_e = j_e/(r_*^2\sqrt{2\pi G\rho_c}) = \tilde{\xi}_e^2\omega_c \tag{11.341}$$

for several polytropic indices n in the state of critical rotation are given in Table 11.9, which is based on calculations of [204, 532, 545]. The ratio of the equatorial and polar radii $\tilde{\xi}_e/\tilde{\xi}_p$ and the non-dimensional parameters for non-rotating polytropes (outer radius ξ_{0n}, mass $\mathscr{M}_{0n}$ and momentum of inertia around the symmetry axis $\mathscr{I}_{0n}$ [204, 269, 545]) are also given.

The viscous flux of specific angular momentum in the accretion disk around a critically rotating star, j_v is therefore negative; near the surface of the star j_v is equal to

$$j_v = j_a - j_e \tag{11.342}$$

as one requires that the total angular momentum flux into the star is equal to $\dot{M}\, j_a$ for self-consistency.

With account of thermal processes, the equations of viscous heat production and heat transfer in the z direction must be added to equilibrium equations (see Problem 2 for the complete system). The total luminosity of the accretion disc is equal to (see (11.254) from Problem 2)

$$L = \int_{r_{in}}^{\infty} 4\pi Q_+ dr = \left(\frac{3}{2} - \xi\right)\dot{M}\frac{GM}{r_{in}}. \tag{11.343}$$

When $\xi = 1$ the star accretes matter with Keplerian angular momentum, and only half of the gravitational energy of the accreted matter is radiated from the disc, according to the virial theorem. For a slowly rotating neutron star there are two possibilities. If the radius of the neutron star r_s is smaller than $3r_g$, then $r_{in} = 3r_g$ ($3r_g = 6GM/c^2$); the disk luminosity is the same as in the case of a black hole, and the remaining gravitational energy, including the part gained during free-fall onto the neutron star surface, is emitted close to the neutron star surface. If $r_s > 3r_g$, then $r_{in} = r_s$, the disc luminosity is $GM\dot{M}/2r_s$ and an almost equal amount of energy is emitted near the stellar surface, where the accreting matter converts its kinetic energy into heat.

The situation changes gradually while the star absorbs angular momentum. Let us consider only the case $r_s > 3r_g$. The luminosity of the disc decreases when the stellar radius increases. The fraction of energy emitted by the neutron star also decreases. It changes from $GM\dot{M}/2r_s$ in the case of a non-rotating star to the difference between the rotational energy of matter in the disc and at the stellar surface.

A rapid change in the efficiency of energy release and disc luminosity occurs when the star rotation reaches its maximum value. The energy emitted near the stellar surface tends towards zero, but the luminosity of the disc suffers a drastic change. The star no longer absorbs all of the angular momentum coming from the disc, but absorbs only the fraction required to maintain the star in a critically rotating state. The specific angular momentum of matter accreted by the star is equal to

Table 11.9 Some parameters of non-rotating and critically rotating polytropic stars

n	$\tilde{\xi}_{n0}$	$\mathscr{M}_{n0}$	$\mathscr{I}_{n0}$	ω_c	$\tilde{\xi}_e$	$\tilde{\xi}_e/\tilde{\xi}_p$	$\mathscr{M}_n$	$\mathscr{I}_n$	ζ_s	$\zeta_a/\zeta_e = \xi$
0.5	2.7528	3.7871		0.389		2.51				
				(0.367)						
0.6				0.362		2.29				
				(0.354)						
0.808				0.326	4.7652	1.917	5.0248	24.43	1.585	0.3304
1.0	π	π	8.104	0.289	4.8265	1.792	4.289	18.73	1.262	0.2805
1.5	3.65375	2.71406	7.413	0.209	5.36	1.626	3.2138	12.18	0.792	0.176
2.0	4.35287	2.41105	7.074	0.147	6.307	1.555	2.6518	9.62	0.533	0.0894
2.5	5.35528	2.18720	7.013	0.09965	7.7623	1.522	2.30563	8.49	0.367	0.0
3.0	6.89685	2.01824	7.234	0.0639	10.123	1.535	2.0743	8.125	0.250	$-\infty$

$(dJ/dM)_c$, and the remaining part is carried away in the disc by viscous stresses. Viscosity carries not only angular momentum, but also energy, so that the energy production and luminosity of the disk rapidly increase, from the value corresponding to $\xi = 1$ to the value corresponding to $\xi = j_a/j_e$ according to (11.343); for polytropic stars, the values of ξ are given in the last column of Table 11.9. This implies a rapid increase in the total luminosity by a factor 2–3. This rapid increase is easy to understand if one remembers that when a star accelerates its rotation, part of the gravitational energy is converted into rotational energy without heat production. When a star reaches the limiting rotation, the growth rate of its rotational energy strongly diminishes and a correspondingly larger fraction of gravitational energy is transformed into heat. The minimum of luminosity of an accreting neutron star is reached when the stellar angular velocity is slightly below the critical value. A rapid increase in luminosity must be accompanied by a corresponding hardening of the emitted spectrum because the energy release is increased, raising the effective temperature in the inner part of the disc. Such events may be expected in objects like LMXB or cataclysmic variables. The young T Tauri stars are born rapidly rotating; so their value of ξ is small from the beginning and close to that of a polytrope $n = 1.5$ with $\xi = 0.176$ given by Table 11.9.

To find the structure of the accretion disc around a rapidly rotating star, we can use the solution in the Problem 2 with

$$\xi = \frac{j_a}{j_e} = \frac{1}{j_e}\left(\frac{\mathrm{d}J}{\mathrm{d}M}\right)_c, \quad \frac{l_{in}}{l} = \xi\sqrt{\frac{r_{in}}{r}}. \tag{11.344}$$

The solution with ξ from (11.344) is valid with good precision up to the very surface of the star, contrary to the case of a black hole, where ξ is close to unity, and the advective term in (11.117) must be taken into account near $r = r_{in}$.

Chapter 12
Dynamic Stability

12.1 Hierarchy of Time Scales

Processes underlying the stellar evolution are characterized by a wide variety of time scales, such as hydrodynamical (τ_h), thermal (τ_{th}), and nuclear (τ_n) times, and a time characterizing the weak interaction rate (τ_β). Throughout most of the evolution, from the phase of the young contracting star to the late evolutionary phases, τ_h remains the smallest of all the time scales. In massive presupernova models, under conditions of nuclear equilibrium, the time scale τ_n is the smallest one (see Sect. 10.1). In an ordinary star there is a rough equilibrium with respect to rapid processes (e.g., static equilibrium), while the evolutionary lifetime is determined by one of the slower processes, as follows:

1. During the stage of gravitational contraction, we have:

$$\tau_h \ll \tau_{th} \ll \tau_\beta, \tau_n. \tag{12.1}$$

   The evolution is determined here by the thermal time τ_{th}, the star is almost in static equilibrium, and the times τ_β and τ_n are so large that the nuclear composition remains frozen (with the exception of the lightest nuclei).
2. The conditions (12.1) hold on the main sequence, but the evolution is determined by the nuclear and weak times τ_n, τ_β, and the star is in a state close to static and thermal equilibrium.
3. After the helium core has formed, a stage of gravitational contraction of central regions and envelope expansion occurs, analogous to stage 1. From the core helium ignition to the final phases, the nuclear time scales remain comparable to thermal ones,

$$\tau_h \ll \tau_{th} \sim \tau_n \leq \tau_\beta, \tag{12.2}$$

   and determine the stellar evolution rate.

   The thermal time in the core is substantially less than the time in the envelope τ_{th}^{env}, similarly for the nuclear time τ_n^{core}, which is minimal for the central regions since in the envelope the nuclear reactions do not proceed at all. For quiet

G.S. Bisnovatyi-Kogan et al., *Stellar Physics: 2: Stellar Evolution and Stability*, Astronomy and Astrophysics Library, DOI 10.1007/978-3-642-14734-0_6,

post-main-sequence evolutionary phases of non-massive stars one has $\tau_{\mathrm{th}}^{\mathrm{env}} \sim \tau_{\mathrm{n}}^{\mathrm{core}}$, that is, the minimum nuclear time of the star is of the same order as the maximum thermal time. During violent events such as helium flashes in degenerate cores or flashes in a helium-burning shell (see Sect. 9.3), the nuclear time in the central regions or in the helium shell reduces to a magnitude comparable to that of the local thermal time. The hydrodynamical time remains minimal in this case and the static equilibrium of the star is not broken.

4. In low- and intermediate-mass stars of $M \leq (6-8)M_{\odot}$ with a core transforming into a white dwarf, the condition of minimum τ_{h} is never broken. Weak interactions cause the neutrino cooling of stars which dominates the photon cooling during the stage of white dwarf formation and determines the thermal time $\tau_{\mathrm{h}} \ll \tau_{\mathrm{th}} \sim \tau_{\beta} \leq \tau_{\mathrm{n}}$. At the end of white dwarf formation, after the envelope ejection, the nuclear reactions cease in the central regions of the dwarf, and only residual hydrogen and helium burning in the envelope is still possible.
5. In more massive stars with a core exceeding, after envelope ejection, the Chandrasekhar limit, a complicated instability develops, incorporating thermal, nuclear, and weak processes. In less massive stars of this range, τ_{n} falls off steeply because of increasing temperature in the degenerate C–O core, so that the condition

$$\tau_{\mathrm{n}} \sim \tau_{\mathrm{th}} \ll \tau_{\mathrm{h}}, \tau_{\beta}, \tag{12.3}$$

is reached, implying the onset of a thermal explosion (see Sect. 10.2).
6. For larger core mass stars, the condition (12.3) never arises, but a high Fermi energy of electrons causes their rate of capture to accelerate until the condition

$$\tau_{\beta} \sim \tau_{\mathrm{h}} \leq \tau_{\mathrm{th}}, \tau_{\mathrm{n}}, \tag{12.4}$$

becomes valid, leading to the onset of collapse due to neutronization (see Sect. 10.3).
7. In the most massive stars, where the degeneracy does not occur, τ_{n} decreases because of increasing temperature, falls below τ_{h}, and nuclear equilibrium sets in. With further energy emission, the star enters a region of dynamical instability which is principally distinct from the above instabilities.

When instabilities are related to thermal, nuclear, or weak processes, their development is accompanied by a rapid decrease in correspondent time scales. If these become of the order of or less than the hydrodynamical time scale, then the instability ends with a thermal explosion or collapse. The dynamic instability arises from changes in the structure of the equilibrium state rather than in the dynamic time scale τ_{h} that vary slightly. Contrary to other types of instability, the dynamic one lends itself to a treatment based on the theory of conservative mechanical systems. Two equivalent methods are basic in the field: the variational principle and the method of small perturbations.

12.2 Variational Principle and Small Perturbations

12.2.1 Variational Principle in General Relativity

We consider the equilibrium of spherically symmetric stars and their stability with the aid of the variational principle. We shall discuss our treatment in the framework of general relativity to have the possibility to apply the results to neutron and supermassive stars. In a metric similar to the Schwarzschild metric [615]

$$ds^2 = -g_{00}\,dt^2 + g_{11}\,dr^2 + r^2(d\theta^2 + \sin^2\theta\,d\varphi^2) \tag{12.5}$$

for an arbitrary spherically symmetric distribution of matter, we may write the total energy of the star ε in the form [615]

$$e \equiv e(r) = 4\pi \int_0^r \tilde{E} n r^2 dr, \quad \varepsilon = e(R), \tag{12.6}$$

where R is the Schwarzschild radius of the star, $\tilde{E}$ is the internal energy per baryon, $\tilde{E} = (\rho_0/n)\,(E + c^2)$. A volume element of the spherical shell $d\nu$ is

$$d\nu = 4\pi r^2 dr \sqrt{g_{11}} = 4\pi r^2 \left(1 - \frac{2Ge}{c^4 r}\right)^{-1/2} dr. \tag{12.7}$$

Similarly, we introduce N, the total number of baryons in the star, and ν, the number of baryons inside the given radius r

$$\nu(r) = 4\pi \int_0^r n r^2 \left(1 - \frac{2Ge}{c^4 r}\right)^{-1/2} dr, \quad N = \nu(R). \tag{12.8}$$

According to the variational principle [449, 1075], the total energy of the star ε has an extremum at the equilibrium point for a given total number of baryons N and fixed entropy distribution over baryons. The total energy ε is, in this case, analogous to the potential energy of a conservative system whose conservatism originates from the invariant character of the entropy distribution. Introducing a Lagrangian coordinate ν from (12.8) so that

$$\frac{de}{d\nu} = \tilde{E}\left(1 - \frac{2Ge}{c^4 r}\right)^{1/2}, \tag{12.9}$$

$$n = \left(1 - \frac{2Ge}{c^4 r}\right)^{1/2} \left(4\pi r^2 \frac{dr}{d\nu}\right)^{-1}. \tag{12.10}$$

Next, we find the variation of the total energy ε as a function of the radius variation $\delta r(\nu)$. Varying (12.10), we obtain

$$\delta n = \frac{\mathrm{Gen}}{c^4 r} \frac{\delta r/r - \delta e/e}{1 - 2Ge/c^4 r} - 2n\frac{\delta r}{r} - \frac{4\pi r^2 n^2 \mathrm{d}(\delta r)/\mathrm{d}\nu}{(1 - 2Ge/c^4 r)^{1/2}}. \tag{12.11}$$

Varying (12.9) using (12.11) and recalling the thermodynamic relation [617] $(\partial \tilde{E}/\partial n)_S = P/n^2$ gives

$$\frac{\mathrm{d}(\delta e)}{\mathrm{d}\nu} = -\frac{P/n + \tilde{E}}{(1 - 2Ge/c^4 r)^{1/2}} \frac{Ge}{c^4 r}\left(\frac{\delta e}{e} - \frac{\delta r}{r}\right) - \frac{2P}{nr}\left(1 - \frac{2Ge}{c^4 r}\right)^{1/2} \delta r - 4\pi r^2 P \frac{\mathrm{d}(\delta r)}{\mathrm{d}\nu}. \tag{12.12}$$

Solving (12.12) for δe as a linear inhomogeneous equation (see [917]) yields

$$\begin{aligned}\delta e(\nu) = \exp&\left\{-\int_0^\nu \left(\frac{P}{n} + \tilde{E}\right)\left(1 - \frac{2Ge}{c^4 r}\right)^{-1/2} \frac{G\,\mathrm{d}\nu}{c^4 r}\right\} \\ &\times \left\{\int_0^\nu \exp\left[\int_0^\nu \left(\frac{P}{n} + \tilde{E}\right)\left(1 - \frac{2Ge}{c^4 r}\right)^{-1/2} \frac{G}{c^4 r}\,\mathrm{d}\nu\right]\right. \\ &\times \left[\left(\frac{P}{n} + \tilde{E}\right)\left(1 - \frac{2Ge}{c^4 r}\right)^{-1/2} \frac{G}{c^4 r}\left(e + 4\pi r^3 P\right)\right. \\ &\left.\left. + \frac{1}{n}\frac{\mathrm{d}P}{\mathrm{d}r}\left(1 - \frac{2Ge}{c^4 r}\right)^{1/2}\right]\delta r\,\mathrm{d}\nu\right\} - 4\pi r^2 P \delta r.\end{aligned} \tag{12.13}$$

Equating the variation of the total energy to zero, $\delta\varepsilon = \delta e(N) = 0$, and using $P(N) = 0$, we have from (12.13) the equilibrium Oppenheimer–Volkov equation (11.2.3) where $\rho c^2 = \tilde{E}n$, $mc^2 = e$. Looking for an extremum of ε from (12.6) at fixed N from (12.8) reduces to an isoperimetric problem of the calculus of variations [918]. The reciprocity principle is valid for this problem, according to which the function $r(\nu)$ determining the extremum of ε at fixed N simultaneously determines the extremum of N at fixed ε.

The stability of a star implies that the second variation of the energy be positive,

$$\delta^2\varepsilon > 0, \tag{12.14}$$

providing the minimum stellar energy for stable equilibrium. If during the evolution $\delta^2\varepsilon$ becomes zero, changing sign from plus to minus, such a state will be critical and correspond to the loss of stability. Deriving the stability condition yields $\delta^2 n$ from (12.11) and $\mathrm{d}(\delta^2 e)/\mathrm{d}\nu$ from (12.12). The latter reduces to a differential equation with respect to $\delta^2 e$ that should be solved analogously to (12.12), giving the following expression for the second variation of energy:

$$\delta^2\varepsilon = \exp\left\{-\int_0^R \frac{\tilde{E}n+P}{1-2Ge/c^4r}\frac{4\pi Gr}{c^4}\mathrm{d}r\right\}$$
$$\times 4\pi\int_0^R \exp\left\{\int_0^r \frac{\tilde{E}n+P}{1-2Ge/c^4r}\frac{4\pi Gr}{c^4}\mathrm{d}r\right\}$$
$$\times\left\{\gamma P\left[2\delta r + r\frac{\mathrm{d}(\delta r)}{\mathrm{d}r} - \frac{Ge}{c^4r}\frac{1+4\pi r^3P/e}{1-2Ge/c^4r}\delta r\right]^2\right.$$
$$-\frac{1}{4}\frac{(P+\tilde{E}n)(1+4\pi r^3P/e)^2}{(1-2Ge/c^4r)^2}\left(\frac{2Ge}{c^4r}\right)^2(\delta r)^2$$
$$\left.-2\frac{P+\tilde{E}n}{1-2Ge/c^4r}\left(1+\frac{1}{2}\frac{4\pi r^3P}{e}\right)\frac{2Ge}{c^4r}(\delta r)^2\right\}\mathrm{d}r, \qquad (12.15)$$

where $\gamma \equiv \gamma_1 = (\partial \ln P/\partial \ln \rho)_S$. If $t_{n,\beta} \gg t_{\mathrm{h}}$, the adiabatic index is evaluated at a fixed chemical composition, if $t_n \ll t_{\mathrm{h}}$, at an equilibrium nuclear composition. The case of $t_n \sim t_{\mathrm{h}}$ or $t_\beta \sim t_{\mathrm{h}}$ is the most difficult. While the relation $t_n \sim t_{\mathrm{h}}$ can hold only in a narrow range of parameters because of an exponential dependence of the reaction rates on the temperature, the relation $t_\beta \sim t_{\mathrm{h}}$ can hold for a sufficiently long period at the beginning of low-mass star collapse (see Sect. 10.3). In such cases, a kinetic description is needed for nuclear and weak reactions, a dissipation of second-viscosity type arises, the entropy increases, and, ultimately, the conditions for using the variational principle break. A phenomenological derivation of the equation of motion has a wider range of applications than the variational one. Eulerian equations in hydrodynamics, for instance, can be derived by either method, while the Navier–Stokes equations including viscosity can be derived only phenomenologically in the hydrodynamics, or from a more general kinetic theory [275]. For a static case at $v=0$ Eulerian and Navier–Stokes equations coincide. The reason for this is that at $v=0$ the viscous dissipation is not present. Using the variational principle requires the absence of dissipation not only at $v=0$ but also at small $v \neq 0$, therefore at $t_\beta \sim t_{\mathrm{h}}$, when the second viscosity ($\sim \mathrm{div}\mathbf{v}$) is at work, the variational principle cannot be used to derive the stability condition. Rather, a perturbation method should be used, which includes the effect of the kinetics of the processes on stability. Both methods yield the same result for conservative systems. In particular, the condition (12.15) has been derived by both the method of small perturbations in [267] and the variational method in [449], see also [223].

12.2.2 Newtonian and Post-Newtonian Limits

We rewrite the variation of energy $\delta\varepsilon$ in a form free from derivatives of thermodynamic functions. Using (12.13) gives

$$
\begin{aligned}
\delta\varepsilon = \exp&\left\{-\int_0^N \left(\frac{P}{n}+\tilde{E}\right)\left(1-\frac{2Ge}{c^4 r}\right)^{-1/2}\frac{G\,\mathrm{d}\nu}{c^4 r}\right\} \\
&\times\left\{\int_0^N \exp\left[\int_0^\nu \left(\frac{P}{n}+\tilde{E}\right)\left(1-\frac{2Ge}{c^4 r}\right)^{-1/2}\frac{G}{c^4 r}\,\mathrm{d}\nu\right]\right. \\
&\times\left[\left(\frac{P}{n}+\tilde{E}\right)\left(1-\frac{2Ge}{c^4 r}\right)^{-1/2}\frac{Ge}{c^4 r^2}\delta r-\frac{2P}{nr}\right. \\
&\left.\left.\times\left(1-\frac{2Ge}{c^4 r}\right)^{1/2}\delta r-4\pi r^2 P\frac{\mathrm{d}(\delta r)}{\mathrm{d}\nu}\right]\mathrm{d}\nu\right\}. \qquad (12.16)
\end{aligned}
$$

In the Newtonian limit, since

$$
\begin{aligned}
&\rho_0 = n m_{\mathrm{u}}, \quad m_0 = \nu m_{\mathrm{u}}, \quad M_0 = N m_{\mathrm{u}}, \\
&\tilde{E}n = \left(E + c^2\right)\rho_0, \\
&\mathrm{d}\nu/n = \mathrm{d}m_0/\rho_0, \qquad (12.17)
\end{aligned}
$$

from (12.6), (12.9), (12.16), and (12.15) we have

$$
\varepsilon = M_0 c^2 + \int_0^{M_0}\left(E - \frac{Gm_0}{r}\right)\mathrm{d}m_0, \qquad (12.18)
$$

$$
\delta\varepsilon = \int_0^{M_0}\left(\frac{Gm_0}{r^2}\delta r - 2\frac{P}{\rho}\frac{\delta r}{r} - \frac{P}{\rho}\frac{\mathrm{d}\delta r}{\mathrm{d}r}\right)\mathrm{d}m_0, \qquad (12.19)
$$

$$
\delta^2\varepsilon = \int_0^{M_0}\left\{\gamma\frac{P}{\rho}\left[2\frac{\delta r}{r}+\frac{\mathrm{d}(\delta r)}{\mathrm{d}r}\right]^2 - 4\frac{Gm}{r}\left(\frac{\delta r}{r}\right)^2\right\}\mathrm{d}m_0. \qquad (12.20)
$$

Here, we take $e = m_0 c^2$ and identify the Schwarzschild radius r with the Newtonian radius r_{N}. The Newtonian equilibrium equation (10.1) follows from the condition $\delta\varepsilon = 0$ in (12.19) upon integrating by parts in the last term.

The post-Newtonian approximation may be obtained in a similar way on retaining the terms $\sim \left(Gm/rc^2\right)^2$ and $P/\rho c^2$. Using (12.17), we therefore obtain from (12.6) and (12.9)

$$
\begin{aligned}
\varepsilon &= \int_0^N \rho_0\left(E+c^2\right)\left(1-\frac{2Ge}{c^4 r}\right)^{1/2}\frac{\mathrm{d}\nu}{n} \\
&= \int_0^{M_0}\rho_0\left(E+c^2\right)\left[1-\frac{Ge}{c^4 r}-\frac{1}{2}\left(\frac{Ge}{c^4 r}\right)^2\right]\frac{\mathrm{d}m_0}{\rho_0} \\
&= M_0 c^2 + \int_0^{M_0}\left(E-\frac{Ge}{c^4 r}\right)\mathrm{d}m_0 - \int_0^{M_0}\frac{GEm_0}{c^2 r}\mathrm{d}m_0 \\
&\quad - \frac{1}{2}\int_0^{M_0}\left(\frac{Gm_0}{rc}\right)^2\mathrm{d}m_0. \qquad (12.21)
\end{aligned}
$$

In the two last terms, we set $e = m_0c^2$. Relativistic corrections to e and r should be included in the second term. Passing to Newtonian descriptions does not alter the physical volume v determined in (12.7) and the Newtonian radius $r_{\mathrm{N}} = (3v/4\pi)^{1/3}$. From (12.9),

$$e_1 = \int_0^{m_0} (E + c^2)\left(1 - \frac{Gm_0}{c^2 r}\right) \mathrm{d}m_0$$
$$= m_0c^2 + \int_0^{m_0} E\, \mathrm{d}m_0 - \int_0^{m_0} \frac{Gm_0\, \mathrm{d}m_0}{r}. \tag{12.22}$$

Equation (12.7) gives

$$\frac{4\pi}{3} r^3 = v - \int_0^v \frac{Gm_0}{c^2 r} \mathrm{d}v, \quad r = r_{\mathrm{N}}\left(1 - \frac{1}{r^3}\int_0^r \frac{Gm_0}{c^2} r\, \mathrm{d}r\right). \tag{12.23}$$

In the last two terms of (12.21) and in the second term (12.23), r and r_{N} may be taken as identical. Substituting (12.22) and (12.23) into (12.21), we find [1080], omitting the index "N"

$$\begin{aligned}\varepsilon = M_0c^2 &+ \int_0^{M_0}\left(E - \frac{Gm_0}{r}\right)\mathrm{d}m_0 - \frac{G}{c^2}\int_0^{M_0}\frac{Em_0}{r}\mathrm{d}m_0\\
&- \frac{G^2}{2c^2}\int_0^{M_0}\frac{m_0^2}{r^2}\mathrm{d}m_0 - \frac{G}{c^2}\int_0^{M_0}\frac{\mathrm{d}m_0}{r}\left(\int_0^{m_0} E\,\mathrm{d}m_0\right)\\
&+ \frac{G^2}{c^2}\int_0^{M_0}\left(\int_0^{m_0}\frac{m_0\,\mathrm{d}m_0}{r}\right)\frac{\mathrm{d}m_0}{r}\\
&- \frac{G^2}{c^2}\int_0^{M_0}\frac{m_0}{r^4}\left(\int_0^r m_0 r\,\mathrm{d}r\right)\mathrm{d}m_0.\end{aligned} \tag{12.24}$$

Consider the post-Newtonian approximation for the case of a mass distribution according to the adiabat $n = 3$ that remains in equilibrium during a uniform expansion or contraction. This corresponds to the linear eigenfunction $\delta r = \alpha r$ in the variations $\delta\varepsilon$ and $\delta^2\varepsilon$. Evaluating the last five integrals in (12.24) which are small corrections accounting for general relativity, we have to take into account the equality $E\rho_0 = 3P$ and the Newtonian equilibrium equation (10.1).[1] Integrating by parts gives

$$\int_0^{M_0}\frac{Em_0}{r}\mathrm{d}m_0 = 3\int_0^{M_0}\frac{P}{\rho_0 r}m_0\,\mathrm{d}m_0 = -3\int_0^R\left(\int_0^{m_0}\frac{m_0\,\mathrm{d}m_0}{\rho_0 r}\right)\frac{\mathrm{d}P}{\mathrm{d}r}\mathrm{d}r$$
$$= 3G\int_0^{M_0}\frac{m_0\,\mathrm{d}m_0}{r^4}\left(\int_0^r m_0 r\,\mathrm{d}r\right),$$

[1] Equation (10.1) implies $m = m_0$ since only the rest mass is gravitating in Newtonian theory.

$$\int_0^{m_0} E\,\mathrm{d}m_0 = 3\int_0^{m_0} \frac{P}{\rho_0}\,\mathrm{d}m_0 = 4\pi P r^3 - 4\pi \int_0^{m_0} r^3 \frac{\mathrm{d}P}{\mathrm{d}r}\,\mathrm{d}r$$

$$= 4\pi P r^3 + G\int_0^{m_0} \frac{m_0\,\mathrm{d}m_0}{r},$$

$$\int_0^{M_0} P r^3 \frac{\mathrm{d}m_0}{r} = \int_0^{M_0} P d\left(\int_0^{m_0} r^2 \mathrm{d}m_0\right)$$

$$= \int_0^{M_0} \frac{\rho G m_0}{r^2}\left(\int_0^{m_0} r^2 \mathrm{d}m_0\right)\mathrm{d}r$$

$$= \int_0^{M_0} \frac{\rho G m_0}{r^2}\left(m_0 r^2 - 2\int_0^{r} m_0 r\,\mathrm{d}r\right)\mathrm{d}r$$

$$= \frac{G}{4\pi}\left[\int_0^{M_0} \frac{m_0^2}{r^2}\,\mathrm{d}m_0 - 2\int_0^{M_0} \frac{m_0\,\mathrm{d}m_0}{r^4}\int_0^{r} m_0 r\,\mathrm{d}r\right]. \tag{12.25}$$

Substituting (12.25) into (12.24) gives

$$\varepsilon = M_0 c^2 + \int_0^{M_0}\left(E - \frac{G m_0}{r}\right)\mathrm{d}m_0$$
$$- \frac{G^2}{c^2}\left[2\int_0^{M_0} \frac{m_0\,\mathrm{d}m_0}{r^4}\left(\int_0^{r} m_0 r\,\mathrm{d}r\right) + \frac{3}{2}\int_0^{M_0} \frac{m_0^2}{r^2}\,\mathrm{d}m_0\right]. \tag{12.26}$$

We shall drop hereafter the subscript "0" for m_0 and M_0. The last term in (12.26) is $\varepsilon_{\mathrm{GR}}$ from (10.1.7) and (10.1.16). Using (10.1.2) and (10.1.8), we have for $n = 3$

$$\varepsilon_{\mathrm{GR}} = -\frac{G^2}{c^2} M^{7/3}\rho_c^{2/3}\left[\frac{(4\pi)^{2/3}}{M_3^{7/3}}\left(\frac{3}{4}\int_0^{\xi_1} \theta^5\xi^2\,\mathrm{d}\xi + \frac{3}{8}\int_0^{\xi_1} \theta^7\xi^4\,\mathrm{d}\xi\right)\right]$$
$$= -0.9183\,\frac{G^2}{c^2} M^{7/3}\rho_c^{2/3}. \tag{12.27}$$

We have used here the integrals J_{52} and J_{74} from the table to the problem from Sect. 10.1. Equation (12.27) has been used in the energetic method in Sect. 10.1 for the treatment of stability to collapse. Using the sum of several profiling functions in the energy functional, instead of just one function in the enegetic method, increases the precision. The Gaterkin method for solving this problem was suggested in [125]. It is equivalent to the spectral method of solving differential equations as applied in [222] for problems related to general relativity. The subsequent post-Newtonian correction to ε has been found in [1011] for nonrotating stars and in [138] for stars with rotation.

12.2.3 Method of Small Perturbations in Newtonian Theory

Let us show the equivalence of the variational principle and the perturbation method on an example of Newtonian theory. We use the equations of hydrodynamics (10.38) and (10.39) for the case of adiabaticity and consider small deviations from the static equilibrium state

$$r = r_0 + r', \quad \rho = \rho_0 + \rho', \quad v_r \text{ is small},$$
$$P = P_0 + P' = P_0 + \gamma_1 \frac{P_0}{\rho_0}\rho'. \tag{12.28}$$

From (10.38) and (10.39), using (12.28), we obtain

$$\frac{\partial r'}{\partial t} = v_r, \quad \frac{\partial v_r}{\partial t} = -4\pi r_0^2 \frac{\partial P'}{\partial m} - 8\pi r_0 r' \frac{\partial P_0}{\partial m} + \frac{2Gm}{r_0^3} r',$$
$$\rho' = -4\pi \rho_0^2 r_0^2 \frac{\partial r'}{\partial m} - 2\rho_0 \frac{r'}{r},$$
$$dm = 4\pi \rho_0 r_0^2 \, dr_0. \tag{12.29}$$

Reducing the system (12.29) to the equation for r' gives

$$\frac{\partial^2 r'}{\partial t^2} = 4\pi r_0^2 \frac{\partial}{\partial m}\left[\gamma_1 \frac{P_0}{\rho_0}\left(4\pi\rho_0^2 r_0^2 \frac{\partial r'}{\partial m} + 2\rho_0 \frac{r'}{r_0}\right)\right]$$
$$- 8\pi r_0^2 \frac{\partial P_0}{\partial m} r' + \frac{2Gm}{r_0^3} r'. \tag{12.30}$$

Solutions to the linear equations with time-independent coefficients are sought in the form

$$(r', \rho', P') = (\overline{r}, \overline{\rho}, \overline{P},)\, \mathrm{e}^{-\mathrm{i}\omega t}. \tag{12.31}$$

Substituting (12.31) into (12.30) and recalling (10.1), we have

$$\omega^2 \overline{r} = 4\pi r_0^2 \frac{\mathrm{d}}{\mathrm{d}m}\left[\gamma_1 \frac{P_0}{\rho_0}\left(\frac{\mathrm{d}\overline{r}}{\mathrm{d}r_0} + 2\frac{\overline{r}}{r_0}\right)\rho_0\right]$$
$$- \frac{4Gm}{r_0^3}\overline{r} \equiv \mathscr{L}(\overline{r}). \tag{12.32}$$

The regularity of the function $\overline{r}/r_0$ at $r_0 = 0$ yields the boundary condition

$$\frac{\mathrm{d}}{\mathrm{d}r_0}\left(\frac{\overline{r}}{r_0}\right) = 0 \quad \text{at } r_0 = 0, \tag{12.33}$$

while the regularity of the function $\overline{P}/P_0$ on the boundary $r = R$ yields for equations of state with $P_0/\rho_0 \to 0$ as $\rho_0 \to 0$ the condition [317]

$$\frac{\mathrm{d}}{\mathrm{d}r_0}\left(\frac{\overline{P}}{P_0}\right) = 0 \quad \text{at } r_0 = R, \tag{12.34}$$

which, on using (12.28) and (12.29), becomes

$$\left(4 - 2\gamma_1 + \frac{\omega^2 r_0^3}{GM}\right)\frac{\overline{r}}{r_0} - \gamma_1 \frac{\mathrm{d}\overline{r}}{\mathrm{d}r_0} = 0 \quad \text{at } r_0 = R. \tag{12.35}$$

The Sturm–Liouville problem [918] for (12.32) at the boundary conditions (12.33) and (12.35) has a finite, physically allowable solution to an arbitrary factor in the form of eigenfunctions $\overline{r}_i(r_0)$ only for eigenfrequencies $\omega = \omega_i$. Real eigenfrequencies ($\omega_i^2 > 0$) correspond to stability. All the eigenvalues of (12.32) are real because the operator $\mathscr{L}(\overline{r})$ is self-conjugate [317]. Consider the normalized eigenfunctions

$$\int_0^M \overline{r}^2 \mathrm{d}m = 1. \tag{12.36}$$

Multiplying (12.32) by $\overline{r}$ and integrating by parts over the star gives

$$\omega^2 = \int_0^M \overline{r}\mathscr{L}(\overline{r})\,\mathrm{d}m = \int_0^M \left[\gamma \frac{P_0}{\rho_0}\left(2\frac{\overline{r}}{r_0} + \frac{\mathrm{d}\overline{r}}{\mathrm{d}r_0}\right)^2 - 4\frac{Gm}{r_0}\left(\frac{\overline{r}}{r_0}\right)^2\right]\mathrm{d}m. \tag{12.37}$$

Obviously, the stability conditions $\omega^2 > 0$ in (12.37) and $\delta^2\varepsilon > 0$ in (12.20) are equivalent if $\overline{r}$ is regarded both as an eigenfunction and arbitrary trial function δr.

Because of the Hermiticity of the operator $\mathscr{L}(\overline{r})$ [317,918], the minimum in ω^2 from (12.37) (or $\delta^2\varepsilon$ from (12.20)) occurs for the eigenfunction $\delta r = \overline{r}$. This means that if any trial function $\delta r(r_0)$ yields a negative $\delta^2\varepsilon$ from (12.20), the corresponding equilibrium state is undoubtedly unstable. On the contrary, a positive $\delta^2\varepsilon$ does not imply the stability of the state for certain. Comparison with the exact static criterion from Sect. 12.3 shows that the linear trial function $\delta r = \alpha r$ in (12.15) determines almost exactly the point of stability loss for a neutron star in general relativity (see Sect. 11.2).

Transforming the last term in (12.20) or (12.37) using the equilibrium equation (10.1), we obtain

$$\delta^2\varepsilon = \int_0^{M_0} \frac{P}{\rho}\left\{\gamma\left[2\frac{\delta r}{r} + \frac{\mathrm{d}(\delta r)}{\mathrm{d}r}\right]^2 - 4\left(\frac{\delta r}{r}\right)^2 - 8\frac{\delta r}{r}\frac{\mathrm{d}(\delta r)}{\mathrm{d}r}\right\}\mathrm{d}m_0, \tag{12.38}$$

and, with $\delta r = \alpha r$, the relation

$$\delta^2\varepsilon = 9\alpha^2 \int_0^{M_0} \frac{P}{\rho}(\gamma - 4/3)\,\mathrm{d}m \tag{12.39}$$

that coincides with the approximate relation (10.20) of the energetic method provided that general relativity effects are ignored and (10.18) is valid.

For highly inhomogeneous models, the eigenfunctions may have a substantially nonlinear form. In the bypolytropic model from Sect. 8.3.1 with $n_2 \gg 1$, for instance, the instability of the static model sets in if the condition (8.45) is satisfied. The eigenfunction $\overline{r}(r_0)$ in this model is close to zero in the core and increases appreciably in the outer layers of the envelope, thereby ensuring a slow, "quasistationary" character of the outflow.

Applying the approximate energetic method from (10.19) and (10.20) based on the homologeneity of contraction leads, for this case, to a considerable error. In particular, at the point of stability loss (8.45) the quantity $\delta^2\varepsilon$ from (10.20) is essentially positive and almost corresponds to $\gamma \simeq \gamma_1, \gamma_1 = 1 + (1/n_1)$ from (8.28) since the mass of the destabilizing envelope with $n = n_2$ is small.

Problem. Derive the stability criterion for a spherical star from the variational principle in Newtonian gravity.

Solution. [1075] The sum of internal and gravitational energies, analogous to the potential energy of the conservative system from (10.8), is equal to

$$\varepsilon = \int_0^M E(\rho, T)\mathrm{d}m - \int_0^M \frac{Gm\mathrm{d}m}{r}. \tag{12.40}$$

Make the first variation, taking into account

$$\delta S = 0, \quad \Delta m = 0 \quad \text{and} \quad \frac{\mathrm{d}r}{\mathrm{d}m} = \frac{1}{4\pi\rho r^2}. \tag{12.41}$$

We then have, using relation $(\partial E/\partial\rho)_S = P/\rho^2$,

$$\delta\varepsilon = \int_0^M \frac{P}{\rho^2}\delta\rho\mathrm{d}m + \int_0^M \frac{Gm\delta r}{r^2}\mathrm{d}m. \tag{12.42}$$

It follows from (12.41)

$$\frac{\delta\rho}{\rho} = -2\frac{\delta r}{r} - \frac{\mathrm{d}\delta r/\mathrm{d}m}{\mathrm{d}r/\mathrm{d}m} = -2\frac{\delta r}{r} - \frac{\mathrm{d}\delta r}{\mathrm{d}r}. \tag{12.43}$$

Taking into account (12.43) in (12.42), we obtain

$$\delta\varepsilon = -\int_0^M \frac{P}{\rho}\left(2\frac{\delta r}{r} + \frac{\mathrm{d}\delta r}{\mathrm{d}r}\right)\mathrm{d}m + \int_0^M \frac{Gm\delta r}{r^2}\mathrm{d}m. \tag{12.44}$$

The second term in the first integral is reduced by partial integration to

$$\int_0^M \frac{P}{\rho}\frac{\mathrm{d}\delta r}{\mathrm{d}r}\mathrm{d}m = 4\pi\int_0^R Pr^2\frac{\mathrm{d}\delta r}{\mathrm{d}r}\mathrm{d}r = -\int_0^M \left(\frac{1}{\rho}\frac{\mathrm{d}P}{\mathrm{d}r} + \frac{2P}{\rho r}\right)\delta r\mathrm{d}m. \tag{12.45}$$

With account of (12.45), the first variation in (12.44) is reduced to

$$\delta\varepsilon = \int_0^M \left(\frac{1}{\rho}\frac{\mathrm{d}P}{\mathrm{d}r} + \frac{Gm}{r^2}\right)\delta r \mathrm{d}m, \tag{12.46}$$

leading to the equilibrium equation (10.1). Variation of (3) with account of $\gamma \equiv \gamma_1$ gives (8.26)

$$\delta^2\varepsilon = \int_0^M \frac{P}{\rho}\left[(\gamma-2)\left(\frac{\delta\rho}{\rho}\right)^2 + \frac{\delta^2\rho}{\rho}\right]\mathrm{d}m - 2\int_0^M \frac{Gm(\delta r)^2}{r^3}\mathrm{d}m. \tag{12.47}$$

Variation of $\delta^2\rho$ is obtained from (12.43) as

$$\frac{\delta^2\rho}{\rho} = \left(2\frac{\delta r}{r} + \frac{\mathrm{d}\delta r}{\mathrm{d}r}\right)^2 + 2\left(\frac{\delta r}{r}\right)^2 - \delta\left(\frac{\mathrm{d}\delta r}{\mathrm{d}r}\right). \tag{12.48}$$

Variation of the last term in (12.48) with account of (12.43) gives

$$\delta\left(\frac{\mathrm{d}\delta r}{\mathrm{d}r}\right) = 4\pi\delta\left(\rho r^2\frac{d\delta r}{dm}\right) = \frac{\delta\rho}{\rho}\frac{\mathrm{d}\delta r}{\mathrm{d}r} + 2\frac{\delta r}{r}\frac{\mathrm{d}\delta r}{\mathrm{d}r} = -\left(\frac{\mathrm{d}\delta r}{\mathrm{d}r}\right)^2. \tag{12.49}$$

Finally, we obtain from (12.47) with account of (12.43), (12.48), and (12.49)

$$\delta^2\varepsilon = \int_0^M \frac{P}{\rho}\left[(\gamma-1)\left(2\frac{\delta r}{r} + \frac{\mathrm{d}\delta r}{\mathrm{d}r}\right)^2 + 2\left(\frac{\delta r}{r}\right)^2 + \left(\frac{\mathrm{d}\delta r}{\mathrm{d}r}\right)^2\right]\mathrm{d}m$$
$$-2\int_0^M \frac{Gm(\delta r)^2}{r^3}\mathrm{d}m. \tag{12.50}$$

The last term in (12.50) with account of the equilibrium equation (10.1) and after partial integration is reduced to

$$\int_0^M \frac{Gm(\delta r)^2}{r^3}\mathrm{d}m = -\int_0^M \frac{1}{\rho}\frac{\mathrm{d}P}{\mathrm{d}r}\frac{\delta r^2}{r}\mathrm{d}m = -4\pi\int_0^R \frac{\mathrm{d}P}{\mathrm{d}r}r\delta r^2 dr$$
$$= \int_0^M \frac{P}{\rho r^2}\frac{\mathrm{d}}{\mathrm{d}r}(r\delta r^2)\mathrm{d}m = \int_0^M \frac{P}{\rho}\left[2\frac{\delta r}{r}\frac{\mathrm{d}\delta r}{\mathrm{d}r} + \left(\frac{\delta r}{r}\right)^2\right]\mathrm{d}m. \tag{12.51}$$

With account of (12.51), we can write (12.50) in three equivalent forms

$$\delta^2\varepsilon = \int_0^M \frac{P}{\rho}\left[\gamma\left(2\frac{\delta r}{r} + \frac{\mathrm{d}\delta r}{\mathrm{d}r}\right)^2 - 2\left(\frac{\delta r}{r}\right)^2 - 4\frac{\delta r}{r}\frac{\mathrm{d}\delta r}{\mathrm{d}r}\right]\mathrm{d}m - 2\int_0^M \frac{Gm(\delta r)^2}{r^3}\mathrm{d}m$$
$$= \int_0^M \frac{P}{\rho}\left[\gamma\left(2\frac{\delta r}{r} + \frac{\mathrm{d}\delta r}{\mathrm{d}r}\right)^2 - 4\left(\frac{\delta r}{r}\right)^2 - 8\frac{\delta r}{r}\frac{\mathrm{d}\delta r}{\mathrm{d}r}\right]\mathrm{d}m$$

$$= \int_0^M \frac{P}{\rho}\gamma\left(2\frac{\delta r}{r} + \frac{\mathrm{d}\delta r}{\mathrm{d}r}\right)^2 - 4\int_0^M \frac{Gm(\delta r)^2}{r^3}\mathrm{d}m. \tag{12.52}$$

The last form in (12.52) coincides with (12.20), and the second one with (12.38).

12.3 Static Criteria for Stability

12.3.1 Non-Rotating Stars

For the case of a nonrotating cold star ($T = 0$, $S = 0$), the stability criterion may be formulated as follows [1075]. Treating pulsations of a steady-state star, we may assume the time dependence of the displacement ξ of a point with Lagrangian radius $\mathbf{r}_0$ to be

$$\xi_n(\mathbf{r}_0, t) = \mathbf{r}(\mathbf{r}_0, t) - \mathbf{r}_0 \tag{12.53}$$

for the n-th normal mode of radial oscillations to take the form

$$\xi_n \sim \mathrm{e}^{-\mathrm{i}\sigma_n t}, \tag{12.54}$$

where all σ_n^2 are real provided that there are no dissipative processes, so that the stability of the n-th mode implies that $\sigma_n^2 > 0$, whereas instability of the n-th mode implies that $\sigma_n^2 < 0$. We next calculate a series of models with a given equation of state

$$P = P(\rho) \tag{12.55}$$

and diverse values of central density ρ_c to obtain the dependence $M(\rho_c)$. Suppose that at $\rho_c = \rho_{c,cr} \equiv \rho_{cc}$, this dependence has an extremum $M(\rho_{cc}) = M_0$. Two solutions exist in this case for the mass M that does not differ greatly from M_0. One of these solutions can be obtained from the other by a slight time-independent displacement. This means that the square of the eigenfrequency σ_n^2 of a certain mode passes through zero at $\rho_c = \rho_{cc}$:

$$\sigma_n^2(\rho_{cc}) = 0. \tag{12.56}$$

An extremum on the curve $M(\rho_c)$ thus always corresponds to a critical point that changes the stability of a certain stellar mode, the fundamental mode always losing its stability at a maximum on the curve $M(\rho_c)$ [1075]. Stable stars are represented only by rising portions on the curve $M(\rho_c)$. This criterion is still valid for isentropic stars with an equal specific entropy S. The point of the loss of stability by the fundamental mode corresponds in this case to a maximum on the curve $M_S(\rho_c)$ [1081]. A more detailed knowledge of the number and stability of unstable modes can be extracted from considering the dependence $M(R)$, where R is the stellar radius [449].

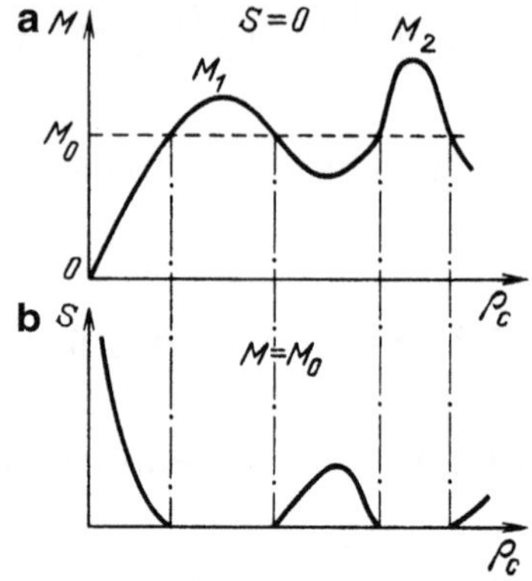

Fig. 12.1 (**a**) The mass M as a function of central density ρ_c, plotted schematically for zero entropy $S = 0$; (**b**) the dependence $S(\rho_c)$ for the mass M_0 below the extrema M_1 and M_2, from [119]

Similar considerations may prove the validity of another static criterion of hot isentropic stars. We shall now fix the stellar mass and calculate models with diverse values of specific entropy. This will give the dependence $S_M(\rho_c)$. It is clear that the extremum in this dependence corresponds to a critical point as well, because the presence of an extremum proves the existence of two close models with an equal entropy S, i.e. $\sigma^2(\rho_{cc}) = 0$ at the extremum. The loss of stability by the fundamental mode now occurs at a minimum on the curve $S_M(\rho_c)$.

It should be noted that calculating a series of models may sometimes lead to an entropy fall-off down to zero, so that in a certain range of ρ_c there will be no static solutions for the given mass. Dips will then be present on the curve $S(\rho_c)$ (Fig. 12.1). This case, nevertheless, does not produce difficulties since at $S = 0$ we can conclude on stability from another static criterion. The presence of a "dip", and of an extremum as well, implies a change of sign for the square of the eigenfrequency of one of an oscillatory mode.

One may thus perceive a certain complementarity in the criteria. On one hand, an extremum on the curve $M_S(\rho_c)$ corresponds to the point of stability change, on the otherhand, an extremum in $S_M(\rho_c)$ does. This is a consequence of the conservation of M and S during adiabatic pulsations. This effect seems to be trivial for spherically symmetric stars, but the situation changes if we proceed to a more interesting case of rotating stars.

It is obvious that all quantities conserved throughout adiabatic pulsations have to be constant near a critical model. This means that if any quantity A conserved throughout pulsations varies along a series of models, the dependence $A(\rho_c)$ has then an extremum at the critical point. The total energy ε of an equilibrium star always increases with increase in its mass or entropy [1074]. As a result, the extrema on the curves $\varepsilon_S(\rho_c)$ and $\varepsilon_M(\rho_c)$ coincide with the extrema on the curves $M_S(\rho_c)$ and $S_M(\rho_c)$ and may be used with a static criterion for stability, while the dependencies $\varepsilon_S(M)$ and $\varepsilon_M(S)$ are not analytical and have cusps.

It should be noted that the static criterion may be applied not only to isentropic stars. We can construct, for example, a series of models of variable mass with specific entropy distribution fixed over the dimensionless Lagrangian coordinate $q = M_r/M$, where M_r is the mass inside r. It is easy to show that the extremum $M_{S(q)}(\rho_c)$ corresponds in this case to a critical point as well.

12.3.2 Criteria for Rotating Stars

It follows from considerations in Sect. 12.3.1 that to formulate a static criterion for stability, it is necessary to indicate quantities $A, B, C, \ldots$, conserved throughout pulsations, chose one of them (say, A) as a varying parameter, and, upon fixing the integral or specific values of the other parameters, independent of A, to construct a series of models. An extremum on the curve $A_{B,C\ldots}(\rho_c)$ will then correspond to the critical point. In the case of rotation under conditions of adiabatic pulsations and zero viscosity, the entropy S, specific angular momentum j of each piece of material, and stellar mass M are conserved. The static criterion for stability could be derived most easily for the case of an isentropic isomomentum star, but for a real case the validity of the condition $j = \text{const.}$ is barely possible. A criterion for real stars with rotation can be derived by calculating, as described above, a series of non-isomomentum models analogous to non-isentropic models.

Consider isentropic models. As we have three conserved quantities, three static criteria are possible. The extrema of the curves

$$1^\circ\ S_{M,j}(\rho);\quad 2^\circ\ J_{M,S}(\rho);\quad 3^\circ\ M_{S,j}(\rho), \tag{12.57}$$

will be critical points.

According to Poincaré's theorem, determining restrictions to the rotational law in barotropic equilibrium rotating stars, the specific angular momentum j and angular velocity Ω in equilibrium are constants at cylinders; therefore fixed are

$$j = j(m), \quad m = \frac{M_b}{M}, \tag{12.58}$$

where M_b is the mass inside cylinder of radius b, for criteria 1° and 3° and

$$j' = \frac{M}{J}\, j(m) \tag{12.59}$$

for criterion 2°.

It may be easily seen that for all three cases the extremum on the curve coincides with a critical point, and we see a full analogy with calculating a series of non-isentropic models.

It is often necessary to examine the stability of models with a solid-body rotation. The specific momentum of each Lagrangian mass varies along a series of such models; therefore, the static criterion should be applied as follows. Let there be a series of models with solid-body rotation. Each model has its own distribution of momentum $j(m)$. Fixing $j(m)$, we can calculate from any model a new "associated" series of models in accordance with (12.57). Critical will be the model with extreme parameters in the associated series.

Fully convective stars, say, supermassive [1080], or cold stars with $S = 0$, may be regarded as isentropic. Degenerate configurations, white dwarfs, and neutron stars may be treated as isothermal because of their high thermal conductivity. For the distribution of angular velocity of a star, see Sect. 8.2. At $\mathrm{d}j/\mathrm{d}b > 0$, the star is stable to turbulence development, but only for axially symmetrical perturbations to the linear approximation. For example, when coaxial cylinders with large Reynolds numbers $\mathrm{Re} = \rho v R/\mu > 10^5$ are rotating, turbulence may develop at $\mathrm{d}j/\mathrm{d}b > 0$ when the outer cylinder rotates more rapidly [614]. The Reynolds numbers are always very large in stars, therefore the possibility for differentially rotating stars with a large velocity to exist is rather questionable. At a low viscosity that is necessary to violate the Thomson conservation-of-circulation theorem,there always exist perturbations sufficiently large to cause turbulence. Instability of non-axisymmetric perturbations with large azimuthal number $m \gg 1$, at $\xi \sim \mathrm{e}^{\mathrm{i}m\varphi}$, also becomes important at large Re.

12.3.3 Removal of Degeneracy of Neutral Oscillatory Modes in Rotating Isentropic Stars

A detailed analysis of static criteria for the stability of rotating stars reveals a strong connection between eigenfunctions corresponding to neutral convective motions of a non-rotating star and neutral eigenfunctions in the presence of rotation, and it also enables us to find a simple relationship between the convective stability of the star and Poincaré's theorem.

An eigenfunction of the neutral perturbation mode in a nonrotating star depends in a critical state on the spherical radius $\xi = \xi(r)$. It represents the difference in radii of two close equilibrium states of the same mass on both sides of the maximum

$$\xi = \xi(q) = r_1(q) - r_2(q), \quad r = r(q), \quad q = \frac{M_\mathrm{r}}{M}, \tag{12.60}$$

in accordance with the static criterion of Sect. 12.3.1. On the other hand, in the case of isentropic stars, for any, including the weakest, rotation

$$\xi = \xi(b), \quad b = b(m). \tag{12.61}$$

Indeed, as the specific angular momentum $j(b) = \Omega b^2$ conserves, we have

$$\frac{\Delta\Omega}{\Omega} = -2\frac{\Delta b}{b}. \tag{12.62}$$

Since Ω, $\Delta\Omega$, and b are functions of b, then $\Delta b = \xi_b(b)$. A jump in the expression for an eigenfunction of the critical state occurring when a weak rotation is added will be easily explained if we recall that in a nonrotating isentropic star not only a radial neutral mode, but also an infinite series of eigenfunctions representing convective motion correspond to the eigenvalue $\omega = 0$, because the whole volume of an isentropic star is in a neutral state relative to convective instability, according to the Schwarzschild criterion (9.4). If several eigenfunctions correspond to the state with a given eigenvalue ω, this state is called degenerate. For an isentropic star without rotation such as the oscillatory mode with $\omega = 0$. The rotation causes neutral convective modes to become either stable or unstable (when $\mathrm{d}j/\mathrm{d}b < 0$); so the degeneracy of the neutral mode is removed.

If a perturbation (small rotation) removes the degeneracy, then a nondegenerate eigenfunction of a perturbed state with the same eigenfrequency will be close to one "regular" superposition of degenerate unperturbed eigenfunctions. An eigenfunction of the neutral mode of a weakly rotating star is represented by a superposition of convective and one radial mode such that the eigenfunction itself depends only on the cylindrical radius b. The angular parts of eigenfunctions of convective modes are Legendre polynomials $P_l(\cos\theta)$ for ξ_r and their derivatives $\mathrm{d}P_l/\mathrm{d}\theta = P_l^1(\cos\theta)$ for ξ_θ with even l [317]. Similar to determination of an eigenfunction of the neutral radial mode for a nonrotating star using a static criterion, for a weakly rotating star, the same procedure enables us to restore the radial parts of all the convective modes [119]. For this purpose, we find the eigenfunction

$$\xi = (\xi_b(b), \xi_z(b,z)), \quad b = r\sin\theta \tag{12.63}$$

in the form of a difference between two close equilibrium models of the same mass. We then evaluate the displacements in a spherical coordinate system

$$\xi_r(b.z) = \sqrt{\xi_b^2 + \xi_z^2}, \quad \xi_\theta(b.z) = \arctan\frac{\xi_z}{\xi_b}, \tag{12.64}$$

and perform the relevant expansions:

$$\begin{aligned}
\xi_r(r,\theta) &= \sum_{l=0}^{\infty} \xi_{rl}(r)\, P_l(\cos\theta),\\
\xi_{rl}(r) &= (l+1/2)\int_{-1}^{1} \xi_r(r,\theta)\, P_l(\cos\theta)\,\mathrm{d}(\cos\theta),\\
\xi_\theta(r,\theta) &= \sum_{l=0}^{\infty} \xi_{\theta l}(r)\, P_l^1(\cos\theta),\\
\xi_{\theta l}(r) &= \frac{(l+1/2)}{l(l+1)}\int_{-1}^{1} \xi_\theta(r,\theta)\, P_l^1(\cos\theta)\,\mathrm{d}(\cos\theta).
\end{aligned} \tag{12.65}$$

The functions ξ_{rl} and $\xi_{\theta l}$ are the radial parts of convective modes of oscillations in a neutral case. This method for calculation of eigenfunctions is much less difficult than the conventional method of small perturbations [317], although it can be applied only to a neutral degenerate mode.

In a star stable to convection, the neutral mode is nondegenerate and its radial dependence remains invariant in slowly rotating stars. This is due to the violation of Poincaré's theorem together with conditions (12.58) and (12.61) for barotropic stars with entropy varying with mass. For an isentrope with $\gamma_1 = 4/3$, the neutral mode is homologous, and for stars with a slow solid-body rotation, a maximum on the curve $M_{\mathrm{S,J}}(\rho_{\mathrm{c}})$ coincides with the point of stability loss.

12.3.4 Numerical Examples [119]

Consider cold white dwarfs with rigid-body rotation determined by an equation of state written with regard for neutronization in the form [853]

$$\rho = Ax^3\left(2 + a_1x + a_2x^2 + a_3x^3\right),$$

$$P = B\left[x\left(2x^2 - 3\right)\left(x^2 + 1\right)^{1/2} + 3\ln\left(x + \sqrt{1+x^2}\right)\right],$$

$$A = 9.82 \times 10^5\ \mathrm{g\ cm^{-3}}, \quad B = 6.01 \times 10^{22}\ \mathrm{erg\ cm^{-3}},$$

$$a_1 = 1.255 \times 10^{-2}, \quad a_2 = 1.755 \times 10^{-5}, \quad a_3 = 1.376 \times 10^{-6}. \qquad (12.66)$$

The stability is established according to the criterion $M_{\mathrm{S,J}}(\rho_{\mathrm{c}}) \equiv M_1(\rho_{\mathrm{c}})$. The self-consistent field method (see Chap. 6, Vol. 1) is used to calculate equilibrium rotating models. First, we calculate a series of models with solid-body rotation and fixed angular momentum J_0 and choose a model near an extremum of the given dependence $M_1(\rho_{\mathrm{c}})$. This model determines a certain distribution of momentum $j(m)$ for which we calculate an associated series of models $M_2(\rho_{\mathrm{c}})$ with the same total momentum J_0. The results of calculations are given in Table 12.1 [119] from which we see that the critical model (an extremum on the curve $M_2(\rho_{\mathrm{c}})$) coincides with an extremum on the curve $M_1(\rho_{\mathrm{c}})$ for models with solid-body rotation. The reason for this is that the equation of state for a white dwarf in the critical state has γ_1 very close to 4/3, and the neutral oscillatory mode is almost homologous. Besides, for a solid-body rotation of a $n = 3$ polytrope, the rotational T to gravitational W energy ratio cannot be large because centrifugal and gravitational forces become rapidly equal at the equator. In the present case, the rotation parameter β, reflecting the relative role of centrifugal forces on the structure of the main body of a star, is $\beta = \Omega^2/8\pi G\rho_{\mathrm{c}} = 6.82 \times 10^{-4}$, which is almost 70% of the terminal value $\beta_{\mathrm{lim}} = 9.83 \times 10^{-4}$ for the polytrope $n = 3$ [545]. For these models also $T/|W| = 5.78 \times 10^{-3}$, and $(T/|W|)_{\mathrm{lim}} = 9.00 \times 10^{-3}$. Such a small value of the kinetic rotational energy shows that the homologous mode is only slightly perturbed, thus ensuring the coincidence of maxima in the series of models with solid-body and differential rotation. The maximum mass in a critical state is 3.5% larger for the limiting rotation than for the nonrotational case. For models from Table 12.1, the increase in mass limit is $\approx 1.5\%$.

Table 12.1 The dependence of central density ρ_c of the mass M_1 of white dwarfs with solid-body rotation at $J = 1.88 \times 10^{49}\ \mathrm{g\,cm^2\,s^{-1}}$, and mass M_2, polar R_p and equatorial R_e radii of differentially rotating white dwarfs with momentum distribution corresponding to a model with solid-body rotation and $x_c = 10.5$. The equation of state is determined in (12.66) from [119]

x_i	ρ_c, g cm^{-3}	$\frac{M_1}{M_\odot}$	$\frac{M_2}{M_\odot}$	R_p, cm	R_e, cm
10.0	2.090(9)	1.28621	1.28618	2.0126(8)	2.1622(8)
10.5	2.427(9)	1.28654	1.28654	1.9339(8)	2.0864(8)
10.6	2.499(9)	1.28654	1.28655	1.9189(8)	2.0720(8)
10.7	2.572(9)	1.28653	1.28654	1.9042(8)	2.0578(8)
10.8	2.646(9)	1.28650	1.28651	1.8896(8)	2.0439(8)
10.9	2.722(9)	1.28644	1.28646	1.8753(8)	2.0302(8)
11.0	2.800(9)	1.28637	1.28640	1.8613(8)	2.0034(8)

If in a polytropic equation of state γ_1 differs sensitively from $4/3$, then the difference between the critical state and the maximum mass on the curve for solid-body rotation may become essential. The equation of state in a parametric form

$$\rho(H) = \frac{1}{2}\rho_1\left[\left(\frac{H}{H_1}\right)^{0.3} + \left(\frac{H}{H_1}\right)^{10}\right],$$

$$P(H) = \frac{1}{2}\rho_1 H_1\left[\frac{(H/H_1)^{1.3}}{1.3} + \frac{(H/H_1)^{11}}{11}\right], \tag{12.67}$$

has been considered in [119]. The value $\gamma_1 = 4.33$ at $\rho = \rho_1$ and falls off steeply when $\rho > \rho_1$. The results of calculations for a model with solid-body rotation and three associated series of models are given in Fig. 12.2. The loss of stability occurs at the point of intersection of the associated curve D with the solid-body one, this point coinciding with the maximum on the curve D. It can be seen from Fig. 12.2 that the point of stability loss differs from the maximum on the solid-body curve by almost 5% with respect to ρ_c. An extension of the static criteria to general relativity and toroidal magnetic fields is made in [119].

12.4 Star Stability in the Presence of a Phase Transition

Under conditions of strong degeneration, the stellar matter may undergo phase transitions of the first kind, i.e. in a certain density range $Z\rho_1 \le \rho \le \rho_2$ the pressure remains constant, $P = P_*$ (Fig. 12.3). Since the pressure decreases monotonically with radius at r_*, so that $P(r_*) = P_*$, a jump in density occurs from ρ_2 to ρ_1. This situation was first encountered in studies of giant planets [5, 821]. Phase transitions due to neutronization may occur in the centre of white dwarfs with a mass close to the Chandrasekhar limit (see Sect. 11.1) and in neutron star envelopes [154]. Phase

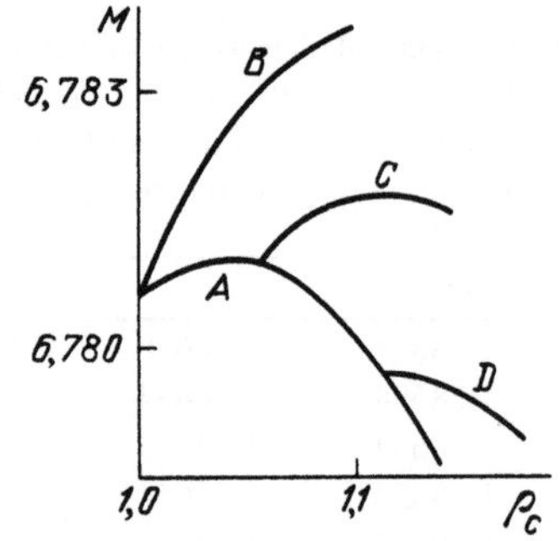

Fig. 12.2 Mass M versus central density ρ_c for models with solid-body (curve A) and differential (curves B, C, and D) rotation with the equation of state (12.67). The distribution of angular momentum over the curves B, C, and D corresponds to the model with solid-body rotation on the curve A at the points of its intersection with curves B, C, and D. The curve C departures from the maximum of the curve A, and the maximum of the curve D coincides with the intersection point which represents just a critical point for models with solid-body rotation. The mass is given in $(4\pi\rho_1)^{-1/2}(H_1/G)^{3/2}$, ρ_c in ρ_1

Fig. 12.3 Pressure versus density in the presence of a phase transition of the first kind

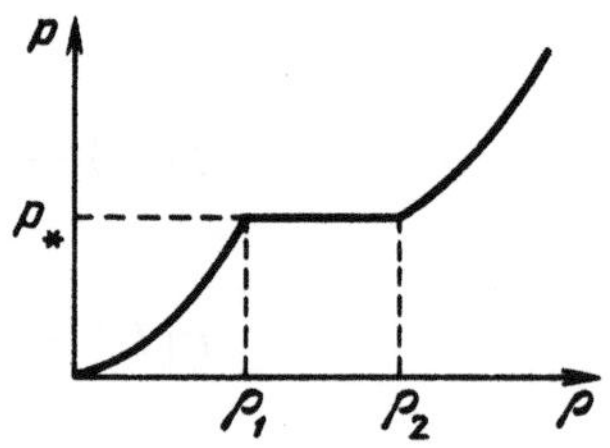

transitions in low-mass white dwarfs [573] due to ionization by pressure are also possible.

A phase transition reduces the stability reserve of the star. At $\rho_2/\rho_1 > 3/2$, the star loses stability immediately after the central pressure has attained $P = P_*$ for any equation of state [638, 893]. The stability of isentropic polytropes with phase transitions at the periphery[2] has been studied in [204, 895] using a static criterion.

The variational method for stability studies presented below has been developed in [169]. Arguments for the validity of the static criterion from Sect. 12.3 for stars with a phase transition are given in [909].

12.4.1 Evaluation of Variations $\delta\varepsilon$ and $\delta^2\varepsilon$

We take for the starting point the Newtonian energy of the star (10.8) without the last term. Let the density in the unperturbed equilibrium state undergo a jump from ρ_2 to ρ_1 at $m = m_*$. Suppose that during perturbation the jump conserves its magnitude

[2] When $S \neq 0$, the phase transitions are assumed to occur at S = const. The validity of this assumption is not obvious, it is only a rough method for describing a phase transition.

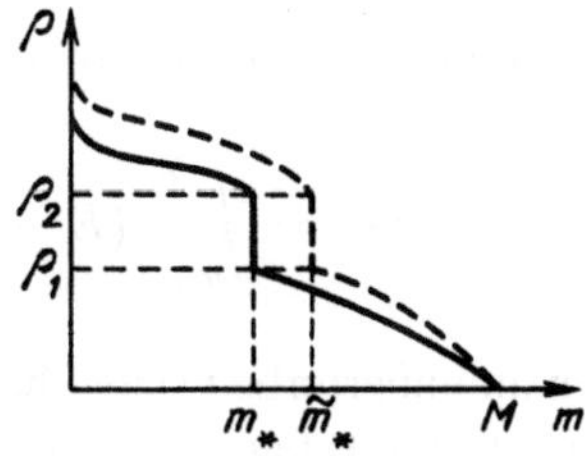

Fig. 12.4 The density profile: *solid line* in equilibrium star, *dashed line* in perturbed star, from [169]

and shifts into a point $\tilde{m}_* > m_*$ (Fig. 12.4). The case of $\tilde{m}_* < m_*$ should be treated analogously. To evaluate the variation of ε, we divide the integral in (10.8) into three parts: ε_2 $(0 \le m \le m_*)$, ε_1 $(\tilde{m}_* \le m \le M)$, and ε_* $(m_* \le m \le \tilde{m}_*)$. Variations $\delta\varepsilon_1$ and $\delta\varepsilon_2$ are calculated similarly to Sect. 12.3. Introducing volume v variation instead of radius r, we obtain:

$$\varepsilon_2 = \int_0^{m_*} E\,\mathrm{d}m - \beta \int_0^{m_*} \frac{m\,\mathrm{d}m}{v^{1/3}}, \quad \beta = \left(\frac{4\pi}{3}\right)^{1/3} G,$$

$$v = \frac{4\pi}{3} r^3, \quad \mathrm{d}v/\mathrm{d}m = 1/\rho; \tag{12.68}$$

$$\delta\varepsilon_2 = -\int_0^{m_*} P \frac{\mathrm{d}(\delta v)}{\mathrm{d}m}\,\mathrm{d}m + \frac{1}{3}\beta \int_0^{m_*} \frac{m\,\delta v}{v^{4/3}}\,\mathrm{d}m$$

$$= -P\,\delta v|_{m_*} + \int_0^{m_*} \left(\frac{\mathrm{d}P}{\mathrm{d}m} + \frac{\beta}{3}\frac{m}{v^{4/3}}\right) \delta v\,\Delta m, \tag{12.69}$$

with

$$P = -\partial E/\partial(1/\rho), \quad \delta(1/\rho) = \mathrm{d}(\delta v)/\mathrm{d}m. \tag{12.70}$$

The parenthesized expression in the integral (12.69) equals zero by virtue of the equilibrium equation, therefore:

$$\delta\varepsilon_2 = -P(m_*)\delta v(m_*). \tag{12.71}$$

Similarly,

$$\delta\varepsilon_1 = P(\tilde{m}_*)\delta v(\tilde{m}_*). \tag{12.72}$$

Evaluating the second variations of the energy, we obtain (cf. (12.20))

$$\delta^2\varepsilon_2 = -\frac{4}{9}\beta \int_0^{m_*} \frac{m(\delta v)^2 \mathrm{d}m}{v^{7/3}} + \int_0^{m_*} \gamma P\rho \left[\frac{\mathrm{d}(\delta v)}{\mathrm{d}m}\right]^2 \mathrm{d}m, \tag{12.73}$$

$$\delta^2\varepsilon_1 = -\frac{4}{9}\beta \int_{\tilde{m}_*}^{M} \frac{m(\delta v)^2 \mathrm{d}m}{v^{7/3}} + \int_{\tilde{m}_*}^{M} \gamma P\rho \left[\frac{\mathrm{d}(\delta v)}{\mathrm{d}m}\right]^2 \mathrm{d}m, \tag{12.74}$$

$$\gamma \equiv \gamma_1 \text{ from (1.1.11).}$$

Evaluate the variation of ε_* with accuracy $(\Delta m_*)^2$, $\Delta m_* = \tilde{m}_* - m_*$:

$$\Delta\varepsilon_* = \int_{m_*}^{\tilde{m}_*} \left(\tilde{E} - E\right) \mathrm{d}m - \beta \int_{m_*}^{\tilde{m}_*} \left(\tilde{v}^{-1/3} - v^{-1/3}\right) m \,\mathrm{d}m. \tag{12.75}$$

In an unperturbed state, the density in the interval $m_* < m < \tilde{m}_*$ is close to ρ_1, in a perturbed state to ρ_2, and $\rho(m_* + 0) = \rho_1$, $\tilde{\rho}\,(\tilde{m}_* - 0) = \rho_2$. So, we have

$$\begin{aligned}
\tilde{E} &= E\left(\frac{1}{\tilde{\rho}}\right) = E\left(\frac{1}{\rho_2}\right) - P_*\left(\frac{1}{\tilde{\rho}} - \frac{1}{\rho_2}\right) + \dots \\
&\approx H_* - \frac{P_*}{\tilde{\rho}} + O\left[(\Delta m)^2\right], \\
E &= E\left(\frac{1}{\rho}\right) = E\left(\frac{1}{\rho_1}\right) - P_*\left(\frac{1}{\rho} - \frac{1}{\rho_1}\right) + \dots \\
&\approx H_* - \frac{P_*}{\rho} + O\left[(\Delta m)^2\right].
\end{aligned}$$

Here, the enthalpy $H_* = E\,(1/\rho_1) + (P_*/\rho_1) = E\,(1/\rho_2) + (P_*/\rho_2)$ is constant in the interval $\rho_1 \le \rho \le \rho_2$. Hence, the first term in the right-hand side of (12.75) is equal to

$$P_* \int_{m_*}^{\tilde{m}_*} \left(\frac{1}{\rho} - \frac{1}{\tilde{\rho}}\right) \mathrm{d}m + \dots \approx P_* \left[v(\tilde{m}_*) - v(m_*) - \tilde{v}(\tilde{m}_*) + \tilde{v}(m_*)\right].$$

Thus,

$$\int_{m_*}^{\tilde{m}_*} \left(\tilde{E} - E\right) \mathrm{d}m \approx P_* \left[\delta v(m_*) - \delta v(\tilde{m}_*)\right]. \tag{12.76}$$

For the second term in (12.75) we use

$$\begin{aligned}
v(m) &\approx v(\tilde{m}_*) + \frac{1}{\rho_1}\,(m - \tilde{m}_*), \\
\tilde{v}(m) &\approx \tilde{v}(\tilde{m}_*) + \frac{1}{\rho_2}\,(m - \tilde{m}_*).
\end{aligned} \tag{12.77}$$

Now

$$\begin{aligned}
\tilde{v}^{-1/3} - v^{-1/3} &\approx -\frac{1}{3}\,v^{-4/3}(\tilde{m}_*)\,(\tilde{v} - v) \\
&\approx -\frac{1}{3}\,v^{-4/3}(\tilde{m}_*)\left[\tilde{v}(\tilde{m}_*) - v(\tilde{m}_*) + (m - \tilde{m}_*)\left(\frac{1}{\rho_2} - \frac{1}{\rho_1}\right)\right].
\end{aligned} \tag{12.78}$$

We then obtain

$$\beta \int_{m_*}^{\tilde{m}_*} (\tilde{v}^{-1/3} - v^{-1/3}) m \,\mathrm{d}m \approx \beta m_* \int_{m_*}^{\tilde{m}_*} (\tilde{v}^{-1/3} - v^{-1/3}) \,\mathrm{d}m$$

$$\approx -\frac{\beta m_* \delta v(\tilde{m}_*)}{3 v^{4/3}(m_*)} \Delta m_* + \frac{\beta m_*}{6 v^{4/3}(m_*)} \left(\frac{1}{\rho_2} - \frac{1}{\rho_1} \right) (\Delta m_*)^2. \quad (12.79)$$

We assume that the increment $\delta v(m)$ for parts 1 and 2 have the same order of smallness as the magnitude of the jump shift $\Delta m_* = \tilde{m}_* - m_*$. Also, we combine (12.71) and (12.72) with the relation

$$P(\tilde{m}_*) \approx P(m_*) - \frac{\beta m_*}{3 v^{4/3}(m_*)} \Delta m_*, \quad (12.80)$$

following from the equilibrium equation. For the total perturbation $\Delta\varepsilon = \delta\varepsilon + \delta^2\varepsilon$, using (12.71–12.74) and (12.76–12.80), we find

$$\Delta\varepsilon = \frac{1}{2} \left(\int_0^{m_*} + \int_{\tilde{m}_*}^{M} \right) \left[\gamma P \rho \left(\frac{\mathrm{d}\,\delta v}{\mathrm{d}m} \right)^2 - \frac{4}{9} \beta \frac{m (\delta v)^2}{v^{7/3}} \right] \mathrm{d}m$$

$$+ \frac{\beta m_*}{6 v^{4/3}(m_*)} \left(\frac{1}{\rho_1} - \frac{1}{\rho_2} \right) (\tilde{m}_* - m_*)^2 . \quad (12.81)$$

Now, we express $(\tilde{m}_* - m_*)$ in terms of δv. By virtue of (12.77)

$$\delta v(\tilde{m}_*) - \delta v(m_*) \approx (\tilde{m}_* - m_*) \left(\frac{1}{\rho_2} - \frac{1}{\rho_1} \right), \quad (12.82)$$

therefore the non-integral term in (12.81) is

$$\frac{1}{2} B \left[\delta v(\tilde{m}_*) - \delta v(m_*) \right]^2,$$

$$B = \frac{\beta m_*}{3 v_*^{4/3}} \left(\frac{1}{\rho_1} - \frac{1}{\rho_2} \right)^{-1}, \quad v_* = v(m_*). \quad (12.83)$$

Take the trial function δv in the form

$$\delta v(m) = \eta \varphi_2(m), \quad 0 \le m \le m_* \quad (\varphi_2(0) = 0)$$

$$= \eta \varphi_1(m), \quad \tilde{m}_* \le m \le M. \quad (12.84)$$

The condition of positivity of $\Delta\varepsilon$ at small η will then be written as

$$\Delta\varepsilon = \int_0^M \left\{ \gamma P\rho \left[\varphi'(m)\right]^2 - \frac{4}{9}\frac{\beta m\varphi^2}{v^{7/3}} \right\} \mathrm{d}m$$

$$+B\left[\varphi_1(m_*) - \varphi_2(m_*)\right]^2 > 0; \tag{12.85}$$

$$\varphi = \varphi_2 \quad 0 \le m \le m_*$$

$$= \varphi_1 \quad m_* \le m \le M.$$

The trial function $\varphi(m)$ may be discontinuous at $m = m_*$, thereby yielding the additional nonintegral term in the stability condition. When several phase transitions occur in a single star, each of them has its own corresponding nonintegral term similar to (12.83).

12.4.2 Other Forms of Stability Criterion

The trial function $\varphi(m)$ in (12.86) is arbitrary, has an arbitrary jump at $m = m_*$, and must only be equal to zero at $m = 0$. We now fix values for the jump $\varphi_2(m_*) = \varphi(m_* - 0)$, $\varphi_1(m_*) = \varphi(m_* + 0)$ and choose $\varphi(m)$ in such a way as to minimize the integral term in (12.86). Euler's equation for such a variational problem with a square functional is linear and coincides with the linearized equilibrium equation

$$\frac{\mathrm{d}}{\mathrm{d}m}\left[\gamma P\rho\varphi'(m)\right] + \frac{4}{9}\beta\frac{m}{v^{7/3}}\varphi = 0. \tag{12.86}$$

Using (12.86), the integral in (12.86) may be expressed in terms of the values at the jump φ_1 and φ_2. The condition (12.86) will then be replaced by

$$-P_*\gamma_2\rho_2\varphi_2'(m_*)\varphi_2(m_*) + B\left[\varphi_1(m_*) - \varphi_2(m_*)\right]^2$$

$$+P_*\gamma_1\rho_1\varphi_1'(m_*)\varphi_1(m_*) \ge 0, \tag{12.87}$$

$$\gamma_2 \equiv \gamma(m_* - 0), \quad \gamma_1 \equiv \gamma(m_* + 0).$$

While the condition $\varphi(0) \equiv \varphi_2(0) = 0$ determines the solution of (12.86) only to a factor, the ratios $\varphi_2'(m_*)/\varphi_2(m_*)$ and $\varphi_1'(m_*)/\varphi_1(m_*)$ are independent of $\varphi_1(m_*)$, $\varphi_2(m_*)$ and are determined only by unperturbed functions. We define

$$A_2 = -P_*\gamma_2\rho_2\frac{\varphi_2'(m_*)}{\varphi_2(m_*)}, \quad A_1 = P_*\gamma_1\rho_1\frac{\varphi_1'(m_*)}{\varphi_1(m_*)}. \tag{12.88}$$

The condition (12.87) then takes the form

$$A_2\varphi_{2*}^2 + B(\varphi_{1*} - \varphi_{2*})^2 + A_1\varphi_{1*}^2 \geq 0 \tag{12.89}$$

for all $\varphi_{1*} = \varphi_1(m_*)$ and $\varphi_{2*} = \varphi_2(m_*)$. Satisfying (12.89) requires that

$$A_1 + B \geq 0, \quad A_2 + B \geq 0, \quad A_1A_2 + B(A_1 + A_2) \geq 0. \tag{12.90}$$

The necessary condition for stability is sometimes useful to be written in the form where only one integral is minimized. Eliminating the integral from 0 to m_* and minimizing with respect to φ_{2*} gives

$$D\,[\varphi_1(m_*)]^2 + \int_{m_*}^{M} \left\{ \gamma P\rho\,[\varphi_1'(m)]^2 - \frac{4}{9}\,\beta\,\frac{[\varphi_1(m)]^2}{v^{7/3}} \right\} dm \geq 0,$$

$$D = A_2B/\,(A_2 + B)\,, \quad A_2 + B \geq 0. \tag{12.91}$$

The condition (12.90) is local by form, but the evaluation of A_1 and A_2 requires solving two Cauchy problems for (12.86): in the interval $0 \leq m \leq m_*$ from the point $m = 0$, and in the interval $m_* \leq m \leq M$ from the point $m = M$. On obtaining these solutions and finding A_1 and A_2 from (12.88), the stability is to be tested by (12.90).

If the conditions (12.86) and (12.89) are taken only with continuous functions $\varphi(m)$, that is, $\varphi_1(m_*) = \varphi_2(m_*)$, the condition (12.89) will give the requirement $A_1 + A_2 \geq 0$. The continuity of $\varphi(m)$ implies perturbations that do not shift the density jump with respect to mass. The same stability condition is valid in the absence of a phase transition and follows from (12.90) as $\rho_2 \to \rho_1$, $B \to \infty$. In (12.91), we can set directly $\rho_1 = \rho_2$ when a phase transition is absent, to obtain $D = A_2$.

12.4.3 Rough Test for Stability

The variational principle (12.86) enables us to perform an approximate stability test upon choosing an appropriate class of trial functions $\varphi(m)$. This method is substantially simpler than solving the Cauchy problem for (12.86) with singularities at $m = 0$ and $m = M$. In the absence of a phase transition, a good result is yielded by applying the linear trial function $\varphi(m) = v(m)$ (see Sect. 11.2 and (11.67) for the case of general relativity). In the presence of a phase transition, a satisfactory accuracy results from using a two-parametric family of trial functions. Replace $\varphi(m)$ by a trial function $\psi(m)$ reducing for the smooth case to $(\delta r/r)$ (see Sect. 12.2):

$$\varphi(m) = 3v(m)\psi(r(m)).$$

The condition (12.86) then becomes

$$\left(\int_0^{r_*} + \int_{r_*}^{R}\right) P r^2 \left[\gamma \left(3\psi + r\frac{d\psi}{dr}\right)^2 - 12\psi^2 - 4r\frac{d\psi^2}{dr}\right] dr$$

$$+K_1(\psi_{1*} - \psi_{2*})^2 + K_2\left(\psi_{2*}^2 - \psi_{1*}^2\right) \geq 0,$$

$$K_1 = Gm_* r_*^2 \left(\frac{1}{\rho_1} - \frac{1}{\rho_2}\right)^{-1}, \quad K_2 = 4P_* r_*^3. \tag{12.92}$$

The condition (12.92) reduces to (12.38) at $\psi_{1*} = \psi_{2*}$. An extension of the linear trial function $\psi = 1$, $\varphi \sim v$ to the case with a phase transition is given by the function

$$\psi = \alpha_2, \quad r < r_*,$$
$$= \alpha_1, \quad r > r_*,$$

which regretfully does not yield a sufficient accuracy for the boundary of stability. The accuracy of the variational criterion has been checked using a static criterion. The choice of the two-parametric family

$$\psi(r) = \alpha_1 \quad r < r_*,$$
$$= \alpha_2 + (\alpha_3/r^2), \quad r > r_*. \tag{12.93}$$

has proved to be more successful. Substituting (12.93) into (12.92), we obtain the quadric form $F(\alpha_1, \alpha_2, \alpha_3)$ required to be positively definite for stability:

$$F(\alpha_1, \alpha_2, \alpha_3) = \sum_{i,k=1}^{3} A_{ik}\alpha_i\alpha_k,$$

$$A_{11} = K_1 + K_2 + 3I_1, \quad A_{12} = A_{21} = -K_1,$$

$$A_{13} = A_{31} = -K_1/r_*^2,$$

$$A_{22} = K_1 - K_2 + 3I_2, \quad A_{23} = A_{32} = (K_1 - K_2)/r_*^2 + I_3,$$

$$A_{33} = (K_1 - K_2)/r_*^4 + I_4. \tag{12.94}$$

Here, K_1 and K_2 are given in (12.92), integrals I_i are

$$I_1 = \int_0^{r_*} (3\gamma - 4) P r^2 dr, \quad I_3 = \int_{r_*}^{R} (3\gamma - 4) P \, dr,$$

$$I_2 = \int_{r_*}^{R} (3\gamma - 4) P r^2 dr, \quad I_4 = \int_{r_*}^{R} (\gamma + 4) \frac{P}{r^2} dr. \tag{12.95}$$

Table 12.2 Approximate values of critical parameters for polytropes with phase transitions according to the criterion (12.92)–(12.95) in comparison with exact values according to the static criterion from Sect. 12.3; ρ_a is the central density at the point of stability loss, ρ_b is the same for the point of stability return (from[169])

	ρ_a/ρ_2		ρ_b/ρ_2		
q	Exact value	Approximate value	Exact value	Approximate value	γ
1.32			stable		
1.33	1.15	–	1.25	–	5/3
1.34	1.10	1.28	1.35	1.35	
1.38	1.03	1.14	1.60	1.6	
1.60	1.00	*	2.63	2.6	
2.00	1.00	*	4.30	4.2	
			stable		
1.09	1.42	–	1.51	–	7/5
1.10	1.19	1.20	2.19	2.18	

* At $\rho_a = \rho_2$ we have $r_* = 0$ so that the integral I_4 from (45.28) diverges.

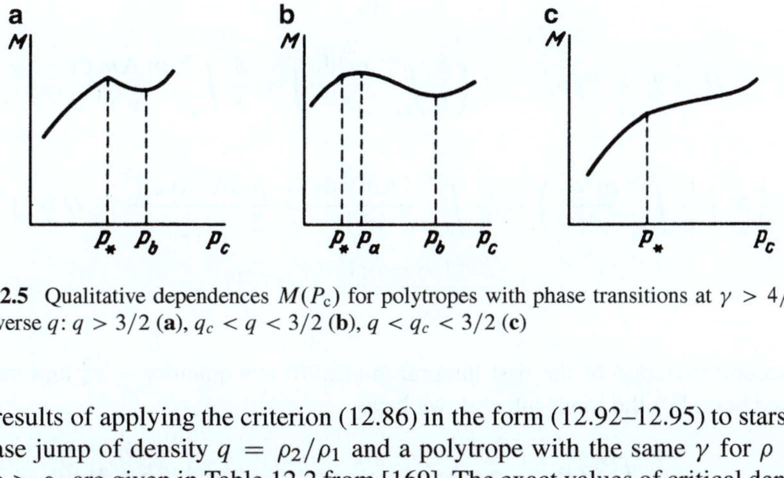

Fig. 12.5 Qualitative dependences $M(P_c)$ for polytropes with phase transitions at $\gamma > 4/3$ and for diverse q: $q > 3/2$ (**a**), $q_c < q < 3/2$ (**b**), $q < q_c < 3/2$ (**c**)

The results of applying the criterion (12.86) in the form (12.92–12.95) to stars with a phase jump of density $q = \rho_2/\rho_1$ and a polytrope with the same γ for $\rho < \rho_1$ and $\rho > \rho_2$ are given in Table 12.2 from [169]. The exact values of critical densities calculated from the static criterion in [204] are given for comparison.

Qualitative dependencies $M(P_c)$ in the neighborhood of the phase transition point in the center of the star are sketched in Fig. 12.5 for diverse γ and q (see [204, 893]). For $q > 3/2$, the stability loss occurs at $P_c = P_*$, but with further increase in P_c and ρ_c for $\gamma > 4/3$ the stability returns at $P_c = P_b$ ($\rho_c = \rho_b$) (Fig. 12.5a). For $q < 3/2$, the stability loss is also possible for a finite new-phase core with $P_c = P_a$, $\rho_c = \rho_a$. For any $\gamma > 4/3$, there exists q_c such that at $q < q_c$ the star is always stable. For $\gamma = 2, 5/3, 3/2, 7/5, 4/3$ the approximate values $q_c = 1.46, 1.33, 1.20, 1.09, 1.00$. It is obvious that the case $\gamma = 4/3$ is degenerate (neutral equilibrium) in Newtonian theory (see (12.39)); therefore, an arbitrary phase transition leads to instability.

12.4.4 Derivation of Stability Condition for a Phase Transition in the Center of Star

Using the Eulerian coordinate v as an independent variable, we write the energy ε from (12.68) in the form

$$\varepsilon = \int_0^\infty E\rho \, \mathrm{d}v - \frac{\beta}{6}\int_0^\infty \frac{m^2 \mathrm{d}v}{v^{4/3}}. \tag{12.96}$$

Let, in an unperturbed star, the central pressure P_c is equal to P_*, and at the phase transition point in the perturbed star $v = v_*$. To evaluate the energy variation, break the integrals in (12.96) into intervals $0 < v < v_*$, $v_* < v < \infty$. We have

$$\begin{aligned} \delta\left(\int_{v_*}^\infty E\rho \, \mathrm{d}v\right) &= \int_{v_*}^\infty H\delta\rho \, \mathrm{d}v = \int_{v_*}^\infty H\frac{\mathrm{d}\,\Delta m}{\mathrm{d}v}\,\mathrm{d}v \\ &= H\,\Delta m\Big|_{v_*}^\infty - \int_{v_*}^\infty \frac{\mathrm{d}H}{\mathrm{d}v}\Delta m \, \mathrm{d}v, \end{aligned} \tag{12.97}$$

$$H = E + P/\rho, \qquad \delta\left(\frac{\beta}{6}\int_{v_*}^\infty \frac{m^2 \mathrm{d}v}{v^{4/3}}\right) = \frac{\beta}{3}\int_{v_*}^\infty \frac{m\,\Delta m\,\mathrm{d}v}{v^{4/3}},$$

$$\begin{aligned} \frac{1}{2}\delta^2\left(\frac{\beta}{6}\int_{v_*}^\infty \frac{m^2 \mathrm{d}v}{v^{4/3}}\right) &= \frac{\beta}{6}\int_{v_*}^\infty \frac{(\Delta m)^2 \mathrm{d}v}{v^{4/3}} = \frac{\beta}{2}\frac{[\Delta m(v_*)]^2}{v_*^{1/3}} + O\left(v_*^2\right) \\ &= \frac{\beta v_*^2(\rho_2 - \rho_1)^2}{2 v_*^{1/3}} + O\left(v_*^2\right). \end{aligned} \tag{12.98}$$

The second variation of the first integral in (12.96) is a quantity $\sim v_*^2$ and may be rejected here. For the inner interval, we have

$$\begin{aligned} \int_0^{v_*} \left(\tilde{E}\tilde{\rho} - E\rho\right)\mathrm{d}v &= \int_0^{v_*} H_*\left(\tilde{\rho} - \rho\right)\mathrm{d}v + O\left(v_*^2\right) \\ &= H_*\Delta m(v_*) + O\left(v_*^2\right), \end{aligned}$$

$$\frac{\beta}{6}\int_0^{v_*} \frac{\mathrm{d}v}{v^{4/3}}\left(\tilde{m}^2 - m^2\right) = \frac{\beta}{10}\left(\rho_2^2 - \rho_1^2\right)v_*^{5/3}. \tag{12.99}$$

Using the equilibrium equation, we obtain in the point v_* of the unperturbed star

$$H(v_*) = H_* - \frac{\beta}{2}\rho_1 v_*^{2/3} + O\left(v_*^{4/3}\right). \tag{12.100}$$

Summing the variations of the energy (12.97–12.99) using the equilibrium equation, and relation (12.100) in the first term of the first relation (12.97), we find

$$\begin{aligned}\Delta\varepsilon &= -\frac{\beta v_*^{5/3}}{10}\left[5(\rho_2-\rho_1)^2-\rho_2^2+\rho_1^2\right]+\left[H_*-H(v_*)\right]\Delta m(v_*)\\ &= \frac{\beta v_*^{5/3}}{10}\left[-5(\rho_2-\rho_1)^2+\rho_2^2-\rho_1^2+5\rho_1(\rho_2-\rho_1)\right]\\ &= \frac{3\beta v_*^{5/3}}{5}(\rho_2-\rho_1)\left(\frac{3}{2}\rho_1-\rho_2\right).\end{aligned} \tag{12.101}$$

Since $\rho_2>\rho_1$, the stability condition $\Delta\varepsilon>0$ reduces to $\rho_2<(3/2)\,\rho_1$. This conclusion was obtained in [638, 893] by a more complicated method. Note that the derivation of the condition (12.101) by direct evaluation of the energy in Lagrangian coordinates failed because the nonintegral terms in (12.86) become of a lower order of smallness and thereby predominant for the result.

12.5 Dynamic Stabilization of NonSpherical Bodies Against Unlimited Collapse

The dynamic stability of spherical stars is determined by an average adiabatic power

$$\gamma=\frac{\partial\log P}{\partial\log\rho}|_S.$$

For a density distribution $\rho=\rho_0\varphi(m/M)$, the star in Newtonian gravity is stable against dynamical collapse when

$$\int_0^R\left(\gamma-\frac{4}{3}\right)P\frac{dm}{\varphi(m/M)}>0,$$

see (12.39). This approximate criterion becomes exact for adiabatic stars with constant γ. The collapse of a spherical star can be stopped only by a stiffening of the equation of state, such as neutron star formation at the late stages of evolution, or formation of a fully ionized stellar core with $\gamma=5/3$ at the collapse of clouds during star formation. Without a stiffening, a spherical star in Newtonian theory would collapse into a point with an infinite density (singularity).

In the presence of a rotation, a star becomes more dynamically stable against collapse. Because of the more rapid increase of the centrifugal force during contraction, in comparison with the Newtonian gravitational force, the collapse of a rotating star will always be stopped at finite density by centrifugal forces. It is shown in [142] that deviation from spherical symmetry in a nonrotating star with zero angular momentum leads to a similar stabilization, and a nonspherical star without dissipative processes will never reach a singularity. Therefore, the collapse to a singularity is connected with a secular type of instability, even without rotation. This conclusion is based on calculations of the dynamical behaviour of a nonspherical, nonrotating star after its loss of linear stability, at the nonlinear stages. An approximate system of dynamic equations derived in [140, 141] is used, describing three degrees of freedom of a uniform self-gravitating compressible ellipsoidal body. The development of instability leads to the formation of a regularly or chaotically oscillating body, in which dynamical motion prevents the formation of the singularity. The regions of chaotic and regular pulsations are found by constructing a Poincaré diagram for different values of the initial eccentricity and initial entropy. The calculations have been done for spheroidal figures with $\gamma = 4/3$ and $\gamma = 6/5$.

12.5.1 Equations of Motion

Let us consider 3-axis ellipsoid with semi-axes $a \neq b \neq c$:

$$\frac{x^2}{a^2} + \frac{y^2}{b^2} + \frac{z^2}{c^2} = 1, \tag{12.102}$$

and uniform density ρ. A mass m of the uniform ellipsoid is written as (V is the volume of the ellipsoid)

$$m = \rho V = \frac{4\pi}{3} \rho abc. \tag{12.103}$$

Let us assume a linear dependence of velocities on coordinates

$$v_x = \frac{\dot{a}x}{a}, \quad v_y = \frac{\dot{b}y}{b}, \quad v_z = \frac{\dot{c}z}{c}. \tag{12.104}$$

The gravitational energy of the uniform ellipsoid is defined as [615]:

$$U_g = -\frac{3Gm^2}{10} \int_0^\infty \frac{du}{\sqrt{(a^2+u)(b^2+u)(c^2+u)}}. \tag{12.105}$$

The equation of state $P = K\rho^\gamma$ is considered here, with $\gamma = 4/3$. A spherical star with $\gamma = 4/3$ collapses to singularity at small enough K, and deviations from a spherical form prevent formation of any singularity. For $\gamma = 4/3$, the thermal energy of the ellipsoid is $E_{th} \sim V^{-1/3} \sim (abc)^{-1/3}$, and the value

$$\varepsilon = E_{th}(abc)^{1/3} = 3\left(\frac{3m}{4\pi}\right)^{1/3} K$$

remains constant in time. A Lagrange function of the ellipsoid is written as

$$L = U_{kin} - U_{pot}, \quad U_{pot} = U_g + E_{th}, \tag{12.106}$$

$$U_{kin} = \frac{1}{2}\rho \int_V \left(v_x^2 + v_y^2 + v_z^2\right) dv = \frac{m}{10}\left(\dot{a}^2 + \dot{b}^2 + \dot{c}^2\right), \tag{12.107}$$

$$E_{th} = \frac{\varepsilon}{(abc)^{1/3}}. \tag{12.108}$$

Equations of motion describing behavior of three semiaxes (a, b, c) is obtained from the Lagrange function (12.106) in the form

$$\ddot{a} = -\frac{3Gm}{2} a \int_0^\infty \frac{\mathrm{d}u}{(a^2+u)\Delta} + \frac{5}{3m}\frac{1}{a}\frac{\varepsilon}{(abc)^{1/3}}, \tag{12.109}$$

$$\ddot{b} = -\frac{3Gm}{2} b \int_0^\infty \frac{\mathrm{d}u}{(b^2+u)\Delta} + \frac{5}{3m}\frac{1}{b}\frac{\varepsilon}{(abc)^{1/3}}, \tag{12.110}$$

$$\ddot{c} = -\frac{3Gm}{2} c \int_0^\infty \frac{\mathrm{d}u}{(c^2+u)\Delta} + \frac{5}{3m}\frac{1}{c}\frac{\varepsilon}{(abc)^{1/3}}, \tag{12.111}$$

$$\Delta^2 = (a^2+u)(b^2+u)(c^2+u).$$

12.5.2 Dimensionless Equations

To obtain a numerical solution of equations, they are written in nondimensional variables

$$\tilde{t} = \frac{t}{t_0},\ \tilde{a} = \frac{a}{a_0},\ \tilde{b} = \frac{b}{a_0},\ \tilde{c} = \frac{c}{a_0},$$

$$\tilde{m} = \frac{m}{m_0},\ \tilde{\rho} = \frac{\rho}{\rho_0},\ \tilde{U} = \frac{U}{U_0},\ \tilde{E}_{\mathrm{th}} = \frac{E_{\mathrm{th}}}{U_0},\ \tilde{\varepsilon} = \frac{\varepsilon}{\varepsilon_0}.$$

The scaling parameters $t_0,\ a_0,\ m_0,\ \rho_0,\ U_0,\ and \varepsilon_0$ are connected by the following relations:

$$t_0^2 = \frac{a_0^3}{Gm_0},\ U_0 = \frac{Gm_0^2}{a_0},\ \rho_0 = \frac{m_0}{a_0^3},\ \varepsilon_0 = U_0 a_0. \tag{12.112}$$

The system of non-dimensional equations is:

$$\ddot{a} = -\frac{3m}{2} a \int_0^\infty \frac{\mathrm{d}u}{(a^2+u)\Delta} + \frac{5}{3m}\frac{1}{a}\frac{\varepsilon}{(abc)^{1/3}}, \tag{12.113}$$

$$\ddot{b} = -\frac{3m}{2} b \int_0^\infty \frac{\mathrm{d}u}{(b^2+u)\Delta} + \frac{5}{3m}\frac{1}{b}\frac{\varepsilon}{(abc)^{1/3}}, \tag{12.114}$$

$$\ddot{c} = -\frac{3m}{2} c \int_0^\infty \frac{du}{(c^2+u)\Delta} + \frac{5}{3m}\frac{1}{c}\frac{\varepsilon}{(abc)^{1/3}},$$

$$\Delta^2 = (a^2+u)(b^2+u)(c^2+u). \tag{12.115}$$

In equations (12.113)–(12.115) only nondimensional variables are used, and "tilde" sign is omitted for simplicity in this section. The nondimensional Hamiltonian (or nondimensional total energy) is:

$$H = U_{\mathrm{kin}} + U_{\mathrm{g}} + E_{\mathrm{th}} = \frac{m}{10}(\dot{a}^2 + \dot{b}^2 + \dot{c}^2)$$
$$- \frac{3m^2}{10}\int_0^\infty \frac{du}{\sqrt{(a^2+u)(b^2+u)(c^2+u)}} + \frac{\varepsilon}{(abc)^{1/3}}. \tag{12.116}$$

In the case of a sphere ($a = b = c$, $\dot{a} = \dot{b} = \dot{c}$), the nondimensional Hamiltonian and nondimensional equations of motion reduce to:

$$H = \frac{3}{10} m\dot{a}^2 - \frac{3}{5a}\left(m^2 - \frac{5}{3}\varepsilon\right), \tag{12.117}$$

$$\ddot{a} = -\frac{1}{ma^2}\left(m^2 - \frac{5}{3}\varepsilon\right). \tag{12.118}$$

As follows from (12.117) and (12.118) for the given mass, there is only one equilibrium value of ε

$$\varepsilon_{\mathrm{eq}} = \frac{3m^2}{5}, \tag{12.119}$$

at which the spherical star has zero total energy, and it may have an arbitrary radius. For smaller $\varepsilon < \varepsilon_{\mathrm{eq}}$, the sphere should contract to singularity, and for $\varepsilon > \varepsilon_{\mathrm{eq}}$ there will be a total disruption of the star with an expansion to infinity. The equations of motion have been solved numerically in [142] for a spheroid with $a = b \neq c$. Using (12.113)–(12.115), these are written for the oblate spheroid with $k = c/a < 1$ as

$$\ddot{a} = \frac{3}{2}\frac{m}{a^2(1-k^2)}\left[k - \frac{\arccos k}{\sqrt{1-k^2}}\right] + \frac{5}{3m}\frac{1}{a}\frac{\varepsilon}{(a^2c)^{1/3}}, \tag{12.120}$$

$$\ddot{c} = -3\frac{m}{a^2(1-k^2)}\left[1 - \frac{k \arccos k}{\sqrt{1-k^2}}\right] + \frac{5}{3m}\frac{1}{c}\frac{\varepsilon}{(a^2c)^{1/3}}; \tag{12.121}$$

and for the prolate spheroid with $k = c/a > 1$ as

$$\ddot{a} = -\frac{3}{2}\frac{m}{a^2(k^2-1)}\left[k - \frac{\cosh^{-1} k}{\sqrt{k^2-1}}\right] + \frac{5}{3m}\frac{1}{a}\frac{\varepsilon}{(a^2c)^{1/3}}, \tag{12.122}$$

$$\ddot{c} = 3\frac{m}{a^2(k^2-1)}\left[1 - \frac{k\cosh^{-1}k}{\sqrt{k^2-1}}\right] + \frac{5}{3m}\frac{1}{c}\frac{\varepsilon}{(a^2c)^{1/3}}. \tag{12.123}$$

It is convenient to introduce variables

$$\varepsilon_* = \frac{5}{3}\frac{\varepsilon}{m^2}, \quad t_* = t\sqrt{m}. \tag{12.124}$$

In these variables, the (12.122)–(12.123), (12.118) are written as (omitting the subscript "*")

$$\ddot{a} = \frac{3}{2a^2(1-k^2)}\left[k - \frac{\arccos k}{\sqrt{1-k^2}}\right] + \frac{1}{a}\frac{\varepsilon}{(a^2c)^{1/3}}, \tag{12.125}$$

$$\ddot{c} = -\frac{3}{a^2(1-k^2)}\left[1 - \frac{k\arccos k}{\sqrt{1-k^2}}\right] + \frac{1}{c}\frac{\varepsilon}{(a^2c)^{1/3}} \tag{12.126}$$

for the oblate spheroid with $k = c/a < 1$, and as

$$\ddot{a} = -\frac{3}{2a^2(k^2-1)}\left[k - \frac{\cosh^{-1}k}{\sqrt{k^2-1}}\right] + \frac{1}{a}\frac{\varepsilon}{(a^2c)^{1/3}}, \tag{12.127}$$

$$\ddot{c} = \frac{3}{a^2(k^2-1)}\left[1 - \frac{k\cosh^{-1}k}{\sqrt{k^2-1}}\right] + \frac{1}{c}\frac{\varepsilon}{(a^2c)^{1/3}} \tag{12.128}$$

for the prolate spheroid with $k = c/a > 1$. For the sphere, where the equilibrium corresponds to $\varepsilon_{eq} = 1$, the (12.122)–(12.123) and (12.118) are reduced to

$$\ddot{a} = -\frac{1-\varepsilon}{a^2}. \tag{12.129}$$

Near the spherical shape, we should use expansions around $k = 1$, which leads to equations of motion valid for both oblate and prolate cases

$$\ddot{a} = -\frac{1-\varepsilon}{a^2} + \left(\frac{\varepsilon}{3} + \frac{3}{5}\right)\frac{1-k}{a^2},$$

$$\ddot{c} = -\frac{1-\varepsilon}{a^2} + \left(\frac{4\varepsilon}{3} - \frac{4}{5}\right)\frac{1-k}{a^2}. \tag{12.130}$$

In these variables, the total energy is written as

$$H_* = \frac{H}{m^2}, \tag{12.131}$$

and omitting "*", we have

$$H = \frac{\dot{a}^2}{5} + \frac{\dot{c}^2}{10} - \frac{3}{5a}\frac{\arccos k}{\sqrt{1-k^2}} + \frac{3}{5}\frac{\varepsilon}{(a^2c)^{1/3}}, \quad \text{(oblate)}$$

$$H = \frac{\dot{a}^2}{5} + \frac{\dot{c}^2}{10} - \frac{3}{5a}\frac{\cosh^{-1} k}{\sqrt{k^2-1}} + \frac{3}{5}\frac{\varepsilon}{(a^2c)^{1/3}}, \quad \text{(prolate)}$$

$$H = \frac{3\dot{a}^2}{10} - \frac{3}{5a}(1-\varepsilon), \quad \text{(sphere)}$$

$$H = \frac{\dot{a}^2}{5} + \frac{\dot{c}^2}{10} - \frac{3}{5a}\left(1 + \frac{\delta}{3} + \frac{2\delta^2}{15}\right) + \frac{3\varepsilon}{5a}\left(1 + \frac{\delta}{3} + \frac{2\delta^2}{9}\right),$$

$$\delta = 1 - k, \quad \text{(around the sphere)}, \quad |\delta| \ll 1. \tag{12.132}$$

Solution of the system of (12.122)–(12.123) was performed in [142] for initial conditions at $t = 0$: $\dot{c}_0 = 0$, different values of initial a_0, $\dot{a}_0$, k_0, and different values of the constant parameter ε. Evidently, at $k_0 = 1$, $\dot{a}_0 = 0$, $\varepsilon < 1$, we have the spherical collapse to singularity. No singularity was reached at $k_0 \neq 1$, for any $\varepsilon > 0$. At $\varepsilon = 0$, a weak singularity is reached during formation of a pancake with infinite volume density and finite gravitational force. At $\varepsilon > 0$, the behavior depends on the value of the total energy H: at $H > 0$, we obtain a full disruption of the body, and at $H < 0$ the oscillatory regime is established at any value of $\varepsilon < 1$. At $\varepsilon \geq 1$, the total energy of spheroid is positive, $H > 0$, determining the full disruption. The case with $H = 0$ is described separately.

At $H < 0$, the type of oscillatory regime depends on initial conditions and may be represented either by regular periodic oscillations or by chaotic behavior. Examples of two types of such oscillations are represented in Figs. 12.6 and 12.7 (regular, periodic), and in Fig. 12.8 (chaotic). For rigorous separation between these kinds of oscillations, a method, developed by Poincaré [637], was used.

12.5.3 Numerical Results for the Case H = 0

In the case of a sphere with zero initial radial velocity, we have an equilibrium state with $\varepsilon = \varepsilon_{eq} = 1$. Let us consider the case of $\dot{a} \neq 0$ and $\varepsilon < 1$. It follows from (12.132) that for non-zero initial velocity $\dot{a}$, entropy function ε is strictly less than unity. The fate of such sphere depends on the sigh of $\dot{a}$. At $\dot{a} < 0$, the star collapses to singularity, and at $\dot{a} > 0$ there total disruption happens, with zero velocity at infinity. For the initial spheroid with $a = b \neq c$, and zero initial velocities $\dot{a} = \dot{c} = 0$, the entropy function is also less than 1. In this case, the value of the entropy function

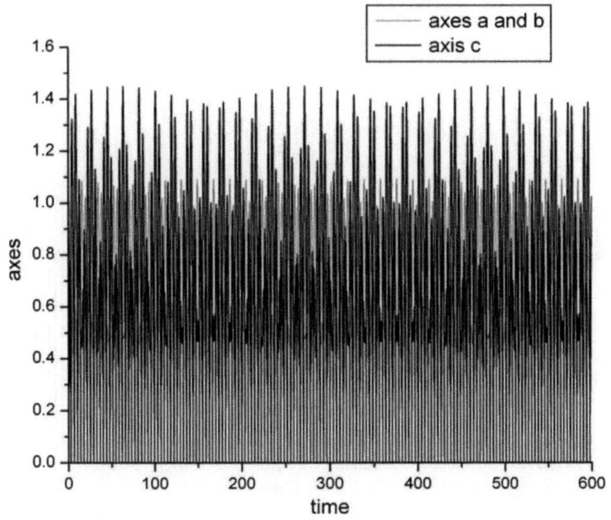

Fig. 12.6 Example of regular motion of spheroid with $\gamma = 4/3$, $H = -1/5$ and $\varepsilon = 2/3$. This motion corresponds to full line on the Poincaré map in Fig. 12.9, from [142]

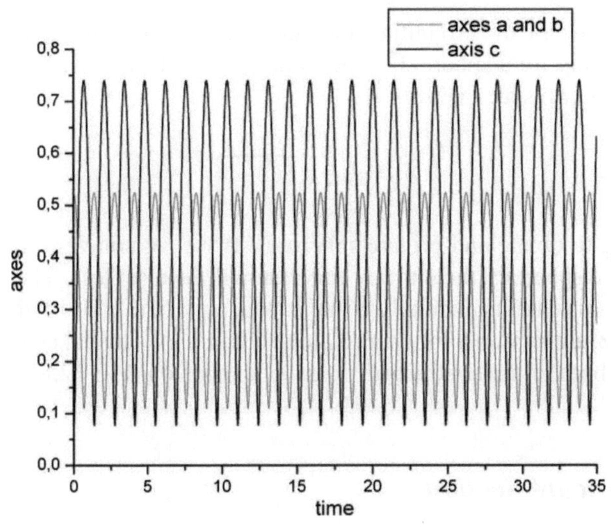

Fig. 12.7 Example of regular motion of spheroid with $\gamma = 4/3$, $H = -1/5$ and $\varepsilon = 2/3$. This motion corresponds to the point inside the regular region on the Poincaré map in Fig. 12.9, from [142]

ε is uniquely determined by the deformation $\delta = (a - c)/a$. This dependence can be found explicitly from (12.132); for small δ, we obtain

$$\varepsilon = 1 - \frac{4}{45}\,\delta^2.$$

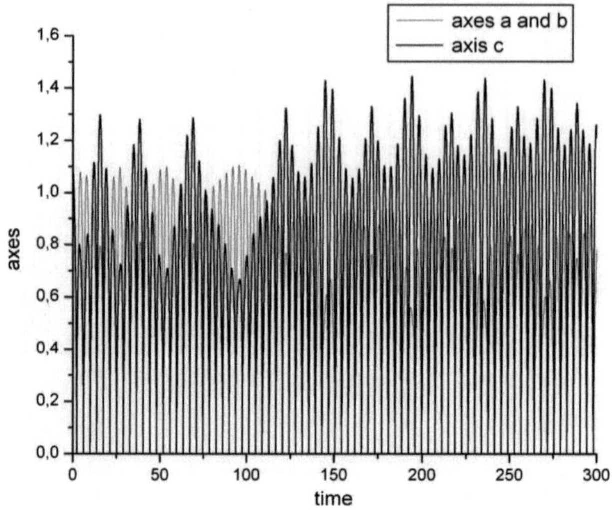

Fig. 12.8 Example of chaotic motion of spheroid with $\gamma = 4/3$, $H = -1/5$ and $\varepsilon = 2/3$. This motion corresponds to grey points on the Poincaré map in Fig. 12.9, from [142]

Thus, even in the case of zero initial velocities, ε is less than unity for a spheroid and reaches unity only for a sphere. For the same ε, the deformation δ of a zero energy body at rest has two states with

$$\delta = \pm\frac{3}{2}\sqrt{5(1-\varepsilon)},$$

corresponding to oblate and prolate spheroids. If we set non-zero initial velocities in the case of a spheroid, the entropy function will be even less. The calculation of the motion in this case leads finally to the expansion of the body, with an oscillating behaviour of δ around zero (oblateprolate oscillations). This takes place under the conditions of both initial contraction or initial expansion (Fig. 12.8).

12.5.4 Poincaré Section

To investigate regular and chaotic dynamics, the method of Poincaré section was used [637], and the Poincare map for different values of the total energy H were obtained. Let us consider a spheroid with semi-axes $a = b \neq c$. This system has two degrees of freedom. Therefore, in this case the phase space is four-dimensional: a, $\dot{a}$, c, $\dot{c}$. If we choose a value of the Hamiltonian H_0, we fix a three-dimensional energy surface $H(a, \dot{a}, c, \dot{c}) = H_0$. During the integration of (12.122) and (12.123), which preserve the constant H, we fix moments t_i, when $\dot{c} = 0$. At these moments, there are only two independent values (i.e. a and $\dot{a}$) because the value of c is determined uniquely from the relation for the Hamiltonian at constant H. At each moment t_i, we put a dot on the plane $(a, \dot{a})$.

For the same values of H and ε, we solve the equations of motion (12.122) and (12.123) at initial $\dot{c} = 0$, and different a and $\dot{a}$. For each integration, we put the points on the plane $(a, \dot{a})$ at the moments t_i. These points are the intersection points of the trajectories on the three-dimensional energy surface with a two-dimensional plane $\dot{c} = 0$, called the Poincaré section.

For each fixed combination of ε and H, we obtain the Poincaré map, represented in Figs. 12.9–12.13. The condition $\dot{c} = 0$ splits into two cases, of a minimum and of a maximum of c. The Poincaré maps are drawn separately, either for the minimum or for the maximum of c, and both maps lead to identical results. The regular oscillations are represented by closed lines on the Poincaré map, and chaotic behaviour fills regions of finite square with dots. These regions are separated from the regions of regular oscillations by a separatrix line.

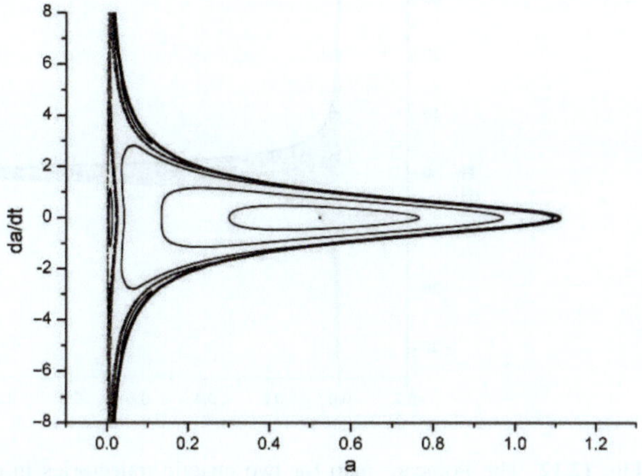

Fig. 12.9 The Poincaré map for five regular and two chaotic trajectories in case of $\gamma = 4/3$, $H = -1/5$, and $\varepsilon = 2/3$. The $(a, \dot{a})$ values are taken in the minimum of c. Full *black line* is the bounding curve. The point inside the regular region corresponds to coherent oscillations with the same period for a and c values, represented in Fig. 12.7

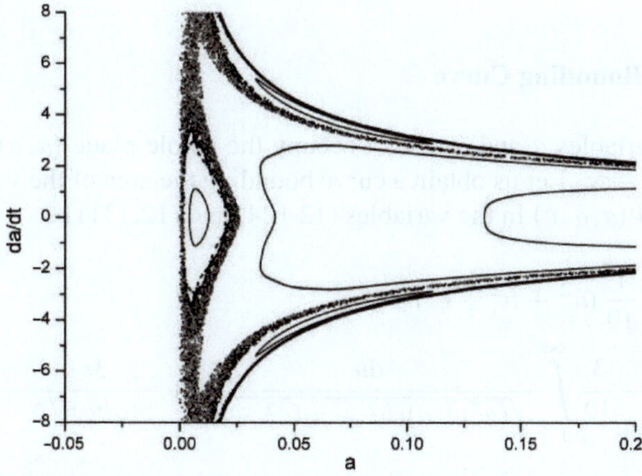

Fig. 12.10 Zoom of Fig. dfig4

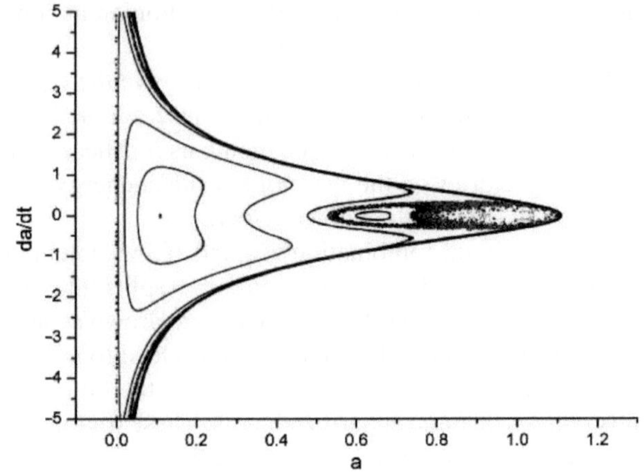

Fig. 12.11 The Poincaré map for five regular and two chaotic trajectories in case of $\gamma = 4/3$, $H = -1/5$, and $\varepsilon = 2/3$. The $(a, \dot{a})$ values are taken in the maximum of c. Full black line is the bounding curve. The point inside the regular region corresponds to coherent oscillations with the same period for a and c values, represented in Fig. 12.7

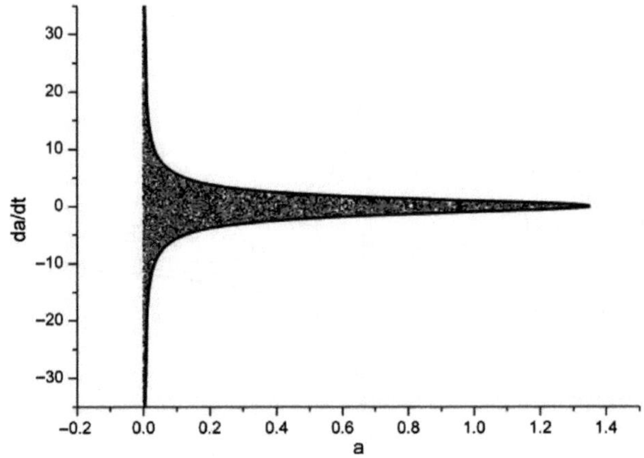

Fig. 12.12 The Poincaré map for two chaotic trajectories in case of $\gamma = 4/3$, $H = -1/2$, $\varepsilon = 1/6$. The $(a, \dot{a})$ values are taken in the minimum of c. Full *black line* is the bounding curve

12.5.4.1 The Bounding Curve

Actually, the variables a and $\dot{a}$ cannot occupy the whole plane $(a, \dot{a})$: $0 < a < \infty, -\infty < \dot{a} < +\infty$. Let us obtain a curve bounding the area of the values a and $\dot{a}$. Let a function $\Phi(a, \dot{a}, c)$ in the variables (12.124) and (12.131) be:

$$\Phi(a, \dot{a}, c) = \frac{1}{10}(\dot{a}^2 + \dot{b}^2 + \dot{c}^2) - \\ - \frac{3}{10}\int_0^\infty \frac{du}{\sqrt{(a^2+u)(b^2+u)(c^2+u)}} + \frac{3\varepsilon}{5(abc)^{1/3}} - H \quad (12.133)$$

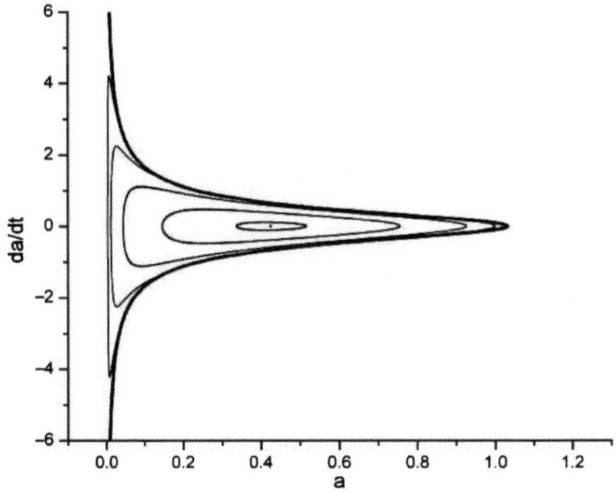

Fig. 12.13 The Poincaré map for six regular trajectories in the case of $\gamma = 4/3$, $H = -3/50$, and $\varepsilon = 9/10$. The $(a, \dot{a})$ values are taken in the minimum of c. Full *black line* is the bounding curve. The point inside the regular region corresponds to coherent oscillations with the same period for a and c values, similar to those represented in Fig. 12.7

at $a = b$, $\dot{a} = \dot{b}$, $\dot{c} = 0$, and fixed value of ε and H. The equation for the bounding curve $f(a, \dot{a}) = 0$ at given ε and H is determined from the following system of equation:

$$\Phi(a, \dot{a}, c) = 0, \quad \frac{\partial}{\partial c}\Phi(a, \dot{a}, c) = 0. \tag{12.134}$$

For $c < a$, this system of equations has the forms

$$\frac{1}{5}\dot{a}^2 - \frac{3}{5}\frac{\arccos(c/a)}{\sqrt{a^2 - c^2}} + \frac{3\varepsilon}{5(a^2 c)^{1/3}} - H = 0, \tag{12.135}$$

$$-\frac{3}{5}c\frac{\arccos(c/a)}{(a^2 - c^2)^{3/2}} + \frac{3}{5}\frac{1}{a^2 - c^2} - \frac{1}{3}\frac{1}{c}\frac{3\varepsilon}{5(a^2 c)^{1/3}} = 0. \tag{12.136}$$

This system is solved numerically. The second equation does not depend on $\dot{a}$. We set a and obtain a corresponding value of c from the second equation. Then we substitute a and c into the first equation and find $\dot{a}$. Thus, we obtain the point $(a, \dot{a})$. Changing a, we obtain numerically the curve $f(a, \dot{a}) = 0$. This bounding curve is shown in Figs. 12.9–12.13 by a bold line.

12.6 General Picture

The main result following from above calculations is the indication of the degenerate nature of the formation of a singularity in unstable Newtonian self-gravitating gaseous bodies. Only pure spherical models can collapse to singularity, but any kind

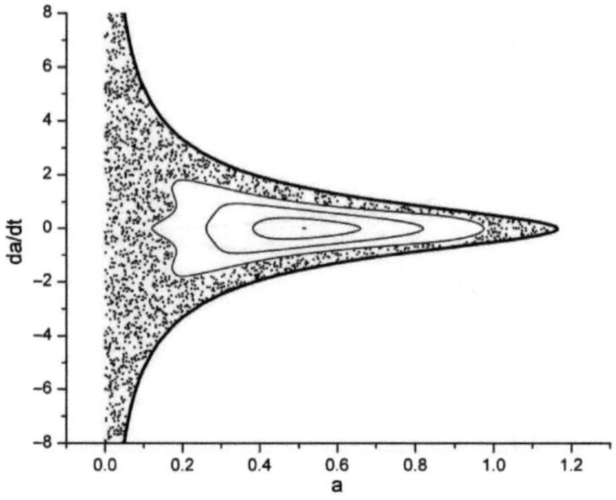

Fig. 12.14 The Poincaré map for one chaotic and four regular trajectories in case of $\gamma = 6/5$, $m = 1$, $\tilde{H} = -3/50$ and $\tilde{\varepsilon} = 27/50$, see (12.137). The $(a, \dot{a})$ values are taken in the minimum of c. Full black line is the bounding curve. The point inside the regular region corresponds to coherent oscillations with the same period for a and c values, similar to those represented in Fig. 12.7

of nonsphericity leads to nonlinear stabilization of the collapse by dynamic motion, and formation of regularly or chaotically oscillating body. This conclusion is valid for all unstable equations of state (i.e., adiabatic with $\gamma < 4/3$). The calculations of the dynamics of the model with $\gamma = 6/5$ gave similar results. The Poincaré map for this case is represented in Fig. 12.14. For $\gamma = 6/5$, we have the entropy function $\varepsilon = E_{\text{th}}(abc)^{1/5}$, and the nondimensional entropy function $\tilde{\varepsilon} = \varepsilon / U_0 a_0^{3/5}$ [see (12.112)]. For a spherical star with $\gamma = 6/5$, the nondimensional Hamiltonian instead of (12.117) is:

$$\tilde{H} = \frac{3m}{10}\dot{a}^2 - \frac{3}{5a}m^2 + \frac{\tilde{\varepsilon}}{a^{3/5}}. \tag{12.137}$$

Note that region of chaotic behaviour on the Poincaré map gradually increases for $\gamma = 4/3$ with decreases in the entropy ε and the total energy H. At $\varepsilon = 9/10$ and $H = -3/50$, only regular oscillations have been found (Fig. 12.13), at $\varepsilon = 2/3$ and $H = -1/5$ both types of oscillations are present (Figs. 12.9–12.11), and only chaotic behaviour is found at $\varepsilon = 1/6$ and $H = -1/2$ (Fig. 12.12). According to [141], this chaotic behavior is connected with development of anisotropic instability, when radial velocities strongly exceed the transversal ones [26, 397].

In reality, the presence of dissipation leads to a damping of these oscillations, and to the final collapse of the nonrotating model, when the total energy of the body is negative. In the case of a core-collapse supernova, the main dissipation is a result of the emission of neutrinos. The time of the neutrino losses is much larger than the characteristic time of the collapse, so we can expect the collapse to lead to the formation of a neutron star where nonspherical modes are excited and exist for several seconds after the collapse. In addition to the damping because of neutrino emission, shock waves will be generated, determining highly variable energy losses during the oscillations. Besides viscosity and radiation, which can damp the ellipsoid-like motions and allow collapse, there is the possibility that inertial interactions with higher-order modes, which must be present in real stratified bodies, may cause an inertial cascade and drain the energy from the second-order modes in a nonsecular way that does not depend on dissipative coefficients. It is tempting to connect chaotic oscillations with the highly variable emission observed in the prompt gamma-ray emission of cosmic gamma-ray bursts [683]. The presence of rotation and a magnetic field greatly complicates the picture of the core-collapse supernova explosion [45]. In reality, a spheroid will become a triaxial ellipsoid during the motion. Variants with triaxial figures have been calculated in [141]. Qualitatively, the same results have been obtained, but a rigorous investigation of the regular and chaotic types of motion by constructing a Poincaré map is impossible, because of a large number of degrees of freedom.

Within the framework of general relativity, dynamic stabilization against collapse by nonlinear nonspherical oscillations is not universal. When the size of the body approaches gravitational radius, no stabilization is possible at any γ. Nevertheless, the nonlinear stabilization may occur at larger radii; so after damping of the oscillations the star would collapse to a black hole. Because of the development of nonspherical oscillations, there is the possibility of the emission of gravitational waves during the collapse of nonrotating stars with an intensity similar to rotating bodies, or even larger.

An account of general relativity will introduce a new nondimensional parameter, which can be written as:

$$p_{\mathrm{g}} = \frac{2Gm_0}{c^2 a_0}.$$

The fate of the gravitating body will depend on the value of this parameter, and we can expect a direct relativistic collapse to a black hole at increasing p_{g}, approaching unity. It is known that a nonrotating black hole is characterized only by its mass [1082]. In the absence of other dissipative processes, the excess of energy, connected with a nonspherical motion, should be emitted by gravitational waves during the formation of a black hole.

Chapter 13
Thermal Stability

13.1 Evolutionary Phases Exhibiting Thermal Instabilities

The development of a dynamical instability in the star causes its transition to a compact state (neutron star or black hole), may sometimes be accompanied by a supernova explosion, and represents the end of its nuclear evolution. Development of thermal instability does not necessarily lead to such a catastrophic result. Some displays of the thermal instability have been considered in Sects. 9.3 and 10.2.

13.1.1 Instability in Degenerate Regions

The instability development under conditions of matter degeneracy, predicted in [689], may be explained quite simply. The temperature increase in this case has almost no effect on pressure. The hydrodynamical mechanism of stabilization is not at work. The nuclear-burning time decreases exponentially with increasing temperature, and there occurs a thermonuclear explosion. In other terms, the thermal instability is related to the positive heat capacity of a degenerate star differing from normal stable stars with negative heat capacity. Such explosions have a large variety of manifestations.

1. Helium flash. This takes place in degenerate helium cores which form in stars with initial masses $M_i < 2.25\, M_\odot$ (Sect. 9.3) after hydrogen has been exhausted in the centre. The helium flash results in removing the degeneracy in the core and transferring the core in a quiescent-burning state (see Fig. 9.40). During helium flash, $\tau_n > \tau_h$ always, and static equilibrium is hardly broken.

2. Carbon flash. In stars with initial mass $M_i = 2.25$–$8 M_\odot$ the degenerate core forms after helium is burnt out in it, and the mixture $^{12}C + ^{16}O$ is formed (Sects. 9.3, 10.2). The fate of such a star depends on the counterbalance between processes of degenerate C–O core growth during helium shell burning and those of matter outflow. As shown in Sect. 9.3, according to observational data, only the most massive stars of this interval succeed in increasing the mass of their cores to $1.39\, M_\odot$ when a thermal instability develops in them to result in an explosion with $\tau_n < \tau_h$

G.S. Bisnovatyi-Kogan et al., *Stellar Physics: 2: Stellar Evolution and Stability*, Astronomy and Astrophysics Library, DOI 10.1007/978-3-642-14734-0_7,

(Sect. 9.3). Such an explosion causes either a complete run-away of the star, or the transition to collapse with subsequent neutron star formation (Sect. 10.2).

3. Neon-oxygen flash. Stars with $M = 8$–$13M_{\odot}$ burn carbon in the core in a non-degenerate state, but the resulting O+Ne+Mg core turns out to be degenerate (Sect. 10.1). For stars with $M = 8$–$10M_{\odot}$ the dynamical instability development results from neutronization of ^{24}Mg leading to a collapse and starts before the thermal instability development in the degenerate core. The oxygen thermal flash in the contraction phase has almost no effect on the collapse. For stars with $M =$ 10–$13M_{\odot}$, the thermal flash develops at the periphery of the degenerate core, removes its degeneracy and leads to subsequent quiescent evolution. Here, as in the helium flash, we have always $\tau_n > \tau_h$.

4. White dwarf envelopes. Degenerate hydrogen-helium shells in white dwarf envelopes result from accretion in binaries. The development of thermal instability of hydrogen thermonuclear burning is thought to be the reason for nova outbursts (Sect. 11.1.5). The value of τ_n then becomes lower than τ_h.

5. Neutron star envelopes. The formation of degenerate hydrogen and helium shells with subsequent thermonuclear explosion is a result of accretion as well. Calculations show [362] that after ignition of hydrogen the neutron star envelopes achieve such high temperatures that helium burning proceeds explosively. Thermal explosions are also possible in a purely helium degenerate envelope. During an explosion, the static condition $\tau_n > \tau_h$ holds. Such explosions are believed to produce X-ray bursters, that is, bursting X-ray sources observed in the spherical component of the Galaxy and in globular clusters in the energy range from tenths to a few tens of keV. Observational data analysis suggests that bursters are very old neutron stars with low magnetic fields incorporated in close binaries together with a low-mass red or white dwarf. Due to a low magnetic field the neutron star may speed its rotation up to very large angular velocities via accretion, transforming the neutron star into a rapid millisecond pulsar.[1]

In calculations with a regular accretion [607], Fig. 13.1, thermal flashes also occur regularly with the time interval between flashes depending on accretion rate $\dot{m}$. Observations reveal a strong irregularity in the occurrence and properties of flashes [958], Fig. 13.2, thus pointing to the possibility of chaotic effects. The emergence of stochasticity in several models of X-ray bursters has been studied in [824].

6. Explosion in the neutron star envelope due to the reaction of fission of superheavy nuclei. In the harder energy range from 20 keV to 20 MeV gamma-ray bursters were detected. Their length, from tenths to tens of seconds, ranges them close to X-ray bursters, but they exhibit a large variety of types (Fig. 13.3 [432]). Contrary to X-ray busters visible on the same irregularly busting sources, gamma-ray bursts are almost never recurrent. The nature of gammy-ray bursts is more ambiguous than that of X-ray busters.

[1] The discovery of the millisecond pulsar in the globular cluster M 28 [658] corroborated this evolutionary scheme considered in [21, 177, 178]. Over forty radio pulsars with rapid rotation were discovered by 2001 in globular clusters [252, 966], see review [641].

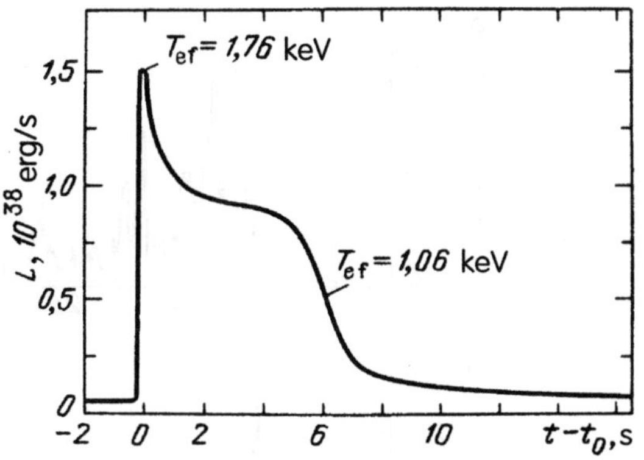

Fig. 13.1 Time dependence of the luminosity during a flash due to hydrogen–helium burning in the neutron star envelope. Shown are effective temperatures at the peak and fall of luminosity (from [607])

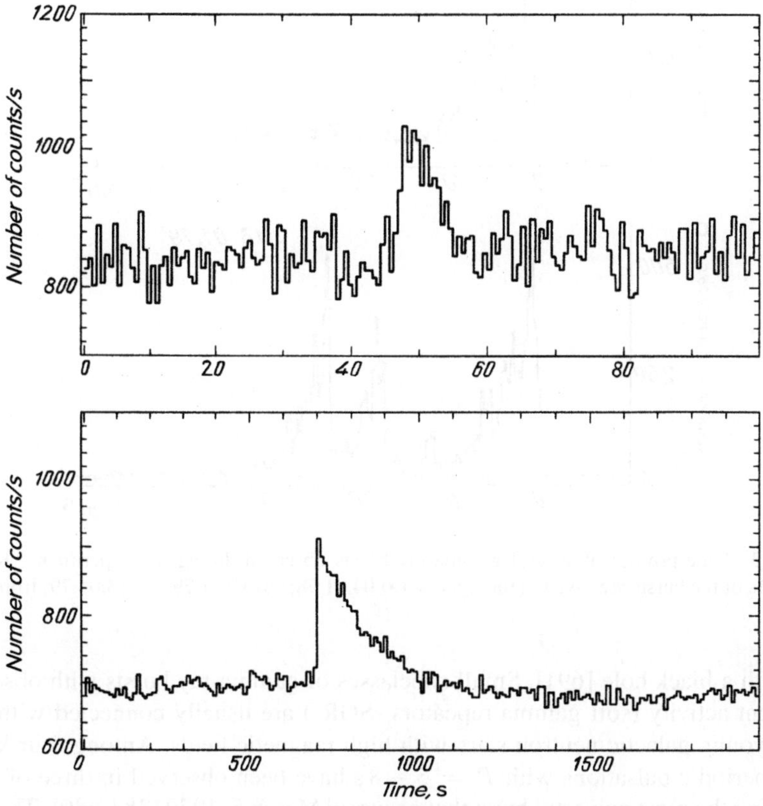

Fig. 13.2 Profiles of two bursts observed at GX 17 + 2 from the EXOSAT satellite 6. September 1984 (*above*) and 20. August 1985 (*below*). The time resolution is 0.63 and 10 s for the upper and lower figures, respectively (from [958])

Some of them are now related to cosmological distances, based mainly on discovery of X-ray and optical afterglows [257, 307, 597]. The necessary huge energy output is usually looked for in collisions between neutron stars, or a neutron

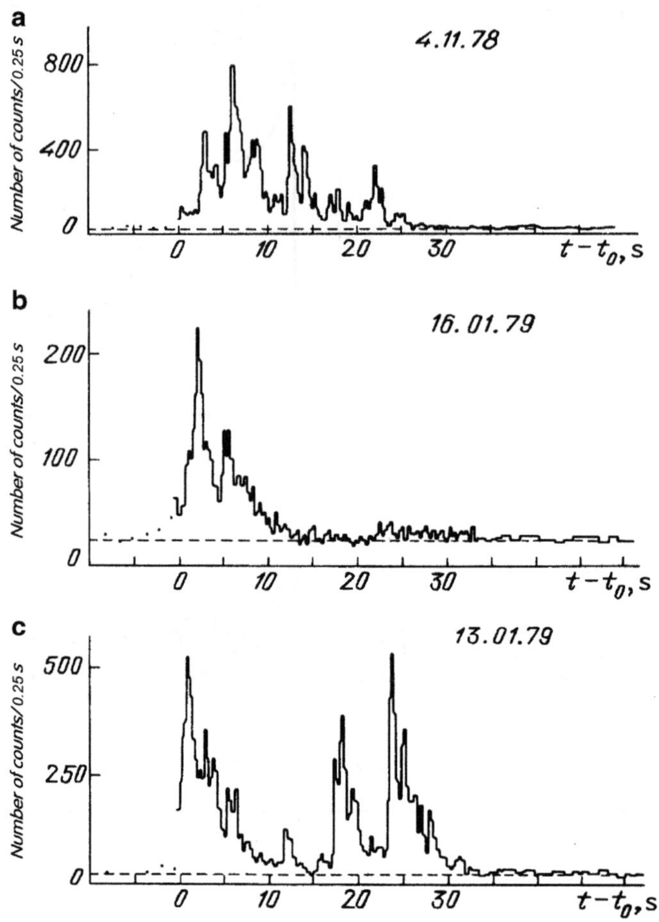

Fig. 13.3 Time profiles of several gamma-ray bursts observed during the experiment "Konus." The dates of the bursts are given in the figures: (**a**) 04.01.78; (**b**) 10.01.79; (**c**) 13.01.79; from [432]

star with a black hole [691]. Small subclasses of gamma-ray bursts with observed, recurrent activity (soft gamma repeators -SGR) are usually connected with relatively young galactic neutron stars with high magnetic fields. Among four known SGR, periodic pulsations with $P = 5 - 8$ s have been observed in three of them, including the most powerful burst that occurred March 5, 1979 [383,680]. The wide variety of properties observed in gamma-ray bursts could be connected with their different origin.

In our opinion, some gamma-ray bursts could occur on old cooled single neutron stars since they have no connection with any stationary optical (or radio) object which might be a normal companion of a neutron star or young neutron star– radio pulsar itself.

The model for gamma-ray burst proposed in [102, 126] and developed in [123, 196] is based on an instability leading to a nuclear explosion due to the reaction of fission of superheavy nuclei in the outer envelope of the neutron star at $\rho \leq 10^9$ g cm^{-3}. Superheavy nuclei form during the neutron star formation and are accumulated in the non-equilibrium layer (see Sects. 1.4.5 and 1.4.6, Vol. 1 [122, 194, 195]). Stable in the non-equilibrium layer at $\rho = 10^{11} - 10^{12}\,\mathrm{g\,cm^{-3}}$, superheavy nuclei become unstable to β-decay after having passed into less dense layers with $\rho \leq 10^9\,\mathrm{g\,cm^{-3}}$. Such a passage may be due to a starquake in the neutron star similar to that causing periodic jumps (glitches) in radio pulsars [670]. An increase in the number of protons in a nucleus makes it unstable to fission after β-decays. The emergence of fast neutrons in spontaneous fission initiates an induced fission which results ultimately in a chain reaction [123] and explosive energy release over a time $\tau_n \sim 10^{-8}$ s well below the hydrodynamical time of the envelope $\tau_{h,\mathrm{env}} = R/v \approx 10^4/10^{10} \approx 10^{-6}$ s. The energy release $Q = 10^{40}$ erg in the neutron star envelope leads to the formation of a shock which, upon reaching the surface, is thought to give the observable gamma-ray burst [166].

The instability resulting in a chain reaction of fission differs in nature from the above effects of thermal instability due to the reduction of the burning time τ_n caused by a temperature increase. The nuclear time of fission depends on the concentration of fast neutrons for a supercritical mass of nuclei undergoing fission, rather than on temperature. A thermal character of this instability consists in that on initial stages of the chain reaction development in degenerate matter the relative increase in pressure is slight, and hydrodynamical motions together with the shock formation occur only after the release of all the fission energy of superheavy nuclei. Such an instability may be called nuclear-thermal. A neutron star starquake may lead to excitation of eigen-oscillations and formation of a high-frequency short-lived radio and gamma pulsar [113].

13.1.2 Instabilities in the Absence of Degeneracy

7. Loops on evolutionary tracks of massive stars in the HR diagram. Evolutionary tracks of stars with $M \geq 3M_\odot$ in the HR diagram turn out to be very sensitive to initial values of parameters and even to the computational scheme (Sects. 9.2 and 2.3). Such an irregular behaviour provides evidence for the presence of a thermal instability having a nature other than the above instabilities. The reduction of τ_n due to temperature variations is of no importance here, moreover, the nuclear-burning rate varies slightly in this case. This instability occurs in stars at the stage of the evolution with core-helium and shell-hydrogen burning phase. From a mathematical standpoint, this means that the time-dependent solutions to the equations of evolution turn out to be unstable, that is, arbitrary small differences in initial conditions lead in time to large differences in solutions. As this instability develops, the relations between the time scales τ_h, τ_{th}, τ_n do not change by orders of magnitude. The presence of an instability is a necessary condition for the development of stochasticity exhibited by numerical computations.

Although evolutionary calculations with loops are numerous, the physical nature of this instability and the forces behind it are still obscure. In [1086], for stars with $M \geq 15M_{\odot}$, this instability is attributed to outgoing of the hydrogen burning-shell beyond the jump in chemical composition at the boundary of the maximum inward penetration of the outer convective zone at previous evolutionary stages. However, the possibility that this penetration itself is a result of this instability cannot be ruled out. For stars of lower masses, there are no hypotheses on the physical nature of their loops.

8. Flashes in non-degenerate helium shells. Instabilities of this type occur in stars of 2.25–8$M_{\odot}$ at the stage when a degenerate C–O core grows because of helium burning in a shell (Sect. 9.3.4). From its effects, this instability is completely analogous to the instability of a nuclear burning under conditions of degeneracy. Here also, the time scale of helium burning τ_{He} decreases strongly with temperature, although never attains τ_h, therefore the static equilibrium is conserved as in cases 1, 3, 5. The dynamical reaction of the helium-burning shell to the temperature increase inside it proves to be weak, but the reason for this is not that the pressure depends weakly on the temperature as in the case of the degenerate matter. Rather, it is related to the response of the star as a whole to the temperature increase in the thin shall, and its positive heat capacity.

13.2 Thermal Instability Development in Non-Degenerate Shells

The prediction of the possibility for thermal instability to develop in thin non-degenerate burning shells owing to peculiarities of the stellar response to perturbations in the burning rate was made in [440]. A mathematical description of this type of instability is given in [884]. We now consider dynamically stable stars.

13.2.1 Stability of a Burning Shell with Constant Thickness

Consider a thermal perturbation, without account of its effect on the shell equilibrium. Rewrite the thermal balance equations for the radiative case (9.24) and (9.94), with the equation of state of a perfect gas in the form

$$L_r = -(4\pi r^2)^2 \frac{4ac}{3} \frac{T^3}{\kappa} \frac{dT}{dm}, \tag{13.1}$$

$$\varepsilon - \frac{dL_r}{dm} = \frac{3}{2} \frac{P}{\rho} \frac{ds}{dt}, \tag{13.2}$$

$$s = \frac{2}{3} \frac{S}{\mathscr{R}} = \ln \frac{P}{\rho^{5/3}} + \text{const.} \quad \mathscr{R} = \frac{k}{\mu m_{\mathrm{u}}}, \tag{13.3}$$

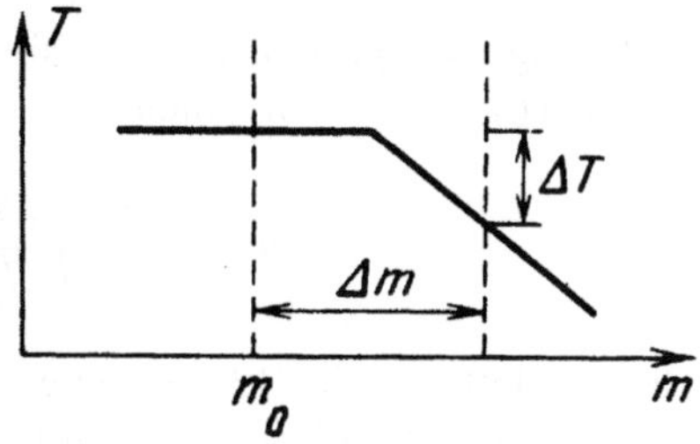

Fig. 13.4 Idealized temperature profile through the burning shell, from [884]

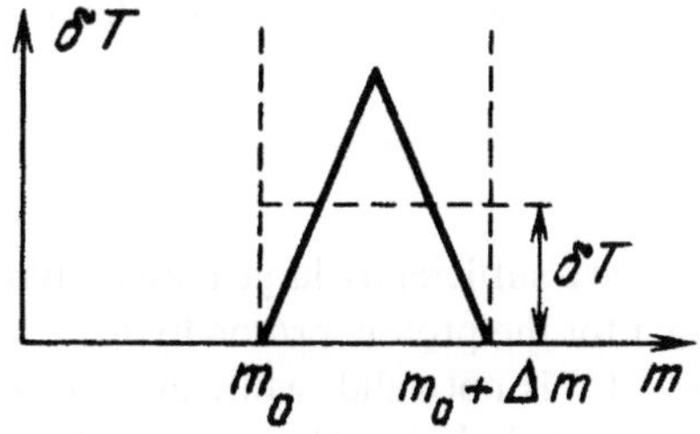

Fig. 13.5 Idealized temperature perturbation profile through the burning shell, from [884]

where S is the entropy of a perfect, fully ionized gas (11.123). Assume that all energy sources are located in a shell of a mass Δm, while the energy release outside this shell is negligible. The idealized temperature distribution in the shell Δm is given in Fig. 13.4 from [884]. Equation (13.1) then gives

$$L_r = 0 \quad \text{for } m = m_0 \tag{13.4}$$

$$L_r = L = \left(4\pi r^2\right)^2 \frac{4acT^3}{3\kappa} \frac{\Delta T}{1/2\Delta m} \quad \text{for } m > m_0 + \frac{\Delta m}{2}. \tag{13.5}$$

Neglecting entropy variations with time gives, for the mean rate of the energy release in the shell

$$\varepsilon = \frac{L}{\Delta m}. \tag{13.6}$$

Let us now perturb the thermal state within the shell [884] as shown in Fig. 13.5. Neglecting the density and temperature perturbations and retaining only the temperature-gradient ones, we have from (13.1)

$$\delta L_r = \mp \left(4\pi r^2\right)^2 \frac{4ac}{3} \frac{T^3}{\kappa} \frac{\delta T}{1/4\Delta m}, \tag{13.7}$$

where "−" refers to the inner, and "+" to the outer surfaces of the shell. The perturbation of the flux divergence is then equal to

$$\delta\left(\frac{dL_r}{dm}\right) = 4\frac{L}{\Delta m}\frac{T}{\Delta T}\frac{\delta T}{T}. \tag{13.8}$$

If the energy release rate is written in the form

$$\varepsilon \sim T^{\nu}, \tag{13.9}$$

neglecting its dependence on density, and only time derivatives of the entropy are included ($d/dt = \partial/\partial t$), then from (13.2), using (13.6), (13.8), and (13.9), we have

$$\left(\nu - 4\frac{T}{\Delta T}\right)\frac{\delta T}{T} = \frac{3}{2}\left(\frac{P}{\rho}\frac{\Delta m}{L}\right)\frac{d(\delta s)}{dt}. \tag{13.10}$$

The first left term in (13.10) represents the gain in heat release, while the second gives the increased heat losses in the shell. A positive left-hand side means that a small increase in shell entropy leads to an increase in the shell temperature, i.e., implies instability. The heat release thus exceeds the heat losses at

$$\frac{\Delta T}{T} > \frac{4}{\nu}, \tag{13.11}$$

i.e., at a sufficiently large temperature difference ΔT in the shell. It is noted in [884] that for the proton–proton hydrogen-burning cycle with $\nu \approx 4$ [878] the condition (47.11) is not valid, while at $\nu \gg 4$ in the case of helium burning [878] (see also Chap. 4, Vol. 1), relatively small temperature differences may lead to instability. To answer definitively the question on the thermal instability of a burning shell, it is necessary to include simultaneously in calculations temperature and density perturbations arising from the hydrostatic adjustment of the whole star to the temperature perturbation in the shell.

13.2.2 Calculations of Density Perturbations [884]

Write the linearized equilibrium equations (10.1) and (9.97) in the form

$$\frac{d(\delta r/r)}{dm} = \frac{1}{4\pi r^3\rho}\left(-3\frac{\delta r}{r} - \frac{3}{5}\frac{\delta P}{P} + \frac{3}{5}\delta s\right), \tag{13.12}$$

$$\frac{d(\delta P/P)}{dm} = \frac{Gm}{4\pi r^4 P}\left(4\frac{\delta r}{r} + \frac{\delta P}{P}\right), \tag{13.13}$$

where

$$\frac{\delta\rho}{\rho} = -\frac{3}{5}\delta s + \frac{3}{5}\frac{\delta P}{P}. \tag{13.14}$$

We look for the solution to the inhomogeneous system of the linear equations (13.12) and (13.13), finite over the whole star, by use of the method of variation of constants. The corresponding homogeneous system (at $\delta s = 0$) has two linearly independent solutions. One of them with subscript "1" is finite in the centre, where the expansions (with the normalization condition $(\delta P/P)_1 = 1$ at $m = 0, r = 0$)

$$(\delta r/r)_1 = -\frac{1}{5} - \frac{3}{250}\frac{\rho G m}{r P} - \dots,$$

$$(\delta P/P)_1 = 1 + \frac{1}{10}\frac{\rho G m}{r P} + \dots \tag{13.15}$$

are valid, while the other, subscripted "2", is finite on the edge of the star, where (with the normalization condition $(\delta P/P)_2 = 1$ at $m = M, r = R$) the expansion

$$\left(\frac{\delta r}{r}\right)_2 = -\frac{1}{4} - \frac{3}{20}\frac{R-r}{R} - \frac{6}{35}\left(\frac{R-r}{R}\right)^2 - \dots,$$

$$\left(\frac{\delta P}{P}\right)_2 = 1 + \frac{3}{7}\left(\frac{R-r}{R}\right) + \frac{10}{21}\left(\frac{R-r}{R}\right)^2 + \dots \tag{13.16}$$

is valid. The solution to the inhomogeneous system which is finite over the star is looked for in the form

$$\delta r/r = C_1(m)(\delta r/r)_1 + C_2(m)(\delta r/r)_2,$$

$$\delta P/P = C_1(m)(\delta P/P)_1 + C_2(m)(\delta P/P)_2. \tag{13.17}$$

Solutions "1" diverge on the boundary, solutions "2" at the centre of the star because everywhere finite non-zero solutions of the homogeneous system are absent. The functions $C_1(m)$ and $C_2(m)$ then have to obey

$$C_1(M) = 0, \quad C_2(0) = 0. \tag{13.18}$$

Substituting (13.17) into (13.12) and (13.13) and solving this system for $C_1(m)$ and $C_2(m)$ with use of (13.18), we obtain the solution to the inhomogeneous system in the form

$$\begin{aligned}\frac{\delta r}{r} &= \left(\frac{\delta r}{r}\right)_1 \int_r^R \frac{1}{\Delta}\left(\frac{\delta P}{P}\right)_2 \frac{3}{5}\delta s\,\frac{dr}{r} \\ &+ \left(\frac{\delta r}{r}\right)_2 \int_0^r \left(\frac{\delta P}{P}\right)_1 \frac{1}{\Delta}\frac{3}{5}\delta s\,\frac{dr}{r}, \\ \frac{\delta P}{P} &= \left(\frac{\delta P}{P}\right)_1 \int_r^R \frac{1}{\Delta}\left(\frac{\delta P}{P}\right)_2 \frac{3}{5}\delta s\,\frac{dr}{r} \\ &+ \left(\frac{\delta P}{P}\right)_2 \int_0^r \left(\frac{\delta P}{P}\right)_1 \frac{3}{5}\frac{\delta s}{\Delta}\frac{dr}{r},\end{aligned} \tag{13.19}$$

where

$$\Delta = \left(\frac{\delta r}{r}\right)_2 \left(\frac{\delta P}{P}\right)_1 - \left(\frac{\delta r}{r}\right)_1 \left(\frac{\delta P}{P}\right)_2. \tag{13.20}$$

The solution (13.19) is written analogously to the solution of the inhomogeneous second-order equation $(d/dx)\,[p(x)y'] - q(x) + f(x) = 0$ in the interval $[a, b]$ with the uniform boundary conditions $\alpha_1 y(a) + \alpha_2 y'(a) = 0$, $\beta_1 y(b) + \beta_2 y'(b) = 0$ by use of the Green function [918]

$$G(x,\xi) = y_1(x)y_2(\xi) \quad (x \le \xi),$$
$$= y_2(x)y_1(\xi) \quad (x \ge \xi),$$

where y_1 satisfies the first, y_2 the second boundary condition. The function $G(x,\xi)$ is the solution to the second-order equation when $f(x) = \delta(x - \xi)$, while the solution with an arbitrary $f(x)$ has the form

$$y = \int_0^b G(x,\xi)\, f(\xi)\, d\xi.$$

For the perturbation δs, localized in a thin shell, we obtain from (13.19) the pressure perturbation in the shell in the form

$$\frac{\delta P}{P} \approx \frac{3}{5}\,Q\,\frac{\Delta r}{r}\,\delta s, \quad Q = \frac{(\delta P/P)_1(\delta P/P)_2}{\Delta}. \tag{13.21}$$

The quantity Q varies slightly throughout the star: from -4 to -8 for a red-giant model with a degenerate carbon core and a helium-burning shell [884]. The negativity of Q causes the entropy increase in the shell to reduce the pressure in it, while the magnitude of the pressure perturbation drops as the shell thickness decreases. From the equation of state for a perfect gas, relations (13.14) and (13.21), we have the temperature perturbation in the shell in the form

$$\frac{\delta T}{T} = \frac{\delta P}{P} - \frac{\delta \rho}{\rho} = \frac{2}{5}\frac{\delta P}{P} + \frac{3}{5}\delta s = \frac{3}{5}\left(1 + \frac{2}{5}\,Q\,\frac{\Delta r}{r}\right)\delta s. \tag{13.22}$$

For wide-burning regions $\Delta r/r \sim 1$ and $Q \approx -6$, a negative heat capacity of the star ($\delta T/T < 0$ for $\delta s > 0$) implying its thermal stability follows from this relation. For sufficiently thin burning shells an increase in temperature with rising entropy is possible if

$$\frac{\Delta r}{r} < \frac{5}{2}\frac{1}{|Q|}. \tag{13.23}$$

The thermal instability development in a burning shell requires that the conditions (13.11) and (13.23) be satisfied. In real burning shells $\Delta T/T \approx \Delta r/r$, so

$$\frac{4}{\nu} < \frac{\Delta r}{r} < \frac{5}{2}\frac{1}{|Q|}, \quad \nu > \frac{8}{5}|Q| \approx 10. \tag{13.24}$$

Helium-burning shells, together with hydrogen shells, where the burning proceeds through a carbon cycle (see Chap. 4, Vol. 1), may thus be thermally unstable. This factor may influence the loop formation in massive star tracks.

13.2.3 A Strict Criterion for Thermal Stability

A strict treatment of the thermal stability of a star requires solving a linearized system of evolution equations with a time dependence

$$\delta L,\ \delta r,\ \delta P,\ \delta T \sim e^{t/\tau}. \tag{13.25}$$

The system comprises (13.12) and (13.13), the linearized equation (13.1)

$$\delta L_r = -\left(4\pi r^2\right)^2 \frac{4ac}{3}\frac{T^3}{\kappa}\left[\frac{d\delta T}{dm} + \frac{dT}{dm}\left(\frac{4\delta r}{r} + \frac{3\delta T}{T} - \frac{\delta\kappa}{\kappa}\right)\right] \tag{13.26}$$

and the linearized equation (13.2) which is the only one to contain a time derivative:

$$\delta\varepsilon - \delta\left(\frac{dL_r}{dm}\right) = \frac{3}{2}\delta\left(\frac{P}{\rho}\right)\frac{ds}{dt} + \frac{3}{2}\frac{P}{\rho}\frac{\delta s}{\tau}. \tag{13.27}$$

The linear homogeneous system of differential equations (13.12), (13.13), (13.26), and (13.27) is solved by writing them in a finite difference form, as in the Henyey method (see Sect. 6.1, Vol. 1). By virtue of the homogeneity of this system, satisfying corresponding linearized boundary conditions, a solution exists only for zero determinant, which provides the single eigenvalue τ. The investigation in [884] of the thermal stability of evolutionary $1M_{\odot}$ models with low content in metals (population II) has revealed the presence of positive eigenvalues: (1) at the phase of the onset of helium flash in a core with $L_{\mathrm{H}} \approx 2600L_{\odot}$, $L_{\mathrm{He}} \approx 1L_{\odot}$, where $\tau \approx 300{,}000$ yr; and (2) at the phase of two non-degenerate burning shells (hydrogen and helium), with a degenerate carbon core at $L_{\mathrm{H}} \approx 122L_{\odot}$, $L_{\mathrm{He}} \approx 118L_{\odot}$, where $\tau \approx 10^6$ yr. In the latter case the temperature and entropy perturbations have one sign inside helium- and opposite signs inside hydrogen-burning shells, thus providing evidence for the instability of helium burning alone. Evolutionary calculations (Sect. 9.3) show that as the helium flash develops in a burning shell in the non-linear regime, τ decreases rapidly. On the boundary of thermal stability $\tau = \infty$, while the determinant of the above linearized system coincides with the Henyey determinant (Sect. 6.1, Vol. 1), therefore its reducing to zero during the evolution implies the onset of thermal instability.

Also, the determinants in the Henyey and Schwarzschild methods become zero when a thermal instability arises, leading to the formation of loops on evolutionary tracks (see [771, 776, 839], Sect. 9.2.1). This also gives rise to an ambiguity in construction of equilibrium stellar models with a fixed distribution of chemical composition.

Chapter 14
Stellar Pulsations and Stability

Being a dynamical system, the star has diverse oscillatory eigenmodes. When one takes thermal processes into account, some of these appear to be unstable. The increment of such an instability γ, equal to the inverse time scale of the amplitude increase by a factor of e, is usually well below the pulsational frequency ω. Other modes can be unstable in the adiabatic approximation as well, as in the case of convective instability (9.4), for $\mathrm{d}S/\mathrm{d}r < 0$.

Stellar pulsations are treated in books and reviews [317, 438, 630, 828, 994]. In this chapter, we consider, in addition, radial oscillations of stars with phase transitions, and the important problem of pulsational stability of massive stars not included in the monographs mentioned above. We will consider first briefly the fundamentals of the theory of stellar pulsation, following [317, 438, 630].

14.1 Eigenmodes

14.1.1 Equations for Small Oscillations

We derive a linearized system of equations of hydrodynamics describing small oscillations of a star. The star is assumed to be non-rotating with $\mathbf{u} = 0$. The Lagrangian perturbations of displacement, velocity, density, pressure and potential are denoted for the given element as

$$\delta\mathbf{r},\ \delta\mathbf{u},\ \delta\rho,\ \delta P,\ \delta\Phi, \tag{14.1}$$

and Eulerian ones at a given point of space as

$$\mathbf{u}'\ \rho',\ P',\ \Phi'. \tag{14.2}$$

The relation between them has the form

$$\delta f = f' + \delta\mathbf{r}\cdot(\nabla f). \tag{14.3}$$

Specifying the time dependence $\sim \mathrm{e}^{-\mathrm{i}\omega t}$, we have from (7.31)

$$\mathbf{u}' = \delta\mathbf{u} = -\mathrm{i}\omega\,\delta\mathbf{r}. \tag{14.4}$$

G.S. Bisnovatyi-Kogan et al., *Stellar Physics: 2: Stellar Evolution and Stability*, Astronomy and Astrophysics Library, DOI 10.1007/978-3-642-14734-0_8,

The continuity equation (7.32) gives:

$$\delta\rho = \rho' + \nabla\rho \cdot \delta\mathbf{r} = -\rho\,(\nabla \cdot \delta\mathbf{r}). \tag{14.5}$$

The equation of motion (7.33) divided by ρ, upon being linearized becomes:

$$-\omega^2\delta\mathbf{r} + \frac{\nabla P'}{\rho} - \frac{\nabla P}{\rho}\frac{\rho'}{\rho} + \nabla\Phi' = 0. \tag{14.6}$$

The Lagrangian perturbations of entropy ΔS are related to δP and δp by:

$$\delta P = \left(\frac{\partial \ln P}{\partial \ln \rho}\right)_S \frac{P}{\rho}\,\delta\rho + \left(\frac{\partial P}{\partial S}\right)_\rho \Delta S = \gamma_1\,\frac{P}{\rho}\,\delta\rho + \rho T\gamma_3\,\Delta S. \tag{14.7}$$

The thermodynamic relations:

$$\left(\frac{\partial P}{\partial S}\right)_\rho = \rho T\left(\frac{\partial \ln T}{\partial \ln \rho}\right)_S = \rho T\gamma_3 \tag{14.8}$$

are used in (14.7), the values γ_1 and γ_3 are given in (8.26) and (11.71). Writing (14.6) in the form

$$-\omega^2\delta\mathbf{r} + \nabla\left(\frac{P'}{\rho} + \Phi'\right) + \frac{P'}{\rho^2}\nabla\rho - \frac{\nabla P}{\rho}\frac{\rho'}{\rho} = 0. \tag{14.9}$$

Using (14.3), we express P' in the third term of (14.9) in terms of δP to obtain:

$$-\omega^2\delta\mathbf{r} + \nabla\left(\frac{P'}{\rho} + \Phi'\right) + \frac{\delta P}{\rho}\frac{\nabla\rho}{\rho} - \frac{(\delta\mathbf{r}\cdot\nabla P)}{\rho}\frac{\nabla\rho}{\rho}$$
$$-\frac{\nabla P}{\rho}\frac{\rho'}{\rho} = 0. \tag{14.10}$$

Since, in a spherical star $\nabla p \parallel \nabla\rho$, the two last terms with use of (14.5) can be unified in a single one:

$$\frac{(\delta\mathbf{r}\cdot\nabla P)}{\rho}\frac{\nabla\rho}{\rho} + \frac{\nabla P}{\rho}\frac{\rho'}{\rho} = \frac{\delta\mathbf{r}\cdot\nabla\rho}{\rho}\cdot\frac{\nabla P}{\rho} + \frac{\rho'}{\rho}\frac{\nabla P}{\rho}$$
$$= \frac{\delta\rho}{\rho}\cdot\frac{\nabla P}{\rho}. \tag{14.11}$$

Substituting (14.7) into the third term of (14.10) and using (14.11) gives:

$$-\omega^2\delta\mathbf{r} + \nabla\left(\frac{P'}{\rho} + \Phi'\right) + \gamma_1\,\frac{P}{\rho}\frac{\nabla\rho}{\rho}\frac{\delta\rho}{\rho} - \frac{\nabla P}{\rho}\frac{\delta\rho}{\rho}$$
$$+T\gamma_3\,\frac{\nabla\rho}{\rho}\,\Delta S = 0. \tag{14.12}$$

Introducing the quantity:

$$A = \frac{1}{\rho}\frac{\mathrm{d}æ}{\mathrm{d}r} - \frac{1}{\gamma_1 P}\frac{\mathrm{d}P}{\mathrm{d}r} \tag{14.13}$$

and expressing $\delta\rho$ in terms of $\delta\mathbf{r}$ from (14.5), we obtain from (14.12):

$$\omega^2\delta\mathbf{r} - \nabla\left(\Phi' + \frac{P'}{\rho}\right) + \gamma_1\frac{P}{\rho}A\,(\nabla\cdot\delta\mathbf{r})\,\frac{\mathbf{r}}{r} = T\gamma_3\frac{\nabla\rho}{\rho}\Delta S. \tag{14.14}$$

In a thermally non-equilibrium star, changes in ΔS over the pulsational period may be important. The energy equation (9.94) gives:

$$-\mathrm{i}\omega T\Delta S = \epsilon - \frac{1}{4\pi\rho r^2}\frac{\mathrm{d}L}{\mathrm{d}r}\,. \tag{14.15}$$

The sign of A from (14.13) is connected with convective stability: in a stable case $A < 0$, in an unstable one $A > 0$. In an isentropic star, neutral with respect to convection, $A = 0$ (see (9.4)). The potential perturbation obeys Poisson's equation (14.7):

$$\nabla^2\Phi' = 4\pi G\rho'\,. \tag{14.16}$$

We consider adiabatic oscillations with $\Delta S = 0$ and have in a spherical coordinate system (r, θ, φ):

$$\delta\mathbf{r} = (\delta r,\ r\sin\theta\delta\varphi,\ r\delta\theta), \tag{14.17}$$

$$\nabla^2\Phi' = \frac{1}{r^2}\frac{\partial}{\partial r}\left(r^2\frac{\partial\Phi'}{\partial r}\right) + \frac{1}{r^2\sin^2\theta}\frac{\partial^2\Phi'}{\partial\varphi^2} + \frac{1}{r^2\sin\theta}\frac{\partial}{\partial\theta}\left(\sin\theta\frac{\partial\Phi'}{\partial\theta}\right), \tag{14.18}$$

$$\nabla\cdot\delta\mathbf{r} = \frac{1}{r^2}\frac{\partial}{\partial r}(r^2\delta r) + \frac{\partial\delta\varphi}{\partial\varphi} + \frac{1}{\sin\theta}\frac{\partial}{\partial\theta}(\sin\theta\,\delta\theta). \tag{14.19}$$

Writing (14.14) in spherical components:

$$\omega^2\delta r - \frac{\partial}{\partial r}\left(\Phi' + \frac{P'}{\rho}\right) + \gamma_1\frac{P}{\rho}A\,(\nabla\cdot\delta\mathbf{r}) = 0, \tag{14.20}$$

$$\omega^2 r\,\delta\theta - \frac{1}{r}\frac{\partial}{\partial\theta}\left(\Phi' + \frac{P'}{\rho}\right) = 0, \tag{14.21}$$

$$\omega^2 r\sin\theta\,\delta\varphi - \frac{1}{r\sin\theta}\frac{\partial}{\partial\varphi}\left(\Phi' + \frac{P'}{\rho}\right) = 0. \tag{14.22}$$

Substituting (14.21) and (14.22) into (14.19) gives:

$$\nabla\cdot\delta\mathbf{r} = \frac{1}{r^2}\frac{\partial}{\partial r}\left(r^2\delta r\right) + \frac{1}{r^2\omega^2}\left\{\frac{1}{\sin^2\theta}\frac{\partial^2\left(\Phi' + \frac{P'}{\rho}\right)}{\partial\varphi^2}\right.$$
$$\left. = +\frac{1}{\sin\theta}\frac{\partial}{\partial\theta}\left[\sin\theta\frac{\partial}{\partial\theta}\left(\Phi' + \frac{P'}{\rho}\right)\right]\right\}. \qquad (14.23)$$

We look for the perturbation eigenfunctions in the form of expansion in spherical harmonics:

$$f = f_{lm}(r)Y_{lm} = f_{lm}P_l^m(\cos\theta)\mathrm{e}^{im\varphi}, \quad l = 0, 1, \ldots,$$
$$-l \le m \le l, \quad f = \delta\mathbf{r}, \rho', P', \Phi'. \qquad (14.24)$$

Here, $P_i^m(\cos\theta)$ are the associated Legendre functions, and spherical harmonics Y_{lm} satisfy the equation:

$$\frac{1}{\sin\theta}\frac{\partial}{\partial\theta}\left(\sin\theta\frac{\partial Y_{lm}}{\partial\theta}\right) + \frac{1}{\sin^2\theta}\frac{\partial^2 Y_{lm}}{\partial\varphi^2} + l(l+1)Y_{lm} = 0. \qquad (14.25)$$

Substituting (14.24) into (14.55), (14.16), and (14.20), using (14.18) (14.23) and (14.25), we obtain for adiabatic oscillations the system of equations for radial dependences of eigenfunctions:

$$\frac{\rho'_{lm}}{\rho} + \frac{1}{\rho}\frac{\mathrm{d}\rho}{\mathrm{d}r}\delta r_{lm} + \frac{1}{r^2}\frac{\mathrm{d}}{\mathrm{d}r}\left(r^2\delta r_{lm}\right)$$
$$- \frac{l(l+1)}{r^2\omega^2}\left(\Phi'_{lm} + \frac{P'_{lm}}{\rho}\right) = 0, \qquad (14.26)$$

$$\omega^2\delta r_{lm} - \frac{\mathrm{d}}{\mathrm{d}r}\left(\Phi'_{lm} + \frac{P'_{lm}}{\rho}\right) + \gamma_1\frac{P}{\rho}A\left[\frac{1}{r^2}\frac{\mathrm{d}}{\mathrm{d}r}\left(r^2\delta r_{lm}\right)\right.$$
$$\left. - \frac{l(l+1)}{r^2\omega^2}\left(\Phi'_{lm} + \frac{P'_{lm}}{\rho}\right)\right] = 0, \qquad (14.27)$$

$$\frac{1}{r^2}\frac{\mathrm{d}}{\mathrm{d}r}\left(r^2\frac{\mathrm{d}\Phi'_{lm}}{\mathrm{d}r}\right) - \frac{l(l+1)}{r^2}\Phi'_{lm} = 4\pi G\rho'_{lm}. \qquad (14.28)$$

Using (14.3), we obtain from (14.7) with $\Delta S = 0$:

$$P'_{lm} + \delta r_{lm}\frac{\mathrm{d}P}{\mathrm{d}r} = \gamma_1\frac{P}{\rho}\left(\rho'_{lm} + \delta r_{lm}\frac{\mathrm{d}\rho}{\mathrm{d}r}\right)$$

and hence, using (14.13) gives:

$$\frac{P'_{lm}}{P} = \gamma_1 \frac{\rho'_{lm}}{\rho} + \gamma_1 \delta r_{lm} A\,. \tag{14.29}$$

The displacement components $\delta\theta_{lm}$ and $\delta\varphi_{lm}$ are obtained by substituting (14.24) into (14.21) and (14.22). We have:

$$\omega^2 r^2 \delta\theta_{lm} = \left(\Phi'_{lm} + \frac{P'_{lm}}{\rho}\right) \frac{\mathrm{d}P_l^m}{\mathrm{d}\theta} \mathrm{e}^{im\varphi}, \tag{14.30}$$

$$\omega^2 r^2 \sin^2\theta\, \delta\varphi_{lm} = \left(\Phi'_{lm} + \frac{P'_{lm}}{\rho}\right) imP_l^m \,\mathrm{e}^{im\varphi}. \tag{14.31}$$

In dimensionless variables [339]:

$$y_1 = \frac{\delta r}{r}, \quad y_2 = \frac{1}{gr}\left(\Phi' + \frac{P'}{\rho}\right), \quad y_3 = \frac{1}{gr}\Phi', \quad y_4 = \frac{1}{g}\frac{\mathrm{d}\Phi'}{\mathrm{d}r}, \tag{14.32}$$

where $g = Gm/r^2$ is the local gravity acceleration, (14.26–14.29) become (the subscripts lm are omitted):

$$r\frac{\mathrm{d}y_1}{\mathrm{d}r} = \left(\frac{g\rho r}{\gamma_1 P} - 3\right) y_1 + \left[g\frac{l(l+1)}{r\omega^2} - \frac{g\rho r}{\gamma_1 P}\right] y_2 + \frac{g\rho r}{\gamma_1 P} y_3, \tag{14.33}$$

$$r\frac{\mathrm{d}y_2}{\mathrm{d}r} = \left(\frac{r\omega^2}{g} + rA\right) y_1 + \left(1 - \frac{r}{m}\frac{\mathrm{d}m}{\mathrm{d}r} - rA\right) y_2 + rAy_3, \tag{14.34}$$

$$r\frac{\mathrm{d}y_3}{\mathrm{d}r} = \left(1 - \frac{r}{m}\frac{\mathrm{d}m}{\mathrm{d}r}\right) y_3 + y_4, \tag{14.35}$$

$$\begin{aligned} r\frac{\mathrm{d}y_4}{\mathrm{d}r} &= -\frac{r}{m}\frac{\mathrm{d}m}{\mathrm{d}r} y_1 + \frac{r}{m}\frac{\mathrm{d}m}{\mathrm{d}r}\frac{g\rho r}{\gamma_1 P} y_2 \\ &+ \left[l(l+1) - \frac{r}{m}\frac{\mathrm{d}m}{\mathrm{d}r}\frac{g\rho r}{\gamma_1 P}\right] y_3 - \frac{r}{m}\frac{\mathrm{d}m}{\mathrm{d}r} y_4. \end{aligned} \tag{14.36}$$

The dimensionless equations (14.33–14.36) are suitable for numerical integration.

14.1.2 Boundary Conditions

Boundary conditions for the equations of small non-radial oscillations follow from the requirement that the eigenfunctions be limited in the centre and at the stellar surface. The requirement of finiteness for solutions in the centre yields the following expansions:

$$\delta r = r^{l-1} \sum_{\nu=0}^{\infty} U_\nu r^\nu, \quad l \geq 1, \tag{14.37}$$

$$P'/\rho = r^l \sum_{\nu=0}^{\infty} Y_\nu r^\nu, \tag{14.38}$$

$$\Phi' = r^l \sum_{\nu=0}^{\infty} \varphi_\nu r^\nu, \tag{14.39}$$

where

$$\omega^2 U_0 = l(Y_0 + \varphi_0) \tag{14.40}$$

and two zero-order coefficients remain free. Recurrence relations for other coefficients are given by substituting the expansions into equations and equating coefficients at equal powers of r [994]. Only coefficients at even powers of r then remain non-zero. Radial adiabatic oscillations have been treated in Sect. 12.2.3. If, in an equilibrium solution, $P/\rho \to 0$ at the surface, then the outer boundary conditions take the form [317]:

$$\frac{\delta P}{P} = \left[l(l+1)\frac{Gm}{\omega^2 r^3} - \frac{\omega^2 r^3}{Gm} - 4 \right] \frac{\delta r}{r}$$
$$+ \left[l(l+1)\frac{Gm}{\omega^2 r^3} - l - 1 \right] \frac{\Phi'}{gr}, \tag{14.41}$$

$$\frac{\mathrm{d}\Phi'}{\mathrm{d}r} + (l+1)\frac{\Phi'}{r} = -4\pi G\rho\,\delta r. \tag{14.42}$$

One of the two free coefficients (14.40) is arbitrary by virtue of the arbitrariness of the normalization condition; therefore, for the boundary conditions (14.41) and (14.42) to be satisfied, the value of ω^2 must be equal to an eigenvalue.

In the case of an isothermal atmosphere with temperature T and sound speed $v_s = \sqrt{\mathcal{R}T}$, there is a boundary frequency [438]:

$$\omega_{c0} \simeq \frac{v_s}{2H}, \quad H^{-1} = -\frac{\mathrm{d}\ln\rho}{\mathrm{d}r} = \frac{g}{\mathcal{R}T}, \quad g = \frac{GM}{R^2}, \quad H \ll R, \tag{14.43}$$

where R is the radius at the base of the isothermal atmosphere, and $\gamma \equiv \gamma_1$ from (8.26). The waves with $\omega > \omega_{c0}$ penetrate into the atmosphere and low-frequency long waves with $\omega < \omega_{c0}$ reflect from it, making a disturbance, exponentially damping with distance.

For long radial waves, the following boundary relation, following from the continuity of $\delta r/r$ and δP, is valid [438] at $r = R$:

$$\frac{\mathrm{d}}{\mathrm{d}r}\left(\frac{\delta r}{r}\right) + k\frac{\delta r}{r} = 0, \tag{14.44}$$

where:

$$k = \frac{1}{2H}\left[1 - \left(1 - \frac{4\omega^2 H^2}{v_s^2}\right)^{1/2}\right]. \tag{14.45}$$

The solution which decreases exponentially to infinity (casual) is chosen here. This condition is used instead of (12.2.31) for a pure polytropic star. The same boundary condition is valid for non-radial oscillations with

$$k = \tilde{k} = \frac{1}{2H}\left\{1 - \left[1 + 4l(l+1)\frac{H^2}{R^2}\left(1 - \frac{N^2}{\omega^2}\right) - \frac{4\omega^2 H^2}{v_s^2}\right]^{1/2}\right\}, \tag{14.46}$$

where, with account of (14.13) $N = g/A$. The potential perturbation Φ' may be approximately taken as zero on the boundary or found using (14.42).

For a sufficiently hot atmosphere with a temperature much higher than on the boundary of the core (like a solar corona), there is an interval of frequencies:

$$\frac{g}{2\sqrt{\mathscr{R}T}} = \omega_{c0} < \omega < \omega_{c1} = \frac{g}{2\sqrt{\mathscr{R}T_1}}, \quad T_1 \ll T, \tag{14.47}$$

where the waves partly penetrate into the hot atmosphere and partly reflect from it. Here, ω_{c1} corresponds to the boundary frequency for the waves, propagating through the isothermal atmosphere with temperature T_1. For the waves from the interval (14.47) with sufficiently small penetration, it is possible to consider approximately the core eigen-oscillations. The boundary condition must be chosen in such a way that in the outer atmosphere with temperature T only the outgoing wave is present [846].

The eigenvalues of the system (14.26–14.29) with the boundary conditions (14.41) and (14.42) are real by virtue of the Hermiticity of the corresponding operator [994].

14.1.3 p-, g- and f-Modes

Solving equations for perturbations and finding eigenvalues and eigenfunctions in simple models have allowed us to study analytically basic properties of stellar oscillations and to establish their classification. In a homogeneous model of a compressive gas, for $n \geq 1$, where n is the number of radial nodes of an eigenfunction, two sets of eigenfrequencies will be written as [317]:

$$\omega_{nl}^2 = \frac{GM}{R^3}\left[2\gamma_1 n^2 + \frac{l(l+1)}{2\gamma_1 n^2}\right] \qquad (p\text{-modes}),$$

$$\omega_{nl}^2 = -\frac{GM}{R^3}\frac{l(l+1)}{2\gamma_1 n^2} \qquad (g\text{-modes}). \tag{14.48}$$

Here, R is the radius of the homogeneous star[1] related to the central pressure P_c by:

$$P_c = \frac{2}{3}\pi G\rho^2 R^2, \quad P = P_c\left(1 - r^2/R^2\right). \tag{14.49}$$

Oscillatory p-modes arise from the counteraction of pressure against inertia and gravity. In the limit of large n they reduce to standing acoustic waves, while for small n and $l = 0$ they represent large-scale radial oscillations of the star. Their frequency

$$\left(\omega_{nl}^2\right)_{p\text{-mode}} \to \infty \quad \text{as} \quad n \to \infty \tag{14.50}$$

just as for acoustic modes ($\omega = 2\pi c_s/\lambda \to \infty$ as $\lambda \to 0$, c_s is the sound velocity). The eigenfunctions of displacements $\delta\mathbf{r}$ for p-modes grow slowly from zero at the centre, oscillating at $n \gg 1$, and rise sharply near the surface, reaching here their maximum values.

In a homogeneous model squares of g-mode, eigenvalues are negative, and hence, unstable. This is due to the convective instability of a homogeneous star, while g-modes themselves represent convective motions in stars. They owe their existence to the counteraction of gravity against buoyancy in matter in the presence of non-uniformities in density. The g-mode frequencies tend to zero as $n \to \infty$, whereas the eigenfunctions of displacements $\delta\mathbf{r}$ grow rapidly from zero at the centre and then fall slowly (oscillating at $n \gg 1$) to a small but finite value at the surface [630]. With increasing l and n, the maximum of the g-mode amplitude moves increasingly closer to the centre, while the maximum of the p-mode moves to the surface. In an adiabatic star, neutral to convection, the g-mode eigenfrequencies become zero (see also Sect. 12.3).

A particular oscillation class is represented by the f-mode. It is the only oscillation type remaining in a figure of a incompressible fluid. The f-mode arises from the tendency of gravity to restore, in the absence of rotation, the spherical form of the figure at any possible perturbation. This mode is absent at $l = 0$ and exists only from $l = 2$ for an incompressible figure and from $l = 1$ for a compressible one. The f-mode eigenfunction grows smoothly from the centre to the surface with no sharp maxima.

Radial g-modes do not exist either since purely radial convective motions are impossible. All radial stellar oscillations are related to p-modes. The value of n for an f-mode in the homogeneous model is equal to zero, and in models of general form they are unity. The f-mode eigenvalues are intermediate between g- and p-modes. The dependence of eigenvalues for various oscillatory modes on n and l is given schematically in Fig. 14.1 from [317].

[1] The equilibrium value of ∇P is achieved owing to a rapid drop in temperature.

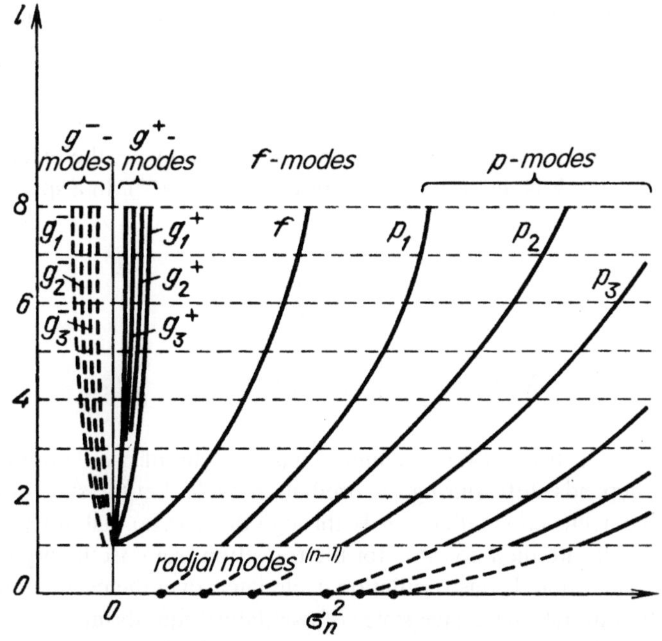

Fig. 14.1 Eigenvalues $\sigma_n^2 = (2\pi/\text{period})^2$ of linear adiabatic non-radial oscillations for various n, the number of nodes along radius, versus l, the number of a spherical harmonic (schematically). Shown are four types of spheroidal modes for non-radial oscillations (p, f, g^+, g^-), in accordance with Cowling's classification. The *dots* on the horizontal axes mark radial oscillations, p_n-mode numbers are equal to $n-1$. Only g^+-modes exist in convectively stable stars with $A < 0$, only g^--modes in convectively unstable stars (see (14.13)), the simultaneous existence in a star g^+– and g^--modes is possible only when somewhere in the star $A < 0$, elsewhere $A > 0$, from [317]

14.1.4 Pulsational Instability

Because of the presence of heat release and heat losses, the pulsation amplitude, constant in the adiabatic case, may increase or decrease according to the resultant effect of heat processes. Consider the conditions for pulsation decay and excitation in the general case, without the assumption of linearity [375]. Divide the total energy of the star $\mathscr{E}$ into pulsation energy W and static energy of the star U. For any element of mass dm, we define the pulsation energy as the value of the kinetic energy at the time of passing through the oscillatory equilibrium state, while the static energy U is the sum of the thermal and gravitational energies of the star in this state. The change ΔW in the energy W over a cycle determines the decay of pulsations for $\Delta W < 0$ or their excitation for $\Delta W > 0$. From the definition, we have:

$$\Delta W = \Delta\mathscr{E} - \Delta U. \tag{14.51}$$

The change in the total energy of the star is the difference between the release and loss over a cycle, or

$$\Delta\mathscr{E} = \int \mathrm{d}m \oint \left(\epsilon - \frac{1}{\rho}\nabla\mathbf{F}\right)\mathrm{d}t. \tag{14.52}$$

The internal integral here is taken over one oscillation cycle, the external integral over the entire star, **F** is the heat flux vector (ergs cm^{-2} s^{-1}). In a spherically symmetric star

$$\mathbf{F} = \left(\frac{L}{4\pi r^2}, 0, 0\right),$$

$$\frac{1}{\rho}\nabla\cdot\mathbf{F} = \frac{1}{\rho r^2}\frac{\mathrm{d}}{\mathrm{d}r}\left(r^2 F_r\right) = \frac{1}{4\pi\rho r^2}\frac{\mathrm{d}L}{\mathrm{d}r}. \tag{14.53}$$

The total energy $\mathscr{E}$ from (14.52) changes, and the state of oscillatory equilibrium on which U depends shifts during the oscillation cycle. Let the state of oscillatory equilibrium be achieved simultaneously through the star, so that any phase shift is absent. Two paths are then possible for the transition from oscillatory equilibrium state I to the same state II one oscillation cycle later: over the oscillation cycle and quasistatically through successive states of oscillatory equilibrium.

In a quasistatic process, the system possesses only the static energy U which changes by (with zero external work)

$$\Delta U = \int \mathrm{d}m \int \mathrm{d}t\, T\frac{\mathrm{d}S}{\mathrm{d}t}. \tag{14.54}$$

As the change in equilibrium location over one period can be taken to be small, the increment γ is much less than the oscillation frequency ω, the relation (14.54) will be written as:

$$\Delta U = \int \mathrm{d}m\, T_\mathrm{e}\,\Delta S, \tag{14.55}$$

where T_e is the equilibrium temperature of the static star whose variations are neglected. When the path between the same states goes through the oscillation cycle, the change in entropy is:

$$\Delta S = \oint \frac{\epsilon - (1/\rho)\nabla\cdot\mathbf{F}}{T}\mathrm{d}t, \tag{14.56}$$

and hence, from (14.55) we have:

$$\Delta U = \int \mathrm{d}m\, T_\mathrm{e}\oint \frac{\epsilon - (1/\rho)\nabla\cdot\mathbf{F}}{T}\mathrm{d}t. \tag{14.57}$$

Substituting (14.52) and (14.57) into (14.51) gives the relation:

$$\Delta W = \int \mathrm{d}m \oint \left(1 - \frac{T_\mathrm{e}}{T}\right)\left(\epsilon - \frac{1}{\rho}\nabla \cdot \mathbf{F}\right)\mathrm{d}t \tag{14.58}$$

determining the pulsation behaviour in star. The condition (14.58) was first obtained by Eddington in 1926 [317].

Regular brightness pulsations have been detected in many stars. Classical cepheids with a period of 1–50 days and RR Lyrae stars with a period of 1.5–25 h are the best known of them. Cepheids have masses 4–14 $M_\odot$, luminosities 300–26,000$L_\odot$ and radii 14–200$R_\odot$. RR Lyrae stars have lower masses, radii and luminosities. Both these types of variables are in post-main-sequence and post-core-hydrogen-burning phases. The mechanism of excitation of radial oscillations in these stars is connected to zones of incomplete helium and, to a lesser extent, hydrogen ionization where the opacity increases with increasing temperature [1071]. Non-radial P-modes of oscillations with periods close to 5 min have been observed on the Sun [260].

14.2 Pulsations in Stars with Phase Transition

The matter of neutron stars at densities close to nuclear may be unstable with respect to π-meson generation [695]. This phenomenon, called π-condensation, results in a van der Waals-like $P(\rho)$ dependence equivalent to a phase transition. The neutronization provides another example of a phase transition. We shall treat pulsations of stars in the presence of a phase transition in the incompressible liquid approximation [190]. Consider stable stars on the rising portion of the curve $M(P_\mathrm{c})$ (see Sect. 11.2). The quantity P_c is convenient to use here as a variable instead of ρ_c for its continuity at jump-like changes of ρ_c.

14.2.1 Equations of Motion in the Presence of a Phase Transition

Let the matter be incompressible everywhere, with the exception of the phase transition, i.e.,

$$\rho = \rho_1 \quad \text{at} \quad P < P_0, \qquad \rho = \rho_2 \quad \text{at} \quad P > P_0. \tag{14.59}$$

From the continuity equation (7.3.2), we have

$$u = u_1(t)\left(\frac{r_2(t)}{r}\right)^2 \quad \text{at} \quad r > r_2, \qquad u = 0 \quad \text{at} \quad r < r_2. \tag{14.60}$$

Here, $r_2(t)$ is the radius of a new phase core with a higher density where matter is at rest; $u_1(t)$ is the matter velocity of the lower density matter at this radius. Substituting (14.60) into the equation of motion (7.3.3) and integrating over radius r from r_2 to R (stellar radius) gives [190, 695]:

$$\dot{u}_1 r_2^2 + 2u_1 r_2 \dot{r}_2^2 - \frac{1}{2} u_1^2 r_2^4 \frac{R^3 + R^2 r_2 + R r_2^2 + r_2^3}{R^3 r_2^3}$$
$$- \frac{P_1}{\rho_1} \frac{R r_2}{R - r_2} + \frac{4\pi G}{3} \left[(\rho_2 - \rho_1) r_2^3 + \frac{\rho_1}{2} R r_2 (R + r_2) \right] = 0. \quad (14.61)$$

To reduce (14.61) to a single function to be found, consider the physical conditions on the discontinuity boundary at $r = r_2$. From the requirement that the mass, momentum, and energy flows be continuous at the jump, similar to the conditions for the shock front [614], we have

$$u_1 = -(\lambda - 1)\dot{r}_2, \quad \lambda = \rho_2/\rho_1, \quad P_2 - P_1 = \rho_2(\lambda - 1)\dot{r}_2^2,$$
$$E_2 - E_1 = \frac{1}{2}(\lambda - 1)^2 \dot{r}_2^2 + \frac{P_1}{\rho_2}(\lambda - 1). \quad (14.62)$$

Here, P_1 and E_1 are the pressure and specific energy on the inner boundary of the old phase, P_2 and E_2 on the boundary of the new phase core. For adiabatic compression:

$$(E_2 - E_1)_{\mathrm{ad}} = \frac{P_0}{\rho_2}(\lambda - 1). \quad (14.63)$$

It is obvious that $P_1 \le P_0$, $P_2 \ge P_0$. Introducing a parameter δ such that

$$P_2 = P_0 + \frac{1+\delta}{2} \rho_2 (\lambda - 1)\dot{r}_2^2,$$

$$P_1 = P_0 - \frac{1-\delta}{2} \rho_2 (\lambda - 1)\dot{r}_2^2, \quad -1 \le \delta \le 1, \quad (14.64)$$

we have for the heat release at the phase jump during contraction

$$q = E_2 - E_1 - (E_2 - E_1)_{\mathrm{ad}}$$
$$= \frac{\delta}{2}(\lambda - 1)^2 \dot{r}_2^2, \quad 0 \le \delta \le 1, \quad (14.65)$$

and during expansion

$$q = E_1 - E_2 - (E_1 - E_2)_{\mathrm{ad}}$$
$$= -\frac{\delta}{2}(\lambda - 1)^2 \dot{r}_2^2, \quad -1 \le \delta \le 0. \quad (14.66)$$

The limitations to δ in (14.65) follow from the condition that the entropy at the jump should not be decreasing. Substituting (14.62–14.65) into (14.61) gives the equation for core radius variations

$$\ddot{r}_2 + \frac{\dot{r}_2^2}{2r_2}\left\{3 + \delta\lambda + \frac{r_2}{R}\left[\lambda\,\frac{\delta - r_2^3/R^3}{1 - r_2/R} - 1 - \frac{r_2}{R} - \frac{r_2^2}{R^2}\right]\right\}$$
$$-\frac{2\pi G\rho_1}{3r_2(1 - r_2/R)(\lambda - 1)}\left[R^2 + (2\lambda - 3)r_2^2 - 2(\lambda - 1)\,\frac{r_2^3}{R}\right]$$
$$+\frac{P_0}{\rho_1 r_2(\lambda - 1)\,(1 - (r_2/R))} = 0. \tag{14.67}$$

Using instead of P_0, equilibrium values of radius R_0 and core radius $r_{2.0}$ for a star of mass M_0

$$M_0 = \frac{4\pi}{3}\left[\rho_1 R_0^3 + (\rho_2 - \rho_1)r_{2.0}^3\right] = \frac{4\pi}{3}\left[\rho_1 R^3 + (\rho_2 - \rho_1)r_2^3\right], \tag{14.68}$$

$$\frac{P_0}{\rho_1} = \frac{2\pi G\rho_1}{3}\left[R_0^2 + (2\lambda - 3)r_{2.0}^2 - 2(\lambda - 1)\,\frac{r_{2.0}^3}{R_0}\right], \tag{14.69}$$

it will be convenient to rewrite (14.67) in the form:

$$\ddot{r}_2 + \frac{\dot{r}_2^2}{2r_2}\left\{3 + \delta\lambda + \frac{r_2}{R}\left[\lambda\,\frac{\delta - r_2^3/R^3}{1 - r_2/R} - 1 - \frac{r_2}{R} - \frac{r_2^2}{R^2}\right]\right\}$$
$$+\frac{2\pi G\rho_1}{3r_2\left(1 - \frac{r_2}{R}\right)}\left\{\frac{2\lambda - 3}{\lambda - 1}\left(r_{2.0}^2 - r_2^2\right) - \frac{r_{2.0}^3 - r_2^3}{R\left(R^2 + RR_0 + R_0^2\right)}\right.$$
$$\left.\times\left[3R^2 + 3RR_0 + 2R_0^2 + 2(\lambda - 1)\,\frac{r_{2.0}^3}{R_0}\right]\right\} = 0. \tag{14.70}$$

Equation (14.70) is valid for pulsations of arbitrary amplitude. For stellar pulsations with small amplitude $|r_2 - r_{2.0}| \ll r_2$, $|R - R_0| \ll r_2$, neglecting square-amplitude terms in (14.70), we obtain the relation:

$$\ddot{r}_2 + \frac{4\pi G\rho_1(r_{2.0} - r_2)}{3\left(1 - \frac{r_{2.0}}{R_0}\right)}\left[\frac{2\lambda - 3}{\lambda - 1} - 4\,\frac{r_{2.0}}{R_0} - (\lambda - 1)\,\frac{r_{2.0}^4}{R_0^4}\right] = 0 \tag{14.71}$$

that at $\omega^2 > 0$ corresponds to harmonic oscillations at a frequency [430, 896]

$$\omega^2 = -\frac{4\pi G\rho_1}{3\,(1 - (r_{2.0}/R_0))}\left[4\,\frac{r_{2.0}}{R_0} + (\lambda - 1)\,\frac{r_{2.0}^4}{R_0^4} - \frac{2\lambda - 3}{\lambda - 1}\right]. \tag{14.72}$$

Obviously, when $\lambda > 3/2$, pulsations are possible only in stars with a finite core [896]. For a star with a small core, $r_{2.0} \ll R_0$, the frequency of small pulsations is:

$$\omega_0 = \left[\frac{4\pi G\rho_1}{3}\,\frac{3-2\lambda}{\lambda-1}\right]^{1/2}. \tag{14.73}$$

If the phase transition occurs in the envelope, $r_{2.0} \approx R_0$, then:

$$\omega^2 = \frac{4\pi G\rho_1}{3}\,\frac{\lambda^2}{\lambda-1}\,\frac{1}{1-(r_{2.0}/R_0)}. \tag{14.74}$$

The non-linear pulsations of a star with a small core ($|r_2 - r_{2.0}| \sim r_{2.0}, r_2; r_{2.0} \ll R_0$) are described by the equation:

$$\ddot{r}_2 + \frac{\dot{r}_2^2}{2r_2}(3+\delta\lambda) + \frac{\omega_0^2}{2r_2}\left(r_2^2 - r_{2.0}^2\right) = 0. \tag{14.75}$$

With $\delta = 1$ and $r_{2.0} = 0$ we obtain the equation examined in [695].

14.2.2 Physical Processes at the Phase Jump

Non-linear decaying pulsations have been studied in [701], where it has been assumed that $\delta = 1$ in the contraction phase, while in the expansion phase there is also dissipation and $\delta = -1$. To substantiate the choice of δ, we shall treat the phase transition as a limiting case of an equation of state where the pressure varies from P_a to P_b with variation of ρ from ρ_1 to ρ_2. With $P_b \to P_0 \leftarrow P_a$ and constant ρ_1 and ρ_2, we thus obtain a phase transition.

In the matter of an intermediate layer, the sound velocity is$a_s \approx \left[(P_a - P_b)/(\rho_2 - \rho_1)\right]^{1/2}$. If the pulsation amplitudes are so small that the motion velocity $v < a_s$, then the oscillations will be adiabatic with $\delta = 0$ at the jump. In the limit $P_b \to P_0 \leftarrow P_a$, we have $a_s \to 0$, hence, in any case there will be $v > a_s$, and the motion in the intermediate layer will become supersonic. The encounter of a flow with supersonic velocity against a wall in the form of a new phase core will lead to the formation of a shock where the kinetic energy converts into heat. It is obvious that the "phase dissipation" of kinetic energy during the star contraction and the growth of the new phase core has the same nature as in the shock. In the limit $P_b \to P_0 \leftarrow P_a$, the value of δ in the condition for jump (14.64) may approach unity, but while $v < a_s$ the value of $\delta = 0$.

The stage of expansion of the star is accompanied by diminishing the new phase core, so it proceeds in another way. It is possible that the supersonic velocity will not result in this case in the shock formation even at a perfectly thin phase jump, since the envelope encounters no obstacle in its motion outward. The pressure on the core boundary exceeds P_0 because of the reaction to the envelope expansion. The expansion phase may thus always remain adiabatic with $\delta = 0$. The pulsation decay in a star with a phase transition occurs in this case only on the stage of star contraction, while in the case of non-ideal, slightly broadened jump there is a possibility of strictly adiabatic oscillations of small but finite amplitude.

14.2.3 Adiabatic Oscillations of Finite Amplitude

The absence of decay at $\delta = 0$ leads to the conservation of the total energy of a pulsating star corresponding to the first integral of (14.67)

$$E = 2\pi\rho_1(\lambda-1)^2\dot{r}_2^2 r_2^3\left(1-\frac{r_2}{R}\right) + \frac{4\pi}{3} r_2^3 P_0(\lambda-1)$$
$$-\frac{16\pi^2 G\rho_1^2 R^5}{15}\left[1+\frac{5}{2}(\lambda-1)\frac{r_2^3}{R^3}+(\lambda-1)\left(\lambda-\frac{3}{2}\right)\frac{r_2^5}{R^5}\right]. \quad (14.76)$$

From (14.76), using (14.68), we find the solution and a relation for the oscillation period in the non-linear case. To avoid cumbersome expressions, we consider non-linear oscillations of a star with a small core, $r_2/R \ll 1$, for which (14.76) combined with (14.69) and (14.73) becomes

$$\dot{x}^2 = \omega_0^2\left(\frac{x^{-3}-x^2}{5} - x_0^2\,\frac{x^{-3}-1}{3}\right). \quad (14.77)$$

Here, $x = r_2/r_{2,\min}$, $x_0 = r_{2.0}/r_{2,\min}$, the initial condition $r_2 = r_{2,\min}$ with $\dot{r}_2 = 0$ is taken into account; $r_{2,\min}$ is the minimum core radius. The non-linear adiabatic oscillation period for the case of a small core is

$$T_{\rm ad} = \frac{2}{\omega_0}\int_0^{x_*}\left[\frac{x^{-3}-x^2}{5} - x_0^2\,\frac{x^{-3}-1}{3}\right]^{-1/2} dx. \quad (14.78)$$

Here, $x_* > x_0 > 1$ corresponds to the maximum core radius $x_* = r_{2,\max}/r_{2,\min}$ and is the root of the denominator in (14.78). The dependence of $T_{\rm ad}\omega_0/2\pi$ on x_0 is given in Fig. 14.2. Oscillations of small amplitude $x = 1+\alpha$, $x_0 = 1+\Delta$, $\alpha, \Delta \ll 1$ are harmonic and (14.78) gives

$$T_0 = \frac{2}{\omega_0}\int_0^{\alpha_{\max}=2\Delta}\left(2\Delta\alpha - \alpha^2\right)^{-1/2} d\alpha = 2\pi/\omega_0,$$

in accordance with (14.73).

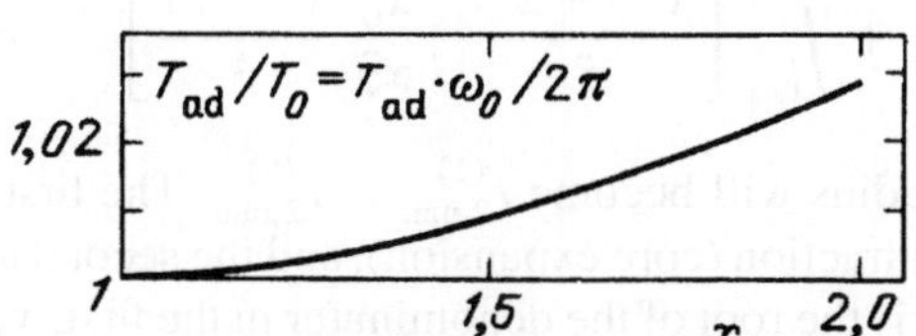

Fig. 14.2 Non-linear adiabatic oscillation period for a star with phase transition $T_{\rm ad}$ versus amplitude for the case of a small core. The abscissa is $x_0 = r_{2,0}/r_{2,\min}$, the ratio of the equilibrium core radius $r_{2,0}$ to the minimum one $r_{2,\min}$; ordinate is $T_{\rm ad}/T_0 = T_{\rm ad}\omega_0/2\pi$, where T_0 and ω_0 are the small harmonic oscillation period and frequency

We now find an estimate for the maximum value of the amplitude of adiabatic oscillations arising from the finite pressure of electrons during pionization. For the two cases examined in [701], we have from (14.72):

$$(1)\quad \lambda = 5, \quad r_2/R = 0.518, \quad \omega = 0.648\,\omega_{00};$$

$$(2)\quad \lambda = 1.2, \quad r_2/R = 0.295, \quad \omega = 1.4\,\omega_{00}, \quad \omega_{00} = \sqrt{4\pi G\rho_1}\,. \tag{14.79}$$

The maximum velocity of the envelope matter motion relative to the jump is $v = \lambda\dot{r}_2 = \lambda\omega r_2\eta$, where η is the relative amplitude of oscillations of the core radius. The sound velocity in the region of the phase transition is $a_s \sim (P_e/\rho)^{1/2}$, where P_e is the electron pressure. For $\rho = 2\times10^{14}$ g cm^{-3}, $P_e = 10^{30}$ dyn, $R = 10$ km, the condition $v < a_s$ gives $\eta \leq 0.003$ and $\eta \leq 0.01$, respectively.

14.2.4 Decaying Finite-Amplitude Oscillations

At sufficiently large oscillation amplitude, or in the case of a perfect phase transition at $P = P_0$, i.e., $a_s = 0$, oscillations decay in the contraction phase and $\delta > 0$. The decay is absent in the expansion phase (when the core diminishes), so we put there $\delta = 0$. The integral of (14.75), similar to (14.77), with $\delta = \text{const.} \neq 0$ becomes:

$$\dot{x}^2 = \omega_0^2\left[\frac{x^{-3-\delta\lambda} - x^2}{5+\delta\lambda} - x_0^2\,\frac{x^{-3-\delta\lambda} - 1}{3+\delta\lambda}\right]. \tag{14.80}$$

If, in the initial state, the minimum core radius is $r_{2,\min}^{(0)}$ and $\dot{r}_2 = 0$, then over a time interval

$$T = \frac{1}{\omega_0}\left\{\int_1^{x_*}\left[\frac{x^{-3-\delta\lambda} - x^2}{5+\delta\lambda} - x_0^2\,\frac{x^{-3-\delta\lambda} - 1}{3+\delta\lambda}\right]^{-1/2}\mathrm{d}x \right.$$
$$\left. + \int_{x_{**}}^{1}\left[\frac{x^{-3} - x^2}{5} - \frac{x_0^2}{x_*^2}\,\frac{x^{-3} - 1}{3}\right]^{-1/2}\mathrm{d}x\right\} \tag{14.81}$$

the minimum core radius will become $r_{2,\min}^{(1)} > r_{2,\min}^{(0)}$. The first integral here corresponds to the star contraction (core expansion), and the second to the star expansion. The quantity $x_* > 1$ is the root of the denominator in the first, $x_{**} < 1$ in the second integral in (14.81). Here:

$$x_0 = r_{2.0}/r_{2,\min}^{(0)}, \quad x_* = r_{2,\max}/r_{2,\min}^{(0)}, \quad x_{**} = r_{2,\min}^{(1)}/r_{2,\max}\,.$$

Hence,

$$r^{(1)}_{2,\min} = r^{(0)}_{2,\min} x_* x_{**}, \quad \text{where} \quad x_* x_{**} > 1$$

by reason of the oscillation decay. We make an expansion in (14.81), setting in the first integral $x = 1 + \alpha$, $x_0 = 1 + \Delta$ in the second $x = 1 - \beta$, $\alpha, \beta, \Delta \ll 1$, and retaining terms $\sim\alpha^3, \beta^3, \Delta^3$ to include the decay in the first non-vanishing term. After simple manipulations, we find

$$x_* = 1 + 2\Delta - \frac{5}{3}\Delta^2 - \frac{2}{3}\delta\Delta^2\lambda,$$

$$x_{**} = 1 - 2\Delta + \frac{17}{3}\Delta^2 + \frac{4}{3}\delta\Delta^2\lambda,$$

$$\frac{r^{(1)}_{2,\min}}{r^{(0)}_{2,\min}} = 1 + \frac{2}{3}\delta\Delta^2\lambda. \tag{14.82}$$

Using the equality $\Delta = (r_{2.0}/r_{2,\min}) - 1$ and the third relation (14.82), taking also the period to be equal to $2\pi/\omega_0$, we obtain the equation for $r_{2,\min}$ variations

$$\frac{\mathrm{d}\Delta}{\mathrm{d}t} = -\frac{\delta\omega_0\lambda}{3\pi}\Delta^2. \tag{14.83}$$

When the initial condition is $\Delta = \Delta_0$, for $t = 0$ we have

$$\frac{\Delta_0 - \Delta}{\Delta\Delta_0} = \frac{\delta\omega_0\lambda}{3\pi}t. \tag{14.84}$$

For a perfect phase transition, we adopt δ = const. > 0 (or perhaps $\delta = 1$, as in [695, 701]). When the transition is broadened, δ smoothly becomes zero near the onset and near the end of the contraction phase, as the ratio v/a_s varies with motion. The qualitative dependence $\delta(t)$ along the oscillation period is shown in Fig. 14.3. In real objects phase transformation rates are finite, and the mechanism of second viscosity is always at work. Quantitative estimates for this mechanism should be made for each type of phase transition separately.

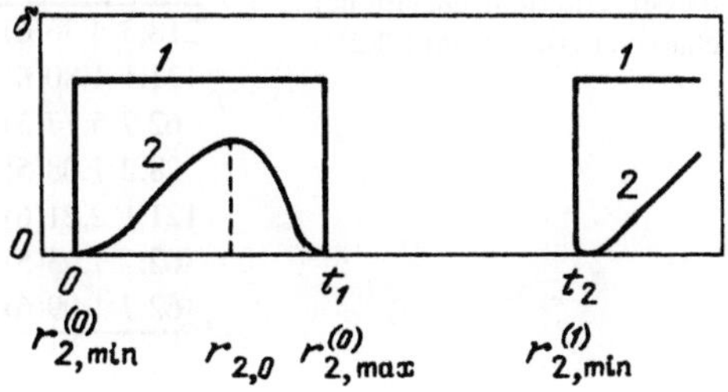

Fig. 14.3 Variations of parameter δ during pulsation (see the relations (14.64), (14.65)), qualitatively: *1* for a perfect phase transition, *2* for a broadened phase transition

14.3 Pulsational Stability of Massive Stars

The upper limit for stellar masses in the Galaxy is not known exactly today. Following the investigation by Ledoux in 1941, several other studies have been carried out, dealing with stability of massive stars to the growth of pulsations in the radial fundamental mode [29, 30, 792, 793, 882, 959, 960, 1084] related to attempts to find theoretically the maximum stellar mass. As the mass increases, the role of radiative pressure becomes more important, and the mean adiabatic index γ_a approaches $4/3$. Since an adiabatic star with $\gamma_1 = 4/3$ is neutral with respect to contraction or expansion (see Sect. 10.1), the effect of destabilizing factors increases with increasing mass because of the reduction of an available pulsational stability resource.

The study of this problem to the linear approximation [882] has shown that in stars of $M > 65M_{\odot}$, the instability to pulsation growth develops quickly and the star may disintegrate. Subsequent calculations have revealed a strong stabilization of these pulsations in the non-linear regime. Stability analyses with new opacity tables have been performed in [421, 422, 936, 940].

14.3.1 The Linear Analysis

Stars with masses 28.2, 62.7, 121.1, 218.3 $M_{\odot}$ and initial chemical composition $x_{\mathrm{H}} = 0.75$, $x_{\mathrm{He}} = 0.22$ have been considered in [882]. Equilibrium models have been calculated by the Schwarzschild method (Sect. 6.1.1, Vol. 1). The radial pulsation period has been calculated by solving equation (12.2.26) with the boundary conditions (12.2.29) and (12.2.31). The opacity was determined by electron scattering (9.2.23a), and the equation of state was defined (11.123) by the sum of the pressure of completely ionized gas and radiation, with γ_1 given in (8.26). Equilibrium model characteristics: the mass M, luminosity L, effective temperature T_{ef}, ages τ, radial pulsation periods P calculated in the linear approximation for the fundamental mode are given in Table 14.1 from [882].

Table 14.1 Equilibrium models and their pulsational characteristics (from [882])

$\frac{M}{M_{\odot}}$	$\frac{L}{L_{\odot}}$	$\frac{R}{R_{\odot}}$	T_{ef}, K	τ, 10^6, yr	P, days	$1/K$, yr
218.3	4.36(6)	20.5	5.82(4)	0	0.546	930
121.1	1.80(6)	14.3	5.58(4)	0	0.383	1800
62.7	5.77(5)	9.71	5.09(4)	0	0.260	44000
28.2	1.08(5)	6.00	4.27(4)	0	0.157	−1400
121.1	2.21(6)	18.75	5.13(4)	1.39	0.502	−2200
62.7	7.93(5)	13.3	4.72(4)	2.08	0.356	−94
62.7	1.09(6)	27.6	3.55(4)	3.68	0.833	−3

Since the characteristic lifetimes of convective elements on the convective core boundary by far exceed the pulsation period P, the convection effect on stability was ignored. The mixing length was assumed to be twice the pressure scale height, $\alpha_{\mathrm{P}} = 2$ in (8.6).

The pulsational stability was tested by a method similar to Sect. 14.1.4 with the inclusion of changes in luminosity, heat release in the core and acoustic wave generation. The following quantity was calculated:

$$K = \frac{1}{2}\frac{L_p}{E_p}, \quad L_p = L_{p\mathrm{N}} - L_{p\mathrm{H}} - L_{ps}, \tag{14.85}$$

where

$$E_p = \left(\frac{2\pi}{P}\right)^2 \int_0^M (\delta r)^2 \,\mathrm{d}m \tag{14.86}$$

is the kinetic energy of the pulsations,

$$L_{p\mathrm{N}} = \int_0^M \delta\epsilon \frac{\delta T}{T} \,\mathrm{d}m \tag{14.87}$$

is the rate of gain of pulsation energy due to nuclear reactions with energy release rate ϵ,

$$L_{p\mathrm{H}} = \int_0^M \frac{\mathrm{d}(\delta L)}{\mathrm{d}m}\frac{\delta T}{T} \,\mathrm{d}m \tag{14.88}$$

is the rate of loss of pulsation energy due to emission from the surface,

$$L_{ps} = 4\pi R^2 \left[\frac{1}{2}\rho_R \left(\frac{2\pi}{P}\right)^2 \left(\frac{\delta r}{r}\bigg|_R R\right)^2\right] \sqrt{\frac{kT_R}{\mu m_{\mathrm{u}}}} \tag{14.89}$$

is the rate of loss of pulsation energy due to acoustic wave generation. Here, the subscript "R" refers to the stellar surface. The term in parenthesis represents the density of acoustic oscillation energy at the photosphere. Photospheric characteristics have been approximately computed from the relations

$$\begin{aligned} T_R &= T_{\mathrm{ef}}, \quad k_R \frac{P_R}{g_R} = \frac{2}{3}, \quad k_R = 0.19\,(1 + x_H), \\ g_R &= \frac{GM}{R^2}, \quad P_R = \frac{aT_R^4}{3} + \frac{kT_{\mathrm{ef}}}{\mu m_{\mathrm{u}}}\rho_R. \end{aligned} \tag{14.90}$$

At a positive K, the star is pulsationally unstable, K representing the reciprocal of the time in which the pulsation amplitude increases by a factor of e. A negative value of K indicates the pulsational stability and $|K|$ is then the inverse time of the pulsation decay. The stability condition (14.85) in the absence of the last term L_{ps} from (14.89) follows from (14.58) if we introduce small sinusoidal oscillations for all quantities in this equation and integrate it over the pulsation period. The quantities δr, δT and δL, in (14.86–14.89) represent the oscillation amplitudes.

The results of stability studies are given in the last column in Table 14.1 from which we see that the star may restore its stability during the evolution. Stars with $M \leq 60M_{\odot} = M_{c0}$ prove to be stable on the main sequence. According to [882], this quantity depends on chemical composition so that:

$$\mu^2 \frac{M_{c0}}{M_{\odot}} = \text{const.} \approx 20, \quad \mu \text{ is a molecular weight}. \tag{14.91}$$

A special investigation of the dependence of M_{c0} on chemical composition has been performed in [1084]. The description of convection and, more importantly, of opacity was different from [882], and the values of M_{c0} turned out to be higher for the same compositions. To calculate equilibrium models, the Henyey method was applied in [1084] (see Sect. 6.1.2, Vol. 1). The values of $\mu^2 M_{c0}/M_{\odot}$ remain approximately constant (=34) for $x_{\rm He}$ varying from 0.28 to 0.4 and the same content of heavy elements $x_{\rm Z} = 0.02$. Variation of $x_{\rm Z}$ including $x_{\rm CNO}$ leads to $\mu^2 M_{c0}/M_{\odot} = 58$ for $x_{\rm He} = 0.384$, $x_{\rm Z} = 0.1$ and $x_{\rm CNO} = 0.06$. The value $M_{c0} = 127M_{\odot}$ is obtained for $x_{\rm He} = 0.184$, $x_{\rm Z} = 0.0201$ and $x_{\rm CNO} = 0.012$.

If we normalize the pulsation amplitude by $(\delta r/r)|_R = 1$, the results of the calculation [882] can be represented as the interpolation formulae:

$$\begin{aligned} \frac{L_p}{L} &= 0.10\left(\frac{M}{M_{\odot}} - 1\right) - 2\,\frac{\tau}{10^6}, \\ \frac{E_p}{L} &= 3750\ \text{yr} \cdot \exp\left[-0.007\left(\frac{M}{M_{\odot}} - 60\right) - 1.2\,\frac{\tau}{10^6}\right], \end{aligned} \tag{14.92}$$

where τ is the evolutionary lifetime in years. The quantity K computed from (14.92) is independent of the normalization condition. It follows from (14.92) that the time $\tau_{\rm cr}$ at which K becomes zero is determined as follows:

$$\tau_{\rm cr} = 0.05\left(\frac{M}{M_{\odot}} - 60\right) \times 10^6\ \text{yr} \tag{14.93}$$

the star will become pulsationally stable. The quantity $\tau_{\rm cr}$ should be compared with $1/K$ from Table 14.1. The stability has time to restore if the gain of pulsation energy is not significant over the time $\tau_{\rm cr}$. The quantity

$$N(\tau) = \int_0^{\tau} K\,\mathrm{d}\tau \tag{14.94}$$

has the sense of the number giving the rate of exponential pulsation energy increase. If we assume that the stability restores when $N(\tau_{\rm cr}) \leq 9$, the estimate for stable star mass will be $M_{\rm cr} = 65M_{\odot}$. It was shown in [882] that $N = 5.7$ for $M = 64M_{\odot}$ and $N = 13.7$ for $M = 66M_{\odot}$.

It was noted in [882] that the quantity M_{cr} obtained by this method turns out to be substantially less than the observed mass of certain stars which attains $\sim 95 M_\odot$ (e.g., ζ^1 Scorpii). To remove this discrepancy, it was suggested in [882] that in 65–95 $M_\odot$ stars, the pulsational instability development does not lead to the complete disruption of the star, but results in an intensive outflow of matter from the surface, thereby giving rise to the formation of stars having spectra of P Cyg type observed in very massive stars.[2] A suggestion is also made in [882] that with increasing pulsation amplitude a non-linear stabilization of oscillations arises, limiting their amplitude with some finite magnitude well before the disruption of the star. A star can lose its mass in a state of quasi-stationary finite-amplitude pulsation.

Investigations of linear stability have been performed for new opacity tables in [512, 513, 837]. Solution of the linearized equations for small radial perturbations was obtained in a full non-adiabatic approach, and stability was determined by the type of eigenvalues of the linearized system. The following values of the critical massed were obtained in [936,940] for main sequence stars depending on the chemical composition:

$$\begin{array}{rclllll} (\mathrm{Y},\mathrm{Z}) & = & (0.27,\ 0.03) & (0.28,\ 0.02) & (0.24,\ 0.004) & (0.24,\ 0.002) & \\ \dfrac{M}{M_\odot} & = & 148 & 121 & 89 & 84 & [936] \\ & & >150 & >150 & 133 & 97 & [940]. \end{array} \tag{14.95}$$

In a similar way, the following critical masses have been obtained for helium-burning main-sequence stars:

$$\begin{array}{rclllll} \dfrac{M}{M_\odot} & = & 20.1, & 18.6, & 14.5, & 14.0 & [936] \\ & & 21.0, & 18.3, & 14.3, & 13.0 & [940] \\ & & & & \text{for} & & \\ \mathrm{Z} & = & 0.03 & 0.02, & 0.004 & 0.002\,. & \end{array} \tag{14.96}$$

Calculations of the same problem also by solving non-adiabatic equations for linear oscillations in [421] gave substantially different results. For main-sequence stars the critical masses are:

$$\begin{array}{rcll} \dfrac{M}{M_\odot} & = & 79, & 92 \\ & & \text{for} & \\ (\mathrm{Y},\mathrm{Z}) & = & (0.27,\ 0.03), & (0.28,\ 0.02)\,. \end{array} \tag{14.97}$$

[2] The spectrum of a star belongs to the type P Cyg if it has lines with absorption in their violet portion, and emission on the red side symmetric with respect to their central frequencies. These are characteristic of stars with a powerful outflow; a typical example for this kind of star is provided by P Cygni [921].

Stability of the evolved stars was investigated in [940]. It was shown that for initial masses of 90–120$M_\odot$ with normal metallic abundances, the instability in the envelope is developed after the end of the central hydrogen burning, which may lead to the strong mass loss and mimic the most basic observed properties of η Carinae and other very luminous blue variables. It is connected with the iron bump in new opacities and its development could lead to the formation of a star with a quasistationary mass outflow, similar to ones considered in Sect. 9.2.4. Stability properties of the evolved stars have been considered in [422].

14.3.2 Non-Linear Oscillations

To study non-linear oscillations of massive stars, the equations of hydrodynamics (10.2.1) have been used in [1084] with an energy equation of the form

$$\frac{\partial E}{\partial t} + P\,\frac{\partial(1/\rho)}{\partial t} = F\left(\epsilon - \frac{\partial L}{\partial m}\right). \tag{14.98}$$

Since the pulsation period (P) is much shorter than the pulsation e-folding time ($\sim 1/K$), it is necessary to get these two time scales close to one another with the aid of some artificial techniques to make possible the numerical investigation of the amplitude growth in the non-linear regime. The factor F in (14.98) is used for this purpose and it must be approximately equal to the ratio $(PK)^{-1}$. Introducing this factor effectively reduces the specific heat of the stellar matter and, accordingly, increases the thermal instability increment. Calculations of non-linear pulsations of a star with mass 100$M_\odot$, $x_{\mathrm{He}} = 0.384$, $x_{\mathrm{Z}} = 0.20$, unstable to linear approximation, have been carried out for $F = 4.133\times 10^5$. It was shown that over some ten periods the surface pulsation amplitude $\delta r/r$ increase, from 0.1 to 4. The pulsations have been stabilized at this amplitude. During the amplitude growth and in the phase of constant-amplitude non-linear oscillations, no mass loss has been obtained. In the non-linear regime, the relative oscillation amplitude decreases rapidly as one moves inward towards the star. It is pointed out in [1084] that the inclusion of the kinetic energy dissipation due to the shock formation in the stellar atmosphere may substantially reduce the amplitude of non-linear oscillations. Using high F strongly reduces the shock-wave dissipation of energy.

To investigate non-linear oscillations, an energy equation in the form (14.98) with a large F has been applied in [959, 960] as well, but in addition to this, the dependence of the stationary amplitude on F has been examined with a more realistic inclusion of the shock-wave dissipation of pulsation energy. Calculations have been performed for 100, 1000 and 10 000$M_\odot$ models. It has been obtained for all cases that a finite-amplitude stability of a pulsational state of limit-cycle type sets in, while the amplitude falls as F decreases. This is due to the increased role of the shock-wave dissipation. For the realistic case of $F = 1$, the pulsation amplitude has been estimated in [960] using the criterion (14.58) where the parameter variations over a non-linear cycle have been approximately specified with the aid of numerical computations. The dissipation due to the shock formation in the atmosphere has been also taken into account. The amplitude of stationary pulsations at the surface for $F = 1$ is $\delta r/r|_R = 0.085, 0.102$ and 0.09 for $M = 100, 1000$ and $10000 M_\odot$, respectively.

The computational technique used in [959,960,1084] has not allowed us to compute the mass loss rate on the stage of non-linear pulsation. Instead, the upper limit has been obtained $\dot{M} < 0.03\, M_\odot/$ yr. Investigation of non-linear oscillations has been made in [29,30] by another method that allows us to include low mass losses. Instead of using the coefficient F in (14.98), the computations have been accelerated by an artificial increase of the oscillation energy at each computational time step, achieved by multiplying the velocity $v(m,t)$ by a factor slightly exceeding unity. The results have been checked by computing the true oscillation energy changes over each period with the aid of a formula similar to (14.58). Physical properties of matter have been adopted in these studies as in [882], and the values of M_{c0} turn out to be almost equal. Investigation of stellar pulsations for stars with masses from 60 to $6{,}000 M_\odot$ has shown that for $M < 100 M_\odot$ the pulsations lead to almost no mass loss. Stars with masses $M = 100 - 200 M_\odot$ do lose mass during pulsations, but this loss is not essential over their lifetime on the main sequence ($\lesssim 3 \times 10^{-5} M_\odot/\text{yr}$). If, on the contrary, the stellar mass exceeds $300 M_\odot$, an essential part of the stellar mass is lost in the hydrogen-burning phase.

In subsequent studies [792, 793], both analytical and direct numerical calculations have given the generation due to the instability of several radial oscillatory modes, which causes the limiting curve $u(t)$ to be fairly complicated. These studies give no mass loss from the surface on the stage of non-linear pulsation for stars with masses from 70 to $210 M_\odot$. It is noted in [793] that the absence of overtone excitation in previous studies is likely to result from the far too large a value of F chosen for numerical computations. There is no mass loss in [793]. The absence of mass loss, similar to [960], seems to be due to the coarse spatial grid near the stellar surface. Linear stability analysis of several radial pulsational modes has been carried out in [421,422].

14.4 On Variable Stars and Stellar Seismology

Observations show that fluxes from all stars are more or less variable; that is, all stars are variables. The discovery of the solar pulsation in the range of periods ~5 min gave rise to a new field in astrophysics–helioseismology. Thousands of solar eigenfrequencies have been observed, starting from the observational discovery of these oscillations in 1975 [322]. The frequencies have been measured to four or five decimal places [28, 1020]. They correspond to light variations of low amplitudes $\Delta m_V = 10^{-5} - 10^{-6}$ and represent high acoustic p-modes with $n = 11 - 33$ and different l varying from zero to a few hundreds. The progress in study of stellar pulsations is connected with observations from satellites. It gave a possibility to observe numerous modes of low-amplitude regular pulsations not only from the Sun, mission SOHO (launched on December 2, 1995), but also from nearby stars, by missions CoRoT (launched on December 2006), and Kepler (launched on March 2009), see i.g. publications [274, 676]. Astero- and helioseismology are extensively treated in the book [7].

For large n, l asymptotic expressions for eigenfrequencies obtained in the Cowling approximation when perturbations of the gravitational potential are neglected may be written in the form [962, 1020]:

$$\nu_{nl} = \frac{\omega_{nl}}{2\pi} \approx \left(n + \frac{l}{2} + \epsilon_p\right) \Delta\nu - \frac{l(l+1) + \delta_p}{n + (l/2) + \epsilon_p} A_p \Delta\nu, \tag{14.99}$$

where

$$\Delta\nu = \left(2 \int_0^R \frac{\mathrm{d}r}{c_s}\right)^{-1}, \quad \epsilon_p, \ \delta_p, \ A_p \tag{14.100}$$

are constants sensitive to stellar structure.

The eigenfrequencies of oscillations with the same $n + l/2$ coincide to a first order. At a fixed l, the frequencies are equidistant along n with an interval $\Delta\nu$. The departure from this degeneracy is represented by the second term in (14.99)

$$\Delta\nu_{nl} = \nu_{nl} - \nu_{n-1,l+2} \approx \frac{2(2l+3)}{n + (l/2) + \epsilon_p} A_p \Delta\nu. \tag{14.101}$$

Determining from observations, the "large" $\Delta\nu$ and "small" $\Delta\nu_{nl}$ of frequency separations provide information about the internal structure of the Sun that can be restored by solving the inverse problem [75, 585, 1020]. It occurs that for the Sun, (14.100) and (14.101) do not give large deviations from the exact frequency values, but for stars with convective cores the error is more considerable [53, 406]. More accurate asymptotic expansions, where perturbations of the gravitational potential are taken into account to higher orders, have been performed in [843–845].

The frequency splitting along an azimuthal number m is due to the solar rotation and magnetic field. Rotational splitting removes the degeneracy with respect to m, making the difference between modes with $(-m)$ and $(+m)$, which is mainly used for reconstruction of the rotation law. The effect of the magnetic field starts from terms $\sim m^{2j}$ with $j \geq 1$, so only degeneracy of modes with different m^2 is removed. There is also an input of rotational terms $\sim \Omega^{2j}$ in the m^{2j} splitting which interferes with the magnetic one, making reconstruction of the magnetic field structure even more complicated than that of the rotation law. The angular velocity distribution of solar interiors was reconstructed in different papers [355, 435, 976] and shows a large discrepancy in results. For a detailed treatment of physics and evolution of rotating stars see [666].

Apart from the Sun, cepheids and RR Lyrae stars, pulsations have been detected in white dwarfs (ZZ Ceti stars) with periods $P = 200$–$1{,}000$ s, δ Scuti variables ($P = 1$–3 h), bright blue stars β Cephei type ($P = 4$–6 h), W Virginis stars

(P = 2–45 days) adjacent to the cepheids and RV Tauri stars (P = 20–150 days) [317]. δ Scuti stars, or dwarf cepheids, are of the same spectral classes as RR Lyrae stars (A2–F3) but of luminosity lower by $2 - 3^m$ than these ($M_{V,\delta\ \mathrm{Scuti}} \approx 2 - 3^m$). The amplitudes of their pulsations are $0.01 - 0.1^m$, and the excitation mechanism is related to incomplete ionization zones, as is the case of cepheids and RR Lyrae stars. The excitation mechanism for pulsations in β Cephei blue giants of spectral classes B1–B2 near the main sequence with masses 10–20$M_\odot$ and luminosities $\sim 10^3 L_\odot$ is not quite clear [672].

Discovery of multiple eigenmodes belonging to one star reveals a possibility to restore their internal structure like that of the Sun, so that helioseismology is extended to stellar seismology. Up to now not more$\sim$10 eigenmodes are known for any particular star apart from the Sun [470], but future observations from space look very promising.

In addition to regularly pulsating stars, there are several types of stars exhibiting non-regular variability. UV Ceti red-bursting low-mass stars are the most numerous among them [921]. A powerful convection in the envelope is the source of their variability and that of more massive young T Tauri stars (see Chap. 8). Non-regular variabilities with time scales 100–700 days are detected on red giants called myrids (after Mira Ceti). Myrids lose mass at a rate $10^{-6} - 10^{-4} M_\odot/$ yr; the outflow is attributed to shocks propagating in their envelopes [368, 981]. Phenomena of non-regularity underlying the variable character of light curves are also detected on RV Tauri stars that are giants less red than the myrids and are already influenced by shock effects. Non-regular variabilities observed on stars of certain types seem to be effects of instabilities that may bring about a chaotic behaviour of the star as a dynamical system.

References

Abbreviations

AsJ	Astronomicheskii Jurnal (Soviet Astronomy)
PAJ	Pis'ma v Astronomicheskii Jurnal (Soviet Astronomy Letters)
A	Astrofizika (Astrophysics)
UFN	Uspekhi Fizicheskikh Nauk (Soviet. Phys. Uspekhi)
JETF	Jurnal Eksperimental'noi i Teoreticheskoi Fiziki (Journal of Experimental and Theoretical Physics, JETP)
NI	Nauchnyie Informatzii Astronomicheskogo Soveta AN SSSR (Scientific Information of the Astonomical Council of the Academy of Sciences of the USSR)
PMTF	Jurnal Prikladnoi Mekhaniki i Tekhnicheskoi Fiziki (Journal of Applied Mechanics and Physics of Engineering)
JVMiMF	Jurnal Vychislitel'noi Matematiki i Matematicheskoi Fiziki (Journal of Applied Mathematics and Mathematical Physics)
M	Moscow
L	Leningrad

References

*1. Abdurashitov, J.N. et al.: Measurement of the solar neutrino capture rate with Gallium metal. Astro-ph/9907113 (1999)

*2. Abdurashitov, J.N. et al.: Measurement of the solar neutrino rate by SAGE and implications for neutrino oscillations in vacuum. Astro-ph/9907131 (1999)

*3. Abramovitz, M., Stegun, I. (eds.): Handbook of Mathematical Functions. Dover, New York (1968)

4. Abramowicz, M.A., Czerny, B., Lasota, J.P., Szuszkiewicz, E.: Slim accretion disks. Astron. J. **332**, 646–658 (1988)

5. Abrikosov, A.A.: On the internal structure of hydrogen planets. Vopr. Kosmogonii **3** 12–19 (1954)

*6. Abrikosov, A.A.: Some properties of highly compressed matter. Jurnal Eksperimental'noi i Teoreticheskoi Fiziki (1960) (J. Exp. Theor. Phys., JETP) **39**, 1797–1805

* References marked with ∗ are cited in Volume 1.

G.S. Bisnovatyi-Kogan et al., *Stellar Physics – 2: Stellar Evolution and Stability*, Astronomy and Astrophysics Library, DOI 10.1007/978-3-642-14734-0,

7. Aerts, C., Christensen-Dalsgaard, J., Kurtz, D.W.: Asteroseismology. Springer, Berlin (2010)
8. Akerlof, C.W., Balsano, R., Barthelemy, S. et al.: Observation of contemporaneous optical radiation from a γ-ray burst. Nature **398**, 400–402 (1999)
*9. Akhiyezer, A.I., Berestetzkii, V.B.: Quantum Electrodynamics Moscow, Nauka (1969)
10. Akiyama, S., Wheeler, J.C., Meier, D.L., Lichtenstadt, I.: The Magnetorotational Instability in Core-Collapse Supernova Explosions. Astrophys. J. **584**, 954–970 (2003)
11. Akmal, A., Pandharipande, V.R., Ravenhall, D.G.: Equation of state of nucleon matter and neutron star structure. Phys. Rev. C **58**, 1804–1828 (1998)
*12. Aksenov, A.G., Blinnikov, S.I.: A Newton iteration method for obtaining equilibria of rapidly rotating stars. Astron. Astrophys. **290**, 674–681 (1994)
13. Aksenov, A.G., Milgrom, M., Usov. V.V.: Pair Winds in Schwarzschild Spacetime with Application to Strange Stars. Proc. 11th Marcel Grossmann Meeting, Berlin, July 2006; (arXiv:astrp-ph/0701235) (2007)
*14. Aleksandrov, A.F., Bogdankevich, L.S., Rukhadze, A.A.: Grounds of plasma electrodynamics Visschaya Schkola, Moscow (1978)
15. Allen, K.U.: Astrophysical Quantities Athlone Press, University of London (1973)
*16. Alfven, G., FältHammar, K.G.: Cosmical electrodynamics. Fundamental principles Oxford, Clarendon (1963)
17. Alcock, Ch., Illarionov, A.F.: The surface chemistry of stars. II. Fractionated accretion of interstellar matter. Astrophys. J. **235**, 541–553 (1980)
18. Alcock, C., Farhi, E., Olinto, A.: Strange stars. Astrophys. J. **310**, 261–272 (1986)
19. Alexander, D.: Low temperature Rosseland opacity tables. Astrophys. J. Suppl. **29**, 363–374 (1975)
20. Alexander, D.R., Ferguson, J.W.: Low-temperature Rosseland opacities. Astrophys. J. **437**, 879–891 (1994)
21. Alpar, M.A., Cheng, A.F., Ruderman, M.A., Shaham, J.: A new class of radio pulsars. Nature **300**, 728–730 (1982)
22. Ambartzumian, V.A.: Problems of the evolution of the universe. Erevan, AN ArmSSR (1968)
23. Ambartzumian, V.A., Fuors. A.: Astron. J. **7**, 557–572 (1971) (Astrophys **7**, 331–339, 1971)
24. Ambartzumian, V.A., Saakyan, G.S.: On degenerate superdense gas of elementary particles. Astron. J. **37**, 193–209 (1960) (Sov. Astron. **4**, 187–201, 1960)
25. Amnuel, P.R., Gusseinov, O.Kh.: Accretion of interstellar matter onto a neutron star with magnetic field. Astron. Tsirk. No.524. (1969)
26. Antonov, V.A.: On the instability of stationary spherical models with merely radial motions. in Omarov G. B., ed., *The Dynamics of Galaxies and Stellar Clusters*. Nauka, Alma-Ata (in Russian), 139–143 (1973)
27. Andersen, J.: Accurate masses and radii of normal stars. Astron. Astrophys. Rev. **3**, 91–126 (1991)
28. Anita, H.M., Basu, S.: High-frequency and high-wave-number Solar oscillations. Astrophys. J. **519**, 400–406 (1999)
29. Appenzeller, I.: The evolution of a vibrationally unstable main sequence star of 130 $M_{\odot}$. Astron. Astrophys. **5**, 355–371 (1970)
30. Appenzeller, I.: Mass loss rates for vibrationally unstable very massive main–sequence stars. Astron Astrophys. **9**, 216–220 (1970)
31. Appenzeller, I., Tscharnuter, W.: The evolution of a massive protostar. Astron Astrophys. **30**, 423–430 (1974)
32. Ardelyan, N.V.: Divergence of difference schemes for two-dimensional equations of acoustics and Maxwell equations. J. Appl. Math. Math. Phys. **23**, 1168–76 (1983)
33. Ardelyan, N.V.: On applying iterational methods for the realization of implicit difference schemes of two-dimensional magnetohydrodynamics. J. Appl. Math. Math. Phys. **23**, 1417–26 (1983)

34. Ardelyan, N.V., Bisnovatyi-Kogan, G.S., Popov, Yu.P.: Investigation of magnetorotational supernova explosion. Cylindrical model. Astron. J. (1979) **56**, 1244–55 (Sov. Astron. **23**, 705–711, 1979)
35. Ardelyan, N.V., Bisnovatyi-Kogan, G.S., Popov, Yu.P., Chernigovskii, S.V.: Calculation of collapse of rotating gas cloud on Lagrangian grid. Astron. J. **64**, 495–508 (1987) (Sov. Astron. **31**, 261–267, 1987)
*36. Ardelyan, N.V., Bisnovatyi-Kogan, G.S., Popov, Yu.P., Chernigovskii, S.V.: Core collapse and the formation of a rapidly rotating neutron star. Astron. J. **64**, 761–772 (1987) (Sov. Astron. **31**, 398–403, 1987)
37. Ardeljan, N.V., Bisnovatyi-Kogan, G.S., Kosmachevskii, K.V., Moiseenko, S.G.: Two-dimensional simulation of the dynamics of the collapse of a rotating core with formation of a neutron star on an adaptive triangular grid in lagrangian coordinates. Astrophys. **47**, 47–64 (2004) (Astrophys. **47**, 37–51, 2004)
38. Ardelyan, N.V., Guzshin, I.S.: On an approach to constructing conservative difference schemes. Vest. MGU, Vych. Mat. i Kib., N 3, pp. 3–10 (1982)
39. Ardelyan, N.V., Chernigovskii, S.V.: On the divergence of difference schemes for two-dimensional equations of gas dynamics with the inclusion of gravity to the acoustical approximation. Differential'nyie Uravneniya **20**, 1119–27 (1984)
40. Ardeljan, N.V, Kosmachevskii, K.V.: Comput. Math. Model. **6**, 209 (1995)
41. Ardeljan, N.V, Kosmachevskii, K.V., Chernigovskii, S.V.: Problems of construction and research of conservative difference schemes for magneto-gas-dynamics. MSU, Moscow (in Russian) (1987)
42. Ardelyan, N.V., Bisnovatyi-Kogan, G.S., Kosmachevskii, K.V., Moiseenko, S.G.: An implicit Lagrangian code for the treatment of non-stationary problems in rotating astrophysical bodies. Astron Astrophys. Suppl. **115**, 573–594 (1996)
43. Ardeljan, N.V., Bisnovatyi-Kogan, G.S., Moiseenko S.G.: 2D calculations of the collapse of rotating magnetized gas cloud. Astrophys. Space Sci. **239**, 1–13 (1996)
44. Ardeljan, N.V., Bisnovatyi-Kogan, G.S., Moiseenko, S.G.: Non-stationary magnetorotational processes in a rotating magnetized cloud. Astron. Astrophys. **355**, 1181–1190 (2000)
45. Ardeljan, N.V., Bisnovatyi-Kogan, G.S., Moiseenko, S.G.: Magnetorotational supernova. Mon. Not. R. Astron. Soc. **359**, 333–344 (2005)
46. Arnett, D.: Mass dependence in gravitational collapse of stellar cores. Canad. J. Phys **45**, 1621–1641 (1967)
47. Arnett, D.: A possible model of supernovae: detonation of ^{12}C. Astrophy. Space Sci. **5**, 180–212 (1969)
*48. Arnett, D.: Neutrino trapping during gravitational collapse of stars. Astrophys. J. **218**, 815–833 (1977)
*49. Arponen, J.: Internal structure of neutron stars. Nucl. Phys. **A191**, 257–284 (1972)
50. Artemova, J.V., Bisnovatyi-Kogan, G.S., Björnsson, G., Novikov, I.D.: Structure of accretion discs with optically thin-thick transitions. Astrophys. J. **456**, 119–123 (1996)
51. Artemova, Y.V., Bisnovatyi-Kogan, G.S., Igumenshchev, I.V., Novikov, I.D.: On the structure of advective accretion disks at high luminosity. Astrophys. J. **549**, 1050–1061 (2001)
52. Artemova, Y.V., Bisnovatyi-Kogan, G.S., Igumenshchev, I.V., Novikov, I.D.: Black hole advective accretion disks with optical depth transition. Astrophys. J. **637**, 968–977 (2006)
53. Audard, N., Provost, J.: Sesmological properties of intermediate-mass stars. Astron Astrophys. **282**, 73–85 (1994)
*54. Audoze, J., Reeves, H.: The Origin of light elements. In: Essays in nuclear astrophysics, pp. 340–358. Cambridge University Press, Cambridge (1982)
55. Auman, J., Bodenheimer, P.: The influence of water-vapor opacity and the efficiency of convection on models of late-tape stars. Astrophys. J. **149**, 641–648 (1967)
56. Baade, W., Zwicky, F.: Supernovae and cosmic rays. Phys. Rev. **45**, 138–139 (1934)
*57. Bahcall, J. (1978): Solar neutrino experiment. Rev. Mod. Phys. **50**, 881–903
*58. Bahcall, J.N.: Neutrino astrophysics. Cambridge University Press, Cambridge (1989)

*59. Bahcall, J.N., Barnes, C.A., Christensen-Dalsgaard, J. et al.: Has a standard model solution to the solar neutrino problem been found? Preprint astro-ph 9404002 (1994)
*60. Bahcall, J., Cleveland, B.T., Davis, R. Jr., Rowley, J.K.: Chlorine and gallium solar neutrino experiment. Astrophys. J. Lett **292**, L79–L82 (1985)
61. Bahcall, J.N., Lande, K., Lanou, R.E., Learned, J.G., Robertson, R.G.H, Wolfenstein, L.: Progress and prospects in neutrino astrophysics. Nature **375**, 29–34 (1995)
*62. Bahcall, J.N., Pinsonneault, M.: Standart solar models with and without helium diffusion and the solar neutrino problem. Rev. Mod. Phys. **64**, 885–926 (1992)
*63. Baker, N., Gough, D.: Pulsation of model RR lyrae stars. Astrophys. J. **234**, 232–244 (1979)
64. Baldo, M., Bombaci, I., Bugio, G.F.: Microscopic nuclear equation of state with three-body forces and neutron star structure. Astron Astrophys. **328**, 274–282 (1997)
*65. Balmforth, N.J., Howard, L.N., Spiegel, E.A.: Equilibria of rapidly rotating polytropes. Mon. Not. R. Astron. Soc. **260**, 253–272 (1993)
*66. Bardeen, J.: Binding energy and stability of spherically symmetric masses in general relativity. Preprint OAP No. 36 (1965)
67. Barkat, Z., Wheeler, J.C.: The convective URCA mechanism. Astrophys. J. **355**, 602–616 (1990)
68. Barkov M.V., Komissarov S.V.: Magnetic acceleration of ultrarelativistic GRB and AGN jets. Int. J. Mod. Phys. D **17**, 1669–1675 (2008)
69. Barkov M.V., Komissarov S.V.: Close binary progenitors of gamma-ray bursts. Mon. Not. R. Astron. Soc. **401**, 1644–1656 (2010)
70. Barkov, M.V., Bisnovatyi-Kogan, G.S.: The afterglow of a dense molecular cloud after the passage of a cosmological gamma-ray burst. Astron. J. **82**, 685-698 (2005) (Astron. Rep. **49**, 611–623)
*71. Basko, M.M.: Two-group approximation in spherically symmetric problems of transfer theory. Preprint ITEP, N 7 (1982)
72. Basko, M.M., Imshennik, V.S.: The stability loss by low-mass stars under conditions of neutronization. Astron. J. **52**, 469–80 (1975) (Sov. Astron. **19**, 286–292, 1976)
73. Basko, M.M., Rudzskii, M.A., Seidov, Z.F.: Low-mass star collapse. Astrophys. **16**, 321–335 (1980) (Astrophys. **16**, 192–201, 1980)
74. Bastrukov, S., Podgainy, D., Yang J., Weber, F.: Magneto Torsional Pulsations of Magnetars. Mem. S.A.It. **73**, 522–527 (2002)
75. Basu, S., Christensen-Dalsgaard, J., Schon, J., Thompson, M.J., Tomczyk, S.: The Sun's hydrostatic structure from LOWL data. Astrophys. J. **460**, 1064–1070 (1996)
76. Baud, B. et al.: High-sensitivity IRAS observations of the Chamaeleon I dark cloud. Astrophys. J. Lett **278**, L53–L55 (1984)
77. Baym, G.: Neutron stars and quark matter. AIP Conf. Proc. **892**, 8–14 (2007)
*78. Baym, G., Bethe, H., Pethick, Ch.: Neutron star matter. Nucl. Phys. A **175**, 255–271 (1971)
*79. Baym, G., Pethick, Ch.: Neutron stars. Ann. Rev. Nucl. Sci. **25**, 27–77 (1975)
80. Baym, G., Pethick, Ch., Sutherland, P.: The groun state of matter at high densities: equation of state and slellar model. Astrophys. J. **170**, 306–315 (1971)
81. Baym, G., Chin, S.A.: Can a neutron star be a giant MIT bag? Phys. Lett. B **62**, 241–244 (1976)
82. Beaudet, G., Petrosian, V., Salpeter, E.E.: Energy losses due to neutrino processes. Astrophys. J. **150**, 979–999 (1967)
*83. Beaudet, G., Salpeter, E.E., Silvestro, M.L.: Rates for URCA neutrino processes. Astrophys. J. **174**, 79–90 (1972)
84. Becker, S., Iben, I.: The asymptotic giant branch evolution of intermediate-mass stars as a function of mass and composition. I. Through the second dredge-up phase. Astrophys. J. **232**, 831–853 (1979)
85. Becker, S., Iben, I., Tuggle, R.: On the frequency–period distribution of Cepheid variables in galaxies in the local group. Astrophys. J. **218**, 633–653 (1977)

86. Beichman, C.A., et al.: The formation of Solar-type stars: IRAS observations of the dark cloud Barnard 5. Astrophys. J. Lett. **278**, L45–L48 (1984)
87. Beichman, C., et al.: Candidate Solar-type protostars in nearby molecular cloud cores. Astrophys. J. **307**, 337–349 (1986)
*88. Berestetzkii, V.B., Lifshitz, E.M., Pitaevskii, A.P. Relativistic quantum theory, part I. Moscow, Nauka (1968)
*89. Berezin, I.S., Zhidkov, N.P.: Calculation methods, vol. 2. Moscow, Fizmat (1960)
90. Berezinsky, V.S., Prilutsky. O.F.: Neutrino-antineutrino annihilation around a collapsar. Astron. Astrophys. **175**, 309–311 (1987)
91. Beskin, G., Karpov, S., Bondar, S. et al.: Central engine activity as seen in Naked Eye Burst prompt emission. Talk on the Workshop "Many faces of GRB phenomena – optics vs high energy". October 12-16, 2009, SAO, Russia (2009)
92. Beskin, V.S.: *Axisymmetric stationary astrophysical flows.* Fizmatlit, Moscow (in Russian) (2005)
93. Bethe, H.: Possible explanation of Solar neutrino puzzle. Phys. Rev. Lett. **56**, 1305–1308 (1986)
94. Bethe, H., Johnson, M.: Dense barion matter calculations with realistic potentials. Nucl. Phys. **A230**, 1–58 (1974)
95. Bethe, H., Wilson, J.: Revival of a stalled supernova shock by neutrino heating. Astrophys. J. **295**, 14–23 (1985)
96. Bethe, H.: The supernova shock. Astrophys. J. **449**, 714–726 (1991)
*97. Bethe, H.A.: Theory of nuclear matter. Ann. Rev. Nucl. Sci. **21**, 93–244 (1971)
*98. Bethe, H.A., Morrison, F.: Elementary nuclear theory New York, Wiley (1956)
99. Bhattacharya, D., van den Heuvel, E.P.J.: Formation and evolution of binary and millisecond radio pulsars. Phys. Rep. **203**, 1–124 (1991)
100. Bhattacharya, D., Wijers, R., Hartman, J., Verbunt, F.: On the decay of the magnetic fields of single radio pulsars. Astron. Astrophys. **254**, 198–212 (1992)
*101. Bisnovatyi-Kogan, G.S.: Stellar envelopes with supercritical luminosity. Astrophys. Space Sci. **22**, 307–320 (1973)
102. Bisnovatyi-Kogan, G. S. (1975): Gamma-ray bursts from neutron stars. Report on COSPAR, Varna. Preprint IKI No. D203
103. Bisnovatyi-Kogan, G.S.: Magnetohydrodynamical processes near compact objects. Nuovo Cimento, Rivista, Serie 3, **2**(1), 1–37 (1979)
104. Bisnovatyi-Kogan, G.S.: Magnetohydrodynamical model of supernovae explosion. Ann. New York Acad. Sci. **336**, 389–394 (1980)
105. Bisnovatyi-Kogan, G.S.: Pre-main-sequence stellar evolution. In: Sugimoto, D., Lamb, D., Schramm, D. (eds.) Proc. Symp. IAU Mo. 93: Fundamental problems of stellar evolution, pp. 87–97. Reidel, Dordrecht (1981)
*106. Bisnovatyi-Kogan, G.S.: Physical processes in stars on late stages of stellar evolution. Astron. Nachrichten **203**, 131–137 (1982)
107. Bisnovatyi-Kogan, G.S.: Angular velocity distribution in convective regions and the origin of solar differential rotation. Solar Phys. **128**, 299–304 (1990)
108. Bisnovatyi-Kogan, G.S.: Recycled pulsars and LMXB. Proc. Coll. IAU No. 128 (Zelena Gora, Pedagog. University Press), 209–212 (1992)
*109. Bisnovatyi-Kogan, G.S.: A self-consistent solution for an accretion disk structure around a rapidly rotating non-magnetized star. Astron. Astrophys. **274**, 796–806 (1993)
110. Bisnovatyi-Kogan, G.S.: Asymmetric neurino emission and formation of rapidly moving pulsars. Astron. Ap. Transact. **3**, 287–294 (1993)
111. Bisnovatyi-Kogan, G.S.: Mechanisms of jet formation. In:Errico, L., Vittone, A.A. (eds.) Stellar Jets and Bipolar Outflows, pp. 369–381. Kluwer, Dordecht (1993)
*112. Bisnovatyi-Kogan, G.S.: Analytical self-consistent solution for the structure of polytropic accretion discs with boundary layers. Mon. Not. R. Astron. Soc. **269**, 557–562 (1994)
113. Bisnovatyi-Kogan, G.S.: Close galactic origin of gamma ray bursts. ApJ Suppl. **97**, 185–187 (1995)

114. Bisnovatyi-Kogan, G.S.: Stellar Oscillations and Stellar Convection in the Presence of an URCA Shell. Mon. Not. R. Astron. Soc. **321**, 315–326 (2001)
115. Bisnovatyi-Kogan, G.S.: Magnetic fields of neutron stars: very low and very high. Proc Vulcano99 Workshop. Mem. Soc. Astron. It. **73**, 318–329 (2002) (*astro-ph/9911275*)
116. Bisnovatyi-Kogan, G.S.: Cosmic γ-ray bursts: observations and modeling. Phys. Part. Nucl. **37**, 647–676 (2006)
117. Bisnovatyi-Kogan, G.S.: Dynamic confinement of jets by magnetotorsional oscillations. Mon. Not. R. Astron. Soc. bf 376, 457-464 (2007)
118. Bisnovatyi-Kogan, G.S., Blinnikov, S.I.: The equilibrium, stability and evolution of a rotating magnetized gaseous disk. Astrophys. Space Sci. **19**, 119–144 (1972)
119. Bisnovatyi-Kogan, G.S., Blinnikov, S.I.: Static criteria for stability of arbitrary rotating stars. Astron. Astrophys. **31**, 391–404 (1974)
120. Bisnovatyi-Kogan, G.S., Blinnikov, S.I.: Disk accretion onto a black hole at subcritical luminosity. Astron. Astrophys. **59**, 111–125 (1977)
121. Bisnovatyi-Kogan, G.S., Blinnikov, S.I.: Spherical accretion onto compact X-ray source with preheating: no thermal limits for the luminosity. Mon. Not. R. Astron. Soc. **191**, 711–719 (1980)
122. Bisnovatyi-Kogan, G.S., Chechetkin, V.M.: Nucleosynthesis in supernova outbursts and the chemical composition of the envelopes of neutron stars. Astrophys. Space Sci. **26**, 25–46 (1974)
123. Bisnovatyi-Kogan, G.S., Chechetkin, V.M.: Nuclear fission in the neutron stars and gamma-ray bursts. Astrophys. Space Sci. **89**, 447–451 (1983)
124. Bisnovatyi-Kogan, G.S., Dorodninsyn, A.V.: On modeling radiation-driven envelopes at arbitrary optical depths. Astron. Astrophys. **344**, 647–654 (1999)
125. Bisnovatyi-Kogan, G.S., Dorodninsyn, A.V.: Application of Galerkin method to the problem of stellar stability, gravitational collapse and black hole formation. Grav. Cosmol. **4**, 174–182 (1998)
126. Bisnovatyi-Kogan, G.S., Imshennik, V.S., Nadyozhin, D.K., Chechetkin, V.M.: Pulsed gamma-ray emission from neutron and collapsing stars and supernovae. Astrophys. Space Sci. **35**, 23–41 (1975)
127. Bisnovatyi-Kogan, G.S., Lovelace, R.V.E.: Particle acceleration in extragalactic jets by intense long-wavelength electromagnetic fields. Astron. Astrophys. **296**, L17–L19 (1995)
128. Bisnovatyi-Kogan, G.S., Lovelace, R.V.E.: Influence of Ohmic heating on advection-dominated accretion flows. Astrophys. J. Lett. **486**, L43–L46 (1997)
129. Bisnovatyi-Kogan, G.S., Lovelace, R.V.E.: Magnetic field limitations on advection dominated flows. Astrophys. J. **529**, 978–984 (2000)
130. Bisnovatyi-Kogan, G.S., Lovelace, R.V.E.: Advective accretion disks and related problems including magnetic fields. New Astron. Rev. **45**, 663–742 (2001)
131. Bisnovatyi-Kogan, G.S., Lovelace, R.V.E.: Large-Scale B-Field in Stationary Accretion Disks. Astrophys. J. **667**, L167–L169 (2007)
132. Bisnovatyi-Kogan, G.S.; Lovelace, R.V.E.; Belinski, V.A.: A cosmic battery reconsidered. Astrophys. J. **580**, 380–388 (2002)
133. Bisnovatyi-Kogan, G.S., Moiseenko, S.G., Ardeljan, N.V.: Different magneto-rotational supernovae. Astron. Zh. **85**, 1109–1121 (Astron. Rep. **52**, 997–1008, 2008) (2008)
134. Bisnovatyi-Kogan, G.S., Nadyozhin, D.K.: The evolution of massive stars with mass loss. Astrophys. Space Sci. **15**, 353–374 (1972)
135. Bisnovatyi-Kogan, G.S., Neishtadt, A.I., Seidov, Z.F., Tsupko, O.Yu., Krivosheev, Yu.M.: Dynamical chaos in the problem of jet collimation. Mon. Not. R. Astron. Soc., submitted (2010)
136. Bisnovatyi-Kogan, G.S., Popov, Yu.P., Samochin, A.A.: The magnetohydrodynamical rotational model of supernovae explosion. Astrophys. Space Sci. **41**, 321–356 (1976)
137. Bisnovatyi-Kogan, G.S., Postnov, K.A.: Geminga braking index and pulsar motion. Nature **366**, 663–665 (1993)
138. Bisnovatyi-Kogan, G.S., Ruzmaikin, A.A.: The stability of rotating supermassive stars. Astron. Astrophys. **27**, 209–221 (1973)

139. Bisnovatyi-Kogan, G.S., Ruzmaikin, A.A.: The accretion of matter by a collapsing star in the presence of a magnetic field. Astrophys. Space Sci. **28**, 45–59 (1974)
140. Bisnovatyi-Kogan, G.S.: A simplified model of the formation of structures in dark matter and a background of very long gravitational waves. Mon. Not. R. Astron. Soc. **347**, 163–172 (2004)
141. Bisnovatyi-Kogan, G.S., Tsupko, O.Yu.: Approximate dynamics of dark matter ellipsoids. Mon. Not. R. Astron. Soc. **364**, 833–842 (2005)
142. Bisnovatyi-Kogan, G.S., Tsupko, O.Yu.: Dynamic stabilization of non-spherical bodies against unlimited collapse. Mon. Not. R. Astron. Soc. **386**, 1398–1403 (2008)
143. Bisnovatyi-Kogan, G.S., Ruzmaikin, A.A.: The accretion of matter by a collapsing star in the presence of a magnetic field. II. Selfconsistent stationary picture. Astrophys. Space Sci. **42**, 401–424 (1976)
144. Bisnovatyi-Kogan, G.S., Vainshtein, S.I.: Generation of magnetic fields in rotating stars and quasars. Astrophys. J. Lett **8**, 151–152 (1971)
145. Bisikalo, D.V., Boyarchuk, A.A., Kuznetzov, O.A., Chechetkin, V.M.: Influence of a common envelope of a binary system on a mass transfer through the inner Lagrange point. Astron. J. **74**, 889–897 (1997) (Sov. Astron. **41**, 794–801, 1997)
146. Bisikalo, D.V., Boyarchuk, A.A., Kuznetzov, O.A., Khruzina, T.S., Cherepashchuk, A.M., Chechetkin, V.M.: Evidences to absence of a shock interaction between jet and disk ("hot spot") in semi-detached binary systems. Astron. J. **75**, 40–53 (1998) (Sov. Astron. **42**, 33–45, 1998)
*147. Bisnovatyi-Kogan, G.S.: Isotropic corrections to Maxwellian distribution functions and the rate of energy exchange. Jurnal Prikladnoi Mekhaniki i Tekhnicheskoi Fiziki (J. Appl. Mechan. Phys. Eng.) No. 3, 74–76 (1965)
148. Bisnovatyi-Kogan, G.S.: Critical mass of hot isothermal white dwarf with the inclusion of general relativity effects. Astron. J. **43**, 89–95 (1966) (Sov. Astron. **10**, 69–74, 1966)
149. Bisnovatyi-Kogan, G.S.: Perfect gas flow in a spherically symmetric gravitational field with the inclusion of radiative heat transfer and radiation pressure. Prikl. Mat. Mech. **31**, 762–769 (1967)
150. Bisnovatyi-Kogan, G.S.: Explosions of massive stars. Astron. J. **45**, 74–80 (1968) (Sov. Astron. **12**, 58–62, 1968)
151. Bisnovatyi-Kogan, G.S.: On the role of radiation in the solar wind formation. Mekh. Jidk. Gaza No 4, 182–183 (1968)
152. Bisnovatyi-Kogan, G.S.: The mass limit of hot superdense stable configurations. Astrophys. **4**, 221–238 (1968) (Astrophys. **4**, 79–86, 1968)
*153. Bisnovatyi-Kogan, G.S.: Late evolutionary stages of stars. Cand. Thes., Inst. Prikl. Mat. (1968)
154. Bisnovatyi-Kogan, G.S.: Pulsar as a neutron star and weak interactions. Radiofiz. **13**, 1868–1872 (1970)
155. Bisnovatyi-Kogan, G.S.: On the mechanism of explosion of rotating star as a supernova. Astron. J. **47**, 813–816 (1970) (Sov. Astron. **14**, 652–655, 1971)
156. Bisnovatyi-Kogan, G.S.: First radiopulsar in the binary system. Priroda, No.3, 100–101 (in Russian) (1975)
157. Bisnovatyi-Kogan, G.S.: Equilibrium and stability of stars and stellar systems. Doct. Thes., Moscow Space Res. Inst. (1977)
158. Bisnovatyi-Kogan, G.S.: X-ray sources in close binaries: theoretical aspects. Abastuman. Obs. Biull. No. 58, 175–210 (1985)
*159. Bisnovatyi-Kogan, G.S.: The cooling of white dwarfs with the inclusion of non-equilibrium beta-processes. PAJ **13**, 1014–1018 (1987) (Sov. Astron. Lett. **13**, 427–429, 1987)
160. Bisnovatyi-Kogan, G.S.: Two generations of low-mass X-ray binaries and recycled radiopulsars. Astrophys. **31**, 567–577; Astrophys. **32**, 193–194 (1989, 1990) (Astrophys. **31**, 751–758, 1989; **32**, 103, 1990)
161. Bisnovatyi-Kogan, G.S.: Binary and recycled pulsars: 30 years after observational discovery. UFN. **176**, 59–75 (2006) (Soviet. Phys. Uspekhi **49**, 53–67, 2006)

162. Bisnovatyi-Kogan, G.S., Blinnikov, S.I.: Hot corona around a disk accreting onto black hole and Cyg X-1 model. PAJ **2**, 489–493 (1976) (Sov. Astron. Lett. **2**, 191–193, 1976)
163. Bisnovatyi-Kogan, G.S., Blinnikov, S.I.: Propagation of waves in medium with high radiative density. Preprint of IKI Pr 421 (1978)
*164. Bisnovatyi-Kogan, G.S., Blinnikov, S.I.: Propagation of waves in a medium with high radiative density. I. Basic equations – homogeneous medium case. Astrophys. **14**, 563–577 (1978) (Astrophys. **14**, 316–325, 1978)
165. Bisnovatyi-Kogan, G.S., Blinnikov, S.I.: Equilibrium of rotating gas disks of a finite thickness. Astron. J. **58**, 312–323 (1981) (Sov. Astron. **25**, 175–181, 1981)
166. Bisnovatyi-Kogan, G.S., Blinnikov, S.I., Zakharov, A.F.: Numerical model of explosion near the neutron star surface and gamma-ray bursts. Astron. J. **61**, 104–111 (1984) (Sov. Astron. **28**, 62–66, 1984)
167. Bisnovatyi-Kogan, G.S., Blinnikov, S.I., Kostyuk, N.D., Fedorova, A.V.: Evolution of rapidly rotating stars on the stage of gravitational contraction. Astron. J. **56**, 770–780 (1979) (Sov. Astron. **23**, 432–437, 1979)
168. Bisnovatyi-Kogan, G.S., Blinnikov, S.I., Fedorova, A.V.: The Role of Rotation in Young Star Evolution. In Early stages of stellar evolution, pp. 40–46. Kiev, Naukova Dumka (1977)
169. Bisnovatyi-Kogan, G.S., Blinnikov, S.I., Schnol', E.E.: Stability of stars in the presence of a phase transition. Astron. J. **52**, 920–929 (1975) (Sov. Astron. **19**, 559–564, 1976)
170. Bisnovatyi-Kogan, G.S., Zel'dovich, Ya.B.: Adiabatic Outflow and equilibrium states with energy excess. Astron. J. **43**, 1200–1206 (1966) (Sov. Astron. **10**, 959–963, 1967)
171. Bisnovatyi-Kogan, G.S., Zel'dovich, Ya.B.: The Matter outflow out of stars under the effect of high opacity in the envelope. Report at 35 Comm. XIII Gen. As. IAU, preprint of IPM (1967)
172. Bisnovatyi-Kogan, G.S., Zel'dovich, Ya.B.: The matter outflow out of stars under the effect of high opacity in the atmosphere. Astron. J. **45**, 241–250 (1968) (Sov. Astron. **12**, 192–198, 1968)
173. Bisnovatyi-Kogan, G.S., Zel'Dovich, Ya.B., Syunyaev, R.A.: Physical Processes in a Low-Density Relativistic Plasma. Astron. J. **48**, 24–31 (1971) (Sov. Astron. **15**, 17–22, 1971)
174. Bisnovatyi-Kogan, G.S., Kazhdan, Ya.M.: Critical parameters of stars. Astron. J. **43**, 761–771 (1966) (Sov. Astron. **10**, 604–612, 1967)
175. Bisnovatyi-Kogan, G.S., Kazhdan, Ya.M., Clypin, A.A., Shakura, N.I.: Accretion onto rapidly moving gravitating centre. Astron. J. **56**, 359–67 (1979) (Sov. Astron. **23**, 201–206, 1979)
176. Bisnovatyi-Kogan, G.S., Komberg, B.V., Fridman, A.M.: A Theory for the Formation and Stability of the Jets in Quasars and Radio Galaxy Nuclei. Astron. J. **46**, 465–471 (1969) (Sov. Astron. **13**, 369–373, 1969)
177. Bisnovatyi-Kogan, G.S., Komberg, B.V.: Pulsars and close binary systems. Astron. J. **51**, 373–381 (1974) (Sov. Astron. **18**, 217–221, 1974)
178. Bisnovatyi-Kogan, G.S., Komberg, B.V.: Binary radiopulsar as an old object with small magnetic field: possible evolutionary scheme. PAJ **2**, 338–342 (1976) (Sov. Astron. Lett. **2**, 130–132, 1976)
179. Bisnovatyi-Kogan, G.S., Lamzin, S.A.: Outflow out of stars on the stage of gravitational contraction. Astron. J. **53**, 742–749 (1976) (Sov. Astron. **20**, 418–422, 1976)
180. Bisnovatyi-Kogan, G.S., Lamzin, S.A.: The matter outflow out of stars on early evolutionary stages. In Early stages of stellar evolution, pp. 107–118. Kiev, Naukova Dumka (1977)
181. Bisnovatyi-Kogan, G.S., Lamzin, S.A.: Models for outflowing envelopes of T Tauri stars. Astron. J. **54**, 1268–1280 (1977) (Sov. Astron. **21**, 720–727, 1977)
182. Bisnovatyi-Kogan, G.S., Lamzin, S.A.: The chromosphere, corona and X-ray emission of RU Lupi, a T Tauri star. PAJ **6**, 34–38 (1980) (Sov. Astron. Lett. **6**, 17–20, 1980)
183. Bisnovatyi-Kogan, G.S., Lamzin, S.A.: Stars with neutron cores. The possibility of the existence of objects with low neutrino luminosity. Astron. J. **61**, 323–332 (1984) (Sov. Astron. **28**, 187–193, 1984)

184. Bisnovatyi-Kogan, G.S., Moiseenko, S.G.: Violation of mirror symmetry of the magnetic field in a rotating star and possible astrophysical manifestations. Astron. J. **69**, 563–571 (1992) (Sov. Astron. **36**, 285–289, 1992)
185. Bisnovatyi-Kogan, G.S., Nadyozhin, D.K.: A Calculation method for evolution of stars with mass loss. Nauchnyie Informatzii Astronomicheskogo Soveta AN SSSR (Scientific Information of the Astonomical Council of the Academy of Sciences of the USSR), Issue 11, 27–39 (1969)
*186. Bisnovatyi-Kogan, G.S., Romanova, M.M.: Diffusion and neutron heat conduction in the neutron star crust. Jurnal Eksperimental'noi i Teoreticheskoi Fiziki (J. Exp. Theor. Phys., JETP) **83**, 449–459 (1982)
*187. Bisnovatyi-Kogan, G.S., Rudzskii, M.A., Seidov, Z.F.: Non-equilibrium beta-processes and the role of excited nuclear states. Jurnal Eksperimental'noi i Teoreticheskoi Fiziki (J. Exp. Theor. Phys., JETP) **67**, 1621–1630 (1974)
188. Bisnovatyi-Kogan, G.S., Seidov, Z.F.: On the relationship between white dwarfs and type I supernovae. Astrophys. **5**, 243–247 (1969) (Astrophysics **5**, 113–114, 1969)
189. Bisnovatyi-Kogan, G.S., Seidov, Z.F.: Non-equilibrium beta-processes as a source of thermal energy of white dwarfs. Astron. J. **47**, 139–144 (1970) (Sov. Astron. **14**, 113–116, 1970)
190. Bisnovatyi-Kogan, G.S., Seidov, Z.F.: On the pulsation of star with a phase transition. Astrophys. **21**, 563–571 (1984) (Astrophys. **21**, 665–670, 1984)
191. Bisnovatyi-Kogan, G.S., Syunyaev, R.A.: Galactic nuclei and qusars as sources of infrared emission. Astron. J. **48**, 881–893 (1971) (Sov. Astron. **15**, 697–707, 1972)
192. Bisnovatyi-Kogan, G.S., Timokhin, A.N.: Optical counterparts of cosmological gamma-ray bursts: Interaction with matter of the parent galaxy. Astron. J. **74**, 483–496 (1997) (Astronomy Reports **41**, 423–435, 1997)
193. Bisnovatyi-Kogan, G.S., Fridman, A.M.: On the mechanism of X-ray emission of a neutron star. Astron. J. **46**, 721–724 (1969) (Sov. Astron. **13**, 566–568, 1970)
194. Bisnovatyi-Kogan, G.S., Chechetkin, V.M.: Non-equilibrium composition of neutron star envelopes and nuclear sources of energy. Pis'ma v Jurnal Eksperimental'noi i Teoreticheskoi Fiziki (J. Exp. Theor. Phys. Lett., JETP Lett) **17**, 622–626 (1973)
195. Bisnovatyi-Kogan, G.S., Chechetkin, V.M.: Non-equilibrium envelopes of neutron stars, their role in supporting X-ray emission and nucleosynthesis. UFN. **127**, 263–296 (1979) (Soviet. Phys. Uspekhi **22**, 89–108, 1979)
196. Bisnovatyi-Kogan, G.S., Chechetkin, V.M.: Gamma-ray bursts as an effect of neutron star activity. Astron. J. **58**, 561–568 (1981) (Sov. Astron. **25**, 320–324, 1981)
197. Black, D.C., Bodenheimer, P.: Evolution of rotating interstellar clouds. I. Numerical techniques. Astrophys. J. **199**, 619–632 (1975)
198. Black, D.C., Bodenheimer, P.: Evolution of rotating interstellar clouds. II. The collapse of protostars of 1, 2, and 5 $M_{\odot}$. Astrophys. J. **206**, 138–149 (1976)
*199. Blake, J., Schramm, D.: A possible alternative to the r-process. Astrophys. J. **209**, 846–849 (1976)
*200. Blake, J., Woosley, S., Weaver, T., Shramm, D.: Nucleosynthesis of neutron-rich heavy niclei during explosive helium burning in massive stars. Astrophys. J. **248**, 315–320 (1981)
201. Blandford, R. D., Znaeck, R. L. (1977): Electromagnetic extraction of energy from Kerr black holes. Mon. Not. R. Astron. Soc. **179**, 433–456
*202. Blatt, J.M., Weiskopf, V.F.: Theoretical Nuclear Physics. New York, Wiley (1952)
203. Blinnikov, S.I., Sasorov, P.V., Woosley S.E.: Self-acceleration of nuclear flames in supernovae. Space Sci. Rev. **74**, 299–311 (1995)
204. Blinnikov, S.I.: Self-consistent field method in the theory of rotating stars. Astron. J. **52**, 243–254 (1975) (Sov. Astron. **19**, 151–156, 1975)
*205. Blinnikov, S.I.: On the thermodynamics of relativistic Fermi-gas. PAJ **13**, 820–823 (1987) (Sov. Astron. Lett. **13**, 346–347, 1987)
*206. Blinnikov, S.I., Dunina-Barkovskaya, N.V., Nadyozhin, D.K.: The equation of state for Fermi-gas: approximations for various degrees of relativism and degeneracy. Astrophys. J. Suppl. **106**, 171–203 (1996)

207. Blinnikov, S.I., Lozinskaya, T.A., Chugai, N.N.: Supernovae and their remnants. Itogi Nauki i Tekhniki, Astronomiya **32**, 142–200 (1987)
208. Blinnikov, S.I., Novikov, I.D., Perevodchikova, T.V., Polnarev, A.G.: Exploding neutron stars in close binaries. PAJ **10**, 422–428 (1984) (Sov. Astron. Lett. **10**, 177–179, 1984)
209. Blinnikov, S.I., Imshennik, V.S., Nadezhin, D.K., Novikov, I.D., Perevodchikova, T.V., Polnarev, A.G.: Explosion of a Low-Mass Neutron Star. Astron. J. **67**, 1181–1194 (1990) (Sov. Astron.**34**, 595–601, 1990)
210. Blinnikov, S.I., Rudzskii, M.A.: Collapse of low-mass iron stellar core. PAJ **10**, 363–369 (1984) (Sov. Astron. Lett. **10**, 152–155, 1984)
*211. Blinnikov, S.I., Rudzskyi, M.A.: Annihilation neutrino losses in dense stellar matter. Astron. J. **66**, 730–736 (1989) (Sov. Astron. **33**, 377–380, 1989)
*212. Blinnikov, S.I., Rudzskii, M.A.: New representations of the thermodynamical functions of Fermi gas. Astrophys. **29**, 644–651 (1988) (Astrophys. **29**, 644–651, 1988)
213. Blinnikov, S.I., Khokhlov, A.M.: Development of detonation in degenerate stars. PAJ **12**, 318–324 (1986) (Sov. Astron. Lett. **12**, 131–133, 1986)
*214. Bludman, S., Van Riper, K.: Diffusion approximation to neutrino transport in dense matter. Astrophys. J. **224**, 631–642 (1978)
*215. Bodansky, D., Clayton, D., Fowler, W.: Nuclear quasi-equilibrium during silicon burning. Astrophys. J. Suppl. **16**, 299–371 (1968)
216. Bodenheimer, P., Ostriken, J.: Rapidly rotating stars. IV. Pre-main sequence evolution of massive stars. Astrophys. J. **161**, 1101–1115 (1970)
217. Bodenheimer, P., Ostriken, J.: Do the pulsars make supernovae? II. Calculations of light curves for type II events. Astrophys. J. **191**, 465–471 (1974)
218. Bodenheimer, P., Tscharnuter, W.: A comparison of two independent calculations of the axysimmetric collapse of a rotating protostar. Astron. Astrophys. **74**, 288–293 (1979)
*219. Bogolyubov, N.N., Shirkov, D.V.: Quantum fields Moscow, Nauka, (1980)
*220. Bohm-Vitenze, E.: Uber die Wasserstoffkonvektionzone in Sternen verschiedener Effectivtemperaturen und Leuchtkrafte. Z. Astrophys. **46**, 108–143 (1958)
221. Bombaci, I.: Quark matter in compact stars: astrophysical implications and possible signature. arXiv:0809.4228 (2008)
222. Bonazzola, S., Gourgoulhon, E., Mark, J.-A.: Spectral methods in general relativistic astrophysics. J. Comput. Appl. Math. **109**, 433–473 (1999)
223. Bonazzola, S. and Gourgoulhon, E.: A virial identity applied to relativistic stellar models. Classical Quant. Grav. **11**, 1775–1784 (1994)
224. Bondi, H.: On spherically symmetrical accretion. Mon. Not. R. Astron. Soc. **112**, 195–204 (1952)
225. Bondi, H., Hoyle, F.: On the mechanism of accretion by stars. Mon. Not. R. Astron. Soc. **104**, 273–282 (1944)
226. Boss, A.P.: Protostellar formation in rotating interstellar cloud. I. Numerical methods and tests. Astrophys. J. **236**, 619–627 (1980)
227. Boss, A.P.: Protostellar formation in rotating interstellar cloud. II. Axially symmetric collapse. Astrophys. J. **237**, 563–573 (1980)
228. Boss, A.P.: Protostellar formation in rotating interstellar cloud. II. Non-axysimmetric collapse. Astrophys. J. **237**, 866–876 (1980)
229. Boss, A.P.: Collapse and equilibrium of rotating adiabatic clouds. Astrophys. J. **242**, 699–709 (1980)
230. Boss, A.P., Haber, J.G.: Axysimmetric collapse of rotating isothermal clouds. Astrophys. J. **255**, 240–244 (1982)
231. Braaten, E., Segel, D.: Neutrino energy loss from the plasma process at all temperatures and densities. Phys. Rev. **D48**, 1478–1491 (1993)
*232. Bray, R.J., Loughhead, R.E.: The Solar Granulation Chapman and Hall, Leningrad (1967)
233. Bruenn, S.W.: Thermal consequences of the convectively driven URCA process. Astrophys. J. **183**, L125–L129 (1973)
*234. Bronshtein, I.N., Semendyaev, K.A.: Handbook on Mathematics. Moscow, Gostekhteorizd (1953)

*235. de Broglie, L.: The magnetic electron. dirac's theory (Khar'kov, Gos. Nauch. -Tekh. Izd. Ukrainy) (1936)

*236. Buchler, J-R., Barkat, Z.: Properties of low density neutron star matter. Phys. Rev. Lett **27**, 48–51 (1971)

*237. Buchler, J-R., Datta, B.: Neutron gas: temperature dependence of the effective interaction. Phys. Rev. **C19**, 494–497 (1979)

*238. Buchler, J-R., Yuen, W.R.: Compton scattering opacities in a partially degenerate electron plasma at high temperatures. Astrophys. J. **210**, 440–446 (1976)

*239. Bugaev, E.V., Bisnovatyi-Kogan, G.S., Rudzskyi, M.A., Seidov, Z.F.: The interaction of intermediate energy neutrinos with nuclei. Nucl. Phys. **A324**, 350–364 (1979)

*240. Bugaev, E.V., Kotov, Yu. D., Rozental, I.L.: Cosmic muons and neutrinos. Moscow, Atomizdat (1970)

*241. Bugaev, E.V., Rudzskii, M.A., Bisnovatyi-Kogan, G.S., Seidov, Z.F.: The interaction of intemediate-energy neutrinos with nuclei. Preprint of IKI AN SSSR Pr-403 (1978)

*242. Burbidge, E.M., Burbidge, G.R., Fowler, W., Hoyle, F.: Synthesis of the elements in stars. Rev. Mod. Phys. **29**, 547–650 (1957)

243. Buras, R., Rampp, M., Janka, H.Th., Kifonidis, K.: Improved models of stellar core collapse and still no explosions: what is missing? Phys. Rev. Lett. **90**, 241101–241105 (2003)

244. Burgay, M., D'Amico, N., Possenti, A., Manchester, R.N., Lyne, A.G., Joshi, B.C., McLaughlin, M.A., Kramer, M., Sarkissian, J.M., Camilo, F., Kalogera, V., Kim, C., Lorimer, D.R.: An increased estimate of the merger rate of double neutron stars from observations of a highly relativistic system. Nature **426**, 531–533 (2003)

245. Burrows, A., Lattimer, J.: The birth of neutron stars. Astrophys. J. **307** 178–196 (1986)

246. Burrows, A.: Convection and the mechanism of Type II supernovae. Astrophys. J. Lett. **318**, L57–L61 (1987)

247. Burrows, A., Fryxell, B.A.: A convective trigger for supernova explosion. Astrophys. J. Lett. **418**, L33–L35 (1993)

248. Burrows, A., Hayes, J., Fryxell, B.A.: On the nature of core collapse supernova explosion. Astrophys. J. **450**, 830–850 (1995)

249. Burrows, A., Livne, E., Dessart, L., Ott, C.D., Murphy, J.: A new mechanism for core-collapse supernova explosions. Astrophys. J. **640**, 878–890 (2006)

250. Burrows, A., Dessart, L., Livne, E., Ott, C.D., Murphy, J.: Simulations of magnetically driven supernova and hypernova explosions in the context of rapid rotation. Astrophys. J. **664**, 416–434 (2007)

251. Cabrit, S., Bertout, C.: CO lines formation in bipolar flows. I. Accelerated outflows. Astrophys. J. **307**, 313–323 (1986)

252. Camilo, F., Lorimer, D.R., Freire, P., Lyne, A.G., Manchester, R.N.: Observations of 20 millisecond pulsars in 47 Tucanae at 20 centimeters. Astrophys. J. **535**, 975–990 (2000)

253. Cannon, R.C., Eggleton, P.P., Zytkow, A.N., Podsiadlowski P.: The structure and evolution of Thorne–Zytkow objects. Astrophys. J. **386**, 206–214 (1992)

254. Carson, T.R., Huebner, W.F., Magee, N.H., Merts, A.L.: Discrepancy in the CNO opacity bump resolved. Astrophys. J. **283**, 466–468 (1984)

*255. Castor, J.L.: Radiative transfer in spherically symmetric flows. Astrophys. J. **178**, 779–792 (1972)

256. Castor, J.I., Abbott, D.C., Klein, R.I.: Radiation-driven winds in Of stars. Astrophys. J. **195**, 157–174 (1974)

257. Castro-Tirado, A.J., Gorosabel, J.: Optical observations of GRB afterglows: GRB 970508 and GRB 980326 revisited. Astron. Astrophys. Suppl. **138**, 449–450 (1999)

258. Caughlan, G.R., Fowler, W.A.: Thermonuclear reaction rates V. Atom. Data Nucl. Data Table **40**, 283–334 (1988)

*259. Caughlan, G., Fowler, W., Harris, M., Zimmerman, B.: Tables of thermonuclear reaction rates for low-mass nuclei ($1 \leq Z \leq 14$). Atom. Data Nucl. Data Table **32**, 197–234 (1985)

260. Cavalerie, A., Isaak, G., et al.: Structure of 5-minute solar oscillations 1976–1980. Sol. Phys. **74**, 51–57 (1981)

*261. Chabrier, G.: Quantum effects in dense Coulombic matter: application to the cooling of white dwarfs. Astrophys. J. **414**, 695–700 (1993)
*262. Chabrier, G., Schatzman, E. (eds): The equation of state in astrophysics. Proc. IAU Coll. Cambridge University Press, Cambridge (1994)
263. Chakrabarti, S.K.: Grand unification of solutions of accretion and winds around black holes and neutron stars. Astrophys. J. **464**, 664–683 (1996)
264. Chandrasekhar, S.: The highly collapsed configuration of a stellar mass. Mon. Not. R. Astron. Soc. **91**, 456–466 (1931)
265. Chandrasekhar, S.: The maximum mass of ideal white dwarfs. Astrophys. J. **74**, 81–82 (1931)
266. Chandrasekhar, S., Fermi, E.: Problems of gravitational stability in the presence of a magnetic field. Astrophys. J. **118**, 116–141 (1953)
267. Chandrasekhar, S.: The dynamical instability of gaseous masses approching the Schwarzschild limit in general relativity. Astrophys. J. **140**, 417–433 (1964)
268. Chandrasekhar, S.: Selected Papers Chicago University Press, Chicago **1**, p. 185 (1981)
269. Chandrasekhar, S.: An introduction to the study of stellar structure. University Chicago Press, Chicago (1939)
*270. Chandrasekhar, S.: Radiative transfer. Clarendon, Oxford (1950)
*271. Chandrasekhar, S.: Ellipsoidal equilibrium figures. Yale University Press, New Haven (1969)
272. Chandrasekhar, S.: The mathematical theory of Black Holes. (1983) Clarendon, Oxford
*273. Chan, K., Sofia, S., Wolff, L.: Turbulent compressible convection in a deep atmosphere. I. Preliminary two-dimensional results. Astrophys. J. **263**, 935–943 (1982)
274. Chaplin, W.J., Appourchaux, T., Elsworth, Y., et al.: The asteroseismic potential of kepler: first results for solar-type stars. Astrophys. J. Lett. **713**, L169–L175 (2010)
275. Chapman, S., Cowling, T.G.: Mathematical theory of inhomogeneous gases. Cambridge University Press, Cambridge (1952)
276. Charbonnel, C., Meynet, G., Maeder, A., Schaller, G., Schaerer, D.: Grids of stellar models. III. From 0. 8 to 120 $M_{\odot}$ at Z=0. 004. Astron. Astrophys. Suppl. **101**, 415–419 (1993)
277. Chechetkin, V.M., Gershtein, S.S., Imshennik, V.S., Ivanova, L.N., Khlopov, M.Yu.: Supernovae of types I and II and the neutrino mechanism of thermonuclear explosion of degenerated carbon-oxygen stellar cores. Astrophys. Space Sci. **67**, 61–98 (1980)
*278. Chechetkin, V.M., Kowalski, M.: Production of heavy elements in nature. Nature **259**, 643–644 (1976)
279. Chen, L.-H., Wang, J.-M.: Slim disks with transition regions and applications to microquasars and narrow-line seyfert 1 galaxies. Astrophys. J. **614**, 101–114 (2004)
280. Cheng, K.S., Zhang, C.M.: Magnetic field evolution of accreting neutron stars. Astron. Astrophys. **337**, 441–446 (1998)
281. Cheng, K.S., Zhang, C.M.: Evolution of magnetic field and spin period in accreting neutron stars. Astron. Astrophys. **361**, 1001–1004 (2000)
282. Chevalier, R., Dwarkadas, V.V.: The presupernova H II region around SN 1987A. Astrophys. J. Lett. **452**, L45–L48 (1995)
283. Cherepashchuk, A.M. (ed.): Catalog of close binary systems at late evolutionary stages. MGU Press, M (1988)
284. Cherepashchuk, A.M.: Masses of black holes in binary systems. UFN. **166**, 809–832 (1996) (Sov. Phys. Uspekhi **39**, 759–780, 1996)
285. Cherepashchuk, A.M.: X-ray nova binary systems. Space Sci. Rev. **93**, 3–4 (2000)
*286. Chechev, V.P., Kramarovski, Ya.M.: Theory of nuclear synthesis in stars: s-process. UFN. **134**, 431–467 (1981) (Sov. Phys. Uspekhi **24**, 566–584, 1981)
*287. Chechetkin, V.M.: Equilibrium state of matter at high temperatures and densities. Astron. J. **46**, 202–206 (1969) (Sov. Astron. **13**, 153–156, 1969)
288. Chechetkin, V.M., Ustyugov, S.D., Gorbunov, A.A., Polezhaev, V.I.: On the neutrino mechanism of supernovae explosion. PAJ **23**, 34–40 (1997) (Sov. Astron. Lett. **23**, 30–36, 1997)

289. Chiosi, C., Stalio, R. (eds.): Effects of mass loss in stellar evolution. In: Proc. IAU Coll. No. 59 Reidel, Dordrecht (1981)
290. Choudhuri A.R., Konar S.: Diamagnetic screening of the magnetic field in accreting neutron stars. Mon. Not. R. Astron. Soc. **332**, 933 (2002)
291. Choudhuri A.R., Konar S.: Why do millisecond pulsars have weaker magnetic fields compared to ordinary pulsars? Curr. Sci. **86** 444–446 (2004)
292. Christodoulou, D.M., Contopoulos, I., Kazanas, D.: Simulations of the Poynting-Robertson cosmic battery in resistive accretion disks. Astrophys. J. **674**, 388–407 (2008)
*293. Christensen-Dalsgaard, J., Däppen, W.: Solar oscillations and the equation of state. Astron. Astrophys. Review **4**, 267–361 (1992)
294. Chugai, N.N.: Neutrino helicity and space velocities of pulsars. PAJ **10**, 210–213 (1984) (Sov. Astron. Lett.**10**, 87–88, 1984)
*295. Clifford, F.E., Tayler, R.J.: The equilibrium distribution of nuclides in matter at high temperatures. Mem. Roy. Astron. Soc. **69**, 21–81 (1965)
296. Cloutman, L., Whitaker, R.: On convective and semiconvective mixing in massive stars. Astrophys. J. **237**, 900–902 (1980)
297. Cohen, M., Kuhi, L.V.: Observational studies of pre-main-sequence evolution. Astrophys. J. Suppl. **41**, 743–843 (1979)
298. Colgate, S.: Hot bubbles drive explosions. Nature **341**, 489–490 (1989)
299. Colgate, S., Petschek, A.G.: Explosive supernova core overturn and mass ejection. Astrophys. J. Lett. **236**, L115–L119 (1980)
300. Colgate, S., Krauss, L.M., Schramm, D.N., Walker, T.P.: Magnetohydrodynamic jets, pulsar formation and SN 1987A. Astrophys. J. Lett. **27**, 411–418 (1990)
301. Colgate, S., Herant, M.E., Benz, W.: Neutron star accretion and the neutrino fireball. Phys. Rep. **227**, 157–174 (1993)
302. Colgate, S., White, R.: The hydrodynamic behavior of supernovae explosions. Astrophys. J. **143**, 626–681 (1966)
303. Colpi, M., Shapiro, S., Teukolsky S.: A hydrodynamical model for the explosion of a neutron star just below the minimum mass. Astrophys. J. **414**, 717–734 (1993)
*304. Commins, P.H., Bucksbaum, P.H.: Weak interactions of Leptons and Quarks Cambridge University Press, Cambridge (1983)
305. Contopoulos, I., Kazanas, D.: A cosmic battery. Astrophys. J. **508**, 859–863 (1998)
306. Contopoulos, I., Kazanas, D., Christodoulou, D.M.: The cosmic battery revisited. Astrophys. J. **652**, 1451–1456 (2006)
307. Costa, E., on behalf of the BEPPOSAX GRB Team: X-ray afterglows of gamma-ray bursts with BeppoSAX. Astron. Astrophys. Suppl. **138**, 425–429 (1999)
308. Cottam, J., Paerels, F., Mendez, M.: Gravitationally redshifted absorption lines in the X-ray burst spectra of a neutron star. Nature, **420**, 51–54 (2002)
*309. Couch, R., Arnett, D.: Advanced evolution of massive stars. I. Secondary nucleothynthesis during helium burning. Astrophys. J. **178**, 771–777 (1972)
310. Couch, R.G., Arnett, W.D.: Carbon ignition and burning in degenerate stellar cores. Astrophys. J. **196**, 791–803 (1975)
311. Couch, R., Schmiedenkamp, A., Arnett, D.: S-process nucleosynthesis in massive stars: core helium burning. Astrophys. J. **190**, 95–100 (1974)
312. Cox, A., Tabor, J.: Radiative opacity tables for 40 stellar mixtures. Astrophys. J. Suppl. **31**, 271–312 (1976)
313. Cox, D.P., Tucker, W.H.: Ionization equilibrium and radiative cooling in a low-density plasma. Astrophys. J. **157**, 1157–1167 (1969)
314. Cowling, T.G.: Magnetohydrodynamics. New York, Interscience (1957)
315. Cox, A.N.: Absorption coefficients and the opacity of stellar matter. In: Aller, L.H. (ed.) Stellar Structure, D. B. McLaughlin. University of Chicago Press, Chicago (1965)
316. Cox, A., Stuart, G.: Radiative absorption and heat conductivity: opacities for 25 stellar mixtures. Nauchnyie Informatzii Astronomicheskogo Soveta AN SSSR (Scientific Information of the Astonomical Council of the Academy of Sciences of the USSR), Issue 15, 3–103 (1969)

317. Cox, J.P.: Stellar pulsation theory. Princeton University Press, Princeton (1980)
318. Däppen, W., Mihalas, D., Hummer, D.G.: The equation of statefor stellar envelopes. III. Thermodynamical quantities. Astrophys. J. **332**, 261–270 (1988)
319. Dar, A., Shaviv, G.: A standard model solution to the solar neutrino problem? Preprint astro-ph 9401043 (1994)
320. Dar, A., Shaviv, G.: Has a standard physics solution to the solar neutrino problem been found?-A response. Preprint astro-ph 9404035 (1994)
*321. De Jager, C.: The brightest stars. Dordrecht, Reidel (1980)
322. Deubner, F.-L.: Observations of low wavenumber nonradial eigenmodes of the sun. Astron. Astrophys. **44**, 371–375 (1975)
323. Dey, M., Bombaci, I., Dey, J., Ray, S., Samanta, B.C.: Strange stars with realistic quark vector interaction and phenomenological density-dependent scalar potential. Phys. Lett. B **438**, 123–128 (1999)
324. Dicus, D.: Stellar energy-loss rate in a convergent theory of a weak and electromagnetic interactions. Phys. Rev. **D6**, 941–949 (1972)
*325. Dolgov, A.D., Zel'dovich, Ya.B.: Cosmology and elementary particles. UFN. (Soviet. Phys. Uspekhi) **130**, 559–614 (1980)
326. Dorodnitsyn, A.V.: Line-driven winds in the presence of strong gravitational fields. Mon. Not. R. Astron. Soc. **339**, 569–576 (2003)
327. Dorodnitsyn, A.V., Novikov, I.D.: On the structure of line-driven winds near black holes. Astrophys. J. **621**, 932–939 (2005)
328. Dorodnitsyn, A.V.: Gravitationally distorted P Cygni profiles from outflows near compact objects Mon. Not. R. Astron. Soc. **393**, 1433–1448 (2009)
329. Dorofeev, O.F., Rodionov, V.N., Ternov, I.M.: Anisotropic emission of neutrinos arising in beta-processes under the effect of a strong magnetic field. PAJ **11**, 302–309 (1985) (Sov. Astron. Lett. **11**, 123–126, 1985)
330. Downes, D., et al.: Outflow of matter in the KL nebula: the role of IRc2. Astrophys. J. **244**, 869–883 (1981)
*331. Dravins, D., Lindegren, L., Nordlund, A.: Solar granulation: influence of convection on spectral line asymmetries and wavelength shifts. Astron. Astrophys. **96**, 345–364 (1981)
332. Drobyshevskii, E.M.: Outward transport of angular momentum by gas convection and the equatorial acceleration of the sun. Sol. Phys. **51**, 473–479 (1977)
*333. Drotleff, H.W., Denker, A., Knee, H., Soine, M., Wolf, G., Hammer, J.W., Greife, U., Rolfs, C., Trautvetter, H.P.: Reaction rates of the s-process neutron sources $^{22}Ne(\alpha,n)^{25}Mg$ and $^{13}C(\alpha,n)^{16}O$. Astrophys. J. **414**, 735–739 (1993)
334. Durney, B.R.: The interaction of rotation with convection. In: Slettebak, A. (ed.) Stellar Rotation, pp. 30–36. Reidel, Dordrecht (1970)
335. Durney, B.R.: On theories of solar rotation. In: Bumba, V., Kleczek, J. (ed.) Basic Mechanisms of Solar Activity, pp. 243–295. Reidel, Dordrecht (1976)
*336. Dungey, J.W.: Cosmic electrodynamics Cambridge, Cambridge University Press (1958)
337. Dyck, H.H., Simon, Th., Zukerman, B.: Discovery of an infrared companion to T Tauri. Astrophys. J. Lett. **255**, L103–L106 (1982)
*338. Dzbelepov, B.S., Peker, L.K.: Schemes of decay of radioactive nuclei. Moscow, Nauka (1966)
339. Dzembowski, W.: Non-radial oscillations of evolved stars. I. Quasiadiabatic approximation. Acta Astron. **21**, 289–306 (1971)
340. Dzytko, H., Turck-Chieze, S., Delburgo-Salvador, P., Lagrange, C.: The screened nuclear reaction rates and the solar neutrino puzzle. Astrophys. J. **447**, 428–442 (1995)
341. Eardley, D.M., Lightman, A.P.: Inverse compton spectra and the spectrum of Cyg X–1. Nature **262**, 196–197 (1976)
*342. Ebeling, W., Kraeft, W.D., Kremp, D.: Theory of Bound States and Ionization Equilibrium in Plasmas and Solids Academic-Verlag, Berlin (1979)
343. Eggleton, P.: The structure of narrow shells in red giants. Mon. Not. R. Astron. Soc. **135**, 243–250 (1967)

*344. Eggleton, P.: The evolution of low mass stars. Mon. Not. R. Astron. Soc. **151**, 351–354 (1971)
*345. Eggleton, P.: Composition changes during stellar evolution. Mon. Not. R. Astron. Soc. **155**, 361–376 (1972)
*346. Eggleton, P.: A numerical treatment of double shell source stars. Mon. Not. R. Astron. Soc. **163**, 279–284 (1973)
347. Eggleton, P.: Towards consistency in simple prescriptions for stellar convection. Mon. Not. R. Astron. Soc. **204**, 449–461 (1983)
*348. Eggleton, P., Faulkner, J., Flannery, B.: An approximate equation of state for stellar material. Astron. Astrophys. **23**, 325–330 (1973)
349. Eich, C., Zimmermann, M.E., Thorne, K.S., Zytkow, A.N.: Giant and supergiant stars with degenerate neutron cores. Astrophys. J. **346**, 277–283 (1989)
350. El Eid, M.: Effect of convective mixing on the red-blue loops in the Hertzsprung–Russell diagram. Mon. Not. R. Astron. Soc. **275**, 983–1002 (1995)
351. El Eid, M., Hartmann D.H.: Stellar models and the brightening of P Cygni. Astrophys. J. **404**, 271–275 (1993)
*352. El Eid, M.F., Hillebrandt, W.: A new equation of state of supernova matter. Astron. Astrophys. **42**, 215–226 (1980)
353. El Eid, M., Langer, N.: The evolution of very luminous stars II. Pair creation supernova in massive Wolf–Rayet stars. Astron. Astrophys. **167**, 274–281 (1986)
*354. Ellis, J.: Particle physics and cosmology. Annu. N. Y. Acad. Sci. **759**, 170–187 (1995)
355. Elsworth, Y., Howe, R., Isaak, G.R., McLeod, C.P., Miller, B.A., New, R., Wheeler, S.J., Gough, D.O.: Slow rotation of the sun's interior. Nature **376**, 669–672 (1995)
356. Emerson, J.P., et al.: IRAS observations near young objects with bipolar outflows: L1551 and HH 46–47. Astrophys. J. Lett. **278**, L49–L52 (1984)
357. Epstein, R.I.: Lepton-driven convection in supernovae. Mon. Not. R. Astron. Soc. **188**, 305–325 (1979)
358. Ergma, E., Paczynski, B.: Carbon burning with convective URCA neutrinos. Acta Astron. **24**, 1–16 (1974)
359. Ergma, E., Tutukov, A.V.: Evolution of carbon–oxygen dwarfs in binary systems. Acta Astron. **26**, 69–76 (1976)
*360. Ergma, E.V.: Models for 4 $M\odot$ and 20 $M\odot$ envelopes. Nauchnyie Informatzii Astronomicheskogo Soveta AN SSSR (Scientific Information of the Astonomical Council of the Academy of Sciences of the USSR), Issue 12 (1969)
*361. Ergma, E.V.: A non-local model of convection for stellar envelopes. Nauchnyie Informatzii Astronomicheskogo Soveta AN SSSR (Scientific Information of the Astonomical Council of the Academy of Sciences of the USSR), Issue 23, 33–46, (1972)
362. Ergma, E.V.: Thermonuclear bursts in neutron star envelopes. Itogi Nauki i Tekhniki, Astronomiya **21**, 130–150 (1982)
363. Ergma, E.V.: Thermonuclear processes in accreting white dwarfs (novae, simbiotic novae and type I supernovae). Itogi Nauki i Tekhniki, Astronomiya **31**, 228–257 (1986)
*364. Eriguchi, Y., Müller, E.: A general method for obtaining equilibria of self-gravitating and rotating gases. Astron. Astrophys. **146**, 260–268 (1985)
*365. Eriguchi, Y., Sugimoto, D.: Another equilibrium sequence of self-gravitating and rotating incompressible fluid. Progr. Theor. Phys. **65**, 1870–1875 (1981)
366. Errico, L., Vittone, A. (eds.): Stellar Jets and Bipolar Outflows. Proc. OAC 6, Capri 1991. Kluwer (1993)
367. Ezer, D., Cameron, A.G.W.: Pre-main-sequence stellar evolution with mass loss. Astrophys. Space Sci. **10**, 52–70 (1971)
368. Fadeev, Yu.A.: On the possibility for dust particles to form in the FG sagittae atmosphere. Nauchnyie Informatzii Astronomicheskogo Soveta AN SSSR (Scientific Information of the Astonomical Council of the Academy of Sciences of the USSR), Issue 50, 3–9 (1981)
369. Farhi, E. and Jaffe, R.L.: Strange matter. Phys. Rev. D **30**, 2379–2390 (1984)
370. Faulkner, J., Roxburgh, I.W., Strittmatter, P.A.: Uniformly rotating main-sequence stars. Astrophys. J. **151**, 203–216 (1968)

*371. Fermi, E.: Nuclear physics. University of Chicago Press, Chicago (1950)
*372. Fermi, E.: Versuch einer Theorie der β-Strahlen. Zt. f. Phys. **88**, 161–171 (1934)
*373. Feygenbaum, M.: Universal behaviour in non-linear systems. Los Alamos Sci. **1**(1), 4–27 (1980)
*374. Fok, V.A.: Fundamentals of quantum mechanics Nauka, Moscow (1976)
375. Frank-Kamenetzkii, D.A.: Physical processes in stellar interiors. Fizmatgiz, Moscow (1959)
*376. Frank-Kamenetzkii, D.A.: Neutronization kinetics at superhigh densities. Jurnal Eksperimental'noi i Teoreticheskoi Fiziki (J. Exp. Theor. Phys., JETP) **42**, 875–879 (1962)
377. Frederiks, D.D., Golenetskii, S.V., Palshin, V.D., Aptekar, R.L., Ilyinskii, V.N., Oleinik, F.P., Mazets, E.P., Cline, T.L.: Giant flare in SGR 1806-20 and its Compton reflection from the Moon. PAJ **33**, 3-21 (2007) (Astron. Lett. **33**, 1–18)
378. Frederiks, D.D., Palshin, V.D., Aptekar, R.L., Golenetskii, S.V., Cline, T.L., Mazets, E.P.: On the possibility of identifying the short hard burst GRB 051103 with a giant flare from a soft gamma repeater in the M81 group of galaxies. PAJ **33**, 22-28 (2007) (Astron. Lett. **33**, 19–24, 2008)
379. Fyodorova, A.V.: The effect of magnetic field on the minimum mass of a main-sequence star. Nauchnyie Informatzii Astronomicheskogo Soveta AN SSSR (Scientific Information of the Astonomical Council of the Academy of Sciences of the USSR), Issue 42, 95–109 (1978)
380. Fyodorova, A.V., Blinnikov, S.I.: The effect of accretion and rotation on the minimum mass of a main-sequence star. Nauchnyie Informatzii Astronomicheskogo Soveta AN SSSR (Scientific Information of the Astonomical Council of the Academy of Sciences of the USSR), Issue 42, 75–94 (1978)
381. Feigelson, E.D., DeCampli, W.M.: Observations of X-ray emission from T Tauri stars. Astrophys. J. Lett. **243**, L89–L94 (1981)
382. Feigenbaum, M.J.: Universal behavior in nonlinear systems. Los Alamos Sci. **1**, No. 1, pp. 4–27 (1980)
383. Feroci, M., Frontera, F., Costa, E., Amati, L., Tavani, M., Rapisarda, M., Orlandini, M.: A giant outburst from SRG 1900+14 observed with the BeppoSAX gamma-ray burst monitor. Astrophys. J. Lett. **515**, L9–L12 (1999)
384. Fillipenko, A.V.: The optical diversity of supernovae. In: Danziger, I.J., Kjär, K., (ed.) Proc. ESO/EIPC Workshop on SN 1987A and other Supernovae, pp. 343–362 (1991)
385. Finzi, A., Wolf, R.: Ejection of mass by radiation pressure in planetary nebulae. Astron. Astrophys. **11**, 418–430 (1971)
386. Fishman, G.J., Meegan, C.A.: Gamma-Ray Bursts. Annu. Rev. Astron. Astrophys. **33**, 415–458 (1995)
*387. Flowers, E., Itoh, N.: Transport properties of dense matter II. Astrophys. J. **230**, 847–858 (1979)
388. Foglizzo, T.: Entropic-acoustic instability of shocked Bondi accretion I. What does perturbed Bondi accretion sound like? Astron. Astrophys. **368**, 311–324 (2001)
*389. Fontaine, G., Graboske, H.C., Jr., van Horn, H.M.: Equation of state for stellar partial ionization zones. Astrophys. J. Suppl. **35**, 293–358 (1977)
390. Forestini, M.: Low-mass stars: pre-main sequence evoluton and nucleosynthesis. Astron. Astrophys. **285**, 473–488 (1994)
*391. Fowler, W.: The solar neutrino problem. Preprint OAP, No. 507 (1977)
392. Fowler, W., Caughlan, G., Zimmerman, B.: Thermonuclear reaction rates. Annu. Rev. Astron. Astrophys. **5**, 525–570 (1967)
393. Fowler, W., Caughlan, G., Zimmerman, B.: Thermonuclear reaction rates II. Annu. Rev. Astron. Astrophys. **13**, 69–112 (1975)
*394. Fowler, W., Engebrecht, C., Woosley, S.: Nuclear partition functions. Astrophys. J. **226**, 984–995 (1978)
*395. Fowler, W., Hoyle, F.: Neutrino processes and pair formation in massive stars and supernovae. Astrophys. J. Suppl. **9**, 201–319 (1964)

*396. Fraley, G.: Supernovae explosions induced by pair-production instability. Astrophys. Space Sci. **2**, 96–114 (1968)

397. Fridman, A.M., Polyachenko, V.L.: Physics of Gravitating Systems. Springer, Berlin (1985)

398. Friedman, B., Pandariphande, V.R.: Hot and cold nuclear and neutron matter. Nucl. Phys. **A361**, 502–520 (1981)

399. Friedman, B., Ipser, J., Parker, L.: Rapidly rotating neutron star models. Astrophys. J. **304**, 115–139 (1986)

*400. Fukasaku, K., Fujita, T.: Reexamination of standard solar model to the solar neutrino problem. Progr. Theor. Phys. **98**, 1251–1260 (1997)

*401. Fukuda, V., et al.: Measurements of the solar neutrino flux from super-kamiokaude. Phys. Rev. Lett. **81**, 1158–1162 (1998)

402. Fuller, G., Fowler, W., Newman, M.: Stellar weak-interaction rates for sd-shell nuclei. I. Nuclear matrix element systematics with application to AL^{26} and selected nuclei of importance to the supernova problems. Astrophys. J. Suppl. **42** 445–473 (1980)

403. Fuller, G., Fowler, W., Newman, M.: Stellar weak-interaction rates for intermediate mass nuclei. II: $A = 21$ to $A = 60$. Astrophys. J. **252**, 715–740 (1982)

404. Fuller, G., Fowler, W., Newman, M.: Stellar weak-interaction rates for intermediate mass nuclei. III: rates tables for the free nucleons and $A = 21$ to $A = 60$. Astrophys. J. Suppl. **48**, 279–295 (1982)

405. Fuller, G., Fowler, W., Newman, M.: Stellar weak-interaction rates for intermediate mass nuclei. IV. Interpolation procedures for rapidly varying lepton capture rates using effective $\lg(ft)$ values Astrophys. J. **293**, 1–16 (1985)

406. Gabriel, M.: The D_{nl} values and the structure of the solar core. Astron. Astrophys. **226**, 278–283 (1989)

407. Gahm, G.F.: X-ray observations of T Tauri stars. Astrophys. J. Lett. **242**, L163–L166 (1980)

408. Gahm, G.F., Fradga, K., Liseau, R., Dravins, D.: The far UV spectrum of T Tauri star RU Lupi. Astron. Astrophys. **73**, L4–L6 (1979)

*409. Galkin, S.A., Denissov, A.A., Drozdov, V.V., Drozdova, O.M.: A finite-difference adaptive grid method for computations of equilibria of self-gravitating rotating barotropic gas. Astron. Astrophys. **269**, 674–681 (1993)

410. Gaustad, J.: The opacity of diffuse cosmic matter and the early of star formation. Astrophys. J. **138**, 1050–1073 (1963)

*411. Gavrichenko, K.V., Nadyozhin, D.K.: Simple formulae for the rates of electron and positron capture by nuclei. Preprint of ITEF No. 123 (1980)

*412. Gayar, M.K., Nikolich, M. (eds.): Weak Interactions (Inst. Nat. de Phys. Nucl. et de Phys. de Particuler, Paris) (1977)

*413. Gershtein, S.S., Zel'dovich, Ya.B.: On meson corrections to the beta-decay theory. Jurnal Eksperimental'noi i Teoreticheskoi Fiziki (J. Exp. Theor. Phys., JETP) **29**, 698–699 (1955)

414. Gershtein, S.S. (2000): Oscillation Structure of Gamma-Ray Bursts and Their Possible Origin. PAJ **26**, 848–854 (Astron. Lett. **26**, 730–735, 2000)

*415. Gershtein, S.S., Zel'dovich, Ya.B.: The rest-mass of the muon neutrino and cosmology. Pis'ma v Jurnal Eksperimental'noi i Teoreticheskoi Fiziki (J. Exp. Theor. Phys. Lett., JETP Lett.) **4**, 174–177 (1966)

416. Geppert, U., Urpin, V.: Accretion driven magnetic field decay in neutron stars. Mon. Not. R. Astron. Soc. **271**, 490–496 (1994)

*417. Ginzburg, V.L.: Theoretical physics and astrophysics Moscow, Nauka (1975)

*418. Ginzburg, V.L., Rukhadze, A.A.: Waves in magnetoactive plasma. Moscow, Nauka (1970)

419. Gillman, R.: Planck mean cross-section for four grain materials. Astrophys. J. Suppl. **28**, 397–403 (1974)

420. Giovanelli, F., Bisnovatyi-Kogan, G.S., Golynskaya, I.M., et al.: Coordinated X-ray, ultraviolet and optical observations of T Tauri stars. In: Oda, M., Giacconi, R. (eds.) Proc. symp. int. X-ray astronomy 84, Bologna, pp. 77–80 (1984)

421. Glatzel, W., Kiriakidis, M.: The stability of massive main-sequence stars. Mon. Not. R. Astron. Soc. **262**, 85–92 (1993)
422. Glatzel, W., Kiriakidis, M.: Stability of massive stars and Humphrey–Davidson limit. Mon. Not. R. Astron. Soc. **263**, 375–384 (1993)
423. Glendenning, N.: First-order phase transitions with more than one conserved charge: consequences for neutron stars. Phys. Rev. D **46**, 1274–1287 (1992)
424. Glendenning, N.K., Weber, F.: Nuclear solid crust on rotating strange quark stars. Astrophys. J. **400**, 647–458 (1992)
425. Glendenning N.: Compact Stars: Nuclear Physics, Particle Physics and General Relativity. (Springer, Heidelberg) (2000)
426. Godnev, I.N.: Calculation of thermodynamic functions on the base of molecular data. Moscow, Gostekhizdat (1956)
*427. Godunov, S.K., Riaben'ki, V.S.: Introduction in the theory of difference schemes. Moscow, Fizmatgiz (1962)
428. Gradshtein, I.S., Ryzhik, I.M.: Tables for integrals, sums, series, productions. Moscow, Nauka (1971)
429. Greenspan, H.P.: The theory of rotating fluids. Cambridge, Cambridge University Press (1968)
430. Grinfel'd, M.A.: Gravitational stability and proper oscillations of liquid bodies with phase transition surfaces. DAN SSSR **258**, 1342–1346 (1981)
431. Goldreich, P., Julian, W.: Pulsar electrodynamics. Astrophys. J. **157**, 869–880 (1969)
432. Golenitskii, S.V., Mazets, E.P., et al.: Annihilation radiation in cosmic gamma-ray bursts. Preprint LFTI im. Ioffe, No. 959 (1985)
433. Golenetskii, S., Aptekar, R., Mazets, E., Pal'Shin, V., Frederiks, D., Cline, T.: Konus-wind observation of GRB 080319B. GCN 7482; http://www.ioffe.rssi.ru/LEA/GRBs/GRB080319_T22370/ (2008)
*434. Gonczi, G., Osaki, Y.: On local theories of time-dependent convection in the stellar pulsation problem. Astron. Astrophys. **84**, 304–310 (1980)
435. Goode, P.R., Dziembowski, W.A., Korzennik, S.G., Rhodes, E.J.: What we know about the sun's internal rotation from solar oscillations. Astrophys. J. **367**, 649–657 (1991)
436. Goodman, J., Dar, A., Nussinov, Sh.: Neutrino annihilation in Type II supernovae. Astrophys. J. Lett. **314**, L7–L10 (1987)
*437. Gough, D.O.: The anelastic approximation for thermal convection. J. Atmos. Sci. **26**, 448–456 (1969)
438. Gough, D.O.: Linear adiabatic stellar pulsation. In: Zahn, J.-P., Zinn-Justin, J. (eds.), Proc. Les Houches 1987, pp. 399–560 (1987)
*439. Gough, D.O., Toomre, J.: Single-mode theory of diffusive layers in thermohaline convection. J. Fluid Mech. **125**, 75–97 (1982)
440. Gurevich, L.E., Libedinskii, A.I.: On the causes of stellar bursts, Proc of 4th Conf. on Cosmogony Problems, Moscow, AN SSSR, pp. 143–171 (1955)
441. Greene, J.M., MacKay, R.S., Vivaldi F., Feigenbaum, M.J: Universal behavior in fam- ilies of area-preserving maps. Physica **3D**, 468–486 (1981)
*442. von Groote, H., Hilf, E.R., Takahashi, K.: A new semiempirical shell correction to the droplet model. Gross theory of nuclear magics. Atom. Data Nucl. Data Tables **17**, 418–427, 476–608 (1976)
443. Grossman, A.: The surface boundary condition and approximate equation of state for low-mass stars. In: Kumar, Sh. (ed.) Proc. Symp. Low-luminosity Stars, pp. 247–254. Gordon and Breach (1969)
444. Grossman, A.: Evolution of low-mass stars. I. Contraction to the main sequence. Astrophys. J. **161**, 619–632 (1970)
445. Grossman, A., Mutschlecner, J., Pauls, T.: Evolution of low-mass stars. II. Effects of premodal deuterium burning and nongray surface condition during pre-main-sequence contraction. Astrophys. J. **162**, 613–619 (1970)

446. Grossman, A., Graboske, H., Jr.: Evolution of low-mass stars. III. Effects of non-ideal thermodynamic properties during the pre-main-sequence contraction. Astrophys. J. **164**, 475–490 (1971)
*447. Grossman, S., Narayan, R., Arnett, D.: A theory of non-local mixing-length convection. I. The moment formalism. Astrophys. J. **407**, 284–315 (1993)
*448. Grossman, S., Narayan, R.: A theory of non-local mixing-length convection. II. Generalized smoothed particle hydrodynamics simulations. Astrophys. J. Suppl. **89**, 361–394 (1993)
449. Harrison, B.K., Thorne, K.S., Wakano, M., Wheeler, J.A.: Gravitational theory and gravitational collapse. University of Chicago Press, Chicago (1965)
*450. Hachisu, I.: A versatile method for obtaining structures of rapidly rotating stars. Astrophys. J. Suppl. **61**, 479–508 (1986)
*451. Hachisu, I.: A versatile method for obtaining structures of rapidly rotating stars. II. Three-dimensional self-consistent field method. Astrophys. J. Suppl. **62**, 461–500 (1986)
452. Hamada, T., Salpeter, E.: Models for zero-temperature stars. Astrophys. J. **134**, 683–698 (1961)
453. Han, Z., Podsiadlowski, P., Eggleton, P.P.: A possible criterion for envelope ejection in AGB or FGB stars. Mon. Not. R. Astron. Soc. **270**, 121–130 (1994)
*454. Hansen, J.P., Torrie, G.M., Viellerfosse, P.: Statistical mechanics of dense ionized matter. VII. Equation of state and phase separation of ionic mixtures in a uniform background. Phys. Rev. **A16**, 2153–2168 (1977)
455. Hanson, R., Jones, B.F., Lin, D.N.C.: The astrometric position of T Tauri and the nature of its companion. Astrophys. J. Lett. **270**, L27–L30 (1983)
456. Harm, R., Schwarzschild, M.: Red giants of population II. IV. Astrophys. J. **145**, 496–504 (1966)
457. Harm, R., Schwarzschild, M.: Transport from a gas giant to a blue nucleus after ejection of a planetary nebula. Astrophys. J. **200**, 324–329 (1975)
458. Harm, R., Schwarzschild, M.: Red giants of population II. III. Astrophys. J. **139**, 594–601 (1964)
459. Harris, M., Fowler, W., Caughlan, G. Zimmerman, B.: Thermonuclear reaction rates III. Annu. Rev. Astron. Astrophys. **21**, 165–176 (1983)
460. Haselgrove, C.B., Hoyle, F.: Main-sequence stars. Mon. Not. R. Astron. Soc. **119**, 112–123 (1959)
461. Hayashi, Ch.: Evolution of protostars. Annu. Rev. Astron. Astrophys. **4**, 171–192 (1966)
462. Hayashi, Ch., Hoshi, R., Sugimoto, D.: Evolution of stars. Progr. Theor. Phys. Suppl. **22**, 1–183 (1962)
*463. Hayashi, Ch., Hoshi, R., Sugimoto, D.: Advanced phases of evolution of population II stars. Growth of the carbon core and shell helium flashes. Progr. Theor. Phys. **34**, 885–911 (1965)
464. Heiselberg, H., Pethick, C.J., and Staubo, E.F.: Quark matter droplets in neutron stars. Phys. Rev. Lett. **70**, 1355–1359 (1993)
*465. Henyey, L.G., Wilets, L., Böhm, K.-H., LeLevier, R., Levee, R.D.: A method for automatic computation of stellar evolution. Astrophys. J. **129**, 628–636 (1959)
466. Herant, M., Benz, W.: Hydrodynamical instabilities and mixing in SN 1987A: two-dimensional simulations of the first 3 months. Astrophys. J. Lett. **370**, L81–L84 (1991)
467. Herant, M., Benz, W., Hix, W.R., Fryer, C.L., Colgate, S.A.: Inside the supernova: a powerful convective engine. Astrophys. J. **435**, 339–361 (1994)
468. Herbig, G.: The widths of absorption lines in T Tauri-like stars. Astrophys. J. **125**, 612–613 (1957)
469. Herbig, G.: Eruptive phenomena in early stellar evolution. Astrophys. J. **217**, 693–715 (1977)
470. Hernández, M.M., Hernández, E.P., Michel, E., Belmonte, J.A., Goupil, M.J., Leberton, Y.: Seismology of δ Scuti stars in the Praesepe cluster. Astron. Astrophys. **338**, 511–520 (1998)

*471. Hillebrandt, W.: The rapid neutron capture process and the synthesis of heavy and neutron-rich elements. Space Sci. Rev. **21**, 639–702 (1978)
472. Hillebrandt, W.: Stellar collapse and supernovae explosion. In: Proc. NATO-ASI High energy phenomena around collapsed stars, Cargese (1985)
473. Hillebrandt, W., Nomoto, K., Wolff, R.: Supernovae explosions of massive stars. The mass range 8 to 10 $M_\odot$. Astron. Astrophys. **133**, 175–184 (1984)
474. Hobbs, G., Lorimer, D.R., Lyne, A.G., Kramer, M.: A statistical study of 233 pulsar proper motions. Mon. Not. R. Astron. Soc. **360**, 974–992 (2005)
*475. Holmes, J., Woosley, S., Fowler, E., Zimmerman, B.: Tables of thermonuclear–reaction–rate date for neutron–induced reaction on heavy nuclei. Atom. Data Nucl. Data Tables **18**, 305–412 (1976)
476. Hoshi, R.: Basis properties of a stationary accretion disk surrounding a black hole. Progr. Theor. Phys. **58**, 1191–1204 (1977)
477. Hoshi, R., Shibazaki, N.: The effect of pressure gradient force on an accretion disk surrounding a black hole. Progr. Theor. Phys. **58**, 1759–1765 (1977)
478. Houck, J.R., et al.: Unidentified point sources in the IRAS minisurvey. Astrophys. J. Lett. **278**, L63–L66 (1984)
479. Hoxie, D.: The structure and evolution of stars of very low mass. Astrophys. J. **161**, 1083–1099 (1970)
480. Hoyle, F., Fowler, W.A.: Nucleosynthesis in supernovae. Astrophys. J. **132**, 565–590 (1960)
481. Hulse, R.A., Taylor, J.H.: Discovery of a pulsar in a binary system. Astrophys. J. Lett. **195**, L51–L53 (1975)
*482. Hummer, D.G., Mihalas, D.: The equation of state for stellar envelopes. I. An occupation probability formalism forr the truncation of internal partition functions. Astrophys. J. **331**, 794–814 (1988)
483. Hunt, R.: A fluid dynamical study of the accretion process. Mon. Not. R. Astron. Soc. **154**, 141–165 (1971)
*484. Hurlburt, N., Toomre, J., Massaguer, J.: Two-dimensional compressible convection extending over multiple scale heights. Astrophys. J. **282**, 557–573 (1984)
485. Iben, I., Jr.: Stellar evolution. I. The approach to the main sequence. Astrophys. J. **141**, 993–1018 (1965)
486. Iben, I.: Stellar evolution. II. The evolution of a 3 $M_\odot$ star from main sequence through core helium burning. Astrophys. J. **142**, 1447–1467 (1965)
487. Iben, I.: Stellar evolution. III. The evolution of a 5 $M_\odot$ star from main sequence through core helium burning. Astrophys. J. **143**, 483–504 (1966)
488. Iben, I.: Stellar evolution. IV. The evolution of a 9 $M_\odot$ star from main sequence through core helium burning. Astrophys. J. **143**, 505–515 (1966)
489. Iben, I.: Stellar evolution. V. The evolution of a 15 $M_\odot$ star through core helium burning from the main sequence. Astrophys. J. **143**, 516–526 (1966)
490. Iben, I.: Stellar evolution. VI. Evolution from the main sequence to the red-giant branch for stars of mass 1 $M_\odot$, 1.25 $M_\odot$ and 1.5 $M_\odot$. Astrophys. J. **147**, 624–649 (1967)
491. Iben, I.: Stellar evolution. VII. Evolution of 2.25 $M_\odot$ star from the main sequence to the helium burning phase. Astrophys. J. **147**, 650–663 (1967)
492. Iben, I.: Stellar evolution within and off the main sequence. Annu. Rev. Astron. Astrophys. **5**, 571–626 (1967)
493. Iben, I.: On the specification of the blue edge of the RR Lyrae instability trip Astrophys. J. **166**, 131–151 (1971)
494. Iben, I.: Post main sequence evolution of single stars. Annu. Rev. Astron. Astrophys. **12**, 215–256 (1974)
495. Iben, I.: Thermal pulses; p-capture, α-capture, s-process nucleosynthesis; and convective mixing in a star of intermediate mass. Astrophys. J. **196**, 525–547 (1975)
496. Iben, I.: Solar oscillations as a guide to solar structure. Astrophys. J. Lett. **204**, L147–L150 (1976)

497. Iben, I.: Futher adventures of a thermally pulsing star. Astrophys. J. **208**, 165–176 (1976)
498. Iben, I., Jr.: URCA neutrino-loss rates under conditions found in the carbon-oxygen cores of intermediate-mass stars. Astrophys. J. **219**, 213–225 (1978)
499. Iben, I., Jr.: More on carbon burning in electron-degenerate matter - within single stars of intermediate mass and within accreting white dwarfs. Astrophys. J. **253**, 248–259 (1982)
500. Iben, I.: Low-mass asymptotic giant branch evolution I. Astrophys. J. **260**, 821–837 (1982)
501. Iben, I.: On the frequency of a planetary nebula nuclei powered by helium burning and on the frequency of white dwarfs with hydrogen-deficient atmospheres. Astrophys. J. **277**, 333–354 (1984)
502. Iben, I.: The life and times of an intermediate mass star – in isolation/in a close binary. Q. J. Roy. Astron. Soc. **26**, 1–39 (1985)
503. Iben, I.: Single and binary star evolution. Astrophys. J. Suppl. **76**, 55–114 (1991)
504. Iben, I., Kaler, J., Truran, J., Renzini, A.: On the evolution of those nuclei of planetary nebulae, that experience a final helium shell flash. Astrophys. J. **264**, 605–612 (1983)
505. Iben, I., Renzini, A.: Asymptotic giant branch evolution and beyond. Annu. Rev. Astron. Astrophys. **21**, 271–342 (1983)
506. Iben, I., Renzini, A.: Single star evolution I. Massive stars and early evolution of low and intermediate mass stars. Phys. Rep. **105**, 329–406 (1984)
507. Iben, I., Rood, R.: Metal-poor stars I. Evolution from the main sequence to the giant branch. Astrophys. J. **159**, 605–617 (1970)
508. Iben, I., Tutukov, A.V.: Cooling of low-mass carbon-oxygen dwarfs from the planetary nucleus stage through the crystallization stage. Astrophys. J. **282**, 615–630 (1984)
509. Ichimaru, S.: Bimodal behavior of accretion disks - theory and application to Cygnus X-1 transitions. Astrophys. J. **214**, 840–855 (1977)
*510. Ichimaru, S.: Strongly coupled plasma: high density classical plasmas and degenerate electron liquids. Rev. Mod. Phys. **54**, 1017–1059 (1982)
*511. Ichimaru, S., Utsumi, K.: Enhancement of thermonuclear reaction rate due to screening by relativistic degenerate electrons long range correlation effect. Astrophys. J. **286**, 363–365 (1984)
512. Iglesias, C.A., Rogers, F.J., Wilson, B.J.: Spin-orbit interaction effects on the Rosseland mean opacity. Astrophys. J. **397**, 711–728 (1992)
513. Iglesias, C.A., Rogers, F.J.: Radiative opacities for carbon and oxygen-rich mixtures. Astrophys. J. **412**, 752–760 (1993)
514. Iglesias, C.A., Wilson, B.J., Rogers, F.J., Goldstein, W.H., Bar-Shalom, A., Oreg, J.: Effects of heavy metals on astrophysical opacities. Astrophys. J. **445**, 855–860 (1995)
515. Igumenshchev, I.V., Abramowicz, M.A., Novikov, I.D.: Slim accretion discs: a model for ADAF-SLE transitions. Mon. Not. R. Astron. Soc. **298**, 1069–1078 (1998)
516. Iida, K., Watanabe, G., Sato, K.: Formation of nuclear pasta in cold neutron star matter. Progr. Theor. Phys. **106**, 551–559 (2001)
517. Illarionov, A.F., Sunyaev, R.A.: Why the number of galactic X-ray stars is so small? Astron. Astrophys. **39**, 185–195 (1975)
518. Imshennik, V.S.: A possible scenario of a supernova explosion as a result of the gravitational collapse of a massive stellar core. PAJ **18**, 445–449 (1992) (Sov. Astron. Lett.**18**, 489–504, 1992)
519. Imshennik, V.S.: Gravitational radiation from approaching neutron stars and the rotational explosion mechanism for collapsing supernovae. PAJ **34**, 412-427 (2008) (Sov. Astron. Lett. **34**, 375–388, 2008)
520. Imshennik, V.S., Kalyanova, N.L., Koldoba, A.V., Chechetkin, V.M.: Is detonation burning possible in a degenerate carbon-oxigen core of pre-supernova ? PAJ **25**, 250–259 (1999) (Sov. Astron. Lett.**25**, 206–214, 1999)
*521. Imshennik, V.S., Kotok, E.V., Nadyozhin, D.K.: Calculation of homogeneous models by the progonka method. Nauchnyie Informatzii Astronomicheskogo Soveta AN SSSR (Scientific Information of the Astonomical Council of the Academy of Sciences of the USSR), Issue 1, 48–54 (1965)

*522. Imshennik, V.S., Morozov, Yu.I.: Relativistically covariant equations for interaction of radiation with matter. Astron. J. **46**, 800–809 (1969) (Sov. Astron. **13**, 628–634, 1970)
523. Imshennik, V.S., Nadyozhin, D.K.: Thermodynamic properties of matter at high temperatures and densities. Astron. J. **42**, 1154–1167 (1965) (Sov. Astron. **9**, 896–906, 1966)
*524. Imshennik, V.S., Nadyozhin, D.K.: Heat conductivity by neutrinos in collapsing stars. Jurnal Eksperimental'noi i Teoreticheskoi Fiziki (J. Exp. Theor. Phys., JETP) **63**, 1548–1561 (1972)
525. Imshennik, V.S., Nadyozhin, D.K.: Final stages of stellar evolution and supernova bursts. Itogi Nauki i Tekhniki, Astronomiya **21**, 63–129 (1982)
526. Imshennik, V.S., Nadyozhin, D.K.: Supernova 1987a in great magellanic cloud: observations and theory. UFN. (Soviet. Phys. Uspekhi) **156**, 561–651 (1988)
*527. Imshennik, V.S., Nadyozhin, D.K., Pinaev, V.S.: The kinetic equilibrium of beta-processes in stellar interiors. Astron. J. **43**, 1215–1225 (1966) (Sov. Astron. **10**, 970–978, 1967)
*528. Imshennik, V.S., Nadyozhin, D.K., Pinaev, V.S.: Neutrino emission of energy in beta-interactions of electrons and positrons with nuclei. Astron. J. **44**, 768–777 (1967) (Sov. Astron. **11**, 617–623, 1968)
*529. Imshennik, V.S., Chechetkin, V.M.: Thermodynamics under conditions of hot neutronization of matter and hydrodynamical stability of stars on late evolutionary stages. Astron. J. **47**, 929–941 (1970) (Sov. Astron. **14**, 747–757, 1971)
*530. Imshennik, V.S. Nadyozhin, D.K.: Neutrino chemical potential and neutrino heat conductivity with allowance for neutrino scattering. Astrophys. Space Sci. **62**, 309–333 (1979)
*531. Ioffe, B.L., Lipatov, L.N., Khoze, V.A.: Deeply inelastic processes. Phenomenology. Quark-Parton Model Moscow, Energoatomizdat (1983)
532. Ipser, J., Managan, R.: On the existence and structure of inhomogeneous analogues of the Dedekind and Jakobi ellipsoids. Astrophys. J. **250**, 362–372 (1981)
533. Istomin Ya.N., Pariev V.I.: Stability of a relativistic rotating electron-positron jet: non-axisymmetric perturbations. Mon. Not. R. Astron. Soc. **281**, 1–26 (1996)
*534. Itoh, N.: Physics of dense plasmas and the enhancement of thermonuclear reaction rates due to strong screening. Progr. Theor. Phys. Suppl. **70**, 132–141 (1981)
*535. Itoh, N., Hayashi, H., Kohyama, Y.: Electrical and thermal conductuvity of dense matter in the crystalline lattice phase. III. Inclusion of lower densities. Astrophys. J. **418**, 405–413 (1993)
*536. Itoh, N., Mutaku, S., Iyetomi, H., Ichimaru, S.: Electrical and thermal conductivities of dense matter in the liquid metal phase I. High temperature results. Astrophys. J. **273**, 774–782 (1983)
537. Itoh, N., Totsuji, H., Ichimaru, S.: Enhancement of thermonuclear reaction rates due to strong screening. Astrophys. J. **218**, 477–483; **220**, 742 (1977–1978)
538. Itoh, N., Totsuji, H., Ichimaru, S., De Witt, H.: Enhancement of thermonuclear reaction rates due to strong screening. II. Ionic mixtures. Astrophys. J. **234**, 1079–1084; **239**, 415 (1979–1980)
539. Ivanenko, D.D., Kurdgelaidze, D.F.: Hypothesis concerning quark stars. Astrophys. **1**, 479–482 (1965) (Astrophys. **1**, 251–252, 1965)
540. Ivanova, L.N., Imshennik, V.S., Nadyozhin, D.K.: A study of the dynamics of supernova explosion. Nauchnyie Informatzii Astronomicheskogo Soveta AN SSSR (Scientific Information of the Astonomical Council of the Academy of Sciences of the USSR), Issue 13, 3–78 (1969)
541. Ivanova, L.N., Imshennik, V.S., Chechetkin, V.M.: Pulsation regime of the thermonuclear explosion of a star's dense carbon core. Astrophys. Space Sci. **31**, 497–514 (1974)
542. Iwakami, W., Kotake, K., Ohnishi, N., Yamada, Sh. and Sawada, K.: Three-dimensional simulations of standing accretion shock instability in core-collapse supernovae. Astrophys. J. **678**, 1207–1222 (2008)
*543. Jackson, S.: Rapidly rotating stars. The coupling of the Henyey and the self-consistent-fluid methods. Astrophys. J. **161**, 579–585 (1970)
544. Jaikumar, P., Gale, C., Page, D., Prakash, M.: Bremsstrahlung photons from the bare surface of a strange quark star. Phys. Rev. D **70**, 023004 (2004)

545. James, R.A.: The structure and stability of rotating gas masses. Astrophys. J. **140**, 552–582 (1964)
546. Janka, H.-T., Müller, E.: Neutron star recoils from anisotropic supernovae. Astron. Astrophys. **290**, 496–502 (1994)
547. Janka, H.-T., Müller, E.: The first second of a Type II supernova: convection, accretion, and shock propagation. Astrophys. J. Lett. **448**, L109–L113 (1995)
548. Janka, H.-Th., Scheck, L., Kifonidis, K., Müller, E., Plewa, T.: Supernova asymmetries and pulsar kicks — views on controversial issues. In: The fate of the Most Massive Stars, ASP Conference Series, **332**, 372–381 (2005)
549. Johnston, S., Hobbs, G., Vigeland, S., Kramer, M., Weisberg, J.M., Lyne, A.G.: Evidence for alignment of the rotation and velocity vectors in pulsars. Mon. Not. R. Astron. Soc. **364**, 1397–1412 (2005)
550. Johnston, S., Kramer, M., Karastergiou, A., Hobbs, G., Ord, S., Wallman, J.: Evidence for alignment of the rotation and velocity vectors in pulsars - II. Further data and emission heights. Mon. Not. R. Astron. Soc. **381**, 1625–1637 (2007)
551. Jorgensen, H.E., Lipunov, V.M., Panchenko, I.E., Postnov, K.A., Prokhorov, M.E.: Evolution of supernova explosion rates in the universe. Astrophys. J. **486**, 110–116 (1997)
552. Juman, C.: Baryon star models. Astrophys. J. **141**, 187–194 (1965)
553. Kamija, Y.: The collapse of rotating gas clouds. Progr. Theor. Phys. **58**, 802–815 (1977)
554. Kaplan, S.A.: 'Superdense' stars. Uchyonye Zapiski L'vovskogo gosuniversiteta, ser. fizmat. **15**, 109–116 (1949)
555. Kaplan, S.A.: Cooling of white dwarfs. Astron. J. (Sov. Astron.) **27**, 31–33 (1950)
556. Kaplan, D.B., Nelson, A.E.: Strange goings on in dense nucleonic matter. Phys. Lett. **B175**, 57–63 (1986)
557. Kato, M., Iben, I.: Self-consistent models of Wolf–Rayet stars as helium stars with optically thick winds. Astrophys. J. **394**, 305–312 (1992)
558. Kardashev, N.S.: Magnetic collapse and the nature of powerful sources of cosmic radio emission. Astron. J. **41**, 807–813 (1964) (Sov. Astron. **8**, 643–648, 1965)
559. Keene, J., et al.: Far-infrared detection of low-luminosity star formation in the Bok globule B335. Astrophys. J. Lett. **274**, L43–L47 (1983)
560. Kellman, S., Gaustad, J.: Rosseland and Planck mean absorption coefficients for particles 1of ice, graphite and silicon dioxide. Astrophys. J. **157**, 1465–1470 (1969)
*561. Khokhlov, A.M.: The structure of detonation waves in supernovae. Mon. Not. R. Astron. Soc. **239**, 785–808 (1989)
562. Khokhlov, A.M.: Mechanisms for the initiation of detonation in degenerate matter of Supernovae. Astron. Astrophys. **246**, 383–396 (1991)
563. Khokhlov, A.M.: Delayed detonation model for type Ia Supernova. Astron. Astrophys. **245**, 114–128 (1991)
564. Kippenhahn, R.: Differential rotation in stars with convective envelopes. Astrophys. J. **137**, 664–678 (1963)
565. Kippenhahn, R., Thomas, H.C., Weigert, A.: Sternentwicklung IV. Zentrales Wasserstoff und Heliumbrenner bei einen Stern von 5 Sonnenmassen. Zeit. Astrophys. **61**, 241–267 (1965)
566. Kippenhahn, R., Thomas, H.C., Weigert, A.: Sternentwicklung V. Der Kohlenstoff-Flash bei einen Stern von 5 Sonnenmassen. Zeit. Astrophys. **64**, 373–394 (1966)
*567. Kippenhahn, R., Thomas, H.C.: Rotation and stellar evolution. In: Sugimoto, D., Lamb, D., Shramm, D. (eds.) Proc. IAU Symp. No. 93, Fundamental Problems of the Theory of Stellar Evolution, pp. 237–256. Reidel, Dordrecht (1981)
568. Kippenhahn, R., Thomas, H.C.: On the compatibility of thermal and hydrostatic equilibrium in thin radiative accretion disks. Astron. Astrophys. **114**, 77–84 (1982)
569. Kippenhahn, R., Weigert, A.: Stellar Structure and Evolution Springer, Berlin (1990)
570. Kippenhahn, R., Weigert, A., Hofmeister, E.: Methods for calculating stellar evolution. Meth. Comput. Phys. **7**, 129–190 (1967)
*571. Kitamura, H., Ichimaru, S.: Pycnonuclear reaction rates in stellar interiors. Astrophys. J. **438**, 300–307 (1995)

*572. Kirzhnits, D.A.: On the internal structure of superdense stars. Jurnal Eksperimental'noi i Teoreticheskoi Fiziki (J. Exp. Theor. Phys., JETP) **38**, 503–508 (1960)

573. Kirzhnits, D.A., Lozovik, Yu.E., Shpatakovskaya, G.V.: A statistic model of matter. UFN. **117**, 3–47 (1975) (Soviet. Phys. Uspekhi **18**, 649–672, 1975)

574. Klebesadel, R.W., Strong, I.B., Olson, R.A.: Observations of gamma-ray bursts of cosmic origin. Astrophys. J. Lett. **182**, L85–L88 (1973)

575. Knunyanz, I.L. (ed.): Deuterium, Chemical encyclopedic dictionary. Sovetskaya Entsiklopediya, Moscow, p. 149 (1983)

576. Kolesnik, I.G.: Gravitational contraction of protostars. I. Bulk losses of energy. Astrometriya i Astrofizika, **18**, 45–58 (1973)

577. Kolesnik, I.G.: The hydrodynamics of protostar collapse. Pre-print of ITF No. 79–44P, Kiev (1979)

578. Kolesnik, I.G.: The numerical simulation of protostar collapse. Physical and mathematical formulation of the problem. I. Basic equations. Astrometriya i Astrofizika No. 40, pp. 3–18; II. Computational algorithms, *Ibid.* No. 41, pp. 40–58 (1980)

579. Koldoba, A.V., Tarasova, E.V., Chechetkin, V.M.: On the instability of the detonation wave in thermonuclear model of supernova. PAJ **20**, 445–449 (1994) (Sov. Astron. Lett. **20**, 377–381, 1994)

*580. Kohijama, Y., Itoh, N., Munakata, H.: Neutrino energy losses in stellar interiors II. Axial-vector contribution to the plasma neutrino energy loss rate. Astrophys. J. **310**, 815–819 (1986)

581. Koide, H., Matzuda, T., Shima, E.: Numerical simulations of axisymmetric accretion flows. Mon. Not. R. Astron. Soc. **252**, 473–481 (1991)

582. Konar, S., Choudhuri, A.R.: Diamagnetic screening of the magnetic field in accreting neutron stars - II. The effect of polar cap widening. Mon. Not. R. Astron. Soc. **348**, 661–668 (2004)

583. Kotake, K., Sato, K., Takahashi, K.: Explosion mechanism, neutrino burst and gravitational wave in core-collapse supernovae. Rep. Prog. Phys. **69**, 971–1143 (2006)

584. Kotake K., Sawai H., Yamada S., Sato K.: Magnetorotational effects on anisotropic neutrino emission and convection in core-collapse supernovae. Astrophys. J. **608**, 391–404 (2004)

585. Kosovichev, A.G., Christensen-Dalsgaard, J., Däppen, W., Dziembovski, W., Gough, D.O., Thompson, M.J.: Sources of uncertainity in direct seismological measurements of the solar helium abundance. Mon. Not. R. Astron. Soc. **259**, 536–547 (1992)

*586. Kopysov, Yu.S.: Neutrinos and the internal structure of the Sun. Preprint of IYaI No. P-0019 (1975)

*587. Kozik, V.S., Liubimov, V.A., Novikov, Ye.G., Nozik, V.Z., Tretyakov, E.F.: On the estimate of n mass by spectrum of beta-decay of Tritium in Valin. Yadernaya Fizika **32**, 301–303 (1980)

588. Kraft, R.P., Greenstein, J.L.: A new method for finding faint members of the Pleiads. In: Low Luminosity Stars, pp. 65–82. Gordon and Breach, New York (1969)

589. Krastev, P.G., Sammaruca, F.: Neutron star properties and the equation of state of neutron-rich matter. Phys. Rev. C **74**, 025808 (2006)

590. Kravtzov, V.A.: Atomic masses and binding energies of nuclei. Moscow, Atomizdat (1974)

591. Krasnov, N.F.: Aerodynamics. Moscow, Vysshaya Shkola (1971)

*592. Krylov, V.I.: The approximate evaluation of integrals. Moscow, Nauka (1967)

*593. Krylov, N.S.: Investigations on the statistical physics foundation. Moscow, Leningrad, AN SSSR (1950)

594. Kuan, P.: Emission envelopes of T Tauri stars. Astrophys. J. **202**, 425–432 (1975)

595. Kuhi, L.V.: Mass loss from T Tauri stars. Astrophys. J. **140**, 1409–1433 (1964)

596. Kulkarni, S.R.: Optical identification of binary pulsars implications for magnetic field decay in neutron stars. Astrophys. J. Lett. **306**, L85–L90 (1986)

597. Kulkarni, S.R., et al.: The afterglow, redshift and extreme energetics of the γ-ray burst of 23 January 1999. Nature **398**, 389–399 (1999)

598. Kundt, W.: Are supernova explosions driven by magnetic strings? Nature **261**, 673–674 (1976)
599. Kutter, G.S., Savedoff, M.P., Schuerman, D.W.: A mechanism for the production of planetary nebulae. Astrophys. Space Sci. **3**, 182–197 (1969)
600. Kutter, G.S., Sparks, W.: Studies of hydrodynamic events in stellar evolution III. Ejection of planetary nebulae. Astrophys. J. **192**, 447–455 (1974)
601. Kwok, S.: From red giants to planetary nebulae. Astrophys. J. **258**, 280–288 (1982)
*602. Lafon, J.-P.J., Beruyer, N.: Mass loss mechanisms in evolved stars. Astron. Astrophys. Rev. **2**, 349–389 (1991)
*603. Lamb, G.: Hydrodynamics. Cambridge University Press, Cambridge (1932)
604. Lamb, D.Q., Pethick, C.J.: Effects of neutrino degeneracy in supernova models. Astrophys. J. Lett. **209**, L77–L82 (1976)
605. Lamb, D.Q., Lattimer, J.M., Pathick, C.J., Ravenhall, D.G.: Hot dense matter and stellar collapse. Phys. Rev. Lett. **41**, 1623–1626 (1978)
606. Lamb, D.Q., Van Horn, H.M.: Evolution of crystallizing pure C^{12} white dwarfs. Astrophys. J. **200**, 306–323 (1975)
607. Lamb, D.Q.: Neutron star binaries, pulsars and burst sources. Preprint (Urbana University) (1981)
608. Lamb, S., Iben, I., Howard, M.: On the evolution of massive stars through the core carbon-burning phase. Astrophys. J. **207**, 209–232 (1976)
*609. Lambert, D.L.: The p-nuclei: abundances and origin. Astron. Astrophys. Rev. **3**, 201–256 (1992)
610. Lamers, H.: The dependence of mass loss on the basic stellar parameters. In: Chiosi, C., Stalio, R. (eds.) Effects of Mass Loss on Stellar Evolution, pp. 19–23. Reidel, Dordrecht (1981)
611. Lamers, H.J.G.L., Leitherer, C.: What are the mass-loss rates of O stars? Astrophys. J. **412**, 771–791 (1993)
*612. Lampe, M. (1968): Transport coefficients of degenerate plasma. Phys. Rev. **170**, 306–319
613. Landau, L.D.: On the theory of stars. Works **I** Nauka, Moscow, pp. 86–89 (1969)
614. Landau, L.D., Lifshitz, E.M.: Mechanics of continuum media. Gostekhteorizdat, Moscow (1953)
615. Landau, L.D., Lifshitz, E.M.: The theory of fields. Nauka, Moscow (1962)
616. Landau, L.D., Lifshitz, E.M.: Quantum mechanics. Nauka, Moscow (1963)
617. Landau, L.D., Lifshitz, E.M.: Statistical physics **I**. Nauka, Moscow (1976)
618. Landau, L.D., Lifshitz, E.M.: Electrodynamics of continua. Nauka, Moscow (1982)
*619. Langanke, K., Wiescher, M., Fowler, W., Gorres, J.: A new estimate of the $Ne^{19}(p,\gamma)Na^{20}$ and $O^{15}(\alpha,\gamma)Ne^{19}$ reaction rates at stellar energies. Astrophys. J. **301**, 629–633 (1986)
*620. Langer, N., El Eid, M.F., Fricke, K.J.: Evolution of massive stars with semiconvective diffusion. Astron. Astrophys. **145**, 179–191 (1985)
621. Langer, N., Hamann, W.-R., Lennon, M., Najarro, F., Pauldrach, A.W.A., Puls, J.: Towards an understanding of very massive stars. A new evolutionary scenario relating O stars, LBVs and Wolf–Rayet stars. Astron. Astrophys. **290**, 819–833 (1994)
622. Larson, R.: The evolution of protostars – theory. Found. Cosm. Phys. **1**, 1–70 (1973)
623. Lattimer, J., Mazurek, T.: Leptonic overturn and shocks in collapsing stellar cores. Astrophys. J. **246**, 955–965 (1981)
*624. Lattimer, J.: The equation of state of hot dense matter and supernovae. Annu. Rev. Nucl. Part. Sci. **31**, 337–374 (1981)
625. Lattimer, J., Mazurek, T.: Stellar implosion shocks and convective overturn. In: Proc. DUMAND-1980 (1980)
626. Lazareff, B.: Thermodynamics of convective URCA cores. Astron. Astrophys. **45**, 141–143 (1975)
*627. Lavrentyev, M.A., Shabat, B.V.: *Methods of the Theory of Functions of a Complex Variable* Nauka, Moscow (1958)
*628. Ledoux, P.: To the conditions of equilibrium in the centre of stars and their evolution. Nuclear Processes in Stars, Proc. of Lirge Coll., pp. 152–162. IL, Moscow, (1957)

629. Le Blank, L.M., Wilson, J.R.: A numerical example of the collapse of a rotating magnetized star. Astrophys. J. **161**, 541–551 (1970)
630. Ledoux, P.: Non-radial oscillations. In: Ledoux, P., Noels, A., Rodgers, A.W. (eds.) Proc. IAU Symp. No. 59, pp. 135–173. Reidel, Dordrecht (1974)
*631. Lewellyn-Smith, C.H.: Neutrino reactions at accelerator energies. Phys. Rep. **3C**, 261–379 (1972)
*632. Lifshitz, E.M., Pitaevskii, L.P.: Physical kinetics. Nauka, Moscow (1979)
633. Li, X.-D., Bombaci, I., Dey, M., Dey, J., van den Heuvel, E.P.J.: Is SAX J1808.4-3658 a Strange Star? Phys. Rev. Lett. **83**, 3776–3779 (1999)
634. Liang, E.P.T.: Accretion disk corona and the nature of X-ray burster 3U 1820-30. Astrophys. J. Lett. **211**, L67–L70 (1977)
635. Liang, E.P.T., Thompson, K.A.: Transonic disk accretion onto black holes. Astrophys. J. **240**, 271–274 (1980)
636. Liang, E.P., Wandel, A.: Luminosity limit for alpha-viscosity accretion disks. Astrophys. J. **376**, 746–749 (1991)
637. Lichtenberg, A.J., Lieberman, M.A.: Regular and Stochastic Motion. Springer, New York (1983)
638. Lighthill, H.J.: On the stability of small planetary cores (II). Mon. Not. R. Astron. Soc. **110**, 339–342 (1950)
*639. de Loore, C.: The influence of mass loss on the evolution of binaries. In: Chiosi, C., Stalio, R. (eds.)Effects of Mass Loss on Stellar Evolution, pp. 405–427. Reidel, Dordrecht (1981)
*640. Lopez, J.L., Nanopoulos, D.V. and Zichichi, A.: The top-quark mass in SU(5)×U(1) supergravity. Preprint ASC-31/94 (1994)
641. Lorrimer, D.R.: Binary and millisecond pulsars at the new millenium. Living reviews in relativity **4**, 5–64; Preprint astro-ph 0104388 (2001)
642. Lorrimer, D.R.: Binary and millisecond pulsars. Living reviews in relativity **8**, 7–84; Preprint astro-ph 0511258 (2005)
643. Lovelace, R.V.E.: Dynamo model of double radio sources. Nature **262**, 649–652 (1976)
644. Lovelace, R.V.E., Bisnovatyi-Kogan, G.S., Romanova, M.M.: Spin-up/spin-down of magnetized stars with accretion discs and outflows. Mon. Not. R. Astron. Soc. **275**, 244–254 (1995)
645. Lovelace, R.V.E., Romanova, M.M., Bisnovatyi-Kogan, G.S.: Screening of the magnetic field of disk accretion stars. Astrophys. J., **625**, 957–965 (2005)
646. Lovelace, R.V.E., Romanova, M.M., Newman, W.I.: Implosive accretion and outbursts of active galactic nuclei. Astrophys. J. **437**, 136–143 (1994)
647. Lovelace, R.V.E., Rothstein, D.M., Bisnovatyi-Kogan, G.S.: Advection/diffusion of large-scale B field in accretion disks. Astrophys. J. **701**, 885–890 (2009)
*648. Loytzianskii, L.G.: Fluid mechanics. Nauka, Moscow (1987)
649. Lozinskaya, T.A.: Supernovae and stellar wind. The interaction with gas of galaxy. Nauka, Moscow (1986)
*650. Lyutostanskii, Yu. S., et al.: A kinetic model for beta-process. Preprint of IPM, No. 95 (1984)
651. Lubow, S.H., Papaloizou, J.C.B., Pringle, J.E.: On the stability of magnetic wind-driven accretion discs. Mon. Not. R. Astron. Soc. **267**, 235–240 (1994)
652. Lucy, L.: Formation of planetary nebulae. Astron. J. **72**, 813 (1967)
653. Lucy, L.B.: Gravity-darkening for stars with convective envelopes. Z. Astrophys. **65**, 89–92 (1967)
654. Lucy, L.B.: Radiatively-driven stellar winds. Preprint No. 419 European Southern Observatory (1986)
655. Luo, C., Liang, E.P.: Stability and oscillations of thermal accretion discs around black-holes. Astrophys. J. **266**, 386–398 (1994)
656. Luyten, W.G.: The white dwarfs: discovery and observation. In: Imshennik, V.S. (ed.) White dwarfs, pp. 13–22. Mir, Moscow (1975)
657. Lynden-Bell, D.: Galactic nuclei as collapsed old quasars. Nature **223**, 690–694 (1969)

658. Lyne, A.G., et al.: The discovery of a millisecond pulsar in the globular cluster M28. Nature **328**, 399–401 (1987)
659. Lyne, A.G., Lorimer, D.R.: High birth velocities of radio pulsars. Nature **369**, 127–129 (1994)
660. Lyne A.G., Burgay M., Kramer M., Possenti A., Manchester R.N., Camilo F., McLaughlin M.A., Lorimer D.R., D'Amico N., Joshi B.C., Reynolds J., Freire P.C.C.: A double-pulsar system: a rare laboratory for relativistic gravity and plasma physics. Science, **303**, 1153–1157 (2004)
661. Mac Donald, J.: The effect of a binary companion on a nova outburst. Mon. Not. R. Astron. Soc. **191**, 933–949 (1980)
662. MacFadyen, A.I., Woosley, S.E.: Collapsars: gamma-ray bursts and explosions in "Failed Supernovae". Astrophys. J. **524**, 262–289 (1999)
663. Maeder, A.: Stellar evolution III: the overshooting from convective cores. Astron. Astrophys. **40**, 303–310 (1975)
664. Maeder, A.: The most massive stars evolving to red supergiants: evolution with mass loss, WR stars, as post-red supergiants and pre-supernovae. Astron. Astrophys. **99**, 97–107 (1981)
665. Maeder, A.: Grid of evolutionary models for upper part of the HR diagram, mass loss and the turning of some red supergiants into WR stars. Astron. Astrophys. **102**, 401–410 (1981)
666. Maeder, A.: Physics, Formation and Evolution of Rotating Stars. Springer, Berlin (2009)
667. Makashima, K., et al.: Simultaneous X-ray and optical observations of GX 339–4 in an X-ray high state. Astrophys. J. **308**, 635–643 (1986)
668. Malone, R., Johnson, M., Bethe, H.: Neutron star model with realistic high-density equations of state. Astrophys. J. **199**, 741–748 (1975)
669. Malov, I.F.: Radiopulsars. M. Nauka (in Russian) (2004)
670. Manchester, R.N., Taylor, G.H.: Pulsars. Mir, Moscow (1980)
*671. Massevich, A.G., Ruben, G.V., Lomnev, S.P., Popova, E.I.: Calculation of homogeneous 4 $M\odot$, 8 $M\odot$, 16 $M\odot$ models for various chemical compositions and various absorption laws. Nauchnyie Informatzii Astronomicheskogo Soveta AN SSSR (Scientific Information of the Astonomical Council of the Academy of Sciences of the USSR), Issue 1, 2–47 (1965)
672. Massevich, A.G., Tutukov, A.V.: The evolution of massive stars and semiconvection problem, Nauchnyie Informatzii Astronomicheskogo Soveta AN SSSR (Scientific Information of the Astonomical Council of the Academy of Sciences of the USSR), Issue 29, 3–26 (1974)
673. Massevich, A.G., Tutukov, A.V.: *Stellar evolution: Theory and Observations* Nauka, Moscow
*674. Massaguer, J.M., Latour, J., Toomre, J., Zahn, J.-P.: Penetrative cellular convection in a stratified atmosphere. Astron. Astrophys. **140**, 1–16 (1984)
*675. Mathews, G., Dietrich, F.: The $N^{13}(p,\gamma)O^{14}$ thermonuclear reaction rate and the hot CNO cycle. Astrophys. J. **287** 969–976 (1984)
676. Mathur, S., Garcia, R.A., Catala, C., et al.: The solar-like CoRoT target HD 170987: spectroscopic and seismic observations. ArXiv e-print (arXiv:1004.4891) (2010)
677. Matsuda, T, Sekino, N., Sawada, K, Shima, E., Livio, M., Anzer, U., Börner, G. On the stability of wind accretion. Astron. Astrophys. **248**, 301–314
678. Matsumoto, R., Kato, S., Fukue, J., Okazaki, A.T.: Viscous transonic flow around the inner edge of geometrically thin accretion disks. Publ. Astron. Soc. Jpn. **36**, 71–85 (1984)
679. Mayle, R., Wilson, J.R., Schramm, D.N.: Neutrinos from gravitational collapse. Astrophys. J. **318**, 288–306 (1987)
680. Mazetz, E.P., et al.: A bursting X-ray pulsar in Dorado. PAJ **5**, 307–312 (1979) (Sov. Astron. Lett. **5**, 163–165, 1979)
681. Mazets, E.P., Cline, T.L., Aptekar', R.L., Butterworth, P.S., Frederiks, D.D., Golenetskii, S.V., Il'Inskii, V.N., Pal'Shin, V.D.: PAJ **25**, 735–749 (1999) (Astron. Lett. **25**, 635–648, 1999)

682. Mazets, E.P., Aptekar, R.L., Cline, T.L., Frederiks, D.D., Goldsten, J.O., Golenetskii, S.V., Hurley, K., von Kienlin, A., Pal'shin, V.D.: A Giant flare from a soft gamma repeater in the andromeda galaxy (M31). Astrophys. J. **680**, 545–549 (2008)
683. Mazets, E.P., Golenetskii, S.V., Il'inskii, V.N., Panov, V.N., Aptekar, R.L., Gur'yan, Y.A., Proskura, M.P., Sokolov, I.A., Sokolova, Z.Ya., Kharitonova, T.V.: Catalog of cosmic gamma-ray bursts from the KONUS experiment data. Astrophys. Space Sci. **80**, 3–83; 85–117; 119–143 (1981)
684. Mazurek, T.: Degeneracy effects of neutrino mass ejection in supernovae. Nature **252**, 287–289 (1974)
*685. Mestel, L.: Meridional circulation in stars, In: Aller, L.H., McLaughlin, D.B. (eds.) Stellar Structure. University of Chicago Press, Chicago (1965)
686. Menard, F., Monin, J.-L., Angelucci, F., Rougan, D.: Disks around pre-main-sequence binary systems: the case of Haro 6-10. Astrophys. J. Lett. **414**, L117–L120 (1993)
*687. Merryfield, W.J.: Hydrodynamics of semiconvection. Astrophys. J. **444**, 318–337 (1995)
688. Mestel, L.: On the theory of white dwarf stars. I. The energy sources of white dwarfs. Mon. Not. R. Astron. Soc. **112**, 583–594 (1952)
689. Mestel, L.: On the theory of white dwarf stars. II. The accretion of interstellar matter by white dwarfs. Mon. Not. R. Astron. Soc. **112**, 598–605 (1952)
*690. Mestel, L., Ruderman, M.A.: The energy content of a white dwarf and its rate of cooling. Mon. Not. R. Astron. Soc. **136**, 27–38 (1967)
691. Meśzáros, P.: Gamma-ray bursts afterglows and their implications. Astron. Astrophys. Suppl. **138**, 533–536 (1999)
692. Mewe, R.: X-ray spectroscopy of stellar coronae. Astron. Astrophys. Rev. **3**, 127–168 (1991)
693. Meier, D.L., Epstein, R.I., Arnett, W.D., Schramm, D.N.: Magnetohydrodynamic phenomena in collapsing stellar cores. Astrophys. J. **204**, 869–878 (1976)
*694. Meyers, W., Swiatecki, W.: Nuclear masses and deformations. Nucl. Phys. **81**, 1–60 (1966)
695. Migdal, A.B.: Fermions and bosons in strong fields. Nauka, Moscow (1978)
*696. Mikhalas, D.: Stellar atmospheres. Parts I and II. Freeman, San Francisco (1978)
697. Mikheevs, S.P., Smirnov, A.Yu.: The resonant enhancement of oscillations in matter and spectroskopy of solar neutrinos. Yadernaya Fizika **42**, 1441–1448 (1985)
698. Mikheevs, S.P., Smirnov, A.Yu.: Neutrino oscillations in a medium with varying density and bursts caused by gravitational collapses of stars. Jurnal Eksperimental'noi i Teoreticheskoi Fiziki (J. Exp. Theor. Phys., JETP) **91**, 7–13 (1986)
699. Michel F.C.: A fast pulsar in supernova 1987A. Mon. Not. R. Astron. Soc. **267**, L4–L6 (1994)
700. Migdal, A.B.: Pion fields in nuclear matter. Rev. Mod. Phys. **50** 107–172 (1978)
701. Migdal, A.B., Chernoutsan, A.I., Mishustin, I.N.: Pion condensation and dynamics of neutron stars. Phys. Lett. **83B**, 158–160 (1979)
*702. Mihalas, D., Däppen, W., Hummer, D.G.: The equation of state for stellar envelopes. II. Algorithm and selected results. Astrophys. J. **331**, 815–825 (1988)
703. Mikami, H., Sato, Y., Matsumoto, T., Hanawa, T.: Three-dimensional magnetohydrodynamical simulations of a core-collapse supernova. Astrophys. J. **683**, 357–374 (2008)
704. Miller, D.S., Wilson, J.R., Mayle, R.W.: Convection above the neutrinosphere in Type II supernovae. Astrophys. J. **415**, 278–285 (1993)
*705. Mitler, H.: Thermonuclear ion–electron screening at all densities. I. Static solution. Astrophys. J. **212**, 513–532 (1977)
706. Mochkovitch, R.: The thermodynamics of the convective URCA process. Astron. Astrophys. **311**, 152–154 (1996)
707. Moiseenko, S.G., Bisnovatyi-Kogan, G.S., Ardeljan, N.V.: A magnetorotational core-collapse model with jets. Mon. Not. R. Astron. Soc. **359**, 501–512 (2006)
708. Moon, D.-S., Cenko, S.B., Adams, J.: GRB041219: confirmed NIR afterglow. GCN Circ. 2876 (2004)
*709. Morozov, Yu.I.: Radiative transfer equations in non-inertial coordinate systems. Jurnal Prikladnoi Mekhaniki i Tekhnicheskoi Fiziki (J. Appl. Mechan. Phys. Eng.) **1**, 3–11 (1970)

*710. Morse, Ph.M., Feschbach, H.: Methods of theoretical physics. McGraw-Hill, New York (1953)

711. Morton, D.: Mass loss from three OB supergiants in Orion. Astrophys. J. **150**, 535–542 (1967)

712. Moss, D.: Models for rapidly rotating pre-main-sequence stars. Mon. Not. R. Astron. Soc. **161**, 225–237 (1973)

*713. Moss, D.: Magnetic star models: toroidal fields and circulation. Mon. Not. R. Astron. Soc. **178**, 51–59 (1977)

*714. Moss, D.: Time-dependent models of rotating magnetic stars. Mon. Not. R. Astron. Soc. **209**, 607–639 (1984)

*715. Mukhin, K.N.: Experimental nuclear physics **1** Physics of atomic nucleus. Energoatomizdat, Moscow (1983)

716. Muchotrzeb, B.: Transonic accretion flow in a thin disk around a black hole II. Acta Astron. **33**, 79–87 (1983)

717. Muchotrzeb, B., Pachinski, B.: Transonic accretion flow in a thin disk around a black hole. Acta Astron. **32**, 1–11 (1982)

718. Müller, E., Hillebrandt, W.: A magnetohydrodynamical supernova model. Astron. Astrophys. **80**, 147–154 (1979)

719. Müller, R.A., Newberg, H.J.N., Pennypacker, C.R., Perlmutter, S., Sasseen, T.P., Smith, C.K.: High rate for type Ic supernovae. Astrophys. J. Lett. **384**, L9–L13 (1992)

720. Munakata, H., Kohyama, Y., Itoh, N.: Neutrino energy loss in stellar interior. Astrophys. J. **296**, 197–203 (1985)

721. Muslimov, A.: The system PSR 1718-19 and a hypothesis about latent magnetism in neutron stars. Astron. Astrophys. **295** L27–L29 (1995)

722. Muslimov, A., Page, D.: Delayed switch-on of pulsars Astrophys. J. Lett. **440**, L7–L80 (1995)

723. Myra, E.S., Bludman, S.A., Hoffman, Y., Lichenstadt, I., Sack, N., van Riper, K.A. The effect of neutrino transport on the collapse of iron stellar cores. Astrophys. J. **318**, 744–759 (1987)

724. Nadyozhin, D.K.: Evolution of a star with M=30 $M\odot$ in hydrogen-burning phase., Nauchnyie Informatzii Astronomicheskogo Soveta AN SSSR (Scientific Information of the Astonomical Council of the Academy of Sciences of the USSR), Issue 4, 37–64 (1966)

*725. Nadyozhin, D.K.: Asymptotic formulae for the equation of state of electron–positron gas. Nauchnyie Informatzii Astronomicheskogo Soveta AN SSSR (Scientific Information of the Astonomical Council of the Academy of Sciences of the USSR), Issue 32, 3–72 (1974)

*726. Nadyozhin, D.K.: Tables for the equation of state of electron–positron gas. Nauchnyie Informatzii Astronomicheskogo Soveta AN SSSR (Scientific Information of the Astonomical Council of the Academy of Sciences of the USSR), Issue 33, 117–142 (1974)

*727. Nadyozhin, D.K., Chechetkin, V.M.: Neutrino emission by URCA-process at high temperatures. Astron. J. **46**, 270–279 (1969) (Sov. Astron. **13**, 213–218, 1969)

728. Nadyozhin, D.K.: The collapse of iron-oxygen stars: physical and mathematical formulation of the problem and computational methods. Astrophys. Space Sci. **49**, 399–425 (1977)

729. Nadyozhin, D.K.: Gravitational collapse of iron cores with masses 2 and 10 $M_\odot$. Astrophys. Space Sci. **51**, 284–302 (1977)

730. Nadyozhin, D.K.: The neutrino radiation for a hot neutron star formation and the envelope outburst problem. Astrophys. Space Sci. **53**, 131–153 (1978)

*731. Nakazawa, K.: Effect of electron capture on temperature and chemical composition in collapsing dense stars. Progr. Theor. Phys. **49**, 1932–1946 (1973)

732. Nakazawa, K., Hayashi, C., Takahara, M.: Isothermal collapse of rotating gas clouds. Progr. Theor. Phys. **56**, 515–530 (1976)

*733. Nakazawa, K., Murai, T., Hoshi, R., Hayashi, C.: Effect of electron capture on the temperature in dense stars. Progr. Theor. Phys. **44**, 829–830 (1970)

734. Narayan, R., Yi, I.: Advection - dominated accretion: underfed black holes and neutron stars. Astrophys. J. **452**, 710–735 (1995)

735. Nayfeh, A.: Perturbation Methods Wiley, New York (1981)
*736. Negele, J., Vautherin, D.: Neutron star matter at sub-nuclear densities. Nucl. Phys. **A207**, 298–320 (1973)
*737. Neuforge, C.: Low temperature Rosseland mean opacities. Astron. Astrophys. **274**, 818–820 (1993)
*738. Neugebauer, G. et al.: The infrared astronomical satellite (IRAS) mission. Astrophys. J. Lett. **278**, L1–L6 (1984)
739. Neuhäuser, R., Sterzik, M.F., Schmitt, J.H.M.M., Wichmann, R., Krautter, J.: ROSAT survey observation of T Tauri stars in Taurus. Astron. Astrophys. **297**, 391–417 (1995)
740. Newman, M.: S-process studies: the exact solution. Astrophys. J. **219**, 676–689 (1978)
741. Ng, C.-Y., Romani, R.W.: Birth kick distributions and the spin-kick correlation of young pulsars. Astrophys. J. **660**, 1357–1374 (2007)
742. Nice, D.J., Splaver, E.M., Stairs, I.H., Löhmer, O., Jessner, A., Kramer M., Cordes, J.M.: A 2.1 $M_{\odot}$ pulsar measured by relativistic orbital decay. Astrophys. J. **634**, 1242–1249 (2005)
743. Niemeyer, J.C., Hillebrandt, W.: Microscopic instabilities of nuclear flames in type Ia supernovae. Astrophys. J. **452**, 779–784 (1995)
744. Niemeyer, J.C., Hillebrandt, W.: Turbulent nuclear flames in type Ia supernovae. Astrophys. J. **452**, 762–778 (1995)
745. Niemeyer, J.C.: Can deflagration-detonation transitions occure in Type Ia supernovae? Astrophys. J. Lett. **523**, L57–L60 (1999)
746. Nomoto, K.: Accreting white dwarf models for type I supernovae II. Off-center detonation supernovae. Astrophys. J. **257**, 780–792 (1982)
747. Nomoto, K.: Neutron star formation in theoretical supernovae – low mass stars and white dwarfs. In: Helfand, D., Huang, J. (eds.) The Origin and Evolution of Neutron Stars. Proc. symp. IAU No. 125 Reidel, Dordrecht (1986)
748. Nomoto, K., Iwamoto, K., Suzuki, T., Yamaoka, H., Hashimoto, M., Höfflich, P., Pols, O.R., van den Heuvel, E.P.J.: Type Ib/Ic/IIb/ II-L supernovae. Preprint No. 95-13, University of Tokyo (1995)
*749. Nomoto, K., Thielemann, F.-K., Wheeler, J.C.: Explosive nucleosynthesis and type I supernovae. Astrophys. J. Lett. **279**, L23–L26 (1980)
*750. Nomoto, K., Thielemann, F.-K., Miyaji, S.: The triple alpha reaction at low temperatures in accreting white dwarfs and neutron stars. Astron. Astrophys. **149**, 238–245 (1985)
751. Nomoto, K., Tsuruta, S.: Cooling of young neutron stars and the Einstein X-ray observations. Astrophys. J. Lett. **250**, L19–L23 (1981)
752. Nomoto, K., Tsuruta, S.: Cooling of neutron stars: effects of finite scale of thermal conduction. Astrophys. J. **312**, 711–726 (1987)
*753. Nordlung, A.: On convection in stellar atmospheres. Astron. Astrophys. **32**, 407–422 (1974)
754. Norman, M.L., Wilson, J.R., Barton, R.T.: A new calculation on rotating protostellar collapse. Astrophys. J. **239**, 968–981 (1980)
755. Novikov, I.D., Thorne, K.S.: Astrophysics of black holes. In: De Witt, B., De Witt, C., (eds.)Black Holes,pp. 343–561. Gordon and Breach, Paris (1973)
756. Novikov, I.D., Frolov, V.P.: Physics of black holes. Nauka, Moscow (1986)
757. Obergaulinger, M., Aloy, M.A., Dimmelmeier, H., Müller, E.: Axisymmetric simulations of magnetorotational core collapse: approximate inclusion of general relativistic effects. Astron. Astrophys. **457**, 209–222 (2006)
758. O'Connell, R.F., Matese, J.J.: Effect of a constant magnetic field on the neutron beta decay rate and its astrophysical implications. Nature **222**, 649–650 (1969)
*759. Ogata, Sh., Ichimaru, S., Van Horn, H.M.: Thermonuclear reaction rates for dense binary–ionic mixtures. Astrophys. J. **417**, 265–272 (1993)
760. Ohnishi, T.: Gravitational collapse of rotating magnetized stars. Tech. Rep. Inst. At. En. Kyoto Univ. No. 198 (1983)

761. Ohnishi, N., Iwakami, W., Kotake, K., Yamada, Sh., Fujioka, S., Takabe, H.: Standing accretion shock instability: numerical simulations of core-collapse supernova. J. Phys. Conf. Ser. **112**, 042018 (2008)
*762. Okun', L.B.: The weak interaction of elementary particles. Nauka, Moscow (1963)
*763. Okun', L.B.: Leptons and quarks. Nauka, Moscow (1981)
*764. Okun', L.B.: The physics of elementary particles. Nauka, Moscow (1984)
765. Oppenheimer, J., Volkoff, G.: On massive neutron cores. Phys. Rev. **55**, 374–381 (1939)
766. Ostriker, J., Gunn, J.: Do pulsars make supernovae? Astrophys. J. Lett. **164**, L95–L104 (1971)
*767. Ostriker, J., Mark, J.: Rapidly rotating stars. I. The self-consistent-fluid method. Astrophys. J. **151**, 1075–1088 (1968)
768. Owocki, S.P., Castor, J.I., Rybicki, G.B.: Time-dependent models of radiatively driven stellar winds. I. Non-linear evolution of instabilities for a pure absorption model. Astrophys. J. **335**, 914–930 (1988)
769. Paczynski, B.: Envelopes of red supergiant. Acta Astron. **19**, 1–22 (1969)
770. Paczynski, B.: Evolution of single stars. I. Stellar evolution from main sequence to white dwarf or carbon ignition. Acta Astron. **20**, 47–58 (1970)
771. Paczynski, B.: Evolution of single stars. II. Core helium burning in population I stars. Acta Astron. **20**, 195–212 (1970)
772. Paczynski, B.: Evolution of single stars. III. Stationary shell source. Acta Astron. **20**, 287–309 (1970)
773. Paczynski, B.: Evolution of single stars. V. Carbon ignition in population I stars. Acta Astron. **21**, 271–288 (1971)
774. Paczynski, B.: Evolution of single stars. VI. Model nuclei of planetary nebulae. Acta Astron. **21**, 471–435 (1971)
775. Paczynski, B.: Carbon ignition in degenerate stellar cores. Astrophys. Lett. **11**, 53–55 (1972)
776. Paczynski, B.: Linear series of stellar models. I. Thermal stability of stars. Acta Astron. **22**, 163–174 (1972)
777. Paczyński, B.: URCA processes in convective stellar cores. Astron. Lett. **15**, 147–149 (1973)
778. Paczynski, B.: Evolution of stars with $M \leq 8\,M_{\odot}$. In: R. Tayler (ed.) Proc. Symp IAU No. 66 Late Stages of Stellar Evolution, pp. 62–69. Reidel, Dordrecht (1974)
779. Paczynski, B.: Helium flush in population I stars. Astrophys. J. **192**, 483–485 (1974)
780. Paczynski, B.: Core mass-interflash period relation for double-shell source stars. Astrophys. J. **202**, 558–560 (1975)
781. Paczynski, B.: Helium shell flashes. Astrophys. J. **214**, 812–818 (1977)
*782. Paczynski, B.: Models of X-ray bursters with radius expansion. Astrophys. J. **267**, 315–321 (1983)
783. Paczynski, B., Bisnovatyi-Kogan, G.S.: A model of a thin accretion disk around a black hole. Acta Astron. **31**, 283–291 (1981)
784. Paczynski, B., Schwarzenberg-Czerny, A.: Disk accretion in U Geminorum. Acta Astron. **30**, 127–141 (1980)
785. Paczynski, B., Wiita, P.: Thick accretion disks and supercritical luminosities. Astron. Astrophys. **88**, 23–31 (1980)
786. Paczynski, B., Ziolkovski, J.: On the origin of planetary nebulae and Mira variables. Acta Astron. **18**, 255–266 (1968)
787. Page, D., Usov, V.V.: Thermal evolution and light curves of young bare strange stars. Phys. Rev. Lett. **89**, 131101 (2002)
788. Palla, F., Stahler, S.W.: The pre-main-sequence evolution of intermediate-mass stars. Astrophys. J. **418**, 414–425 (1993)
789. Pallavicini, R.: X-ray emission from stellar coronae. Astron. Astrophys. Rev. **1**, 177–207 (1989)
*790. Pandharipande, V.: Dense neutron matter with realistic interaction. Nucl. Phys. A **A174**, 641–656 (1971)

791. Pandharipande, V., Pines, D., Smith, R.: Neutron star structure: theory, observation and speculation. Astrophys. J. **208**, 550–566 (1976)
792. Papaloizou, J.C.B.: Non-linear pulsations of upper main sequence stars. I. A perturbation approach. Mon. Not. R. Astron. Soc. **162**, 143–168 (1973)
793. Papaloizou, J.C.B.: Non-linear pulsations of upper main sequence stars. II. Direct numerical investigations. Mon. Not. R. Astron. Soc. **162**, 169–187 (1973)
794. Papaloizou, J.C.B., Whelan, J.A.J.: The structure of rotating stars: the J^2 method and results for uniform rotation. Mon. Not. R. Astron. Soc. **164**, 1–10 (1973)
795. Park, M.-G.: Self-consistent models of spherical accretion onto black holes. I. One-temperature solutions. Astrophys. J. **354**, 64–82 (1990)
796. Park, M.-G.: Self-consistent models of spherical accretion ontoblack holes. II. Two-temperature solutions with pairs. Astrophys. J. **354**, 83–97 (1990)
797. Parker, E.N.:The generation of magnetic fields in astrophysical bodies. VII. Dynamical considerations. Astrophys. J. **168**, 239–249 (1971)
798. Parker, E.N.: Interplanetary dynamical processes. Interscience, New York (1963)
799. Patterson, J.: The evolution of cataclysmic and low-mass X-ray binaries. Astrophys. J. Suppl. **54**, 443–493 (1984)
800. Petrov, P.P.: T Tauri stars: modern observational data. In: Early Stages of Stellar Evolution, pp. 66–100. Naukova Dumka, Kiev (1977)
*801. Petrosian, V., Beaudet, G., Salpeter, E.E.: Photoneutrino energy loss rates. Phys. Rev. **154**, 1445–1454 (1967)
802. Pieper, S.C., Wiringa, R.B., Carlson, J.: Quantum Monte Carlo calculations of excited states in A=6–8 nuclei. Phys. Rev. C **70**, 054325 (2004)
803. Piro, L., Garmire, G., Garcia, M.R., et al.: The X-Ray afterglow of GRB 000926 observed by BeppoSAX and Chandra: a mildly collimated fireball in a dense medium? Astrophys. J. **558**, 442–447 (2001)
*804. Pikel'ner, S.B.: Fundamentals of cosmic electrodynamics. Nauka, Moscow (1961)
805. Poe, C.H., Owocki, S.P., Castor, J.I.: The steady state solutions of radiatively driven stellar winds for a non-Sobolev, pure absorption model. Astrophys. J. **358**, 199–213 (1990)
806. Pogorelov, N.V., Ohsugi, Y., Matsuda, T.: Towards steady-state solutions for supersonic wind accretion on to gravitating objects. Mon. Not. R. Astron. Soc. **313**, 198–208 (2000)
*807. Pollock, E.L., Hansen, J.P.: Statisticsl mechanics of dense ionized matter. II. Equilibrium properties and melting transition of the crystallized one-component plasma. Phys. Rev. **8A**, 3110–3122 (1973)
808. Podsiadlowski, P., Cannon, R.C., Rees, M.J.: The evolution and final fate of massive Thorne–Zytkow objects. Mon. Not. R. Astron. Soc. **274**, 485–490 (1995)
809. Pols, O.R., Tout, Ch.A., Eggleton, P.P., Han, Zh.: Approximate input physics for stellar modeling. Mon. Not. R. Astron. Soc. **274**, 964–974 (1995)
810. Pontecorvo, B., Bilenky, S.: Neutrino today. Preprint JINR, Dubna, No. E1, 2-87-567 (1987)
811. Pozanenko, A., Loznikov, V., Beskin, G., Karpov, S. Bondar, S., Ivanov, E., Katkova, E., Biryukov, A., Zolotukhin, I., Hurley, K., Rumyantsev, V. (2004): A Search for Prompt Optical Emission from GRBs. Proc. 22nd Texas Symp. on Relativ. Ap. at Stanford Univ., 2219
*812. Porfiryev, V.V., Redkoborodyi, Yu.N.: The electron screening effect in nuclear reactions at high densities. Astrophys. **5**, 393–413 (1969) (Astrophys. **5**, 184–191, 1969)
813. Pskovskii, Yu. P.: Novae and supernovae stars. Nauka, Moscow (1985)
*814. Ptitzyn, D.A., Chechetkin, V.M.: To the problem of formation of elements beyond the iron peak in supernovae. PAJ **8**, 600–606 (1982) (Sov. Astron. Lett. **8**, 322–326, 1982)
815. Prilutsky, O.F., Usov, V.V.: On the nature of gamma-ray bursts. Astrophys. Space Sci. **34**, 395–401 (1975)
816. Pringle, J.E., Rees, M.J.: Accretion disc models for compact X-ray sources. Astron. Astrophys. **21**, 1–9 (1972)
*817. Radtzig, A.A., Smirnov, B.M.: Handbook on Atomic and Molecular Physics. Atomizdat, Moscow (1980)

818. Racusin, J.L., Karpov, S.V., Sokolowski, M., et al.: Broadband observations of the naked-eye γ-ray burst GRB080319B. Nature **455**, 183–188 (2008)
819. Radhakrishnan, V.: On the nature of pulsars. Contemp. Phys. **23**, 207–231 (1982)
*820. Raikh, M.E., Yakovlev, D.G.: Thermal and electrical conductivities of crystals in neutron stars and degenerate dwarfs. Astrophys. Space Sci. **87**, 193–203 (1982)
821. Ramsey, W.H.: On the stability of small planetary cores (I). Mon. Not. R. Astron. Soc. **110**, 325–338 (1950)
*822. Ravenhall, D., Bennett, C., Pechick, C.: Nuclear surface energy and neutron-star matter. Phys. Rev. Lett. **28**, 978–981 (1972)
*823. Redkoborodyi, Yu.N.: The quantum theory of screening effects in thermonuclear reactions. I. Relativistic electron plasma. Astrophys. **12**, 495–510 (1976) (Astrophys. **12**, 325–336, 1976)
824. Regev, O., Lilio, M.: X-ray bursters – hot way to chaos. In: Perdang, J.M., Buchler, J.R., Spiegel, E.A. (eds.) Chaos in Astrophysics. Reidel, Dordrecht (1985)
825. Regev, O.: The disk-star boundary layer and its effect on the accretion disk structure. Astron. Astrophys. **126**, 146–151 (1983)
826. Reimers, D.: Winds in red giants. In: Iben, I., Renzini, A. (eds.) Physical Processes in Red Giants. Reidel, Dordrecht, pp. 269–284 (1981)
*827. Reeves, H.: Sources of stellar energy in stellar structure. In: Aller, L.H., McLaughlin, D.B. (eds.) Stellar Structure. University of Chicago Press, Chicago (1965)
828. Rosseland, S.: Pulsation theory of variable stars. Clarendon, Oxford (1949)
829. Rublyov, S.V., Cherepazshuk, A.M.: [Wolf–Rayet Stars]. In: Non-stationary Phenomena and Stellar Evolution, pp. 47–124. Nauka, Moscow (1974)
*830. Ruderman, M.: *Astrophysical Neutrino*, Reports on Progress in Physics **61**, 411–462 (1965)
*831. Rudzskii, M.A., Seidov, Z.F.: Thermal effects in beta-processes. Izv. AN Azerb. SSR, ser. fiz. mat. i tekh. Nauk **4**, 98–106 (1974)
*832. Rudzskii, M.A., Seidov, Z.F.: Low-mass star collapse and heating by beta-processes. Astron. J. **51**, 936–939 (1974) (Sov. Astron. **18**, 551–553, 1975)
833. Riffert, H., Herold, H.: Relativistic accretion disk structure revisited. Astrophys. J. **408**, 508–511 (1995)
834. Robertson, H.P.: Dynamical effects of radiation in the solar system. Astrophys. J. **97**, 423–438 (1937)
*835. Robnik, M., Kundt, W.: Hydrogen at high pressures and temperatures. Astron. Astrophys. **120**, 227–233 (1983)
836. Rosenfeld, L.: Astrophysics and gravitation. In: Proceedings of the 16th Solvay Conference on Physics, Universite de Bruxells, Bruxells, p. 174 (1974)
837. Rogers, F.J., Iglesias, C.A.: Rosseland mean opacities for variable compositions. Astrophys. J. **401**, 361–366 (1992)
838. Rose, W., Smith, R. (1970): Final evolution of a low-mass star I. Astrophys. J. **159**, 903–912
839. Roth, M., Weigert, A.: Example of multiple solutions for equilibrium stars with helium cores. Astron. Astrophys. **20**, 13–18 (1972)
*840. Roxburgh, I.W.: Integral constraints on convective overshooting. Astron. Astrophys. **211**, 361–364 (1989)
*841. Roxburgh, I.W.: Limits on convective penetration from stellar cores. Astron. Astrophys. **266**, 291–293 (1992)
*842. Roxburgh, I.W., Simmons, J.: Numerical studies of convective penetration in plane parallel layers and the integral constraint. Astron. Astrophys. **277**, 93–102 (1993)
843. Roxburgh, I.W., Vorontsov, S.V.: Seismology of the stellar cores: a simple theoretical discription of the 'small frequency separations'. Mon. Not. R. Astron. Soc. **267**, 297–302 (1994)
844. Roxburgh, I.W., Vorontsov, S.V.: The asymptotic theory of stellar acoustic oscillations: fourth-order approximation for low-degree modes. Mon. Not. R. Astron. Soc. **268**, 143–158 (1994)

845. Roxburgh, I.W., Vorontsov, S.V.: An asymptotic description of solar acoustic oscillations of low and intermediate degree. Mon. Not. R. Astron. Soc. **278**, 940–946 (1996)
846. Roxburgh, I.W., Vorontsov, S.V.: An asymptotic description of solar acoustic oscillations with an elementary exitation source. Mon. Not. R. Astron. Soc. **272**, 850–858 (1995)
*847. Rubbia, C., Jacob, M.: The Z^0. Am. Sci. **78**, 502–519 (1990)
848. Ruffert, M., Janka, H.-Th.: Colliding neutron stars. Gravitational waves, neutrino emission, and gamma-ray bursts. Astron. Astrophys. **338**, 535–555 (1998)
849. Ruffert, M., Janka, H.-Th.: Gamma-ray bursts from accreting black holes in neutron star mergers. Astron. Astrophys. **344**, 573–606 (1999)
850. Ruffini, R., Salmonson, J.D., Wilson, J.R., Xue, S.-S.: On the pair-electromagnetic pulse from an electromagnetic black hole surrounded by a baryonic remnant. Astron. Astrophys. **359**, 855–864 (2000)
851. Rüdiger, G.: Differential rotation and stellar convection, sun and solar-type stars. Akademie-Verlag, Berlin (1989)
852. Rybicki, G.V., Owocki, S.P., Castor, J.I.: Instabilities in line-driven stellar winds. IV. Linear perturbations in three dimensions. Astrophys. J. **349**, 274–285 (1990)
853. Saakyan, G.S.: Equilibrium configurations of degenerate gas masses. Nauka, Moscow (1972)
854. Sager, R., Piskunov, A.E., Myakutin, V.I., Joshi, U.C.: Mass and age distributions of stars in young open clusters. Mon. Not. R. Astron. Soc. **220**, 383–403 (1986)
*855. Sakashita, S., Hayashi, C.: Internal structure and evolution of very massive stars. Progr. Theor. Phys. **22**, 830–834 (1959)
856. Salgado, M., Bonazzola, S., Gourgoulhon, E., Haensel, P.: High precision rotating neutron star models. Astron. Astrophys. **291**, 155–170 (1994)
857. Salpeter, E.E.: Zero temperature plasma. Astrophys. J. **134**, 669–682 (1961)
858. Salpeter, E.E.: Accretion of interstellar matter by massive objects. Astrophys. J. **140**, 796–799 (1964)
859. Salpeter, E.E., Van Horn, H.M.: Nuclear reaction rates at high densities. Astrophys. J. **155**, 183–202 (1965)
*860. Salpeter, E.E., Zapolsky, H.S.: Theoretical high-pressure equations of state including correlation energy. Phys. Rev. **158**, 876–886 (1967)
861. Samarskii, A.A., Popov, Yu.P.: Difference schemes in the gas dynamics. Nauka, Moscow (1975)
862. Samarskii, A.A., Popov, Ju.P.: Difference methods for the solution of problems of gas dynamics. Moscow, Nauka, (in Russian) (1992)
*863. Sampson, D.H.: The opacity at high temperatures due to Compton scattering. Astrophys. J. **129**, 734–751 (1959)
*864. Sato, K.: Nuclear composition in the inner crust of neutron stars. Progr. Theor. Phys. **62**, 957–968 (1979)
*865. Sawai, H., Kotake, K., Yamada, S.: Numerical simulations of equatorially asymmetric magnetized supernovae: formation of magnetars and their kicks. Astrophys. J. **672**, 465–478 (2008)
866. Scalo, J.: Observations and theories of mixing in red giants. In: Iben, I., Renzini, A. (eds.) Physical Processes in Red Giants, pp. 77–114. Reidel, Dordrecht (1981) Progr. Theor. Phys. **62**, 957–968
867. Schaller, G., Schaerer, G., Meynet, G., Maeder, A.: New grids of stellar models from 0. 8 to 120 $M_\odot$ at Z=0. 02 and Z=0. 001. Astron. Astrophys. Suppl. **96**, 269–331 (1992)
*868. Schatzman, E.: Transport of angular momentum and diffusion by the action of internal waves. Astron. Astrophys. **279**, 431–446 (1993)
*869. Schatzman, E.: Solar neutrino and transport processes. In: Roxburgh, I.W., Masnou, J.-L. (eds.) Proc. Symp. Physical Processes in Astrophysics. Springer, Berlin, pp. 171–184 (1995)
870. Schatzman, E., Praderie, F.: The Stars. Springer, Berlin (1993)

871. Scheck, L., Kifonidis, K., Janka, H.-Th., Müller, E.: Multidimensional supernova simulations with approximative neutrino transport I. Neutron star kicks and the anisotropy of neutrino-driven explosions in two spatial dimensions. Astron. Astrophys. **457**, 963-986 (2006)

*872. Schinder, P. et al.: Neutrino emission by the pair, plasma and photoprocesses in the Weinberg–Salam model. Astrophys. J. **313**, 531–542 (1987)

*873. Schmidt-Kaler, Th.: Physical parameters of the stars. Landholt-Börnstein. Astron. Astrophys. **2**, 1–35; 449–456 (1982)

874. Schonberner, D.: Asymptotic giant branch evolution with steady mass loss. Astron. Astrophys. **79**, 108–114 (1979)

875. Schonberner, D.: Late stages of stellar evolution: central stars of planetary nebulae. Astron. Astrophys. **103**, 119–130 (1981)

876. Schonberner, D.: Late stages of stellar evolution II. Mass loss and the transition of asymptotic giant branch into hot remnant. Astrophys. J. **272**, 708–714 (1981)

877. Schonberner, D.: Late stages of stellar evolution III. The observed evolution of central stars of planetary nebulae. Astron. Astrophys. **169**, 189–193 (1986)

878. Schwarzschild, M.: Structure and Evolution of the Stars. Princeton University Press, Princeton (1958)

*879. Schramm, D., Wagoner, R.: Element production in the early Universe. Annu. Rev. Nucl. Sci. **27**, 37–74 (1977)

880. Schwarzschild, B.: Anomalous cosmic-ray data suggest oscillation between neutrino flavors. Phys. Today **No. 10**, 22–24 (1994)

881. Schwarzschild, B.: Chromium surrogate sun confirms that solar neutrinos really are missing. Phys. Today **No. 4**, 19–21 (1995)

882. Schwarzschild, M., Harm, R.: On the maximum mass of stable stars. Astrophys. J. **129**, 637–646 (1959)

883. Schwarzschild, M., Harm, R.: Red giants of population II. II. Astrophys. J. **139**, 158–165 (1962)

884. Schwarzschild, M., Harm, R.: Thermal instability in non-degenerate stars. Astrophys. J. **142**, 855–867 (1965)

885. Schwarzschild, M., Harm, R.: Hydrogen mixing by helium-shell flashes. Astrophys. J. **150**, 961–970 (1967)

886. Schwarzschild, M., Harm, R.: Stability of the sun against spherical thermal perturbations. Astrophys. J. **184**, 5–8 (1973)

887. Schwarzschild, M., Sebberg, H.: Red giants of population II. I. Astrophys. J. **136**, 150–157 (1962)

*888. Schweber, S.S.: An introduction to relativistic quantum field. Peterson and Co., Evanston, Row (1961)

*889. Schlichting, H.: The appearance of turbulence. IL, Moscow (1962)

*890. Sears, R.L., Brownlee, R.R.: The Evolution of Stars and Determination of Their Age. In: Aller, L.H., McLaughlin, D.B. (eds.) Stellar Structure, University of Chicago Press, Chicago (1965)

891. Sedrakyan, D.N., Shakhbazyan, K.M., Movsisyan, A.G.: Magnetic momenta of neutron stars of real baryon gas. Astrophys. **21**, 547–561 (1984) (Astrophysics **21**, 656–664, 1984)

*892. Seeger, P., Fowler, W., Clayton, D.: Nucleosynthesis of heavy elements by neutron capture. Astrophys. J. Suppl. **11**, 121–166 (1965)

893. Seidov, Z.F.: Equilibrium of a star with a phase transition. Astrophys. **3**, 189–201 (1967) (Astrophysics **3**, 90–95, 1967)

894. Seidov, Z.F.: Stars With a Phase Transition. (PhD Thesis, Erevan) (1970)

895. Seidov, Z.F.: Polytropes with a phase transition. II., The polytrope $n = 1$. Soobzsh. Shemakh. Obs., Issue 5, 58–69 (1971)

896. Seidov, Z.F.: The equilibrium, stability and pulsation of stars with a phase transition. Preprint of IKI, No. Pr-889 (1984)

897. Shakura, N.I.: Accretion disk model around relativistic star in a close binary system. Astron. J. **49**, 921–929 (1972) (Sov. Astron. **16**, 756–762, 1973)

898. Shakura, N.I., Sunyaev, R.A.: Black holes in binary systems. Observational appearance. Astron. Astrophys. **24**, 337–355 (1973)
899. Shapiro, S.L., Teukolsky, S.A.: Black holes, white dwarfs, and neutron stars: The physics of compact objects. Wiley-Interscience, NY (1983)
900. Shapiro, S.L., Lightman, A.P., Eardley, D.M.: A two-temperature accretion disc model for Cygnus X-1: structure and spectrum. Astrophys. J. **204**, 187–199 (1976)
901. Shaviv, G., Salpeter, E.E.: Convective overshooting in stellar interior models. Astrophys. J. **184**, 191–200 (1973)
*902. Shekhter, V.M.: Weak interaction with neutral currents. UFN. **119**, 593–632 (1976) (Soviet. Phys. Uspekhi **19**, 645–665, 1976)
*903. Shirokov, Yu.M., Yudin, N.P.: Nuclear physics Nauka, Moscow (1980)
904. Shklovski, I.S.: On the nature of planetary nebulae and their nuclei. Astron. J. (Sov. Astron.) **33**, 315–329 (1956)
905. Shklovski, I.S.: Supernovae stars. Nauka, Moscow (1976)
*906. Shi, X., Schramm, D.N., Dearborn, D.S.P.: On the solar model solution to the solar neutrino problem. Phys. Rev. **D50**, 2414–2420 (1994)
907. Shima, E., Matsuda, T., Takeda, H., Sawada, K.: Hydrodynamic calculations on axisymmetric accretion flow. Mon. Not. R. Astron. Soc. **217**, 367–386 (1985)
908. Shimizu, T., Yamada, S., Sato, K.: Axisymmetric neutrino radiation and the mechanism of supernova explosions. Astrophys. J. Lett. **432**, L119–L122 (1994)
909. Shnol', E.E.: On the stability of stars with a density jump caused by phase transition. Preprint of IPM, No. 93 (1974)
910. Shustov, B.M.: An algorithm for calculation of gasodynamic evolution of the envelope of a massive protostar. Nauchnyie Informatzii Astronomicheskogo Soveta AN SSSR (Scientific Information of the Astonomical Council of the Academy of Sciences of the USSR), Issue 42, 60–74 (1978)
911. Shustov, B.M.: The evolution of protostar envelopes I. The stage of cocoons. Nauchnyie Informatzii Astronomicheskogo Soveta AN SSSR (Scientific Information of the Astonomical Council of the Academy of Sciences of the USSR), Issue 46, 63–92 (1979)
912. Shustov, B.M.: The evolution of protostar envelopes II. spectra of outgoing radiation of protostars and compact zones N II. Nauchnyie Informatzii Astronomicheskogo Soveta AN SSSR (Scientific Information of the Astonomical Council of the Academy of Sciences of the USSR), Issue 46, 93–110 (1979)
913. Shvartsman, V.F.: Halos around black holes. Astron. J. **48**, 479–488 (1971) (Sov. Astron. **15**, 377–384, 1971)
*914. Slattery, W.L., Doolen, G.D., De Witt, H.E.: N dependence on the classical one-component plasma Monte-Carlo calculations. Phys. Rev. A **A26**, 2255–2258 (1982)
915. Smak, J.: Eruptive binaries VI. rediscussion of U Geminorum. Acta Astron. **26**, 277–300 (1976)
916. Smak, J.: Accretion in cataclysmic binaries IV. Accretion disks in dwarf novae. Acta Astron. **34**, 161–189 (1984)
917. Smirnov, V.I.: Course of high mathematics, vol. 2 Nauka, Moscow (1962)
918. Smirnov, V.I.: Course of high mathematics, vol. 4 Gostekhizdat, Moscow (1962)
919. Smirnova, T.V., Shishov, V.I., Malofeev, V.M.: The spatial structure of pulsar emission sources determined using interstellar scintillation. Astrophys. J. **462**, 289–295 (1996)
920. Sobolev, V.V.: Moving envelopes of stars. Harvard University Press, Cambridge (1960)
921. Sobolev, V.V.: Course of theoretical astrophysics. Nauka, Moscow (1967)
*922. Sofia, S., Chau, K.: Turbulent compressible convection in a deep atmosphere II. Two-dimensional results for main sequence A5 and F0 type envelopes. Astrophys. J. **282**, 550–556 (1984)
923. Sokolov, V.V. (2001): Evidence for a supernova in the Ic band light curve of the optical transient of GRB 970508. ESO astrophysics symposia. ISBN 3-540-42771-6. Costa, E., Frontera, F., Hjorth, J. (eds.) Springer-Verlag, pp. 136–138
*924. Spiegel, E.A.: A generalization of the mixing-length theory of turbulent convection. Astrophys. J. **138**, 216–225 (1963)

*925. Spiegel, E.A., Zahn, J.-P. (eds.): Problems of stellar convection. Proc. IAU Coll. **No. 38** Springer, Berlin (1976)
926. Spruit, H.: The rate of mixing in semiconvective zones. Astron. Astrophys. **253**, 131–138 (1992)
927. Spruit H.C.: Dynamo action by differential rotation in a stably stratified stellar interior. Astron. Astrophys., **381**, 923–932(2002)
928. Spruit, H.C., Uzdensky, D.A.: Magnetic flux captured by an accretion disk. Astrophys. J. **629**, 960–968 2005
929. Sramek, R., Panagia, N., Weiler, K.: Radio emission from type I supernova SN 1983. 51 in NGC 5236. Astrophys. J. Lett. **285**, L59–L62 (1984)
930. Stahler, S.: The birthline for low-mass stars. Astrophys. J. **274**, 822–829 (1983)
931. Stahler, S., Shu, F., Taam, R.: The envelope of protostars. I. Global formulation and results. Astrophys. J. **241**, 637–654 (1980)
932. Stahler, S., Shu, F., Taam, R.: The envelope of protostars. II. The hydrostatic core. Astrophys. J. **242**, 226–241 (1980)
933. Stahler, S., Shu, F., Taam, R.: The envelope of protostars. III. The accretion envelope. Astrophys. J. **248**, 727–737 (1981)
934. Stanek, K.Z., Matheson, T., Garnavich, P.M., et al.: Spectroscopic discovery of the supernova 2003dh associated with GRB 030329. Astrophys. J. Lett. **591**, L17–L20 (2003)
935. Stein, J., Barkat, Z., Wheeler, J.C.: The role of kinetic energy flux in the convective URCA process. Astrophys. J. **523**, 381–385 (1999)
936. Stothers, R.B.: Upper limit to the mass of pulsationally stable stars with uniform chemical composition. Astrophys. J. **392**, 706–709 (1992)
937. Stothers, R., Chin, C.: Stellar evolution at high mass based on the Ledoux criterion for convection. Astrophys. J. **179**, 555–568 (1973)
938. Stothers, R., Chin, C.: Stellar evolution at high masses including the effects of a stellar wind. Astrophys. J. **233**, 267–279 (1979)
939. Stothers, R., Chin, C.: Stellar evolution at high mass with convective core overshooting. Astrophys. J. **292**, 222–227 (1985)
940. Stothers, R.B., Chin, C.: Dynamical instability as the cause of the massive outburst in Eta Carinae and other luminous blue variables. Astrophys. J. Lett. **408**, L85–L88 (1993)
941. Stothers, R., Chin, C.: Iron and molecular opacities and the evolution of population I stars. Astrophys. J. **412**, 294–300 (1993)
942. Stothers, R., Chin, C.: Galactic stars applied to tests of the criterion for convection and semiconvection in an inhomogeneous star. Astrophys. J. **431**, 797–805 (1994)
943. Strom, S.E., Strom, K., Rood, R.T., Iben, I.: On the evolutionary status of stars above the horizontal branch in globular clusters. Astron. Astrophys. **8**, 243–250 (1970)
*944. Sugimoto, D.: Helium flash in less massive stars. Progr. Theor. Phys. **32**, 703–725 (1964)
*945. Sugimoto, D.: On the numerical stability of computation of stellar evolution. Astrophys. J. **159**, 619–628 (1970)
946. Sugimoto, D., Nomoto, K.: Presupernova models and supernovae. Space Sci. Rev. **25**, 155–227 (1980)
*947. Sugimoto, D., Nomoto, K., Eriguchi, Y.: Stable numerical method of computations of stellar evolution. Progr. Theor. Phys. Suppl. **70**, 115–131 (1981)
*948. Sugimoto, D., Yamamoto, Y.: Second helium flash and an origin of carbon stars. Progr. Theor. Phys. **36**, 17–36 (1966)
949. Sunyaev, R.A., Titarchuk, L.G.: Comptonization of X-rays in plasma clouds. Typical radiation spectra. Astron. Astrophys. **86**, 121–138 (1980)
950. Supernova: in *Proc. workshop ESO*, July 1987 (1987A)
951. Suwa, Y., Takiwaki, T., Kotake, K., Sato, K.: Magnetorotational Collapse of Population III Stars. Publ. Astron. Soc. Jpn. **59**, 771–785 (2007)
952. Suwa, Y., Kotake, K., Takiwaki, T., Whitehouse, Liebendörfer, M., Sato, K.: Explosion geometry of a rotating 13 $M_\odot$ star driven by the sasi-aided neutrino-heating supernova mechanism. *astro-ph 0912.1157v1* (2009)

953. Sweigart, A.: A method for suppression of the thermal instability in helium-shell burning stars Astrophys. J. **168**, 79–97 (1971)
954. Sweigart, A.: Initial asymptotic branch evolution of population II stars. Astron. Astrophys. **24**, 459–464 (1973)
955. Sweigart, A., Mengel, J., Demarque, P.: On the origin of the blue halo stars. Astron. Astrophys. **30**, 13–19 (1974)
*956. Swinney, H.L., Gollub, J.P. (eds.):Hydrodynamic instabilities and the transition to turbulence. Springer, Berlin (1981)
957. Symbalisty, E.M.D.: Magnetorotational iron core collapse. Astrophys. J. **285**, 729–746 (1984)
958. Sztajno, M., et al.: X-ray bursts from GX 17+2, a new approach. Mon. Not. R. Astron. Soc. **222**, 499–511 (1986)
959. Talbot, R.J.: Non-linear pulsations of unstable massive main-sequence stars I. Small-amplitude tests of an approximation technique. Astrophys. J. **163**, 17–27 (1971)
960. Talbot, R.J. Non-linear pulsations of unstable massive main-sequence stars II. Finite-amplitude stability. Astrophys. J. **165**, 121–138 (1971)
961. Tanvir, N. at IofA U.Cambridge (2009): GRB 090423: VLT/ISAAC spectroscopy. GCN circular number: 9219
962. Tassoul, M.: Asymptotic approximations for stellar non-radial pulsations. Astrophys. J. Suppl. **43**, 469–490 (1980)
*963. Tassoul, J.-L., Tassoul, M.: Meridional circulation in rotating stars VIII. The solar spin-down problem. Astrophys. J. **286**, 350–358 (1984)
964. Tassoul, J.L.: Theory of rotating stars. Princeton University Press, Princeton (1978)
965. Tayler, R.J.: The adiabatic stability of stars containing magnetic fields-I. Toroidal fields. Mon. Not. R. Astron. Soc. **161**, 365–380(1973)
966. Taylor, J.H.: Pulsar timing and relativistic gravity. Philos. Trans. R. Soc. Lond. A. **341**, 117–134 (1992)
967. Taylor, J.H., Hulse, R.A., Fowler, L.A., Gullahorn, G.E., Rankin J.M.: Further observations of the binary pulsar PSR 1913+16. Astrophys. J. Lett **206**, L53–L58 (1976)
968. Taylor, J.H., Manchester, R.N., Lyne, A.G.: Catalog of 558 pulsars. Astrophys. J. Suppl. **88**, 529–568 (1993)
969. Taylor, J.H., Weisberg, J.M.: A new test of general relativity: gravitational radiation and the binary pulsar PSR 1913+16. Astrophys. J. **253**, 908–920 (1982)
970. Taylor, J.H., Weisberg, J.M.: Further experimantal tests of relativistic gravity using the binary pulsar PSR 1913+16. Astrophys. J. **349**, 434–450 (1989)
971. Tchekhovskoy, A., McKinney, J.C., Narayan, R.: Simulations of ultrarelativistic magnetodynamic jets from gamma-ray burst engines. Mon. Not. R. Astron. Soc. **388**, 551–572 (2008)
972. Terebey, S., Chandler, C.J., Andre, P.: The contribution of disks and envelopes to the millimeter continuum emission from very low-mass stars. Astrophys. J. **414**, 759–772 (1993)
973. Thomson, R.C., Mackay, C.D., Wright, A.E.: Internal structure and polarization of the optical jet of the quasar 3C273. Nature **365**, 133–135 (1993)
974. Thorne, K.S., Zytkow, A.N.: Stars with degenerate neutron cores. I. Structure of equilibrium models. Astrophys. J. **212**, 832–858 (1977)
975. Tohlins, J.E. Ring formation in rotating protostellar clouds. Astrophys. J. **236**, 160–171 (1980)
976. Tomczuk, S., Schou, J., Thompson, M.J.: Measurement of the rotational rate in deep solar interior. Astrophys. J. Lett. **448**, L57–L60 (1995)
*977. Tomonaga, S.: Innere Reibung und Warmeleitfahigkeit der Kernmaterie. Z. Phys. **110**, 573–604 (1938)
*978. Toomre, J.: Thermal convection and penetration. In: Zahn, J.-P., Zinn-Justin, J. (eds.) Proc. Les Houches 1987 (1993)
*979. Tooper, R.F.: On the equation of state of a relativistic Fermi-Dirac gas at high temperatures. Astrophys. J, **156**, 1075–1100 (1969)

980. Tsuruta, S., Cameron, A.G.W.: URCA shells in dense stellar interiors. Astrophys. Space. Sci. **7**, 374–395 (1970)
981. Tutukov, A.V., Fadeev, Yu.A.: The formation of an extended envelope around pulsating star. Nauchnyie Informatzii Astronomicheskogo Soveta AN SSSR (Scientific Information of the Astonomical Council of the Academy of Sciences of the USSR), Issue 49, 48–63 (1981)
982. Tutukov A.V., Cherepashchuk, A.M.: Wolf-Rayet Stars, Black Holes, and Gamma-Ray Bursters in Close Binaries. Astron. J. **80**, 419435 (2003) (Astron. Rep. **47**, 386–400, 2003)
*983. Trautvetter, H., et al.: The Ne-Na cycle and the $C^{12} + \alpha$ reaction. Proc. Second Workshop on Nuclear Astrophysics. Preprint MPA, No. 90, pp 24–33 (1983)
*984. Trimble, V. The origin and abundances of chemical elements. Rev. Mod. Phys. **47**, 877–976 (1975)
*985. Trimble, V.: The origin and abundances of the chemical elements revisited. Astron. Astrophys. Rev. **3**, 1–46 (1991)
986. Tscharnuter, W.: On the collapse of rotating protostars. Astron. Astrophys. **39**, 207–212 (1975)
*987. Uehling, E.A., Uhlenbeck, G.E.: Transport phenomena in Einstein–Bose and Fermi–Dirac gases. I. Phys. Rev. **43**, 552–561 (1933); II. Phys. Rev. **46**, 917–929 (1934)
*988. Ulrich, R.: A non-local mixing-length theory of convection for use in numerical calculations. Astrophys. J. **207**, 564–573 (1976)
*989. Ulrich, R.K.: S-Process. In: Essays in nuclear astrophysics. Cambridge University Press, Cambridge (1982)
990. Umeda, H., Nomoto, K., Tsuruta, S., Muto, T., Tatsumi, T.: Neutron star cooling and pion condensation. Astrophys. J. **431**, 309–320 (1994)
*991. Unno, W.: Development of the stellar convection theory. Progr. Theor. Phys. Suppl. No. 70, 101–114 (1981)
*992. Unno, W., Kondo, M.: The Eddington approximation generalized for radiative transfer in spherically symmetric systems I. Basic method. Publ. Astron. Soc. Japan **28**, 347–354 (1976)
*993. Unno, W., Kondo, M.: The Eddington approximation generalized for radiative transfer in spherically symmetric systems II. Non-gray extended dust-shell models. Publ. Astron. Soc. Japan **29**, 693–710 (1977)
994. Unno, W., Osaki, Y., Ando, H., Shibahashi, H.: *Non-radial Oscillations of Stars* Tokyo University Press, Tokyo (1989)
995. Upton, I.K.L., Little, S.J., Dworetsky, M.M.: Dynamical stability in pre-main-sequence stars. Astrophys. J. **154**, 597–611 (1968)
*996. Urpin, V.A., Yakovlev, D.G.: The heat conductivity caused by inter-electron collisions in degenerate relativistic electron gas. Astron. J. **57**, 213–215 (1980) (Sov. Astron. **24**, 126–127, 1980)
*997. Urpin, V.A., Yakovlev, D.G.: Thermogalvanomagnetic phenomena in white dwarfs and neutron stars. Astron. J. **57**, 738–748 (1980) (Sov. Astron. **24**, 425–431, 1980)
998. Usov, V.V.: Millisecond pulsars with extremely strong magnetic fields as a cosmological source of gamma-ray bursts Nature **357**, 472–474 (1992)
999. Usov, V.V.: Bare quark matter surfaces of strange stars and e^+e^- emission. Phys. Rev. Lett. **80**, 230–233 (1998)
1000. Usov, V.V.: Thermal emission from bare quark matter surfaces of hot strange stars. Astrophys. J. Lett. **550**, L179–L182 (2001)
1001. Usov, V.V.: Low-mass normal-matter atmospheres of strange stars and their radiation. Astrophys. J. Lett. **481**, L107–L110 (1997)
1002. Ustyugova, G.V., Koldoba, A.V., Romanova, M.M., Chechetkin, V.M., Lovelace, R.V.E.: Magnetocentrifugally driven winds: comparison of MHD simulations with theory. Astrophys. J. **516**, 221–235 (1999)
1003. Ustyugov, S.D., Chechetkin, V.M.: Supernova explosion in presence of large scale convective instability in rotating protoneutron star. Astron. J. **76**, 816–824 (1999) (Sov. Astron. **43**, 718–726, 1999)

1004. Uus, U.: Calculations of the structure of thin nuclear-burning shells in stellar models. Nauchnyie Informatzii Astronomicheskogo Soveta AN SSSR (Scientific Information of the Astonomical Council of the Academy of Sciences of the USSR), Issue 13, 126–144 (1969)
1005. Uus, U.: The evolution of 1. 5, 2. 3 and 5 $M\odot$ stars on the stage of carbon core growth. Nauchnyie Informatzii Astronomicheskogo Soveta AN SSSR (Scientific Information of the Astonomical Council of the Academy of Sciences of the USSR), Issue 17, 3–24 (1970)
1006. Uus, U.: The penetration of the convective envelope of a star in the nuclear-burning zone. Nauchnyie Informatzii Astronomicheskogo Soveta AN SSSR (Scientific Information of the Astonomical Council of the Academy of Sciences of the USSR), Issue 20, 60–63 (1971)
*1007. Uus, U.: On the possibility of a consistent treatment of stellar turbulent convection. Publ. of Tartus. Astrofis. Obs. im. V. Ya. Struve **47**, 103–119 (1979)
*1008. Vaisenberg, A.O.: Mu-Meson. Moscow, Nauka (1964)
1009. Vandakurov, Yu.V.: Theory for the stability of a star with a toroidal magnetic field. Astron. J. **49**, 324–333 (1972) (Sov. Astron. **16**, 265–272, 1972)
1010. Vandakurov, Yu.V.: Convection in the Sun and 11-year cycle. Leningrad, Nauka (1976)
1011. Vartanyan, Yu.L., Ovsepyan, A.V., Adzhjyan, G.S.: On the stability and radial pulsations of rotating neutron stars. Astron. J. **50**, 48–59 (1973) (Sov. Astron. **17**, 30–37, 1973)
*1012. Vartanyan, Yu.L., Ovakimova, N.K.: Neutron-rich nuclei in Fermi-gas. Astron. J. **49**, 306–315 (1972) (Sov. Astron. **16**, 250–258, 1972)
*1013. Vartanyan, Yu.L., Ovakimova, N.K.: Cold neutron evaporartion from nuclei in a superdense matter. Soobzsh. Burokan Obs. **49**, 87–95 (1976)
1014. Vartanyan, Yu.L., Ovsepyan, A.V.: Evolution and radial pulsations of isothermal white dwarfs with the inclusion of rotation, neutronisation and GR effects. Astrophys. **7**, 107–119 (1971) (Astrophys. **7**, 62–69, 1971)
1015. Vartanyan, Yu.L.: Superdense stars containing strange baryons. Astrophys. **53**, 25–40 (2010) (Astrophys. **53**, 18–31, 2010)
*1016. Varshavskii, V.I.: The Schwarzschild method with logarithmic variables. Nauchnyie Informatzii Astronomicheskogo Soveta AN SSSR (Scientific Information of the Astonomical Council of the Academy of Sciences of the USSR), Issue 16, 77–82 (1970)
1017. Varshavskii, V.I.: Evolution of massive stars with the inclusion of semiconvection. Nauchnyie Informatzii Astronomicheskogo Soveta AN SSSR (Scientific Information of the Astonomical Council of the Academy of Sciences of the USSR), Issue 21, 25–45 (1972)
1018. Varshavskii, V.I., Tutukov, A.V.: Evolution of massive stars. Astron. J. **52**, 227–233 (1975) (Sov. Astron. **19**, 142–145, 1975)
*1019. Vlasiuk, M.P., Polezhaev, V.I.: Numerical study of convective motions in a horizontal gas layer heated from below. Preprint of IPM AN SSSR 1–74 (1970)
1020. Vorontzov, S.V., Zharkov, V.N.: Helioseismology. Itogi Nauki i Tekhniki, Astron. **38**, 253–338 (1988)
1021. Vanbeveren, D.: Evolution with mass loss: massive stars, massive binaries. PhD thesis, Brussels University (1980)
1022. Van den Hulst, J., et al.: Radio discovery of a young supernova. Nature **306**, 566–568 (1983)
*1023. Van Horn, H.M.: Crystallization of white dwarfs. Astrophys. J. **151**, 227–238 (1968)
1024. van Putten M.H.P.M.: Gamma-ray bursts: LIGO//VIRGO sources of gravitational radiation. Phys. Rep. **345**, 1–59 (2001)
1025. Vardya, M.S.: Hydrogen–Helium adiabats for late type stars. Astrophys. J. Suppl. **4**, 281–336 (1960)
1026. Vardya, M.S.: Thermodynamics of a solar composition gaseous mixture. MNRAS **129**, 205–213 (1965)
1027. Wagenhuber, J., Weiss, A.: Termination of AGB-evolution by hydrogen recombination. Astron. Astrophys. **290**, 807–814 (1994)
*1028. Wagoner, R.: Synthesis of the elements within objects exploding from very high temperatures. Astrophys. J. Suppl. **18**, 247–296 (1969)

*1029. Wallace, R.K., Woosley, S.E.: Explosive hydrogen burning. Astrophys. J. Suppl. **45**, 389–420 (1981)

1030. Wandel, A., Liang, E.P.: Hybrid accretion disks in active galactic nuclei. I - Structure and spectra. Astrophys. J. **380**, 84–97 (1991)

1031. Wang, J.C.L., Sulkanen, M.E., Lovelace, R.V.E.: Intrinsically asymmetric astrophysical jets. Astrophys. J. **390**, 46–65 (1992)

1032. Weaver, T., Woosley, S.: Evolution and explosion of massive stars. Ann. N. Y. Acad. Sci. **336**, 335–357 (1980)

1033. Weaver, T., Zimmerman, G., Woosley, S.: Presupernova evolution of massive stars. Astrophys. J. **225**, 1021–1029 (1978)

1034. Weigert, A.: Sternentwicklung VI. Entwicklung mit neutrinoverlusten und thermische pulse der Helium-Schalenquelle bei einem Stern von 5 sonnenmassen. Z. Astrophys. **64**, 395–425 (1966)

1035. Weir, A.D.: Axisymmetric convection in rotating sphere. Part I. Stress-free surface. J. Fluid Mech. **75**, 49–79 (1976)

1036. Wendell, C.E., Van Horn, H.M., Sargent, D.: Magnetic field evolution in white dwarfs. Astrophys. J. **313**, 284–297 (1987)

1037. Westbrook, Ch., Tarter, B.: On protostellar evolution. Astrophys. J. **200**, 48–60 (1975)

1038. Wiedemann, V.: The initial/final mass relation for stellar evolution with mass loss. In: Chiosi, C., Stalio, R. (ed.) Effects of Mass Loss on Stellar Evolution, pp. 339–349. Reidel, Dordrecht (1981)

1039. Wilson, L.A.: Fe I fluorescence in T Tauri stars II. Clues to the velocity field in the circumstellar envelopes. Astrophys. J. **197**, pp. 365–370 (1975)

1040. Wilson, J.R., Mayle, R., Woosley, S., Weaver, T.: Stellar core collapse and supernova. Ann. N.Y. Acad. Sci. **470**, 267–293 (1986)

*1041. Wilson, J.R., Mayle, R.W.: Report on the progress of supernova research by the Livermore group. Phys. Rep. **227**, 97–111 (1993)

1042. Wiringa, R.B., Ficks, V., Fabrocini, A.: Equation of state for dense nucleon matter. Phys. Rev. C **38**, 1010–1037 (1988)

1043. Witten, E.: Cosmic separation of phases. Phys. Rev. D **30**, 272–285 (1984)

1044. Wolfram, S.: The Mathematica Book Cambridge University Press, Cambridge (1996)

1045. Wolszczan, A.: A nearby 39.7 ms radio pulsar in a relativistic binary system. Nature **350**, 688–690 (1991)

1046. Wood, P.R.: Dynamical models of asymptotic-giant-branch stars. In: Ledoux, P. et al. (ed.) Proc. IAU Symp. No. 59: Stellar Instability and Evolution, pp. 101–102. (Reidel, Dordrecht) (1974)

1047. Wood, P.R.: Pulsation and mass loss in Mira variables. Astrophys. J. **227**, 220–231 (1979)

1048. Wood, P.R., Faulkner, D.J.: Hydrostatic evolutionary sequences for the nuclei of planetary nebula. Astrophys. J. **307**, 659–674 (1986)

1049. Woodward, P.: Theoretical models of star formation. Annu. Rev. Astron. Astrophys. **16**, 555–584 (1978)

*1050. Woosley, S., Fowler, W., Holmes, J., Zimmerman, B.: Semiempirical thermonuclear reaction rate data for intermediate mass nuclei. Atom. Data Nucl. Data Tables **22**, 371–441 (1978)

*1051. Woosley, S., Fowler, W., Holmes, J., Zimmerman, B.: Tables of thermonuclear reaction rate data for intermediate mass nuclei. Preprint OAP, No. 422, pp. 1-1-5, A1–A179 (1975)

1052. Woosley, S.E., Langer, N., Weaver, T.A: The presupernova evolution and explosion of helium stars that experience mass loss. Astrophys. J. **448**, 315–338 (1995)

1053. Woosley, S., Weaver, T.: Theoretical models for type I and type II supernovae. In: Audouze, J., Mathieu N. (eds.) Nucleosynthesis and Its Implications on Nuclear and Particle Physics, pp. 145–166. Reidel, Dordrecht (1986)

*1054. Wu, C.S., Moszkowski, S.A.: Beta-Decay Interscience, New York (1966)

1055. Wynn-Williams, C. (1982): The search for infrared protostar. Annu. Rev. Astron. Astrophys. **20**, 587–618

*1056. Yakovlev, D.G.: Phenomena of Heat and Charge Transfer in Neutron Stars and White Dwarfs (PhD Thesis, Leningrad, LFTI) (1980)
*1057. Yakovlev, D.G., Urpin, V.A.: On the heat and charge conductivity in neutron stars and white dwarfs. Astron. J. **57**, 526–536 (1980) (Sov. Astron. **24**, 303–310, 1980)
*1058. Yakovlev, D.G., Shalybkov, D.A.: The effect of electron screening on thermonuclear reaction rates. PAJ **13**, 730–736 (1987) (Sov. Astron. Lett. **13**, 308–310, 1987)
*1059. Yakovlev, D.G., Shalybkov, D.A.: The degenerate cores of white dwarfs and neutron star envelopes: thermodynamics and plasma screening in thermonuclear reactions. Itogi Nauki i Tekhniki, Astronomiya **38**, 191–252 (1988)
1060. Yamada, S., Sawai, H.: Numerical study on the rotational collapse of strongly magnetized cores of massive stars. Astrophys. J. **608**, 907–924 (2004)
1061. Yorke, H.: The evolution of protostellar envelopes of masses 3 $M_\odot$ and 10 $M_\odot$. I. Structure and hydrodynamic evolution. Astron. Astrophys. **80**, 308–316 (1979)
1062. Yorke, H.: The evolution of protostelar envelopes of masses 3 $M_\odot$ and 10 $M_\odot$. II. Radiation transfer and spectral apppearance. Astron. Astrophys. **85**, 215–220 (1979)
1063. Yorke, H.: Protostars and their evolution. In: Proc. ESO Conf. on Scientific importance of high angular resolution of infrared and optical wavelength (Garching), pp. 319–340 (1981)
1064. Yorke, H., Krugel, H.: The dynamical evolution of massive protostellar clouds. Astron. Astrophys. **54**, 183–194 (1977)
1065. Yorke, H., Shustov, B.M.: The spectral appearance of dusty protostellar envelopes. Astron. Astrophys. **98**, 125–132 (1981)
1066. Young, E.J., Chanmugam, G.: Postaccretion magnetic field evolution of neutron stars. Astrophys. J. Lett. **442**, L53–L56 (1995)
1067. Zahn, J.-P.: Convective penetration in stellar interiors. Astron. Astrophys. **252**, 179–188 (1991)
*1068. Zahn, J.-P.: Circulation and turbulence in rotating stars. Astron. Astrophys. **265**, 115–132 (1992)
1069. Zapolsky, H.S., Salpeter, E.E.: The mass-radius relations for cold spheres of low mass. Astrophys. J. **158**, 809–813 (1969)
1070. Zenkevich, O.C., Morgan, K.: Finite Elements and Approximation. Wiley, New York (1983)
1071. Zhevakin, S.A.: Physical basis of the pulsation theory of variable stars. Annu. Rev. Astron. Astrophys. **1**, 367–400 (1963)
*1072. Zel'dovich, Ya.B.: On nuclear reactions in a superdense cold gas. Jurnal Eksperimental'noi i Teoreticheskoi Fiziki (J. Exp. Theor. Phys., JETP) **33**, 991–993 (1957)
1073. Zel'dovich, Ya.B.: Equation of state at a superhigh density and relativistic restrictions. Jurnal Eksperimental'noi i Teoreticheskoi Fiziki (J. Exp. Theor. Phys., JETP) **41**, 1609–1615 (1961)
1074. Zel'dovich, Ya.B.: Static solutions with energy excess in general relativity. Jurnal Eksperimental'noi i Teoreticheskoi Fiziki (J. Exp. Theor. Phys., JETP) **42**, 1667–1671 (1962)
1075. Zel'dovich, Ya.B.: Hydrodynamical stability of star. Vopr. Kosmog. **9**, 157–170 (1963)
1076. Zel'dovich, Ya.B.: Chemical physics and hydrodynamics, selected works. Moscow, Nauka 279–280 (1984)
1077. Zel'dovich, Ya.B., Blinnikov, S.I., Shakura, N.I.: Physical grounds of stellar structure and evolution. Moscow, MSU (1981)
*1078. Zel'dovich, Ya.B., Gusseinov, O.Kh.: The neutronization of matter during collapse and neutrino spectrum. DAN SSSR **162**, 791–793 (1965)
1079. Zel'dovich, Ya.B., Ivanova, L.N., Nadyozhin, D.K.: Non-stationary hydrodynamical accretion onto neutron star. Astron. J. **49**, 253–264 (1972) (Sov. Astron. **16**, 209–218, 1972)
1080. Zel'dovich, Ya.B., Novikov, I.D.: Relativistic astrophysics. II. UFN. **86**, 447–536 (1965) (Soviet. Phys. Uspekhi **8**, 522–577, 1966)
1081. Zel'dovich, Ya.B., Novikov, I.D.: Gravitation theory and stellar evolution. Moscow, Nauka (1971)

1082. Zel'dovich, Ya.B., Novikov, I.D.: Relativistic astrophysics. Moscow, Nauka (1967)
1083. Zhou, Sh., Evans, N.J., Kömpe, C., Walmsley, C.M.: Evidence for protostellar collapse in B335. Astrophys. J. **404**, 232–246 (1993)
1084. Ziebarth, K.: On the upper mass limit for main sequence stars. Astrophys. J. **162**, 947–962 (1970)
*1085. Zimmerman, B., Fowler, W., Caughlan, G.: Tables of thermonuclear reaction rates. Preprint OAP, No. 399, pp. 1–35 (1975)
1086. Ziolkowski, J.: Evolution of massive stars. Acta Astron. **22**, 327–374 (1972)
1087. Źytkov, A.: On the stationary mass outflow from stars I. The computational method and results for 1 $M_\odot$ star. Acta Astron. **22**, 103–139 (1972)
1088. Źytkov, A.: On the stationary mass outflow from stars II. The results for 30 $M_\odot$ star. Acta Astron. **23**, 121–134 (1973)

List of Symbols and Abbreviations

Here, only global symbols used throughout the book are indicated. Some of the notations may be used for other variables in different parts of the book, where they are defined.

Latin Symbols

A, A_i:	Atomic mass of a nucleus
$\mathbf{A}, A_i$:	Vector potential of electromagnetic field
a:	Constant of radiation energy density
a:	Average distance between ions
a_0:	Bohr radius
a_Z:	Atomic radius in the Thomas–Fermi model
$\mathbf{B}, B_i$:	Vector of the magnetic field strength
$B_c = m_e^2 c^3/e\hbar$:	Critical value of the magnetic field
$B_{A,Z}$:	Binding energy of a nucleus with atomic mass A and charge Z
B_n:	Binding energy per nucleon
$B(T)$:	Energy density of an equilibrium radiation
$B_\nu(T)$:	Spectral intensity of equilibrium Planck radiation
c:	Speed of light in vacuum
c_p, C_p:	Heat capacity at constant pressure
c_v, C_v:	Heat capacity at constant volume
d, d_{over}:	Overshooting length
d_{pc}:	Distance to a star in parsecs
ds:	Interval of space-time
dV, dv:	Element of a volume
$d\Omega$:	Element of a solid angle
$\mathscr{D}(x)$:	Debye function
E:	Specific energy of matter
E_b:	Binding energy per nucleon

E_e, E_{e^-}:	Specific energy of electrons in matter
E_{e^+}:	Specific energy of positrons in matter
E_n:	Specific energy of neutrons in matter
E_N:	Specific energy of nuclei in matter
E_{conv}:	Specific energy of convective motion
E_{zp}:	Zero-point energy of three-dimensional oscillator
E_{ν_e}:	Specific energy of electron neutrinos in thermodynamic equilibrium
$E_{\tilde{\nu}_e}$:	Specific energy of electron antineutrinos in thermodynamic equilibrium
$\mathbf{E}$, E_i:	Electrical field strength
E_M:	Magnetic energy per unit volume
$\tilde{E}$:	Internal energy per baryon
e:	Electrical charge of the electron
$e \equiv e(r)$:	Energy of a star within radius r
F:	Specific free energy of matter
F_i:	Vector of radiative energy flux density
F_{conv}:	Convective energy flux density
F_{rad}:	Radiative energy flux density
$F_0(u)$:	Fermi function of beta decay
f:	Particle distribution function
f_e:	Electron distribution function
f_ν:	Photon distribution function
f_ν:	Rate of neutrino emission losses per unit mass
G:	Gravitational constant
G_W:	Coupling constant of the weak interaction
$G_V = G_W$:	Constant of weak vector-type interaction
G_A:	Constant of weak axial-type interaction
$g_A = G_V/G_A$:	Relative constant of axial weak interaction
g:	Gravitational acceleration
$g_\odot$:	Gravitational acceleration at the surface of the Sun
$g_{\mathbf{ef}}$:	Effective gravitational acceleration, including centrifugal acceleration
g_{ik}:	Metric tensor
$g_{A,Z}$:	Statistical weight of a nucleus with atomic mass number A and charge Z
g_0, g_1:	Statistical weights of nuclei entering a nuclear reaction
g_2, g_3:	Statistical weights of nuclei resulting from a nuclear reaction
g_{ij}:	Statistical weight of an element i in the ionization state j
g_n:	Statistical weight of a neutron
g_p:	Statistical weight of a proton
g_{bf}:	Gaunt factor for bound–free transitions
g_{ff}:	Gaunt factor for free–free transitions

H:	Matrix element of nuclear transformation
H:	Specific enthalpy of matter
H_β:	Matrix element of beta reaction
H_μ:	Matrix element of muon decay
H_{n}:	Matrix element of a neutron decay
H_p:	Pressure scale height
H_ρ:	Density scale height
$\hbar$:	Planck constant divided by 2π
h:	Planck constant
h:	Thickness of a layer; half-thickness of a disk
I:	Nuclear spin
I_{ij}:	Ionization energy (potential) of the j-th electron of element i
I_ν:	Spectral intensity of radiation
$I_{\nu\epsilon}$:	Spectral intensity of neutrino radiation
I_ϕ:	Azimuthal component of a surface electric current density
J_0:	Total angular momentum of a star
j:	Specific angular momentum of matter
j_i:	Vector of electrical current density
j_ν:	Spectral coefficient of emission per unit mass
K, K_1, K_2:	Coefficients in a polytropic equation of state
k:	Boltzmann's constant
k:	Wavenumber of a turbulent vortex
k_s:	Diffusion coefficient in an ion binary gas mixture
k_T:	Coefficient of temperature diffusion
L:	Stellar luminosity
L:	Maximal scale of a turbulent vortex
L_B:	Magneto-bremstrahlung luminosity
L_{opt}:	Photon luminosity of a star
L_{cr}, L_c:	Eddington critical stellar luminosity
L_{RG}:	Luminosity of a red giant star
L_r:	Radial energy flux from a star
L_t:	Stellar luminosity at the turning point off a main sequence
L_{k}:	Flux of kinetic energy from a star
L_m:	Stellar luminosity in a peak of a helium shell flash
L_ν:	Neutrino luminosity of a star
L_r^{rad}:	Radial heat flux due to radiation heat conductivity
L_{th}:	Thermal heat flux due to heat conductivity
$L_{\mathrm{conv}}, L_r^{\mathrm{conv}}$:	Radial energy flux due to convection
$L_\odot$:	Luminosity of the Sun
L_{H}:	Stellar luminosity due to hydrogen burning
L_{He}:	Stellar luminosity due to helium burning
L_{g}:	Stellar luminosity due to gravitational energy production

l: Mixing length of a convective element
l: Specific angular momentum of matter
l: Current scale of a turbulent vortex
l_i: Unit vector in the direction of photon motion
l_{in}: Specific angular momentum on the inner boundary of an accretion disk
l_T: Mean free path of a neutrino
l_{WZ}: Radius of a Wigner–Seitz spherical cell
M: Stellar mass
M: Absolute stellar magnitude
M_0: Rest mass of a star
M_i: Initial mass of a star
M_e: Mass of the hydrogen envelope of a star
M_n: Mass of a neutron star
M_n: Non-dimensional Lane–Emden stellar mass for polytropic index n
$M_\odot$: Mass of the Sun
M_c, M_{core}: Mass of the stellar core
$M_{C\text{He}}$: Mass of the helium core of a star
M_{CO}: Mass of the carbon–oxygen stellar core
M_{Fe}: Mass of the iron stellar core
M_{Ch}: Chandrasekhar limit of the stellar mass
$\dot{M}$: Mass loss rate from a star, or mass flux into a star
M_{bol}: Bolometric absolute stellar magnitude
M_0, M_1: Masses of nuclei entering a nuclear reaction
M_2, M_3: Masses of nuclei emerging from a nuclear reaction
M_{n}: Matrix element of neutron decay
M_Z: Matrix element of beta decay of a nucleus (A,Z)
m: Lagrangian mass coordinate in a spherically symmetric star
$m = M/M_\odot$: Non-dimensional stellar mass
m_0: Rest mass of a star within radius r
m: Visual stellar magnitude
m_{ph}: Photo-visual stellar magnitude
m_B: Visual stellar magnitude in a filter B
m_U: Visual stellar magnitude in a filter U
m_V: Visual stellar magnitude in a filter V
$m_{A,Z}$: Mass of a nucleus with atomic number Z and atomic mass A
m_{e}: Electron mass
m_{n}: Neutron mass
m_{p}: Proton mass
m_i: Mass of an atomic nuclei with mass number $A_i (\approx A_i m_{\text{u}})$

m_{u}:	Atomic mass unit = 1/12 times the mass of the isotope ^{12}C
m_{μ}:	Muon mass
$m_{\nu_e 0}, m_{\nu_e}$:	Rest mass of the electron neutrino
$m_{\nu_\mu 0}, m_{\nu_\mu}$:	Rest mass of the muon neutrino
$\dot{m} = \dot{M}c^2/L_c$:	Non-dimensional accretion mass flux
N:	Number of baryons in a star
N_{b}:	Total number of baryons in a neutron star
N_A:	Avogadro number
N_{n}:	Total number density of neutrons (free and bound in nuclei)
N_{p}:	Total number density of protons (free and bound in nuclei)
n:	Number density of baryons
n:	Landau level
n, n_1, n_2:	Polytropic indices
n_0, n_1:	Number densities of nuclei entering a reaction
n_2, n_3:	Number densities of nuclei emerging from a reaction
n_A:	Number density of atoms
$n_{A,Z}$:	Number density of nuclei with atomic number Z and atomic mass A
n_{b}:	Number density of baryons
$n_{\mathrm{e}}, n_{\mathrm{e}^-}, n_-$:	Number density of electrons
n_{e^+}, n_+:	Number density of positrons
n_{ij}:	Number density of ions of element i in ionization state j
n_{n}:	Number density of neutrons
n_{p}:	Number density of protons
P:	Pressure of matter
$P_{\mathrm{e}}, P_{\mathrm{e}^-}$:	Electron pressure
P_{e^+}:	Positron pressure
P_{n}:	Neutron pressure
P_g:	Gas pressure
P_r:	Radiation pressure
P_N:	Nucleus pressure
P_{ik}:	Pressure tensor of radiation field
P_{01}:	Reaction rate per unit volume
p:	Electron momentum
p_z:	z Component of electron momentum
$p_{\mathrm{e}i}$:	Electron four-vector
$p_{\mu i}$:	Muon four-vector
p_{Fe}:	Fermi momentum of electrons
p_{Fn}:	Fermi momentum of neutrons
$\mathscr{P}$:	Pressure integrated over the accretion disk thickness
Q:	Energy release per nuclear reaction

Q_6: Energy obtained as heat per nuclear reaction, expressed in MeV
Q_{pair}: Energy loss rate per unit volume by neutrino emission produced due to electron–positron pair annihilation
Q_{tot}: Energy release per nuclear reaction, including the energy of free outflowing neutrinos, expressed in MeV
Q_n: Energy of a neutron strip from a nucleus
Q_p: Energy of a proton strip from a nucleus
$Q_{\mathrm{CN}}, Q_{\mathrm{CNO}}$: Heat produced during the formation of a helium nucleus in the CNO cycle of hydrogen burning
$Q_{3\alpha}$: Heat produced during the formation of a ^{12}C nucleus in the 3α reaction of helium burning
$Q_{^{12}\mathrm{C}\alpha}$: Heat produced during the formation of a ^{16}O nucleus in the α-capture reaction by ^{12}C
$Q_{^{16}\mathrm{O}\alpha}$: Heat produced during the formation of a ^{20}Mg nucleus in the α-capture reaction by ^{16}O
q_i: Heat flux density
R: Radius of a star
$R_g, r_g = 2GM/c^2$: Stellar gravitational radius
$R_\odot$: Radius of the Sun
Re: Non-dimensional Reynolds number
$R_{\mathrm{cr}}, r_{\mathrm{cr}}$: Critical radius where the flow velocity equals the local sound velocity
r: Radial coordinate
r_i: Radius of an isothermal core
r_{in}: Inner radius of an accretion disk
$r_{\mathscr{D}\mathrm{e}}$: Debye radius for electron screening
$r_{\mathscr{D}\mathrm{i}}$: Debye radius for ion screening
$\mathscr{R}$: Gas constant
S: Specific entropy of matter
S: Total density of radiation energy
S_ν: Spectral density of radiation energy
$S_{\mathrm{e}}, S_{\mathrm{e}^-}$: Entropy of electrons
S_{e^+}: Entropy of positrons
$s = 2S/3\mathscr{R}$: Non-dimensional entropy
T: Thermodynamic temperature
T: Rotational energy of a star
T_c: Central temperature of a star
T_9: Temperature expressed in 10^9 K
T_i: Temperature of an isothermal core
$T_{\mathrm{ef}}, T_{\mathrm{e}}$: Effective stellar temperature
T_{cr}: Critical temperature where the flow velocity equals the local sound velocity
T_m: Temperature of crystal melting

t_{ms}:	Time needed for a contracting star to reach the main sequence
t_{RG}:	Lifetime of a star during the red giant stage
$t_{r\phi}$:	$r\phi$ component of the viscosity stress tensor
t_t:	Lifetime of a star prior to the turning point off the main sequence
t_n:	Characteristic time of nuclear reactions
t_β:	Characteristic time of beta processes
$u, \mathbf{u}$:	Velocity
u_s:	Sound velocity in a gas
u_{cr}:	Critical velocity of a flow, equal to the local sound velocity
$u = \Delta/m_e c^2$:	Non-dimensional energy of beta decay
v, v_i:	Velocity
v_e:	Electron velocity
v_n:	Velocity of a neutron star
v_s:	Same as u_s
$\langle v_i \rangle$:	Average (diffusive) electron velocity
W:	Gravitational energy of a star
W_n:	Probability of neutron decay
W_e:	Probability of electron capture by a nucleus
$W_\beta, W_{A,Z}, W^+_{A,Z}, W^-_{A,Z}$:	Probabilities of different beta reactions with nuclei
W_μ:	Probability of muon decay
X_H, x_H:	Mass abundance of hydrogen
X_α, x_{He}:	Mass abundance of helium
$x = pc/kT$:	Non-dimensional electron momentum
$x = rc^2/GM$:	Non-dimensional current radius of an accretion disk
x_A, x_Z:	Mass abundance of an element heavier than He
x_i:	Mass abundance of an element with atomic number i
$x_{^{12}C}$:	Mass abundance of carbon
x_0, x_1:	Mass abundances of nuclei entering a reaction
x_2, x_3:	Mass abundances of nuclei emerging from a reaction
Y_e:	Ratio of the total number of protons to the total number of baryons in nuclear matter
Y_l:	Lepton charge per baryon in nuclear matter
$y = p_{Fe}/m_e c$:	Non-dimensional Fermi momentum of electrons
$y_n = p_{Fn}/m_n c$:	Non-dimensional Fermi momentum of neutrons
y_{ij}:	Fraction of the i-th element ionized to the j-th state
Z:	Electrical charge of a nucleus (nuclear number)

Greek Symbols

α:	Fine-structure constant
$\alpha = m_e c^2/kT$:	Non-dimensional inverse temperature

α: Coefficient connecting the $r\phi$ components of the viscous stress tensor with pressure
α: Coefficient connecting the mixing length of a convective element and the characteristic scale height
α_p, α: Coefficient connecting the mixing length of a convective element and the pressure scale height
α_P: Planck-averaged absorption coefficient per unit mass
α_T: Thermal expansion coefficient
α_ν: Spectral coefficient of absorption per unit mass
α_ν^*: Spectral coefficient of absorption per unit mass taking into account stimulated transitions
α_ν^{bf}: Spectral coefficient of bound–free absorption per unit mass taking into account stimulated transitions
α_ν^{ff}: Spectral coefficient of free–free absorption per unit mass taking into account stimulated transitions
α_ρ, α: Coefficient connecting the mixing length of a convective element and the density scale height
$\beta = \mu_{te}/kT$: Non-dimensional chemical potential of electrons
β_g: Ratio of gas pressure to total pressure
Γ: Relativistic adiabatic index
$\Gamma = Z^2e^2/kT\, l_{WZ}$: Non-dimensional gas parameter
$\Gamma_Z = Z^2e^2/kT\, a$: Non-dimensional gas parameter
$\Gamma; \Gamma_1, \Gamma_2$: Total and partial widths of a Breit–Wigner resonance
γ: Polytropic or adiabatic power index
$\gamma_1, \gamma_2, \gamma_3$: Adiabatic power indices
γ_i: Dirac matrices, $i = 1, 2, 3, 4, 5$
$\gamma_{\text{rad}} = d\ln T/d\ln P$: Logarithmic derivative along the radius of a radiative star
Δ: Total energy of beta decay
$\Delta\nabla T$: Excess of a temperature gradient in star over the corresponding adiabatic gradient
$\Delta\nabla\rho$: Excess of a density gradient in star over the corresponding adiabatic gradient
Δt_{ThF}: Duration of a thermal helium shell flash
$\delta = \Delta/m_ec^2$: Non-dimensional total energy of beta decay
δ_{ij}: Kronecker symbol
ϵ: Total energy of a star (with or without its rest mass)
ϵ, ϵ_e: Total energy of an electron
ϵ_G: Newtonian gravitational energy of a star
ϵ_{GR}: First-order correction to the energy of a star connected with general relativity
ϵ_i: Internal energy of a star
ϵ_{eq}: Total energy of a star in static equilibrium (without rest mass)
ϵ_M: Magnetic energy of a star
ϵ_N: Total energy of a Newtonian star (without rest mass)

ϵ_{01}:	Energy per unit mass released in a reaction between nuclei 0 and 1
$\epsilon_{2\gamma}$:	Energy produced during photo-disintegration of nucleus 2
ϵ_{ff}:	Total free–free emission rate of electrons per unit mass
ϵ_B:	Total emission rate of electron magneto-bremstrahlung per unit mass
ϵ_{Fe}:	Fermi energy of electrons
ϵ_{Fn}:	Fermi energy of neutrons
ϵ_β:	Kinetic energy of particles produced in beta decay (absorbed in beta capture)
ϵ_ν:	Energy loss by a star due to neutrino emission per unit mass
ϵ_n:	Rate of nuclear energy production in a star per unit mass
ϵ_{gr}:	Rate of gravitational energy production in a star per unit mass
ϵ_{CNO}:	Rate of heat production per unit mass in the CNO cycle of hydrogen burning
$\epsilon_{3\alpha}$:	Rate of heat production per unit mass in the 3α reaction of helium burning
$\epsilon_{^{12}\mathrm{C}\alpha}$:	Rate of heat production per unit mass in the α-capture reaction by ^{12}C
$\epsilon_{^{16}\mathrm{O}\alpha}$:	Rate of heat production per unit mass in the α-capture reaction by ^{16}O
η:	Coefficient of (dynamic) viscosity
η_T:	Coefficient of turbulent viscosity
θ:	Debye temperature of a Coulomb lattice
θ:	Lane–Emden function
θ:	Angle of $e\nu$ or $e\tau$ neutrino mixing in vacuum
θ_W:	Weinberg angle
θ_m:	Angle of $e\nu$ or $e\tau$ neutrino mixing in matter
κ:	Rosseland opacity
κ_{bf}:	Rosseland opacity for bound–free absorption
κ_e:	Opacity connected with electron heat conductivity
κ_{ff}:	Rosseland opacity for free–free absorption
κ_B:	Rosseland opacity for magneto-bremsstrahlung absorption
κ_{se}^r:	Rosseland opacity for scattering on free relativistic electrons
$\kappa_\nu = \alpha_\nu + \sigma_\nu$:	Spectral coefficient of combined absorption and scattering
κ_ν:	Neutrino opacity
κ_L:	Normalized monochromatic line absorption coefficient
$\kappa_{\mathscr{D}}$:	Reciprocal radius of Debye screening
Λ:	Coulomb logarithm
λ:	Wavelength of a photon
λ:	Coefficient of heat conductivity
λ_{e}:	Coefficient of electron heat conductivity

λ_{n}: Coefficient of neutron heat conductivity in the presence of nuclei

λ_{nn}: Coefficient of neutron heat conductivity of a pure neutron gas

$\lambda_1(0)$: Probability of a nuclear reaction between nucleus 0 and nucleus 1

μ: Molecular weight ≡ number of nucleons per particle

$\mu_{A,Z}$: Chemical potential of nuclei with atomic number Z and atomic mass A

μ_N: Average number of nucleons per nucleus

μ_Z: Number of nucleons per electron

μ_{n}: Chemical potential of neutrons

μ_{p}: Chemical potential of protons

μ_{te}, μ_{te^-}: Chemical potential of electrons

$\mu_{te^+} = -\mu_{te^-}$: Chemical potential of positrons

μ_{ν_e}: Chemical potential of electronic neutrinos

$\mu_{\tilde{\nu}_e} = -\mu_{\nu_e}$: Chemical potential of electronic antineutrinos

ν: Photon frequency

ν: Coefficient of kinematic viscosity

$\nu \equiv \nu(r)$: Number of baryons within radius r

ν_e: Total frequency of electron collisions

ν_{ee}: Frequency of collisions between electrons

ν_{ei}: Frequency of collisions between electrons and ions

ξ: Non-dimensional Lane–Emden radius

Π: Pressure term connected with artificial viscosity

ρ: Matter density

ρ_c, ρ_{c0}: Central density of a star

ρ_0: Rest mass density

$\rho_{c,\mathrm{cr}}$: Central density of a star at a point of loss of stability

$\bar{\rho}$: Average density of a star

ρ_{n}: Density of neutrons

ρ_{nd}: Neutron drip line density

ρ_{cr}: Critical density, where the flow velocity equals the local sound velocity

Σ: Surface matter density of a disk

σ: Cross-section of a nuclear reaction

σ: Coefficient of electro-conductivity

σ_{eff}: Cross-section of free–free emission

σ_{aff}: Cross-section of free–free absorption

σ_{abf}: Cross-section of bound–free absorption

σ_{efb}: Cross-section of free–bound emission (recombination cross-section)

σ^*_{abf}: Cross-section of bound–free absorption taking into account stimulated transitions

σ^*_{aff}: cross-section of free–free absorption taking into account stimulated transitions
σ_e: Total cross-section of scattering by free electrons (Thomson scattering)
σ_e: Electron spin
σ_{er}: Total cross-section of scattering by free relativistic electrons
σ_T: Thomson coefficient of scattering by free electrons per unit mass
σ_T: Coefficient of turbulent electro-conductivity
σ_ν: Spectral coefficient of scattering per unit mass
σ: Vector of Pauli matrices
τ: Optical depth
τ: Decay time of a stellar magnetic field
τ_{H}: Characteristic time of hydrogen burning in a star
τ_{He}: Characteristic time of helium burning in a star
τ_h: Characteristic hydrodynamic time
τ_n: Characteristic time of nuclear reactions
τ_{ph}: Optical depth at the level of a photosphere
τ_{th}: Characteristic time of thermal processes
τ_β: Characteristic time of beta processes
τ_ν: Cooling time of a star due to neutrino emission
$\tau_{1/2}$: Half-life in beta decay
$\tau_1(0)$: Average lifetime of nucleus 0 until a reaction with nucleus 1
$\tau_\gamma(2)$: Time of photo-disruption of nucleus 2
τ_μ: Lifetime of a muon
Φ, ϕ_G: Gravitational potential
ψ: Particle psi-function (bispinor)
$\bar{\psi}$: Dirac conjugate psi-function
ψ_{e}: Electron psi-function
ψ_{n}: Neutron psi-function
ψ_{p}: Proton psi-function
ψ_μ: Muon psi-function
ψ_ν: Neutrino psi-function
Ω, ω: Angular velocity of matter
Ω_K: Keplerian angular velocity
ω: Circular frequency of a photon
ω, ω_{nl}: Frequency of stellar oscillations
ω_1, ω_2: Terms with artificial viscosity
ω_B: Larmor frequency
ω_{i}: Circular frequency of ion oscillations in a crystal
$\omega_{p\mathrm{i}}$: Plasma frequency of ions
$\nabla = d\ln T/d\ln\rho$: As γ_{rad}
$\langle 01\rangle \equiv \langle\sigma v\rangle_{01}$: Averaging over reacting nuclei, with relative velocity v

List of Abbreviations

1-D:	One-dimensional
2-D:	Two-dimensional
AGB:	Asymptotic giant branch
CAK:	Castor–Abbott–Klein
CHF:	Core helium flash
CP:	Charge conjugation-spatial parity
CVC:	Conservation of vector current
D:	Degenerate
D:	Dipole
DH:	Debye–Hückel
GR:	General relativity
HB:	Horizontal branch
HR:	Herzsprung–Russel
IMS:	Initial main sequence
IRZ:	Intermediate regime zone
LD:	Landau–Darreus
LI:	Low-intermediate(-mass stars)
LTE:	Local thermodynamic equilibrium
MES:	Minimum energy state
MHD:	Magneto-hydrodynamical
MHD:	Mihalas–Hummer–Däppen
MRE:	Magneto-rotational explosion
MS:	Main sequence
ND:	Non-degenerate
NM:	Nuclear matter
P:	Pole
PCAC:	Partial conservation of axial current
PN:	Planetary nebula
PNN:	Planetary nebula nuclei
PSR:	Pulsar
RGB:	Red giant branch
RT:	Rayleigh–Taylor
SFC:	Self-consistent
SN:	Supernova
SNI:	Supernova of type I
SN Ia,b,c:	Supernova of types Ia, Ib, Ic
SNII, SN II:	Supernova of type II
SNU:	Aolar neutrino unit
SPH:	Smooth particle hydrodynamics
TF:	Thomas–Fermi
ThF:	Thermal flash
TFDH:	Thomas–Fermi Debye–Hückel
UHB:	Upper horizontal branch

UWI:	Universal weak interaction
WD:	White dwarf
WKB:	Wentzel–Kramers–Brillouin
WR:	Wolf–Rayet
WS:	Wigner–Seitz
ZAMS:	Zero-age main sequence
ZAHB:	Zero-age horizontal branch

Some Important Constants

$\pi = 3.1415926536$
$e = 2.7182818285$
$\log e = 0.4342944819$
$1 \text{ radian} = 57.2957795131°$

Physical Constants

Light velocity:	$c = 2.997925 \times 10^{10}$ cm·s^{-1}
Gravitational constant:	$G = 6.67 \times 10^{-8}$ dyn· cm^2·g^{-2}
Plank constant divided by 2π:	$\hbar = 1.05459 \times 10^{-27}$ ergs·s
Electron charge:	$\mathrm{e} = 4.80325 \times 10^{-10}$ CGSE units
Electron mass:	$m_{\mathrm{e}} = 9.10956 \times 10^{-28}$ g, $m_{\mathrm{e}}c^2 = 0.511003$ MeV $= k \cdot 5.93013 \times 10^9$K
Physical mass unit:	$m_{\mathrm{u}} = (1/12) m_{^{12}\mathrm{C}} = 1.660531 \times 10^{-24}$ g, $m_{\mathrm{u}}c^2 = 931.481$ MeV
Proton mass:	$m_{\mathrm{p}} = 1.672661 \times 10^{-24}$ g $= 1.00727\, m_{\mathrm{u}}$
Neutron mass:	$m_{\mathrm{n}} = 1.674911 \times 10^{-24}$ g
Muon mass:	$m_{\mu} = 1.88357 \times 10^{-25}$ g
Boltzmann constant:	$k = 1.38062 \times 10^{-16}$ erg·K^{-1}
Fine structure constant:	$\alpha = \mathrm{e}^2/\hbar c = (137.036)^{-1}$
Classical electron radius:	$l_{\mathrm{e}} = \mathrm{e}^2/m_{\mathrm{e}}c^2 = 2.81794 \times 10^{-13}$ cm
Compton electron wavelength:	$\lambda_{\mathrm{e}} = \hbar/m_{\mathrm{e}}c = 3.861592 \times 10^{-11}$ cm
Photon wavelength of the energy 1 eV:	λ (1 eV) $= 12398.54 \times 10^{-8}$ cm
Photon frequency of the energy 1 eV:	ν(1 eV) $= 2.417965 \times 10^{14}$ c^{-1}
Energy corresponding to 1 eV:	E_0(1 eV) $= 1.602192 \times 10^{-12}$ ergs
Temperature corresponding to 1 eV:	T(1 eV) $= 11604.8K = E_0/k$, (E_0/k) $\log e = 5.039.9$K
Radiation energy density const.:	$a = \frac{\pi^2 k^4}{15 c^3 \hbar^3} = 7.56464 \times 10^{-15}$erg·cm^{-3}·K^{-4}
Stephan-Boltzmann const.:	$\sigma = ac/4 = 5.66956 \times 10^{-5}$erg· cm^{-2}·K^{-4}·c^{-1}

Astronomical Constants

1 astronomical unit:	a.u. $= 1.495979 \times 10^{13}$cm
1 parsec:	pc $= 3.085678 \times 10^{18}$cm
1 light	year $= 9.460530 \times 10^{17}$cm
Solar mass:	$M_{\odot} = 1.989 \times 10^{33}$g
Solar radius:	$R_{\odot} = 6.9599 \times 10^{10}$cm
Solar luminosity:	$L_{\odot} = 3.826 \times 10^{33}$erg $\cdot$s^{-1}
Earth mass:	$M_{\oplus} = 5.976 \times 10^{27}$g
Earth equatorial radius:	$R_{\oplus,e} = 6378.164$km
Jupiter mass:	$M_{\rm Jup} = 317.83 M_{\oplus}$
Jupiter equatorial radius:	$R_{\rm Jup,e} = 71300$km
Tropical year (from equinox to equinox):	1 year $= 3.1556926 \times 10^7$s
Connection between absolute bolometric stellar magnitude and full luminosity:	$M_{bol} = 4.74 - 2.5 \log(L/L_{\odot})$
Connection between absolute (M) and visual (m) stellar magnitudes:	$M = m + 5 \log d_{\rm pc} - A_{\rm absorption}$; $d_{\rm pc}$ is a distance to the star in parsec.

Subject Index

MIX
Papier aus verantwortungsvollen Quellen
Paper from responsible sources
FSC® C105338

If you have any concerns about our products,
you can contact us on
ProductSafety@springernature.com

In case Publisher is established outside the EU,
the EU authorized representative is:
Springer Nature Customer Service Center GmbH
Europaplatz 3, 69115 Heidelberg, Germany

Printed by Libri Plureos GmbH
in Hamburg, Germany